TECHNIQUES OF CHEMISTRY

WILLIAM H. SAUNDERS, JR., *Series Editor*
ARNOLD WEISSBERGER, *Founding Editor*

VOLUME XXI

SOLUBILITY BEHAVIOR OF ORGANIC COMPOUNDS

TECHNIQUES OF CHEMISTRY

VOLUME XXI

SOLUBILITY BEHAVIOR OF ORGANIC COMPOUNDS

DAVID J. W. GRANT

Professor of Pharmaceutics
University of Minnesota
Minneapolis, Minnesota

and

TAKERU HIGUCHI

Late Regents Distinguished
Professor of Pharmacy
and of Chemistry
University of Kansas
Lawrence, Kansas

WILEY

A WILEY-INTERSCIENCE PUBLICATION
JOHN WILEY & SONS
New York • Chichester • Brisbane • Toronto • Singapore

The Library of Congress has cataloged this serial Publication as follows:

Techniques of chemistry.—   —New York:
    Wiley,
        v.
    Irregular.
    Began in 1971.
    Description based on: Vol. 4, pt. 1.
    Photo-offset reprint separately cataloged and classified in LC.
    Editor: A. Weissberger.
    Merger of: Technique of inorganic chemistry; and: Technique of organic
chemistry.
    ISSN 0082-2531 = Techniques of chemistry.
    1. Chemistry—Manipulation—Collected works. I. Weissberger, Arnold,
1898– . II. Title: Techniques of chemistry
    [DNLM: W1 TE197K]
    QD61.T4   542—dc19   85-649508
    ISBN 0-471-61314-2   AACR 2 MARC-S

    Library of Congress   [8610]

Printed in the United States of America

10 9 8 7 6 5 4 3 2 1

INTRODUCTION TO THE SERIES

Techniques of Chemistry is the successor to Technique of Organic Chemistry and its companion, Technique of Inorganic Chemistry. The newer series reflects the fact that many modern techniques are applicable over a wide area of chemical science. All of these series were originated by Arnold Weissberger and edited by him for many years.

Following in Dr. Weissberger's footsteps is no easy task, but every effort will be made to uphold the high standards he set. The aim remains the same: the comprehensive presentation of important techniques. At the same time, authors will be encouraged to illustrate what can be done with a technique rather than cataloging all known applications. It is hoped in this way to keep individual volumes to a reasonable size. Readers can help with advice and comments. Suggestions of topics for new volumes will be particularly welcome.

WILLIAM H. SAUNDERS, JR.

Department of Chemistry
University of Rochester
Rochester, New York

FOREWORD

In recent years the solubility of organic compounds has been the subject of a number of excellent books, both monographs and multiauthored volumes. Another attempt in this field may at first glance appear to be superfluous. It is apparent, however, in the earlier works that the role of specific molecular interactions in influencing solubility behavior has not been fully treated. It is our view that these specific interactions play major roles in controlling the solubility behavior of most of the bioactive organic species present in everyday life, such as drugs, herbicides, pesticides, disinfectants, cosmetics, food additives, and environmental contaminants. Moreover, specific interactions are not only largely responsible for the solubility behavior of these substances, but can be used to elicit preferential solubility and often to permit rational selection of solvents. We have undertaken to discuss solubility phenomena in this book broadly in terms of various widely accepted concepts. We have, however, placed greatest emphasis on the influence of specific interactions, primarily hydrogen bonding.

Specific interactions in the present work are largely expressed in terms of their corresponding association (complexation) constants. The thrust of the book has been directed toward attempts to predict quantitatively the values of these constants and other relevant physicochemical quantities for given solute–solvent systems in order to predict solubility behavior. Much attention has been paid to solubility and specific interactions occurring in environments of low polarity. However, emphasis has also been placed on the explanation and prediction of solubility phenomena in self-associated solvents such as alcohols and water. The treatment of solubility phenomena in aqueous environments is based primarily on various group contribution approaches.

Even though it would not normally be expected to be included under the title of this book, some material on the kinetics of diffusion and dissolution has been incorporated. In view of the importance of these rate processes in the behavior of the named groups of bioactive substances in solution, and in view of the strong relationships between these kinetic phenomena and the equilibrium solubility phenomena, we hope that the inclusion of these topics will enhance the value of the book.

We hope that the reader will forgive us for the use of data and examples which have been drawn primarily from the research of the authors and of the authors' collaborators. In many cases the material has been chosen on account of familiarity, convenience, and its direct relationship to concepts arising from specific interactions.

The concepts of solubility are now so broad and deep that it is exceedingly difficult, if not impossible, to avoid presenting a specific picture of a very fundamental physicochemical phenomenon. We hope this volume will be of value to scientists working with organic chemicals and, in certain cases, with biological systems. The level of the text is appropriate for graduates with some knowledge of physical and organic chemistry, such as chemists, pharmacists, biologists, and engineers. The text should be of particular value to graduate students, postdoctoral scientists, and other graduates who are involved in research and development work that requires knowledge of solubility phenomena and interactions in solution.

On March 24, 1987, at a time when the preparation of this manuscript had reached an advanced stage, my co-author, Dr. Takeru Higuchi, died and the pharmaceutical world was shaken by the enormity of a great loss. Known as Tak to his many friends and colleagues, Dr. Higuchi has been widely acknowledged as the foremost creator of the applied science of pharmaceutics and as the father of physical pharmacy, one of its major fields. The highest research award in pharmaceutical sciences and technology, the Higuchi Research Prize, had been established in his honor four years previously. Tak's seminal researches into the optimization of drug design, targeting, and delivery are outstanding examples of the rigorous development of applied physical chemistry. However, Tak never lost sight of the beauty and structure of the basic discipline which he applied so skillfully. He was fascinated by intermolecular interactions, thermodynamics, and kinetics, threads of which run through this book.

Having benefitted from numerous discussions with Tak before and during the preparation of the manuscript, and having enjoyed his legendary generosity, hospitality, and friendship, I ask the reader's forgiveness for briefly stating his scientific contributions to the subject matter of this book. The reader can be sure that any shortcomings of this book are due to me and not to Tak!

In the field of solubility and partitioning of organic compounds, Dr. Takeru Higuchi has made three particularly important contributions. First, he developed phase solubility analysis and highlighted the role of intermolecular complexation in increasing the solubility of organic molecules in various solvents. Second, and related to the former contribution, Dr. Higuchi has demonstrated the importance of specific intermolecular interactions, such as hydrogen bonding, in interpreting and predicting solubility and partitioning behavior. He called attention to the important role of specific interactions in solubility some 60 years after Van Laar had prevailed over Dolezalek in emphasizing purely nonspecific influences, termed van der Waals forces. By studying the solubility of a number of polar solutes in various solvent media, Dr. Higuchi demonstrated that regular solution theory in its simplest form fails in polar solvents, because it assumes only nonspecific interactions. The errors are compounded when definite solute–solvent hydrogen bonds are formed. Third, Dr. Higuchi helped to demonstrate the advantages and limitations of group contribution approaches and linear-free-energy relationships for correlating and predicting partition coefficients, Henry's law constants, and in certain cases complexation constants and activity coefficients. Dr. Higuchi's work with drug substances has no doubt kindled the interest of other scientists in the solubility behavior of these polyfunctional compounds.

We gratefully acknowledge the following for their various contributions and assistance during the preparation of the manuscript: Dr. S. Lindenbaum, the late Dr. T. Mikkelson, Dr. J. H. Rytting, Dr. V. Stella, Dr. W. E. Acree, Jr., and many graduate students of the Department of Pharmaceutical Chemistry, University of Kansas, for useful and stimulating discussions; Professor A. T. Florence, Professor E. G. Rippie, and Professor L. Saunders for reviewing the manuscript and for their valuable suggestions; Mrs. Nancy Harmony, Mr. Jeffrey Kuehnhoff, and Ms. Maure Dillsaver for their untiring secretarial and administrative support; Mr. Andrew Dallas for preparing the indexes; Mrs. Jean Marr, Ms. Ann Musser, and Mr. Leen-How Soh for artwork and drawings; and Dr. R. T. Borchardt, Summerfield Distinguished Professor and Chairman, Department of Pharmaceutical Chemistry, University of Kansas, for his enthusiastic support and encouragement.

DAVID J. W. GRANT

Minneapolis, Minnesota
June, 1989

CONTENTS

LIST OF ABBREVIATIONS

Abbreviation	Description	Chapter
ABDAC	alkylbenzyldimethylammonium chloride	9
ASOG	analytical solution(s) of groups	7
BTPC	benzyltriphenylphosphonium chloride	9
BL	γ-butyrolactone	3
^{13}C	carbon-13 isotope	8
^{13}C-NMR	carbon-13 nuclear magnetic resonance spectroscopy	4
$C_{14}BDAC$	tetradecylbenzyldimethylammonium chloride	9
CIN	cinnamaldehyde	3
CMC	critical micelle concentration	9
CMR	Carter, Murrell, and Rosch	5
DMPC	l-α-dimyristoyl phosphatidoyl choline	7
EA	ethyl acetate	3
ESR	electron spin resonance spectroscopy	9
$^{19}FNMR$	fluorine-19 nuclear magnetic resonance spectroscopy	5
h	hydrocarbon	8
hc	hydrocarbon	7
HPLC	high performance liquid chromatography	7
IR	infrared spectroscopy	9
IR	infrared spectroscopy	4
IR	infrared spectroscopy	6
ISO	isooctane	3
L	ligand	10
l-PTB	l-S-ethyl-O-ethylphenylphosphonothionate	5
LFER	linear free-energy relationship(s)	5
LLP	liquid liquid partitioning	5
LSER	linear solvation energy relationship	2
MEP	2-methyl-3-pentanol	3
MHz	megahertz	6
mmHg	millimeters of mercury	7
MSGA	molecular and group surface area approach	7
NMR	nuclear magnetic resonance spectroscopy	4
OH	hydroxyl group	7
OMe	methoxy group	7
ORD	optical rotatory dispersion	5

Abbreviation	Description	Chapter
P	polar group(s)	8
PEG	polyethylene glycol	8
pK_a	$-\log_{10}$ (dissociation constant)	11
PMR	proton magnetic resonance spectroscopy	4
QSAR	quantitative structure-activity relationship(s)	3
RMRI	molecular redundancy index	8
S	substrate	10
s.d.	standard deviation	2
SCG	sodium cromoglycate	9
SCL	supercooled liquid	8
SDDS	sodium dodecylsulfate	9
SOLV	polarizable solvent	3
SSO	saturated solution in l-octanol	8
SSW	saturated solution in water	8
TBP	tri-*n*-butylphosphate	5
TMPPT	1,3,7,9-tetramethylpyrimido(5,4-g)pteridine-2,4,6,8 (1H, 3H, 7H, 9H)-tetrone	10
TOL	toluene	3
UNIFAC	UNIQUAC functional-group activity coefficient	7
UNIFAC	UNIQUAC functional-group activity coefficient	1
UNIFAC	UNIQUAC functional-group activity coefficient	8
UNIQUAC	universal quasi-chemical equation or model	7
UNIQUAC	universal quasi-chemical equation or model	8
UNIQUAC	universal quasi-chemical equation or model	1
UV	ultraviolet absorption spectroscopy	5
UV	ultraviolet-visible spectroscopy	9
UV	ultraviolet-visible spectroscopy	6
W	water	7

LIST OF SYMBOLS

This list of symbols is presented in the following order: All Latin letters are alphabetized before Greek letters. Superscripts and subscripts are defined at the end of this list.

Symbol	Description	Section
A	concentration of the dissolved reactant 7-acetyltheophylline	11.8.2
A	hydrogen-bond (proton) acceptor	5.2.1
A	measured peak area resulting from a solute in gas chromatography	6.3.2
A	relative surface area of a group relative to the *tert*-butyl group	8.8
A	surface area of an individual group or molecule	7.2.4
A	surface area of a solid available for dissolution	8.9.3, 11.1
A	wetted surface area of a solid in dissolution	11.1, 11.7.2
$Å$	Angstrom unit $= 0.1$ nm	2.11.5
A^*	measured peak area of the pure solute vapor in gas chromatography	6.3.2
$[A]$	activity of hydrogen bond acceptor	6.7
$[A]$	concentration of solute A	1.2.4
$[A]$	molar concentration of the free hydrogen-bond acceptor, i.e. the complexing agent	5.2.1
A_B	surface area of a solute molecule B	8.3
A_L	linear rise in $[S]_t$ with increasing $[L]_t$ in the formation of a soluble complex	10.4
A_N	negative curvature in a plot of $[S]_t$ *vs.* $[L]_t$ with the formation of a soluble complex	10.4
A_P	positive curvature in a plot of $[S]_t$ *vs.* $[L]_t$ in the formation of a soluble complex	10.4
A_B^h	surface area of a hydrocarbon group in the solute molecule B	8.3
A_B^p	surface area of a polar group in the solute molecule B	8.3
A_{wk}	van der Waals group surface-area parameter	7.5.2
$[A^-]$	molar concentration of the ionized species of a weak acid HA	11.3
$[A_1]$	molar concentration of monomer, e.g. an alcohol	6.5.3, 6.6.1
$[A]_o$	molar concentration of free, uncomplexed hydrogen bond acceptor	5.7.5, 6.7

Symbol	Description	Section
$[A]_t$	molar concentration of a hydrogen bond acceptor functional group	4.4.3
$[A]_t$	total molar concentration of a hydrogen-bond (proton) acceptor (or an alcohol)	6.6.1, 6.7
$[A_nB]$	concentration of the complex A_nB	1.2.4
$[AD_n]$	molar concentration of a complex of stoichiometric number n	6.7
$[A^-]_0$ as $[HA]_0$; $[A^-]$ at $x = 0$		11.4.1, 11.5.1
$[A^-]_h$ as $[HA]_h$; $[A^-]$ at $x = h$		11.4.1, 11.5.1
a	molecular area	11.2.5
a	thermodynamic activity	2.1
a^{c*}	thermodynamic activity of a pure species, the standard state of unit activity being a hypothetical 1 mol dm^{-3} solution which behaves as if it were infinitely dilute, in a paraffinic solvent	3.10
a^s	thermodynamic activity of a solid solute	2.6, 2.11.3, 8.3
a_1	thermodynamic activity of the solvent	8.5
a_2	thermodynamic activity of the solute	8.5
a_A	thermodynamic activity of the proton acceptor	5.2.1
a_A, a_B	thermodynamic activity of species A or B	2.6
a_D	thermodynamic activity of the proton donor	5.2.1
a_{DA}	therymodynamic activity of a hydrogen-bonded donor–acceptor complex	5.2.1
a_i	thermodynamic activity of species i	11.3
a_{mn}	group interaction parameter between groups m and n in UNIFAC	7.5.2
a_{nm}	group interaction parameters between groups n and m in UNIFAC	7.5.2
$a_{(X)}$	number of occurrences of fragment X in a molecule	7.2.2
a_2^*	activity of the pure solute	3.4
a_1^c	activity of the monomer species in solution, the standard state being a hypothetical 1 mol dm^{-3} solution which behaves as if it were infinitely dilute	6.3.2
a_2^c	activity of the solute 2 based on the Henry's law-molarity scale	3.3
a_2^{c*}	activity of the pure solute 2 for which the standard state of unit activity is a hypothetical 1 mol dm^{-3} solution which behaves as if it were infinitely dilute	3.4

Symbol	Description	Section
$a_2^{c}*$	molarity-based activity of the pure solute for which the standard state of unit activity is a 1 mol dm^{-3} solution which behaves as if it were infinitely dilute	3.4
a_2^{m}	activity of solute 2 based on the Henry's law-molality scale	3.3
a_2^{R}	activity of the pure solute 2, based on Raoult's law, for which the standard state is the pure substance	3.5
a_2^{x}	activity of the solute 2 based on the Henry's law mole fraction scale	3.3
$a_2^{x}*$	mole fraction-based activity of the pure solute 2 for which the standard state is a hypothetical solution of unit mole fraction which behaves as if it were infinitely dilute	3.4
B	base	11.3
B'	empirical methylene group contribution to log (solubility of a solid)	8.6
[B]	concentration of solute B	1.2.4
B_I	[S]$_t$ vs. [L]$_t$ plot with the formation of an insoluble complex	10.4
B_o	molar concentration of a base, e.g. hydroxide ion, at the solid–liquid interface	11.8.2
B_S	[S]$_t$ vs. [L]$_t$ plot with the formation of complexes with limited solubility	10.4
[B]$_h$ as [HA]$_h$; [B] at $x = h$		11.5.1
[B$^-$]$_h$ as [HA]$_h$; [B$^-$] at $x = h$		11.4.1
[B$^-$]$_0$ as [HA]$_0$; [B$^-$] at $x = 0$		11.4.1
[B]$_0$ as [HA]$_0$; [B] at $x = 0$		11.5.1
[BH$^+$]$_0$ as [HA]$_0$; [BH$^+$] at $x = 0$		11.5.1
[BH$^+$]$_h$ as [HA]$_h$; [BH$^+$] at $x = h$		11.5.1
[BTPC]	molar concentration of BTPC	9.5
$b_{(\xi)}$	number of times structural factor ξ occurs in the molecule	7.2.2
C	cell constant in measurements of electrical conductivity of solutions	9.5
C	number of components in the system	10.2
C_1, C_2, C_3	concentration of compound 1, 2, or 3	10.2
C_A	susceptibility of an acid to form a covalent bond	2.11.7
C_L	concentration of ligand in complexation solubility analysis	10.4

Symbol	Description	Section
C_S	concentration of substrate in complexation solubility analysis	10.4
$C_a A^-$	concentration of the anion A^- in the stirred aqueous phase (in ion pair phase transfer)	9.3.4
$C_{ai} A^-$	concentration of the anion A^- immediately at the aqueous side of the interface (in ion pair phase transfer)	9.3.4
$C_{ai} M^+$	concentration of the cation M^+ immediately at the aqueous side of the interface (in ion pair phase transfer)	9.3.4
$C_a M^+$	concentration of the cation M^+ in the stirred aqueous phase (in ion pair phase transfer)	9.3.4
$C_o M^+ A^-$	concentration of the ion pair $M^+ A^-$ in the stirred organic phase (in ion pair phase transfer)	9.3
$C_{oi} M^+ A^-$	concentration of the ion pair $M^+ A^-$ immediately at the organic side of the interface (in ion pair phase transfer)	9.3.4
c	concentration	6.10.2
c	concentration of a solid solute dissolved at time t	8.9.3
c	curvature correction factor in the MGSA approach	8.3
c	molar concentration	8.11
c''	empirical constant	5.6.1
c_s	equilibrium solubility of a dissolving solid	11.1
c^s	molar solubility, i.e. solubility in mol dm^{-3} = mol/L	2.9
$c^s(\text{calc})$	calculated molar solubility	2.9
$c^s(\text{obs})$	observed molar solubility	2.9
c_1	molar concentration of the solute monomer, e.g. an alcohol	6.3.2
c_f	final total molarity of the solute after dilution	6.3.1
c_B	molar concentration of salt B	8.11
C_B	susceptibility of a base to form a covalent bond	2.11.7
c_b	analyzed concentration of a dissolved solid in the bulk of the dissolution medium	11.1
c_i	initial total molarity of the solute	6.3.1
c_o^s	molar solubility of the unionized species of a weak acid or weak base	11.3
c_n^s	solubility of a compound containing n methylene groups	8.6

Symbol	Description	Section
c_t^s	total molar solubility of a weak acid or weak base in both the ionized and unionized form	11.3
c_2^s	molar solubility of organic solute 2 in an aqueous solution of salt B of molar concentration c_B	8.11
$c_{2,A}^s$	molar solubility of solute 2 in pure A	8.11
$c_{2,B}^s$	molar solubility of solute 2 in pure B	8.11
$c_{2,0}^s$	molar solubility of organic solute 2 in an aqueous solution in the absence of salt B	8.11
$c_{2,w}^s$	solubility of a solute 2 in water	8.5
c_o^s	solubility of a compound containing zero methylene groups	8.6
c_w^{sat}	solubility of the solute in water	8.7.2
c_{hc}^{sat}	solublity of a solute in a hydrocarbon solvent	8.7.3
D	dielectric constant (i.e. relative permittivity) of a medium or solvent	2.11.2, 2.11.5, 9.2.2
D	diffusivity of a solute, i.e. diffusion coefficient	6.10.1, 8.9.3
D	distribution ratio; apparent partition coefficient	9.3.1.2
D	hydrogen-bond (proton) donor	5.1
D	observed distribution coefficient of a solute expressed as the ratio of its molar concentration in an organic phase to that in an aqueous phase	7.2.4, 9.3.1.2, 9.4
$[D]$	molar concentration of the fatty acid monomer	6.7
D_0	distribution ratio of an ion pair solute (solvent/water) in the absence of a specific solvating agent M	9.4
D_1	diffusion coefficient of the monomer of phenol	6.10.2
$[D_2]$	molar concentration of the fatty acid dimer	6.7
D_5	diffusion coefficient of the pentamer of phenol	6.10.2
D_o	limiting diffusivity at low concentrations of the solute	6.10.1
D_S	diffusivity of species S	11.4.1
D_X	diffusivity of molecule X	11.7.2
$[D]_o$	molar solubility of the proton donor in the absence of acceptor	5.2.1
$[D]_o$	concentration of uncomplexed proton donor	5.7.5
$[D]_t$	total molar solubility of the hydrogen-bond (proton) donor	5.2.1

Symbol	Description	Section
$[D]_t$	total molar concentration of the hydrogen-bond (proton) donor	6.7
$[DA]$	molar concentration of the one-to-one hydrogen-bonded donor–acceptor complex	5.2.1
d	mean hydraulic diameter	11.2.5
dc/dt	rate of increase in the concentration of a dissolving solid solute	8.9.3
dc/dx	concentration gradient	11.2.3
dm/dt	dissolution rate of a dissolving solid	11.1
$(dQ/dt)_x$	rate at which solute A crosses unit area of any plane of constant distance x within the diffusion layer	11.8.2
E	electric field	2.11.4
$E(r)$	potential energy of an ion pair as a function of the distance r of separation of the ions	9.2.2
E^*	conditional extraction constant	9.3.3
E_A	susceptibility of an acid to undergo electrostatic interaction	2.11.7
E_B	susceptibility of a base to undergo electrostatic interaction	2.11.7
E_C	absorbance or extinction of a charge-transfer complex	5.3.5
E_s	Taft steric parameter	5.5.2
F	number of degrees of freedom	10.2
F_t	volume increase factor during a conductometric titration	9.5
F_{CH_2}	ratio by which the introduction of a methylene group increases the partition coefficient (organic solvent/water)	3.2
$F(X)$	ratio by which substituent X changes the partition coefficient	7.2.2
f	fugacity	3.4
f	refractive index function, i.e. $(n^2 - 1)/(n^2 + 2)$	2.11.2, 2.11.6, 3.11
f_1	fugacity of the solute monomer, e.g. an alcohol	6.3.2
f_A	formal molarity, i.e. total molarity of an alcohol	6.8.2
$f_{(R)}, f_{(H)}, f_{(X)}$	fragmentation constants of group R, H, and X	7.2.2
f_2^*	fugacity of the pure solute 2	3.4

Symbol	Description	Section
$f_2^{c}*$	molarity-based fugacity of the solute 2 in the pure state	3.4
f_2^{θ}	fugacity of the solute 2 in the standard state	3.4
$f_2^{c\theta}$	molarity-based fugacity of the solute 2 in the (hypothetical) standard state	3.3
$f_2^{x}*$	mole fraction-based fugacity of the solute 2 in the pure state	3.4
$f_2^{x\theta}$	mole fraction-based fugacity of the solute 2 in the (hypothetical) standard state	3.3
G	Gibbs free energy	1.2.6
$\overline{G}$	partial molar Gibbs free energy of a solute	3.8
$\overline{G}^{\theta}$	partial molar Gibbs free energy of a solute in its standard state in solution	3.8
G_g	molar free energy of a solute vapor	3.8
G_i	initial conductance of the solution in a conductometric titration	9.5
G_t	conductance of the solution at the time t during a conductometric titration	9.5
G_g^{θ}	molar free energy of a solute vapor in its standard state of 1 atm	3.8
H	enthalpy, i.e. heat content	1.2.6
HA	weak acid	11.3
$[H^+]$	hydrogen-ion concentration	9.3.1.3
$[H^+]_0$ as $[HA]_0$; $[H^+]$ at x = 0		11.5.1
$[H^+]_h$ as $[HA]_h$; $[H^+]$ at x = h		11.5.1
$[HA]_0$	molar concentration of HA at x = 0, i.e., immediately next to the surface of the dissolving solid at the start of the diffusion layer	11.4.1, 11.5.1
$[HA]_h$	molar concentration of HA at x = h, i.e., in the bulk solution at the end of the diffusion layer of the dissolving solid	11.4.1, 11.5.1
$[HB]_0$ as $[HA]_0$; $[HB]$ at x = 0		11.4.1
$[HB]_h$ as $[HA]_h$; $[HB]$ at x = h		11.4.1
h	Henry's law constant	3.2
h	thickness of the diffusion layer next to a dissolving solid	8.9.3, 11.2.3
h_2	Henry's law constant of solute 2	3.11
h^c	Henry's law constant based on molarity	3.2
h^m	Henry's law constant based on molality	3.2
h^x	Henry's law constant based on mole fraction	3.2

Symbol	Description	Section
h_A	parameter representing the effect of an acceptor on the hydrogen bonding equilibrium in paraffinic solvents at 25°C ($h_A = 1$ for tri-n-butylphosphate)	5.6.1
$\bar{h}_D$	mean value of h_D	5.6.5
h_D	parameter representing the effect of a donor on the hydrogen bonding equilibrium in paraffinic solvents at 25°C ($h_D = 0$ for phenol)	5.6.1
h_o	Henry's law constant of a solute in an organic solvent	7.1
h_w	Henry's law constant of a solute in water	7.1, 8.7.2
h_A^*	h_A based on an alternative solvent (carbon tetrachloride at 30°C)	5.6.5
h_D^*	h_D based on an alternative solvent (carbon tetrachloride at 30°C)	5.6.5
h_1^c	Henry's law constant of the solute monomer based on molarity	6.3.2
h_B^c	molarity-based Henry's law constant of species B	3.3
h_B^m	molality-based Henry's law constant of species B	3.3
h_2^x	Henry's law constant of the solute 2 based on mole fraction	3.2
h_B^x	mole fraction-based Henry's law constant of species B	3.3
h_w^x	Henry's law constant based on mole fraction in dilute aqueous solution	7.1
h_{hc}^x	Henry's law constant based on mole fraction in dilute solution in a paraffinic hydrocarbon	7.1
I	ionic strength	8.11
I_i	ionization energy of molecular species i	2.6
[IP]	concentration of the nonconducting ion pair	9.5
J	flux; dissolution rate per unit surface area, i.e. intrinisic dissolution rate	11.1, 11.4.2
J_A	flux of species A; rate of dissolution of A per unit surface area	11.8.2
J_S	flux of species S	11.4.3
J_t	total flux, i.e. intrinsic dissolution rate of a solid	11.4.3

Symbol	Description	Section
K	apparent equilibrium constant for the reaction between HA and B^-	11.4.1
K	conductivity of a solution	9.5
K	equilibrium constant of the ion-pair association reaction	9.4
K	thermodynamic ionization constant of a weak acid HA	11.3
K	equilibrium constant, e.g. stability constant of a complex	1.2.4, 1.2.6, 10.6
K	stability constant of ion-pair solvation by a specific solvating agent M	9.4
K'	equilibrium constant of a second reaction	5.5.1
K'	true stability constant of the complex in terms of activities	5.3.2
K_ω	stepwise equilibrium constant for self-association	6.8.1
K_a	apparent ionization constant of a weak acid based on molarity	9.3.1.3, 11.3, 11.4.3
K_c	stability constant of a charge transfer complex, molarity based	5.3.5
K_D	thermodynamic dimerization constant	6.7
K_d	dimerization constant for self-association of a solute in the organic phase, based on molarity	5.3.4, 6.7
K_e	ion-pair extraction constant	7.2.4, 9.3.1.2
K_m	conductivity of a mixture of ion pairs and salts in solution	9.5
K_n	overall stability constant of the complex DA_n or SA_n based on molarity	4.4.1, 6.6.2
K_0	molarity-based extraction constant with a specific solvating agent M	9.3.1.2
K_0	stablity constant with a reference donor in a paraffinic solvent	5.6.1
k_R	reaction rate constant (first order) for the interfacial reaction in the dissolution of a solid	11.2.1
K_s	apparent solubility product based on molarity	9.5
K_0'	K_0 at unit molar concentration of solvating agent	9.3.1.2
K_1'	apparent equilibrium constant for a weak acid HA reacting with hydroxide ion	11.4.3
K_a'	apparent equilibrium constant for a weak acid HA reacting with water	11.4.3

Symbol	Description	Section
K''_g	log (K''_g) represents the net solution interaction energy after correcting for dispersion interactions	2.7
K'_n	apparent stability constant of the complex AD_n based on molarity	6.7
K'_w	apparent equilibrium constant for the reaction between H_3O^+ and OH^-	11.4.3
K^{eq}	equilibrium constant, stability constant	5.2.1
$K_{1:1}$	molarity based stability constant of a 1:1 complex	4.3.1, 10.6
$K_{1:2}$	molarity based stability constant of a 1:2 complex	10.6
$K_{1,2}$	molarity based dimerization constant	6.10.1
$K_{1,5}$	molarity based equilibrium constant of pentamer formation	6.3.2
$K_{1:n}$	molarity based stability constant of a 1:n complex	4.3.1
$K_{1,n}$	molarity based equilibrium constant for self-associated species of stoichiometry n, i.e. n-mer	6.3.1
K_{B-H}	Benesi–Hildebrand equilibrium constant, molarity based	5.7.5
K_{DA}	dimensionless formation constant of a one-to-one complex between donor D and acceptor A	4.4.1
K_{DA}	formation constant of a one-to-n complex between donor D and acceptor A	4.4.1
K_{eq}	thermodynamic equilibrium constant, usually for complexation	6.7
K_{HA}	apparent ionization constant of the weak acid HA	11.4.1
K_{HB}	apparent ionization constant of the weak acid HB	11.4.1
K_{IP}	ion-pair association constant	9.5
K'_{DA}	molarity-based equilibrium constant for a donor-acceptor complex obtained experimentally in the presence of an inert solvent, such as a paraffin	5.7.5
K''_{DA}	apparent molarity-based equilibrium constant determined experimentally in the presence of an interfering solvent	5.7.5
K'_{SA}	molarity-based equilibrium constant for a solvent-acceptor complex	5.7.5
K'_{SD}	molarity-based equilibrium constant for a solvent-donor complex	5.7.5

Symbol	Description	Section
K_{DA}^{eq}	thermodynamic stability constant (interaction constant) of a hydrogen-bonded donor-acceptor complex	5.2.1
$(K_c)_{B-H}$	stability constant generated by the Benesi-Hildebrand equation in terms of molarity	5.3.5
K_{CMR}	Carter, Murrell and Rosch equilibrium constant, molarity based	5.7.5
K_{ass}	self-association constant, based on molarity	6.6.2
K_{SDDS}	conductivity of SDDS in water	9.5
k	Boltzmann constant, 1.3806×10^{-23} J K^{-1}	2.11.2
k	chromotographic capacity factor	7.2.3
k	functional group in UNIFAC	7.5.2
k_1^*	first-order rate constant for the dissolution of a solid	11.2.1
k_2	salt-effect parameter	8.11
k_A	parameter representative of the hydrogen-bond accepting capability of a hydrogen-bond acceptor	5.6.1
k_B	second-order rate constant for reaction between dissolved solute A and component B in solution	11.8.2
k_D	parameter representative of the hydrogen-bond donating capability of a hydrogen-bond donor	5.6.1
k_T	rate constant for transport of the dissolved solute across the diffusion layer in the dissolution of a solid	11.2.1
$k_{2,A}$	constant which reflects the sensitivity of the solubility of a solute 2 to the volume fraction of an organic cosolvent A	8.11
$k_{2,B}$	constant which reflects the sensitivity of the solubility of a solute 2 to the volume fraction of an organic cosolvent B	8.11
k_{NW}	first-order (Noyes Whitney) rate constant for the dissolution of a solid	11.1
k_{app}	apparent first-order hydrolytic rate constant	11.8.1
L	ligand, interactive cosolvent	5.1
L	Avogadro constant, 6.0222×10^{23} mol^{-1}	8.3
[L]	molar concentration of the ligand (or cosolvent) in a defined situation	4.3.1
L^{θ}	partial molar enthalpy of solution at infinite dilution	5.3.8

Symbol	Description	Section
$[L]_f$	molar concentration of the free uncomplexed ligand	6.6.1
$\{L\}_f$	activity of the dissolved uncomplexed ligand	4.3.3
$[L]_h$	molar concentration of the ligand in the bulk solution, i.e. at $x = h$	11.7.1
$[L]_o$	molar concentration of the ligand at the solid-liquid interface, i.e. at $x = 0$	11.7.1
$[L]_t$	total molar concentration of ligand	10.4
l	film thickness for diffusion of a solid	6.10.2
l	path length of a UV-visible spectrophotometric sample cell	5.3.5
M	solvating agent for an ion par	7.2.4
M	molecular weight	2.11.6
M	total number of components in the mixture in UNIQUAC	7.5.1
[M]	molar concentration of the ion-pair solvating agent (e.g., chloroform) in the organic phase	7.2.4, 9.3.1.2
M_S, M_X	molecular weight of a molecule S or X	11.7.2
m	mass of a solid dissolved at time t	11.1
m	molality	3.3
m	stoichiometric number of S in the complex S_mL_n	10.5
m''	empirical constant	5.6.1
m_B	molality of species B	3.3
m_B	molality of salt B	8.11
m/n	stoichiometric ratio for the complex S_mL_n	10.5
N	total number of functional groups in a liquid mixture in UNIFAC	7.5.2
n	effective molecularity of the solvating agent in the ion pair	9.3.1.2
n	number of data points	2.12
n	refractive index	2.11.6
n	stoichiometry of the complex A_nB	1.2.4
n	stoichiometric number of alcohol molecules in the complex SA_n	6.6.2
n	stoichiometric number of L in the complex S_mL_n	10.5
n	stoichiometric number for solvation of an ion pair	7.2.4

Symbol	Description	Section
n_D	refractive index for visible light using the sodium D line at 589 nm	2.11.6
n_D^{25}	refractive index at 25° using the sodium D line at 589 nm	3.11
n_{25}^D	refractive index at 25° with respect to the sodium D line at 589 nm	2.11.2
$[OH^-]_0$ as $[HA]_0$; $[OH^-]$ at $x = 0$		11.5.1
$[OH^-]_h$ as $[HA]_h$; $[OH^-]$ at x $= h$		11.5.1
P	monomer of phenol	6.10.2
P	number of phases in the system	10.2
P	partition coefficient	2.9
P'	solvent polarity parameter of Rohrschneider	2.7
P^c	partition coefficient based on molar concentration, i.e. molarity	2.11.3
P^x	partition coefficient of a solute based on mole fraction	2.9
P_5	pentamer of phenol	6.10.2
P_E	electronic polarization	2.11.6
$P_{\alpha/\beta}$	partition coefficient of a solute between phases α and β	3.6
$P_{g/w}$	partition coefficient of a solute between the gas phase and a solution in water	2.12
$P_{o/w}$	partition coefficient of a solute between an organic solvent (often 1-octanol) and water	2.12, 7.1, 7.2.3
$P_{p/w}$	partition coeffceint of a solute between phospholipid vesicles and an aqueous solution	7.2.3
$P_{o/w}^c$	partition coefficint (1-octanol/water) based on molarity	2.9
$P_{o/w}^c$	partition coefficient (organic solvent/water) based on molarity	7.2.2
$P_{(RH)}^c$	molarity-based partition coefficient of compound RH	7.2.2
$P_{(RX)}^c$	molarity-based partition coefficient of compound RX	7.2.2
$P_{hc/w}^c$	partition coefficient, based on the molar concentration scale, of a solute between a paraffinic hydrocarbon and water	7.3.2
$P_{\alpha/\beta}^x$	partition coefficient based on mole fraction of a solute between liquid phase α and liquid phase β	2.9
$P(RH)$	partition coefficient of compound RH	7.2.2
$P(RX)$	partition coefficient of compound RX	7.2.2

Symbol	Description	Section
p	vapor pressure, partial vapor pressure	3.8, 6.3.2
p^*	saturated vapor pressure of a pure substance	2.1.2, 6.3.2, 8.7.2
p_1	partial vapor pressure of the solute monomer	6.3.2
p_A, p_B	partial vapor pressure of species A or B	3.3
P_2^*	saturated vapor pressure of the pure solute 2	3.5.2, 3.9
Q	enthalpy change associated with the formation of n moles of a complex	5.3.8
q_1, q_2	charge on an ionic species 1 or 2	2.11.2
q_h	van der Waals surface area of the molecule h in the pure liquid in UNIQUAC	7.5.1
R	gas constant, 8.3143 J K^{-1} mol^{-1} or 1.987 cal K^{-1} mol^{-1}	2.6
R_D	molar refractivity using the sodium D line at 589 nm	2.11.6
Re	Reynolds number	11.2.4
R_F	ratio of eluent movement to solvent movement in open-bed chromatography	7.2.3
R_M	retention parameter in open-bed chromatography	7.2.3
R_m	molar refractivity	2.11.2
$R(\lambda = \infty)$	molar refractivity for light of infinite wavelength	2.11.6
r	correlation coefficient	2.12, 5.6.5
r	distance of separation of the ions in an ion pair	9.2.2
r	intermolecular distance, i.e. distance between two ionic or molecular centers	2.11.2, 2.11.3, 3.9
r	mean radius of a diffusing particle	6.10.1, 11.2.5
r	radius of rotating disc	11.2.5
r_d	summation of the interatomic distances of the molecule	8.5
r_h	van der Waals volume of the molecule h in the pure liquid in UNIFAC	7.5.1
S	entropy	2.2.1
S	substrate, dissolving solid, solute	5.1
[S]	molar concentration of a solute species in a defined situation	4.3.1, 6.5.3

Symbol	Description	Section
S_1, S_2, S_3	solubility of compound 1, 2, or 3	10.2
$\overline{S}_B$	partial molar entropy of component B in a mixture	2.10
S_g	molar concentration of the pure solute when saturating the gas phase	2.12
S_w	molar solubility in water	2.12
$\{S\}_o$	activity of the dissolved uncomplexed solute	4.3.3
$[S]_o$	equilibrium molar solubility of the substrate S in the absence of the ligand L	10.4
$[S]_o$	molar concentration of uncomplexed solvent	5.7.5
$[S]_t$	total molar concentration of the dissolved substrate S, i.e. the apparent solubility of the substrate	4.3.1, 6.6.1, 10.4
$[S]_t$	total molar concentration of the dissolved solute in the bulk solution during the dissolution of a solid S in the presence of a complexing agent L	11.7.1
S_B^*	molar entropy of pure component B	2.10
$\{SL\}$	activity of the dissolved one-to-one complex	4.3.3
$[SL]_h$	molar concentration of the complex SL in the bulk solution, i.e., at $x = h$	11.7.1
$[SL]_o$	molar concentration of the complex SL at the solid-liquid interface, i.e., at $x = 0$	11.7.1
$[SDDS]$	molar concentration of SDDS	9.5
s_o	solubility in an organic solvent	8.8
s_w	solubility in water	8.8
T	absolute temperature	2.6
T_1	^{13}C spin-lattice relaxation time	8.2.4
T_b	absolute boiling point of a liquid	3.4
T_c	phase transition temperature	7.2.3
T_m	absolute melting point temperature	8.5
t	corrected retention time of an eluent in HPLC	7.2.3
t	time	2.5, 11.1
t_L	lag time	6.10.2
t_B	burst time	6.10.2
t_o	corrected retention time of an eluent slightly enriched with water in HPLC	7.2.3
U_{12}	solvent–solute pair-potential interaction energy	3.11

Symbol	Description	Section
U_{AB}	molar internal energy of interaction between two molecules A and B	2.11.8
U_{mn}	energy of interaction between groups m and n in UNIFAC	7.5.2
u	linear flow rate, i.e. distance/time	11.2.5
u(at.)	pair-potential energy of attraction between two molecules	2.11.5
u(dispersion)	pair-potential energy of interaction due to London dispersion forces between two molecules	2.11.4, 2.11.6
u(fixed dipoles)	pair-potential energy of interaction between two fixed dipoles	2.11.3
u(induction)	pair-potential energy of interaction between a rotating dipolar molecule and a rotating polarizable molecule (Debye forces)	2.11.4
u(orientation)	pair-potential energy of interaction between two rotating dipoles averaged over all possible orientations (Keesom forces)	2.11.3
u_{AA}	solvent–solvent pair-potential interaction energy	2.6
u_{AB}	solvent–solute pair-potential interaction energy	2.6
u_{AB}	pair-potential energy of interaction between two molecules A and B	2.11.8
u_{BB}	solute–solute pair-potential interaction energy	2.6
u_{hh}	pair potential energy between two h molecules in UNIQUAC	7.5.1
u_{hj}	pair potential energy between one h and one j molecule in UNIQUAC	7.5.1
u_{jh}	pair potential energy between one j and one h molecule in UNIQUAC	7.5.1
u_{jj}	pair potential energy between two j molecules in UNIQUAC	7.5.1
V	molar volume	6.10.1
V	partial molar volume	11.2.5
V	total volume of a solution	5.3.8
V	volume of the dissolution medium	11.1
V	volume of the solvent in a batch-type dissolution rate determination	8.9.3, 11.1
V_2	molar volume of the solute 2 in the liquid state	8.5

Symbol	Description	Section
V_B	partial molar volume of species B, usually the solute	2.8
V_i	molar volume of species i	2.6
V_o	molar volume of the organic solvent, e.g. 1-octanol	2.9, 7.2.2
V_t	volume of titrant added at time t during a conductometric titration	9.5
V_w	molar volume of water	2.9, 7.2.2
V_{wk}	van der Waals group volume parameter	7.5.2
v	molecular volume	2.11.6, 11.2.5
W	energy correction term to allow for solvent–solute non-dispersion interactions in an extended Hildebrand solubility parameter approach	2.7
W	rotation speed of a disc, in revolutions per second, i.e. Hz	11.2.4
W_A^c, W_B^c	work of cohesion of species A or B	2.8
W_{AA}	work of interaction between molecules of species A	2.8
W_{AB}	work of interaction between molecules of two different species, A and B	2.8
W_{BB}	work of interaction between molecules of species B	2.8
W_{ij}	reversible work of interaction between molecules of species i and j	2.8
W_{AB}^a	work of adhesion between molecules of two different species, A and B	2.8
w_1, w_2, w_3	mass fraction of compound 1, 2, or 3	10.2
$\{X\}$	activity of substance X	6.7
$[X^-]$	concentration of a free anion X^- in the aqueous phase	7.2.4, 9.3.1.2
x	distance	6.10.2
x	distance of a point from the dissolving solid surface within the diffusion layer, or film, of thickness h	6.10.2, 11.4.1, 11.8.2
x^s	solubility of a solid expressed as mole fraction	8.3

Symbol	Description	Section
x^s	solubility expressed as mole fraction	2.9, 7.1
x_h	mole fraction of h molecules in the liquid in UNIQUAC	7.5.1
x_s	mole fraction of the solvent in a mixture or solution	5.7.5
x_w^l	mole fraction solubility of a liquid in water	2.9
x_A, x_B	mole fraction of component A or B	2.6, 2.9
Y	generalized equilibrium constant for phase transfer	1.2.6
Y	generalized equilibrium property	7.1
$\underline{Y}$	logarithm of the stability constant K_{DA}	5.6.5
$\overline{Y}$	mean value of Y	5.6.5
Y_A, Y_B	physical property for a solution of A or B alone	5.3.6
y^c	activity coefficient based on Henry's law with the molarity scale	3.3, 9.5
y^x	activity coefficient based on Henry's law mole fraction scale	3.3, 3.8
y_1^c	activity coefficient of the monomer in solution, the standard state being a hypothetical 1 mol dm^{-3} solution which behaves as if it were infinitely dilute	6.3.2
y_2^c	molarity-based activity coefficient of solute 2 for which the standard state of unit activity is a hypothetical 1 mol dm^{-3} solution which behaves as if it were infinitely dilute	3.4
y_2^m	activity coefficient of solute 2 based on Henry's law with the molality scale	3.3
y_2^x	activity coefficient of solute 2 based on Henry's law with the mole fraction scale	3.3
z	number of nearest neighbors in the solution at the molecular level in UNIFAC	7.5.1
z	valence of an ion	8.11
α	degree of dissociation of an electrolyte	9.5
α	dimensionless relative concentration of A at a given point x in the diffusion layer	11.8.3
α	fraction of monomers that are associated	6.8.3
α	molecular polarizability	2.11.2
α	ratio of the monomer molarity to the total molarity of a self-associating solute	6.3.2

Symbol	Description	Section
α	solvatochromic acidity parameter denoting hydrogen-bond donating ability	2.12
α	solvent phase in partitioning (numerator)	3.6
α	thermal expansivity (i.e., coefficient of thermal expansion) of the solvent or solution	6.5.3
α_m	monomer value for the solvatochromic measure of hydrogen-bond donating ability	2.12
α/v	polarizability of a molecule per unit volume	3.9
β	dimensionless relative concentration of B at a given point x in the diffusion layer	11.8.3
β	isothermal compressibility	9.4
β	solvatochromic basicity parameter denoting hydrogen-bond accepting ability	2.12
β	solvent phase in partitioning (denominator)	3.6
β_m	monomer value for the solvatochromic measure of hydrogen-bond accepting ability	2.12
χ_d	proton-donor selectivity parameter with dioxane	2.7
χ_e	proton-acceptor selectivity parameter with ethanol	2.7
χ_n	dipole selectivity parameter with nitromethane	2.7
ΔA	Hemholtz free energy change	2.8
ΔC_p	change in heat capacity at constant pressure	8.5
$\Delta(C_p)_2$	partial molar heat capacity of solute 2	8.2.4
ΔG^E	molar excess Gibbs free energy of mixing	8.2.1
ΔG^s	solvatochromic standard molar Gibbs free energy of solution	2.12
ΔG^θ	standard Gibbs free-energy change	5.5.1, 9.4
ΔG_h	Gibbs free energy of hydration	2.5
ΔG_{mix}	Gibbs free energy of mixing	2.12
$\Delta G_t(w \rightarrow o)$	Gibbs free energy of transfer of a solute from water to an organic solvent	2.12
ΔG_B^E	partial molar excess Gibbs free energy of species B	2.10
$(\Delta G^s)_{cav}$	solvatochromic cavity formation contribution to the molar free energy of solution	2.12
$(\Delta G^s)_{DP}$	solvatochromic dipolarity/polarizability contribution to the molar free energy of solution	2.12
$(\Delta G^s)_{HB}$	solvatochromic hydrogen bonding contribution to the molar free energy of solution	2.12

Symbol	Description	Section
$(\Delta G^s)_{reorg}$	solvatochromic reorganizational contribution to the molar free energy of solution	2.12
$(\Delta G^s)_{total}$	total solvatochromic molar free energy of solution	2.12
ΔG_e^Θ	standard Gibbs free energy of extraction of an ion-pair species from an aqueous phase to an organic phase	7.2.4
$\Delta G_{1,5}^\Theta$	standard Gibbs free energy of self-association to the pentamer	6.3.2
ΔG_s^Θ	standard molar Gibbs free energy of solution	8.2.1
ΔG_t^Θ	standard molar Gibbs free energy of transfer	7.1
ΔG_p^Θ	contribution of polar groups to the standard Gibbs free energy of transfer	7.2.2
ΔG_{hc}^θ	contribution of hydrocarbon groups to the standard Gibbs free energy of transfer	7.2.2
$\Delta G_{v \to s}^\theta$	standard Gibbs free energy of transfer of a species from the vapor state to the solution state	5.7.9
$\Delta_{alc}^{iso} G^\theta$	standard Gibbs free energy of transfer of an alcohol from the pure liquid alcohol to an infinitely dilute solution in isooctane	6.5.1
$\Delta_{pure}^{iso} G^\theta$	standard molar Gibbs free energy of transfer of a solute from the pure solid or liquid to the Henry's law-mole fraction standard state in isooctane	3.8
$\Delta_s^{ssw} G_2^\theta$	standard molar Gibbs free energy of transfer of pure solid solute 2 to the saturated solution in water	8.5
$\Delta_{alc}^{vap} G^\theta$	standard Gibbs free energy of transfer of an alcohol from the pure liquid alcohol to the vapor state	6.5.3
$\Delta_{iso}^{vap} G^\theta$	standard Gibbs free energy of transfer of a solute from an infinitely dilute solution in isooctane to the vapor state	6.5.1
$\Delta_{iso}^{vap} G^\theta$	standard molar Gibbs free energy of transfer of a solute from the Henry's law-mole fraction standard state in isooctane solution to the vapor standard state at 1 atm	3.8
$\Delta_{pure}^{vap} G^\theta$	standard Gibbs free energy of transfer of a solute from the pure condensed state to the vapor standard state at 1 atm	3.8

Symbol	Description	Section
$\Delta_s^{sol} G_2^\theta$	standard molar Gibbs free energy of transfer of the pure solid solute 2 to the supercooled liquid state	8.5
$\Delta_{sso}^{ssw} G_2^\theta$	standard molar Gibbs free-energy of transfer of solute 2 from a saturated solution in 1-octanol to a saturated solution in water	8.5
$\Delta_{iso}^{solv} G_2^\theta$	standard molar Gibbs free energy of transfer of a solute from a 1 mol dm^{-3} solution of isooctane to a 1 mol dm^{-3} solution in another solvent, both solutions behaving as if they were infinitely dilute	3.11
$\Delta_{scl}^{sso} G_2^\theta$	standard molar Gibbs free energy of transfer of solute 2 from the supercooled liquid state to the saturated solution in 1-octanol	8.5
ΔH	change in enthalpy	6.3.1
$\Delta H^\theta / n$	standard enthalpy of association per mole of monomer associated	6.8.3
$\Delta_{soln}^{vap} H$	molar enthalpy of transfer of a solute from a solution to the vapor state	3.4
$\Delta H(\text{ns})$	contribution of nonspecific interactions to the enthalpy change in the pure-base method	5.3.8
ΔH^E	excess enthalpy	2.10
ΔH^v	enthalpy of vaporization	7.2.1
ΔH^θ	standard enthalpy change	9.4
$\Delta \overline{H}_2$	partial molar enthalpy of solution of a solute 2	8.2
ΔH_f	enthalpy of fusion	2.9
ΔH_f	standard enthalpy of complex formation at infinite dilution	5.3.8
ΔH_h	enthalpy of hydration	2.5
ΔH_s	enthalpy of solution	8.9.1
ΔH_{mix}	enthalpy of mixing	2.6
ΔH_f^T	molar enthalphy of fusion at temperature T	2.2
ΔH_e^θ	standard enthalpy of extraction of an ion-pair species from an aqueous phase to an organic phase	7.2.4
ΔH_s^θ	standard molar enthalpy of solution	8.2.1
ΔH_t^θ	standard molar enthalpy of transfer	7.1
$\Delta H_{1,n}^\theta$	standard molar enthalpy of self-association of stoichiometry n	6.3.1
ΔH_{mix}^{id}	enthalpy of mixing to form an ideal solution	2.10

Symbol	Description	Section
ΔH_{vap}^{bpt}	molar enthalpy of vaporization at the boiling point	8.2.4
$\Delta_{alc}^{iso} H^{\theta}$	standard enthalpy of transfer of an alcohol from the pure liquid alcohol to an infinitely dilute solution in isooctane	6.5.3
$\Delta_{iso}^{vap} H^{\theta}$	standard enthalpy of transfer of a solute from isooctane solution to the vapor state	6.5.3
$\Delta_{alc}^{vap} H^{\theta}$	standard enthalpy of transfer of an alcohol from the pure liquid alcohol to the vapor state	6.5.3
ΔK_m	change in the conductivity of a solution as a result of ion-pair formation	9.5
ΔR_m	group contributions to R_m	7.2.3
ΔS^E	excess entropy	2.10
ΔS^{θ}	standard entropy change	2.11.7
$\Delta \overline{S}_2$	partial molar entropy of solution of a solute 2	8.2.1
ΔS_f	entropy of fusion	8.5
ΔS_h	entropy of hydration	2.5
ΔS_{mix}	entropy of mixing	2.6
ΔS_B^M	partial molar entropy of mixing of component B	2.10
ΔS_e^{θ}	standard entropy of extraction of an ion-pair species from an aqueous phase to an organic phase	7.2.4
ΔS_s^{θ}	standard molar entropy of solution	8.2.1
ΔS_t^{θ}	standard molar entropy of transfer	7.1
$\Delta S_{H_2O}^{\theta}$	difference in entropy between two anions in water	9.4
$\Delta_{iso}^{vap} S^{\theta}$	standard entropy of transfer of a solute from an infinitely dilute solution in isooctane to the vapor state	6.5.3
ΔS_{mix}^{id}	entropy of mixing to make an ideal solution	2.6
ΔU	internal energy change	2.6, 2.8
ΔU^v	energy of vaporization	2.11.6
ΔU_{mix}	internal energy of mixing	2.6
ΔU_i^v	molar energy of vaporization of compound i	2.6

Symbol	Description	Section
$\Delta U^{\theta}_{v \to s}$	standard internal energy of transfer of a species from the vapor state to the solution state	5.7.10
Δu	overall change in pair-potential interaction energy	2.6
Δu_{mix}	overall change in pair potential energy on mixing	2.6
ΔV_{mix}	volume change on mixing	2.6
ΔX_a	thickness of the aqueous diffusion layer	9.3.4
ΔX_o	thickness of the organic diffusion layer	9.3.4
ΔY	difference between a measured property and that expected if no interaction takes place	5.3.6
$\Delta \mu^{E}_{B}$	partial molar excess chemical potential (i.e., excess free energy) of component B in a mixture	2.10
$\Delta \nu$	change in the frequency of an infrared stretching absorption band	2.11.7
$\Delta(\Delta G)$	group contribution to the Gibbs free energy change	7.1
$\Delta \Delta G(R)$	group contribution of a given hydrocarbon group in a molecule	7.2.4
$\Delta \Delta G_t$	group contribution to the standard Gibbs free energy of transfer	1.2.6
$\Delta \Delta G^{\theta}_t$ (CH_2)	methylene group contribution to the standard Gibbs free energy of transfer	7.2.1
$\Delta \Delta G^{\theta}_t$ (H$\cdots$OH)	group contribution to the standard Gibbs free energy of transfer for the hydroxyl group together with that for a terminal methyl hydrogen	7.2.1
$\Delta \Delta G^{\theta}_t$ (NH_2)	amino group contribution to the standard Gibbs free energy of tranfer	7.2.1
$\Delta \Delta G^{\theta}_t$ (OH)	hydroxyl group contribution to the standard Gibbs free energy of transfer	7.2.1
$\Delta \Delta G^{\theta}_t$ (X)	group contribution of group X to the standard Gibbs free energy of transfer	8.4
$\Delta(\Delta^{vap}_{alc} G^{\theta})$	difference between the standard Gibbs free energy of transfer of an alcohol from pure liquid alcohol to the vapor state	6.5.3
$\Delta(\Delta H)$	group contribution to the enthalpy change	7.1

Symbol	Description	Section
$\Delta\Delta H_t^\theta$ (CH$_2$)	methylene group contribution to the standard enthalpy of tranfer	7.3.1
$\Delta\Delta H_t^\theta$ (X)	group contribution of group X to the standard enthalpy change	7.2.1
$\Delta\Delta H_t^\theta$ (X)	group contribution of group X to the standard enthalpy of transfer	8.4
$\Delta(\Delta S)$	group contribution to the entropy change	7.1
$\Delta\Delta S_{ext}$	difference in the standard entropy change of extraction ΔS_{ext}^θ	9.4
$\Delta\Delta S_t^\theta$ (CH$_2$)	methylene group contribution to the standard entropy of transfer	7.3.1
$\Delta\Delta S_t^\theta$ (X)	group contribution of group X to the standard entropy of transfer	8.4
$\Delta\Delta S_t^\theta$ (X)	group contribution of group X to the standard entropy of transfer	7.2.1
δ^a	partial solubility parameter to account for the contribution of the acidic properties of a compound to its overall solubility parameter	2.7
δ^b	partial solubility parameter to account for the contribution of the basic properties of a compound to its overall solubility parameter of a compound	2.7
δ^d	partial solubility parameter to account for the dispersion force contribution of a compound to its overall solubility parameter	2.7
δ^h	partial solubility parameter to account for the hydrogen-bonding contribution to the overall solubility parameter of a compound	2.7
δ^p	partial solubility parameter to account for the polar contribution to the overall solubility parameter of a compound	2.7
δ_2	Hildebrand solubility parameter of the solute	8.5
δ_A	chemical shift of the acidic protons in the uncomplexed hydrogen-bond donor A in PMR	5.3.7
δ	actual thickness of the diffusion layer	11.2.4
δ	Hildebrand solubility parameter	2.11.6
δ_1	Hildebrand solubility parameter of the solvent	2.12
δ^2	cohesive energy density	2.12, 9.3.3
$\delta_A, \delta_B, \delta_i$	solubility parameter of species A, B, or i	2.6
δ_C	chemical shift of the acidic protons, hydrogen-bonded in the complexed form C, in PMR	5.3.7

Symbol	Description	Section
δ_{obs}	observed chemical shift of acidic protons in PMR as a result of the formation of a hydrogen-bonded complex	5.3.7
δ_1^2	cohesive energy density of the solvent	2.12
δ_2^2	cohesive energy density of the solute	2.12
ε	permittivity of a medium or solvent	2.11.2
ε_c	molar absorptivity (i.e., extinction coefficient) of a complex	5.3.5
ε_o	permittivity of a vacuum	2.11.2
η	dynamic viscosity	11.2.4
Γ	surface excess concentration	9.5
Γ_k	residual activity coefficient of group k in a solution (UNIFAC)	7.5.2
$\Gamma_k^{(i)}$	residual activity coefficient of group k in a reference solution containing only molecules of type i (UNIFAC)	7.5.2
γ	activity coefficient	2.6
γ^∞	limiting activity coefficient of a solute at infinite dilution	7.1
γ_B	activity coefficient of component B in a mixture	2.8, 8.3
γ_h	activity coefficient of component h in a liquid mixture (UNIQUAC)	7.5.1
γ_s	activity coefficient of species s	11.3
γ_X	molarity based activity coefficient of substance X	6.7
$(\gamma^\alpha)^\infty$	activity coefficient of a solute at infinite dilution in phase α	2.9
$(\gamma^\beta)^\infty$	activity coefficient of a solute at infinite dilution in phase β	2.9
γ_h^C	combinatorial contribution to the activity coefficient of the molecule h in UNIQUAC	7.5.1
γ_i^C	combinatorial contribution to the activity coefficient of the group i in UNIFAC	7.5.2
γ_h^R	residual contribution to the activity coefficient of the molecule h in UNIQUAC	7.5.1
γ_i^R	residual contribution to the activity coefficient of the group i in UNIFAC	7.5.2
γ_o^∞	limiting activity coefficient of a solute at infinite dilution in an organic solvent	7.2.2
γ_w^∞	activity of a solute at infinite dilution in water	2.9

Symbol	Description	Section
γ_w^∞	limiting activity coefficient of a solute at infinite dilution in aqueous solution	7.2.2
$\gamma_{2,0}$	activity coefficient of organic solute 2 in aqueous solution in the absence of salt B	8.11
$\gamma_{2,A}$	activity coefficient of solute 2 in pure A	8.11
$\gamma_{2,B}$	activity coefficient of solute 2 in pure B	8.11
γ_B^{th}	temperature-dependent (thermal) contribution to the activity coefficient of component B	2.10
γ_B^{ath}	temperature-independent (athermal) contribution to the activity coefficient of component B	2.10
κ	the HPLC analog of R_m, i.e. log of the capacity κ factor	7.2.3
Λ_j	molar conductivity of a salt j in solution	9.5
Λ_o	molar conductivity of a salt at infinite dilution	9.5
Λ_{BTPC}	molar conductivity of BTPC in solution	9.5
Λ_{SDDS}	molar conductivity of SDDS in solution	9.5
λ_i	molar conductivity of an ion i at infinite dilution	9.5
λ_{max}	wavelength of maximum absorption	10.8
μ_i	induced dipole moment	2.11.4
μ_1 or μ_2	dipole moment of molecule 1 or 2	2.11.3
ν	kinematic viscosity of the fluid medium	11.2.4
ν_i	chemical shift of the monomer in PMR	6.8.2
ν_n	chemical shift of the n-mer in PMR	6.8.2
$\nu_k^{(i)}$	number of groups of kind k in one molecule of i in UNIFAC	7.5.2
$\nu_m^{(j)}$	number of groups of kind m in one molecule of j in UNIFAC	7.5.2
ν_{obs}	observed chemical shift in PMR	6.8.2
ω	rate of angular rotation of a disc in radians per second	11.2.4
ϕ_A, ϕ_B, ϕ_i	volume fraction of components A, B, or i	2.6
ϕ_c	volume fraction of the cosolvent in a solvent mixture	8.3
ϕ_h	molecular volume fraction of the molecule h in a liquid mixture on the UNIQUAC model	7.5.1
ϕ_L	relative apparent molar enthalpy of dilution of a solute from a solution of a given molarity to infinite dilution	6.3.1, 6.8.3

Symbol	Description	Section
ϕ_S	rate of neutralization of species S	11.4.1
ϕ_w	volume fraction of the main solvent in a solvent mixture	8.3
π^*	solvatochromic parameter representing dipolarity and polarizability of a molecule	2.12
$\pi_{(X)}$	contribution of substituent group X to the partition coefficient of the molecule RX; i.e., Hansch hydrophobic group substituent constant	7.2.2
ρ	density of the pure substance, usually a liquid	2.11.6, 6.3.1, 11.2.4
ρ	equilibrium weight of total dissolved solutes per unit quantity of solvent or solution	10.2
ρ	Hammett polar reaction constant	5.5.2
ρ^*	Taft polar reaction constant	5.5.2
Ψ_{mn}	parameter for the energetic interaction between two groups m and n in a liquid mixture on the UNIFAC model	7.5.2
σ	Hammett polar substituent constant	5.5.2
σ	residual standard deviation	6.3.1
σ	surface tension of a liquid	8.2.1
σ	weight of solid sample added in a phase solubility analysis	10.2
σ^+	Taft polar substituent constant for an electron-supplying group	5.5.2
σ^-	Taft polar substituent constant for an electron-withdrawing group	5.5.2
σ^*	Taft polar substituent constant	5.5.2
σ_c	molecular surface energy parameter of the cosolvent in a solvent mixture	8.3
σ_i	surface free energy (surface tension) of liquid i	2.8
σ_m	molecular surface energy parameter of a solvent mixture	8.3
σ_w	molecular surface energy parameter of the main solvent in a solvent mixture	8.3
σ_1^d	contribution of dispersion forces to the surface tension of a liquid solvent 1	8.2.1
σ_w^h	interfacial tension between a hydrocarbon group of a solute and water or other solvent	8.3

Symbol	Description	Section
σ_{12}	interfacial tension between liquid 1, i.e. water, and liquid 2, i.e. a saturated aliphatic hydrocarbon	8.2.1
σ_{AB}	interfacial free energy (interfacial tension) between two liquids, A and B, or two molecules, A and B, e.g., solvent and solute	2.8, 8.3
σ_{AB}^h	microscopic interfacial free energy of interaction (interfacial tension) of a hydrocarbon group in the solute molecule B and the solvent A (water)	8.3
σ_{AB}^P	microscopic interfacial free energy of interaction (interfacial tension) between a polar group in the solute molecule B and the solvent A (water)	8.3
τ_{hj}	interaction parameter between h and j molecules in UNIQUAC	7.5.1
τ_{jh}	interaction parameter between j and h molecules in UNIQUAC	7.5.1
Θ	temperature in °C	2.9
θ_n	molecular surface area fraction of the molecule h	7.5.1
θ_m	group surface area fraction of group m in a mixture in UNIFAC	7.5.2
θ_m	melting point temperature in °C	8.5
ξ	dimensionless distance at point x within the diffusion layer with respect to the total thickness h, i.e. x/h	11.8.3

SUPERSCRIPTS (STATE)

a   -acidic (i.e. hydrogen bond donor) partial solubility parameter
     -adhesion
ath   -athermal, i.e. temperature dependent, contribution to the activity coefficient
b   -basic (i.e. hydrogen bond acceptor) partial solubility parameter
     -normal boiling point (at 1 atm pressure)
bpt   -normal boiling point (at 1 atm pressure)

C — combinational contribution to the activity coefficient in UNIQUAC and UNIFAC

c — based on molar concentration (i.e. molarity, mol/dm^3 of solution), the standard state of unit activity being a hypothetical 1 mol dm^{-3} solution which behaves as if it were infinitely dilute
— cohesion

D — sodium D line at $\lambda = 589$ nm

d — dispersion force contribution in the dispersion partial solubility parameter
— dispersion contribution to interfacial tension

E — excess thermodynamic function

eq — equilibrium

h — hydrocarbon group contribution, e.g. to solubility parameter and to molecular surface area

i — molecular species in UNIQUAC and UNIFAC

id — ideal solution value

iso — in isooctane solution (usually at infinitely dilution)

j — molecular species in UNIQUAC and UNIFAC

M — mixing (rarely)

m — based on molality (i.e. mol/kg of solvent), the standard state of unit activity being a hypothetical 1 mol kg^{-1} solution which behaves as if it were infinitely dilute

p — polar group contribution, e.g. to solubility parameter and to molecular surface area

R — based on Raoult's law, the standard state being the pure liquid, supercooled and hypothetical, if necessary
— residual contribution to the activity coefficient in UNIQUAC or UNIFAC

s — solid
— solubility
— solution

sat — saturated solution

scl — supercooled liquid as the final state

solv — solvent

sso — saturated solution in 1-octanol as the final state

ssw — saturated solution in water as the final state

T — at absolute temperature T

th — thermal (i.e. temperature dependent) contribution to the activity coefficient

v — vaporization

vap — vapor state

x — based on the mole fraction, the standard state being a hypothetical solution in which the solute is present at unit mole fraction but which behaves as if it were infinitely dilute

α — of phase "alpha"

β	-of phase "beta"
θ	-standard state
*	-alternative reference interaction when applied to h_A^* or h_D^*
	-pure state (commonly)
	-solvatochromic dipolar/polarizable parameter, when applied to π^*
	-Taft polar reaction constant, when applied to ρ^*
	-Taft polar substituent constant, when applied to σ^*
+	-Taft polar substituent constant for electron supplying groups, when applied to σ^+
−	-Taft polar subtituent constant for electron withdrawing groups, when applied to σ^-
25	-at 25°, i.e. 298.15 K
∞	-infinity
	-at infinite dilution

SUBSCRIPTS (SPECIES, PROCESS, INITIAL STATE)

A	-acid (rarely)
	-component of a mixture
	-hydrogen-bond acceptor
a	-aqueous
	-acidity
AA	-interaction between two A molecules
AB	-interaction or complex between A and B molecules
ai	-aqueous interface
alc	-pure liquid alcohol
app	-apparent
aq	-aqueous
ass	-self-association
B	-base
	-burst (time)
	-component of a mixture
b	-bulk (concentration)
	-normal boiling point (at 1 atm pressure)
BB	-interaction between two B molecules
B–H	-Benesi–Hildebrand
C	-complex
c	-complex
	-cosolvent
	-critical
calc	-calculated
cav	-cavity
CMR	-Carter, Murrell, and Rosch value
D	-dimerization based on activity
	-hydrogen bond donor
	-sodium D line at $\lambda = 589$ nm

d	-dimerization based on molarity
	-distribution
	-hydrogen-bond donation to dioxane in χ_d
DA	-donor-acceptor hydrogen-bonded complex
D_2A	-donor$_2$-acceptor hydrogen-bonded complex
DA_2	-donor-acceptor$_2$ hydrogen-bonded complex
DP	-dipolar/polarizability
E	-electronic
e	-extraction
	-hydrogen-bond acceptance from ethanol in χ_e
eq	-equilibrium
exp	-experimental
ext	-extraction
F	-factor as in R_F
f	-final
	-free, i.e. uncomplexed, species
	-fusion
fm	-formation
g	-gas or vapor
g/w	-gas or vapor /water
h	-at $\chi = h$, i.e., in the bulk solution at the end of the diffusion layer of a dissolving solid
	-hydration
	-molecular species in UNIQUAC
HB	-hydrogen bonding
hc	-hydrocarbon
hc/w	-hydrocarbon/water
i	-induced
	-initial
	-interface
	-molecular species (in general)
iso	-in isooctane solution (usually at infinite dilution)
j	-molecular species (in UNIQUAC)
k	-functional group in UNIFAC
L	-lag (time)
	-ligand or cosolvent which forms a complex with a solute S
m	-functional group in UNIFAC
	-melting
	-mixed solvent system, water + cosolvent
	-molar
	-monomer
	-stoichiometric number for self-association (or complexation, rarely)
max	-maximum
mix	-mixing

n	-functional group in UNIFAC
	-number of methylene groups (rarely)
	-stoichiometric number for complexation or self-association
	-dipolar/polarizability interaction with nitromethane in χ_n
NW	-Noyers–Whitney
o	-absence of interacting species
	-at $\chi = O$, i.e., immediately next to the surface of the dissolving solid at the start of the diffusion layer
	-1-octanol
	-organic phase or solution
	-organic solvent
	-reference compound
	-uncomplexed species in solution
obs	-observed
oi	-organic interface
org	-organic phase
o/w	-organic solvent or 1-octanol /water
p	-constant pressure
	-polar
pure	-in the pure state
p/w	-phospholipid/water
R	-reaction (rate constant)
reorg	-reorganization
RH	-unsubstituted species (e.g. alkane)
RX	-substituted species (H replaced by group X)
S	-solute undergoing dissolution
s	-solid
	-solubility (in solubility product K_s)
	-solution
	-solvent (rarely)
	-species
	-steric
SA	-solvent-hydrogen bond acceptor complex
SD	-solvent-hydrogen bond donor complex
scl	-supercooled liquid as the initial state
soln	-in solution (usually at infinite dilution)
sso	-saturated solution in 1-octanol as the initial state
ssw	-saturated solution in water as the initial state
T	-transport (rate constant)
t	-transfer
v	-vaporization
vap	-vapor
v→s	-transfer from the vapor state to the solution state (usually at infinite dilution)
w	-water
wk	-van der Waals parameter of functional group k in UNIFAC

w→o   -transfer from a solution in water to a solution in an organic solvent, e.g. 1-octanol

X   -functional group (rarely)

       -molecule (rarely)

α -liquid phase "alpha"

β -liquid phase "beta"

1   -first order (rarely)

       -monomer

       -solvent of a solution (commonly)

       -solvent 1

1:n   -1:n complex, e.g. SL_n or DA_n

1:1   -1:1 complex, e.g. SL or DA

1:2   -1:2 complex, e.g. SL_2 or DA_2

1,n   -self-associated n-mer species, e.g. A_n

1,5   -self-associated pentameric species, e.g P_5

2   -second order (rarely)

       -dimer

       -solute in a solution (commonly)

       -solvent 2

2:1   -2:1 complex, e.g S_2L or D_2A

5   -pentamer, e.g of phenol

25   -at 25°C, i.e. 298.15K

∞ -infinity

       -stepwise equilibrium with stoichiometry increasing from (n-1) to n as in K_∞

Chapter **I**

INTRODUCTION AND SURVEY

1.1 GENERAL CONSIDERATIONS

In routine chemical procedures and in many fields of research, a knowledge of solubility values has a broad application. Solubility information is particularly vital in areas such as chemical processing, the development and use of analytical methods, the prediction of the ecological impact of chemicals, and the assessment of transport and distribution problems with organic species. This book is primarily concerned with understanding and predicting the solubility of organic species in various solvents, including water.

Much of the current research in the pharmaceutical and biological sciences requires knowledge of solubility behavior of bioactive compounds such as drugs, herbicides, pesticides, disinfectants, toiletries, cosmetics, food additives, and environmental contaminants, all of which fall within the scope of this book. This work, however, does not discuss the solubility of inorganic substances or of polymeric or macromolecular materials, since these areas are not within the scientific domain of the authors.

In many instances, chemists need to know how much of a particular organic substance will dissolve in a given quantity of a certain solvent at a given temperature, or in what proportions a certain solute will distribute or partition itself between two immiscible solvents under equilibrium conditions. Guidelines are needed which will allow solubility and partition behavior of organic compounds to be estimated without recourse to experiments. The major concern of this book is in developing such guidelines for the prediction of solubility.

Solubility behavior and methods for predicting solubilities have been previously studied by many laboratory workers and theoreticians. The reader is encouraged to consult these contributions, which are referenced throughout the book. A historical survey of the early work on solubility, including that of polymers, has been made by Mellan (1968). One of the earliest rules of solubility introduced by medieval alchemists was stated in the Latin phrase "similia similibus solvuntur," which is often translated as "like dissolves like." Although this rule has been used in the past as a guide by many workers and is embodied in the concepts of ideal and regular solutions, we shall see that for most real systems it has serious deficiencies in its practical applications. An excellent text, which includes both recent and earlier concepts; has been written by James (1986).

In most scientific texts it is necessary or at least desirable to make some value judgment concerning opposing scientific theories. In this work our

1

conclusions are guided by the philosophical teachings of William of Occam (1343), as embodied in the Latin sentence "Entia non sunt multiplicanda prater necessitatem." This famous dictum, known as "Occam's Razor," may be translated as "Entities are not to be multiplied without necessity." In the modern scientific domain this statement may be paraphrased as follows: the theory with the fewest unproven assumptions is the most acceptable. This concept may be supported by statistics, in which it is known that the addition of more adjustable parameters into an equation, while providing a better fit to experimental data, may result in some of the parameters becoming statistically insignificant.

In this work we have tried to cover in particular the elements of the thermodynamics of solubility, and to present in an organized fashion some of the concepts and experimental results concerning solubility which have been developed in the senior author's laboratories during the past four decades. In this chapter we will introduce the reader to the general scope of the material to be covered in this book.

1.2 OUTLINE OF SUBSEQUENT CHAPTERS

1.2.1 Chapter 2

Chapter 2 describes the classical concepts in the field of solubility derived from the efforts of such early pioneers as Raoult, Henry, van't Hoff, and others, who defined the properties of ideal and dilute solutions. The influence of inter-molecular forces on solubility are reviewed. Deviations from ideality resulting from imbalances of intermolecular interactions as set forth in the regular solution theory are discussed, particularly as expressed in the writings of Scatchard (1931) and Hildebrand, Scott, and co-workers (1950, 1962, and 1970). Extensions of regular solution theory including concepts of three-dimensional solubility parameters and an introduction to the recent molecular and group-surface area approach of Yalkowsky, Amidon, and Valvani are described. The final section in Chapter 2 describes the approach of Taft and co-workers (Taft et al., 1985a, b) to predict solubilities from molecular parameters that are derived from considerations of the underlying molecular forces. The basic threads of thermo-dynamic theory in solution chemistry introduced by Gibbs (see Lewis and Randall, 1961; Koltz, 1964; and Prausnitz, 1969) are presented throughout the chapter in our attempt to tie together the various topics outlined above.

Also in Chapter 2, solubility is treated as an equilibrium phenomenon. Solubility is under the same fundamental thermodynamic control as each of the following equilibrium quantities to which it is intimately related:

1. Thermodynamic activity and activity coefficient.
2. Partition coefficient and osmotic pressure in liquid–liquid phase equilibria.
3. Surface phenomena, such as absorption, which represent equilibria at interfaces.

4. Melting-point and solid–liquid phase equilibria which reflect the stability of a crystal lattice.

5. Vapor pressure and liquid–vapor or solid–vapor phase equilibria which reflect fugacity (i.e., escaping tendency).

6. Chromatographic constants, such as R_F value and retention time, which depend on partition, adsorption, or vapor pressure.

1.2.2   Chapter 3

In a thermodynamic treatment of solubility the energetics of a substance in a dissolved state is compared with that in a defined reference state or "standard state" for that substance, as discussed at length in Chapter 3. Since current laboratory research necessitates comparisons of the solubility and thermo-dynamic properties of a wide variety of organic compounds, a thermodynamic standard state is generally defined for all substances. A frequent choice for the standard state of a solute is pure liquid at 25°C. For substances that are solid under ambient conditions, this standard state is a supercooled or hypothetical state. Unfortunately, the intermolecular interactions are often different and unpredictable for different substances in this state. Consequently, the prediction of solubility in interactive solvents is difficult.

To circumvent this problem, we define the standard state of the solute in terms of an infinitely dilute solution in a paraffinic hydrocarbon solvent in which the only interactions experienced by the solute(s) are nonspecific and often predictable. In Chapter 3 the reasons for this choice of a standard state and its consequences are discussed.

A fine control can be exercised on the solubility of a substance by adding a second solvent, or cosolvent, to the first. The solubility of organic compounds in simple binary solvents has been examined, thermodynamically for example, by Bertrand, Acree, and co-workers (Acree and Bertrand, 1981; Acree, 1984; Judy et al., 1987), using a theoretical treatment known as the nearly ideal binary solvent (NIBS) approach. It should be emphasized that the NIBS theory often deals with fine differences in solubility and does not appear to be appropriate for predicting solubilities in interactive solvents with strong solute–solvent inter-actions. For this reason, detail of the NIBS approach has been omitted. In the solubility predictions attempted throughout this book, the level of accuracy for interactive solvents is within a factor of 2 or 3 of the actual solubility.

1.2.3   Chapter 4

Chapter 4 introduces the role of specific interactions in solubility behavior, such as the solubility of polar solutes in various organic solvents. Perhaps because the concept is relatively old, it has had few supporters in recent years. The regular solution theory, which assumes that the main forces acting on solutions are the nonspecific London dispersion forces, is shown to be incapable of predicting solubility data obtained experimentally. It is further shown that the solubility behavior of polar solutes in some pure solvents, and in binary mixtures of polar and nonpolar solvents in most practical situations, can be explained and predicted satisfactorily in terms of specific interactions—in particular, hydrogen bonding between polar species.

To select the most appropriate solvent for a particular problem involving solubility or separation, we shall see that specific interactions between the solute and solvent in question often offer the most challenging opportunities. This approach has been given the major emphasis in this book.

1.2.4   Chapter 5

Chapter 5 discusses in greater detail the enormous influence of hydrogen bonding and resulting complexation between a polar solute and a polar cosolvent on solubility in the presence of a nonpolar solvent. Hydrogen bonding can contribute about $-4\,\text{kcal}\,\text{mol}^{-1}$ net ($-17\,\text{kJ}\,\text{mol}^{-1}$) to the enthalpy of inter- action between solute and cosolvent, which is much greater than that derivable from dispersion forces. Hydrogen bonding is also associated with an appreciable steric requirement for the interacting molecules. Although the resulting reduction in entropy partially offsets the reduction in enthalpy, the overall free-energy change is appreciably exergonic. As a result, hydrogen bonding occurs whenever possible and increases the solubility by a factor of about 10^3–10^4 times that expected from dispersion interactions only. Equations 1.1–1.3 are used in Chapter 5 to explain how the stoichiometry n of complexation and the stability constant (formation constant or K value) can be calculated when the concen- tration of the cosolvent A and the solubility of the solute B are known:

$$nA + B \rightleftharpoons A_nB \qquad (1.1)$$

$$K \cong \frac{[A_nB]}{[A]^n \cdot [B]} \qquad (1.2)$$

$$\begin{aligned} \text{solubility of solute B} = &\text{ solubility of free B} \\ &+ \text{concentration of complexed B} \end{aligned} \qquad (1.3)$$

Conversely, if the stoichiometry and stability constant of the complex are known, the solubility of a hydrogen-bonding solute B in the presence of a given concentration of hydrogen-bonding cosolvent A can be predicted. The stability constants of hydrogen-bonded complexes are shown to be related to molecular structure by means of the linear free-energy relationships of Brønsted, Hammett, Taft, Hansch, and others. Hence, appropriate substituent parameters and reaction constants may be used to predict the stability constant of the complex. Consequently, the solubility of a substance in a solvent of given composition can often be determined.

When the interacting solvent molecule consists of a functional group attached to the end of an alkyl chain, it is frequently found that the length of the hydro- carbon chain exerts little or no influence on the stability constant of the complex or on the solubility of the solute in the presence of a given concentration of cosolvent. Furthermore, the complexation constant appreciably increases when there is a good steric fit, corresponding to a "lock-and-key" relationship between the putative interacting functional groups of the solute and the solvent or cosolvent. These facts emphasize the overwhelming role of the specific interaction between solute and cosolvent on the solubility of the solute in such systems, and examples are provided in Chapters 4 and 5.

1.2.5   Chapter 6

Compounds such as alcohols and phenols, which act as both donors and acceptors of hydrogen bonds, are in common use and are capable of self-association. As discussed in Chapter 6, the structure of the solvent, particularly that determined by self-association, exerts a profound influence on the solubility and related properties of solutes in solution. Knowledge of the stoichiometry and equilibrium constant for self-association is crucial in order to understand and predict the influence of self-association on these properties.

Several stoichiometries of self-association (self-association models) have been proposed in the literature for a given cosolvent system; some of the relevant models are discussed critically in Chapter 6. Methods for determining the stoichiometry and equilibrium constants for self-association are also described and examples of the prediction of solubility of polar solutes in the presence of common self-association solvents such as alcohols are included.

The diffusional transport of bioactive substances in solution is a kinetic phenomenon which is related somewhat indirectly to the equilibrium phenomenon of solubility. However, both phenomena depend on intermolecular interactions in solution and are of the greatest relevance in accounting for the observed effects of bioactive substances in chemical and biological systems. For these reasons, coupled with the particular interests of the authors, a section on the influence of specific intermolecular interactions on diffusional transport has been included at the end of Chapter 6.

1.2.6   Chapter 7

The solubility of solutes in water and their partition coefficients (organic solvent/water) are difficult to predict theoretically, although elegant and heroic attempts have been made, as outlined in Chapter 8. Partly to lay the groundwork for Chapter 8, Chapter 7 presents and discusses a semiempirical method, known as the "group-contribution" approach, which has a wide practical application and predictive value for various phase-transfer processes.

The concept of group contributions is indebted to the pre-1940 pioneering work of Langmuir (1925) and Butler (1962). The group-contribution approach is based on the assumption that each of the individual groups in a molecule is associated with a specific constant increment, $\Delta\Delta G_t$, in the standard free energy of transfer, ΔG_t^θ, from one phase to another. For example, in the case of octanol, $CH_3(CH_2)_7OH$,

$$\Delta G_t^\theta = \Delta\Delta G_t(CH_3) + 7\Delta\Delta G_t(CH_2) + \Delta\Delta G_t(OH) \tag{1.4}$$

Since

$$\Delta G_t^\theta = -2.303RT\log Y \tag{1.5}$$

where Y is the equilibrium constant for phase transfer, such as partition coefficient, solubility, or Henry's constant, then,

$$\log Y = \Delta \log Y(CH_3) + 7\Delta \log Y(CH_2) + \Delta \log Y(OH) \tag{1.6}$$

where each $\Delta \log Y$ term is the respective group contribution to Y (in the logarithmic form). This example illustrates how group contributions, in the form ΔG_t^θ or $\log Y$, may be used to predict solubilities and partition coefficients. It follows from Eq. 1.5 and 1.6 that

$$\Delta\Delta G_t = -2.303RT\Delta \log Y \tag{1.7}$$

Because of the intimate relationships (Eqs. 1.5 and 1.7) between free energy and $\log Y$, Eqs. 1.4 and 1.6, used in group contributions, are examples of "linear free-energy relationships," which find wide application in medicinal and biological chemistry. The group-contribution concept also applies to enthalpies (ΔH) and entropies (ΔS) of transfer as well as to free energies. In real life, of course, the various functional groups often interact with each other and do not always act independently.

To allow for the mutual interactions of the functional groups and limitations to their independence, various attempts have been made to assign two or more types of group contribution to each group. Examples include the UNIQUAC and UNIFAC approaches of Fredenslund et al. (1977b), Gmehling et al. (1978), Prausnitz, and others. Because a spectrum of mutual interactions is considered, the predictions agree very well with experiment. Considerable computational effort may, however, be required. These developments receive only cursory attention here. Readers interested in these powerful but more complicated extensions of group contributions will wish to refer to the works cited and to a publication by Ochsner and Sokoloski (1985), who are applying the procedure to the prediction of the solubility of pharmaceuticals.

The temperature dependence of equilibrium constants K or Y (or of standard free-energy changes) provides enthalpies of the process, be it reaction or phase transfer, achieved by applying the well-known van't Hoff isochore equation, which may be written in the following form at constant atmospheric pressure p:

$$\left(\frac{\partial \ln(K \text{ or } Y)}{\partial T}\right)_p = \frac{\Delta H}{RT^2} \tag{1.8}$$

If ΔH is assumed to be constant, which will be a good approximation over a relatively small range of temperature, this equation may be integrated to give

$$\ln(K \text{ or } Y) = -\Delta H/RT + I \tag{1.9}$$

where I is an integration constant. Then ΔH may be calculated from the slope of the plot of $\ln(K$ or $Y)$ against $1/T$. The appreciable standard errors in the slopes of many of the van't Hoff plots lead to corresponding errors in ΔH. Consequently, we are led to treat with some skepticism ΔH values so determined and to favor calorimetric values of ΔH, which usually have much lower standard deviations.

1.2.7   Chapter 8

In biological and pharmaceutical literature lipoidal materials such as hydrocarbons, oils, fats, and other substances consisting of long alkyl chains are frequently described, in relation to their solubility in water, as being lipophilic or hydrophobic, words which give the impression that these species have a strong affinity for each other, but a weak affinity for water. On the contrary, the intermolecular interactions between two lipoidal molecules are actually weaker than the weak interactions between lipoidal and water molecules. Therefore, the word lipophilic is somewhat misleading. The consequences of these bonding forces and the role of enthalpy and entropy in the overall free-energy changes responsible for the hydrophobic interactions are included in Chapter 8.

The importance of the predictive value of the group-contribution approach in the estimation of the aqueous solubility of organic compounds is brought out in Chapter 8. To give one possible line of argument, Henry's law constants may be predicted, especially for compounds in paraffinic hydrocarbons or water, by applying the group-contribution approach to free energies of transfer of solutes from a solution to the vapor state. From the predicted or measured Henry's law constant of a compound in a paraffinic hydrocarbon, together with knowledge of the vapor pressure of the compound, the escaping tendencies of the compound in the pure state relative to the chosen standard may be calculated. From this activity value, the solubility of the compound in the saturated hydrocarbon may be predicted. Similarly, from the predicted Henry's constant of a compound in water, together with a knowledge of the vapor pressure of the compound, the solubility of the compound in water may be predicted. Since this method of solubility prediction requires knowledge of the vapor pressure of the solute, it is only applicable to relatively volatile compounds.

As indicated above, partition coefficients may also be predicted by applying the group-contribution approach to free energies of transfer from water to a given organic solvent. Partition coefficients have been predicted for a wide variety of solutes and organic solvents and are particularly important quantities for chemists concerned with organic compounds. Since the most commonly used partition coefficients are those between water and organic solvents, their prediction is specifically considered in Chapters 7 and 8.

For steroids and other substances which are sparingly soluble in water, the direct experimental measurement of the aqueous solubility may not be feasible. By dividing the experimental or predicted value of the solubility in the paraffinic hydrocarbon by the predicted partition coefficient, a predicted value has been found to agree well with the experimental value of the aqueous solubility in those cases where the latter is accessible. Even when it is possible to measure directly the aqueous solubility of solutes, a number of practical problems arise which are also discussed in Chapter 8 together with methods of overcoming them.

The partition or the distribution of solutes between immiscible solvents is analogous to solubility. The first deals with the solution–solution equilibrium, while the second concerns the solid solute–solution or liquid solute–solution equilibrium. Partition or distribution also give rise to further methods of analysis, separation, or purification, of which the most widely used are chromatographic techniques. In the pharmaceutical sciences, the partition coefficient and distribution of a drug between the aqueous and lipoidal phases within the body are vital for understanding drug absorption and distribution.

1.2.8   Chapter 9

Under certain conditions one or more electrolytes can be converted into ion pairs which behave as unionized molecules. These unionized molecules may be polar or effectively nonpolar, depending on the nature of the interacting ions. The presence or lack of polarity may have important practical application to solubility, solution properties, and transport behavior of the interacting substances. Ion pairs can also be exploited theoretically for deducing the group contributions to solubility and partitioning.

The changes in solubility and phase-transfer properties which result from ion pairing are considered in Chapter 9. Cationic drugs and anionic surface-active agents, which are more soluble in polar solvents such as water than in nonpolar solvents, interact with each other to give effectively nonpolar species. These nonpolar species are conversely more soluble in nonpolar solvents than in polar solvents. Also, solubility of an ion pair consisting of a large organic cation with a small inorganic anion, such as an amine or quaternary ammonium salt, in a nonpolar solvent is enhanced by the presence of a suitable electrophilic solvating agent such as chloroform. On the other hand, an ion pair consisting of a large organic anion with a small inorganic cation, such as a Grignard reagent or other organometallic compound, experiences an enhanced solubility in the presence of a nucleophilic solvent such as diethyl ether.

1.2.9   Chapter 10

The term "phase-solubility analysis" denotes procedures for determining the purity of a substance and also for determining relative amounts, solution properties, interactions, and complexation of solutes in mixtures. Phase-solubility analysis, discussed in Chapter 10, is a direct application of solubility phenomena which, in its simplest form, can be used for the quantitative analysis of a mixture.

Also included in Chapter 10 is a treatment of the phenomenon of complexation between the solute and ligand, particularly in aqueous solutions. A ligand represents a third component or cosolvent dissolved in the solvent. Complexation in water relates directly to the more theoretical material discussed in Chapters 4 and 5. The phenomenon of complexation examined by phase-solubility analysis may be used to control the apparent solubility of a solute by the addition of an appropriate quantity of a ligand. Phase-complexation analysis can also be used to determine the stoichiometry and equilibrium constants of a complex.

1.2.10   Chapter 11

In science and technology, particularly in the field of pharmaceutics, it is sometimes more important to understand the kinetics of dissolution than the equilibrium solubility. Chapter 11 deals with the kinetics and mechanism of dissolution rates and the factors which affect them. Of course, dissolution will only occur if the thermodynamic conditions for a spontaneous process (i.e., reaction moving toward an equilibrium state represented by a saturated solution) are satisfied. Thus the dissolution rate depends on solubility and is therefore treated after solubility has been considered.

In the pharmaceutical sciences the dissolution rate of an orally administered drug (Chapter 11) and the permeation rate of a topically administered drug across the skin barrier (end of Chapter 3) are of vital importance. The dissolution and permeation rates, if rate-limiting, can control the absorption rate of a drug, and hence can determine the therapeutic efficacy and potential toxicity of the drug.

In practice, the dissolution rate of a solid is usually controlled by diffusion across a thin film in contact with the solid surface. In Chapter 11 particular attention is focused on the influence of pH, acids, bases, and buffers on the dissolution rates of weak acids (and, by implication, of weak bases too). The dissolution of weak acids is found to be most influenced by the intrinsic solubility and pK_a of the acid and by the pK_a and concentration of the base. Surprisingly perhaps, the bulk pH itself usually plays a subordinate role in a buffered medium. Also considered are the possible influence of slow reversible ionization of a carbon acid on its dissolution behavior and the effects of reversible molecular complexation and irreversible reaction on dissolution rates. The thermodynamic background is covered throughout the book, particularly in Chapters 2, 3, 6, 8, and 9.

1.3   BOUNDS TO THE TEXT

A number of aspects of solubility have been omitted from the book since their inclusion would have yielded a volume of unreasonable size or would have necessitated a briefer discussion of topics deemed of great importance by the authors. Omissions were made after careful consideration and justification. Statistical mechanical calculations have been omitted since they do not readily lend themselves to the simple predictions of solubility in practical situations. Those interested in statistical thermodynamic treatments may wish to consult texts by Prigogine (1957), Kirkwood (1968), and Ben Naim (1980) and a review by Pierotti (1976).

Regular solutions and solubility parameters are featured less strongly in this book than in recent monographs and research papers. Regular solution theory assumes that the intermolecular interactions in solution are nonspecific in nature and consist essentially of London dispersion forces. However, this assumption greatly restricts the utility of the theory for our purposes, since many organic compounds are highly polar and are therefore capable of undergoing specific interactions in solution. Such interactions include hydrogen bonding or charge-transfer interactions in addition to dipole-induced forces.

Specific interactions are considered particularly important because they can be exploited in choosing the correct solvent for a particular substance. Of course, specific interactions due to hydrogen bonding and polarity can be allowed for by introducing additional parameters, such as three-dimensional solubility parameters. However, these additional parameters stretch regular solution theory much further than the originators intended. This point is discussed further in Chapter 2. Texts on regular solution theory include the classical ones by Hildebrand and co-workers (1950, 1962, 1970) and a monograph by Shinoda (1978). Extensions of solubility parameter theory have been reviewed by Barton (1975) and Snyder (1978 and 1980) and include the work of Martin et al. (1981).

A third omission is the solubility behavior of high polymers. Such polymer solutions are unusual in that the solute molecules are much larger than the solvent molecules. Consequently, the entropy of mixing behaves differently from that when both solute and solvent molecules are of similar size (Flory, 1941, 1942; Huggins, 1941, 1942). For this reason high-polymer solubility has been omitted from this book. Polymer solutions have been treated in the classical text by Flory (1953); in texts by Harris and Seymour (1977), Eisenberg (1976), and Morawetz (1975); in elementary monographs by Shinoda (1978) and Blackadder (1975); and in a review by Bagley and Scigliano (1976).

The solubility behavior of surface-active agents has also been omitted. Surfactants give rise to association colloids, which include micelles and liquid crystalline phases, and they stabilize emulsions and suspensions. It therefore seems appropriate to leave the consideration of the solubility of solutes in surfactant solutions to texts on solubilization and surface activity. The interested reader may wish to refer to Elworthy et al. (1968), Shinoda (1978), and Attwood and Florence (1983).

The solubility of gases has been only lightly reviewed in this text, since the vast majority of organic compounds are solids or liquids, whereas only a small number are gases at room temperature (e.g., methane, formaldehyde, and ethyl chloride). The solubility of gases is considered explicitly in connection with Henry's law, which applies to the vapor–solution equilibrium and was discovered from experiments dealing with the solubility of gases in water.

The interested reader may wish to refer to Gerrard (1976), reviews by Wilhelm and Battino (1973), Clever and Battino (1976) and Wilhelm et al. (1977), and the text dealing with regular solutions mentioned above. Since most gas molecules experience only very weak intermolecular forces, it is generally expected that they will not undergo specific interactions with most solvents. Obvious exceptions include aqueous solutions of formaldehyde, ammonia, and hydrogen chloride, all of which have molecules that form much stronger hydrogen bonds with the water molecules than with each other. In general, with the exceptions noted above, the solubility of gases in liquids can be adequately treated on the basis of regular solution theory.

A sixth omission is the phenomenon of solid solutions, that is, the solubility of solids, liquids, and gases in an intimate mixture of individual molecules in the solid state. Solid solutions often have properties which are desirable for the pharmaceutical sciences and for technologies dealing with plastics and metals. Although solid solutions can have useful properties, these solid phases are frequently in metastable states. Particular molecular arrangements and inter- actions which are stable at a higher temperature are thermodynamically unstable at lower temperatures. The higher activation energies involved in molecular rearrangements within the rigid structure characteristic of the solid state kinetically stabilize the metastable state such that the equilibrium solid solubility is often exceeded.

Those wishing to study solid solutions are referred to textbooks on phase equilibria, such as the classical text by Bowden (1938), which represents an elementary treatment, and more recent books by Findlay et al. (1951) and Reisman (1970). The pharmaceutical applications of solid solutions and eutectics, which are collectively referred to as dispersion systems, have been reviewed by Chiou and Reigelman (1971).

Although certain aspects of solubility have been omitted, this book does include topics which have not hitherto been included in monographs on solubility, such as phase-solubility analysis, complexation analysis, self-asso- ciation of solutes in solution, ion pairs, group contributions to phase transfer and to solubility in water, and dissolution rates. The authors hope that the inclusion of these topics will draw the reader's attention to the importance of these aspects of solubility and lead to further investigations in these areas.

Chapter II

SOLUBILITY, INTERMOLECULAR FORCES, AND THERMODYNAMICS

2.1 SOLUBILITY AS AN EQUILIBRIUM PROCESS

In this chapter the thermodynamic basis of the solubility equilibrium of solids is treated. This approach most logically constitutes the starting point for any rational treatment of solubility behavior. However, its practical utility is severely limited. Predictions of solubility based on the classical thermodynamics of ideal and regular solutions are compared in this chapter with some experimentally observed values. The solubility of polymorphs is discussed from the same viewpoint. We have also included the possible impact of various intermolecular interactions on the energetics of solvent–solvent and solute–solvent interactions.

The concept of solubility implies that the process of solution (or dissolution) has reached an equilibrium state such that the solution has become saturated. The dissolution process is associated with changes in thermodynamic quantities such as energy, degrees of disorder (i.e., entropy), and volume. The subsequent sections of this chapter present certain principles of equilibrium thermodynamics and other important concepts which are necessary for the prediction of solubility.

Solubility may be expressed in any appropriate units, depending on the composition of the saturated solution at a given temperature, and at a given partial pressure in the case of a gas. Typical quantities used to define the composition, strength, concentration, or solubility are

(a) mole fraction, x; mass or weight fraction, w; volume fraction, ϕ; or the corresponding percentages;

(b) the ratio of quantity of solute to solvent, such as the mole or mass ratio;

(c) mass concentration, ρ, units: $g\,dm^{-3}$ or $g\,L^{-1}$ of solution;

(d) molar concentration or molarity, c, units: $mol\,dm^{-3}$ or $mol\,L^{-1}$ of solution, or simply $mol\,L^{-1}$ or M; because the material in this book has been abstracted from diverse sources, we hope the reader will forgive us for using all of these symbols, since their meanings are identical;

(e) molarity, m, units: $mol\,kg^{-1}$ of solvent; and

(f) normality, N units: equivalents per liter of solution.

For sparingly soluble solutes, the density of the solution may be assumed to be equal to that of the solvent such that the volume of the solution and of the solvent may be interchanged. This is done when the solubility is expressed in milligrams percent, that is, milligram of solute per 100 mL of solution or solvent at a given temperature. In some pharmacopoeias, for example, in the *U.S. Pharmacopeia* (1985), earlier editions of the *British Pharmacopoeia*, and in *Martindale's Extra Pharmacopoeia* (1983), solubility is expressed as a *dilution*, that is, in a reciprocal fashion, namely as the number of parts (mL or cm^3) of solvent which will form a saturated solution with one part of solute (1 mL of liquid or 1 g of a solid) at 25°C (or 20°C).

Extensive solubility data are available from the following sources: Seidell (1958), Yalkowsky et al. (1987), Stephen and Stephen (1963, 1964), and a continuing series of IUPAC publications. The compilers have initially examined each literature source before presenting the data. The latter works (Stephen and Stephen 1963, 1964; IUPAC Publication on solubility) are the result of the deliberations of expert committees (Akademiia Nauk SSSR, Chief Coordinator V. V. Kafarov, and IUPAC).

Detailed data on the physical properties of solvents have been compiled by Riddick and Bunger (1970), Riddick et al. (1986), Mellan (1977), and Durrans (1971). Additional physicochemical data for all types of substances, including potential solutes and solvents, are available in the *Merck Index* (Windholz et al., 1983) and in the *CRC Handbook of Chemistry and Physics* (Weast et al., 1989).

Solubility and partition behavior are inverse reflections of the escaping tendency of the solute from the solvent, whereas vapor pressure is related directly to escaping tendency. Lewis and Randall (1961) introduced the term fugacity, *f*, as a measure of escaping tendency. Fugacity, as defined below, is dimensionless:

$$\mu_A = RT \ln f_A; \qquad \mu_B = RT \ln f_B \qquad (2.1)$$

where f_A and f_B are the fugacities and μ_A and μ_B are the chemical potentials of components A and B. When applied to gases, fugacity usually represents that pressure which would be exerted by a gas if the gas behaved ideally. In the pharmaceutical and biological sciences, the vapor pressures of interest are usually sufficiently low that deviations from the ideal gas laws are negligible and so the approximations $p_B \cong f_B$ and $p_B^\theta \cong f_B^\theta$ are valid.

The fugacity, or escaping tendency, of a substance in a given state is more easily compared with that of a substance in another state than determined absolutely. Accordingly, the term activity *a* or relative activity (IUPAC, 1979) is classically defined as the ratio of the fugacity *f* of the substance to that in a defined standard state, f^θ. Thus, for two substances A and B,

$$a_A = f_A/f_A^\theta; \qquad a_B = f_B/f_B^\theta \qquad (2.2)$$

A solution is said to be saturated when the chemical potential (or the thermodynamic activity) of the dissolved solute is equal to that of the undissolved solute. The problem which we are addressing in this book is the prediction of the conditions under which these activities are equal. In general,

$$\text{activity} = \text{concentration} \times \text{activity coefficient} \qquad (2.3)$$

For a saturated solution:

$$
\begin{aligned}
\text{activity of undissolved solute} &= \text{activity of dissolved solute} \\
&= \text{solubility} \times \text{activity coefficient of the dissolved solute}
\end{aligned}
\qquad (2.4)
$$

therefore,

$$\text{solubility} = \frac{\text{activity of undissolved solute}}{\text{activity coefficient}} \qquad (2.5)$$

Solubility prediction, therefore, amounts to the prediction of both (1) the activity of the undissolved solute (or pure solute if it does not take up a significant amount of solvent) and (2) the activity coefficient of the dissolved solute.

Equation 2.5 shows that solubility is an inverse reflection of the activity coefficient. Thus, if the activity of the pure solute remains constant, a change which causes the activity coefficient to be doubled will cause the solubility to be halved. Since the activity coefficient of the dissolved solute is a complicated function of the intermolecular interactions in solution, particularly of solute–solvent interactions, there is usually no simple, convenient way of predicting activity coefficients.

By definition the activity of any substance is unity when the substance is in its standard state (see Eq. 2.2). Unfortunately, no single standard state is universally accepted for comparison purposes. Instead, a number of standard states are used, each of which is admirably suited to a particular theory or experimental procedure. The choice of a reference or standard state is discussed further in Chapter 3.

2.2 SOLUBILITY IN IDEAL SYSTEMS

2.2.1 Definition of an Ideal Solution

The prediction of the solubility of a solute in a solvent with which it forms an ideal solution is particularly simple because the activity coefficient is constant. A simple, practical definition of an ideal solution is one in which the activity coefficient γ_B of any component B is equal to unity when the general equation (Eq. 2.6) is written in the form

$$x_B = a_B / \gamma_B \qquad (2.6)$$

where x_B is the mole fraction and a_B is the activity, which for this purpose is usually defined in terms of the liquid state of the pure substance B as the standard state of unit activity. Thus, when this component is present at mole fraction x_B^{id} in an ideal solution,

$$x_B^{id} = a_B \tag{2.7}$$

For an ideal liquid mixture, the components are mutually soluble in all proportions and their vapor pressures obey Raoult's law.

The simplest theoretical definition of an ideal mixture or solution is a mixture in which the interactions between unlike molecules A and B are identical in strength and nature to those between like molecules (A with A, B with B) in the same state. Thus, an interchange of any molecule, A or B, between any position in the mixture or in the pure state leaves the internal energy and the volume of the system unchanged. In view of this definition, the energy of mixing, ΔU_{mix}^{id}, the volume of mixing, ΔV_{mix}^{id}, and the enthalpy of mixing, ΔH_{mix}^{id}, are all zero. Since the various molecular sites are equally available to molecules of A and B, the ideal entropy of mixing, ΔS_{mix}^{id}, can be calculated statistically according to the equation

$$\Delta S_{mix}^{id} = - R(n_A \ln x_A + n_B \ln x_B) \tag{2.8}$$

where n_A and n_B are the number of moles of A and B, and x_A and x_B are the respective mole fractions in the random mixture.

Ideal liquid mixtures arise when the components are similar, such as two molecules (1) which differ in isotopic substitution, (2) which are adjacent or close members of the same homologous series, (3) which are closely related isomers, or (4) for which a difference in one thermodynamic quantity is compensated by difference in another quantity, so that the ideal equations (Eqs. 2.7 and 2.8) are still obeyed (e.g., chlorine and carbon tetrachloride, chlorobenzene and naphthalene).

The ideal system is not a highly useful concept because in most instances the solute–solvent and solvent–solvent interactions are quite different both qualitatively and quantitatively. The phenomenon of limited solubility arises from the very departure from ideality with which we are largely concerned in most practical situations. We have included this treatment of ideal solutions to provide some indication of its applicability, usefulness, and limitations.

2.2.2 Solid Solutes

Most solutes with which we are concerned are solids and, according to Eq. 2.7, prediction of their ideal solubility is equivalent simply to the prediction of the activity of the pure solid, a^s. For this section (and indeed for the whole of Chapter 2), we find it convenient to continue to use as our standard state the pure liquid. For a stable solid solute, this standard state is the pure supercooled liquid which is an unstable state that is not usually attainable in practice. The activity a^s of a stable solid is, therefore, less than unity. Chapter 3 provides a more extensive discussion of this concept. The activity a_B^s of the solid solute B can be predicted according to the arguments that follow.

The dissolution of a solid in a solvent is equivalent to the melting of the solid followed by mixing of the resulting liquid solute with the solvent. The enthalpy of solution ΔH_s of a solid is therefore given by

$$\Delta H_s = \Delta H_f^T + \Delta H_{mix} \tag{2.9}$$

where ΔH_f^T is the enthalpy of fusion at the absolute temperature T of the solution, which is invariably below the normal melting point T_m of the solid, and ΔH_{mix} is the enthalpy of liquid–liquid mixing. Since for an ideal solution $\Delta H_{mix} = 0$, the ideal enthalpy of solution of a solid, ΔH_s^{id}, is given by

$$\Delta H_s^{id} = \Delta H_f^T \tag{2.10}$$

For any solute, classical thermodynamics shows that

$$\left(\frac{\partial \ln a}{\partial T}\right)_p = \frac{\Delta H_s}{RT^2} \tag{2.11}$$

and

$$\left(\frac{\partial \ln a}{\partial (1/T)}\right)_p = -\frac{\Delta H_s}{R} \tag{2.12}$$

For an ideal saturated solution of a solid solute B in a liquid solvent A, $a_B^{s,sat}$ may be replaced by $x_B^{id,sat}$, ΔH_s by ΔH_f^T, and ΔS_s by ΔS_f^T, where ΔS_f^T is the entropy of fusion of the solid solute at temperature T. Ideality of solutions of solids in liquids can, therefore, be tested by comparing the value of ΔH_s obtained from the gradient of the plot of $\ln x_B$ against $1/T$ with the value of ΔH_f^T, if ΔH_s is assumed, as a first approximation, to be independent of T.

If the enthalpy of fusion of solid B is assumed to be independent of temperature, $\Delta H_f^T = \Delta H_f$, the enthalpy of fusion at the melting point T_m of B and Eq. 2.11 or 2.12 can be integrated to give

$$\ln x_B^{id} = \ln a_B \cong I - \frac{\Delta H_f}{R}\frac{1}{T} \tag{2.13}$$

where I is the integration constant.

This equation indicates that ideal mole fraction solubility x_B^{id} increases with increasing temperature until, when $T = T_m$, the solid solute melts, that is, forms a liquid in the absence of solvent, so $x_B^{id} = 1$. Therefore, $I = \Delta H_f/RT_m$, such that

$$\ln x_B^{id} \cong \ln a_B = -\frac{\Delta H_f}{R}\left(\frac{1}{T} - \frac{1}{T_m}\right); \qquad \Delta H_f \cong \Delta H_s^{id} \tag{2.14}$$

or

$$\ln x_B^{id} = \ln a_B \cong -\frac{\Delta H_f}{R}\frac{T_m - T}{TT_m}; \qquad \Delta H_f \cong \Delta H_s^{id} \tag{2.15}$$

2.2.3 Prediction of Solid Solubility in Ideal Systems

Figure 2.1 shows that Eq. 2.14 predicts quite accurately the solubility of naphthalene in the aromatic hydrocarbons benzene and xylene, indicating (a) that the solutions behave ideally, as might be expected because of the similarity between the solute and each solvent, and (b) that the enthalpy of fusion of naphthalene $(\Delta H_f = 19.1 \, \text{kJ} \, \text{mol}^{-1} = 4.56 \, \text{kcal} \, \text{mol}^{-1}$ at the melting point 353.4 K) is not very temperature-dependent.

Ideal behavior is readily explicable on account of the similarity of molecular properties of the solute naphthalene and the solvents benzene and xylene. All are polarizable, but without significant dipole moments, and so their intermolecular interactions are almost entirely due to London dispersion forces and ΔU_{mix} is very small. All the molecules have quite similar molar volumes (solid naphthalene, $112 \, \text{cm}^3 \, \text{mol}^{-1}$; liquid benzene, $89.4 \, \text{cm}^3 \, \text{mol}^{-1}$; toluene, $107 \, \text{cm}^{-3}$ mol^{-1}; p-xylene, $124 \, \text{cm}^3 \, \text{mol}^{-1}$; all at 25°C), so ΔV_{max} is also likely to be small.

However, the solubility of naphthalene in ethanol is much less than the ideal value, largely because of the hydrogen-bonded structure of ethanol, although

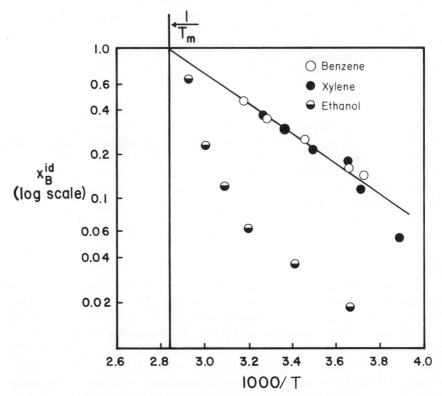

Figure 2.1 Van't Hoff plot of log mole fraction solubility of solid naphthalene, B, in several solvents as a function of the reciprocal of the absolute temperature T. The solid line corresponds to Eq. 2.14 for the ideal solubility of solids.

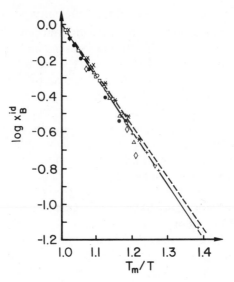

Figure. 2.2 Near-ideal solubility behavior of the following aromatic solids in benzene: □, naphthalene; ○, biphenyl; ▽, fluorene; ◇, acenaphthene; ×, o-terphenyl; +, pyrene; △, fluoranthene; ●, m-terphenyl; ∗, phenanthrene (for individual structures, see Scheme 2.1). Ideal solubility line, − − −. x_B^{id} is the mole fraction solubility at absolute temperature T, and T_m is the melting point of the solute. Reproduced with permission of the copyright owner, the Royal Society of Chemistry, from McLaughlin, E. and Zainal, H. A. (1959). *J. Chem. Soc.*, **1959**, 863–867.

the polarity of the molecule ($\mu = 1.66$ D), the low molar volume of the liquid (58.7 cm^3 mol^{-1} at 25°C), and the high polarizability of the aromatic solute undoubtedly play a part. Later chapters show that hydrogen bonding in particular exerts a profound influence on solubility.

Since $\Delta H_f/T_m = \Delta S_f$, the entropy of fusion of the solid at its true melting point, Eq. 2.15 may be stated in the following entropy form:

$$\ln x_B^{id} = \ln a_B \cong -\frac{\Delta S_f}{R}\left(\frac{T_m}{T} - 1\right) \tag{2.16}$$

This equation indicates that the magnitude of the ideal mole fraction solubility is determined by the ratio T_m/T and by the entropy of fusion of the solid solute. Figure 2.2 shows that Eq. 2.16 gives a satisfactory prediction of solubility at temperatures near the melting point ($T_m/T < 1.1$) for a number of solid aromatic hydrocarbons dissolved in benzene (McLaughlin and Zainal, 1959). The molecular structures of the compounds are presented in Scheme 2.1. The entropy of fusion is much the same for these hydrocarbons, the mean value being 54.4 J K^{-1} mol^{-1} ($= 13.0$ cal K^{-1} mol^{-1}), and this value was employed to construct the ideal solubility line in Figure 2.2.

The ideal solubility represents, in most instances, an upper limit of solubility.

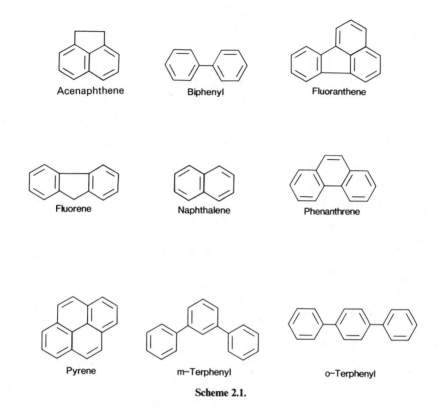

Acenaphthene   Biphenyl   Fluoranthene

Fluorene   Naphthalene   Phenanthrene

Pyrene   m–Terphenyl   o–Terphenyl

Scheme 2.1.

Most of the solutes in Figure 2.2, particularly the bulky molecules acenaphthene and *m*-terphenyl, whose structures are shown in Seheme 2.1, have significantly lower solubilities than the ideal (McLaughlin and Zainal, 1959). As in Figure 2.1, the solute and solvent molecules have intermolecular forces of similar strength and nature. The deviations from ideality here are more likely to be due to differences in molar volume between solute and solvent (McLaughlin and Zainal, 1959) than to the approximation introduced in deriving Eqs. 2.13–2.16.

The prediction of the solubility of a polar solute such as benzoic acid, in terms of ideal behavior (Eq. 2.14), generally fails unless the solute and solvent exhibit the same polar interactions. Thus, Figure 2.3 shows that benzoic acid ($\mu = 1.65$ D in benzene; McClellan, 1963) apparently forms a nearly ideal solution in acetophenone, which is also polar ($\mu = 2.92$ D in benzene; McClellan, 1963) and possesses similar molecular size and structure. The nearly ideal behavior may result from a fortuitous mutual cancellation of intermolecular effects, such as the tendency of benzoic acid to form an associated species with acetophenone, a proton acceptor, or with another molecule of benzoic acid in a dimer ($\mu = 0.97$ D in benzene; McClellan, 1963). Furthermore, the presence of the benzene ring gives similar polarizabilities to the molecules and hence similar Debye and London forces. Thus, ΔU_{mix} is close to zero.

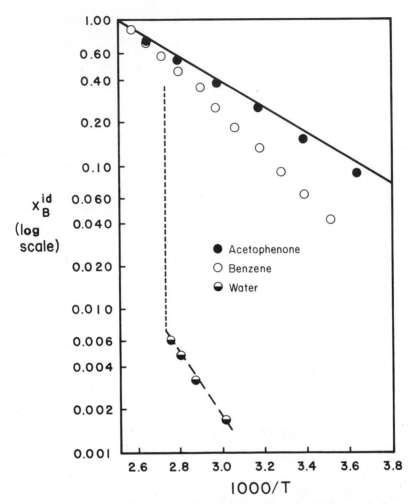

Figure 2.3 Van't Hoff plot of log mole fraction solubility of benzoic acid in several solvents as a function of the reciprocal of the absolute temperature T. The solid line corresponds to the idealized relationship, Eq. 2.14.

Benzoic acid, like other carboxylic acids, forms hydrogen-bonded dimers in nonpolar solvents such as benzene. Consequently, the solute becomes more ordered and this negative influence reduces ΔS_{mix} below the ideal value. This effect more than compensates for the enhancement caused by the contributions of both dimer and monomer. As a result, the benzoic acid solubility plot in benzene has a steeper slope than ideal, giving solubilities less than ideal (Fig. 2.3). Also, ΔU_{mix} may deviate from zero. Mixtures of excess benzoic acid and water form two liquid phases between 89.7 and 117.2°C; the latter is the upper initial solution temperature at which the two liquid phases coalesce to give a homogeneous solution containing 32.34% (w/w) benzoic acid and 67.66% (w/w) water, corresponding to $x_B = 0.0659$ (*Merck Index*, Windholz et al., 1983; Ward and Cooper, 1930). These solutions deviate considerably from ideality and contain hydrogen-bonded structures involving both water and benzoic acid.

In the case of polymorphism, if one polymorph forms an ideal solution in a certain solvent, the other polymorph(s) will do so as well, because ideality is determined by the behavior of the liquid solute and the liquid solvent toward each other. Since the polymorphs which are indicated by subscripts 1 and 2 in Figure 2.4 have different crystalline lattices, they have different activities, a^s, which are reflected in their different enthalpies of fusion, ΔH_f, and different melting points, T_m. These differences give rise to different slopes and intercepts, respectively, of the ideal solubility lines predicted by Eq. 2.14 and depicted by Figure 2.4.

At a given temperature the polymorph with the lower activity, and hence mole fraction solubility, is the more stable. If the two lines cross, as in Figure 2.4, the temperature at the point of intersection is the transition point T_t, and the two polymorphs are enantiotropes. If the two lines do not intersect, the polymorphism is monotropic. Further discussion of the influence of polymorphism on solubility will be presented in Section 2.3 and will not be restricted to ideal system. Solvates will be discussed in Section 2.4.

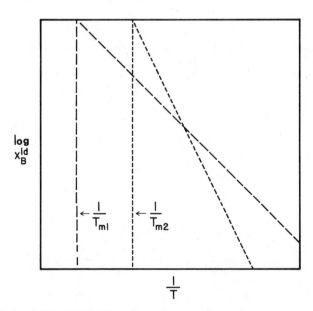

Figure 2.4 Ideal mole fraction solubility x_B^{id} of two polymorphic forms which are enantiotropes of melting point T_{m1} and T_{m2}, respectively.

2.2.4 Thermodynamics of Solid Solubility

Some years ago it was true to state that the properties (e.g., activity) of the crystalline phase could not be predicted from the structure of the solute. By treating the nonbonded interactions of atoms in certain very simple organic molecules by complex computer averaging techniques, it is now possible to predict with some accuracy the structure and some thermodynamic properties of their crystals (Kitaigorodsky, 1978). We are still some distance, however, from predicting precisely the structure and properties of crystals of drug molecules, including the relative stability of their different forms (polymorphs or solvates) under various conditions. This is because most organic molecules have low symmetries, irregular shapes, and complex intermolecular interactions which are often dominated by hydrogen bonding. The resulting crystal lattices have low symmetries and appreciable interstices.

Some organic compounds can form a metastable amorphous form on quenching a melt. Moreover, the conditions of crystallization usually introduce imperfections such as point defects and dislocations into the crystal lattice (Hüttenrauch, 1978). Some defects result from lattice vacancies and others from the presence of foreign molecules, which give impurity defects. All these subtle imperfections contribute to batch-to-batch variations in the physical properties, thermodynamic characteristics (e.g., activity and solubility), and rates of interaction (e.g., dissolution rate and stability) of organic raw materials such as drugs and excipients (Jones, 1981; York, 1983; Chow et al., 1984, 1985).

It is possible to predict approximately the activities of such solids by making estimates of their entropies of fusion, ΔS_f. The enthalpy of fusion ΔH_f may be estimated from $\Delta H_f = T_m \Delta S_f$, if the melting point is known. Walden (1908) was the first to note that the entropy of fusion for most organic compounds is close to $13 \, \text{cal} \, K^{-1} \, \text{mol}^{-1}$. At 25°C (298.15 K),

$$(T_m/T) - 1 = (T_m - T)/T = (\Theta_m - 25)/298.15 \qquad (2.17)$$

where Θ_m is the Celsius melting point.

Using Walden's (1908) approximation for the entropy of fusion and Eq. 2.17, Irmann (1965) wrote Eq. 2.16 in the form

$$\log \frac{s^s}{s^o} = \log x^{id} = \log a^s = -0.0095(\Theta_m - 25) \qquad (2.18)$$

and used it to calculate the solubility of the solid s^s from the solubility of the supercooled liquid, s^o.

Yalkowsky (1979) estimated the entropy of fusion, ΔS_f, of organic compounds using the additive relationship

$$\Delta S_f = \Delta S_{exp} + \Delta S_{pos} + \Delta S_{rot} + \Delta S_{int}$$

where

$$\Delta S_{exp} + \Delta S_{pos} = \Delta S_{trans}$$

$$(2.19)$$

and where ΔS_{exp} is the entropy component of expansion as the solid melts; ΔS_{pos} is the positional entropy component of the molecules due to the crystal lattice, both of which comprise ΔS_{trans}, the translational entropy change; ΔS_{rot} is the rotational entropy component of the molecules in the liquid; and ΔS_{int} is the internal entropy component, which represents the increase in the number of degrees of freedom of the groups in flexible molecules as they become released from the constraints of the crystal lattice. Yalkowsky (1979) proposed the most likely values and the upper and lower limits of these components of the entropies of fusion, as shown in Table 2.1. For spherical molecules, he has suggested that

$$\Delta S_f = \Delta S_{trans} \sim 3.5\, cal\, K^{-1}\, mol^{-1} = 14.6\, J\, K^{-1}\, mol^{-1} \qquad (2.20)$$

and for rigid molecules,

$$\Delta S_f = 13.5\, cal\, K^{-1}\, mol^{-1} = 56.5\, J\, K^{-1}\, mol^{-1} \qquad (2.21)$$

whereas for flexible molecules,

$$\begin{aligned} \Delta S_f &= 13.5 + 2.5(n-5)\, cal\, K^{-1}\, mol^{-1} \\ &= 56.5 + 10.5(n-5)\, J\, K^{-1}\, mol^{-1} \end{aligned} \qquad (2.22)$$

where n is the number of flexible links in the chain and only enters into the calculation if $n > 5$.

It turns out that the entropy of fusion of compounds which have nonhydrogen bonding groups (e.g., CH_3, Cl, Br, NO_2) are not significantly different from those of compounds having one or two hydrogen bonding groups (e.g., NH_2, OH, COOH). There is also little systematic difference between the ΔS_f values

Table 2.1 Component Entropies of Fusion

Type of Entropy	Most Likely Values $(cal\, K^{-1}\, mol^{-1})$	Normal Range of Values $(cal\, K^{-1}\, mol^{-1})$	
		Low	High
Expansional	2	1	3
Positional	2.5	2	3
Rotational	9	7	11
Total (rigid molecules)	13.5	10	17
Internal[a]	$2.5(n-5)$ (for $n > 5$)	(2.3 to 2.7)	$(n-3)$ to $(n-6)$
Total (flexible molecules)	$13.5 + 2.5(n-5)$		

[a] n is the number of flexible links in the chain and only exerts a significant effect on the entropy of fusion when n exceeds 5.

of ortho, meta, and para isomers of aromatic compounds. Yalkowsky pointed out that the treatment described above provides a simple means of estimating approximately the ideal solubility from the structure and the melting point, but cannot be expected to provide highly accurate estimates for all compounds because it is based on many assumptions and approximations. For the purposes of prediction, Eq. 2.16 and 2.17 can be written in the form

$$\log x^{id} = \log a^s = -\frac{\Delta S_f}{2.303\,R\,298.15}(\Theta_m - 25) \qquad (2.23)$$

If these assumptions are permissible, Eq. 2.23 can be written as

$$\log x^{id} = \log a^s = -k(\Theta_m - 25) \qquad (2.24)$$

where k is a constant which is approximately equal to 0.0099 for rigid molecules. Figure 2.5 illustrates a practical situation for which

$$a^s = x^s \gamma^s = k's \qquad (2.25)$$

where s is the molar solubility of each of the various solid 3-acyloxymethyl-5, 5-diphenylhydantoins (Scheme 2.2) in ethyl oleate (Yamaoka et al., 1983) and k' is a function of both the factor for converting mole fraction to molar solubility

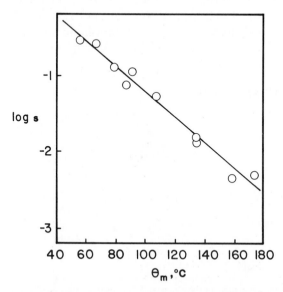

Figure 2.5 Plot of the log molar solubility s in ethyl oleate against the melting point Θ_m in °C (according to Eq. 2.26) for various 3-acyloxymethyl-5,5-diphenylhydantoins. Reproduced with permission of the copyright owner, the American Pharmaceutical Association, from Yamaoka, Y., Roberts, R. D., and Stella, V. J. (1983). *J. Pharm. Sci.*, **72**, 400–405.

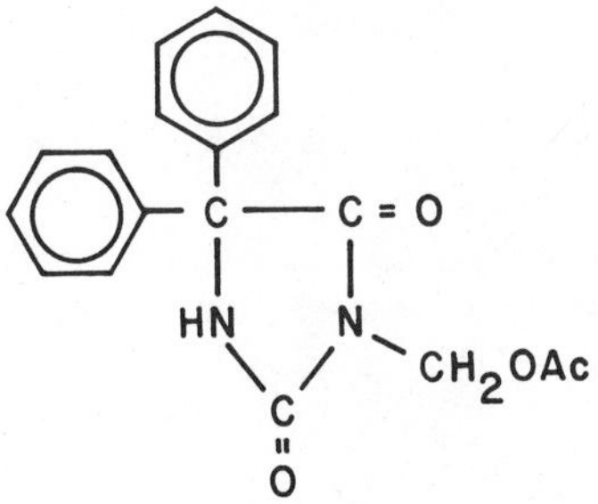

Scheme 2.2.

and of the activity coefficient which reflects the intermolecular interactions in solution. The latter is assumed to be independent of the nature of the acyl group. Combining Eqs. 2.24 and 2.25 leads to

$$\log s = k'' - k\Theta_m \qquad (2.26)$$

where

$$k'' = 25k - \log k'$$

Application of this equation to the plot in Figure 2.5 shows that $k'' = 0.4606$ and $k = 0.0167$ with a correlation coefficient $r = -0.988$. The deviation of the slope, k, from the value of -0.0099 results from (a) nonconstancy and enhancement of ΔS_f, which mainly reflects the lack of rigidity in the molecules, and (b) nonconstancy of the activity coefficients, which reflects the influence of the acyl group on solute–solvent interactions in solution. If Yalkowsky's approximation in Table 2.1 is used to calculate ΔS_f, and if the mean value of $n = 8.6$ is assumed for the sidechains in the series of solutes in Scheme 2.2, we obtain the mean value $\Delta S_f = 22.5\,\text{cal K}^{-1}\,\text{mol}^{-1}$ ($94.1\,\text{J K}^{-1}\,\text{mol}^{-1}$). This value compares well with $\Delta S_f = 22.8\,\text{cal K}^{-1}\,\text{mol}^{-1}$ ($95.3\,\text{J K}^{-1}\,\text{mol}^{-1}$) derived from the experimental value of k. The experimental intercept, $k'' = 0.4606$, is appreciably smaller than that calculated assuming $\gamma = 1$, that is, $25\,k - \log k' = 0.8659$, mainly because of the reduced activity coefficients of the polar solute molecules in the polar solvent, ethyl oleate. Furthermore, the assumed proportionality between mole fraction and molarity breaks down at higher concentrations.

This method (Yalkowsky, 1979) of estimating the entropy of fusion is based on many assumptions and approximations and cannot, therefore, be expected to provide accurate predictions of the activity and ideal solubility of all organic solids. The treatment does, however, give useful estimates of these quantities from a knowledge of the molecular structure and the melting point of a particular compound. The melting temperature must, of course, correspond to that of the crystalline form of the compound whose solubility is being estimated. In many instances polymorphic or desolvation transitions occur during the determination of the melting point.

This approach has been extended to provide approximate predictions of aqueous solubility (Yalkowsky and Valvani, 1980) as discussed in Sections 2.8

Table 2.2 Influence of Substituent Groups on the Melting Point and Molar Solubility of Pteridine in Water at 20°C[a,b]

Substituent Group	2-Position		4-Position		6-Position		7-Position	
	m.p. (°C)	Solubility (mol dm^{-3})	m.p. (°C)	Solubility (mol dm^{-3})	m.p. (°C)	Solubility (mol dm^{-3})	m.p. (°C)	Solubility (mol dm^{-3})
Cl	106	0.200	140	—	146	0.0751	95	0.133
CH$_3$	146	0.806	151	0.342	—	—	134	0.14
OCH$_3$	150	0.0772	195	0.0772	124	0.0726	130	0.123
SCH$_3$	136	0.0176	191	0.00432	—	—	143	0.0281
NMe$_2$	125	2.29	165	0.0952	212	—	204	0.952
NH$_2$	275(d)	0.00504	305(d)	0.00486	300(d)	0.00454	>320(d)	0.00486
OH	240(d)	0.0113	>350(d)	0.0338	248(d)	0.00193	230(d)	0.00751
SH	>205(d)	0.00235	>290(d)	0.00226	—	—	>260(d)	0.00197

[a] 1 mol dm^{-3} = 1 mol L^{-1} = 1 M.
[b] Unsubstituted pteridine: m.p. = 140°C; solubility = 1.05 mol dm^{-3}; d = incipient decomposition.

Reproduced with permission of the copyright owner, the Royal Society of Chemistry, from Brown, D. J. (1954). The monosubstituted pteridines. In *Ciba Foundation Symposium on Chemistry and Biology of Pteridines*. Wolstenholme, G. E. W. and Cameron, M. P. (Eds.). Little, Brown, Boston.

and 2.9. It is apparent from the equations given above that the melting point can provide useful information concerning the activity or ideal solubility of a solid. For example, Eq. 2.24 suggests that a $100°C$ increase in T_m would decrease $a^s(= x^{id})$ by a factor of 1000, if other factors are equal.

Unfortunately, there are virtually no dependable procedures for accurately calculating melting points or enthalpies of fusion. This is because of the complications arising from the specific intermolecular interactions in the solid state and the complexities of the entropic changes which occur during fusion. Yalkowsky, however, noted the following general guidelines which can be used in a qualitative way:

(a) Gross molecular geometry exerts a powerful influence on the melting point (Ubbelohde, 1965; Kitaigorodski, 1973). The melting point increases with increasing compactness of the molecular packing, with increasing molecular size, rigidity, and symmetry, and with increasing alkyl chain length when the molecular size has exceeded a certain value. A decrease in melting point to a minimum with increasing alkyl chain length for short chains, as observed among the alkylbenzenes and alkylnaphthalenes, and an odd–even alternation, as observed among the alkanes and other homologous series, can be explained in terms of efficiency in packing of the molecules in the crystal lattice (Ubbelohde, 1965; Kitaigorodski, 1973).

(b) The addition of a polar group to a molecule introduces Keesom and Debye forces (see Section 2.11) and thereby increases the melting point and reduces the solubility, provided that the efficiency of molecular packing in the crystal is not greatly reduced. Table 2.2 shows that these considerations often apply, for example, to the addition of the methyl, methoxy, thiomethyl, and dimethylamine group to pteridine. Substitution at the 4 position shows the effect best because it does not seriously disturb the crystal packing, whereas the 2 and 7 positions are more sensitive to packing effects, which explains why the melting point is sometimes reduced. As expected, the increase in melting point often parallels the dipole moment of the added groups.

(c) The presence of hydrogen-bond donor and acceptor groups strengthens the intermolecular interactions and so increases the melting point. The latter increases with increasing number of intermolecular hydrogen bonds per molecule, which is itself equal to twice the minimum number of either donor or acceptor sites that the molecule possesses.

2.2.5   Influence of Hydrogen Bonding on Molecular Packing in the Solid State and Solubility

Primary amides ($RCONH_2$) have three acceptor sites and two protons for donation and can therefore form a lattice in which each molecule takes part in up to four hydrogen bonds. Carboxylic acids ($RCOOH$) have four acceptor sites but only one donor proton and can therefore form a lattice in which there are an average of only two hydrogen bonds per molecule. Similarly, alcohols

(ROH), primary amines, and secondary amines form lattices with an average of two hydrogen bonds per molecule.

Amino carboxylic acid molecules can donate three protons and have five electron pairs which can accept protons and therefore can form six intermolecular hydrogen bonds. Dicarboxylic acids and diamines, by the same argument, can only form four hydrogen bonds and so generally have lower melting points than the amino carboxylic acids. Groups containing lone pairs of electrons, such as halogen atoms, ether groups, and carbonyl groups, act as proton acceptors and thereby increase the melting point of amines. When, however, the molecular packing is upset, the reverse effect may be observed, as is illustrated in Table 2.2.

Heterocyclic bases such as pyridine, pyrimidine, pteridine, and quinoline have only proton acceptor groups and so are not particularly high melting compounds, in spite of their polarity and planarity. Addition of hydroxyl, amino, or thiol groups, which can also accept protons, permits intermolecular hydrogen bonding and so greatly increases the melting point. The increase in intermolecular interactions in the crystalline state is much stronger than the increase in solute–solvent interactions in aqueous solution (which reduces the activity coefficient), so the aqueous solubility is reduced. Table 2.2 illustrates these effects with the pteridine molecule. Yalkowsky suggested that, whereas the melting point depends on twice the minimum number of *either* donor or acceptor sites on the molecule, activity coefficients in a hydrogen-bonded solvent such as water correlate with the *sum* of the number of donor and acceptor sites.

Certain groups, such as saccharides and semicarbazones, and ionizable structures, such as phosphates and succinates, may increase the melting point without greatly reducing the solubility simply because they interact very strongly with water, reducing the activity coefficient much more than they reduce a^s.

Intramolecular hydrogen bonding may occur between neighboring hydrogen bonding groups within the same molecule, an effect which greatly reduces the extent of intermolecular hydrogen bonding and thereby reduces the melting point. For example, among the isomeric aminobenzoic acids, hydroxybenzoic acids, diaminobenzenes, dicarboxylbenzenes, and dihydroxybenzenes, the ortho isomer has a lower melting point than the meta or para isomer. The meta isomer has a lower melting point than the para isomer for different reasons, namely, less efficient packing and lower overall dipole moment.

To increase the solubility of a solid drug substance, it may be expedient to reduce a^s by reducing the efficiency of molecular packing reflected in the melting point. This may be achieved by introduction of a suitably shaped bulky group such as trimethylacetyl (pivaloyl), or a straight-chain group such as pentanoyl. For example, the conversion of diphenylhydantoin (phenytoin) to certain acyl derivatives (e.g., pentanoyl) greatly reduces the melting point and at the same time increases its solubility in cyclohexane, triolein, trioctanoin, tributyrin, and ethyl oleate (Fig. 2.5) (Yamaoka et al., 1983). Although such a procedure probably changes the activity coefficient, perhaps in an unfavorable direction, this stratagem for reducing a^s frequently results in a higher solubility.

Aryl substituents which confer efficient molecular packing and intermolecular hydrogen bonding have been employed by Morozowich et al. (1979) to prepare prostaglandin aryl esters. The melting points of the aryl esters parallel those of the corresponding phenols. The aryl esters are, however, poorly soluble in water because of their high a^s values, which are reflected in their high melting points.

2.2.6 Influence of Temperature on Enthalpy of Fusion

These methods of predicting ΔH_f or ΔS_f assume, however, that the enthalpy of fusion at the melting point, ΔH_f^m, is equal to the enthalpy of fusion, ΔH_f^T, at a lower temperature T, to which the activity of the solid and ideal solubility refers. This is tantamount to assuming that the heat capacity of the supercooled liquid at T is the same as that of the solid at T_m. This assumption can lead to significant errors, because for most substances the molar heat capacity of the supercooled liquid, C_p^*, is greater than that of the solid, C_p, at a given temperature T. The finite difference ΔC_p, causes ΔH_f to vary with temperature according to the equation

$$\left(\frac{\partial(\Delta H_f)}{\partial T}\right)_p = \left(\frac{\partial(H^* - H)}{\partial T}\right)_p = C_p^* - C_p = \Delta C_p \tag{2.27}$$

If proper allowance is made for this effect, the ideal solubility of the solute B can be predicted accurately from ΔH_f at any temperature T below the melting point T_m.

If the second-order approximation is made such that ΔC_p is independent of temperature, we can use the Kirchhoff equation:

$$\Delta H_f^T = \Delta H_f^m - \Delta C_p(T_m - T) \tag{2.28}$$

where ΔH_f^m is the enthalpy of fusion at the melting point T_m. Introducing this into Eq. 2.11 and integrating affords

$$\ln x_B^{id} = \ln a_B = \frac{-\Delta H_f^m}{R}\frac{T_m - T}{TT_m} + \frac{\Delta C_p}{R}\frac{T_m - T}{T} - \frac{\Delta C_p}{R}\ln\frac{T_m}{T} \tag{2.29}$$

Hildebrand and Scott (1950) have further discussed the heat capacity correction. James and Roberts (1968) have discussed other approximations involving ΔC_p and have described its experimental measurement using differential scanning calorimetry.

Hildebrand and Scott (1962) suggested another approximate equation for ideal solubility, which assumes that $\Delta H_f^T \sim T\Delta S_f^m \sim T\Delta C_p$. This is at least as good an approximation as was the assumption that $\Delta C_p = 0$ in eqs. 2.14–2.16. The result is

$$\left(\frac{\partial \ln x_B^{id}}{\partial \ln T}\right)_p = \left(\frac{\partial \ln a_B}{\partial \ln T}\right)_p \cong \frac{\Delta H_f^m}{RT_m} = \frac{\Delta S_f^m}{R} \tag{2.30}$$

Integrating, with the boundary conditions $x_B = 1$ when $T = T_m$, we find:

$$\ln x_B^{id} = \ln a_B \cong \frac{\Delta S_f^m}{R} \ln \frac{T}{T_m}; \qquad \Delta S_f^m \sim \Delta S_s^{id} \qquad (2.31)$$

Equations 2.30 and 2.31 indicate that ideal solubility can be tested by plotting $\ln x_B$ against $\ln T$ at least as well as by plotting $\log x_B$ against $1/T$ according to Eq. 2.14. The plot of $\ln x_B$ for iodine in a number of solvents against $\ln T$ gives absolutely straight lines in the region where $x_B < 0.1$, whereas plots of $\log x_B$ against $1/T$ give curved lines throughout (Hildebrand and Scott, 1962). If the logarithm in Eq. 2.29 is expressed as an infinite series, $\ln y = 1 - 1/y + \cdots$ for $y > 1/2$, the equation can be written:

$$\ln x_B^{id} = \ln a_B \cong \frac{\Delta S_f^m}{R} \left(1 - \frac{T_m}{T} \right) \qquad (2.32)$$

which is identical with Eq. 2.16.

The temperature dependence of the solubility of polycyclic aromatic hydro-carbons in benzene conforms more closely to the van't Hoff equation (Eq. 2.13) than to the Hildebrand equation (Eq. 2.29) (Yalkowsky, 1981). This finding implies that ΔC_p is closer to zero than it is to ΔS_f, and appears to agree with data for the aqueous solubility of many varied organic nonelectrolytes (Yalkowsky and Valvani, 1980).

In nonideal systems the enthalpy of solution $\Delta H_s \neq \Delta H_f^T$ because $\Delta H_m \neq 0$ in Eq. 2.9, while the corresponding $\Delta C_p \neq 0$. In general, over a wide range of temperature (e.g., 0–60°C), log (solubility) is *not* a linear function either of $1/T$ (Eq. 2.13) or of $\log T$ (Eq. 2.31), particularly in the case of polar compounds dissolved in water; this is shown by Figure 2.6. Over a wide range of temperature the following equation provides a good representation of the temperature dependence of the mole fraction solubility of polar organic compounds in water (Grant et al., 1984):

$$\ln x_B = -\frac{a}{R} \frac{1}{T} + \frac{b}{R} \ln T + c \qquad (2.33)$$

Equation 2.33 contains three adjustable parameters ($a, b,$ and c) and may have wide application to most solute–solvent combinations; it may be derived by integrating Eq. 2.11, making the following assumptions:

$$\Delta H_s^* = a + bT \qquad (2.34)$$

and

$$\left(\frac{\partial \ln x_B}{\partial \ln a_B} \right)_T = \text{constant} \qquad \text{(unity only for ideal solutions)} \qquad (2.35)$$

where ΔH_s^* is the apparent enthalpy of solution which is related to the actual

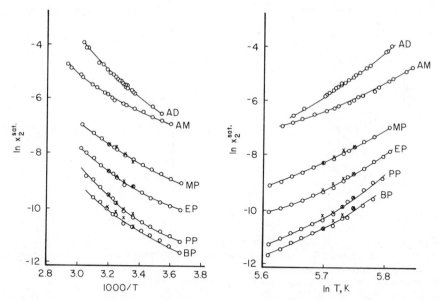

Figure 2.6 Van't Hoff plots (left) according to Eq. 2.13 and Hildebrand plots (right) according to Eq. 2.31, representing the influence of absolute temperature T on the aqueous mole fraction solubility x_2^{sat} of adipic acid (AD), acetaminophen (AM), methyl p-hydroxybenzoate (MP), ethyl p-hydroxybenzoate (EP), propyl p-hydroxybenzoate (PP), and butyl p-hydroxybenzoate (BP). $\bigcirc$, data of Grant et al. (1984); $\times$, data of Alexander et al. (1978). Reproduced with permission of the copyright owner, Elsevier Science Publishing Co., Inc., from Grant, D. J. W., Mehdizadeh, M., Chow, A. H.-L, and Fairbrother, J. E. (1984). *Int. J. Pharm.*, **18**, 25–38.

enthalpy of solution ΔH_s, as (Hollenbeck, 1980)

$$\Delta H_s^* = \Delta H_s \left(\frac{\partial \ln x_B}{\partial \ln a_B} \right)_T \tag{2.36}$$

For nonideal solutions, $\Delta H_s^* \neq \Delta H_s$ because the partial derivative (i.e., partial differential) is not equal to unity. Equation 2.33 contains two temperature-dependent terms, one of which is similar to that in the classical van't Hoff equation (Eq. 2.13), while the other is analogous to Equation 2.31 (Hilderbrand-Scott, 1962). Equations 2.33 and 2.34 are analogous to the more rigorous Eqs. 2.28 and 2.29 for ideal solubility, since b represents a ΔC_p term (the difference between the molar heat capacity of the solute in solution and that of the pure solute) and a represents a reduced ΔH term (the hypothetical value of ΔH_s^* at $T = 0 \, K$).

If the pure solid solute B is sufficiently volatile, it is sometimes possible to measure its vapor pressure p_B^s at the temperature of interest, T, and to calculate a_B from the equation

$$a_B = p_B^s / p_B^* \tag{2.37}$$

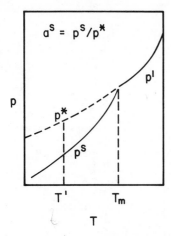
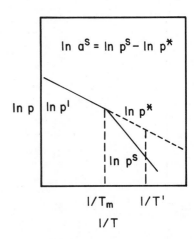

Figure 2.7   Influence of temperature T on the vapor pressure of the solid, p^s, and of the liquid, p^1. At the temperature of interest, T', p^* is the vapor pressure of the supercooled liquid; a^s is the activity of the solid with respect to the pure supercooled liquid; T_m is the melting point. Adapted from Shinoda, K., in *Principles of Solution and Solubility*, Marcel Dekker, New York, 1978.

(p_B^s is also equal to the vapor pressure of B above its saturated solution, since the pure solid is in equilibrium with the solution). Here p_B^* is the vapor pressure of the pure (hypothetical) supercooled liquid at temperature T, and is evaluated by extrapolating the vapor pressure of liquid B (determined at temperatures above the freezing point) down to the temperature of interest, T', as shown in Figure 2.7. The extrapolation may be carried out by means of plots of $\ln p$ against $1/T$, which according to the Clapeyron–Clausius equation (Eq. 3.15) will be linear if the enthalpy of vaporization of B is independent of temperature. Very sensitive methods may have to be used to determine the vapor pressure, particularly of the solid. Such methods include analysis of a sample of the vapor by means of gas chromatography or mass spectrometry.

The activity of solids using the pure supercooled liquid as the standard state in this chapter involves an extrapolation (of ΔH_f or p^s) down to the temperature of interest, a most questionable matter for complex molecules. In Chapter 3 a different standard state is used that does not require extrapolations.

Since the preceding discussion has been concerned with the solid state, it is appropriate at this point to discuss the influence of polymorphism and solvate formation on solubility. The following section is generally applicable and is not restricted to ideal solutions.

2.3   SOLUBILITY OF POLYMORPHS

If a given chemical substance is capable of existing in two (or more) different crystalline forms, such as polymorphs or solvates, their chemical potentials, activities, and escaping tendencies are likely to be different. At a given temperature T, the polymorphic form which has the higher activity will also have the higher solubility and lower stability. For the more stable form,

$$G_I = H_I - TS_I \tag{2.38}$$

For the less stable form,

$$G_{II} = H_{II} - TS_{II} \tag{2.39}$$

The standard state of a solid is often taken to be a fluid phase, which may be either the pure hypothetical supercooled liquid (Hildebrand and Scott, 1950), an infinitely dilute solution in an inert solvent such as isooctane (Rytting et al., 1972), or even the gas or vapor phase (Butler, 1937, 1962). Since each of these standard states represents complete breakdown of the various crystalline forms, it will be independent of crystalline form and we can write

$$G^\theta = H^\theta - TS^\theta \tag{2.40}$$

Subtracting Eq. 2.40 from Eq. 2.38 for constant temperature T, we obtain for form I:

$$G_I - G^\theta = (H_I - H^\theta) - T(S_I - S^\theta) \tag{2.41}$$

$$= \Delta G_I^\theta = \Delta H_I^\theta - T\Delta S_I^\theta \tag{2.42}$$

$$= RT \ln(a_I/1) \tag{2.43}$$

Similarly, using Eq. 2.40 and 2.39, we obtain for form II:

$$G_{II} - G^\theta = (H_{II} - H^\theta) - T(S_{II} - S^\theta) \tag{2.44}$$

$$= \Delta G_{II}^\theta = \Delta H_{II}^\theta - T\Delta S_{II}^\theta \tag{2.45}$$

$$= RT \ln(a_{II}/1) \tag{2.46}$$

Subtracting Eq. 2.43 from Eq. 2.46 (or Eq. 2.38 from Eq. 2.39), we obtain

$$(G_{II} - G_I) = (H_{II} - H_I) - T(S_{II} - S_I)$$

That is,

$$\Delta G(I \to II) = \Delta H(I \to II) - T\Delta S(I \to II) \tag{2.47}$$

$$= RT \ln(a_{II}/a_I) \tag{2.48}$$

The activities a_I and a_{II} are, of course, proportional to the escaping tendencies or fugacities f_I and f_{II} and to the vapor pressures p_I and p_{II} and to the molalities

m_I and m_{II}, if Henry's law is obeyed.

$$\frac{a_{II}}{a_I} = \frac{f_{II}}{f_I} = \frac{p_{II}}{p_I} = \frac{y_{II}^m m_{II}}{y_I^m m_I} \sim \frac{m_{II}}{m_I} \tag{2.49}$$

Equations 2.48 and 2.49 are particularly useful in relating molecular interactions in the crystals and escaping tendencies of the polymorphs.

If the standard state is taken to be a hypothetical solution of unit molality which acts as if it were infinitely dilute, then $y^\infty = 1$ and $a = y^m m$, where m is the molality ($mol\,kg^{-1}$) and y^m is the molal activity coefficient. If m_I and m_{II} represent the molal solubilities of the respective polymorphs, the standard molar enthalpies of solution are, respectively:

$$H^\theta - H_I = \Delta H_I \qquad \text{and} \qquad H^\theta - H_{II} = \Delta H_{II} \tag{2.50}$$

while the standard molar entropies of solution are, respectively:

$$S^\theta - S_I = \Delta S_I \qquad \text{and} \qquad S^\theta - S_{II} = \Delta S_{II} \tag{2.51}$$

Application of the van't Hoff isochore to these situations gives:

$$\left(\frac{\partial \ln a_I}{\partial (1/T)}\right)_p = \left(\frac{\partial \ln (y_I^m m_I)}{\partial (1/T)}\right)_p = \frac{-(H_I - H^\theta)}{R} = \frac{\Delta H_I}{R} \tag{2.52}$$

$$\left(\frac{\partial \ln a_{II}}{\partial (1/T)}\right)_p = \left(\frac{\partial \ln (y_{II}^m m_{II})}{\partial (1/T)}\right)_p = \frac{-(H_{II} - H^\theta)}{R} = \frac{\Delta H_{II}}{R} \tag{2.53}$$

$$\left(\frac{\partial \ln (a_{II}/a_I)}{\partial (1/T)}\right)_p = \left(\frac{\partial \ln (y_{II}^m m_{II}/y_I^m m_I)}{\partial (1/T)}\right)_p = \frac{-(H_{II} - H_I)}{R} = \frac{\Delta H_{II} - \Delta H_I}{R} \tag{2.54}$$

If Henry's law is obeyed because of the very low solubility of the polymorphs in the solvent, then $y_I^m = y_{II}^m = 1$,

$$\left(\frac{\partial \ln (m_{II}/m_I)}{\partial (1/T)}\right)_p = \frac{\Delta H_{II} - H_I}{R} \tag{2.55}$$

where

$$\Delta H_{II} - \Delta H_I = H_I - H_{II} = -\Delta H(I \to II), \tag{2.56}$$

$$\left(\frac{\partial \ln (m_{II}/m_I)}{\partial (1/T)}\right)_p = \frac{-\Delta H(I \to II)}{R} \tag{2.57}$$

This equation indicates that the solubility ratio (or ratio of escaping tendency) is independent of the solvent, as long as Henry's law is obeyed.

At some temperature T_t, known as the transition temperature, the two forms may be in equilibrium, that is,

$$0 = \Delta G(\text{I} \rightarrow \text{II}) = \Delta H(\text{I} \rightarrow \text{II}) - T_t \Delta S(\text{I} \rightarrow \text{II}) \tag{2.58}$$

or

$$\Delta H(\text{I} \rightarrow \text{II}) = T_t \Delta S(\text{I} \rightarrow \text{II}) \tag{2.59}$$

and the activities, fugacities, vapor pressures, and solubilities will be equal, that is,

$$m_\text{I} = m_\text{II} \tag{2.60}$$

When the two polymorphs are not in equilibrium, Eqs. 2.48 and 2.49 indicate that polymorph I will change spontaneously to polymorph II if polymorph I has the higher activity, fugacity, escaping tendency, vapor pressure, and solubility. The converse is also true. In other words, the less stable polymorph has the higher relative activity, absolute activity, escaping tendency, fugacity, vapor pressure, or solubility.

These thermodynamic relationships are helpful for characterizing and evaluating polymorphic systems. In particular, Eqs. 2.57 and 2.58 show that by means of solubility measurements at different temperatures it is possible to obtain both enthalpy and entropy differences between polymorphic forms of crystals.

The van't Hoff plots of log (solubility) in n-decanol against $1/T$ for methylprednisolone forms I and II (Higuchi et al., 1963), according to Eqs. 2.52

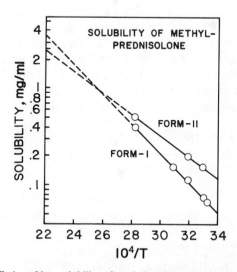

Figure 2.8 Van't Hoff plot of log solubility of methylprednisolone in water as a function of the reciprocal of the absolute temperature T; ○ refers to data for form I and form II. Reproduced with permission of the copyright owner, the American Pharmaceutical Association, from Higuchi, W. I., Lau, P. K., Higuchi, T., and Shell, J. W. (1963). *J. Pharm. Sci.*, **52**, 150–153.

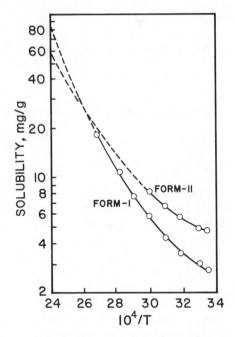

Figure 2.9 Van't Hoff plot of log solubility of the two forms of methylprednisolone in decanol as a function of the reciprocal of the absolute temperature. Reproduced with permission of the copyright owner, the American Pharmaceutical Association, from Higuchi, W. I., Lau, P. K., Higuchi, T., and Shell, J. W. (1963). *J. Pharm. Sci.*, **52**, 150–153.

and 2.53, are shown in Figures 2.8 and 2.9. Since form II is the more soluble form (i.e., is metastable), its solubility is not the equilibrium value. Examination of the linearity of the plots shows that the standard enthalpies of solution, ΔH_I and ΔH_{II}, are independent of temperature in water, but are dependent on temperature in n-decanol. Then enthalpies of solution are all positive (endothermic), ΔH_I being greater than ΔH_{II}. The two curves intersect at the transition temperature T_t (118°C). The van't Hoff plots of $\log(m_{II}/m_I)$ (that is, log ratio of the solubilities in three different solvents: n-dodecanol, n-decanol, and water) against $1/T$ according to Eq. 2.57 show (Fig. 2.10) a single straight line whose slope gives:

$$\Delta H_{(I \to II)} = \Delta H_I - \Delta H_{II} = 1600 \text{ cal mol}^{-1} = 6700 \text{ J mol}^{-1}$$

when $m_{II}/m_I = 1$, $T = T_t = 391 \text{ K} = 118°C$. The deviations at low temperatures are attributable to deviations from Henry's law, that is, $y_{II} \neq y_I \neq 1$. Application of Eq. 2.59 affords:

$$\Delta S_{(I \to II)} = \Delta S_I - \Delta S_{II} = 4.1 \text{ cal K}^{-1} \text{mol}^{-1} = 17.1 \text{ J K}^{-1} \text{mol}^{-1}$$

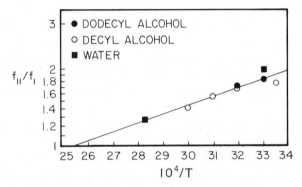

Figure 2.10 A plot of the logarithm of the fugacity or solubility ratio for the two forms of methylprednisolone as a function of solvent and the reciprocal of the absolute temperature T. This plot shows that fugacity, that is, the solubility ratio, is independent of the solvent. Reproduced with permission of the copyright owner, the American Pharmaceutical Association, from Higuchi, W. I., Lau, P. K., Higuchi, T., and Shell, J. W. (1963). *J. Pharm. Sci.*, **52**, 150–153.

Thus, the ratio of solubilities, m_{II}/m_I, the transition temperature, and the enthalpy and entropy of transition are independent of the solvent in accordance with Eqs. 2.57 and 2.59. The entropy difference may be a result of greater localization of the functional groups in the sidechain of form I, resulting from intermolecular and intramolecular interactions. Loss of freedom or rotation about the carbon–carbon single bond contributes about ($R \ln 3$) per methylene group to the entropy of fusion for long paraffin chains. Thus, the effects of two or more linkages in the sidechain may account for the observed molar entropy of transition.

The results for two polymorphic forms of methylprednisolone demonstrate that significant differences in solubilities and other thermodynamic behavior may exist among polymorphs. At room temperature, form II has the higher solubility (a factor of about 1.8 greater than that of I) and is, therefore, the less stable, more active form. The lower limits of the solubility of amorphous or supercooled liquid methylprednisolone in Figure 2.8 at various temperatures were determined as the concentration of a steroid in aqueous solution necessary for rapid appearance of turbidity when a concentrated solution of the steroid in dimethylformamide was added dropwise to water. This polymorphic form is about 20 times more soluble than I. From X-ray diffraction and other structural methods, it is always found that the amorphous, vitreous, glassy, or supercooled liquid form has a noncrystalline structure similar to that of the liquid state, in which long-range order is lost, but short-range order is present. The amorphous form always has a higher thermodynamic activity and solubility than the normal crystalline form (I in the case of methylprednisolone).

As discussed in Section 3.9 and 3.10, the thermodynamic activity of a pure drug controls the chemical, biological, and pharmaceutical properties of the drug. It is a general truth that the more unstable physical form of the pure drug has higher values of the following properties: free energy, chemical potential, thermodynamic activity, fugacity, escaping tendency, vapor pressure, solubility in any solvent, dissolution rate in any solvent, rate of any chemical reaction including decomposition, and rate and extent of absorption and bioavailability in the body. Different physical forms include different polymorphs and pseudo-polymorphs. Solvates present a somewhat different situation, owing to the presence of molecules of solvent in the crystal lattice of the main component.

2.4   THERMODYNAMIC THEORY OF SOLUBILITY OF SOLVATES

The solubility of a nonsolvated form of a crystalline nondissociating organic compound A in water is proportional to the equilibrium constant K_s in the equilibrium

$$A(solid) \overset{K_s}{\rightleftharpoons} A(aqueous) \tag{2.61}$$

The solubility depends on temperature, pressure, and polymorphic form and is approximately proportional to the activity of the latter. The solubility of a hydrate of A, that is, $A \cdot mH_2O$, in water is similarly proportional to the equilibrium constant K'_s in the equilibrium

$$A \cdot mH_2O(solid) \overset{K'_s}{\rightleftharpoons} A(aqueous) + mH_2O \tag{2.62}$$

The formation of hydrated crystals from anhydrous crystals is represented by the equilibrium

$$A(solid) + mH_2O \overset{K_h}{\rightleftharpoons} A \cdot mH_2O(solid) \tag{2.63}$$

$$K_h = \frac{a[A \cdot mH_2O(solid)]}{a[A(solid)]a[H_2O]^m} \tag{2.64}$$

The hydrated solid $A \cdot mH_2O(solid)$ will be more stable than the anhydrous solid $A(solid)$ when $K_h > 1$, that is, when $a(H_2O) > [a(A \cdot mH_2O(solid)]/ \{a[A(solid)]K_h\}]^{1/m}$. The anhydrous solid will be the more stable form in the inverse situation when $K_h < 1$. The corresponding relations hold not only for hydrates, but for other solvates and indeed for any situation in which a solid complex is formed. The activity of water may be reduced by dilution of the liquid with a suitable cosolvent such as ethanol, acetone, or dioxan.

A rule applying to solubility behavior is that solid solvates are always less soluble in the solvent forming the solvate than is the original solid. Thus, hydrated crystals are less soluble in water than the corresponding anhydrous solid. Solvates formed from other solvents, if the solvent is water-miscible, are more soluble in water than the corresponding nonsolvated form. In effect, in these situations the free energy of solution of the solvent into the water acts to increase the apparent solubility of the solvate. For example, caffeine hydrate is much less soluble in water than anhydrous caffeine, but the hydrate is very much more soluble in ethanol than anhydrous caffeine.

Since Eq. 2.61 is given by adding Eqs. 2.62 and 2.63,

$$K_h = \frac{K_s}{K_s'} \tag{2.65}$$

and

$$\Delta G_h^\theta = -RT \ln K_h = -RT \ln \left(\frac{K_s}{K_s'} \right) \tag{2.66}$$

where ΔG_h^θ is the standard free energy of hydration which is represented by Eq. 2.63. The ΔG_h^θ can be calculated from Eq. 2.66 by inserting the solubilities of the anhydrous and hydrated forms in place of K_s and K_s', respectively. The corresponding standard enthalpy and entropy of hydration, ΔH_h^θ and ΔS_h^θ, can be calculated from the temperature dependence of the solubility ratio (K_s/K_s') using the van't Hoff isochore (Eq. 1.9).

The solubility product K_{sp} in water of a crystalline solvate of A ($A \cdot nB$ where B is the solvating solvent, usually organic) is determined from an equation analogous to Eq. 2.62, thus:

$$A \cdot nB \, (\text{solid}) \overset{K_{sp}}{\rightleftharpoons} A \, (\text{aqueous}) + nB \, (\text{aqueous}) \tag{2.67}$$

in which

$$K_{sp} = [A][B]^n \tag{2.68}$$

Since Eq. 2.67 can be derived by combining a series of chemical equilibria (e.g., Eq. 2.61), K_{sp} and the standard free energy change can be evaluated from (1) the solubility of solvated A in pure B, (2) the solubility of unsolvated A in B, (3) the solubility of the same unsolvated A in water, and (4) the activity coefficient of B in water or the reciprocal of the solubility of B in water if low (Eq. 2.104). An example of this concept is provided by cefamandole sodium methanol solvate (Pikal et al., 1983). Equation 2.67 usually represents a far greater free energy difference than Eqs. 2.61 and 2.62.

A theoretical and experimental study of the solubility of solvates was made by Shefter and Higuchi (1963). Subsequently, the pharmaceutical aspects of solvates and inclusion compounds were reviewed by Haleblian (1975), Frank (1975), Byrn (1982), while more general aspects have been treated by MacNicol et al. (1978), Atwood et al. (1984), and Atwood and Davies (1987).

2.5   SOLUBILITY AND DISSOLUTION RATE OF SOLVATES

The thrust of this section is to show how the intrinsic solubility of a substance depends on the particular solid phase (solvate or anhydrate) that is present. This may, however, change with time, owing to interaction of the solid phase with the liquid dissolution medium. Since the intrinsic solubility controls the dissolution rate of the solid, this section considers those aspects of dissolution kinetics that are relevant to solvates. A detailed treatment of dissolution rates is, however, postponed until Chapter 11, since it draws on material covered throughout this book.

If dissolution is a diffusion-controlled process, the intrinsic dissolution rate J of a solid A, which is the rate of dissolution per unit surface area (see Chapter 11, Eq. 11.8), is given by

$$J = \frac{dm_A}{dt}\frac{1}{A_s} = \frac{dc_A}{dt}\frac{V}{A_s} = \frac{D_A}{h}(c_A^* - c_A) \tag{2.69}$$

where m_A is the mass of A dissolved and c_A is the concentration of A dissolved in time t; dm_A/dt is the dissolution rate of A; A_s is the surface area of the dissolving solid; V is the volume of the dissolution medium; D_A is the diffusion coefficient of A; c_A^* is the concentration of A at the immediate vicinity of the crystal surface, which is normally equal to the solubility of the solid in the dissolution medium; and h is the thickness of the diffusion layer, which depends on the geometry of the system and agitation conditions.

If the solid dissolving is the solvate or adduct, $A \cdot nB$ of a drug A with an organic solvent, then the concentration c_B of dissolved B is given by

$$c_B = nc_A \tag{2.70}$$

where n is the stoichiometric number and c_A is the concentration of dissolved A. Writing Eq. 2.69 in tems of c_B, we obtain

$$J = \frac{dc_A}{dt}\frac{V}{A_s} = \frac{1}{n}\frac{dc_B}{dt}\frac{V}{A_s} = \frac{D_B}{nh}(c_B^* - c_B) \tag{2.71}$$

where D_B is the diffusion coefficient of B and c_B^* is the concentration of B at the immediate vicinity of the crystal surface, which is normally equal to n times the solubility of solid $A \cdot nB$ in the dissolution medium (cf. Eq. 2.70).

The case $n = 1$ applies to many solvates or adducts containing large molecules of organic solvents. The case $D_A = D_B$ is also a good approximation for many organic molecules in water. Under these circumstances, $c_A = c_B$ and Eq. 2.68 affords $c_B^* = K_{sp}/c_A^*$, so that Eq. 2.71 can be written

$$J = \frac{dc_A}{dt}\frac{V}{A_s} = \frac{D_A}{h}\left(\frac{K_{sp}}{c_A^*} - c_A\right) \tag{2.72}$$

In view of the factors that influence K_{sp} discussed above, $(K_{sp})^{1/2}$ for the solid adduct may well exceed the equilibrium solubility of unsolvated A.

Other simple relationships can be derived from the equations shown above when $n = 1$ but $D_A \neq D_B$. Thus, from Eqs. 2.69 and 2.68 for the solubility product, we obtain

$$J = \frac{D_A}{h}\left(\frac{K_{sp}}{c_B^*} - c_A\right) \tag{2.73}$$

and since

$$J = \frac{D_B}{h}(c_B^* - c_B) \tag{2.74}$$

$$c_B^* = J\frac{h}{D_B} + c_B \tag{2.75}$$

By substituting Eq. 2.75 into Eq. 2.73, the following quadratic equation is obtained:

$$J^2 + J\left(D_B\frac{c_B}{h} + D_A\frac{c_A}{h}\right) - D_B D_A\frac{K_{sp}}{h^2} + D_B D_A c_B\frac{c_A}{h^2} = 0 \tag{2.76}$$

When $c_A = c_B$, the quadratic is easily solved to yield

$$J = \frac{dc_A}{dt}\frac{V}{A_s} = \left\{\frac{c_A}{2h}\left((D_B - D_A)^2 + \frac{4K_{sp}D_A D_B}{c_A^2}\right)^{1/2} - (D_A + D_B)\right\} \tag{2.77}$$

This equation leads to a zero rate only when $c_A^2 = K_{sp}$, and does not represent a first-order approach to saturation as in the case of Eqs. 2.69 and 2.71.

If $n = 1$, $D_A \neq D_B$ and c_B is increased much more than c_A by addition of the solvating solvent to the dissolution media. Equation 2.76 can be solved and the solution simplified by a binomial expansion to give

$$J = \frac{D_A K_{sp} - D_A c_A}{h c_B} \tag{2.78}$$

This indicates that the addition of the solvent B to the solution will, by increasing c_B, decrease the dissolution rate of the solvate. The solvates of higher order than $n = 1$ lead to equations more complicated than Eq. 2.76, but the dissolution rates of these solvates also depend on their K_{sp} values.

As pointed out earlier, the solubility and dissolution rate in water of the anhydrous form are greater than those of the hydrated form at room temperature. This is because the hydrated solid form, if more soluble, cannot be prepared in a stable form, since it would spontaneously yield free water and the less soluble anhydrous form. Figure 2.11 illustrates this by comparing the dissolution curves of anhydrous and hydrated crystalline forms of theophylline. Other examples include the hydrated and anhydrous forms of cholesterol,

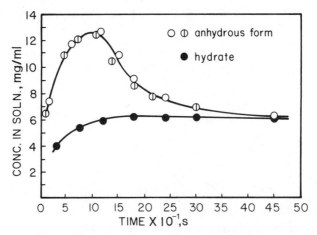

Figure 2.11 The dissolution–time curves for the anhydrous and hydrated crystalline forms of theophylline in water at 25°C. The two types of circles for the anhydrous form represent successive experimental runs. Reproduced with permission of the copyright owner, the American Pharmaceutical Association, from Shefter, E. and Higuchi, T. (1963). *J. Pharm. Sci.*, **52**, 781–791.

caffeine, and glutethimide. The maximum concentration observed may in some instances correspond to the solubility of the anhydrous crystalline phase, and in others may represent a temporary steady state in which the rates of dissolution of the metastable anhydrous form and the rate of crystallization of the stable hydrate are equal. The decreasing concentration represents crystallization of the stable hydrate from a solution supersaturated with respect to this species.

Succinylsulfathiazole forms several hydrates and polymorphs and a pentanol solvate. As expected from the discussion above and as shown in Figure 2.12,

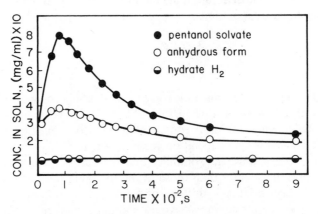

Figure 2.12 The dissolution–time behavior of anhydrous, hydrated (H_2), and pentanol–solvated forms of succinylsulfathiazole in a sulfuric acid solution ($0.5 \, mmol \, dm^{-3}$) at 25°C. Reproduced with permission of the copyright owner, the American Pharmaceutical Association, from Shefter, E. and Higuchi, T. (1963). *J. Pharm. Sci.*, **52**, 781–791.

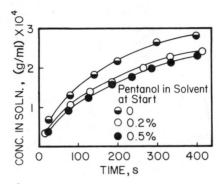

Figure 2.13 The influence of the addition of pentanol to the 15% aqueous ethanol solution on the dissolution of the pentanol solvate of fludrocortisone acetate at 20°C. Reproduced with permission of the copyright owner, the American Pharmaceutical Associaton, from Shefter, E. and Higuchi, T. (1963). *J. Pharm. Sci.*, **52**, 781–791.

the dissolution rate and solubility (maximum value) of the anhydrous form are higher than those of the hydrated form, H_2. On storage, however, the concentration of each of the solutions decreases from an apparently stable level to a lower, more stable level, each corresponding to a different crystalline hydrated form. The lowest level corresponds to the solubility of the most stable hydrate, H_1. The pentanol solvate gives a higher dissolution rate and solubility (maximum value) than the anhydrous form because the negative free energy of mixing of the organic solvent with water makes an additional contribution to the negative free energy of solution. Similarly, the solvates of other organic solvents that are appreciably soluble in water will have a higher dissolution rate and momentary solubility than the anhydrous form. The addition of the solvating organic solvent (e.g., pentanol) to the aqueous dissolution medium lowers the dissolution rate of the solvate (e.g., the pentanol solvate of fludrocortisone acetate) as shown in Figure 2.13, because the negative free energy of mixing of the released organic solvent with the dissolution medium is reduced, thereby reducing the contribution to the negative free energy of solution. Analogous behavior is expected for other solvates of organic compounds.

If the maximum concentration of solute released in the dissolution experiments (e.g., Fig. 2.11) corresponds to the solubility of the dissolving species, the initial increase in concentration with time obeys first-order kinetics according to the integrated form of Eq. 2.69, provided that the exposed crystalline surface area does not change significantly. This applies to glutethimide, theophylline, and succinylsulfathiazole. Van't Hoff plots of log (solubility of the anhydrous form) againt $1/T$, shown in Figures 2.14 and 2.15, give linear relationships, the slopes of which afford the standard enthalpy of solution ΔH_s^θ. The standard free energy of solution, ΔG_s^θ is given by

$$\Delta G_s^\theta = -RT \ln (\text{solubility}) \qquad (2.79)$$

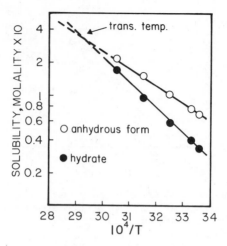

Figure 2.14 Van't Hoff-type plot of molal solubility (mol kg^{-1}, log scale) against the reciprocal of the absolute temperature for the anhydrous (N_1) and hydrated (H_1) forms of theophylline in water. Reproduced with permission of the copyright owner, the American Pharmaceutical Associaton, from Shefter, E. and Higuchi, T. (1963). *J. Pharm. Sci.*, **52**, 781–791.

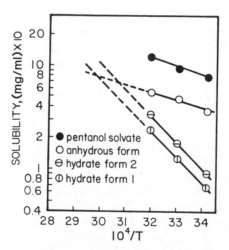

Figure 2.15 Vant' Hoff-type plot of solubility for the anhydrous form, hydrated form 1, and hydrated form 2 of succinylsulfathiazole in a $\cong 0.001$-N sulfuric acid solution (0.5 mmol dm^{-3}). The maximal concentration of the sulfa drug attained in solution with the pentanol solvate is included. Reproduced with permission of the copyright owner, the American Pharmaceutical Associaton, from Shefter, E. and Higuchi, T. (1963). *J. Pharm. Sci.*, **52**, 781–791.

and the standard entropy of solution is

$$\Delta S_s^\theta = \frac{\Delta H_s^\theta - \Delta G_s^\theta}{T} \tag{2.80}$$

Tables 2.3 and 2.4 show these thermodynamic quantities for solution of various anhydrous and hydrated organic compounds.

By analogy with the van't Hoff plots for different polymorphs, the temperature at the point of intersection of the van't Hoff line of each solvate with that of the anhydrous form represents the transition temperature at which the solvate and anhydrous form have equal solubilities, free energies, and stabilities. This is illustrated by Figures 2.14 and 2.15. The transition temperatures, shown in Tables 2.3 and 2.4, agree with the values determined for glutethimide and

Table 2.3 Thermodynamic Values Calculated for Anhydrous–Hydrated Systems of Glutethimide and Theophylline (1 cal = 4.184 J)

Compound	Transition Temperature (°C)	ΔH_s^θ (cal mol^{-1}) Hydrate	ΔH_s^θ (cal mol^{-1}) Anhydrous	ΔH_h (cal mol^{-1})	$\Delta G_h^{\theta a}$ (cal mol^{-1})	$\Delta S_h^{\theta a}$ (cal K^{-1} mol^{-1})	ΔS_h^b (cal K^{-1} mol^{-1})
Glutethimide	52	11,700	9,700	−2,000	−280	−5.8	−6.1
Theophylline	73	10,700	7,400	−3,300	−410	−10.0	−9.5

^{a}at 298 K.
^{b}at transition temperature.

Reproduced with permission of the copyright owner, the American Pharmaceutical Association, from Shefter, E. and Higuchi, T. (1963). *J. Pharm. Sci.*, **52**, 781–791.

Table 2.4 Thermodynamic Values Calculated for Anhydrous–Hydrated System of Succinyl Sulfathiazole (1 cal = 4.184 J)

Crystal Form	ΔH_s (cal mol^{-1})	ΔG_h^θ (298 K) (cal mol^{-1})	ΔS^θ (298 K)^a (cal K^{-1} mol^{-1})	Transition Temperature with Anhydrous Form (°C)
Form 1 hydrate	12,000	—	—	60
Form 2 hydrate	11,000	−210	Small + Value	51
Anhydrous form (N$_1$)	3,400	−880	−26	—

^{a}Calculated for the conversion to the most stable hydrate, form 1.

Reproduced with permission of the copyright owner, the American Pharmaceutical Association, from Shefter, E. and Higuchi, T. (1963). *J. Pharm. Sci.*, **52**, 781–791.

theophylline in a melting-point apparatus. Entropies of transition were calculated from Eq. 2.59 and are quoted in Table 2.3.

The enthalpy of hydration, ΔH_h, of an anhydrous organic compound is the enthalpy change in Eq. 2.63, and according to Hess's law it is equal to the enthalpy of solution of the anhydrous form (ΔH for Eq. 2.61) minus the enthalpy of solution of the hydrated form (ΔH for Eq. 2.62) (see Tables 2.3 and 2.4).

The free energy of hydration, ΔG_h, is the free energy change in Eq. 2.66, and at constant temperature and pressure is given by Eq. 2.66 in the form

$$\Delta G_h = -RT \ln \frac{\text{solubility of anhydrous form}}{\text{solubility of the hydrate}} \tag{2.81}$$

Some values of ΔG_h are listed in Tables 2.3 and 2.4.

The entropy of hydration, ΔS_h, can be calculated from the equation

$$\Delta S_h = \frac{\Delta H_h - \Delta G_h}{T} \tag{2.82}$$

The values of ΔS_h for glutethimide and theophylline which are listed in Table 2.3 are quite close to the molar entropy of freezing of water ($-6 \, \text{cal K}^{-1} \, \text{mol}^{-1} = -25 \, \text{J K}^{-1} \, \text{mol}^{-1}$ at 25°C).

This suggests that the energy involved in the transformation of the dehydrated form of these compounds to the hydrate is related mainly to the decrease in entropy of the water molecules in the hydrate structure. Throughout the lattice of the theophylline hydrate, the water molecules form a chain network in which one water molecule is hydrogen-bonded to two other water molecules and to one theophylline molecule. This type of structure could be responsible for the larger decrease in entropy associated with the hydration of theophylline.

From the foregoing discussion it can be seen that the structure of the solid state profoundly affects the dissolution rate and momentary solubility, which may be many times greater for an unstable physical form or polymorph, a nonsolvate, or solvate of an organic compound than for the stable physical form, polymorph or hydrate. This may be exploited in chemical processing and in pharmaceutical technology. Complexes (e.g., solvates) between the compound in question and a ligand (e.g., solvent) have higher dissolution rates and momentary solubilities in water than do the nonsolvate or hydrate. This is because the system utilizes, in effect, the free energy of dilution of the complexing agent to raise the solubility of the compound in question.

2.6 QUALITATIVE PREDICTION BASED ON SOLUBILITY PARAMETERS AND RELATED CONCEPTS

The simplicity of solubility predictions for ideal systems arises from the constancy of the activity coefficients, that is, $\gamma = 1$, if the standard state is the pure liquid form. For nonideal systems, γ is not constant but depends on the solvent, temperature, and other conditions. Many attempts at solubility prediction for nonideal systems use (1) Eq. 2.6, for which the pure (supercooled) liquid is the standard state, (2) the prediction of a_B as described above for an ideal solution, and (3) the prediction of γ_B by making special assumptions concerning the nature of the intermolecular interactions, to which we shall return later.

For the purpose of clarifying the changes in intermolecular interaction energies, the solution process was divided by Hildebrand and Scott (1950) and by Scatchard (1931) into three steps, as shown in Figure 2.16.

In the first stage, one molecule of solute surrounded by other solute molecules is transferred into the vapor state. If the potential energy between a pair of molecules is u_{BB}, the energy absorbed in breaking the interaction between adjacent molecules is $2u_{BB}$. When the molecules enter the vapor phase, the hole left in the solute closes and the pair-potential energy u_{BB} is liberated. The increase in potential energy in breaking a solute–solute, B–B, interaction in Stage 1 is therefore $+u_{BB}$ (positive).

In the second stage, a hole is created in the solvent just large enough to accept the solute molecule. If the potential energy between a pair of solvent molecules is u_{AA}, the energy absorbed in separating the members of a pair by breaking a solvent–solvent, A–A, interaction to produce a hole is $+u_{AA}$ (positive).

Finally, the free solute molecule in the vapor phase is transferred to fill the hole in the solvent, which is just big enough to receive it. If the potential energy of interaction between a solute molecule and a solvent molecule is u_{AB}, the liberated

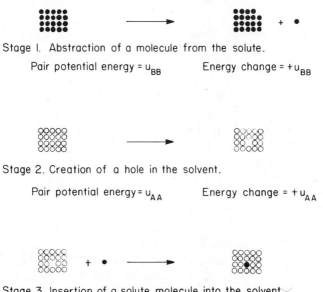

Stage 1. Abstraction of a molecule from the solute.

Pair potential energy = u_{BB} Energy change = $+u_{BB}$

Stage 2. Creation of a hole in the solvent.

Pair potential energy = u_{AA} Energy change = $+u_{AA}$

Stage 3. Insertion of a solute molecule into the solvent.

Pair potential energy = u_{AB} Energy change = $-2u_{AB}$

Figure 2.16   Hypothetical stages in the solution process.

energy is $2u_{AB}$, since two new pair interactions, A–B and B–A, are created on filling the hole. The energy liberated in creating the solute–solute interaction in Stage 3 is therefore $-2u_{AB}$ (negative).

This scheme is highly simplified since it involved only molecular pair potentials. The overall change in the pair-potential energy Δu is given by

$$\Delta u = u_{AA} + u_{BB} - 2u_{AB} \tag{2.83}$$

The intermolecular forces ensure that the hole created in the solvent has the same size as the solute molecule. The potential energies (u_{AA}, u_{BB}, and u_{AB}) can be expressed either in terms of the energy per unit *volume* of the solute molecule or in terms of the energy per unit *surface area* of the solute molecule. The molecular volume approach was originally developed by Hildebrand and Scott (1950), Scatchard (1931), and others in regular solution theory and in solubility parameter theory. The alternative surface area concept was developed by Yalkowsky et al. (1972, 1975, 1976), Amidon et al. (1974, 1975), and Valvani et al. (1976) as the molecular and group surface area (MGSA) approach.

Hildebrand (1929) stated, "A regular solution is one involving no entropy change when a small amount of one of its components is transferred to it from an ideal solution of the same composition, the total volume remaining unchanged," that is, $\Delta S_{mix} = \Delta S_{mix}^{id}$, $\Delta V_{mix} = 0$. No restriction is placed on ΔU_{mix} or ΔH_{mix}.

The energy or enthalpy of interaction between two different molecules, A and B, in a regular solution is assumed to be given by the geometric mean of the values between two identical molecules A with A and B with B, respectively. The theoretical justification for this is easiest to visualize from the point of view of London dispersion forces, which lead to the following intermolecular pair potential between two different molecules:

$$u_{AB} = \frac{2(I_A I_B)^{1/2}}{I_A + I_B}(u_{AA} u_{BB})^{1/2} \tag{2.84}$$

Since the values of the ionization energies I_A and I_B are fairly constant for many organic molecules (ca. 10.4 ± 1.3 eV) and much the same (ca. 11.9 ± 1.4 eV) for simple inorganic molecules, such as halogens, covalent halides, and water (Shinoda, 1978), the term $2(I_A I_B)^{1/2}/(I_A + I_B)$ is close to unity, so

$$u_{AB} \cong (u_{AA} u_{BB})^{1/2} \tag{2.85}$$

This is the geometric mean rule. If u_{AB} were amended to be given by an arithmetic mean rule, that is, $u_{AB} = (u_{AA} + u_{BB})/2$, Eq. 2.83 would become $\Delta u_{mix} = 0$ and the mixture would behave ideally. Assuming the geometric mean rule, Eq. 2.83 becomes

$$\Delta u_{mix} = u_{AA} + u_{BB} - 2(u_{AA} u_{BB})^{1/2}$$
$$= \{u_{AA}^{1/2} - u_{BB}^{1/2}\}^2 \tag{2.86}$$

Since the geometric mean of two quantities is always smaller than the arithmetic mean, if u_{AA} and u_{BB} differ, Δu_{mix} is always positive. The solvent–solvent interaction energy u_{AA}, when summed over all the molecues, is represented by the cohesive energy density, that is, the molar energy of vaporization, ΔU_A^v, divided by the molar volume V_A; this quantity is represented by the square of the so-called "solubility parameter" δ_A, thus

$$u_{AA} = \frac{\Delta U_A^v}{V_A} = \delta_A^2 \quad \text{or} \quad u_{BB} = \frac{\Delta U_B^v}{V_B} = \delta_B^2 \tag{2.87}$$

Equation 2.86 can be written as

$$\Delta U_{mix} = (\delta_A - \delta_B)^2 \tag{2.88}$$

from which the following equation for the acitivity coefficient can be derived:

$$\ln \gamma_B = \frac{V_B \phi_A^2 (\delta_A - \delta_B)^2}{RT} \tag{2.89}$$

where V_B is the molar volume of the solute and ϕ_A is the volume fraction of the solvent. Combination with Eq. 2.5 affords

$$\ln x_B = \ln a_B - \frac{V_B \phi_A^2 (\delta_A - \delta_B)^2}{RT} \tag{2.90}$$

This is known as the Hildebrand solubility equation and enables solubilities in regular solutions to be predicted. The equation requies knowledge of the solubility parameters of the solvent δ_A, and of the solute, δ_B. Solubility parameters may be calculated from the molar volume at the temperature of interest and the molar energy of vaporization, which may itself be estimated from the boiling point using the Hildebrand rule (1962) or other methods.

Solubility parameters have been the subject of several reviews (Barton, 1975, 1983; Kumar and Prausnitz, 1975; Snyder, 1978–1980) as well as being discussed extensively in the books by Hildebrand and Scott (1950, 1962) and Hildebrand et al. (1970). Values of δ have been tabulated in various reference works and some examples are quoted in Table 2.5. The square of the solubility parameter (i.e., the cohesive energy density) can be equated for nonpolar liquids with the internal pressure $(\partial U / \partial V)_T$. The stronger the London dispersion forces, the greater the value of δ and of related quantities.

The standard state of unit activity for regular solutions, as stated earlier, is generally taken to be the pure liquid state, which is a hypothetical supercooled state for solids, so that a^s can be calculated using the same equations and procedures as for ideal solutions. Regular solution theory predicts that the smaller the difference between δ_A and δ_B, the smaller is the activity coefficient

Table 2.5 Classification of Solvent Properties for Selected Pure Liquids

Solvent	Hildebrand	Hansen's Values[b]				Karger's Values[c]				Rohrschneider's[d] and Snyder's[e] System				
	δ	δ	(δd)	(δp)	(δh)	(δd)	(δp)	(δa)	(δb)	(P')[d]	(χe)[e]	(χd)[e]	(χn)[e]	(e)[f]
Isooctane	7.0					7.0	0.0	0.0	0.0	0.1				—[g]
n-Hexane	7.3	7.24	7.23	0.0	0.0	7.3	0.0	0.0	0.0	0.1				—[g]
Ethyl ether	7.4	7.62	7.05	1.4	2.5	6.7	2.4	0.0	3.0	2.8	0.55	0.11	0.34	I
Triethyl amine	7.5	7.4				7.5	0.0	0.0	4.5	1.9	0.61	0.07	0.32	I
Cyclohexane	8.2	8.18	8.18	0.0	0.0	8.2	0.0	0.0	0.0	0.2				—[g]
Carbon tetrachloride	8.6	8.65	8.65	0.0	0.0	8.6	0.0	0.0	0.0	1.6	0.30	0.38	0.32	—[g]
Ethyl acetate	8.9	9.10	7.44	2.6	4.5	7.0	4.0	0.0	2.7	4.4	0.34	0.25	0.42	VI
Tetrahydrofuran	9.1	9.52	8.22	2.8	3.9	7.6	3.5	0.0	3.7	4.0	0.41	0.19	0.40	III
Benzene	9.2	9.15	8.95	0.5	1.0	9.2	0.0	0.0	0.0	2.7	0.29	0.28	0.43	VII
Chloroform	9.3	9.21	8.65	1.5	2.8	8.1	3.0	6.5	0.5	4.1	0.28	0.39	0.33	VIII
Methylene chloride	9.6	9.93	8.91	3.1	3.0	6.4	3.0	0.0	3.0	3.4	0.34	0.17	0.49	V
Acetone	9.7	9.77	7.58	5.1	3.4	9.1	5.1	0.0	3.0	5.1	0.36	0.24	0.40	VI
Carbon disulfide	10.2	9.97	9.96	0.0	0.0	10.2	0.0	0.0	0.0	0.3				—[g]

Solvent														
Dioxane	10.1	10.0	9.30	0.9	0.3	7.8	5.2	0.0	4.6	4.8	0.38	0.21	0.41	VI
Dimethlformamide	11.8	12.14	8.52	6.7	5.5	7.9				6.4	0.41	0.21	0.38	III
Propanol	12.0	11.97	7.75	3.3	8.5	7.2	2.6	5.3	6.3	3.9	0.53	0.21	0.26	II
Dimethyl sulfoxide	12.0	12.93	9.00	8.0	5.0	8.4	6.1	0.0	5.2	7.2	0.35	0.27	0.38	III
Acetonitrile	12.1	11.9	7.50	8.8	3.0	6.5	8.2	0.0	3.8	5.8	0.33	0.26	0.41	VI
Ethanol	12.7	12.92	7.73	4.3	9.5	6.8	3.4	6.9	6.9	4.3	0.51	0.21	0.28	II
Ethylene glycol	14.3					8.2				6.9	0.47	0.23	0.30	IV
Methanol	14.5	14.28	7.42	6.0	10.9	6.2	4.9	8.3	8.3	5.1	0.51	0.19	0.30	II
Formamide	19.2	17.8	8.4	12.8	9.3	8.3				9.6	0.40	0.28	0.32	IV
Water	23.4	23.5	8.65	0.5	1.5	6.3				10.2	0.40	0.34	0.26	VIII

[a] The superscripts *inside* the parentheses are symbols used in the literature. The superscripts *outside* the parentheses refer to the footnotes. The units of all the δ and P' values are cal$^{1/2}$ cm$^{-3/2}$. χ_e, χ_d, and χ_n are dimensionless quantities. 1 cal = 4.184 J.

[b] Burrell (1975); Hoy and Martin (1975); Hansen and Beerbower (1971).

[c] Karger et al. (1976, 1978); Snyder (1980).

[d] Rohrschneider (1973).

[e] Snyder (1979, 1980).

[f] Selectivity group.

[g] Nonpolar, that is, no polar selectivity.

Adapted with permission from Snyder, L. (1980). *Chemtech*, March 1980, 188–193. Copyright © 1980, American Chemical Society.

(Eq. 2.89) and the higher is the solubility (Eq. 2.90). When $\delta_A = \delta_B$, γ_B reaches a minimum value of unity, the solubility is maximal, and the mixture behaves ideally. Thus, the theory does not account for situations in which γ_B is less than unity, corresponding to negative deviations from Raoult's law and to appreciable affinity between solute and solvent.

Figure 2.17 illustrates these points for the practical case of testosterone propionate studied by James and Roberts (1968), who found that the solubility is closest to the ideal value when the solubility parameter of the solvent is equal to that of the solute. Although the two straight lines in Figure 2.17 should really be replaced by a single continuous smooth curve, this replacement will not significantly influence the value of the solubility parameter that corresponds to the maximum solubility. Hildebrand and Scott (1962) consider that regular solutions are formed by esters only when the dipole is well buried within the molecule.

In the case of mixed-solvent systems, regular solution theory indicates that the solubility parameter δ of the mixture should be well represented by the equation

$$\delta = \phi_A \delta_A + \phi_B \delta_B + \cdots \tag{2.91}$$

where A and B refer to individual solvents, ϕ_A and ϕ_B are their volume fractions in the solvent mixture, and δ_A and δ_B are their solubility parameters.

The Hildebrand solubility equation (Eq. 2.90) provides good predictions of the solubility of nonpolar solutes in nonpolar solvents, that is, when the primarily intermolecular forces are London dispersion forces. Such solute–solvent combinations may therefore be classed as regular solutions, and include mixtures of fluorocarbons with hydrocarbons. Because of the assumptions made in the derivation of Eq. 2.90, it cannot be expected to provide good predictions of

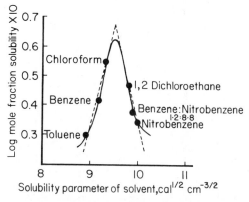

Figure 2.17 Dependence of the solubility of testosterone propionate on the solubility parameter of the solvent. Reproduced with permission of the copyright, owner, the Pharmaceutical Society of Great Britain, from James, K. C. and Roberts, M. (1968). *J. Pharm. Pharmacol.*, **20**, 709–714.

solubility under the following circumstances: (1) when the solute and solvent molecules are polar, that is, when dipole–dipole, dipole–induced-dipole, charge-transfer, or hydrogen-bonding interactions occur, so that the geometric mean rule breaks down; (2) when specific molecular orientation effects occur in solution, often as a result of hydrogen bonding, so that the entropy of mixing is no longer ideal; and (3) when the solute and solvent molecules have rather different sizes, so that the volume of mixing is not zero.

Although these points have been made by Hildebrand and co-workers, many authors still attempt to use solubility parameters and the Hildebrand equation (Eq. 2.90) for solubility prediction in these unsuitable cases, because of the relative simplicity of the regular solution concept and because the errors involved in making the various assumptions often mutually cancel.

Regular solution theory usually fails to predict solubility when dipoles are present, that is, for nonpolar or polar solutes in polar solvents, such as methylene chloride or nitromethane. This lack of predictability is particularly marked for systems containing hydrogen-bonded species, such as water or an alcohol, mainly because hydrogen-bonding interactions are stronger than London dispersion forces and are specifically oriented. The simple theory also fails for solutions of iodine in polar solvents owing to specific charge-transfer interactions which also change the absorption spectrum (i.e., color).

The experimentally determined solubility parameters for some common solvents are often significantly larger than the values calculated from dispersion interactions alone. Regular solution theory partially corrects for this effect by incorporating these extra interactions into the final value of δ. These extra interactions will depend on the other species present and will vary as these change. For example, carbon disulfide and nitrobenzene have similar solubility parameters, $\sim 10 \, cal^{1/2} \, cm^{-3/2}$, but quite different solvency properties. Carbon disulfide readily dissolves aromatic hydrocarbons ($\delta = 9$ to $10 \, cal^{1/2} \, cm^{-3/2}$) and phosphorus ($\delta = 13.1 \, cal^{1/2} \, cm^{-3/2}$), owing to the overwhelming predominance of dispersion forces. Nitrobenzene permits a lower solubility of these compounds, but is a good solvent for compounds with appreciable dipole moments or polar groups.

Thus, if the molecules are polar or are undergoing specific interactions, they will take up preferred orientations, lowering both the energy (and enthalpy) and the entropy of the system. The geometric mean rule and the basic assumption that the entropy of mixing is ideal will then not be valid, and regular solution theory is not equipped to cope with this situation. However, in certain special cases it can be argued that the lowering of both ΔH and ΔS from regularity could cancel each other in the equation $\Delta G = \Delta H - T\Delta S$, leaving the solubility much the same as that predicted by theory. To take another example, the polar hydrogen-bonded liquid water is often assigned (Hildebrand and Scott, 1962) a δ value of $23.4 \, cal^{1/2} \, cm^{-3/2}$, which accounts quite well for its solubility in saturated hydrocarbon solvents (Black et al., 1948) ($x_{H_2O} \cong 10^{-4}$ to 10^{-3}) at 25°C, partly because the solutions are so dilute that the solute–solvent interactions are mainly London dispersion forces. However, $\delta = 15.65 \, cal^{1/2} \, cm^{-3/2}$ for water (Davis et al., 1972) may better fit the solubility behavior of hydrocarbons (and of the methylene group of other organic compounds) in water at 25°C.

2.7 EXTENSIONS OF REGULAR SOLUTION THEORY

Attempts have been made to modify or "patch up" (Taft et al., 1969; Rohrschneider, 1973) the Hildebrand solubility parameter treatment to improve its accuracy, applicability and predictability. One attempt is from Hansen and co-workers (Hansen and Beerbower, 1971; Burrell, 1975; Hoy and Martin, 1975), who introduced partial or three-dimensional solubility parameters (δ^d, δ^p, δ^h) to predict the solubilities of polymers in various solvents. The overall solubility parameter δ is then given by

$$\delta^2 = (\delta^d)^2 + (\delta^p)^2 + (\delta^h)^2 \qquad (2.92)$$

where δ^d, δ^p, and δ^h refer to dispersion, polar, and hydrogen-bonding interactions, respectively. Karger et al. (1976, 1978) extended this approach by subdividing the hydrogen-bonding interactions into mutual acid–base interactions between the molecule A as an acid (partial solubility parameter δ^a_A) and the molecule B as a base (partial solubility parameter δ^b_B) and vice versa, thus,

$$\delta^h_{AB} = \delta^a_A \delta^b_B + \delta^b_A \delta^a_B \qquad (2.93)$$

This concept has been criticized by Snyder (1978–1980), mainly because of its unsatisfactory assumptions and complexity.

Another attempt at extending solubility parameters has been proposed by Rohrschneider (1973), who represented each type of solute interaction by means of model solute, such as ethanol as a hydrogen bond donor, dioxane as a hydrogen bond acceptor, and nitromethane as a dipolar molecule. The relative solubility of these model solutes in the solvent in question was determined and corrected for dispersion forces by comparing the result with an equivalent nonpolar species known as a homomorph.

After correction for dispersion interactions (Snyder, 1978), a net solution energy for each solute in a given solvent was obtained: $\log(K''_g)$ethanol, $\log(K''_g)$dioxane, and $\log(K''_g)$nitromethane. These three terms were assumed to give an overall solvent polarity value P', which can be compared with the solubility parameter δ. Values of P' for common solvents are listed in Table 2.5. Plots of P' against δ generally give a good positive curvilinear correlation. However, solvents which are strong electron donors (i.e., proton acceptors), but have no proton-donor ability (e.g., diethyl ether and triethylamine) fall on a different curve displaced toward lower δ values. These solvents are moderately polar on the Rohrscheider scale (P' between 1.8 and 2.9), but their δ values are similar to the alkanes (7–8 cal$^{1/2}$ cm$^{-3/2}$). This is because solubility parameters are based on the properties of the pure solvent in which the strong proton acceptor or electron donor properties are not apparent. The P' values of diethyl ether and trimethylamine more accurately reflect their interactions than do values of δ.

Snyder (1978–1980) calculated the individual fractional contributions of proton-acceptor interactions, χ_e, proton-donor interactions, χ_d, and polar interactions, χ_n, by expressing the individual $\log(K'')$ terms as fractions of P' in the following manner:

$$\chi_e = \frac{\log(K_g'')\,\text{ethanol}}{P'} \tag{2.94}$$

where χ_e is the proton-acceptor selectivity parameter. The proton-donor selectivity parameter χ_d and the dipole selectivity parameter χ_n are similarly defined. These solvent selectivity parameters are somewhat analogous to Karger's δ^b, δ^a, and δ^p values. When Snyder's selectivity parameters are plotted on triangular coordinates, as in Figure 2.18, solvents within the same homologous series and with the same functional groups are represented by points which are close together and which can be circumscribed by circles representative of the respective chemical characteristics summarized in Table 2.6.

Snyder's classification of solvents is shown in Table 2.6. It appears from this that dispersion interactions play an insignificant role in solvent selectivity. This is an important point which is emphasized in this book and which is relevant to the chemical, biological and pharmaceutical sciences. Snyder (1978–1980) also proposed a useful empirical scheme for choosing the best solvent for maximizing solubility of a particular compound or maximizing the ratio of solubilities in two different solvents for separation. First, a suitable solvent polarity is selected. Second, a suitable solvent selectivity group is chosen. The solubility of a solute in

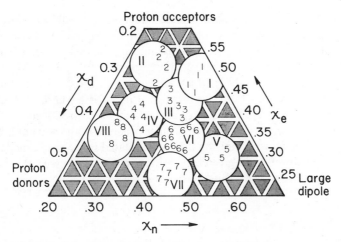

Figure 2.18 Grouping of pure solvents by selectivity, in terms of fractional polar-selectivity values χ_e, χ_n, and χ_d. Reprinted with permission from Snyder, L. (1980). *Chemtech*, March 1980, 188–193. Copyright © 1980 by the American Chemical Society.

Table 2.6   Classification of Solvent Selectivity in Figure 2.18

Group	Solvents
I	Aliphatic ethers, tetramethylguanidine, hexamethyl phosphoric acid amide, trialkylamines
II	Aliphatic alcohols
III	Pyridine derivatives, tetrahydrofuran, amides (except formamide), glycol ethers, sulfoxides
IV	Glycols, benzyl alcohol, acetic acid, formamide
V	Methylene chloride, ethylene chloride
VI	(a) Tricresyl phosphate, aliphatic ketones and esters, polyethers, dioxane
	(b) Sulfones, nitriles, propylene carbonate
VII	Aromatic hydrocarbons, halosubstituted aromatic hydrocarbons, nitro compounds, aromatic ethers
VIII	Fluoroalkanols, m-cresol, water, chloroform

Reproduced with permission from Snyder, L. (1980). *Chemtech*, March 1980, 188–193. Copyright © 1980, American Chemical Society.

solvent of the same selective group (Figure 2.18 and Table 2.6) will tend to approach that for ideal solutions, as discussed in Section 2.2. The solubility of a solute in a given solvent will tend to increase with increasing propensity for specific interaction between the solute and the solvent, for example, Group VIII solutes, being good proton donors, tend to be highly soluble in solvents of Groups I, II, and III, which are good proton acceptors. This concept is introduced in Section 2.11.8 and is discussed fully in Chapters 4 and 5. Finally, the actual solvent is selected to provide the most appropriate "balance" of properties.

Martin and co-workers (1980, 1981) and Adjei et al. (1980) have proposed an extended Hildebrand solubility parameter approach to correlate the solubility of polar solutes, such as caffeine and theophylline, in polar solvents, such as water, ethanol, dioxane, glycerol, and polyethylene glycols. In essence, Martin et al. expressed the Δu term (Eq. 2.83) which appears as $(\delta_A^2 - \delta_B^2)$ in the Hildebrand solubility equation in the alternative form $(\delta_A^2 + \delta_B^2 - 2W)$, where W is an energy correction term which represents $2u_{AB}$ and which was introduced to allow for nondispersion interactions between molecules of solute A and solvent B. Such an approach would be more promising if W were related to polar characteristics of the molecule. However, for a given solute, Martin et al. expressed W as a power series in δ_A for each solvent, thus,

$$W = a + b\delta_A + c\delta_A^2 + d\delta_A^3 + \cdots \tag{2.95}$$

where a, b, and c are adjustable parameters to which numerical values were assigned by a computer to give the best prediction of the solubility of the solute in the series of solvents. The fit was, of course, excellent, since it is possible to fit any data, including solubility, with a sufficient number of adjustable parameters. Although this approach can be used to correlate and possibly to condense experimental data, its predictive capability is doubtful.

In the extensions of regular solution theory, the Hildebrand solubility equation (Eq. 2.90) is playing a purely empirical role, since the original assumptions of regular solution theory do not apply. Extend solubility parameters and associated equations and concepts have been successfully applied by technologists working with polar solvents in a number of industries (Kumar and Prausnitz, 1975), including the paint and polymer industries. But as we have seen, these approaches are essentially empirical.

2.8   THE MOLECULAR AND GROUP-SURFACE-AREA (MGSA) APPROACH

The molecular and group-surface-area (MGSA) approach is an important variation and extension of the Scatchard–Hildebrand treatment of interactions in solution which has been developed by Yalkowsky et al. (1972, 1975, 1976), Amidon et al. (1974, 1975), and Valvani et al. (1976). In this treatment, a two dimensional analogy of Eq. 2.83 and 2.89 is derived. Reversible work W, which is effectively a Helmholtz free-energy change ΔA, expresses the molecular pair interactions in solution instead of potential energy, which is effectively an internal energy change ΔU. The use of reversible work indicates that entropy changes are being included as well as energy changes.

This treatment goes beyond the regular solution concept, which assumes ideal entropy of mixing, and is suitable for considering the solubility of nonelectrolytes in water and in other polar solvents whose high degree of order is altered by the presence of the solute. Clearly, in such solvents dispersion forces play only a minor role. Consequently, the geometric mean approximation in Eq. 2.85 is invalid, and Eqs. 2.86 and 2.89 do not apply. However, Eq. 2.83 is valid and so Eq. 2.89 can be written in the modified form

$$\ln \gamma_B = (W_{AA} + W_{BB} - 2W_{AB}) \frac{V_B \phi_A^2}{RT} \qquad (2.96)$$

in which reversible work W replaces potential energy U. As before, V_B is the partial molar volume of the solute and ϕ_A is the volume fraction of the solvent, which for dilute solution is close to unity. If W_{AB} could be measured directly without recourse to any form of averaging, this equation would be most useful for predicting solubilities and a distinct advantage over the solubility parameter treatment.

In the two-dimensional analogy of Eq. 2.96, initially proposed by Yalkowsky et al. (1976), molar surface area A replaces molar volume V. The terms W_{AA} and W_{BB} are replaced by the surface free energies (surface tensions) σ_A and σ_B of the pure liquids A and B, while W_{AB} is replaced by the interfacial free energy (interfacial tension) σ_{AB} between the two liquids.

Referring to Figure 2.16, the work required to remove a solute molecule from a bulk phase in Stage 1 is equal to the surface area created multiplied by the surface tension of the liquid. Since removal of a single molecule from the bulk liquid phase does not significantly change the bulk surface area, only half the work of cohesion, W_B^c, is involved in the process, so that

$$W_{BB} = 0.5\,W_B^c = \sigma_B A_B \qquad (2.97)$$

Similarly, in Stage 2 the work done on the system to create a cavity in the solvent phase just big enough to accommodate the solute is given by half the work of cohesion of the solvent W_A^c, so that

$$W_{AA} = 0.5\,W_A^c = \sigma_A A_B \qquad (2.98)$$

Lastly, in Stage 3 the work done by the system in inserting the solute molecule into the solvent cavity is equal to the work of adhesion W_{AB}^a between the two species, so that

$$2W_{AB} = W_{AB}^a = +(\sigma_A + \sigma_B - \sigma_{AB})A_B \qquad (2.99)$$

Introducing the expressions of the last three equations into Eq. 2.96 gives us

$$\ln \gamma_B = \frac{\sigma_{AB} A_B}{kT} \qquad (2.100)$$

The derivation of this equation is analogous to the derivation of Eq. 2.89. The interfacial tension σ_{AB} is easily measured for substance of different polarity when solubility parameters cannot be used, but becomes very difficult to measure when the polarities are low and very similar, a condition to which the geometric mean rule and solubility parameters may be applied. Thus, the two approaches are complementary. The application of the MGSA approach to the quantitation and prediction of the solubility behavior of organic compounds in water is developed in Section 8.30.

2.9 PREDICTION OF SOLUBILITY IN WATER FROM PARTITION COEFFICIENTS

As explained in Chapter 1, some attention is given to partition coefficients because they are valuable quantities in themselves and can be used to predict solubility.

The partition coefficient of a solute between two phases α and β is related to activity coefficients at infinite dilution as follows:

$$P_{\alpha/\beta}^x = \frac{(\gamma^\beta)^\infty}{(\gamma^\alpha)^\infty} \qquad (2.101)$$

the concentration scales being in mole fraction units.

Yalkowsky and Valvani (1980) suggested that for most drugs γ_w^∞ (in water) can be equated with $P_{o/w}^x$ ($= x$ in octanol/x in water) because the solubility parameters of most drugs are similar to that of octanol; that is, in octanol γ^∞ is assumed to be unity, meaning that the solution is assumed to be ideal. In fact, $\delta/\text{cal}^{1/2}\,\text{cm}^{-3/2}$ = 10.3 for n-octanol and lies between about 8 and 12 for most drugs.

A great advantage of this approach is that these partition coefficients are simply related, by Eq. 7.26 in Section 7.2.2, to the partition coefficients calculated from molarity in n-octanol/molarity in water, which have been determined or can be predicted (Hansch and Leo, 1979) for the purpose of developing structure–activity relationships in medicinal chemistry (Hansch and Fujita, 1964). As we shall see in Chapter 8, activity coefficients in water, solubility in water, and partitioning between water and organic liquids can be predicted rather success-fully using the group contribution approach.

Implicit in this treatment are a number of linear free-energy relationships, including the following:

$$\log \gamma^\infty = d \log P^x + e \tag{2.102}$$

where P^x is the partition coefficient based on mole fraction, while d and e are constants. The assumption that $\gamma_w^\infty = P_{o/w}^x$ corresponds to the assumption that $d = 1$ and $e = 0$. Yalkowsky and Valvani (1980) and Yalkowsky et al. (1983) tested these assumptions by means of empirical correlations. In order to convert partition coefficients based on molarities, $P_{o/w}^c$, which are readily available (Hansch and Leo, 1979), to partition coefficients based on mole fraction, $P_{o/w}^x$, Eq. 7.28 may be employed and, when combined with Eq. 2.102, affords

$$\log \gamma_w^\infty = \log P_{o/w}^x = \log P_{o/w}^c + \log \frac{V_o}{V_w} \tag{2.103}$$

where $P_{o/w}^c$ = (molarity in n-octanol)/(molarity in water) and $\log (V_o/V_w) = 0.94$. V_o and V_w are the molar volumes of n-octanol and water. The value of this equation is further discussed in Sections 7.2.2, 8.7.3, and 8.7.5.

If the standard state of unit activity is assumed to be the pure liquid state or, in other words, is based on Raoult's law, then for a sparingly soluble liquid solute B, which is in equilibrium with its saturated solution, $a_B^R \cong 1$ Eq. 2.6 then gives

$$x_B (\text{sparingly soluble liquid}) = \frac{1}{\gamma_B^\infty} \tag{2.104}$$

The activity coefficient γ_B of the solute B in the saturated but very dilute solution is designated γ_B^∞, where the superscript ∞ indicates that the solution is infinitely dilute. Equation 2.104 assumes that the pure liquid solute takes up a negligible amount of solvent from the saturated solution, a circumstance which implies negligible mutual solubility.

For many non-self-associating liquids as solutes, the following empirical correlation was obtained:

$$\log x_w^1 = -\log \gamma_w = -1.08 \log P_{o/w}^c - 1.04 \tag{2.105}$$

$$n = 417, \qquad r = 0.973, \qquad \text{s.d.} = 0.356$$

where x_w^1 and γ_w are respectively the mole fraction solubility and activity coefficient of each liquid in water, n is the number of observations, r is the correlation coefficient, and s.d. is the residual standard deviation corresponding to scatter about the regression line. The slope and intercept are close to the predicted values of -1.0 and -0.94, respectively.

The molar solubilities c^s and mole fraction solubilities x^s are related as follows:

$$c^s = \frac{1000 \rho x^s}{M_1 + (M_2 - M_1)x^s} \tag{2.106}$$

where ρ is the density of the saturated solution and M_1 and M_2 are the molecular weight of the solvent and solute, respectively. For dilute solutions such as are given by sparingly soluble solutes in water, $\rho = 1\,\text{g cm}^{-3}$, $M_1 = 18.02$, and $M_1 \gg (M_2 - M_1)x^s$. Therefore,

$$c^s/\text{mol dm}^{-3} \cong 55.5x^s \tag{2.107}$$

therefore,

$$\log c^s \cong \log x^s + 1.74 \tag{2.108}$$

For saturated solutions of solids in water at 25°C, Eqs. 2.6, 2.103 and 2.23 can be combined to give

$$\log x_w^s = -\log P_{o/w}^c - 0.94 - \frac{\Delta S_f(\Theta_m - 25)}{2.303\,R\,298.15} \tag{2.109}$$

Combining Eqs. 2.108 and 2.109 leads to the following generalized equation for nonelectrolytes:

$$\log c^s \cong -\log P_{o/w}^c - \frac{\Delta S_f(\Theta_m - 25)}{2.303\,R\,298.15} + 0.80 \tag{2.110}$$

For rigid and short molecules, $\Delta S_f = 13.5\,\text{cal K}^{-1}\,\text{mol}^{-1}$ and Eq. 2.110 affords

$$\log c^s \cong -\log P_{o/w}^c - 0.01\,\Theta_m + 1.05 \tag{2.111}$$

Yalkowsky and Valvani (1980) tested the last two equations by linear regression analysis between the observed molar solubility $c^s(\text{obs})$ and the calculated molar solubility $c^s(\text{calc})$ at 25°C using the equation

$$\log c^s(\text{obs}) = m \log c^s(\text{calc}) + c \tag{2.112}$$

For polycyclic aromatic hydrocarbons, halobenzenes, alkyl *p*-aminobenzoates (crystalline consisting of rigid and flexible molecules), aliphatic alcohols (liquid or crystalline, rigid or flexible), m was close to unity; for steroid hormones, however, m was 0.847. The major variation between the groups was in c, which would be zero if the predictions were correct. The mean linear regression for the five series of liquid and crystalline solutes is

$$\log c^s(\text{obs}) \cong \log c^s(\text{calc}) - 0.5 \qquad (2.113)$$

When Yalkowsky and Valvani (1980) fitted the complete set of data at their disposal by multiple linear regression to a function of $\log P_{o/w}^c$ and $\Delta S_f(\Theta_m - 25)$, they obtained the semiempirical equation

$$\log c^s = -1.00 \log P_{o/w}^c - 1.11 \cdot \frac{\Delta S_f(\Theta_m - 25)}{2.303\, R\, 298.15} + 0.54 \qquad (2.114)$$

$$n = 167, \quad r = 0.995, \quad \text{s.d.} = 0.242$$

This equation estimated the solubility of 95% of the 167 solutes to within a factor of 3 (0.5 log unit). Although the solubilities spanned nine orders of magnitude, the error never reached a factor of 10. The linear regression for the rigid molecules gave

$$\log c^s = -1.05 \log P_{o/w} - 0.012 \Theta_m + 0.87 \qquad (2.115)$$

$$n = 155, \quad r = 0.989, \quad \text{s.d.} = 0.308$$

The coefficient of $\log P_{o/w}^c$ agrees well with the value obtained by Hansch et al. (1968) for 140 liquids. Deviations from the predicted equations arise because (1) ΔH_f and ΔS_f in any equation for predicting a^s vary with temperature owing to the effect of ΔC_p, (2) solubility and partitioning may show slightly different dependence on specific interactions in solutions, and (3) there may be experimental errors in c_s, $P_{o/w}$, and Θ_m.

These equations enable the aqueous solubility to be estimated from one single physical measurement, the melting point. Partition coefficients or $\log \gamma_w^\infty$ can readily be estimated from group contributions (Hansch and Leo, 1979). Although the equations are essentially empirical, they facilitate an appreciation of the influence of structural modifications on aqueous solubility.

2.10 ATHERMAL SOLUTIONS

The deviations of athermal solutions from ideality are attributed entirely to the entropy of mixing arising from the differences in the size of the molecules of the substances in solution. For athermal solutions, $\Delta S_{mix} < \Delta S_{mix}^{id}$. (i.e., the excess entropy ΔS^E is negative), whereas $\Delta H_{mix} = \Delta H_{mix}^{id} = 0$ (i.e., the excess enthalpy ΔH^E is zero). Athermal solutions can be treated theoretically by a statistical mechanical analysis of mixtures of two kinds of molecule arranged on a lattice. This treatment is particularly relevant to solutions of polymers.

The entropy of mixing of long-chain molecules was studied by Flory (1941 and 1942) and Huggins (1941 and 1942), who independently arrived at the following expression for the partial molar entropy of mixing (of component B with A):

$$\Delta S_B^M = \bar{S}_B - S_B^* = -R\left[\ln \phi_B + \phi_A\left(1 - \frac{V_B}{V_A}\right)\right] \tag{2.116}$$

where $\bar{S}_B$ is the partial molar entropy of B in the mixture, S_B^* is the molar entropy of pure B, ϕ_A and ϕ_B are the volume fractions and V_A and V_B are the molar volumes of A and B. The Flory–Huggins entropy, Eq. 2.116, can also be derived by assuming that the partial molar volumes of the components are proportional to their molar volumes.

For mixtures of liquids or solutions of small molecules, such as drugs, the actual partial molar entropy of mixing is usually better approximated by the ideal value (Eq. 2.8) than by the Flory–Huggins entropy (Eq. 2.116), which is usually too high (Shinoda, 1978).

The deviations from Raoult's law by mixtures of nonpolar molecules of unequal size can be simply treated (Ashworth and Everett, 1960; Denbigh, 1971) by assuming that the partial molar excess free energy of a given component B,

$$\Delta G_B^E = \Delta \mu_B^E = RT\ln \gamma_B \tag{2.117}$$

consists of two contributions which are mutually approximately independent, thus,

$$\ln \gamma_B = \ln \gamma_B^{ath} + \ln \gamma_B^{th} \qquad (\text{or } \gamma_B = \gamma_B^{ath}\gamma_B^{th}) \tag{2.118}$$

In this equation $\ln \gamma_B^{ath}$ is a temperature-independent (athermal) term associated with statistical effects due to differences in volume of solute and solvent molecules and $\ln \gamma_B^{th}$ is a temperature-dependent (thermal) term associated with disparities in the intermolecular energies of interaction between solvent–solvent, solute–solute, and solvent–solute. The next section considers the various intermolecular energies of interactions which are associated with the thermal contribution. The $\ln \gamma_B^{ath}$ term can be calculated from the Flory–Huggins approach discussed above, and $\ln \gamma_B^{th}$ can be calculated from the regular solution theory of Hildebrand and Scatchard discussed earlier in this chapter. The use of Eq. 2.118 to predict solute activity coefficients in any solvent of high molecular weight has been described by Martire (1967).

2.11   INTERMOLECULAR INTERACTIONS AND THEIR INFLUENCE ON SOLUBILITY

2.11.1   Introduction to Intermolecular Interactions

The development and selection of methods for solubility prediction require knowledge of the nature and relative importance of the intermolecular forces:

solute–solute, solvent–solvent, and solute–solvent. Because of their importance in understanding solubility behavior, a relatively brief discussion of some of the more important forces is presented in this section. The influence of the most important member, the hydrogen-bonding interaction on solubility, is developed in Chapters 4–6.

2.11.2 Coulombic Interactions

The simplest type of intermolecular force is the coulombic (i.e., interionic) interaction, which is important when considering ion pairs, as in Chapter 9. This is a valence force and is one of the strongest known chemical interactions, the bond energy of which (~ 100–200 kcal mol^{-1}, i.e., 400–800 kJ mol^{-1}) is comparable to that of a covalent bond (~ 50–150 kcal mol^{-1} or 200–600 kJ mol^{-1}). The energy of interaction u(ion–ion) between two ions of charge q_1 and q_2, respectively, and separated by distance r is given by

$$u(\text{ion–ion}) = \frac{q_1 q_2}{rD} \tag{2.119}$$

where D is the dielectric constant of the medium separating and surrounding the charges and is dimensionless, being exactly unity for a vacuum and 80.36 for water at 25°C. The equilibrium distance between the oppositely charged centers may be relatively large, since the energy of attraction is inversely proportional to the first power of the interionic separation.

In S. I. units Eq. 2.119 is written

$$u(\text{ion–ion}) = \frac{q_1 q_2}{r 4\pi\varepsilon} \tag{2.120}$$

where ε is the permittivity of the medium in C^2 m^{-2} J. In the S.I. system the dielectric constant D is termed the relative permittivity and is equal to the permittivity of the medium, ε, divided by the permittivity of a vacuum, ε_0, that is,

$$D = \frac{\varepsilon}{\varepsilon_0} \tag{2.121}$$

where $\varepsilon_0 = 8.8542 \times 10^{-12}$ C^2 m^{-1} J. This concept is discussed by McGlashan (1973).

Values of dielectric constant D of various liquids are quoted by Riddick and Bunger (1970) and by Mellan (1977). Table 2.7 also lists a few values. The magnitude of the coulombic energy of attraction between a cation and an anion is greatest in a crystal of a salt, D being unity in free space, but becomes weaker as D increases. This effect and the formation of ion–dipole interactions explains why salts tend to be soluble in water ($D = 80$), rather less soluble in less polar solvents such as ethanol, chloroform, and nitromethane, but virtually insoluble in nonpolar solvents such as hydrocarbons. Certain organic ions which are

Table 2.7 Values of Dielectric Constants[a] (D), Dipole Moments[a] (μ), Principal Molecular Polarizabilities[b] (α), Refractive Indices[a] (n_{25}^D), Refractive Index Function (f), Molar Volumes (V_m), Molar Refractivities (R_m), Boiling Points (b.p. at 760 mm Hg), and Solubility Parameters[c] (δ) of Some Common Organic Compounds (1 cal = 4.184 J)

Compound	D	μ (D)	α_1	α_2	α_3	Mean α	n_{25}^D	$\dfrac{f}{(n^2-1)/(n^2+2)}$	V_m (cm³ mol⁻¹)	R_m (cm³ mol⁻¹)	b.p. (°C)	δ (cal$^{1/2}$ cm$^{-3/2}$)
				(10⁻²⁴ cm³)								
CH$_4$	Gas	0	2.58	2.58	2.58	2.58	—	—	—	—	−161.5	5.4
CS$_2$	2.641	0.06	13.08	5.58	5.58	8.08	1.62409	0.35312	60.645	21.415	46.2	10.0
CCl$_4$	2.238	0	10.26	10.26	10.26	10.26	1.45739	0.27255	97.087	26.461	76.8	8.6
CH$_3$Cl	Gas	1.87	5.09	4.11	4.11	4.44	1.3344	0.20648	29.209	6.031	−23.8	9.7
CH$_3$Br	Gas	1.80	6.56	4.99	4.99	5.51	1.4130	0.24936	54.821	13.670	3.6	9.6
CH$_3$I	7.00	1.48	8.72	6.57	6.57	7.29	1.5270	0.30744	62.669	19.267	42.4	10.2
CH$_3$CN	37.5	3.44	5.43	3.70	3.70	4.28	1.34163	0.21052	52.862	11.129	81.6	10.5
CHCl$_3$	4.806	1.15	6.73	9.01	9.01	8.25	1.44293	0.26507	80.667	21.383	61.2	9.3
CHBr$_3$	4.39	0.99	9.11	12.58	12.58	11.42	1.5956	0.34007	87.887	29.888	149.6	10.5
CHI$_3$	Solid	—	15.90	17.85	17.85	17.20	—	—	—	—	(m.p. 120)	—
C$_2$Cl$_6$	Solid	—	16.57	16.57	13.61	15.58	—	—	—	—	(m.p.187)	—
Benzene	2.275	0	11.20	11.20	7.36	9.92	1.49792	0.29308	89.407	26.203	80.1	9.2
Chlorobenzene	5.621	1.54	14.78	12.55	8.21	11.85	1.52214	0.30506	102.24	31.190	131.7	9.5
Bromobenzene	5.40	1.55	16.83	13.01	8.92	12.92	1.55709	0.32196	105.507	33.969	155.9	9.9
Iodobenzene	Solid	—	19.71	15.88	9.96	15.18	1.61645	0.34965	126.215	44.131	189	10.1
Benzonitrile	25.2	4.05	16.16	11.60	8.15	11.97	1.52547	0.30669	103.063	31.608	191.1	8.4
Thiophene	2.705	0.52	10.15	10.14	6.70	9.00	1.52572	0.30681	79.464	24.381	84.2	9.8
Pyridine	12.4	2.37	10.72	10.43	6.45	9.20	1.50745	0.29782	80.862	24.082	115.3	10.7
CH$_2$=CCl$_2$	—	1.30	8.96	8.79	5.57	7.83	1.42119	0.25390	82.527	20.954	31.6	9.1
cis-CHCl=CHCl	9.20	1.76	7.80	9.46	6.08	7.78	1.44611	0.26672	75.993	20.269	60.6	9.1

[a] Riddick and Bunger (1970), Riddick et al. (1986).
[b] Le Fèvre (1965).
[c] Barton (1975); Burrell (1975); Shinoda (1978).

separated from their counterions by water tend to form ion pairs in solvents of dielectric constant lower than about 30, as discussed in Chapter 9.

2.11.3   Dipole Moments

Although most of the molecules with which we are concerned are uncharged, many will be polar, possessing a permanent electric dipole moment μ, and all will be polarizable to an extent represented by the polarizability α.

The magnitude of the dipole moment is given by the product of the electric charge associated with either end of the dipole and the distance between the charges. Although the S.I. unit of dipole moment is the C m, the Debye unit D is still in general use where $1 D = 3.336 \times 10^{-30} C m$. Dipole moment is a vector since it has direction as well as magnitude. The overall dipole moment of a molecule is therefore the vectorial sum of the dipole moments of all its bonds and lone pairs (or its groups).

In solubility phenomena and solution thermodynamics the dipole moments of bonds or groups are of greater significance than the dipole moment of the whole molecule. If the latter determined the degree of polarity in solution, the three disubstituted benzenes shown in Figure 2.19 would have very different solubilities and partition coefficients P between two given solvents. This is not so because the neighboring molecules of the solvent are small enough to respond almost identically to the presence of the two polar groups in these molecules. In fact, the $\log P^c$ (octanol/water) values for o-, m-, and p-dichlorbenzene are virtually identical, being 3.39, 3.38, and 3.38, respectively, and even the values for o-, m-, and p-dinitrobenzene are fairly constant, being 1.49, 1.58, and 1.49, respectively (Hansch and Leo, 1979).

Another example of this effect is given by the mole fraction solubilities of m-, o-, and p-dinitrobenzene in benzene at 50°C of 0.376, 0.175, and 0.031, respectively. These values do not reflect the overall dipole moments of the molecules, but instead are inversely related to the melting points (m.p. 90, 116, and 170°C, respectively) (Hildebrand and Scott, 1950), which reflect the thermodynamic activity of the solid, a^s, and the molecular packing in the crystal.

Between two fixed permanent dipoles with dipole moments of μ_1 and μ_2,

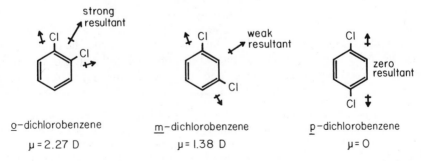

Figure 2.19   Dipole moments of symmetrical dichlorobenzenes measured at 24°C in benzene solution.

whose axes are parallel and whose centers are separated by a distance r, the potential energy of interaction u(fixed dipoles) is given by

$$u(\text{fixed dipoles}) = \frac{-2\mu_1\mu_2}{r^3} \tag{2.122}$$

Unless otherwise constrained, the dipoles will align themselves such that the positive end of one dipole is oriented toward the negative end of the other dipole. The interaction energy will then have a minimum negative value, corresponding to an attraction. If many dipolar molecules were arranged in this fashion they would tend to form a regularly repeating structure representing a crystal lattice. This interaction contributes to the stability, that is, low thermodynamic activity, of many organic crystals.

In the liquid state the dipoles would be free to rotate, creating additional degrees of freedom which would contribute to the entropy of fusion, as discussed in Section 2.2.4. Certain orientations of the rotating dipoles are more probable on account of their lower potential energy. Over all possible orientations the potential energy of interaction u(orientation), is given by

$$u(\text{orientation}) = \frac{-2\mu_1^2\mu_2^2}{3kTr^6} \tag{2.123}$$

where T is the absolute temperature, k is the Boltzmann constant, and the negative sign indicates that the force is always attractive. This force is known as the *orientation effect*, the (rotating) *dipole–dipole* force, or the *Keesom force*.

2.11.4 Induced Dipole Moments and Polarizability

As mentioned before, polarizability is important in solubility prediction because it makes possible certain other intermolecular forces. The higher the polarizability α of a molecule, the greater is the strength of the dipole moment μ_i, induced by a given electric field E. This field may be produced by the dipole of a neighboring molecule or by the charge of a neighboring ion, that is,[†]

$$\mu_i = \alpha E \tag{2.124}$$

The polarizability represents the ability of the electrons and the nuclei of the molecule to be displaced from the mean positions and is the sum of the mean molecular electronic polarizability and the mean vibrational (or atomic)

[†] If the quantities in this equation are expressed in the c.g.s., e.s.u. system of units in the Gaussian or symmetrical version, α has the units cm^3. Sometimes α is expressed as $A^{\circ 3}$ ($1\,A^{\circ 3} = 10^{-24}\,cm^3$). If the quantity in the previous equation are expressed in the S.I. system of units, α is expressed in $C.m.^2\,V^{-1}$. McGlashan (1973) discusses the interconversion of these units. For the numerical values, we note that

$$\alpha/C.m.^2\,V^{-1} (\text{in S.I.}) = \alpha/cm^3 (\text{in c.g.s. Gaussian}) \times 1.113 \times 10^{-16} \tag{2.125}$$

polarizability. The electronic polarizability represents the ability of the electrons to be displaced by unit electric field from their mean positions relative to the nuclei, or simply the deformability of the electron clouds. The vibrational polarizability represents the ability of the nuclear skeleton of the molecule to be deformed by unit electric field, and is about one tenth of the electronic polarizability of a molecule. Molecules and atoms possessing high polarizability have delocalized electrons, such as in aromatic or conjugated systems, or have large molar volumes, such as iodine, iodide ions, and their compounds. Atoms or ions of small molar volume (e.g., Na^+, F^-) and compact molecules containing few lone pairs and with σ rather than π bonds, such as the alkanes, have low polarizabilities. Molar refraction is related to polarizability and both are nearly independent of temperature (Le Fèvre, 1965).

Molecular polarizabilities and related quantities have been discussed by Le Fèvre (1965), and some values are quoted in Table 2.7. The polarizability of a molecule depends on its orientation, unless it is a sphere (e.g., I^-) or a regular tetrahedron (e.g., CCl_4). In the liquid or gaseous states molecules rotate rapidly, so it is possible to represent the polarizability by a mean value. The principal polarizabilities of a molecule can be analyzed in terms of bond polarizabilities (Le Fèvre, 1965) (e.g., longitudinal $\alpha/10^{-24}$ cm^3 for C—C 0.99, for C=C 2.8, for C≡C 3.5; transverse $\alpha/10^{-24}$ cm^3 for C—C 0.27, C=C 0.73, for C≡C 1.3). This shows that the π electrons, being exposed, are very polarizable. A second-order effect, known as hyperpolarizability (Buckingham and Orr, 1967), can usually be neglected.

As a result of its polarizability α, a given molecule may participate in an attractive interaction with a neighboring polar molecule of dipole moment μ_p, separated from it by a distance r. The energy of interaction u(induction) is given by

$$u(\text{induction}) = -\frac{2\alpha\mu_p^2}{r^6} \tag{2.126}$$

Both molecules are here assumed to be polar molecules of identical type, such that each permanent dipole creates an identical induced dipole in the other. The energy is attractive, independent of temperature, and can exist provided that at least one of the molecules is polar. This force is known as the *dipole–induced-dipole* force or the *induction effect*. It is also termed the *Debye force* after the originator of the concept.

The induction effect increases the attraction between already polar molecules. It also gives rise to an attraction between a polar molecule and a polarizable nonpolar molecule.

The dipole–induced-dipole interaction explains the formation of complex crystal lattices containing both the very polar molecules of picric acid (2,4,6-trinitrophenol) and the polarizable molecules of a polynuclear aromatic hydrocarbon such as anthracene.

Quantum mechanical considerations indicate that all molecules, polar or nonpolar, attract each other by *induced-dipole–induced-dipole* interactions, which are also known as *dispersion forces* or *London forces*. To a first approximation the potential energy of attraction between two similar molecules is given by

$$u(\text{dispersion}) \cong \frac{-3\alpha^2 I}{4r^6} \tag{2.127}$$

where α is the polarizability, I is the ionization potential, and r is the distance of separation of the molecules. The dispersion energy between the dissimilar molecules, subscripts 1 and 2, is given by

$$u(\text{dispersion}) \cong -\frac{3\alpha_1\alpha_2}{2r^6}\left(\frac{I_1 I_2}{I_1 + I_2}\right) \tag{2.128}$$

The units of α^2/r^6 and $\alpha_1\alpha_2/r^6$ can easily be made to cancel by appropriate choice of units (e.g., α in cm^3, r in cm or α in A$^{\circ 3}$, r in A$^\circ$), in which case $\mu(\text{dispersion})$ will be expressed in the same units as the ionization potentials.

Because of their ubiquity and their similarity in strength from molecule to molecule, as shown in Table 2.8, London dispersion forces cannot be exploited to provide selectivity in solubility phenomena. Since the regular solution theory in its simplest form, which is due to Hildebrand, Scatchard, and co-workers, assumes that dispersion forces are the only forces acting, this theory cannot be utilized to account for or to achieve selectivity in solubility behavior.

Table 2.8   Relative Magnitude of the Potential Energies of Intermolecular Attraction Between Pairs of Various Molecules Whose Centers are Separated by the Same Distance

Molecule	μ (D)	α (10^{-24} cm^3)	I(eV)	$u(orientation)r^6$ (10^{-67} J cm^6)	$u(induction)r^6$ (10^{-67} J cm^6)	$u(dispersion)r^6$ (10^{-67} J cm^6)
H$_2$	0	0.81	16.4	0	0	11.3
Ar	0	1.63	16.5	0	0	57
N$_2$	0	1.74	17.1	0	0	62
CH$_4$	0	2.58	14.5	0	0	117
Cl$_2$	0	4.60	18.2	0	0	461
CO	0.12	1.99	14.3	0.0034	0.057	67.5
HI	0.38	5.4	12	0.35	1.68	382
HBr	0.78	3.58	13.3	6.2	4.05	176
HCl	1.03	2.63	13.7	18.6	5.4	105
NH$_3$	1.5	2.21	16	84	10	93
H$_2$O	1.84	1.48	18	190	10	47

2.11.5   Relative Importance of the Various Intermolecular Forces

The dipole–dipole, dipole–induced-dipole, and dispersion forces are all attractive and are collectively known as *van der Waals forces*. Their relative magnitudes for an isolated pair of similar molecules are indicated in Table 2.8, and calculated values for methylene groups in various solvents are shown in Table 2.9. Nonpolar molecules can only interact by dispersion forces (e.g., CH_4 in Table 2.8; $-CH_2-$ in cyclohexane and CCl_4 in Table 2.9). Even between polar molecules, dispersion forces may predominate over the other attractions (e.g., HCl in Table 2.8).

The potential energies of all three interactions are proportional to r^{-6} and so fall off rapidly with increasing intermolecular separation. The combined potential energy of attraction $u(at.)$ may therefore be written

$$u(at.) = -Ar^{-6} \tag{2.129}$$

It should be kept in mind, however, that certain very weak interactions in solution could result from dipole–quadrupole, quadrupole–quadrupole, and dipole–octopole interactions, but that the energies of attraction of these interactions fall off much more rapidly with increasing distance than do the usual van der Waals forces, and therefore can usually be ignored.

The importance of Debye forces (i.e., dipole–induced-dipole interactions) in separation science has not been fully appreciated. In fact, the literature is replete with statements that contributions from these induction forces to the total intermolecular energy of attraction are very small and unimportant compared with those from electrostatic forces (e.g., dipole–dipole or Keesom forces) and London dispersion forces (Buckingham, 1967; Hirschfelder, 1967; Maitland et al., 1981). Such statements have been qualified by words such as "often" (Buckingham, 1967), "usually" (Hirschfelder, 1967), or even "always" (Maitland et al., 1981).

This generalization is based on the relative magnitudes of the pair-potential energies, which may be readily calculated from the physical properties of the substances under consideration by means of classical expressions (Eq. 2.123 and Eqs. 2.126–2.128) for each of the Keesom, Debye, and London interactions. This generalization is often valid in pure, moderately polar liquids, for example, esters, aldehydes, ketones, and amides. However, Debye interactions are often stronger than London dispersion forces for highly polar molecules, such as γ-butyrolactone, in solvents of low polarity, such as isooctane and toluene. Moreover, the strength of Debye forces rivals that of Keesom forces for polar molecules in the weakly polar aromatic solvents such as toluene.

Despite such statements propounding their unimportance, Debye interactions may be very important in explaining and promoting selectivity in solubility phenomena and in separation science, in the petroleum industry, for example. Thus, the Debye force explains why certain polar and polarizable molecules are mutually soluble. For example, the polar compound benzil, $C_6H_5 \cdot CO \cdot CO \cdot C_6H_5$, is 14 times more soluble in the nonpolar but polarizable

Table 2.9 Calculated Enthalpies of Intermolecular Attraction[a] of —CH_2— Groups[b] of an Aliphatic Alcohol Dissolved in Various Solvents[c]

Interacting Solvent	Dielectric Constant[d] (D)	Dipole Moment[e] (μ_1/D)	Dipole–Dipole for CH_2[f]	Intermolecular Interactions of CH_2 with Solvent			
				Dipole–Induced-Dipole[g]	Dispersion Term[h]	Cavity Term[i]	Dipole–Dipole[j]
Diethyl ether	4.34	1.36	0 (0)	−0.02 (−0.08)	−1.9 (−7.9)	+1.0 (+4.2)	−0.1 (−0.4)
Cyclohexane	2.02	0	0 (0)	0 (0)	−2.1 (−8.8)	+1.2 (+5.0)	0 (0)
CCl_4	2.24	0	0 (0)	0 (0)	−2.1 (−8.8)	+1.3 (+5.4)	0 (0)
$CHCl_3$	4.81	1.15	0 (0)	−0.02 (−0.08)	−2.3 (−9.6)	+1.5 (+6.3)	−0.07 (−0.3)
Tetrahydrofuran	7.58	1.63	0 (0)	−0.04 (−0.2)	−2.3 (−9.6)	+1.6 (+6.7)	−0.2 (−0.8)
Pyridine	12.4	2.37	0 (0)	−0.07 (−0.3)	−2.6 (−10.9)	+2.0 (+8.4)	−0.3 (−1.3)
$Me_2N\cdot CHO$	36.7	3.82	0 (0)	−0.2 (−0.8)	−2.9 (−12.1)	+2.5 (+10.5)	−0.8 (−3.3)
Me_2SO	46.7	4.49	0 (0)	−0.2 (−0.8)	−3.2 (−13.4)	+3.0 (+12.6)	−0.9 (−3.8)

[a] In kcal mol⁻¹. Value in parentheses are in kJ mol⁻¹.

[b] $\mu = 0$; $\alpha = 1.83 \times 10^{-24}$ cm³.

[c] Temperature, 25°C. 1 cal = 4.184 J.

[d] Riddick and Bunger (1970).

[e] McClellan (1963).

[f] Zero because $\mu = 0$ for CH_2 (and C—H).

[g] Calculated from Eq. 2.126.

[h] Calculated (Spencer et al., 1979).

[i] The enthalpy change required to create a cavity in the solvent to enable a CH_2 group to be introduced (Spencer et al., 1979).

[j] Calculated from Eq. 2.123 for a hypothetical solute of $\mu_2 = 1$, D, r = 4 Å.

Adapted with permission from Spencer, J. N., Gleim, J. E., Blevins, C. H., Garrett, R. C., and Mayer, F. J. (1979). *J. Phys. Chem.* **83**, 1249–1255. Copyright © 1979, American Chemical Society.

solvent carbon tetrachloride than in the nonpolar but poorly polarizable solvent *n*-hexane (Acree and Bertrand, 1980). Furthermore, the polarizable aromatic compounds are much more readily extracted from crude oil by the polar solvents sulfolane ($\mu = 4.81$ D) or *N*-methyl-2-pyrrolidinone ($\mu = 4.09$ D) than are the poorly polarizable aliphatic compounds which occur with them (e.g., Prokic et al., 1970).

N-methyl-2-pyrrolidinone   sulfolane

This type of selectivity is primarily the result of the relative strengths of the Debye interactions; thus, the Debye force may be used to increase the solubility of a polarizable solute, either per se or relative to a less polarizable solute, in which case a solvent with a high dipole moment μ can be selected on the basis of tabulated properties (Riddick and Bunger, 1970). The Debye force can also be used to increase the solubility of a polar solute either per se or relative to a less polarizable solute, in which case the solvent with a high polarizability α can be selected.

A demonstration of the relative importance of the Debye interaction has been presented by Rytting et al. (1986). This hinges on a comparison of the "escaping tendency" of polar compounds from solvent media of different polarizabilities into the vapor state, as measured by the Henry's law constants. Since these concepts are developed throughout Chapter 3, further discussion of the importance of the dipole–induced-dipole forces is postponed until Section 3.10. The dipole–induced-dipole interaction has to be considered when explaining solvation enthalpies, especially those between nonpolar solutes in polar solvents (Krishnan and Friedman, 1971; Spencer et al., 1979).

2.11.6 Polarizability and Intermolecular Forces

The polarizability α of neighboring molecules is fundamental in accounting for and predicting the strength of both the Debye and London forces between them (Eqs. 2.126–2.128). Values of α have been determined for various molecular orientations of a number of molecules, but not for all solvents (and solutes) which may be of interest to us. For our purpose we need a simple means of assessing the relative strength of a potentially useful Debye interaction from an approximate value of the mean polarizability α, which may be defined as

$$\alpha = \frac{\alpha_1 + \alpha_2 + \alpha_3}{3}$$

(2.130)

Electromagnetic theory (Le Fèvre, 1965) shows that

$$R(\lambda = \infty) = P_E = \frac{4\pi L\alpha}{3} \tag{2.131}$$

where L is Avogadro's constant $(6.022 \times 10^{23} \, mol^{-1})$ and P_E is the electronic polarization of the molecule whose value is given by the molar refractivity R for light of infinite wavelength $(\lambda = \infty)$. Evaluation of the constants affords

$$\alpha = 0.3964 \times 10^{-24} R(\lambda = \infty) \tag{2.132}$$

R is given by

$$R = \frac{n^2 - 1}{n^2 + 2} V_m \tag{2.133}$$

where n is the refractive index and V_m is the molar volume, which is given by the molecular weight M divided by the density ρ of the pure liquid. Equation 2.133 may be applied to the refractive index for visible light (e.g., the sodium D line at 25°C, R_D, n_D^{25}) provided the wavelength is not close to an absorption band.

In columns 2–4 of Table 2.10, values of mean polarizability obtained by different methods are compared. Since the measured values, such as those in columns 2 and 3, are available for relatively few compounds, indirect methods

Table 2.10 Comparison of Calculated and Measured Values of Mean Polarizabilities $(\alpha, 10^{-24} \, cm^3)$ of Some Molecules

Molecule	Mean of Measured Values[a]	α Values[b]	α Values[c]	α Values[d]	$\alpha/v =$ $\alpha L\rho/M^e$	$\left(\dfrac{n^2-1}{n^2+2}\right)\left(\dfrac{3}{4\pi}\right)^f$
2-Methylbutane	—	9.72	10.02	10.04	0.051	0.051
Chloroform	8.25	8.29	8.38	8.48.	0.063	0.063
Benzene	9.92	9.92	10.46	10.39	0.070	0.070
Carbon tetrachloride	10.26	10.25	10.29	10.49	0.064	0.065
Benzonitrile	11.97	11.79[d]	12.22	12.54	0.072	0.073
Iodobenzene	15.18	15.18	15.58	15.51	0.082	0.083
Griseofulvin	—	—	32.93	(solid)	0.079	—

[a] Le Fèvre (1965).
[b] Calculated from bond polarizabilities (Eq. 2.130) (Le Fèvre, 1965).
[c] Calculated from atom and group contribution to R_D^{20} and Eq. 2.132 (Vogel, 1948).
[d] Calculated from n_D^{25} and V_m data (Eqs. 2.132 and 2.133) (Riddick and Bunger, 1970).
[e] From column 4.
[f] From n_D^{25} data.
[g] Bond polarizabilities of $-C\equiv N$ were assumed to be equal to those of $-C\equiv C-$

used to obtain the figures in columns 4 and 5 usually have to be used. These indirect methods merely involve substitution of literature values into Eqs. 2.132 and 2.133, and for the sodium D line at 25°C give only small, fairly consistent errors, about 3% too high. Riddick and Bunger (1970) give extensive data, including n_D^{25}, ρ, and M for organic solvents which readily enable R_D and hence α to be calculated. Fortunately, molar refractivity is an additive property of a molecule. For solid compounds, including newly synthesized molecules, R_D (and n_D) can readily be calculated from atomic, group, and structural refractivities for which many values have been stated in literature (Le Fèvre, 1965; Vogel, 1948). The effect of temperature differences of a few degrees, say 20 or 25°C, on n_D or R_D is well within the errors quoted above. Taking as an example the drug griseofulvin, whose structure is shown in Scheme 6.2, R_D is calculated to be 83.1 cm^3 mol^{-1} from Vogel's (1948) atomic, group, and structural refractivities, and so $\alpha = 83.1 \times 0.3964 \times 10^{-24}$ cm$^3 = 32.93 \times 10^{-24}$ cm^3.

It can be seen from Eq. 2.126 and 2.127 that the energies of the Debye and London interactions are increased by increasing the polarizability, but decreased by increasing molecular radius or volume. Although increasing size (and molecular weight) will certainly increase the molar refractivity and the polarizability, it will also increase the distance of closest approach of the other interacting molecule and will tend to cancel out the increase in energy of attraction.

It has been suggested that a useful measure of the potential for Debye and London interactions is the mean polarizability α of the molecule divided by the mean volume v of the molecule. The latter is taken to be equal to the molar volume V_m of the liquid (calculated as the molecular weight divided by the density) divided by Avogadro's constant L. For a solid substance, the density and hence the molar volume of the solid form would be a useful first approximation. The sixth and seventh columns of Table 2.10 present α/v values, for which the rank order seems intuitively to be correct. We propose that the Debye–London interaction parameter α/v is given by

$$\frac{\alpha}{v} = \frac{\alpha \rho L}{M} \frac{3}{4\pi} \frac{R}{V_m} = \frac{3}{4\pi} \left(\frac{n^2 - 1}{n^2 + 2} \right) \tag{2.134}$$

For simplicity we propose that α and v be expressed in the same units (e.g., volume per molecule, cm^3), so that α/v is a dimensionless parameter. We note that R and V_m both have the dimension volume per mole (usually cm^3 mol^{-1}), while refractive index is dimensionless, and that α/v can be readily calculated from published values either of refractive index or of atom, group, or structural refractivities.

The London dispersion pair-potential energy between similar molecules is given by Eq. 2.127. For many organic molecules the ionization energy is very similar (ca. 11.9 ± 1.3 eV) (Shinoda, 1978; see Section 2.6) and can be assumed to be constant. For a number of liquids the free volume can also be assumed constant (Shinoda, 1978), such that the molecular volume is proportional to the cube of the intermolecular distance, that is, $v \propto r^3$. Equation 2.127 becomes

$$u(\text{dispersion}) \propto \left(\frac{\alpha}{v}\right)^2 \qquad (2.135)$$

For molecules without significant dipole moments, Eq. 2.87 of regular solution theory shows that the cohesive energy density $\Delta U^v / V_m$ (i.e., the square of the Hildebrand solubility parameter δ) is proportional to the London pair-potential energy. From Eq. 2.135,

$$\frac{\alpha}{v} \propto u(\text{dispersion})^{1/2} \propto \left(\frac{\Delta U^v}{V_m}\right)^{1/2} = \delta \qquad (2.136)$$

Comparison with Eq. 2.134 indicates that

$$\delta = \left(\frac{\Delta U^v}{V_m}\right)^{1/2} \propto \frac{n^2 - 1}{n^2 + 2} = f \qquad (2.137)$$

Thus, for nonpolar but polarizable liquids having similar ionization energies and molar free volumes and interacting only by London dispersion forces, δ is proportional to the refractive index function f. This has been demonstrated empirically (Sewell, 1966; Lawson and Ingham, 1969; Keller et al., 1970; Karger et al., 1976; Grant and Abougela, 1983), as shown in Figure 2.20, using literature values (Riddick and Bunger, 1970) of ΔU^v, V_m, and n_D for liquids of low polarity ($\mu < 0.4$ D; $f \leq 0.28$; $n_D \leq 1.472$; and $\delta < 8.67$ cal$^{1/2}$ cm$^{-3/2}$ at 20–25°C).

Although the extrapolated line passes through the origin, actual organic liquids have $f > 0.2$, $n_D > 1.32$, or $\delta > 6.1$ cal$^{1/2}$ cm$^{-3/2}$. With increasing dipole moments of the liquids, the points deviate from the line in the direction of greater cohesive energy density, as expected. The deviation may be small or negligible if the dipole is buried within the molecule. Examples include (1) tertiary amines which are in a state of oscillation, such as triethylamine, $\mu = 0.87$ D, and (2) organic iodides, owing to the size of the iodine atoms. In these cases the dipole–dipole forces are weakened by the increased distance between the dipoles of two neighboring molecules.

Hydrogen bond donors which can also act as hydrogen acceptors deviate greatly in the direction of increasing cohesive energy density because of the additional hydrogen-bond interaction. Examples are alcohols, phenols, acids, amines (primary and secondary), and amides (primary and secondary).

To summarize, the Debye interactions are more important than are generally realized and their exploitation offers selectivity in solvency, extraction, and solubility behavior. Furthermore, the parameter $(n^2 - 1)/(n^2 + 2)$, or simply the refractive index itself, gives a good "rule of thumb" measure of the potential of a molecule to undergo London and Debye polarizability interaction. Tables of refractive index suggest the following scale:

Compounds	n_D^{20-25}	
CH_2I_2	1.7	
CS_2	1.6	
Monoiodo compounds	1.5	
Dibromo compounds	1.5	
Aromatic hydrocarbons	1.5	Increasing polarizability,
Polychloro paraffins	1.4_5	London and Debye
Amides	1.4_3	interactions, and
Monobromo compounds	1.4_2	refractive index
Monochloro compounds	1.4	
Olefins	1.4	
Esters	1.3_8	
Paraffins	1.3_8	
Ethers	1.3_6	

This list merely indicates trends. There is considerable overlap between representatives.

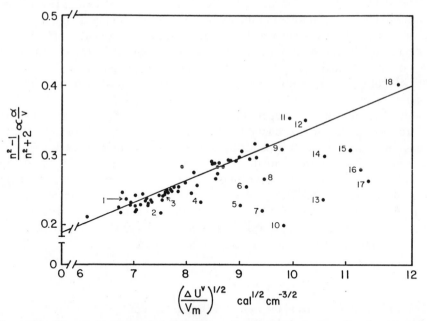

Figure 2.20 Plot of electronic polarizability per unit volume expressed as the refractive index function, $(n^2 - 1)/(n^2 + 2)$, against (cohesive energy density)$^{1/2}$. 1, 2,2,4-trimethylpentane (isooctane, $\mu = 0$); 2, diethyl ether ($\mu = 1.15$ D); 3, triethylamine ($\mu = 0.87$ D); 4, hexafluorobenzene ($\mu = 0.33$ D); 5, ethyl acetate ($\mu = 1.88$ D); 6, furan ($\mu = 0.71$ D); 7, acetone ($\mu = 2.69$ D); 8, chloroform ($\mu = 1.15$ D); 9, iodomethane (methyl iodide, $\mu = 1.48$ D); 10, dichloromethane (methylene chloride, $\mu = 1.14$ D); 11, carbon disulfide ($\mu = 0$); 12, iodobenzene ($\mu = 1.30$ D), from McClellan (1963); 13, tert-butanol (1,1-dimethylethanol, $\mu = 1.66$ D); 14, pyridine ($\mu = 2.37$ D); 15, benzonitrile ($\mu = 4.05$ D); 16, N-methyl-2-pyrrolidinone ($\mu = 4.09$ D); 17, N,N-dimethylacetamide ($\mu = 3.72$ D); 18, diiodomethane (methylene iodide, $\mu = 1.08$ D). Dipole moment μ in benzene solution at 20–30°C, from Riddick and Bunger (1970).

Forces of attraction are opposed by forces of repulsion which arise from the interactions of the electron clouds as the molecules or atoms approach each other. The potential energy of repulsion is by definition positive and increases very steeply with decreasing separation. At the equilibrium distance of separation between two molecules, corresponding to the sum of the van der Waals radii, van der Waals forces have energies of interaction which are one or two orders of magnitude less than those of conventional valency bonds and are not directional or oriented in space as are covalent, dative, or hydrogen bonds. Furthermore, van der Waals radii are greater than covalent radii. Consequently, van der Waals forces are classed as nonspecific interactions and do not give clear-cut chemical entities nor true complexes in solution (Section 2.11.8). Van der Waals interactions account for, or contribute to, the enthalpy of vaporization of organic liquids and crystals. For example, ΔH_v for solid carbon dioxide has the value 15.3 kJ mol^{-1} ($3.66 \text{ kcal mol}^{-1}$), which is largely attributed to London dispersion forces.

2.11.7 Hydrogen Bonds

The hydrogen bond (Pimentel and McClellan, 1960; Speakman, 1975) is an intermolecular or intramolecular interaction which has a strength similar to that of van der Waals forces but which is directional like a covalent bond, so that under suitable circumstances it can form discrete recognizable units consisting of two or more single molecules. The enthalpy of vaporization of water (40.7 kJ mol^{-1}; $9.72 \text{ kcal mol}^{-1}$) is composed of contributions from the dispersion force (about 24%), the dipole–dipole force (about 39%), the induction effect (about 3%), and the hydrogen bond (about 34%). The allocation of the experimentally determined energy among the various hypothetical forces of attraction is arbitrary and depends upon the simplified models used.

Hydrogen bonds were originally postulated by Latimer and Rodebush (1920) to explain, among other effects, the high boiling points of compounds containing the groups —OH, —NH$_2$, or > NH as compared with isomeric molecules with no hydrogen directly attached to the oxygen or nitrogen atom (see Table 2.11). The former compounds are associated because of hydrogen bonding in the liquid and solid states, but much less so in the gaseous state.

Table 2.11 Boiling Points of the Isomers of C_2H_6O and of C_3H_9N at 1 atm Pressure

Isomer of C_2H_6O	Boiling Point (°C)	Isomer of C_3H_9N	Boiling Point (°C)
C_2H_5OH	78	$CH_3CH_2CH_2NH_2$	49
CH_3OCH_3	−25	$(CH_3)_2CHNH_2$	32
		$C_2H_5NHCH_3$	34
		$(CH_3)_3N$	3

The hydrogen bond is represented by the broken line in Figure 2.21a. Significant bonding will occur only if atoms A and B are strongly electronegative and small, and in practice A and B must be either fluorine, oxygen, or nitrogen (Speakman, 1975). Some other elements, such as chlorine or sulphur, can sometimes form weaker hydrogen bonds. A and B may be single atoms or attached to other atoms or groups as in Figure 2.21b and c. Mixed hydrogen bonds such as O—H⋯N or N—H⋯O are common, the latter being found in proteins and nulceic acids.

The high electronegativities of A and B indicate that these atoms are electron-rich, as shown in Figure 2.21. Since the hydrogen atom is positively polarized and very small, it is specially endowed with the property of allowing B to approach as closely as possible. This is facilitated if B is at the negative end of a dipole in a

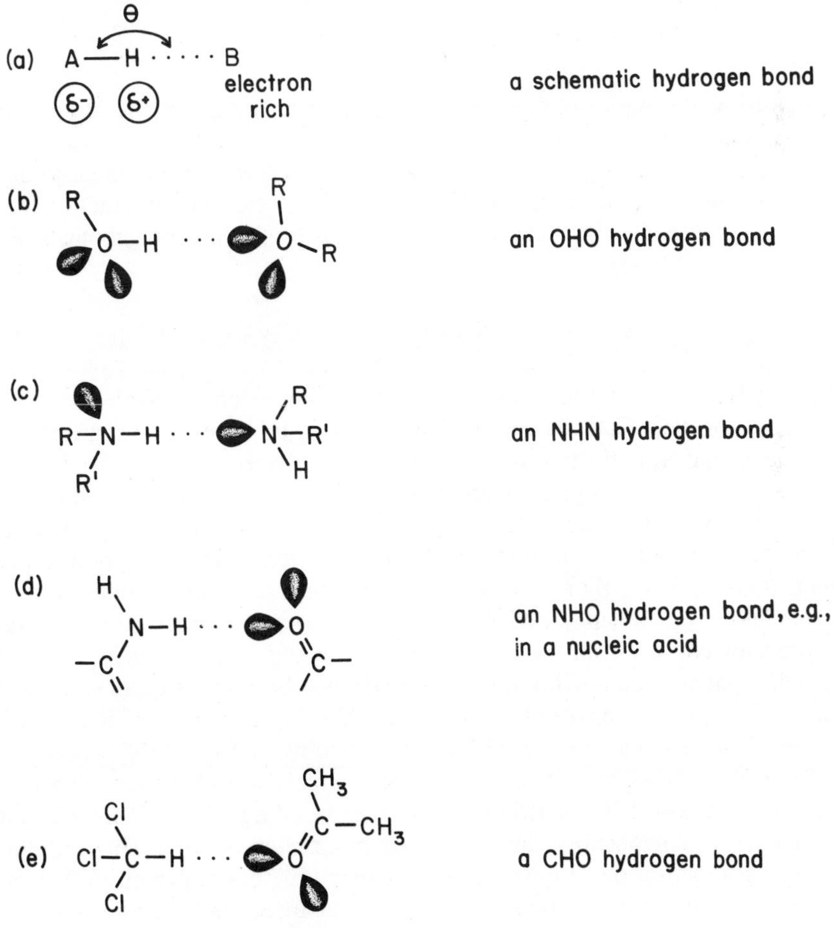

Figure 2.21   Hydrogen bonding.

larger molecule. Thus, the hydrogen bond has definite electrical components of the fixed dipole–dipole and the fixed-dipole–induced-dipole types. This is not all however. Quantum mechanics indicates that there must be mutual overlap (Kollman and Allen, 1972; Morokuma, 1977) of the σ^* (antibonding) orbital of the proton donor H—A, and lone pair n (nonbonding) orbital of the proton acceptor B. H—A must therefore be acidic and B must be basic. For this reason a positively polarized or acidic hydrogen, when attached to a carbon atom (Green, 1974) as in $CHCl_3$, can often form a weak hydrogen bond with traditional hydrogen bond acceptors (F, O, or N atoms). In chloroform, the three electronegative chlorine atoms increase the acidity of the proton considerably as compared with the relatively nonpolar C—H bond in a hydrocarbon group.

The concept of orbital overlap can be considered alternatively as electron delocalization involving resonance between two canonical structures (Speakman, 1975).

$$A—H\cdots B \longleftrightarrow A\cdots H—B$$

The electrostatic repulsions of nonbonding electrons prevent the atoms from approaching too closely.

Up to a point, hydrogen bonding becomes more favored as the acidic strength of H—A and the basic strength of B increase. If the acid–base interaction is too strong, however, B will take complete control of the proton, giving rise to ionization, thus (Speakman, 1975),

$$A—H + B \rightarrow A—H\cdots B \rightarrow A^- + H—B^+$$

The stronger the hydrogen bond, the more important orbital overlap or delocalization becomes, the closer is the hydrogn bond atom to the midpoint between A and B, and the shorter the distance between A and B.

The necessity for orbital overlap accounts for the directional nature of hydrogen bonds. The closer the angle θ is to 180° in Figure 2.21, the stronger the hydrogen bond. For strong bonds θ is rarely less than 165°, and for weak bonds θ is rarely less than 140°. Genuine hydrogen bonding becomes minimal at smaller angles. When A and B are parts of larger molecules, stereochemical factors may reduce θ from 180°. As indicated by examining the data in Table 2.12 for bulky phenyl and *tert*-butyl groups, hydrogen bonding is quite sensitive to steric effects, being weakened by bulky neighboring groups (e.g., R and R' in Figure 2.21*b* and *c*) as a result of bond lengthening and bond distortion.

Table 2.12 shows that hydrogen bonding involves both negative enthalpy and entropy changes. Thus, although hydrogen bonding is favored energetically, it is disfavored entropically, the system becoming more ordered. This enthalpy–entropy compensation (Lumry and Rajender, 1970; Tomlinson, 1983) is a characteristic of hydrogen-bonded systems and causes the Gibbs free energy of interaction to be relatively small, of the order of 1 kcal mol^{-1} (4 kJ mol^{-1}). For

Table 2.12 Thermodynamics of Hydrogen Bond Formation and Infrared Shifts of the A—H Stretching Frequency. Phenol, Alcohols, and Aniline in CCl_4 Solutions

Acid	Base	$-\Delta G^{\theta}$ (kcal mol^{-1})	$-\Delta H^{\theta}$ (kcal mol^{-1})	$-\Delta S^{\theta}$ (cal K^{-1} mol^{-1})	$\Delta \nu$ (cm^{-1})
Phenol	Diphenyl ether $(C_6H_5)_2O$	−0.146	2.06	7.4	130
	Tetrahydrofuran $(CH_2)_4O$	1.66	5.30	12.2	284
	Diethyl ether $(C_2H_5)_2O$	1.30	5.41	13.8	278
	Di-tert-butyl ether $(t\text{-}C_4H_9)_2O$	0.751	7.31	22.0	330
	Di-n-butyl selenide $(n\text{-}C_4H_9)_2Se$	−0.752	3.69	14.9	240
	Di-n-butyl sulfide $(n\text{-}C_4H_9)_2S$	0.165	4.19	13.5	254
	Propionaldehyde C_2H_5CHO	0.821	4.25	11.5	170
	Methyl acetate CH_3COOCH_3	1.18	4.49	11.1	170
	Acetone $(CH_3)_2CO$	−, 1.44	4.94, 4.6	−, 10.6	193
	N,N-Dimethyl formamide $HCON(CH_3)_2$	2.46	6.1	12.2	294
	Triethylamine $(C_2H_5)_3N$	−, 2.52	9.1, 7.8	−, 17.7	550
	Acetonitrile CH_3CN	0.991, 1.34	4.3, 4.65	11.1	155
	Cyclohexene C_6H_{10}	−0.864	3.40	14.3	97
	Benzene C_6H_6	−0.766	1.56	7.8	47
	Dimethyl sulfoxide $(CH_3)_2SO$	3.08	8.00	16.5	350
Methanol	Ethyl acetate $CH_3COOC_2H_5$	0.194	2.52	7.8	84
	Acetone $(CH_3)_2CO$	0.344	2.52	7.3	112
	N,N-Dimethyl formamide $HCON(CH_3)_2$	1.01	3.72	9.1	160
	Pyridine C_5H_5N	0.660	3.88	10.8	286
Ethanol	Diethyl ether $(C_2H_5)_2O$	−0.181	2.92	10.4	144
	Dioxane $O(CH_2)_4O$	0.019	3.09	10.3	123
	Ethyl acetate $CH_3COOC_2H_5$	0.0044	2.33	7.8	80
	Acetone $(CH_3)_2CO$	0.121	3.46	11.2	109
	N,N-Dimethyl formamide $HCON(CH_3)_2$	0.749	3.88	10.5	155
	Pyridine C_5H_5N	0.529	3.66	10.5	276
Aniline	Tetrahydrofuran $(CH_2)_4O$	0.0585	3.04	10.0	
	Ethyl acetate $CH_3COOC_2H_5$	0.238	3.13	9.7	
	N,N-Dimethyl acetamide $CH_3CON(CH_3)_2$	1.30	4.85	11.9	

Reproduced with permission from the *Annual Review of Physical Chemistry*, Vol. 22, 1971 by Annual Reviews, Inc., from Pimentel, G. C. and McClellan, A. L. (1971). *Ann. Rev. Phys. Chem.*, **22**, 347–385.

most hydrogen bonds the bonding enthalpy is of the order $12–30\,kJ\,mol^{-1}$ $(3–8\,kcal\,mol^{-1})$. The thermodynamics of hydrogen bonding will be considered in greater detail in Chapter 5 in connection with complex formation.

To summarize its nature, a hydrogen bond is a spatially oriented donor–acceptor (i.e., charge-transfer) interaction which owes its uniqueness to the fact that it involves a moderately polar, short, and strong A—H bond as a proton donor. Electrostatic forces are generally the greatest contributors to bonding, but are not strong enough to allow bond stretching and variation of geometry which would result in greater electron–electron repulsion and an unfavorable energy balance (Pimentel and McClellan, 1971; Kollman and Allen, 1972; Kollman, 1977; Morokuma, 1977). In any hydrogen bond, Debye and London forces will supplement the forces of electrostatic attraction and electron delocalization (or orbital overlap) discussed above.

Numerous attempts have been made to correlate acid–base behavior using multiparameter equations (Spencer et al., 1979). For the empirical treatment of donor–acceptor interactions in the gas phase and solvents which do not undergo specific interactions with either species A or B, the following equation from Drago and Wayland (1965) and Drago et al. (1971, 1977) is perhaps the most useful:

$$\Delta H = E_A E_B + C_A C_B \qquad (2.138)$$

where ΔH is the enthalpy of interaction, E_A and E_B are the parameters representing the susceptibility of the acid and the base to undergo electrostatic interaction, and C_A and C_B are parameters representing the susceptibility of the acid and the base to form a covalent bond. The equation can be used for predicting many enthalpies of interactions of proton donors and acceptors (according to the Brønsted–Lowry concept) and electron pair donors and acceptors (according to the Lewis concept) since the parameters E_A, E_B, C_A, and C_B have been extensively tabulated (Drago and Wayland, 1965; Drago et al., 1971).

Careful inspection of Table 2.12 shows a limited correlation between the enthalpy of hydrogen bonding, ΔH, and the infrared spectral displacement, $\Delta \nu_{OH}$, of the O—H stretching vibrations. This correlation is known as the Badger–Bauer correlation and is somewhat limited (Ghersetti and Lusa, 1965; Arnett et al., 1970, 1974; Pimentel and McClellan, 1971; Kollman and Allen, 1972). Ghersetti and Lusa (1965), working with a series of phenols interacting with different acceptors, found that $\Delta \nu_{OH}$ values may be used to predict ΔH values only for a given acceptor and for a very homogeneous series of hydrogen donors. They tentatively suggested that the substituent effects in the substituted phenols were themselves affected by the strengths of the acceptors.

2.11.8   Interactions and Solubility

We have seen that the various intermolecular interactions profoundly influence solubility. In this connection, two classical contributions are worthy of particular mention, namely, those from van Laar's group in the Netherlands, who followed the traditions of van der Waals and van't Hoff, and those from Dolezalek's school in Charlottenburg. In addition to their fundamental contributions, the two schools generated a scientific controversy which acted as a psychological catalyst for further advances in the field of the physical chemistry of solutions. The deviations of actual solutions of nonelectrolytes from ideality represented by Raoult's law were attributed by the van Laar school to physical interactions, namely, van der Waals forces. On the other hand, chemical interactions involving the formation of new chemical species were advocated by the school of Dolazalek (1908, 1910) in their explanation of deviations from ideal, Raoult's law, behavior. For example, the Dolezalek school postulated an intermolecular compound or complex between chloroform and acetone to explain the negative deviations of this system from ideal behavior. For this particular example, Dolezalek's interpretation is largely correct, since certain spectroscopic evidence (Green, 1974) strongly suggests that a hydrogen-bonded complex is formed between chloroform and acetone, according to Figure 2.21e.

However, the Dolezalek school attributed positive deviations from Raoult's law to the formation of associated species of one of the original components. While self-association of alcohols and phenols may partly explain the positive deviations of their solution in, for example, hydrocarbon solvents, it cannot be the correct explanation in systems such as carbon tetrachloride + benzene and methyl salicylate + diethyl ether. The postulate of species such as $(CCl_4)_2$ and methyl salicylate dimers was, and still is, clearly unacceptable.

The scientific papers during the period of the World War I record great hostility between the two schools of thought. The controversy was summarized by Timmermanns (1921), Scatchard (1931), and Hildebrand and Scott (1950). The physical interpretation prevailed as the explanation for the behavior of most mixtures of nonelectrolytes, and when the equations of van Laar were freed of the restrictions of the van der Waals equation by Scatchard (1931) and Hildebrand (1950), the foundations of regular solution theory were laid.

Physical interactions between solute and solvent involve van der Waals forces, especially London dispersion forces, which are nonspecific in nature. Chemical interactions between solute and solvent in solution, on the other hand, are necessarily specific, but were difficult to envisage at the time of the controversy since all the valencies of the atoms were believed to be accounted for. When in 1920 the hydrogen bond was proposed by Latimer and Rodebush (1920), the theory of chemical or specific interaction had lost credibility.

The view prevailing at the end of the classical period of solution thermodynamics is summarized in the *Transactions of the Faraday Society*, Volume 33 (1937). At that time some workers had begun to postulate specific interactions in solution based upon intermolecular hydrogen bonding. The charge-transfer interaction, which is also specific, was proposed by Hassel and Hroslef in 1954, and may account for the enhanced solubility of anhydrides in ethers (Joris et al., 1972).

From a survey of recent literature, it is apparent that specific interactions possess the following features:

(a) A relatively high pair-potential energy, $u_{AB} \gg kT$, or in other words, a high molar internal energy of interaction, $U_{AB} \gg RT$ ($= 592\,\text{cal mol}^{-1}$ $= 2480\,\text{J mol}^{-1}$ at 25°C).

(b) A fairly high steric requirement or, in other words, an appreciable negative entropy change.

The former criterion permits a solute–solvent bond to persist for many molecular vibrations, whereas the second criterion does not allow the bonded molecules to spin and jeopardize the bond. According to these criteria hydrogen bonding and certain charge-transfer interactions are classed as specific interactions, whereas van der Waals forces (dipole–dipole, dipole–induced-dipole, and dispersion forces) are classed as nonspecific interactions. Many hydrogen bonds have $-\Delta H$ values of $> 3\,\text{kcal mol}^{-1}$ and a strict steric requirement, whereas London forces and other nonspecific interactions, although they may be relatively strong, lack the steric requirement.

In this book we do not wish to resurrect the old controversy involving specific versus nonspecific interactions by implying that most interactions in solution are specific in nature. We would like, however, to stress that specific interactions can be most readily exploited and that is our reason for weighting the book toward specific interactions in solution, as will be apparent in Chapters 4–6.

2.12 PREDICTION OF SOLUBILITY FROM DERIVED MOLECULAR PARAMETERS: THE SOLVATOCHROMIC APPROACH

Prediction of solubilities and partition coefficients from the known physical properties of the solute and solvents is our primary objective and must inevitably be preceded by an understanding of the relevant intermolecular forces. In this chapter we have examined in a preliminary way the influence of (a) the interactions in the solid state, (b) enthalpic contributions based on cavity formation and London forces (regular solution theory), (c) entropic contributions based on differences in molecular size (Flory–Huggins entropy and athermal solutions), (d) the relationship between solubilities and partition coefficients, and (e) intermolecular forces and their influences.

Recently these concepts have been extended and developed quantitatively by Taft, Abraham, Doherty, Kamlet, and co-workers (see Taft et al., 1985a, b). Over a number of years Taft, Abraham, and Kamlet have expressed the various intermolecular forces in solution in terms of bulk solvation. These workers have introduced "solvatochromic parameters," calculable from known physical properties of the relevant substances, to calculate the solvation energies. This approach has led to the following conceptualization and prediction of the solubility of liquids and partition coefficients.

In principle the solubility s of a given liquid solute is expressed in terms of the standard molar free energy of solvation, ΔG^s, thus,

$$\Delta G^s = E - RT \ln s \qquad (2.139)$$

where E is a constant which depends on the definition of the standard state of the solute. We note that the molar free energy of solvation, ΔG^s, is essentially equivalent to the molar free energy of mixing, ΔG_{mix}, and this differs from the molar free energy of solution, ΔG_s, according to the following equation, which is analogous to Equation 2.9, thus,

$$\Delta G^s = \Delta G_{mix} = \Delta G_s - \Delta G_f^T \qquad (2.140)$$

Here ΔG_f^T is the molar energy of fusion at the temperature T and is obviously zero for liquids, but is positive for solids below the melting point.

Solvation of a solute molecule is considered to be composed of the following distinct processes, each with a corresponding contribution to ΔG^s: (a) formation of a cavity in the solvent of volume equal to that of the incoming solute molecule, according to Figure 2.16, corresponding to a molar free-energy change $(\Delta G^s)_{cav}$; (b) reorganization of solvent molecules around the cavity and reorganization of the conformation of the solute within the cavity, corresponding to $(\Delta G^s)_{reorg}$; (c) formation of Keesom and Debye interactions between the solute molecue and the neighboring solvent molecules on account of the dipole moments (dipolarity) and polarizability of the respective molecules (see Section 2.11) corresponding to $(\Delta G^s)_{DP}$; and (d) formation of hydrogen bonds, where possible, between the solute molecule and the neighboring solvent molecules (see Sectio 2.11) corresponding to $(\Delta G^s)_{HB}$. Each of these molar free-energy terms is determined by a certain physical property or parameter for the solute and one for the solvent, as described in the next four paragraphs.

The cavity term is derived from regular solution theory (cf. Eq. 2.89), thus,

$$(\Delta G^s)_{cav} = A\delta_1^2 \frac{V_2}{100} \qquad (2.141)$$

where V_2 is the molar volume of the solute (i.e., molecular weight divided by density of the liquid), δ_1 is the Hildebrand solubility parameter of the solvent (δ_1^2 being the cohesive energy density of the solvent according to Eq. 2.87), and A is a constant. Division by 100 is introduced so that the magnitude of $V_2/100$ is comparable to those of the solvatochromic parameters considered below. A is positive because the cavity term is endoergonic, disfavoring the solution process.

The reorganization term, $(\Delta G^s)_{reorg}$, is either too small to be considered, or may be included in the entropy component of the other terms.

The dipolar term is given by

$$(\Delta G^s)_{DP} = B\pi_1^*\pi_2^* \qquad (2.142)$$

where π_1^* is the solvatochromic parameter representing both the dipolarity and polarizability of the solvent molecules, π_2^* is that of the solute molecules, and B is a constant. Here B is negative because the dipolar term is exoergonic and a high value of π_1^* or π_2^* will encourage the solution process.

The term for hydrogen bonding consists of two complementary donor–acceptor interactions, thus,

$$(\Delta G^s)_{HB} = C\alpha_1\beta_2 + D\alpha_2\beta_1 \tag{2.143}$$

where α_1 is the solvatochromic parameter scaling for hydrogen-bond donating ability (acidity) of the solvent (and α_2 is that of the solute), while β_1 is the solvatochromic parameter scaling for hydrogen-bond accepting ability (basicity) of the sovent (and β_2 is that of the solute). Both C and D are constants which are negative, because hydrogen bonding, if it can occur, is exoergonic and a large value of α or β will encourage the solution process.

The total molar free energy of solvation in Eq. 2.139 is assumed by Taft et al. (1985a, b) to be equal to the sum of the molar free-energy terms represented by Eq. 2.141–2.143, thus,

$$(\Delta G^s)_{total} = (\Delta G^s)_{cav} + (\Delta G^s)_{reorg} + (\Delta G^s)_{DP} + (\Delta G^s)_{HB}$$
$$= E - RT \ln s \tag{2.144}$$

Because of the assumption in Eq. 2.144, the derived equations (Eqs. 2.145, 2.146, 2.152, 2.153) and the fitted equations (Eqs. 2.150, 2.154) are termed by Taft et al. (1985a, b) linear solvation energy relationshps (i.e., LSERs). Substituting the individual expressions (Eqs. 2.141–2.143) we obtain

$$(\Delta G^s)_{total} = A\delta_1^2 \frac{V_2}{100} + B\pi_1^*\pi_2^* + C\alpha_1\beta_2 + D\alpha_2\beta_1$$
$$= E - RT \ln s \tag{2.145}$$

When the solubility of a range of solutes in a given solvent is considered, the quantities δ, π_1^*, α_1, and β_1 are constants, in which case Eq. 2.145 reduces to

$$\log s = e + a\frac{V_2}{100} + b\pi_2^* + c\beta_2 + d\alpha_2 \tag{2.146}$$

Table 2.13 lists values of $V_2/100$, π_2^*, β_2, and $\log s_w$, where s_w is the molar solubility in water at 25°C for a number of liquid solutes that do not donate hydrogen bonds significantly in water, but may nevertheless act as hydrogen-bond acceptors. In the case of liquid solutes that take up appreciable quantities of water so as to be miscible in all proportions, the s_w values in Table 2.13 were calculated from

$$s_w = \frac{s_g}{P_{g/w}} \tag{2.147}$$

Table 2.13 Experimental and Predicted Values of Molar Solubilities in Water at 25°C and Partition Coefficients (Octanol/Water)[a]

Solute	$V_2/100$ ($cm^3\,mol^{-1}$)	π_2^*	β_2	$\log s_w$ ($mol\,dm^{-3}$)		$\log P_{o/w}^c$	
				Exp.	Eq. 2.150	Exp.	Eq. 2.154
n-C$_7$H$_{16}$	1.465	-0.08	0.00	-4.53	-4.41	—	—
(C$_2$H$_5$)$_3$N	1.401	0.14	0.71	-0.14	-0.38	1.45	1.43
CH$_3$COC$_2$H$_5$	0.895	0.67	0.48	0.49	0.36	0.29	0.36
(C$_2$H$_5$)$_2$O	1.046	0.27	0.47	-0.13	-0.38	0.89	1.18
C$_2$H$_5$COOCH$_3$	0.963	0.55	0.45	-0.14	-0.08	—	—
C$_2$H$_5$NO$_2$	0.721	0.82	0.25	-0.24	-0.19	0.18	0.55
HCON(CH$_3$)$_2$	0.770	0.88	0.69	(1.90)	1.98	-1.01	-0.90
CH$_3$COCH$_3$	0.734	0.71	0.48	(0.92)	0.92	-0.24	-0.12
CH$_3$CON(CH$_3$)$_2$	0.924	0.88	0.76	(2.15)	1.83	-0.77	-0.73
(CH$_3$)$_2$SO	0.710	1.00	0.76	(2.59)	2.60	-1.35	-1.43
CH$_3$COOC$_5$H$_{11}$-n	1.487	0.53	0.45	-1.74	-1.85	—	—
CH$_3$OH	0.405	0.60	0.40	(1.35)	1.56	-0.65	-0.64
(CH$_3$)$_2$CHOH	0.765	0.48	0.51	(0.83)	0.87	0.05	0.07
n-C$_7$H$_{15}$OH	1.414	0.43	0.45	-1.83	-1.65	—	—
2-C$_6$H$_{13}$OH	1.255	0.44	0.51	-0.82	-0.80	—	—
t-C$_5$H$_{11}$OH	1.094	0.41	0.57	0.09	0.04	0.89	0.83

[a]The predicted values were calculated using Eq. 2.150 or Eq. 2.154 from the molar volume V_2 and the solvatochromic parameters π_2^* and β_2 of the solute. The unit mol dm^{-3} is identical with mol L^{-1}, that is, molarity. Solubilities in parentheses were calculated from Eq. 2.147.

Adapted with permission from *Nature* and the *Journal of Pharmaceutical Sciences*, from Taft, R. W., Abraham, M. H., Doherty, R. M. and Kamlet, M. J. (1985). *Nature*, **313**, 384–386. Taft, R. W. (Abraham, M. H., Famini, G. R., Doherty, R. M., Abboud, J.-L. M., and Kamlet, M. J. (1985). *J. Pharm. Sci.*, **74**, 807–814.

where s_g is the molar concentration of the pure solute when saturating the gas phase and $P_{g/w}$ is the partition coefficient (gas phase/solution in water). If the pressures are well below 1 atm, the ideal gas law may be employed to calculate $P_{g/w}$ from the Henry's law constant in water, h_w, and s_g from the saturated vapor pressure p^*, thus:

$$P_{g/w} = h_w/24.46 \text{ atm dm}^3 \text{ mol}^{-1} \tag{2.148}$$

$$s_g = p^*/24.46 \text{ atm dm}^3 \text{ mol}^{-1} \tag{2.149}$$

Taft, Abraham, Doherty, and Kamlet have taken a step-by-step empirical approach toward solubility prediction guided by relevant theory. Thus, each constant A, B, C, and D is an adjustable parameter (or "exploratory variable"). With the inclusion of each adjustable parameter into the general solubility equation (Eq. 2.145), the goodness of fit between the experimental and calculated solubility must either improve or remain the same. Statistical procedures based on Student's t test or F statistic were employed to demonstrate whether or not the inclusion of the new adjustable parameter was statistically justified.

An essential requirement for the correlations is that the independent parameters V_2, π_2^*, β_2, and α_2 in Eq. 2.146 do not exhibit colinearity, as shown by low values of the coefficient of correlation betwen each pair. For 93 liquid, nonhydrogen-bond-donating solutes (i.e., $\alpha_2 = 0$) and water as the solvent at 25°C, Eq. 2.146 was tested by Taft et al. (1985a) using the values of $V_2/100$, π_2^*, β_2, and log s_w, some of which are listed in Table 2.13. The best fit was given by

$$\log (s_w, \text{mol dm}^{-3}) = (0.55 \pm 0.09) - (3.36 \pm 0.06)\frac{V_2}{100}$$

$$+ (0.46 \pm 0.09)\pi_2^* + (5.23 \pm 0.09)\beta_2 \tag{2.150}$$

for which $n = 93$, $r = 0.9943$, and s.d. $= 0.144$.

Weak hydrogen-bond donor solutes, such as the alkanols, could be included in Eq. 2.146 if the β value corresponded to that of the non-self associated "monomeric" form (β_m values), thus: methanol, 0.41; other primary alcohols, 0.45; all secondary alcohols, 0.47; all tertiary alcohols, 0.57. These values are close to the corresponding β_m values determined by Abboud et al.(1985) from the association constants of the 1:1 complexes of the monomeric alcohols with 3,4-dinitrophenol in cyclohexane: methanol, 0.415; ethanol, 0.47; 2-propanol, 0.51; tert-butanol, 0.56.

Although the unsubstituted alkanols are amphiprotic, they appear to act only as hydrogen-bond acceptors, not donors, in aqueous solution. This seems reasonable if, for the alkanol monomer solutes, $a_m \cong 0.3$ and $\beta_m = 0.40-0.57$, and for water-solvent clusters, $\alpha = 1.17$ and $\beta \cong 0.2$. For stronger hydrogen-bond donor solutes, such as 2,2,2-trifluoroethanol or di(trifluoromethyl)methanol, the values of $\log (s_g/P_{g/w})$ are 0.80 and 0.74, respectively, which are greater than the values predicted by Eq. 2.150 by 2.07 and 3.56, respectively, (Taft et al., 1985a). This suggests the need for including the α_2 term (final terms in Eqs. 2.145 and 2.146) in order to correlate and predict the solubilities of stronger hydrogen-bond donating solutes.

Since the parameters $V_2/100$, π_2^*, and β_2 in Eq. 2.150 have been scaled to span roughly the same range, the magnitudes of their coefficients indicate the relative importance of the various terms. These show that the major factors influencing the solubilities of organic nonelectrolytes in water are the exergonic term ($c\beta_2$ in Eq. 2.146 or $C\alpha_1\beta_2$ in Eq. 2.145) for water-donor–solute-acceptor hydrogen bonding and the opposing endergonic cavity term ($aV_2/100$ in Eq. 2.146 or $A\delta_1^2 V_2/100$ in Eq. 2.145). The exergonic term ($b\pi_2^*$ in Eq. 2.146 or $B\pi_1^*\pi_2^*$ in Eq. 2.145) for solute–water Keesom and Debye interactions appears to exert a much smaller influence. It should be mentioned that a term representing the endergonic separation of a single solute molecule from the bulk liquid solute has been omitted from the LSER (Stage 1 in Fig. 2.16). However, since this term is probably highly correlated by $V_2/100$ and π_2^*, good correlations can be obtained without its explicit inclusion (Taft et al., 1985b).

The free energy of transfer $\Delta G_t(\text{w} \rightarrow \text{o})$, and the partition coefficient of a given solute between two relatively immiscible solvents, such as a suitable organic solvent (subscript o, e.g., 1-octanol) and water (subscript w), is given by subtraction of Eq. 2.139, when applied to water, from Eq. 2.139, when applied to the organic solvent, thus,

$$\Delta G_t(\text{w} \rightarrow \text{o}) = \Delta G_o^s - \Delta G_w^s = -RT \ln \frac{S_o}{S_w} = -RT \ln P_{\text{o/w}} \qquad (2.151)$$

where $P_{\text{o/w}}$ is the partition coefficient (organic solvent/water). If Eq. 2.145 is applied to each solvent in turn and is substituted into Eq. 2.151, we obtain

$$\Delta G_t(\text{w} \rightarrow \text{o}) = -RT \ln P_{\text{o/w}} = E_w - E_o$$

$$+ A\left(\frac{V_2}{100}\right)(\delta_o^2 - \delta_w^2) + B\pi_2^*(\pi_o^* - \pi_w^*)$$

$$+ C\beta_2(\alpha_o - \alpha_w) + D\alpha_2(\beta_o - \beta_w) \qquad (2.152)$$

For a given pair of solvents, o and w, their parameters are constant, so that Eq. 2.152 reduces to

$$\log P_{\text{o/w}} = e' + a'\left(\frac{V_2}{100}\right) + b'\pi_2^* + c'\beta_2 + d'\alpha_2 \qquad (2.153)$$

Kamlet et al. (1984) and Taft et al. (1985b) fitted this equation to the partition coefficients (octanol/water) of 102 solutes which do not donate hydrogen bonds significantly, but which may nevertheless act as hydrogen-bond acceptors, that is, $\alpha_2 \cong 0$ while β_2 is significant. The solutes comprised aliphatic and aromatic compounds including some alkanols, and the empirical equation of best fit was

$$\log P_{o/w} = (0.20 \pm 0.07) + (2.74 \pm 0.05)\frac{V_2}{100}$$
$$- (0.92 \pm 0.08)\pi_2^* - (3.49 \pm 0.09)\beta_2 \tag{2.154}$$

for which $n = 102$, $r = 0.989$, and s.d. $= 0.17$.

The signs of a', b', and c' in Eq. 2.153 are $+$ve, $-$ve, and $-$ve, respectively, in Eq. 2.154. These signs are intuitively expected from examination of the solvent parameters in Eq. 2.152. Since $\delta_1/\text{cal}^{1/2} \text{ cm}^{-3/2}$ for water (15.65–23.4) is larger than that for 1-octanol (10.2), a greater V_2 will encourage partitioning into 1-octanol. Furthermore, since π_2^* for water (1.09) is greater than that for 1-octanol (~ 0.4), a larger π_2^* will encourage partitioning into water. Moreover, since α_1 for water (1.17) is higher than that for 1-octanol (~ 0.6), a greater β_2 will also direct the solute into water (Taft et al., 1985b). To accommodate in the correlation a wider range of aromatic solutes than the rather restricted range included in Eq. 2.154, Taft et al. (1985b) devised a rather cumbersome set of "ground rules" which represent correction terms to the "input parameters" V_2, π_2^*, and β_2 for the aromatic segment of the solute molecule. Although these ground rules, that is, corrections, can be rationalized on the basis of the intermolecular interactions and physical organic chemistry, they appear to weaken the usefulness of this treatment of solubility and partitioning at its present stage of development.

Yalkowsky et al. (1988) recently published a detailed critique of the solvatochromic approach. They showed that the contribution of the π^* term is uncertain and that it can be dropped from the correlation without substantially affecting the degree of fit. Yalkowsky et al. (1988) also compared the prediction of water solubility from the solvatochromic method with that from their own relationship between aqueous solubility and the octanol–water partition coefficient (Section 2.9). They demonstrated that the latter offers superior predictions. However, the two approaches have different philosophies for solubility prediction. Yalkowsky's method is based on readily acessible experimental quantities, such as octanol–water partition coefficient and melting point. The solvatochromic method of Taft and co-workers bases its predictions on parameters for the actual underlying intermolecular interactions, such as cavity formation, nonspecific polar and polarizabilty effects, and specific donor–acceptor interactions. Each approach therefore has different strengths and weaknesses. Other generally applicable methods of solubility prediction are presented and discussed in Chapters 3–10.

Chapter **III**

ACTIVITIES OF SOLUTES, SELECTION OF STANDARD STATE AND HENRY'S LAW CONSTANTS

3.1 INTRODUCTION

3.1.1 Thermodynamic Activity

We have already seen in Eqs. 2.2–2.5 and Sections 2.2.2–2.2.6 that the thermodynamic activity of solute molecules in the pure solute phase is a major determinant in solubility behavior. In this chapter we consider directly the concept of activity and the fundamental relationships between the equilibrium concentrations of chemical species in the vapor and solution phases. Classical thermodynamics is concerned with activities or fugacities (i.e., escaping tendencies), concepts which are extremely useful not only in consideration of solubility behavior but also in many other aspects of applied chemistry. A particular problem is the selection of an appropriate reference state or standard state of unit activity.

From the thermodynamic point of view, the solubility of a solid crystalline substance is equal to that concentration of the substance at which the activity of the substance in solution is equal to the activity of the pure substance. By convention, in classical equilibrium thermodynamics this activity is designated as equal to one. However, this definition of the standard state is not always the most useful when comparing the solubilities of different or even closely related solute species. In addition, if a pure solid form is chosen as the standard state, interpretation of polymorphic behavior greatly complicates the situation. The consequences of polymorphism are considered in Fig. 2.4 and Section 2.3.

The stable pure state, solid or liquid, has been used by Ferguson (1939) as the standard state for relating the biological activity to the thermodynamic activity for narcotic agents and other toxic compounds. The reason for this choice is partly one of convenience, since the pure stable solute standard state (a) is usually physically attainable, (b) is independent of the solvent, and (c) has physical properties which are often easily characterized, such as energy and entropy of vaporization, molar volume, solubility parameter, and vapor pressure. However, when the pure solid or the pure liquid solute is chosen as the standard state, prediction of relative solution properties of different solutes is complicated by the fact that the molecules of each solute do not have the same environments in the reference state. In other words, the reference state is actually different for each solute at the molecular level.

3.1.2 Standard States

The standard state used for solute(s) throughout most of Chapter 2 is the pure liquid (supercooled in the case of solids) at an appropriate temperature and pressure. This reference state is usually preferred by most workers when considering ideal solutions, regular solutions (Hildebrand and Scott, 1950, 1962; Hildebrand et al., 1970), and liquid mixtures (Rowlinson, 1969). However, this choice of standard state has one serious disadvantage. In the case of solid solutes, the pure liquid form at the defined temperature and pressure (e.g., 25°C, 1 atm) may not be physically attainable. The standard state is then usually taken as the hypothetical supercooled liquid with consequences discussed in Sections 2.2.1–2.2.3. This standard state is also less appropriate for predicting the relative escaping tendencies of a series of closely related substances since the molecular constituents of each pure reference state are again exposed to different environments. Therefore, prediction of solubility is also difficult to make since solubility behavior is directly related to escaping tendency.

Theoretically, any set of conditions may be chosen to define the standard state. However, in practice, only a few such sets are sufficiently convenient to have gained general acceptance (Hildebrand and Scott, 1950; Lewis and Randall, 1961; Klotz, 1964; Prausnitz, 1969). Rytting et al. (1972) have given careful consideration to this point and have proposed appropriate standard states for drugs and other organic substances.

Any alternative reference state must be widely applicable and permit thermodynamic comparisons. Infinitely dilute solutions are often used as the reference state for most aqueous solutions of electrolytes when comparisons of chemical potentials and activities are made. For biological and pharmaceutical systems, a standard state based on a solution is particularly appropriate since most biochemical processes occur in solution. Also, drug molecules react in the immediate environment of their receptor sites, where their concentration is often low.

Before a solution standard state can be selected, a suitable solvent must be proposed. Although water is the most common solvent for many electrolytes and biological systems, it is not the most convenient solvent for a reference state because of its complicated and ordered structure. In water the solvent–solvent interactions are strong and are affected by the solute molecules. Furthermore, the solute–solvent interactions vary greatly, depending on the nature and structure of the solute molecules. These variable interactions greatly complicate interpretation of thermodynamic and biological data in aqueous solution.

Hansch (1968) used 1-octanol and water as solvents for the measurement of activity coefficients in quantitative structure–activity relationships (QSAR) involving the action of drugs on biological systems. The reasons for this choice were essentially practical ones, since 1-octanol has solvent properties said to be similar to those of cellular lipid. Other alcohol–water systems appear to give comparable results (Scholtan, 1968). Alcohols as reference solvents for thermodynamic studies have the same disadvantages as water; they are highly structured because of hydrogen bonding and they undergo variable interactions which depend largely on the nature and structure of the solute.

3.2 RECOMMENDED STANDARD STATE FOR DISSOLVED SOLUTES

Rytting et al. (1972) have proposed the following universal standard state for organic molecules: a hypothetical solution of unit molarity, molality, or mole fraction, which acts as if it were infinitely dilute, and for which the solvent is a suitable paraffin (i.e., a saturated aliphatic hydrocarbon) such as cyclohexane or isooctane. The reference state of a drug would then be the drug in a hydrocarbon solvent at an infinite dilution, and Henry's law, referred to in Section 2.12, would be the limiting law for the system. This state is similar to that used by Deal et al. (1962) and Christian et al. (1970). It is sometimes necessary to refer to later chapters in this book in order to justify the chemical advantages of such a standard state, which are as follows:

(a) This standard state is attainable in practice by extrapolation to infinite dilution or by measurement of very low solubility.

(b) The substance (solute) in the standard state is too dilute to interact with itself so solute–solute interactions are absent even for strongly polar and hydrogen-bonding solutes. Comparisons between different solutes is then possible since each solute has essentially the same environment. This situation contrasts with the classical standard states of regular solution theory in which the molecules have different environments in each of the pure (supercooled) liquids. In addition, deviations from Henry's law at moderate concentrations of a polar or hydrogen-bonded solute can provide information concerning the solute–solute interactions and molecular association of the solute (Anderson et al., 1978; Rytting et al., 1978a), as will be discussed in Section 6.3.2 and Figures 6.2–6.8.

(c) Since the solvent chosen for the reference state is nonpolar, the solute–solvent interactions which dominate the thermodynamic behavior of the system at infinite dilution are simple, being essentially London dispersion forces for a nonpolar solute and both London and Debye for a polar solute. The energies of these interactions can be calculated and compared relatively easily. The choice of a nonpolar solvent eliminates Keesom forces, hydrogen bonding, and charge-transfer interaction in the standard state.

As will be discussed in Chapter 5, the chemical potentials of substances such as phenols, steroids, carboxylic acids, esters, ketones, ethers, amines, amides, and organic ion pairs are drastically altered by hydrogen bonding with the solvent (Nakano et al., 1967, Higuchi et al., 1969, Michaelis and Higuchi, 1969). In solvents capable of hydrogen bonding, charge transfer, and other specific interactions, deviations of physical properties from ideal behavior (in a hydrocarbon) provide information which enables these interactions to be quantified (see Chapters 5 and 6).

Table 3.1 Henry's Law Constants (h_2^x, torr) for Several Organic Solutes in Various Aliphatic Hydrocarbon Solvents[a]

Solvents	Solutes			
	Hexane	Heptane	Methyl Ethyl Ketone	Acetone
Hexane	—	—	490[e]	—
Cyclohexane	137[b]	—	—	—
Heptane	—	—	470[e]	1628[e]
2,3,4-Trimethylpentane	133[c]	38[c]	—	—
Decane	—	—	475[e]	1430[e]
Dodecane	117[b]	—	—	—
Hexadecane	109[b]	31[b]	416[e]	1342[e]
Heptadecane	120[d]	38[d]	—	—

[a] These values were calculated from activity coefficients (mole fraction concentration scale) extrapolated to infinite dilution. These activity coefficients were then multiplied by the vapor pressure of the pure solute to yield Henry's constants (i.e., $h_2^x = \gamma_2^\infty p_2^*$, Eq. 3.25).
[b] Rowlinson, J. S., *Liquids and Liquid Mixtures*, Butterworths, London, England, 1969. $T = 20°C$.
[c] Rescan, R. E. and Martin, J. J., *Anal. Chem.*, **38**, 1661 (1966). $T = 20°C$.
[d] Matire, D. E. and Riedl, P., *J. Phys. Chem.*, **72**, 3478 (1968). $T = 22.5°C$.
[e] Pierotte, G. J., Deal, C. H., and Derr, E. L., *Am. Doc. Inst.*, Document No. 5782, Library of Congress, 1958, pp. 1–53. $T = 25°C$.

Reproduced with permission of the copyright owner, the American Pharmaceutical Association, from Rytting, J. H., Davis, S. S., and Higuchi, T. (1972). *J. Pharm. Sci.*, **61**, 816–818.

(d) Since the proposed saturated hydrocarbon (i.e., paraffinic) solvents have negligible polarity and relatively low polarizabilities, they form nearly ideal solutions with each other, making it possible to convert chemical potentials in one solvent (e.g., cyclohexane) to those in another (e.g., isooctane). Thus, provided the solvents have similar molar volumes, the Henry's law constants for a given solute in various hydrocarbon solvents should be essentially the same, as shown in Table 3.1.

The Henry's law constants listed in Table 3.1 show remarkable agreement considering the diverse sources of data and the particularly large differences in molar volume of the solvents (e.g., for cyclohexane $V_m = 108.8 \, \text{cm}^3 \, \text{mol}^{-1}$ at 20°C; for hexadecane $V_m = 344.6 \, \text{cm}^3 \, \text{mol}^{-1}$ at 20°C). A similar constancy is observed when the solubilities of various polar substances in hydrocarbon solvents are modified to allow for differences in molar volume of the solvent by expressing solubility in units of molarity (e.g., Table 3.4).

These data indicate that, within the limits of dilute solutions, the solubility of a given solute is relatively independent of the paraffinic solvent and is essentially a function of the crystalline structure and the intermolecular forces of the solid solute. In unsaturated or aromatic hydrocarbon solvents such as toluene or in carbon tetrachloride, which are more polarizable then paraffins although still essentially nonpolar, the stronger Debye interactions and London dispersion forces will reduce the Henry's constants of most solutes (Figure 3.5 and Table 3.12).

(e) The group contributions of methylene ($\log F_{CH_2}$) (Davis et al., 1972) and other groups to the partition coefficient between water and various nonpolar organic solvents can be calculated as described in Chapter 7. ($-2.303RT \log F_{CH_2}$ corresponds to the free-energy transfer of a methylene group included in a molecule at infinite dilution from water to the organic solvent.) Since the methylene group contribution is largely unaffected by the choice of the nonpolar organic solvent, as shown in Tables 7.5 and 7.8, the value of the proposed standard state is emphasized.

A biological advantage of the proposed standard state concerns the application of partition coefficients to QSARs. The saturated aliphatic hydrocarbons chosen as model solvents simulate hydrocarbon polymethylene chains inside biological membranes and fatty (triglyceride) tissues and may reflect the lipoidal characteristic of such tissues (Beckett and Moffatt, 1969). Although the analogy between the liquid aliphatic hydrocarbons and biological membranes is weakened because the molecular chains of the former are arranged randomly, whereas those of the latter are relatively ordered, it may be difficult to find more convenient model solvents.

3.3   STANDARD STATES BASED ON HENRY'S LAW

When the reference state is an infinitely dilute solution, the limiting law is not Raoult's law but Henry's law, as shown in Figure 3.1. (For this reference state, the components are usually represented by alphabetical or numerical subscripts in this book.) Henry's law states that the partial vapor pressure of a solute above a dilute solution at a given temperature is proportional to its concentration in that solution. The proportionality constant is known as the Henry's law constant h and is proportional to the partition coefficient gas phase/solution phase (e.g., Eq. 2.148). Since Henry's law is limiting at very dilute solutions, any appropriate concentration scale can be chosen for the horizontal axis in Figure 3.1. However, the value of the Henry's law constant h_B will depend on the concentration scale. Thus,

$$p_B = h_B^x x_B = h_B^m m_B = h_B^c c_B \qquad (3.1)$$

where p_B is the partial vapor pressure of B, x_B is the mole fraction of B, m_B is the molality of B, c_B is the molarity of B, and h_B^x, h_B^m, and h_B^c are the respective Henry's constants. In dilute solutions.

$$\lim_{x_B \to 0} (m_B) = \frac{1000}{M_A} x_B \qquad (3.2)$$

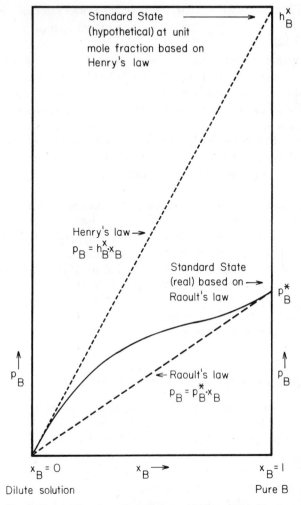

Figure 3.1   Standard states based on Raoult's law and Henry's law (Davis et al., 1972).

$$\lim_{x_B \to \infty} (c_B) = \frac{1000\,\rho_A}{M_A} x_B = \frac{1000}{V_A} x_B \tag{3.3}$$

where M_A, ρ_A, and V_A are the molecular weight, density, and molar volume, respectively, of the solvent A. The relationships between the various Henry's constants are

$$h_B^m = \frac{M_A}{1000} h_B^x \tag{3.4}$$

$$h_B^c = \frac{M_A}{1000\,\rho_A} h_B^x = \frac{V_A}{1000} h_B^x \tag{3.5}$$

In this chapter we wish to distinguish between the solvent (subscript 1, replacing A) and the solute (subscript 2, replacing B). Based on Henry's law, we can define a standard state of unit activity ($a_2^\theta = 1$) of the solute 2 as a hypothetical solution of *either* unit mole fraction ($x_2 = 1$), *or* unit molarity ($c_2 = 1 \, \text{mol dm}^{-3}$), *or* unit molality ($m_2 = 1 \, \text{mol kg}^{-1}$), which acts as if it were infinitely dilute. These standard states are, of course, hypothetical, since $x_2 = 1$ corresponds to the pure solute while $c_2 = 1 \, \text{mol dm}^{-3}$ and $m_2 = 1 \, \text{mol kg}^{-1}$ represents the solute in situations far removed from an infinitely dilute solution. The respective relationships for the solute are *either*

$$a_2^x = y_2^x x_2, \qquad y_2^x \to 1 \quad \text{as } x_2 \to 0 \tag{3.6}$$

or

$$a_2^c = y_2^c c_2, \qquad y_2^c \to 1 \quad \text{as } c_2 \to 0 \tag{3.7}$$

or

$$a_2^m = y_2^m m_2, \qquad y_2^m \to 1 \quad \text{as } m_2 \to 0 \tag{3.8}$$

where y_2^x, y_2^c, and y_2^m are the respective activity coefficients of the solute 2. The corresponding relationship for the solvent 1 is based on Raoult's law. Thus,

$$a_1 = \gamma_1 x_1, \qquad \gamma_1 \to 1 \quad \text{as } x_1 \to 1 \tag{3.9}$$

Figure 3.2 corresponds to the most common situation in which the deviations from Raoult's law are so great that the solute (subscript 2) can reach a solubility limit in the solvent (subscript 1). The vapor is assumed to behave as an ideal gas, thus justifying the usual assumption that the partial vapor pressure p_2 is equal to the fugacity f_2. This assumption is a very good approximation for organic compounds such as drugs, for which the boiling point is above room temperature. The curve corresponds to the concentration dependence of vapor pressure of the solute.

The two concentration scales, molarity and mole fraction, are related by Eq. 3.3 under the limiting condition corresponding to the Henry's law constant. This constant is equal to the vapor pressure of the solute in the hypothetical standard state, since the standard state corresponds to unit concentration. Thus,

$$p_2^{x\theta} = h_2^x; \qquad p_2^{c\theta} = h_2^c \cdot 1 \, \text{mol dm}^{-3} \tag{3.10}$$

Since the dimension of $p_2^{c\theta}$ is pressure, and the dimension of h_2^c is pressure divided by molar concentration (e.g., $\text{torr dm}^3 \, \text{mol}^{-1}$), the inclusion of $1 \, \text{mol dm}^{-3}$ is necessary for dimensional correctness. Analogous considerations apply to the molal scale.

At any point on the curve in Figure 3.2 the activity with respect to any standard state is given by Eq. 2.2, into which we may insert the Henry's law constants given by Eq. 3.10 and the good approximation $f_2 = p_2$.

Since we are now making a distinction between the solvent (subscript 1,

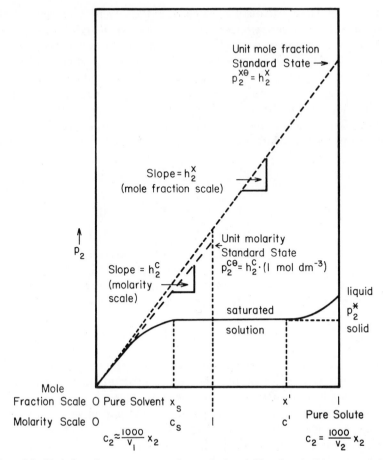

Figure 3.2   Variation of vapor pressure of a sparingly soluble solute 2 with concentration.

replacing A) and the solute (subscript 2, replacing B), Eq. 2.2 may be written

$$a_2^x = \frac{f_2}{f_2^{x\theta}} = \frac{p_2}{h_2^x} \text{ (mole fraction scale)} \tag{3.11}$$

or

$$a_2^c = \frac{f_2}{f_2^{c\theta}} = \frac{p_2}{h_2^c \cdot 1 \text{ mol dm}^{-3}} \text{ (molarity scale)} \tag{3.12}$$

Equations analogous to Eqs. 3.10 and 3.12 can be written with respect to the molal scale.

In Figure 3.2 the solubility of the solute in the solvent is given by c_s or x_s, whereas c' and x' correspond to the solute composition for a saturated solution of the solvent dissolved in the solute. The latter situation is realistic if the solute is a liquid (e.g., aniline in hexane), but is not attainable for most solid solutes (e.g., NaCl in water) unless the crystal lattice can expand to accommodate solvent molecules, thereby producing a solvate or inclusion compound.

The phase rule demands that p_2 is constant as long as two phases are present, such as a saturated solution and the pure solute (or vice versa). A rise in p_2 will occur between c' or x' and the pure solute if the solute is capable of dissolving some of the solvent, (e.g., n-butanol–H_2O, benzoic acid–H_2O) p. 20. For solid solutes which are unable to dissolved the solvent (e.g., acetaminophen–H_2O) or to incorporate any more solvent (e.g., ampicillin trihydrate–H_2O), c' or x' will have no meaning and p_2 will be constant.

3.4 ACTIVITY OF THE PURE SOLUTE FROM THE HENRY'S LAW CONSTANT

The activity of the pure solute, a_2^*, is given by the fugacity of the pure solute, f_2^*, divided by the fugacity of the solute in the standard state, f_2^θ, whatever that might be. This rule applies to both solids and liquids. Taking as our reference state an infinitely dilute solution, that is, one that obeys Henry's law, application of Eqs. 3.11 and 3.12 to the pure solute gives

$$a_2^{x*} = \frac{f_2^*}{f_2^{x\theta}} = \frac{p_2^*}{h_2^x} \quad \text{(mole fraction scale)} \tag{3.13}$$

or

$$a_2^{c*} = \frac{f_2^*}{f_2^{c\theta}} = \frac{p_2^*}{h_2^c \cdot 1 \, \text{mol dm}^{-3}} \quad \text{(molarity scale)} \tag{3.14}$$

An equation analogous to Eq. 3.14 applies to the molal scale.

Equation 3.14 is particularly useful for estimating the activity a_2^{c*} of pure, relatively volatile organic compounds based on the molarity scale, since only a knowledge of the Henry's law constant h_2^c in a paraffinic solvent and the saturated vapor pressure p_2^* is required. Table 3.2 lists values of the activities a_2^{c*} so calculated. If the Henry's law constant in a paraffinic solvent is not available in the literature or is not accessible experimentally, it may be estimated quite reliably from a knowledge of the molecular structure of the solute using the group-contribution approach, provided that the molecules are relatively inflexible, as discussed in Sections 7.1 and 7.2.1. The saturated vapor pressures p_2^* of relatively volatile organic compounds are usually available in the literature (Hala et al., 1968; Riddick and Bunger, 1970; Boublík et al., 1984; Wichterle et al., 1973, 1976; Weast et al. 1986; Windholz et al., 1983).

If the saturated vapor pressure p^* is known at a given temperature T, the value $p^{*'}$ at the temperature of interest T' may be evaluated by means of the empirical equations stated in the relevant literature or may be calculated from the appropriate integrated form of the Clapeyron–Clausius equation, thus,

$$\ln p^* = -\frac{\Delta H_v}{R} \frac{1}{T} + C \tag{3.15}$$

Table 3.2 Approximate Thermodynamic Activities a_2^{c*} of Some Pure Organic Compounds[a]

Substances	h_2^c (torr dm^3 mol^{-1})	p_2^* (torr)[b]	a_2^{c*}
	Hydrocarbons		
Pentane	99.3[c]	512.5	5.16
Hexane	27.2[c]	151.3	5.56
Hexane (20°C)	22.0[d]	121.2	5.52
Heptane (20°C)	6.29[d]	35.4	5.65
Benzene	8.5[e]	95.2	11.2
	Chlorinated Hydrocarbons		
Chloroform	15.7[e]	194.8	12.4
	Normal Alkanols		
Methanol	1250[e]	125.03	0.1
Ethanol	460[f]	59.77	0.130
1-Propanol	127[f]	20.85	0.164
1-Butanol	33.3[f]	6.18	0.186
1-Pentanol	11.8[f]	2.35	0.199
1-Hexanol	3.76[f]	0.82	0.218
1-Heptanol	1.00[f]	0.232	0.232
1-Octanol	0.293[f]	0.075	0.256
	Branched-Chain Alcohols		
2-Methylpropanol (isobutanol)	46.8[g]	10.22	0.218
2-Butanol (*sec*-butanol)	62.4[g]	18.29	0.293
2-Methyl-2-propanol (*tert*-butanol)	119[g]	41.98	0.353
2,4-Dimethyl-3-pentanol	3.04[h]	3.407	1.121
2-Methyl-3-heptanol	0.881[h]	0.777	0.882
2,4-Dimethyl-3-hexanol	9.10[h]	—	—
	Normal Butanol and tert-Butanol: Temperature Dependence		
n-Butanol (in isooctane) (15°C)	20.9[g]	2.79	0.133
(25°C)	33.3[g]	6.18	0.186
(35°C)	50.2[g]	12.7	0.253
tert-Butanol (in isooctane) (15°C)	80.5[g]	21.3	0.265
(25°C)	119[g]	41.98	0.353
(35°C)	173[g]	77.9	0.450
tert-Butanol (35°C) (in hexadecane)	288[i]	77.9	0.270
	Phenols		
Phenol (in isooctane, 25°C)	4.6[j]	0.41	0.089
(cyclohexane, 25°C)	2.8[j]	0.41	0.146

Table 3.2 (Continued)

Substances	h_2^c (torr dm^3 mol^{-1})	p_2^* (torr)[b]	a_2^{c*}
Normal Alkanals			
1-Propanal	184[c]	314	1.71
1-Butanal	50.3[c]	118	2.35
1-Pentanal	13.9[c]	42.0	3.04
1-Hexanal	5.67[c]	13.6	2.40
1-Heptanal	1.48[c]	3.76	2.54
Ketones			
Acetone (in heptane, 25°C)	240[d]	74.0	1.04
2-Butanone (in heptane, 25°C)	69.3[d]	90.2	1.31
Esters			
Methyl acetate	316[k]	216.2[l]	0.684
Ethyl acetate	104.5[k]	94.63[l]	0.906
1-Propyl acetate	21.9[k]	33.65[l]	1.54
1-Butyl acetate	5.42[k]	11.47[l]	2.17
1-Pentyl acetate	1.09[k]	3.82[l]	3.50
1-Hexyl acetate	0.280[k]	1.00[l]	3.57
2-Propyl acetate (isopropyl acetate)	43.4[k]	59.4[l]	1.37
2-Methylpropyl acetate (isobutyl acetate)	10.4[k]	17.9[l]	1.72
2-Methyl-2-propyl acetate (*tert*-butyl acetate)	26.6[k]	41.4[l]	1.56
Lactones			
γ-Butyrolactone	4.03[m]	0.3127	0.0776
Ethers			
Diethyl ether	55.49[e]	534.2	9.55
Amides			
N,N-Dimethylacetamide	2.90[m]	1.3[o]	0.45
1-Methyl-2-pyrrolidinone	0.831[m]	0.39[p]	0.47
Aliphatic Amines			
1-Butylamine (in heptane, 25°C)	19.5[n]	104[b]	5.33
1-Pentylamine (in heptane, 25°C)	6.16[n]	34.5[b]	5.60
1-Hexylamine (in heptane, 25°C)	1.91[n]	11.4[b]	5.97
1-Heptylamine (in heptane, 25°C)	0.56[n]	3.95[b]	7.05
1-Octylamine (in heptane, 25°C)	0.17[n]	1.34[b]	7.88

Table 3.2   (Continued)

Substances	h_2^c (torr dm^3 mol^{-1})	p_2^* (torr)^b	a_2^{c*}
	Pyridines		
Pyridine	23.6^k	20.7^l	0.877
2-Methylpyridine	6.70^k	11.2^l	1.67
3-Methylpyridine	4.26^k	6.00^l	1.41
4-Methylpyridine	—	5.68^l	—
2,6-Dimethylpyridine	2.44^k	5.59^l	2.29

[a]The standard state is a $1\,\mathrm{mol\,dm}^{-3}$ (i.e., $1\,\mathrm{mol\,L}^{-1}$) solution, acting as if it were infinitely dilute, in isooctane at 25°C, unless otherwise stated. a_2^{c*} was calculated from the molarity-based Henry's constant h_2^c and from the saturated vapor pressure p^*.
[b]Riddick and Bunger (1970).
[c]Schreiber (1979).
[d]Rytting et al. (1972).
[e]Higuchi (1977).
[f]Anderson et al. (1978).
[g]Rytting et al. (1978).
[h]McHan et al. (1980).
[i]Tucker and Becker (1973).
[j]Anderson et al. (1979).
[k]Warycha et al. (1980).
[l]Boublik et al. (1973).
[m]Grant et al. (1984).
[n]Rytting et al. (1978).
[o]Bogoslavskii et al. (1972).
[p]Gustin and Renon (1974).

or

$$\ln \frac{p^{*'}}{p^*} = \frac{\Delta H_v}{R} \frac{T' - T}{T T'} \tag{3.16}$$

where ΔH_v is the enthalpy of vaporization, which is assumed to be constant, R is the gas constant ($8.314\,\mathrm{J\,K^{-1}\,mol^{-1}}$), and C is a constant of integration.

If only one data point (p^* at T) is available, ΔH_v for unassociated liquids may be estimated approximately at the boiling point using Trouton's rule, thus,

$$\Delta H_v / T_b = \Delta S_v \cong 21\,\mathrm{cal\,K^{-1}\,mol^{-1}} \cong 88\,\mathrm{J\,K^{-1}\,mol^{-1}} \tag{3.17}$$

In the case of nonpolar liquids, it may be estimated at 25°C using the Hildebrand rule (Hildebrand, 1915, 1918; Hildebrand and Scott, 1950) in the following form:

$$\Delta H_v / \mathrm{cal\,mol^{-1}} = -2950 + 23.7 T_b + 0.020 T_b^2 \tag{3.18}$$

where in both equations T_b is the absolute boiling point of the liquid. Equation 3.16 may then be applied to calculate the saturated vapor pressure at the relevant temperature (e.g., 25°C). This is illustrated in the following example.

Example 3.1 Estimate the saturated vapor pressure of guaiacol (2-methoxyphenol) at 25°C (298.15 K) given that its boiling point is 204–206°C (at 1 atm).

Since guaiacol is a highly polar compound, application of Hildebrand's rule is inappropriate. However, because guaiacol is an orthosubstituted phenol, it undergoes intramolecular hydrogen bonding, indicating the absence of intermolecular self-association and therefore allowing Trouton's rule to be applied.

At the boiling point, $T_b = 478$ K and $\Delta H = 88 \times 478$ J mol^{-1} = 42064 J mol^{-1}. Given that $p^* = 760$ torr at $T = 478$ K, we wish to calculate $p^{*\prime}$ at $T' = 298.15$ K. Substitution into Eq. 3.16 affords

$$\ln \frac{p^{*\prime}}{760 \text{ torr}} = \frac{42064}{8.314} \left(\frac{298.15 - 478}{478 \times 298.15} \right)$$

whence $p^{*\prime} = 1.28$ torr.

Thus, the saturated vapor pressure of guaiacol is approximately 1 torr at 25°C. This method of calculating p^* using Trouton's rule can only indicate the order of magnitude.

The prediction of the enthalpy of vaporization is particularly difficult for associated liquids such as alcohols and amines. It is, in any case, more accurate to estimate the saturated vapor pressure at a given temperature from the respective values at two or more other temperatures, using the linear regression expressed by Eq. 3.15. In this case ΔH is implicit in the data and is not assigned a specific value, but is assumed to be constant over the temperature range of the data.

Example 3.2 Estimate the saturated vapor pressure of guaiacol at 25°C (298.15 K) given that its boiling point is 204–206°C at 760 torr and 53–55°C at 4 torr Windhole et al., 1983).

Substitution of the data into Eq. 3.16 affords

$$\ln \frac{760}{4} = \frac{\Delta H_v / \text{J mol}^{-1}}{8.314} \frac{151}{478 \times 327}$$

whence $\Delta H_v = 45190$ J mol^{-1}.

Inserting this value and one pair of (p^*, T) data into Eq. 3.15 gives

$$p^{*\prime} = 0.7947 \text{ torr} \quad \text{at} \quad T' = 298.15 \text{ K}$$

Thus, the extrapolated saturated vapor pressure of guaiacol is 0.79 torr at 25°C. Improved accuracy would be achieved if more pairs of (p^*, T) data were used and if the temperatures were closer to 25°C, so that constancy of ΔH_v would become a more reliable assumption. A linear regression based on Eq. 3.15 could then be used.

The Henry's law constant h can similarly be calculated at a relevant temperature, say 25°C, if its values at two or more other temperatures are known. For this purpose Eq. 3.15 or Eq. 3.16 may be used, in which p^* is replaced by h and ΔH_v by the molar enthalpy of vaporization of the solute from the solution, $\Delta_{soln}^{vap} H$. The latter quantity, like h itself, may be predicted using the group-contribution approach, as discussed in Chapter 7. This approach is not generally applicable to p^*, ΔH_v, or a^*, thus reflecting our present inability to predict simply the properties and behavior of the pure state of polar organic compounds.

The experimental determination of Henry's constant demands great sensitivity in the instrumentation, because to obtain an accurate value of the limiting slope in Figure 3.2, the vapor pressure of the solute must be measured above solutions as dilute as $10^{-4}-10^{-2}$ mol dm^{-3} ($10^{-5}-10^{-3}$ mole fraction). Suitable techniques include quantitative gas chromatographic or mass spectrometric analysis of $0.1-1$ cm^3 of vapor withdrawn from the head space above the solution by means of an evacuated injection loop. The details and results of these techniques will be further discussed in Section 6.3.2.

Table 3.1 lists values of h_2^x while Table 3.3 presents values of a_2^*

Table 3.3 Activity of Several Pure Organic Solutes on the Henry's Law-Based Mole Fraction Scale[a]

	Solutes			
Solvents	Hexane	Heptane	Methyl Butyl Ketone	Acetone
Hexane	—	—	0.1849[e]	—
Cyclohexane	0.885[b]	—	—	—
Heptane	—	—	0.1928[e]	0.1413[e]
2, 3, 4-Trimethylpentane	0.911[c]	0.933[c]	—	—
Decane	—	—	0.1907[e]	0.1608[e]
Dodecane	1.036[b]	—	—	—
Hexadecane	1.112[b]	1.143[b]	0.2178[e]	0.1714[e]
Heptadecane	1.130[d]	1.061[d]	—	—

[a] Activity of the pure (liquid) solute. a_2^x, on the Henry's law-based mole fraction scale was calculated using Eq. 3.13, $a_2^x = p_2^*/h_2^x$, where p_2^* is the vapor pressure of the pure solute from Riddick and Bunger (1970) and h_2^x is the mole fraction-based Henry's law constant from Table 3.1.

[b,c] At 20°C, p_2^*, torr: hexane, 121.2; heptane, 35.4.

[d] At 22.5°C, p_2^*, torr: hexane, 135.6; heptane, 40.6.

[e] At 25°C, p_2^*, torr: methyl butyl ketone, 90.6; acetone, 223.0.

calculated from h_2^x and from the saturated vapor pressure of the solute (Riddick and Bunger, 1970) using Eq. 3.13. The pure saturated aliphatic hydrocarbons hexane and heptane have activities close to 1.0 (1.026 ± 0.103) with respect to the infinitely dilute standard state on the Henry's law-based mole fraction scale in each of five other saturated aliphatic hydrocarbons. Since such mixtures are expected to behave ideally, they will obey Raoult's law over the entire mole fraction range and will consequently also obey Henry's law over the whole range. The two limiting laws are indistinguishable for ideal solutions. Therefore, the activities based on each are also indistinguishable. The activities of the two pure ketones, with respect to the infinitely dilute standard state on the Henry's law-based mole fraction scale in saturated aliphatic hydrocarbons, are about 0.2. This value is considerably less than for the paraffins, since the intermolecular attraction in the pure ketones is enhanced by their carbonyl dipoles.

3.5   ACTIVITY OF THE PURE SOLUTE FROM ACTIVITY COEFFICIENTS OR FROM THE SOLUBILITY IN PARAFFINIC SOLVENTS

3.5.1   Introduction

We have seen in Chapter 2 that prediction of the activity of the undissolved (usually pure) solute is an essential process in solubility prediction (see Eq. 2.5). Consequently, further equations for deriving the activity of the pure solute are developed and discussed in this section.

3.5.2   Relationships Between Activity Coefficients

The activity coefficient of the solute 2 on the Henry's law-based mole fraction scale is given by combining Eqs. 3.6 and 3.11, thus,

$$y_2^x = p_2/h_2^x x_2 \qquad (3.19)$$

The analogous equation on the Henry's law-based molarity scale is obtained by combining Eqs. 3.7 and 3.12 in the form

$$y_2^c = p_2/h_2^c c_2 \qquad (3.20)$$

An analogous equation applies on the molal scale. These equations apply to both solid and liquid solutes.

When the standard state of component 2, the solute, is based on Raoult's law and is defined as the pure liquid (supercooled or not), $a_2^\theta = a_2^* = 1$ and the activity of the solute is given by

$$a_2^R = \gamma_2 x_2 \qquad (3.21)$$

By definition of the standard state, when $x_2 \to 1$, then $a_2^R \to 1$, provided component 2 is in the liquid state. At the same time, $a_2^R \rightarrow x_2$, and the activity coefficient $\gamma_2 \rightarrow 1$ and Raoult's law is obeyed more and more closely, as shown in Figure 3.1.

$$a_2^R = p_2/p_2^* \tag{3.22}$$

where p_2^* is the saturated vapor pressure of the pure liquid solute. Combining the last two equations gives

$$\gamma_2 = p_2/p_2^* x_2 \tag{3.23}$$

Dividing Eq. 3.23 by Eq. 3.19 yields

$$\gamma_2/y_2^x = h_2^x/p_2^* \tag{3.24}$$

where p_2^* refers to the pure liquid. This useful and simple equation relates, for any given composition of the system, the activity coefficient based on the Raoult's law–mole fraction scale to that based on the Henry's law–mole fraction scale.

For infinitely dilute solutions, $x_2 \rightarrow 0$ and $p_2/x_2 \rightarrow h_2^x$ (see Figure 3.2) and Eq. 3.19 shows that $y_2^x \rightarrow 1$. At the same time γ_2 tends toward a limiting value known as γ_2^∞. Under this condition Eq. 3.24 gives

$$\gamma_2^\infty = h_2^x/p_2^* \tag{3.25}$$

where p_2^* refers to the pure liquid state.

For compositions approaching the pure liquid solute, $x_2 \rightarrow 1$, $\gamma_2 \rightarrow 1$, and $p_2 \rightarrow p_2^*$, whereas y_2^x tends towards a limiting value known as $(y_2^x)^\infty$. Application of Eq. 3.19 to this condition yields

$$(y_2^x)^\infty = p_2^*/h_2^x \tag{3.26}$$

Similarly, as $c_2 \rightarrow 1$, $y_2^c \rightarrow (y_2^c)^\infty$, and Eq. 3.20 becomes

$$(y_2^c)^\infty = p_2^*/h_2^c \tag{3.27}$$

where p_2^* still refers to the pure liquid state. An analogous equation also applies to the molality scale.

Comparison of Eqs. 3.25 and 3.26 shows that

$$(y_2^x)^\infty = 1/\gamma_2^\infty \tag{3.28}$$

This equation applies to all solutes (solids, liquids, and gases) and indicates that the Henry–mole fraction and Raoult–mole fraction activity coefficients, at the extreme compositions opposite to the respective limiting conditions on which they are based, are reciprocally related.

Equation 3.25 is particularly important for the following reasons:

(a)   The Henry's law constant h_2^x of the solute can be calculated from the saturated vapor pressure p_2^* of the pure liquid solute and from the Raoult's law-based activity coefficient γ_2 of the solute which has been extrapolated to infinite dilution (γ_2^∞). (The values of h_2^x in Table 3.1 were calculated by Rytting et al., 1972, using this method.) The equation also enables γ_2^∞ to be calculated if the other two quantities are known.

(b)   For a sparingly soluble liquid solute, γ_2^∞ is the reciprocal of the mole fraction solubility, provided it is low and subject to the conditions discussed in Section 2.9. Equation 3.25, therefore, enables Henry's constant to be calculated from low mole fraction solubility and vice versa.

3.5.3   Activity of the Pure Solute From Henry's Law Constant or Solubility in a Paraffinic Solvent

For a saturated solution of a solute, whether solid or liquid, which does not take up the solvent,

$$\text{activity of pure solute} = \text{activity of dissolved solute}$$
$$= \text{solubility} \times \text{activity coefficient}$$

as stated in Eq. 2.4.

With respect to any Henry's law-based standard state, the activity coefficient is unity for a very dilute solution, as given by a sparingly soluble solute. Therefore, the activity of any pure solute, solid or liquid, relative to any Henry's law-based concentration scale, is equal to its solubility in a hydrocarbon solvent, provided the solubility is expressed in the same units and is also low (less than about 0.01 mole fraction or 0.1 mol dm^{-3}). Combining this fact with Eq. 3.13, 3.25, and 3.26 leads to

$$a_2^{x*} = p_2^*/h_2^x \quad \text{(useful if the solute is sufficiently volatile)} \tag{3.13}$$

$$= \text{(low) mole fraction solubility for any solute, solid or liquid,}$$
$$\text{in a paraffinic solvent} \tag{3.29}$$

$$= (y_2^x)^\infty = 1/\gamma_2^\infty \quad \text{for a liquid solute in a paraffin} \tag{3.30}$$

Combining the former fact with Eqs. 3.14 and 3.27 in an analogous manner leads to

$$a_2^{c*} = p_2^*/h_2^c \quad \text{(useful if the solute is sufficiently volatile)} \tag{3.14}$$

$$= \text{(low) molar solubility for any solute, solid or liquid,}$$
$$\text{in a paraffinic solvent} \tag{3.31}$$

$$= (y_2^c)^\infty \quad \text{for a liquid solute in a paraffin} \tag{3.32}$$

Table 3.4 Approximate Thermodynamic Activities a_2^{c*} of Some Pure Organic Compounds[a]

Substances	Solvent and Temperature (°C)	Melting Point[b,c]	Boiling Point[b,c]	a_2^{c*}
Aliphatic Hydrocarbons				
Dotriacontane $C_{32}H_{66}$	Isooctane[d] 25	69.7	>310	0.044
	Heptane[d] 28.4			(0.1)
Aromatic Hydrocarbons				
Biphenyl	Heptane[d] 26.5	70	256	0.009
	Isooctane[d] 25			0.0066
Anthracene	Hexane[d] 25	218	342	0.03
Phenanthrene	Hexane[d] 25	100	340	(0.7)
	Hexane[d] 25			(0.3)
	Cyclohexane[d] 25			(0.3)
Halogenated Hydrocarbons				
1,1,1-Trichloro-2,2-bis-(p-chlorophenyl ethane) (DDT)	Petroleum ether[d] 25	108.5–108.9		(0.2)
Phenols				
4-Phenylphenol	Isooctane[e] 25	164–165	305–8	0.00124
4-Iodophenol	Isooctane[e] 25	93–94	Decomposes	0.0150
2,4,6-Triiodophenol	Isooctane[e] 25	156–158	Decomposes	0.0050
4-Nitrophenol	Isooctane[e] 25	114	279 decomposes	0.00033

Compound	Solvent	mp (°C)	bp (°C)	Temp (°C)	Solubility
2,4,6-Trinitrophenol (picric acid)	Cyclohexane[f]	122–123		25	0.00041
	Hexane[f]			25	0.00030
	Heptane[f]			25	0.00033
	Isooctane[f]			25	0.00025
	Decane[f]			25	0.00042
	Dodecane[f]			25	0.00033
	Hexadecane[f]			25	0.00039
Phenolphthalein	Petroleum ether[f]	261–262		25	0.0008
Carboxylic Acids					
Benzoic acid	Ligroin[d]	122.4	249.2	15	0.06
	Ligroin[d]			25	0.1
	Petroleum ether[d]			26	0.08
	Hexane[d]			25	0.1
o-Chlorobenzoic acid	Heptane[d]	142	Sublimes	25	0.0091[†]
	Ligroin[d]			15	0.004
m-Chlorobenzoic acid	Heptane[d]	158	Sublimes	25	0.0094[†]
	Ligroin[d]			15	0.005
o-Hydroxybenzoic acid	Cyclohexane[f]	159	211	25	0.0040
	Hexane[f]			25	0.0034
	Heptane[f]			25	0.0033
	Isooctane[f]			25	0.0023
	Decane[f]			25	0.0032
	Dodecane[f]			25	0.0028
	Hexadecane[f]			25	0.0030
	Heptane[d]			29	0.0074
2-Hydroxy-5-methylbenzoic acid	Heptane[d]	152.5		25	0.0067
Cinnamic acid	Petroleum ether[d]	135–136	300	26	0.006
(±)-3-Phenyl-2,3-dibromopropanoic acid	Petroleum ether[d]	204	Sublimes	26	0.002
Camphoric acid	Ligroin[d]	188		25	0.0005
Retinoic acid	Petroleum ether[d]	181		25	0.0002

Table 3.4 (Continued)

Substances	Solvent and Temperature (°C)		Melting Point[b,c]	Boiling Point[b,c]	a_2^*
Diols and Polyols					
1,2-Ethanediol	Cyclohexane[g]	25	−13.2	198	0.0011†
	Heptane[g]	25			0.00092†
1,3-Propanediol	Cyclohexane[g]	25		213.5	0.0011†
	Heptane[g]	25			0.00067†
1,4-Butanediol	Cyclohexane[g]	25	19	235	0.00011†
	Heptane[g]	25			0.00079†
1,5-Pentanediol	Cyclohexane[g]	25	−18	260	0.0011†
	Heptane[g]	25			0.00077†
1,6-Hexanediol	Cyclohexane[g]	25	43	250	0.0013†
	Heptane[g]	25			0.00083†
Diethylene glycol	Heptane[h]	25	−10.5	245	0.0028†
Andromedotoxin I (grayanotoxin I, $C_{31}H_{51}O_{10}$)	Petroleum ether[d]	12	267–270		9.1×10^{-5}
Triglycerides					
Glyceryl trioctadecanoate (α-polymorph)	Hexane[a]	27.8	53.5		0.098
Carboxylic Acid Anhydrides					
Acetic anhydride	Petroleum ether[d]	27.6	−73.1	136.4	(1)
Phthalic anhydride	Cyclohexane[f]	25	130.8	284.5	0.0064
	Hexane[f]	25			0.0050
	Heptane[f]	25			0.0049
	Isooctane[f]	25			0.0042
	Decane[f]	25			0.0049
	Dodecane[f]	25			0.0048
	Hexadecane[f]	25			0.0047

Unsaturated Mixed-Functional Ethers, Ketones, Lactones

Compound	Solvent	Temperature	m.p.	Notes	Solubility
Anthraquinone	Isooctane[e]	25	286	379.8	0.00028
Griseofulvin	Hexane[i]	25	220		1.0×10^{-5}
Santonin	Petroleum ether[d]	25	173	Sublimes	0.0009
1,3-Diphenyl-2-bromo-3-methoxy-2-propene-1-one	Ligroin[d]	19	120 71–72	(Geometric isomers)	(0.12) 0.09
1,3,3-Triphenyl-3-methoxy-1-propyne	Petroleum ether[d]	16	124		0.05
	Petroleum ether[d]	16	51–52		(1)
	Petroleum ether[d]	16	42–43		(3)

Steroids

Compound	Solvent	Temperature	m.p.	Notes	Solubility
Hydrocortisone	Isooctane[j]	25	217–220		$<2 \times 10^{-6}$
Hydrocortisone acetate	Isooctane[j]	25	223 decomposes		$<2 \times 10^{-6}$
17-Methyltestosterone	Isooctane[j,k]	25	176–177[j] 165–166[k]		0.0053[j] 0.0013[k]
Norethindrone	Isooctane[k]	25	203–204		6.9×10^{-5}
Norethindrone acetate	Isooctane[k]	25	161–162		0.0025
Norethindrone enanthate	Isooctane[k]	25			0.0235

Compounds Containing Sulfur

Compound	Solvent	Temperature	m.p.	Notes	Solubility
3-Pentyl-3-propanesulfone ($C_8H_{16}O_3S$)	Petroleum ether[d]	20	129		0.02
2,2-bis-Ethylsulphonyl butane (Trional)	Petroleum ether[d]	25	75.7	Decomposes	0.03
Sulfamethoxazole	Petroleum ether[d]	25	167		0.0008

Table 3.4 (Continued)

Substances	Solvent and Temperature (°C)		Melting Point[b,c]	Boiling Point[b,c]	a_2^c*
	Amines				
Carbazole	Cyclohexane[f]	25	247	355	0.0017
	Hexane[f]	25			0.0013
	Heptane[f]	25			0.0016
	Isooctane[f]	25			0.0011
	Decane[f]	25			0.0015
	Dodecane[f]	25			0.0014
	Hexadecane[f]	25			0.0017
1,2-Diaminoethane	Cyclohexane[d]	25	8.5	116.5	0.05[†]
	Hexane[d]	25			(0.4)
Perfluorotributylamine	Methylcyclohexane[d]	19.2		179	0.028
	Hexane[d]	25.5			(0.2)
	Amides				
Acetanilide	Isooctane[f]	25	113–114		0.00075
	Heptane[f]	25			0.00093
	Isooctane[f]	25			0.00093
	Decane[f]	25			0.00093
	Dodecane[f]	25			0.00078
	Hexadecane[f]	25			0.00083
Acetophenetidin (phenacetin)	Petroleum ether[d]	20	134–135		0.0008
	Miscellaneous Nitrogen-Containing Compounds				
p-Nitrobenzyl chloride	2-Methylbutane[d]	25	71		0.05
	Hexane[d]	25			0.05

Compound	Solvent	Temp	mp (°C)	Solubility
α-Phenylglyoxal-phenylhydrazone	Cyclohexane[d]	25	114–117	(0.11)
	Heptane[d]	25		(0.14)
β-Phenylglyoxal-phenylhydrazone	Cyclohexane[d]	25	138	0.036
	Heptane[d]	25		0.091
Trimethoprim	Petroleum ether	25	200	0.0007

Alkaloids[1]

Compound	Solvent	Temp	mp (°C)	Solubility
Aconitine	Petroleum ether	25	204	0.0007
Brucine	Petroleum ether	20	178	0.003
Cinchonidine	Petroleum ether	20	210	0.002
Cinchonine	Petroleum ether	20	265	0.002
Cocaine	Petroleum ether	20	98	0.1
Codeine	Petroleum ether	25	157–158.5	0.01
Colchicine	Petroleum ether	20	155–157	0.002
Cytisine	Petroleum ether	15	154.5	0.09
dl-Hyoscine	Petroleum ether	20	82	0.04
Hydrastine	Petroleum ether	20	135	0.003
Hyosayamine	Petroleum ether	20	108.5	0.005
Morphine	Petroleum ether	20	254 decomposes	0.005
Noscapine	Petroleum ether	25	174–176	0.0008
Quinidine	Petroleum ether	20	174–175	0.001
Quinine	Petroleum ether	20	177	0.001
Quinine hydrate	Petroleum ether	20	57	0.0005
Strychnine	Petroleum ether	20	268–290	0.0004

Organic Salts

Compound	Solvent	Temp	mp (°C)	Solubility
Magnesium hydrocarpate $Mg(C_{16}H_{27}O_2)_2$	Petroleum ether[d]	30		0.00002
Magnesium chaulmoograte $Mg(C_{18}H_{31}O_2)_2$	Petroleum ether[d]	30		0.00001

Table 3.4 (Continued)

Substances	Solvent and Temperature (°C)		Melting Point[b,c]	Boiling Point[b,c]	a_2^{c*}
Inorganic Substances (Covalent)					
Iodine I_2	Isooctane[d]	24.9	113.5	184.35	0.075
White P_4	Heptane[d]	25	44.1	280	(0.2)
Rhombic S_8	Hexane[d]	20	112.8	444.6	0.015
	Heptane[d]	25			0.021
Indium iodide (InI_3)	Petroleum ether[d]	20	210		0.01
Stannic iodide (SnI_4)	Heptane[d]	25	144.5	364.5	0.079
Water	Pentane[m]	20	0	100	0.0031
	Hexane[m]	20			0.0041
	Heptane[m]	20			0.0050
	Octane[m]	20			0.0055
	2-Methylbutane[m]	20			0.0031
	Cyclohexane[m]	20			0.0043

[a] The standard state is a $1\ mol\ dm^{-3}$ (i.e., $1\ mol\ L^{-1}$) solution, acting as if it were infinitely dilute, in isooctane at 25°C, unless otherwise stated. a_2^{c*} was calculated from the molar solubility in the stated saturated aliphatic hydrocarbon solvent at the stated temperature. Values in parentheses indicate the order of magnitude only. Values in petroleum ether or ligroin should be regarded as approximate, owing to the presence of traces of aromatic hydrocarbons. The symbol † indicates that several values were interprolated or extrapolated to 25°C using the van't Hoff isochore (Eq. 1.9).
[b] Windholz et al. (1983).
[c] Weast et al. (1986).
[d] Stephen and Stephen (1963, 1964).
[e] Anderson et al. (1980).
[f] Fung and Higuchi (1971).
[g] Staveley and Milward (1957).
[h] Johnson and Francis (1968).
[i] Elworthy and Lipscomb (1968).
[j] Higuchi (1977).
[k] Higuchi et al. (1979).
[l] Müller (1903).
[m] Black et al. (1948).
† Extrapolated value using the van't Hoff isochore (Eq. 1.9).

An analogous set of equations applies to the molal scale. The last six equations summarize the methods of determining the activity of a pure solute with respect to the preferred reference state, that is, at infinite dilution in a paraffinic solvent such as isooctane.

Table 3.4 lists values of these activities on the molar scale (i.e., a_2^{c*}) which, according to Eq. 3.31, are given by the molar solubility of the substance in a saturated aliphatic hydrocarbon solvent. Most of these substances have low volatilities, low escaping tendencies, low solubilities (i.e., $< 0.1 \, mol \, dm^{-3}$), and low activities. The solubility must be small in order for Eq. 3.31 to be valid. Unless otherwise stated, the preferred solvent in Tables 3.2 and 3.4 is isooctane at a temperature of 25°C. When other saturated aliphatic hydrocarbon (paraffin) solvents and temperatures were used, as in Table 3.4, the solvent and the temperature are stated in separate columns.

The influence of temperature on solubility, and hence on a_2^*, can be readily predicted and allowed for by means of the van't Hoff isochore (Eq. 1.9) or Eqs. 2.33 and 2.34, provided that the molar enthalpy of solution in the solvent is known at the temperature of interest. Solubilities were extrapolated to 25°C by means of Eq. 2.52 in the case of the α, ω-diols (Staveley and Milward 1957), that is, polymethylene glycols, which are indicated in Table 3.4.

The solvent is normally expected to exert a profound influence on the solubility, on Henry's constant, and on a_2^*. If, however, the solvents are restricted to saturated aliphatic hydrocarbons, such as cyclohexane, hexane, heptane, isooctane, decane, dodecane, and hexadecane, little difference is observed in Tables 3.3 and 3.4, especially if the molarity scale is used (Fung and Higuchi, 1971).

The lack of variability is due to the relatively constant dispersion energy (Eq. 2.128) of solute–solvent interaction per unit volume of solution, as discussed in Chapter 5. Although ligroin and petroleum ether are usually classed as mixtures of saturated aliphatic hydrocarbons, these organic compounds may contain traces of aromatic hydrocarbons whose presence increases the solubility of polar solutes because of the introduction of Debye interactions. Consequently, ligroin or petroleum ether may lead to abnormally high values of a_2^*, particularly of polar solutes. This possibility should be remembered when referring to Table 3.4.

3.6 SOLUBILITY PREDICTION FROM THE ACTIVITY OF THE PURE SOLUTE AND THE PARTITION COEFFICIENT

Since the partition coefficient relates the concentration or solubility in one solvent to that in another, it may be used to predict the solubility in a solvent of interest, such as water, from the solubility in a paraffinic solvent such as isooctane. As we have seen, the latter solubility is equal to the activity a^* of the pure solute with respect to our recommended reference state based on Henry's law.

When a solute species distributes itself between two immiscible solvents at constant temperature, its activity at equilibrium in one solvent is equal to that in the other. This gives rise to the partition law (i.e., distribution law), which states that the concentration ratio, known as the partition coefficient $P_{\alpha/\beta}$, is a constant provided that the solutions are dilute, thus,

$$P_{\alpha/\beta} \cong \frac{\text{concentration in solvent } \alpha \text{ (if low)}}{\text{concentration in solvent } \beta \text{ (if low)}} \tag{3.33}$$

Since the activities in the two solvents are equal, Eq. 2.3 affords

$$P_{\alpha/\beta} = \frac{\text{activity coefficient in solvent } \beta \text{ at infinite dilution}}{\text{activity coefficient in solvent } \alpha \text{ at infinite dilution}} \tag{3.34}$$

Comparison with Eq. 3.25 shows that

$$P_{\alpha/\beta} = \frac{\text{Henry's law constant in solvent } \beta}{\text{Henry's law constant in solvent } \alpha} \tag{3.35}$$

This equation is derived from standard free energies of transfer (Eqs. 7.15 and 7.16). Since the activities of the solute in the two solvents are equal, Eqs. 3.33 and 2.4 yield

$$P_{\alpha/\beta} \cong \frac{\text{solubility in solvent } \alpha \text{ (if low)}}{\text{solubility in solvent } \beta \text{ (if low)}} \tag{3.36}$$

The units in the numerator and denominator of each must, of course, be the same. The influence of self-association of the solute on its apparent distribution coefficient is discussed in Chapter 6.

If β is taken to be the solvent in which the solubility is to be predicted and α is a paraffinic solvent such as isooctane, in which, as we have seen, the solubility is equal to a^*, then Eq. 3.36 leads to

solubility in the solvent of interest $= a^*/P$ (paraffin/solvent)   (3.37)

Furthermore, in our recommended Henry's law reference state in a paraffinic solvent, the activity coefficient is unity, so Eq. 3.34 becomes

activity coefficient in solvent of interest $= P$ (paraffin/solvent)   (3.38)

Substitution into the fundamental Eq. 2.5 for solubility prediction leads once again to Eq. 3.37. Thus, Eq. 3.37, which has been derived in two slightly different ways, may be regarded as an alternative form of Eq. 2.5 that is particularly suited for general solubility predictions using the activity of the pure solute based on our recommended reference state. It should be stressed that the concentration scales which are used to express P and the Henry's law standard defining a^* must be the same (i.e., either mole fraction, molarity, or molality). Moreover, the calculated solubility in the solvent in question is also expressed in terms of this scale. The principles developed in this section will now be illustrated with examples.

Example 3.3 The partition coefficient (isooctane/water) of 17-methyltestosterone acetate ($C_{22}H_{33}O_2 = 344.5$) is 3300 at 25°C on a molar concentration basis (Higuchi et al., 1979). Using the activity value stated in Table 3.4, calculate the aqueous solubility of the steroid at 25°C.

The data, P^c (isooctane/water) = 3300 and $a^{c*} = 0.0053$, may be substituted directly into Eq. 3.37, since they are expressed in terms of the molarity scale, whence

$$\text{solubility in water} = (0.0053/3300)\,\text{mol dm}^{-3}\ (\text{i.e., mol/L})$$

$$= 1.606\,\mu\text{mol dm}^{-3} = 0.553\,\text{mg dm}^{-3}\ (\text{i.e., mg/L})$$

Since the aqueous solubility is very low (< 1 ppm), the value determined in this way is probably more accurate than that measured directly owing to (1) slow equilibration, (2) the influence of impurities, and (3) heterogeneity in the energy content of the solid (Higuchi et al., 1979), as discussed in Secton 8.9.2 and 8.9.3.

Example 3.4 Isobutyl acetate (2-methylpropyl acetate, $C_6H_{12}O_2 = 116.16$) is reported (Riddick and Bunger, 1970) to be soluble in 180 parts of water at room temperature ($\sim 25°C$). Using the data in Table 3.2, calculate the partition coefficient between isooctane and water, the Henry's law constant in water, and the Henry's law-based activity coefficient $(y^c)^\infty$ in water.

Solubility of isobutyl acetate = 1 g in $180\,\text{g} = (1 \times 1000)/(180 \times 116.16)\,\text{mol}$ $\text{kg}^{-1} = 0.0478\,\text{mol kg}^{-1} \cong 0.0478\,\text{mol dm}^{-3}$ (i.e., mol L^{-1}) since the solution is dilute in water at ambient temperature. This solubility is sufficiently low ($< 0.1\,\text{mol dm}^{-3}$) for Eq. 3.37 to be applied. Therefore, P^c (isooctane/water) = a^{c*}/molar solubility in water.

From Table 3.2 $a^{c*} = 1.72$. Therefore, P^c (isooctane/water) = 1.72/0.0478 = 35.96. Thus, the partition coefficient on the molar scale at 25°C is approximately 36.

The Henry's law constant in water is given by Eq. 3.35 in the form: h^c (in water) = h^c (in isooctane) P^c (isooctane/water). From Table 3.2 $h^c = 10.4$ torr $\text{dm}^3\,\text{mol}^{-1}$. Therefore, h^c (in water) = $10.4 \times 35.96 = 374$ torr $\text{dm}^{-3}\,\text{mol}^{-1}$.

The Henry's law-based activity coefficient in water is equal to the partition coefficient between the reference solvent and water (Eq. 3.38). Using the molarity scale, $(y^c)^\infty_{\text{water}} = 36$.

3.7 ACTIVITY AND SOLUBILITY OF BIFUNCTIONAL COMPOUNDS: GLYCOLS

The solubilities of the first five α, ω-diols (i.e., polymethylene glycols $HO \cdot (CH_2)_n \cdot OH$, in cyclohexane, heptane, and benzene) were studied at various temperatures by Staveley and Milward (1957) and serve to illustrate various points made so far in this book. The concentrations are sufficiently low to ensure that the activity coefficients are unity and to reduce the tendency of the solute to undergo self-association by intermolecular hydrogen bonding, discussed in Chapter 6. Furthermore, the solubilities of the solvents in the glycols are sufficiently low to ensure that the activity of glycol in the glycol-rich phase is virtually the same as that of the pure glycol, so that Eqs. 3.13 and 3.31–3.36 may be applied. The mole fraction solubilities in Table 3.5 can therefore be regarded as the activities of the pure glycols with respect to the Henry's law–mole fraction based standard state and are seen to be very similar in heptane and cyclohexane as solvents.

The thermodynamic quantities in Tables 3.6 and 3.7 were calculated from the temperature dependence of the solubility (Eq. 1.9) and were applied to the

Table 3.5 Interpolated Solubilities of Glycols in Benzene, Heptane, and Cyclohexane at 39.4°C, Expressed as the Mole Fraction of Glycol $\times 10^4$

Glycol	Benzene	Heptane	Cyclohexane
Ethylene glycol	36.8	2.86	2.66
Trimethylene glycol	47.9	2.86	2.91
Tetramethylene glycol	50.9	2.83	2.70
Pentamethylene glycol	47.0	2.74	2.75
Hexamethylene glycol	56.9	3.37	3.25

Reproduced with permission of the copyright owner, the Royal Society of Chemistry, from Staveley, L. A. K. and Milward, G. L. (1957). *J. Chem. Soc.*, **1957**, 4369–4375.

Table 3.6 Enthalpies of Solution (ΔH, kcal mol^{-1}) and Standard Entropies of Solution[a]

Glycol	Benzene		Heptane		Cyclohexane	
	ΔH	ΔS^θ	ΔH	ΔS^θ	ΔH	ΔS^θ
Ethylene glycol	7.07	16.70	9.60	19.29	10.14	21.18
Trimethylene glycol	7.08	17.26	10.62	22.62	10.32	21.93
Tetramethylene glycol	7.96	20.18	10.65	22.29	10.69	22.94
Pentamethylene glycol	8.30	21.12	11.15	24.14	10.69	23.01
Hexamethylene glycol	9.00	23.72	11.15	24.57	11.02	24.38

[a] ΔS^θ, cal K^{-1} mol^{-1}. 1 cal = 4.184 J.

Reproduced with permission of the copyright owner, the Royal Society of Chemistry, from Staveley, L. A. K. and Milward, G. L. (1957). *J. Chem. Soc.*, **1957**, 4369–4375.

Table 3.7 Comparison of the Molar Enthalpies of Solution (ΔH, kcal mol^{-1}) of Primary Alcohols and Glycols[a]

Compound	Benzene	Cyclohexane	Glycol in Heptane
$2\Delta H$ for CH_3OH	7.0	12.4	11.4
ΔH for $(CH_2OH)_2$	7.1	10.1	9.6
$2\Delta H$ for C_2H_5OH	7.4	11.6	11.4
ΔH for $(C_2H_4OH)_2$	8.0	10.7	10.6
$2\Delta H$ for C_3H_7OH	7.7	—	11.3
ΔH for $(C_3H_6OH)_2$	9.0	11.0	11.2

[a]The values for the alcohols refer to infinitely dilute solution. 1 cal = 4.184 J.

Reproduced with permission of the copyright owner, the Royal Society of Chemistry, from Staveley, L. A. K. and Milward, G. L. (1957). *J. Chem. Soc.*, **1957**, 4369–4375.

transfer of 1 mole of glycol to infinite dilution in the solvent, the standard state being a hypothetical solution of unit molality which obeys Henry's law. The values of ΔH for the glycols with even numbers of carbon atoms are approximately double those for the primary alcohols with half the number of carbon atoms (Table 3.7). From this Staveley and Milward suggest that the two hydroxyl groups in the bifunctional glycols are acting as independent entities. This conclusion must be tempered by the fact that the bifunctional solute molecules may undergo intramolecular and/or intermolecular hydrogen bonding, resulting in significant cyclization and/or self-association in the

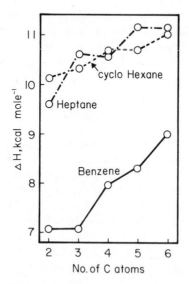

Figure 3.3 Dependence on enthalpy of solution ΔH of glycols ($\alpha\omega$-diols) on the number of carbon atoms. 1 cal = 4.184 J. Reproduced with permission of the copyright owner, the Royal Society of Chemistry from Staveley, L. A. K. and Milward, G. L. (1957). *J. Chem. Soc.*, **1957**, 4367–4375.

hydrocarbon phase. A more thorough study is required to elucidate the solution chemistry of these systems.

In benzene the higher solubilities (Table 3.5) and lower enthalpies (Tables 3.6 and 3.7) than in heptane or cyclohexane suggest a stronger interaction of the polar (hydroxyl) groups with the more polarizable solvent (benzene). Furthermore, ΔS^θ is smaller in benzene, indicating a greater ordering of the solvent molecules around the solute molecules. Essentially the same behavior has been found for the solubility of water in aromatic and olefinic hydrocarbons as compared with saturated hydrocarbons (Bent, 1976), for which the same explanation can be offered. The enhanced polarizability of the π-electron systems in olefinic and aromatic solvents gives rise to stronger Debye and dispersion forces and possibly to significant charge-transfer complexation with the polar and acidic hydroxyl group.

The enthalpy of solution increases with increasing chain length of the glycols and shows an alternation (Fig. 3.3). This behavior differs from that of the alcohols and may result either from cyclization of the solute or from the constraining influence of the hydrocarbon chain on the interactions of the hydroxyl groups of the glycols with the solvent (Staveley and Milward, 1957).

3.8   RELEVANCE OF THE GAS PHASE AND GIBBS FREE ENERGIES OF TRANSFER

In the reference state proposed and discussed earlier in this chapter, the solute interacts with the solvent. The preferred reference state for the solute could reasonably be considered to be one in which all intermolecular interactions are absent, as occurs in the gas phase or vapor state. Accordingly, Butler (1937, 1962) proposed the vapor state (at a defined fugacity or partial pressure of 1 atm = 760 torr = 101,325 Pa and at a defined temperature of usually 25°C) as a reference state for solutes. Thus, the molar free energy G_g of the solute vapor at a partial pressure p is given by

$$G_g = G_g^\theta + RT \ln(p/\text{atm}) \tag{3.39}$$

where G_g^θ is the molar free energy of the vapor in its standard state. The notation (p/atm) is employed in this section to emphasize the required units for p. If p^* is the saturated vapor pressure expressed in atmosphere and G^* is the molar Gibbs free energy of the pure solute, $(G_g^\theta - G^*)$ represents the standard molar Gibbs free energy of transfer of the solute from the pure condensed state to the vapor standard state. This is shown in Scheme 3.1 and in the following equation

$$\Delta_{\text{pure}}^{\text{vap}} G^\theta = G_g^\theta - G^* = -RT \ln(p^*/\text{atm}) \tag{3.40}$$

With the aid of vapor-pressure tables (Jordan, 1954; Riddick and Bunger, 1970; Boublik et al., 1984), p^* values can be found and $\Delta_{\text{pure}}^{\text{vap}} G^\theta$ can readily be calculated. Table 3.2 presents some values of p^* expressed in torr ($= \text{mm Hg}$).

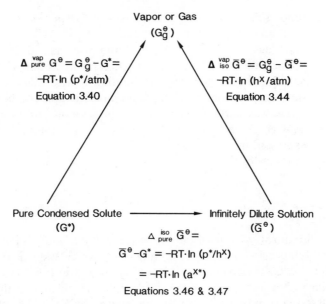

Scheme 3.1 Partial molar free-energy differences for a solute substance (solid or liquid) between the gas, condensed, and solution phases.

The partial molar Gibbs free energy $\bar{G}$ of the solute at activity a in solution is given by

$$\bar{G} = \bar{G}^{\theta} + RT \ln a \qquad (3.41)$$

where $\bar{G}^{\theta}$ is the standard partial molar Gibbs free energy of the solute in the solution.

Butler defined the standard state of unit activity of the solute in solution such that in infinitely dilute solution $a = x$. This standard state corresponds to the Henry's law-based mole fraction scale, for which a is defined by Eq. 3.9. Equation 3.41 then leads to

$$\bar{G} = \bar{G}^{\theta} + RT \ln(y^x x) \qquad (3.42)$$

where y^x is the activity coefficient based on the Henry's law–mole fraction scale. If the solute vapor at a pressure p is in equilibrium with the solute in solution, $G_g = \bar{G}$, and Eqs. 3.39 and 3.42 may be combined to give $(G_g^{\theta} - \bar{G}^{\theta})$, which is the standard molar Gibbs free energy of transfer of the solute from the infinitely dilute standard state in solution (e.g., in isooctane) to the defined vapor state. This relation is shown in Scheme 3.1 and in the following equations:

$$\Delta_{iso}^{vap} \bar{G}^{\theta} = G_g^{\theta} - \bar{G}^{\theta} = -RT \ln(p/y^x x) \qquad (3.43)$$

$$= -RT \ln(h^x/atm) \qquad (3.44)$$

Again, the notation (h^x/atm) is used to emphasize the required units for h^x.

Since $p/y^x x = h^x$ according to Eq. 3.19, the quantity $(\bar{G}^\theta - G_g^\theta)$, which is equal to $RT \ln h^x$, is in fact $-\Delta_{iso}^{vap} \bar{G}^\theta$ ($= \Delta_{vap}^{iso} \bar{G}^\theta$), and was named by Butler as the (standard molar Gibbs) free energy of solvation (or hydration, if the solvent is water). In fact, Butler assumed that the solution was so dilute that $y^x = 1$. His reference state requires that h^x be based on mole fraction and be expressed in atmosphere. (He was particularly interested in alcohols and other organic compounds dissolved in water.) Table 3.2 presents some values of h^c based on molarity and expressed in torr.

Equation 3.5 indicates that

$$\frac{h^x}{atm} = \frac{1}{760} \frac{1000}{V_1} \frac{h^c}{torr\, dm^3\, mol^{-1}} \tag{3.45}$$

where V_1 is the molar volume of the solvent ($166.08\ cm^3\, mol^{-1}$) for isooctane (Riddick and Bunger, 1970) or ($18.07\ cm^3\, mol^{-1}$) for water (Riddick and Bunger, 1970), both at 25°C. The required units of h^c are stated in Eq. 3.45.

The molar Gibbs free energy of the pure solute, G^*, relative to the Henry's law–mole fraction standard rate in solution is given by Eq. 3.41 in the form

$$G^* = \bar{G}^\theta + RT \ln a^{x*} \tag{3.46}$$

$$= \bar{G}^\theta + RT \ln(p^*/h^x) \tag{3.47}$$

Since $a^{x*} = p^*/h^x$ according to Eq. 3.13, this point of view is taken in presenting the activities in Table 3.2. The standard molar Gibbs free energy of transfer of the solute from the pure solid or liquid to the Henry's law–mole fraction standard state in (isooctane) solutions is $(\bar{G}^\theta - G^*)$. This is shown in Scheme 3.1 and in the following equation:

$$\Delta_{pure}^{iso} \bar{G}^\theta = \bar{G}^\theta - G^* = -RT \ln(p^*/h^x) = -RT \ln a^{x*} \tag{3.48}$$

Elimination of h^x and h^c from Eqs. 3.5, 3.13, and 3.14 leads to

$$a^{x*} = \frac{V_1}{1000} a^{c*} \tag{3.49}$$

Equations 3.45 and 3.49 provide simple means of interconverting thermodynamic values based on the Henry's law–mole fraction scale and those based on the Henry's law–molarity scale. Tables 3.2 and 3.4 present some values of x^{c*} based on molarity.

Equation 3.48 can also be derived by subtracting Eq. 3.44 from Eq. 3.40. In fact, any one of the Eqs. 3.40, 3.44, and 3.48 can be derived from the other two, since each of these equations expresses the difference in the partial molar Gibbs free energy (chemical potential) of the solute substance between two of the three phases in Scheme 3.1.

Any one of these three phases can be used as the defined reference state for an organic compound. As discussed at the beginning of this chapter, many workers have previously chosen the pure condensed solute (Hildebrand et al., 1970) (liquid or solid) as the reference state, whereas an infinitely dilute solution in isooctane is recommended as the reference state in this book. Butler (1937, 1962), on the other hand, selected the pure vapor as the reference state, which has the advantage that there are no solute–solvent interactions to consider, thus simplifying the theory.

However, when dealing with relatively complicated organic molecules such as drugs, the gas phase has little practical significance, since most organic compounds are solids (or liquids) whose behavior in solution is of particular interest. Therefore, the more practical reference state appears to be the compound in infinitely dilute solution, and, as we have seen, we can use this as the appropriate standard state for calculating the (relative) activity of the pure solid (or liquid). Furthermore, the conformations of chain-like molecules in the vapor state may differ markedly from those in solution, thus further complicating the interpretations of data for drug molecules.

Molecules containing hydrocarbon chains of about 20 or more carbon atoms coil up in the vapor under the influence of dispersion forces to give a "droplet," whereas in hydrocarbon solution solute–solvent dispersion forces ensure relative freedom in the flexibility of the chains. The solubility of a solute in various solvents is in any case more closely related to the solubility in isooctane or other paraffins (which is governed by the defined activity of the solute) than to the behavior in the vapor state. It must be kept in mind, however, that no perfect reference state exists and we should select that which is most useful for a given purpose.

3.9 INFLUENCE OF MOLECULAR STRUCTURE ON THE ACTIVITY OF PURE ORGANIC COMPOUNDS

Tables 3.2 and 3.4 list the activity a^{c*} of a number of pure organic compounds, both solids and liquids, the standard state being a hypothetical 1 mol dm^{-3} (i.e., 1 mol L^{-1} solution), acting as if it were infinitely dilute, in isooctane at 25°C unless otherwise stated. For volatile solutes, particularly liquids listed in Table 3.2, a^{c*} was calculated as p_2^*/h_2^c using Eq. 3.14. For involatile solutes, particularly solids listed in Table 3.4, a_2^{c*} was given by the molar solubility (provided it was low) in isooctane at 25°C. For volatile solutes, h_2^c and p_2^* show similar trends indicative of partial compensation. This effect is illustrated by the free-energy diagram in Scheme 3.1, in which any increased escaping tendency is paralleled by a more negative partial molar free-energy change.

The thermodynamic activity a_2^{c*} represents the escaping tendency (fugacity) of solute molecules from the pure state to the solution; this is directly related to a high p_2^*, which represents the fugacity of solute molecules from the pure state to the vapor phase. However, the fugacity from the pure state to the solvent is negatively reflected by a high Henry's constant, which represents the fugacity of solute molecules from the solution to the vapor phase.

Since the saturated aliphatic hydrocarbons are expected to form nearly ideal solutions with isooctane and with each other, $a^{x*} \cong 1$, as shown in Table 3.3. Conversion of the concentration scales according to Eq. 3.3 gives $a_2^{c*} \cong 1000/V_1$. Since, for isooctane, $V_1 = 166.08 \, cm^3 \, mol^{-1}$ at 25°C (Riddick and Bunger, 1970), $a_2^{c*} \cong 6.02$, when the solute forms a nearly ideal solution in this solvent. The values of a_2^{c*} for pentane, hexane, and heptane in Table 3.2 are indeed close to this figure.

It is of great practical and theoretical importance to be able to relate a^{c*} to molecular structure and to molecular properties. Although Henry's constants can be related to molecular structure, the saturated vapor pressure of a pure substance is difficult to predict since it depends on the precise structure of the liquid or crystalline state of the substance. However, the term $\bar{G}^\theta - G^*$ in Scheme 3.1 and the activity of the pure solute would be expected to reflect the intensities of molecular interactions (such as intermolecular hydrogen bonding, Debye, Keesom and London forces, etc.) in the pure and solution states and the entropy difference between those states. Quantitative relationships of this type should be forthcoming in the near future.

The following qualitative features can be discerned from Table 3.2 and 3.4. (a) Increased strength or contribution of intermolecular hydrogen bonding, as in the alcohol series, reduces the activity of the pure state. (b) Increased molecular asymmetry, irregularity, and chain branching (e.g., among the butanols) increases the thermodynamic activity of the pure state. (c) Within a homologous series of polar compounds (e.g., among the aliphatic aldehydes, ketones, and esters), increasing hydrocarbon character on ascending the homologous series increases a^{c*}, presumably as a result of decreasing strengths of the intermolecular Keesom and Debye forces. (d) Pure liquids and solids whose constituent molecules or ions have strong mutual attractions exhibit low values of a^{c*}. This is especially pronounced for ionic solids, including organic salts, which are held together by long-range coulombic forces, and for molecular solids or liquids containing intermolecular hydrogen bonds, for example, the diols and polyols in Table 3.4. The early literature has related the strengths of these attractive forces to melting and boiling point. Consequently, Table 3.4 shows a rough inverse relationship between these temperatures of disruption and thermodynamic activity. Examples include anthracene and phenanthrene, 3-ethers of 1, 3, 3-triphenyl-1-hydroxy-1-propyne, steroids, and alkaloids.

The importance of molecular conformation may be seen by comparing the properties of the cyclic molecule γ-butyrolactone with those of its acyclic homomorph ethyl acetate (Tables 3.2 and 3.8) (Grant et al., 1984). Figure 3.4 shows that γ-butyrolactone has only one conformation, which is necessarily cyclic and is essentially planar. Consequently, its dipole moment is fixed at a high value of 4.12 D dictated by the ·CO·O· group. On the other hand, ethyl acetate has at least two conformations, one of high dipole moment and one of low dipole moment, which results in a lower mean value of 1.88 D (Marsden and Sutton,

Table 3.8 Physical Properties of Some Pure Liquids at 25°C

Liquid	$T_b^{a,g}$	$P*^{b,g}$	$V_m^{c,g}$	$\mu^{d,h}$	$n_D^{25e,g}$	I^f
γ-Butyrolactone	479	0.3127	76.50	4.12	1.4348	10
Ethyl acetate	350	94.53	98.49	1.88	1.3698	10.11
N-Methylpyrrolidinone	475	0.39	96.44	4.09	1.4680	10
N,N-Dimethylacetamide	439	1.3	93.02	3.72	1.4356	8.81
2-Methyl-3-pentanol	400	6.0	124.51	1.70	1.4148	10
Cinnamaldehyde	526	0.0292	125.9	3.63	1.6195	10
Toluene	384	28.5	106.85	0.31	1.4941	8.5
Isooctane	372	49	166.08	0	1.3890	9.86

[a] Boiling point (K) at 760 torr.
[b] Vapor pressure in torr at 25°C.
[c] Molar volume at 25°C in $cm^3 \, mol^{-1}$.
[d] Determined in benzene, units are Debye.
[e] Refractive index, sodium D line at 25°C.
[f] Ionization energy in eV from Weast and Astel, 1980; unavailable values of I are assumed to approximate to 10 eV.
[g] Riddick and Bunder (1970), Weast and Astle (1980), Windholz et al. (1983), Gustin and Renon (1974), Hovorka et al. (1940), Davis et al. (1974).
[h] Riddick and Bunger (1970), Bentley et al. (1949) for 2-methyl-3-pentanol; μ is assumed to approximate to 1.70 D, as for most secondary alcohols (McClellan, 1963).

Reproduced with Permission of the copyright owner, Plenum Publishing Corp., from Grant, D. J. W., Higuchi, T., Hwang, Y. T., and Rytting, J. H. (1984). *J. Soln. Chem.*, **13**, 297–311. Adapted from Grant et al. (1984a) and Rytting et al. (1986).

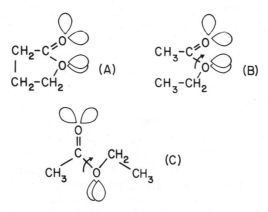

Figure 3.4 The conformations and dipoles of: (A) γ-butyrolactone, $\mu = 4.12$ D (Riddick and Bunger, 1970); (B) ethyl acetate, cis conformation $\mu(calc) \cong 3.5–4$ D (Gustin and Renon, 1974); (C) ethyl acetate, *trans* conformation $\mu(calc) \cong 1.5$ D (Marsden and Sutton, 1936). The experimental μ for ethyl acetate = 1.88 D (Riddick and Bunger, 1970). The orientations of the lone pairs contribute greatly to the overall dipole moment (Grant et al., 1984). Reproduced with permission of the copyright owner, Plenum Publishing Corp., from Grant, D. J. W., Higuchi, T., Hwang, Y. T., and Rytting, J. H. (1984). *J. Soln. Chem.*, **13**, 297–311.

1936). If differences of intermolecular separation are ignored, Eqs. 2.123 and 2.126 indicate that the Keesom and Debye interaction energies of γ-butyrolactone are, respectively, about $(4.12/1.88)^4 = 23.1$ times and $(4.12/1.88)^2 = 4.8$ times as strong as for ethyl acetate.

At 25°C the refractive index of γ-butyrolactone is 1.4348, while that of ethyl acetate is 1.3698 (Table 3.8). Consequently, Eq. 2.134 indicates that the interaction parameter α/v, which measures the approximate strength of London dispersion forces, is only 1.15 times greater for γ-butyrolactone than for ethyl acetate. It is apparent that the Keesom and Debye interactions exert the strongest differential influences in the pure liquid state, whereas the Debye interaction exerts a pronounced differential effect in a hydrocarbon solvent such as isooctane or toluene. Consequently, the escaping tendencies into less condensed phases (as measured by the activity of the pure liquid, the Henry's law constant, and the saturated vapor pressure) are much smaller for γ-butyrolactone than for ethyl acetate, as shown in Table 3.2. The greatest difference is in p_2^*, which represents the escaping tendency from the most-condensed to the least-condensed phase. More detailed discussion is provided by Grant et al. (1984).

As a result of the greater intermolecular attractions with γ-butyrolactone than with ethyl acetate, the molecules of γ-butyrolactone are pulled together more tightly in the liquid state than are the molecules of ethyl acetate, so the molar volume (and molar free volume) of the former are smaller than those of the latter (Table 3.8). Owing to the shorter intermolecular distance r for γ-butyrolactone than for ethyl acetate, the actual van der Waals energies of attractions (Eqs. 2.123, 2.126, and 2.127) are greater for the former by factors larger than those predicted above (Table 3.9). The much greater intermolecular energies of attraction in liquid γ-butyrolactone cause the boiling point, density, cohesive energy density, and solubility parameter (204°C; $1.125 \, \text{g cm}^{-3}$ at 25°C; $161.0 \, \text{cal cm}^{-3}$; $12.7 \, \text{cal}^{1/2} \, \text{cm}^{-3/2}$) to be appreciably greater than those of ethyl acetate (77.1°C; $0.8945 \, \text{g cm}^{-3}$ at 25°C; $81.1 \, \text{cal cm}^{-3}$; $9.0 \, \text{cal}^{1/2} \, \text{cm}^{-3/2}$).

The relative contributions of the various conformations of a flexible molecule such as ethyl acetate will depend on the solvent and the temperature and so will the mean dipole moment. This will cause the following quantities to have a considerable dependence on the nature of the solvent; Keesom and Debye interaction energies, Henry's law constant, deviations from Henry's law, cohesive energy density, and solubility parameter. These possibilities should always be kept in mind when considering polar flexible molecules such as esters. In the case of relatively inflexible molecules such as γ-butyrolactone or N-methyl-2-pyrrolidinone, the dependence of the dipole moment on the solvent will be relatively small.

N, N-Dimethylacetamide has physical properties (Tables 3.8 and 3.9) and intermolecular interaction energies (Table 3.9) which are very similar to those of N-methyl-2-pyrrolidinone. This is because resonance ensures that the amide group can exist in only one conformation, which is planar, so the dipole moments of amides are relatively insensitive to cyclization or to the attachment of nonpolar groups. For example, in benzene at 25 or 30°C, nine common amides have a mean

Table 3.9 Approximate Pair-Potential Energies[a] of Intermolecular Interactions in the Pure Liquids and in Dilute Solutions in Isooctane and Toluene[b]

Compound		Calculated Interaction Energies		
	$-u_{1,2}$ (Keesom) (Eq. 2.123)	$-u_{1,2}$ (Debye) (Eq. 2.126)	$-u_{1,2}$ (London) (Eqs. 2.127 or 2.128)	$-u_{1,2}$ (Total)
Pure Liquid		*Interaction Energies in Pure Liquids*		
γ-Butyrolactone	793	4.56	128	966
Ethyl acetate	20.7	6.40	97	124
N-Methylpyrrolidinone	485	38.0	145	668
N,N-Dimethylacetamide	357	31.0	113	500
2-Methyl-3-pentanol	8.68	4.58	118	131
Cinnamaldehyde	176	29.0	231	437
Toluene	0.013	0.206	135	136
Isooctane	0	0	104	104
Solute		*Interaction Energies in Isooctane*		
γ-Butyrolactone	0	20.7	115	136
Ethyl acetate	0	3.35	100	104
N-Methylpyrrolidinone	0	16.2	123	139
N,N-Dimethylacetamide	0	13.9	108	122
2-Methyl-3-pentanol	0	2.16	110	113
Cinnamaldehyde	0	9.76	155	165
Toluene	0	0.084	118	118
Isooctane	0	0	104	104
Solute		*Interaction Energies in Toluene*		
γ-Butyrolactone	3.21	25.6	131	160
Ethyl acetate	0.520	4.20	114	119
N-Methylpyrrolidinone	2.51	20.0	139	161
N,N-Dimethylacetamide	2.16	17.2	124	143
2-Methyl-3-pentanol	0.34	2.75	126	129
Cinnamaldehyde	1.52	12.1	176	190
Toluene	0.013	0.206	135	136
Isooctane	0	0.084	118	118

[a]Units are 10^{-23} J. The temperature is 25°C and the fundamental physical qualities are taken from Table 3.8.
[b]From Grant et al. (1984) and Rytting et al. (1986).

Reproduced with permission of the copyright owner, Plenum Publishing Corp., from Grant, D. J. W., Higuchi, T., Hwang, Y. T., and Rytting, J. H. (1984). *J. Soln. Chem.*, **13**, 297–311.

dipole moment of 3.69 D, the standard deviation being 0.24 (Riddick and Bunger, 1970).

Isomerism can profoundly affect the intermolecular and/or the intramolecular interactions and hence the activity of the pure state and the solubility behavior. Examples which are of particular relevance to the biological and pharmaceutical

Table 3.10   Influence of Isomerism on the Melting Point, Boiling Point,[a,b] Density at 25°C,[a,b] Molar Solubility,[a,c] and Partition Coefficients (α) of o-and p-Hydroxybenzoic Acids and their Methyl Esters

	Salicylic Acid	p-Hydrox-benzoic Acid	Methyl Salicylate	Methyl p-Hydroxy-benzoate
Melting point (°C)	159	214	−7	131
Boiling point (°C)	211	—	223	270–280
Density (g cm^{-3})	1.443	1.46	1.174	—
Solubility (mol dm^{-3} or mol L^{-1} at 25°C unless otherwise stated)				
Water	0.016	0.045	0.0079	0.016 (20°C)
Ethanol	4	—	∞	2
Acetone	4	1.6 (23°C)	—	2
Diethyl ether	3	0.68 (17°C)	Soluble	0.6
Benzene	0.00036	0.00022	—	—
Partition Coefficient (organic solvent/water at 25°C)				
1-Octanol	170	38	—	90
Diethyl ether	280	16	—	—
Benzene	3	0.085	—	—
Carbon tetrachloride	0.71	0.042	—	—
Cyclohexane	0.22	0.017	—	0.058

[a] Windholz et al. (1980).
[b] Weast and Astle (1980).
[c] Seidell (1958).
[d] Hansch and Leo (1979).

sciences are presented in Tables 3.10 and 3.11. p-Hydroxybenzoic acid and methyl p-hydroxybenzoate in Table 3.10 undergo intermolecular hydrogen bonding which strengthens their crystal lattices, thereby increasing the melting points, reducing their escaping tendencies and the activity of the solids, and reducing the solubilities in most solvents (except in water) as compared with their respective ortho isomers. Salicylic acid and methyl salicylate, on the other hand, undergo intramolecular hydrogen bonding for steric reasons, which results in reduced intermolecular interactions and weaker crystal lattices. This is reflected in a reduced melting point and in a higher escaping tendency, activity, and solubility in most solvents (with the exception of water), as compared with the corresponding para isomers mentioned above.

The rank order of the aqueous solubility of the hydroxybenzoic acids and their esters is dominated by the strength of the interaction with water. The ortho-hydroxy group forms a weaker hydrogen bond with water, due to the intramolecular hydrogen bonding which arises from the proximity of the neighboring substituent group. Consequently, the ortho-hydroxy compound is

Table 3.11 Influence of Pentose Configuration on the Melting Point, Density at 20°C, and Molar Solubility in Water at 25°C[a]

	Adenosine (9-β-D-Ribofuranosyl 9H-Purine-6-Amine)	*Vidarabine* (9-β-Arabinofuranosyl- 9H-Purine-6-Amine)
Melting point (°C)	234–235 (monohydrate)	257–257.5 (anhydrate)
Density (g cm^{-3})	1.54 (anhydrate)	1.576 (anhydrate)
Solubility (mol dm^{-3} in water at 20°C)	~0.03 (monohydrate)	~0.009 (anhydrate)

[a]Windholz et al. (1983).

less soluble in water than the corresponding para compound. This emphasizes that the solubility of a solute in an interactive solvent such as water depends not only on the activity of the pure solute but also on the extent of solute–solvent interaction, which is represented by the partition coefficient (in Eq. 3.37) or the activity coefficient (in Eq. 2.4). In a highly interactive system, such as a hydroxy compound dissolving in water, the latter effects tends to exert a greater influence than the former. Indeed, Table 3.10 shows that the partition coefficient (organic solvent/water) is much less for the para isomer than for the corresponding ortho isomer.

When adenosine (adenine riboside) is compared with vidarabine (Ara-A, vira-A, spongoadenosine, adenine arabinoside) as shown in Table 3.11, the latter has the higher melting point and lower escaping tendency, activity, and solubility in various solvents. This difference arises because an adenosine molecule, which differs from a vidarabine molecule in the configuration of a single hydroxyl group about one carbon atom of the pentose moiety, packs more efficiently into its crystal lattice than the less symmetrical vidarabine (Bunick and Voet, 1974). Vidarabine forms a more compact crystal lattice of higher density than adenosine, because the orientation of the hydroxyl groups alternate in the vidarabine molecule, thereby promoting more compact hydrogen bonding.

3.10   THERMODYNAMIC ACTIVITY IN TRANSDERMAL DRUG DELIVERY

Higuchi (1977) proposed the following theorem which emphasizes the importance of a^{c*} to percutaneous (i.e., transdermal) drug delivery: "All other factors being equal, the maximal rate of transdermal transport of any chemical species is directly proportional to the thermodynamic activity of the pure form of the species defined in terms given above" (i.e., a^{c*}). Direct experimental evidence of this is available in the literature (e.g., Barry et al., 1985). To quote other examples, norethindrone acetate is expected to be delivered approximately 36 times faster than norethindrone itself (an alcohol), and chloroform roughly 100 times faster than ethanol. Thus, the discomfort and skin damage elicited by chloroform can be attributed to its high a^{c*} value, shown in Table 3.2. Furthermore, the high a^{c*} value for benzene indicates that all skin contact with this very toxic substance should be avoided.

The following empirical rules for percutaneous absorption clearly illustrate the direct relationship with thermodynamic activity emobdied in the Higuchi (1977) theorem. (a) Nonpolar substances are better absorbed than polar substances. (b) Lipoidal agents are more easily delivered through the skin than nonlipids. (c) Solids of low melting point are more easily absorbed than those that melt at a high temperature. (d) Organic liquids are better absorbed than solids.

Lipophilicity is a term used to denote the affinity of a substance for lipoidal materials which, in their simplest form, are reproduced by hydrocarbon chains. According to this definition, Eq. 3.31 indicates that the lipophilicity of a substance can be represented by its activity in the pure state with respect to an infinitely dilute solution in a paraffinic sovent as the reference state. Since the skin presents a lipoidal barrier to drug molecules, it is reasonable to suppose that the ability of a drug to penetrate the skin is reflected in this thermodynamic activity, which represents the lipophilicity of the drug.

3.11   INFLUENCE OF DIPOLE–INDUCED-DIPOLE INTERACTIONS ON HENRY'S LAW CONSTANTS AND SOLUBILITY

We have seen in Section 2.11 that dipole–induced-dipole interactions (Debye forces) between polar molecules and nonpolar, but polarizable, molecules may be exploited to achieve selectivity in extraction, solvency, and solubility behavior. These observations are contrary to statements in the literature, quoted in Section 2.11.5, that these induction forces are unimportant in comparison with dipole–dipole and dispersion forces. Sufficient material has now been covered to facilitate the presentation of quantitative experimental proof of the relative importance of Debye interactions and to demonstrate their influence on escaping tendencies, as measured by Henry's law constants, vapor pressures and activity, and consequently on solubility. For this purpose, all quantitative comparisons will be made in infinitely dilute solution so that the solute–solute interactions are negligible and only the solute–solvent interactions are considered.

We shall compare the molar Gibbs free energy of a polar solute in two nonpolar solvents of different polarizabilities per unit volume, as measured by the refractive index function f (Eq. 2.137). The solute may be any polar molecule; however, the greater the dipole moment, the stronger will be the Debye interaction (Eq. 2.126). The solvents should ideally be nonpolar (i.e., $\mu = 0$) so that the dipole–dipole interactions (i.e., Keesom forces, Eq. 2.123) are zero, but one solvent should be more polarizable than the other to facilitate comparison. Although increasing polarizability of the solvent strengthens both the London dispersion forces (Eq. 2.128) and the Debye forces (Eq. 2.126), the Debye forces increase with increasing dipole moment of the solute, while the London forces show comparatively little variation (Table 3.9).

The influence of the Debye interaction can be expressed in a quantitative empirical manner as the standard molar Gibbs free energy of transfer of a polar solute (e.g., γ-butyrolactone, ethyl acetate, or griseofulvin) from an infinitely dilute solution in a paraffinic hydrocarbon (e.g., isooctane, indicated by iso), to an infinitely dilute solution in a nonpolar but polarizable solvent (indicated by solv) (e.g., benzene or toluene). This free energy of transfer may be represented by $\Delta_{\text{iso}}^{\text{solv}} G_2^\theta$ and may be calculated either (a) from the ratio of the Henry's law constants of the solute in the two solvents, if the solute is sufficiently volatile, or (b) from the ratio of the solubilities of the solute in the two solvents, if the solute is a solid of low volatility.

Expressing the required free energy of transfer in terms of the Henry's law constant [case $a \equiv (\text{isooctane} \rightarrow \text{solvent}) \equiv (\text{isooctane} \rightarrow \text{vapor}) - (\text{solvent} \rightarrow \text{vapor}]$ and applying Eq. 3.44 in Scheme 3.1,

$$\Delta_{\text{iso}}^{\text{solv}} G_2^\theta = \Delta_{\text{iso}}^{\text{vap}} G_2^\theta - \Delta_{\text{solv}}^{\text{vap}} G_2^\theta \tag{3.50}$$

$$= -RT \ln h_2(\text{in iso}) + RT \ln h_2(\text{in solv}) \tag{3.51}$$

If the standard state of the solute dissolved in isooctane and in the solvent is in each case an infinitely dilute solution (molarity scale), while that in the vapor state is a pressure of 1 atm = 760 torr,

$$\Delta_{\text{iso}}^{\text{solv}} G_2^\theta = -RT \ln (h_2^c \text{ in iso}/h_2^c \text{ in solv}) \tag{3.52}$$

We note that this equation is independent of the standard state of the vapor.

Using various polar solutes, Rytting et al. (1986) examined the influence of the polarizability of the solvent of h_2^c, h_2^x, and γ_2^∞ (Eq. 3.25) and on the molar Gibbs free energies of transfer given above (Table 3.12). In the following discussion and in Table 3.12 and Figure 3.5, the following abbreviations will be used: isooctane (ISO), toluene (TOL), γ-butyrolactone (BL), ethyl acetate (EA), 2-methyl-3-pentanol (MEP), and cinnamaldehyde (CIN). TOL in ISO ($\gamma_2^\infty = 2.05$) and ISO in TOL ($\gamma_2^\infty = 3.30$) show only modest departures from ideality (Table 3.12). This

Table 3.12 Henry's Law Constants, Activity Coefficients, and Standard Molar Gibbs Free Energies of Transfer of Some Liquid Solutes Dissolved in Various Solvents at 25°C[a]

Solute 2	Solvent 1	h_2^{cb}	h_2^{xc}	$\gamma_2^{\infty d}$	$\Delta_{sol}^{vap}G_2^{\theta e}$	$\Delta_{iso}^{sol}G_2^{\theta e}$
ISO	ISO	8.14	49	1.00[e]	11.25	0[g]
TOL	ISO	9.7	58.4	2.05	10.81	0[g]
MEP	ISO	9.0	54.2	9.03	11.00	0[g]
EA	ISO	121	729	7.71	4.56	0[g]
BL	ISO	3.86	23.2	74.3	13.10	0[g]
ISO	TOL	17.3	162	3.30	9.38	1.87
TOL	TOL	3.05	28.5	1.00[f]	13.68	−2.87
MEP	TOL	2.34	21.9	3.65	14.34	−3.34
EA	TOL	13.1	123	1.30	10.07	−5.51
BL	TOL	0.023	0.215	0.688	25.79	−12.60
MEP	MEP	0.747	6.0	1.00[f]	17.17	−6.17
EA	EA	9.31	94.53	1.00[f]	10.91	−6.36
ISO	BL	311.6	4073	83.1	2.21	9.04
TOL	BL	128.0	1673	58.7	4.42	6.40
BL	BL	0.0239	0.3127	1.000[f]	25.70	−12.60
ISO	CIN	145.7	1157	23.6	4.09	7.15
TOL	CIN	95.8	761	26.7	5.13	5.68
EA	CIN	22.4	178	1.88	8.74	−4.18
BL	CIN	0.033	0.262	0.838	24.90	−11.80
CIN	CIN	0.0292	0.232	1.00[f]	25.20	—

[a] Abbreviations for solutes and solvents are as follows: isooctane (ISO); toluene (TOL); γ-butyrolactone (BL); cinnamaldehyde (CIN); 3-methyl-2-pentanol (MEP); and ethyl acetate (EA).
[b] Henry's law constant (torr dm^3 mol^{-1}).
[c] Henry's law constant (torr) based on mole fraction of the solute.
[d] Activity coefficient of the solute at infinite dilution calculated from Eq. 3.25.
[e] Units are kJ mol^{-1}. 1 cal = 4.184 J.
[f] Unity, by definition.
[g] Zero, by definition

Adapted from Rytting et al., (1986). Reproduced with permission of the copyright owner, Plenum Publishing Corp., from Rytting, J. H., McHan,D. R., Higuchi, T., and Grant, D. J. W. (1986). *J. Soln. Chem.*, **15**, 693–703.

suggests that the appreciably greater polarizability of TOL than ISO impinges only slightly on the interactions between these essentially nonpolar molecules. Indeed, Table 3.9 shows that London dispersion forces are the predominant interactions between TOL and ISO. Consequently, mixtures of TOL + ISO are assumed to constitute a solvent system for which the polarizability may be varied continuously from a relatively low value in pure ISO to an appreciably higher value in pure TOL, as in the horizontal axis in Figure 3.5. The polarizability employed here is the polarizability per unit volume, which is measured according to Eq. 2.137 by the function $(n^2 - 1)/(n^2 + 2)$, where n is the refractive index, specifically n_D^{25}, at 25°C using the sodium D line (Table 3.8).

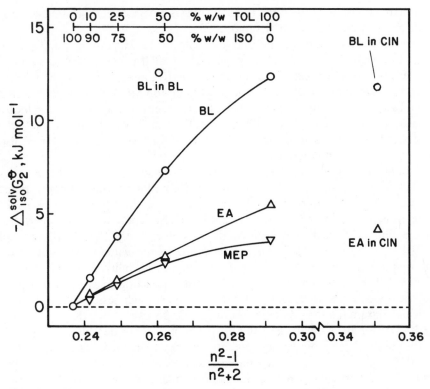

Figure 3.5 Influence of the solvent medium (of refractive index n) and polarizability per unit volume represented by $(n^2 - 1)/(n^2 + 2)$ on the standard molar Gibbs free energy of transfer of the solute BL ($\bigcirc$), EA ($\triangle$), or MEP ($\triangledown$) from dilute solution in ISO to dilute solution in the solvent medium ISO + TOL, TOL, CIN, or BL. 1 cal = 4.184 J. Adapted from Rytting et al. (1986). Reproduced with permission of the copyright owner, Plenum Publishing Corp., from Rytting, J. H., McHan, D. R., Higuchi, T., and Grant, D. J. W. (1986). *J. Soln. Chem.*, **15**, 693–703.

Strictly speaking, one should take into account the statistical enrichment of the more polarizable solvent component, which would occur in the immediate vicinity of any dipolar solute molecule, and the effect this would have on the effective average polarizability. This, however, would be a second-order phenomenon and we have chosen to ignore it.

Figure 3.5 shows that increasing polarizability of the solvent mdeium, as defined above, progressively favors the transfer of polar solutes from ISO to that solvent medium, as measured by the increasing negative values of $\Delta_{iso}^{solv} G_2^\theta$. The slope of each curve at any point reflects the degree of preference of that polar solute for the more polarizable solvent medium and this increases with increasing dipole moment of the solute molecules, that is, MEP < EA < BL (Table 3.8). This reflects increasing strength of the Debye interaction (Table 3.9). The polarizability of the solvent medium was increased by increasing the proportion of TOL. Since TOL possesses a small dipole moment (0.31 D), it can participate in Keesom interactions (Table 3.9). Although the use of benzene ($\mu = 0$) would have eliminated Keesom interactions, this solvent was not used because of the hazards to health presented by its toxicological effects.

Just as escaping tendency (h_2^c and h_2^x, Table 3.12) of each of the polar solutes (BL, EA, and MEP) from TOL is less than that from ISO, so the escaping tendency of TOL from each of the polar solvents (BL and CIN) is less than that of ISO. For example, h_2^c of TOL from BL is 128 torr dm^3 mol^{-3}, while that of ISO from BL is 312 torr dm^3 mol^{-1} (Table 3.12). Furthermore, $\Delta_{iso}^{solv} G_2^\theta$ is in each case smaller into TOL than it is into ISO. These effects may again be attributed to the stronger Debye interactions in TOL than in ISO, although Keesom forces also contribute to the interactions in TOL, as discussed above.

The much larger value of $-\Delta_{iso}^{solv} G_2^\theta$ for BL to TOL (12.7 kJ mol^{-1}) than for EA to TOL (5.15 kJ mol^{-1}) arises from stronger Debye forces which result from the greater dipole moment of the BL for the reasons explained previously (Fig. 3.5). The value of 12.7 kJ mol^{-1} (i.e., 5.12 RT) for BL approaches the energy of interaction of a weak hydrogen bond. However, $-\Delta_{iso}^{solv} G_2^\theta$ for BL to itself (12.6 kJ mol^{-1}) is only 5.7 kJ mol^{-1} in excess of the value (6.9 kJ mol^{-1}) for transfer to a mixture of TOL + ISO of the same polarizability, that is, $(n^2 - 1)/(n^2 + 2) = 0.26084$. This suggests that the Keesom interactions may be contributing to the overall intermolecular interactions to a much lesser extent than is calculated from the classical expressions [Eq. 2.123 gives $-u_{1,2}$ (Keesom) $= 801 \times 10^{-19}$ J for BL with BL in Table 3.9]. This casts considerable doubt on the values of the dipole–dipole pair-potential energy calculated from Eq. 2.213.

CIN is a liquid of unusually high refractive index ($n_D^{25} = 1.6195$, Table 3.8), indicating a particularly high polarizability arising, no doubt, from extensive conjugation. When the solvent medium of BL is changed from TOL to CIN, $-\Delta_{iso}^{solv} G_2^\theta$ continues the general trend shown in Figure 3.5, although at a lower rate of increase, which is somewhat hidden by the contraction in the horizontal scale. In view of the high dipole moment of CIN (3.63 D, Table 3.8), Keesom forces must contribute to its interactions with polar solutes such as EA and BL. However, $-\Delta_{iso}^{solv} G_2^\theta$ for BL increases by only 2.3 kJ mol^{-1} as the solvent is changed from TOL to CIN. Presumably; strong solvent–solvent interactions in CIN partially oppose the strong solute–solvent interactions (Table 3.9). This effect, together with conformational differences discussed in Section 3.9, may explain why $-\Delta_{iso}^{solv} G_2^\theta$ for EA decreases (by 1.3 kJ mol^{-1}) as the solvent is changed from TOL to CIN.

As mentioned earlier in this section, the free energy of transfer that is proposed for the assessment of the Debye interaction can be calculated from the ratio of solubilities [case b $\equiv$ (isooctane $\rightarrow$ solvent) $\equiv$ (pure solute $\rightarrow$ solvent) $-$ (pure solute $\rightarrow$ isooctane)]. Applying the equation for solubility at the upper right of Scheme 1.1, we obtain

$$\Delta_{iso}^{solv} G_2^\theta = \Delta_{pure}^{solv} G_2^\theta - \Delta_{pure}^{iso} G_2^\theta \tag{3.53}$$

$$= -RT \ln (\text{solubility in solvent}) + RT \ln (\text{solubility in isooctane}) \tag{3.54}$$

If the standard state of the solute dissolved in isooctane and in the solvent is in

Table 3.13 Micromolar Solubilities of Griseofulvin and the Standard Molar Gibbs Free Energy of Transfer of Griseofulvin from Isooctane to Various Nonpolar Solvents of Different Polarizability as Measured by $(n^2 - 1)/(n^2 + 2)^a$

Solvent 1	n_D^{25b}	$\dfrac{n^2 - 1}{n^2 + 2}$	μ^a (D)	c_2^s ($\mu\,mol\,dm^{-3}$)	$\Delta_{iso}^{solv} G_2$ ($kJ\,mol^{-1}$)
Isooctane	1.38898	0.23650	0	9.36^c	0^g
n-Heptane	1.38511	0.23441	0.085	10.12^d	−0.195
Benzene	1.49792	0.29308	0	33300^d	−20.27
Toluene	1.49413	0.29119	0.31	1250^e	−12.13
Carbon tetrachloride	1.45739	0.27255	0	5680^f	−15.88

[a] Temperature, 25°C. 1 cal = 4.184 J.
[b] Riddick and Bunger (1970).
[c] Mehdizadeh and Grant (1984).
[d] Elworthy and Lipscombe (1968).
[e] Cook (1978).
[f] Townley (1979).
[g] Zero, by definition.

each case an infinitely dilute solution (molarity scale) as before,

$$\Delta_{iso}^{solv} G_2^\theta = - RT \ln (c_2^s \text{ in solv}/c_2^s \text{ in iso}) \tag{3.55}$$

The solid phase in equilibrium with each saturated solution must, of course, be unchanged (see Sections 2.3 and 2.4) and the solubilities must be sufficiently low that the laws of dilute solution apply.

Table 3.13 shows the application of this concept to griseofulvin, which can be classed as a very polar molecule, since it has two keto groups, four ether groups, and a chloro group on a benzene ring (Scheme 6.1). The solubilities of griseofulvin in the polarizable but nonpolar solvents benzene and carbon tetrachloride are several orders of magnitude higher than in the paraffinic solvents isooctane and heptane. The solubility in toluene is notably less than in benzene and carbon tetrachloride despite the small dipole moment of toluene. Thus, Keesom forces play an insignificant role. The free energy of transfer of griseofulvin from isooctane to toluene (and the other polarizable solvents in Table 3.13) is similar in magnitude to that of γ-butyrolactone from isooctane to toluene (and of cinnamaldehyde in Table 3.12), being about $- 12$ to $- 20\,kJ\,mol^{-1}$. These values correspond to about 5–8 RT, which indicates that the Debye interaction can be comparable in strength to a hydrogen bond.

The material discussed above, which is based on Henry's law constants and solubilities, demonstrates that Debye interactions play a significant and sometimes important role in accounting for the behavior of polar solutes in polarizable solvents. Dipole–induced-dipole interactions are not only responsible for the extractive separation of aromatics from aliphatics in petroleum products by addition of a highly polar species, but may also provide a useful degree of selectivity in solubility phenomena.

Chapter IV

SPECIFIC INTERACTIONS IN SOLUBILITY PHENOMENA

4.1 INTRODUCTION

The purpose of this chapter is to demonstrate to the reader that specific solute–solvent interactions are overwhelmingly important in controlling the solubility behavior of polar solutes in the presence of nonaqueous polar solvents. The arguments in support of this thesis are as follows:

(a) Regular solution theory involving solubility parameters provides an insufficient or inadequate explanation which lacks predictive capability.

(b) The athermal contribution, that is, the Flory–Huggins entropy, can only account for the relative solubilities of sparingly soluble polar solutes in different paraffinic solvents.

(c) The relative solubilities of a given polar solute in hydrocarbons and polar solvents (e.g., chloroform, ethers, esters, amides, alcohols, and carboxylic acids) can be explained in terms of the presence and nature of acid–base interactions.

(d) The stoichiometry of the specific solute–cosolvent interactions, derived by an analysis of the solubility data in mixtures of paraffins and a polar cosolvent, accords with the number and the nature of the functional groups in the solute and cosolvent molecules.

(e) The equilibrium constant for the solute–cosolvent interaction of a given solute in mixtures of a paraffin (or carbon tetrachloride) and a cosolvent is determined by the functional group of the cosolvent and not by the hydrocarbon chain; this is shown by homologous series of cosolvents, for example, alcohols, amides, or esters.

4.2 LIMITATIONS OF REGULAR SOLUTION THEORY AND THE ATHERMAL TREATMENT

As discussed in Chapter 2, regular solution theory does not take explicit account of solute–solvent interactions of a specific nature, such as hydrogen bonding or certain types of charge-transfer interactions. These interactions are inevitable in solutions of polar molecules such as drugs, especially when the solvent is also polar, and are essentially responsible for their solubility behavior.

Such interactions are characterized by (a) an absolute requirement for a definite steric relationship between the solute molecule and the solvent molecule, and (b) an energy of interaction or bond strength (e.g., $4\,\text{kcal mol}^{-1} = 17\,\text{kJ mol}^{-1}$) considerably greater than kT per molecule or RT per mole ($= 0.592\,\text{kcal mol}^{-1} = 2.48\,\text{kJ mol}^{-1}$ at 25°C), which permits the attachment to persist for many molecular vibrations. However, these attached donor–acceptor pairs may separate and reform in a dynamic equilibrium which obeys the simple mass action relationship.

Conditions (a) and (b) are frequently met by polar solutes in polar solvents, especially when one is a hydrogen-bond donor and the other is an acceptor. It is therefore not surprising that regular solution theory cannot effectively predict the solubility of polar solutes in polar solvents, for example, (a) sulfonamides (proton donors) in water, ethanol, propylene glycol, glycerol, and dimethyl-acetamide (proton acceptors) (Sunwoo and Eisen, 1971); (b) testosterone esters (proton acceptors) in chloroform (Bowden and James, 1970) and in other proton-donating solvents; (c) benzoic acid and esters of p-hydroxybenzoic acid in n-alkanols (Restaino and Martin, 1964). Numerous other examples of the failure of regular solution theory have been found.

Higuchi et al. (1971b) pointed out that, although straight-line correlations can sometimes be obtained for irregular systems using solubility parameters, many attempted correlations with regular solution theory often go far beyond the expectations of its originators, Scatchard (1931) and Hildebrand and Scott (1950). In some instances the solvent has been treated as a continuum. For example, attempts have been made to correlate solubility properties of non-aqueous systems with a single parameter such as dielectric constant (Sorby et al., 1963; Paruta et al., 1964; Paruta and Irani, 1965). This approach also has severe limitations. Only in specific instances has it been possible to relate solubility of substances to the bulk properties of the solvent. Table 4.1 shows a typical case in which relative solubility appears to bear no obvious relation to dielectric constant D, dipole moment μ, solubility parameters δ, or molar volume V_m.

To explain merely the rank order of solubilities, thoughts turn naturally to similarities and differences between functional groups and the possibilities of specific interactions between solute and solvent. For most organic substances and drugs, this seems to be the only satisfactory method for understanding solubility behavior. A rigorous and consistent explanation of solubility behavior on the basis of solute–solvent (and solute–cosolvent) interactions has been developed by Higuchi and Connors (1965) and Higuchi et al. (1969). This will be discussed in depth in Chapters 5 and 10. At this point we present the philosophy of this approach in a broader perspective.

Fung and Higuchi (1971) studied the solubility of various nonelectrolytes such as acetanilide, carbazole, picric acid, salicylic acid, and phthalic anhydride in organic solvents ranging from nonpolar hydrocarbons to polar solvents such as diethyl ether, chloroform, and acetic acid. They analyzed the data in terms of (1) Flory–Huggins entropy (Section 2.10), (2) regular solution theory

Table 4.1   Molar[a] and Mole Fraction Solubilities of Phthalic Anhydride in Solvents with Nonzero Dipole Moments and in a Paraffinic Solvent, n-Heptane, at 25°C and Physical Properties of the Solvents

Solvent	D^b	μ^c	$V_m{}^d$	δ^e	Molar Solubility (mol dm^{-3})^a	Solubility (mole fraction)
n-Heptane	1.92	0	148	7.4	0.0049	0.00072
Diethyl ether	4.34	1.15	105	7.4	0.116	0.012
Di-n-propyl ether	3.4	1.3	138	6.9^h	0.04	0.0055
Chloroform	4.8	1.02	81	9.2	0.78	0.063
$tert$-Butyl alcohol	10.9^f	1.0^g	95	10.6	0.04	0.0068
Acetic acid	6.15	1.74	57	14.4^i	0.27	0.015
Tetrahydrofuran	7.4	1.7	81	8.9^i	1.2	0.097
Pyridine	12.3	2.2	81	10.7	2.9	0.23
Acetone	20.7	2.89	74	10.0^h	1.15	0.085

^{a}1 mol dm^{-3} = 1 mol L^{-1} = 1 M.
^{b}Dielectric constant, values obtained from *Handbook of Chemistry and Physics*, 48th ed., Chemical Rubber Co., Cleveland, Ohio, 1967.
^{c}Dipole moment D from same source as in footnote b.
^{d}Molar volume, cm^3 mol^{-1}; (= molecular weight/density) at 25°C, from same source as in footnote b.
^{e}Solubility parameter, cal$^{1/2}$ cm$^{-3/2}$, from Hildebrand and Scott (1962). 1 cal = 4.184 J.
^{f}30°C.
^{g}From same source as in footnote e.
^{h}Crowley, J. D., Teague, Jr., G. S., and Lowe, Jr., J. W. *J. Paint Technol.*, **38**, 272 (1966).
^{i}Estimated from heat of vaporization through Eq. 2.87.

Reproduced with permission of the copyright owner, the American Pharmaceutical Association, from Fung, H. L. and Higuchi, T. (1971). *J. Pharm. Sci.*, **60**, 1782–1788.

(Section 2.6), and (3) specific interactions. The relative importance of each theory was then assessed. For poorly soluble polar solutes in a series of hydrocarbon solvents, the entropic correction term appeared to account for most of the deviation from ideal behavior, whereas the regular solution correction appeared to be unnecessary. To explain the solubility of polar solutes in solvents possessing an acid–base character, specific interactions were found to be the dominant factors. The evidence for these specific interactions will now be considered. Concentration scales based on mole fraction and molarity will also be compared. The following general features are apparent.

(a)   Table 4.2 shows that the *mole fraction* solubilities of the polar solutes in inert saturated hydrocarbon solvents generally increase with increasing molar volume of the solvent, but the differences in solubility from solvent to solvent are quite small.

(b)   Table 4.4 shows that the *molar* solubilities of the solutes in inert hydrocarbon solvents appear to be almost identical within experimental error, and comparison with Table 4.3 shows that they are independent of the molar volume V_m and the solubility parameter δ of the solvent for these limited cases. The

Table 4.2 Mole-Fraction Solubilities and Molar Volumes (V_m, $cm^3\ mol^{-1}$) of Several Polar Substances in Nonpolar Solvents at 25°C Unless Otherwise Stated

Solvent	V_m	Acetanilide	Carbazole	Solute Picric Acid	Salicylic Acid	Phthalic Anhydride
Cyclohexane	109	8.1×10^{-5}	1.8×10^{-4}	4.4×10^{-5}	4.3×10^{-4}	6.9×10^{-4}
n-Hexane	132	—	1.7×10^{-4}	3.9×10^{-5}	4.8×10^{-4}	6.5×10^{-4}
n-Heptane	148	1.4×10^{-4}	2.3×10^{-4}	4.8×10^{-5}	4.8×10^{-4}	7.2×10^{-4}
Isooctane	166	1.2×10^{-4}	1.8×10^{-4}	4.1×10^{-5}	3.8×10^{-4}	7.0×10^{-4}
Decane	192	1.8×10^{-4}	2.9×10^{-4}	8.2×10^{-5}	6.2×10^{-4}	9.6×10^{-4}
Dodecane	222	1.8×10^{-4}	3.2×10^{-4}	7.5×10^{-5}	6.4×10^{-4}	10.9×10^{-4}
Hexadecane	294	2.9×10^{-4}	5.9×10^{-4}	13.5×10^{-5}	10.4×10^{-4}	16.2×10^{-4}
Chloroform	81	$1.8 \times 10^{-1\,a}$	3.7×10^{-3}	$4.4 \times 10^{-3\,b}$	$1.3 \times 10^{-2\,a}$	6.3×10^{-2}
Diethyl ether	105	$1.9 \times 10^{-2\,a}$	1.3×10^{-2}	$6.4 \times 10^{-3\,b}$	$2.5 \times 10^{-1\,c}$	1.2×10^{-2}

[a] 30°C, estimated from values from Seidell, A., *Solubilities of Organic Compounds*, Vol. II, 3rd ed., Van Nostrand, New York, 1941.
[b] 20°C, from same source as in footnote a.
[c] Estimated from *The Merck Index*, 8th ed., Merck & Co., Rahway, NJ, 1968.

Reproduced with permission of the copyright owner, the American Pharmaceutical Association, from Fung, H. L. and Higuchi, T. (1971). *J. Pharm. Sci.*, **60**, 1782–1788.

Table 4.3 Physical Properties of Solvents Used for Solubility Studies of Polar Solutes at 25°C

Solvent	D^a	μ^b	V_m^c	δ^d
Cyclohexane	2.02	0	109	8.2
n-Hexane	2.0	0	132	7.3
n-Heptane	1.92	0	148	7.4
Isooctane	1.94	0	166	6.9
Decane	2.0	0	192^e	7.8^f
Dodecane	2.0	0	222^e	7.9^f
Hexadecane	—	0	294	8.0
Chloroform	4.8	1.02	81	9.2
Diethyl ether	4.34	1.15	105	7.4

^{a}Dielectric constants, values obtained from *Handbook of Chemistry and Physics*, 48th ed., Chemical Rubber Co., Cleveland, OH, 1967.

^{b}Dipole moment, D, from same source as in footnote a.

^{c}Molar volume (cm^3 mol^{-1}) at 25°C (Riddick and Bunger, 1970).

^{d}Solubility parameter, cal$^{1/2}$ cm$^{-3/2}$ (Riddick and Bunger, 1970). 1 cal = 4.184 J.

^{e}Estimated by assuming 15 cm^3/CH$_2$ unit in alkanes.

^{f}Estimated from heat of vaporization through Eq. 2.87.

Reproduced with permission of the copyright owner, the American Pharmaceutical Association, from Fung, H. L. and Higuchi, T. (1971). *J. Pharm. Sci.*, **60**, 1782–1788.

Table 4.4 Molar Solubilities (mol dm^{-3})^a of Several Polar Substances in Nonpolar Solvents

Solvent	Solute				
	Acetanilide	Carbazole	Picric Acid	Salicylic Acid	Phthalic Anhydride
Cyclohexane	7.5×10^{-4}	1.7×10^{-4}	4.1×10^{-4}	4.0×10^{-3}	6.4×10^{-3}
n-Hexane	—	1.3×10^{-3}	3.0×10^{-4}	3.4×10^{-3}	5.0×10^{-3}
n-Heptane	9.3×10^{-4}	1.6×10^{-3}	3.3×10^{-4}	3.3×10^{-3}	4.9×10^{-3}
Isooctane	9.3×10^{-4}	1.1×10^{-3}	2.5×10^{-4}	2.3×10^{-3}	4.22×10^{-3}
Decane	9.3×10^{-4}	1.5×10^{-3}	4.2×10^{-4}	3.2×10^{-3}	4.9×10^{-3}
Dodecane	7.8×10^{-4}	1.4×10^{-3}	3.3×10^{-4}	2.8×10^{-3}	4.8×10^{-3}
Hexadecane	8.3×10^{-4}	1.7×10^{-3}	3.9×10^{-4}	3.0×10^{-3}	4.7×10^{-3}
Chloroform	2.23^b	4.5×10^{-2}	5.5×10^{-2c}	1.6×10^{-1b}	6.3×10^{-2}
Diethyl ether	1.5×10^{-1b}	1.25×10^{-1}	6.1×10^{-2c}	2.4^d	1.2×10^{-2}

^{a}1 mol dm^{-3} = 1 mol L^{-1} = 1 M.

^{b}30°C, estimated from values from Seidell, A., *Solubilities of Organic Compounds*, Vol. II, 3rd ed., Van Nostrand, New York, 1941.

^{c}20°C, from same source as in footnote b.

^{d}Estimated from *The Merck Index*, 8th ed., Merck & Co., Rahway, NJ, 1968.

Reproduced with permission of the copyright owner, the American Pharmaceutical Association, from Fung, H. L. and Higuchi, T. (1971). *J. Pharm. Sci.*, **60**, 1782–1788.

choice of molarity as the unit for expressing solubilities has the effect of correcting for (1) differences in molar volume between solvents and (2) nonspecific solute–solvent interactions. The importance of the choice of concentration units for expressing solubilities and for calculating the parameters of weak complexes (Kuntz et al., 1968; Christian and Lane, 1976) will be discussed in Section 5.7.

(c)   Tables 4.2 and 4.4. show that the solubilities are much higher in chloroform and ether than in the hydrocarbons and appear to be strongly dependent on the nature of the solute.

These observations can be explained by considering the relative contributions of nonspecific (physical) and specific (chemical) interactions between solute and solvent.

In saturated aliphatic hydrocarbons specific solute–solvent interaction is assumed to be largely absent. The solubility of a poorly soluble solute in these solvents is determined by entropic (athermal) and enthalpic (regular solution) contributions. The Flory–Huggins equation (Eq. 2.116) (Flory, 1941, 1942; Huggins, 1941, 1942) for calculating the partial molar entropy of mixing of molecules of different sizes can be used (Longuett-Higgins, 1953; Ashworth and Everett, 1960) to calculate the statistical contribution to the activity coefficient γ_B of the sparingly soluble solute, the standard state being the pure supercooled liquid ($a_B^* = a_B^\theta = 1$), thus,

$$\ln \gamma_B^{ath} = \ln \frac{V_B}{V_A} + \left(1 - \frac{V_B}{V_A} \right) \tag{4.1}$$

where V_A and V_B are the molar volumes of the solvent and solute, respectively. If the entropic factor is mainly responsible for deviations from ideality, $\gamma_B^{ath} = \gamma_B$. When the solute is only very slightly soluble, $a_B = $ constant, but $x_B = a_B/\gamma_B$ (Eq. 2.6). Therefore,

$$\ln \frac{a_B}{x_B} = \ln \gamma_B = \ln \frac{V_B}{V_A} + \left(1 - \frac{V_B}{V_A} \right) \tag{4.2}$$

This equation may be applied to any solvent, s = A, and again to a reference solvent cyclohexane, c = A. Subtraction of the second result from the first leads to

$$\ln \left(\frac{x_{B,c}}{x_{B,s}} \right)^{ath} = \ln \left(\frac{\gamma_{B,s}}{\gamma_{B,c}} \right)^{ath} = \ln \frac{V_c}{V_s} + V_B \left(\frac{1}{V_c} - \frac{1}{V_s} \right) \tag{4.3}$$

The activity of the solid solute a_B then cancels from Eq. 4.2. For a solid solute, V_B is the molar volume of the supercooled liquid, values of which are not available. However, an approximate value of $100 \, cm^3 \, mol^{-1}$ can be assigned to V_B for salicylic acid and phthalic anhydride, based upon published values for organic solutes (Wakabayashi, 1967; Sunwoo and Eisen, 1971).

In Table 4.5 the logarithm of the ratio of the observed mole-fraction solubility

Table 4.5 Comparison of Athermal Contribution with Observed Solubilities of Phthalic Anhydride and Salicylic Acid[a]

Solvent (S)	$\log(x_{B,c}/x_{B,s})^{ath}$	$\log(x_{B,c}/x_{B,s})_{obs}$		$\log(x_{B,c}/x_{B,s})_{obs} - \log(x_{B,c}/x_{B,s})^{ath}$	
		Phthalic Anhydride	*Salicylic Acid*	*Phthalic Anhydride*	*Salicylic Acid*
n-Hexane	−0.013	0.0261	−0.010	0.0391	0.003
n-Heptane	−0.028	−0.018	−0.048	0.010	−0.020
Isooctane	−0.046	−0.006	−0.0539	0.040	0.008
Decane	−0.074	−0.143	−0.159	−0.069	−0.085
Dodecane	−0.106	−0.199	−0.173	−0.093	−0.067
Hexadecane	−0.18	−0.374	−0.383	−0.194	−0.203
Chloroform	−0.009	−1.960	−1.480	−1.951	−1.471
Diethyl ether	+0.0009	−1.240	−2.764	−1.239	−2.765

[a]Symbols are defined in the text.

Reproduced with permission of the copyright owner, the American Pharmaceutical Association, from Fung, H. L. and Higuchi, T. (1971). *J. Pharm. Sci.*, **60**, 1782–1788.

of each of these solutes B in cyclo-hexane, c, to that in each of the solvents s [i.e., $\ln(x_{B,c}/x_{B,s})$], is compared with the analogous quantity calculated from Eq. 4.3 [i.e., $\ln(x_{B,c}/x_{B,s})^{ath}$] assuming entropic (athermal) dominance. The difference between these quantities, stated in the last two columns of Table 4.5, is nearly zero for the hydrocarbon solvents, indicating that the Flory–Huggins entropy (athermal) correction can be used to account for virtually all of the differences in solubility of two highly polar solutes in seven saturated aliphatic hydrocarbons. This conclusion is consistent with the finding that the molar concentration solubilities (Table 4.4) are essentially independent of the properties of the hydrocarbon solvents (Table 4.3).

The enthalpic (regular solution) contribution to the solubility, can, in principle, be estimated from the Hildebrand solubility equation (Eq. 2.90), the major problem being the assignment of a suitable value to the solubility parameter of the solute. In principle, solubility parameters can be calculated from the energy of vaporization and the molar volume using Eq. 2.87. The former can be estimated from the normal boiling point (Riddick and Bunger, 1970; Riddick et al., 1986; Weast et al., 1986; Windholz et al., 1983) using the Hildebrand rule (Eq. 3.18), but for calculation of the latter, knowledge of the density of the supercooled liquid is required. Assuming, as before, a molar volume of $100 \, cm^3 \, mol^{-1}$, the solubility parameters of the solutes in $cal^{1/2} \, cm^{-3/2}$ can be estimated: acetanilide, 13.0; carbazole, 13.9; picric acid > 12.9; salicylic acid, 11.2; phthalic anhydride, 12.7. These values are probably fair approximations, since the solubility parameters δ for the sulfonamides (Sunwoo and Eisen, 1971) lie between 11 and $17 \, cal^{1/2} \, cm^{-3/2}$, as do those of many organic solutes. By means of the Hildebrand solubility equation, solubility ratios x_s/x_c were predicted for a range of δ_B values in representative solvents of various δ_A, as seen in Table 4.6.

Table 4.6 Predicted Relative Mole-Fraction Solubility of a Hypothetical Polar Solute in Various Solvents[a] for Varying Solubility Parameter Values of the Solute δ_2 and of the Solvent δ_1

Solubility Parameter of Solute, δ_2 (cal$^{1/2}$ cm$^{-3/2}$)	x_s/x_e			
	n-Hexane ($\delta_1 = 7.3$)	Heptane + Ether ($\delta_1 = 7.4$)	Isooctane ($\delta_1 = 6.9$)	Chloroform ($\delta_1 = 9.2$)
11	0.651	0.687	0.518	1.402
12	0.517	0.611	0.428	1.623
13	0.500	0.543	0.357	1.879
14	0.438	0.483	0.292	2.176
15	0.384	0.430	0.242	2.250
16	0.337	0.382	0.200	2.917
17	0.295	0.340	0.165	3.378

[a]The reference solvent in cyclohexane. 1 cal = 4.184 J.

Reproduced with persmission of the copyright owner, the American Pharmaceutical Association, from Fung, H. L. and Higuchi, T. (1971). *J. Pharm. Sci.*, **60**, 1782–1788.

Thus, regular solution theory predicts that (1) solubility will be greatest in cyclohexane among the hydrocarbons, (2) solubility in heptane will be identical to that in ether, since both solvents have the same solubility parameter, and (3) solubility in chloroform will not be more than about 3–4 times higher than that in cyclohexane. Examination of Table 4.2 shows that all these predictions are incorrect; solubilities in cyclohexane are generally the lowest among the hydrocarbons, but in ether and chloroform are much higher than predicted.

Thus, the regular solution treatment of solubility of the polar solutes in noninteractive solvents (saturated aliphatic hydrocarbons) is at most only marginally valid, whereas in interactive solvents (chloroform and ether), it is particularly unsuccessful.

Regular solution theory is found generally to fail for polar solutes in interactive solvents. Thus, Anderson et al. (1980) found that the solubility of the hydrogen donor solute carbazole ($\delta = 10$ cal$^{1/2}$ cm$^{-3/2}$, $V_m = 150$ cm^3 mol^{-1}) is 20–100 times higher in hydrogen-accepting solvents, such as ethers and esters, than in relatively inert alkane solvents which have the same solubility parameters as the ethers and esters. A simple comparison of the solubility values in Table 4.7 provides a qualitative indication of the importance of hydrogen bonding in determining relative solubilities. The molar solubility of the strong hydrogen donor p-nitrophenol is nearly 4000 times greater in the hydrogen-acceptor solvent dibutyl ether than in isooctane, whereas for the weaker hydrogen donor carbazole, the molar solubility is increased by a factor of only about 38 by the same change in solvent.

In the case of anthracene, which does not act as a hydrogen-bond donor, the solubility is increased only about twofold. The differences are also large when the solubilities are calculated as mole fractions, because the molecular

Table 4.7 Molar Solubilities of Various Organic Solutes in Several Organic Solvents at 25°C[a]

Solute	Structure	Isooctane	Butyl Butyrate	Butyl Ether	Pentyl Ether	Chloroform
p-Nitrophenol		3.3×10^{-4}	—	1.27	—	—
p-Iodophenol		1.5×10^{-2}	—	3.39	—	—
p-Phenylphenol		1.24×10^{-3}	0.62	0.21	0.12	0.10
2,4,6-Triiodophenol		5.0×10^{-3}	—	0.10	—	—
Carbazole		7.7×10^{-4}	0.099	0.029	0.018	0.047
Anthraquinone		2.8×10^{-4}	0.0038	—	—	0.044
Anthracene		6.6×10^{-3}	—	0.021	0.016	—
Solvent δ_1 (cal$^{1/2}$ cm$^{-3/2}$)		6.9	8.0	7.6	7.9	9.2

[a] $1 \, \mathrm{mol\,dm^{-3}} = 1 \, \mathrm{mol\,L^{-1}} = 1 \, M$. $1 \, \mathrm{cal} = 4.184 \, \mathrm{J}$.

Reproduced with permission of the copyright owner, the American Pharmaceutical Association, from Anderson, B. D., Rytting, J. H., and Higuchi, T. (1980). *J. Pharm. Sci.,* **69**, 676–680.

weight of dibutyl ether (130.2) is quite similar to that of isooctane (114.2). Regular solution theory cannot account for the large difference in the solubilities since the solubility parameter of dibutyl ether ($\delta = 7.6\,\text{cal}^{1/2}\,\text{cm}^{-3/2}$) is quite similar to that of isooctane ($\delta = 6.9\,\text{cal}^{1/2}\,\text{cm}^{-3/2}$). However, the large differences in solubility are quite consistent with specific interaction theory.

A further test of the usefulness of simple solubility parameters is to determine whether they can reliably predict the solubility of a solute in mixed solvents using Eq. 2.91. By means of this equation, the solubility of anthraquinone ($V_m = 150\,\text{cm}^3\,\text{mol}^{-1}$) was calculated in mixtures of chloroform and isooctane, assuming for anthraquinone that $\delta = 11.8\,\text{cal}^{1/2}\,\text{cm}^{-3/2}$, which was deduced from the fact that the mole-fraction solubility of this solute is 77 times higher in chloroform than in isooctane. Figure 4.1 shows that the solubility parameter approach fails to predict the shape of the solubility profile in the mixed-solvent system in which polar effects and/or specific interactions predominate.

In attempts to overcome the inadequacy of simple one-component solubility parameters, we have seen in Section 2.7 that various more or less empirical multicomponent solubility parameters have been proposed which incorporate the effects of specific interactions (Crowley et al., 1966; Hansen and Beerbower, 1971; Hoy and Martin, 1975; Barton, 1975, 1983; Burrell, 1975; Snyder, 1978, 1980). Although such attempts lack a firm theoretical foundation, they do nevertheless demonstrate the great importance of specific interactions.

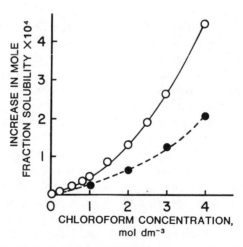

Figure 4.1 Increase in mole-fraction solubility of anthraquinone versus molarity of chloroform (○) in isooctane at 25°C. The solid line was calculated from the equilibrium constants in Table 4.10. The dashed line represents predicted solubilities using solubility parameters described from solubility data in the pure solvents. Reproduced with permission of the copyright owner, The American Pharmaceutical Association, from Anderson, B. D., Rytting, J. H., and Higuchi, T. (1980). *J. Pharm. Sci.*, **69**, 676–680.

4.3   IMPORTANCE OF SPECIFIC INTERACTIONS WITH SOLVENTS

4.3.1   Theoretical Considerations

The early criticism of the importance of specific (or chemical) interactions was largely based on the difficulty of proving the existence of new chemical species from thermodynamic and solubility data alone (Hildebrand and Scott, 1950). Abundant spectroscopic data now prove the existence of new stoichiometric species formed between the solute and the solvent or cosolvent. For example, IR frequency shifts (Chamberlain and Drago, 1976), proton NMR studies (Wiley and Miller, 1972; Martire et al., 1976), and ^{13}C-NMR data (Nakashima et al., 1974) support the existence of molecular complexes between various hydrogen donors such as phenols (Gramstad and Mundheim, 1972) and chloroform (Creswell and Allred, 1962), and hydrogen acceptors, including such weak acceptors as benzene (Baker and Wilson, 1970).

In view of the wealth of spectroscopic evidence, including that mentioned above, it would be unreasonable to ignore the important contribution made by hydrogen-bonding interactions between hydrogen-donor solutes (e.g., phenols, amines, and carboxylic acids) and acceptor solvents or cosolvents (e.g., ethers, ketones, and esters), and between hydrogen-acceptor solutes (e.g., acid anhydrides and ketones) and hydrogen-donor solvents or cosolvents (e.g., chloroform).

The solubilities of the polar solutes studied by Fung and Higuchi (1971) were found to be much higher in chloroform (a proton-donor solvent) and ether (a proton-acceptor and electron donor) than in paraffinic hydrocarbon solvents. These increases are too large to be explained by the molar volume (Table 4.5) or by the solubility parameter (Table 4.6) of the solvent. The measured solubility is very dependent on the chemical structure of both the solute and solvent. Furthermore, phthalic anhydride solubility (Table 4.1) does not correlate simply with the physical properties of the solvent. For example, the solubility in pyridine is higher than in acetone, although acetone has the higher dielectric constant and dipole moment. The solubility is at least five times higher in chloroform than in ether, despite the similar dielectric constants and dipole moments of these solvents.

Information about specific solvation in slightly polar organic solvents can be obtained from solubility data in mixed solvents consisting of an interactive component, such as chloroform (a proton-donor solvent) or ether (a proton acceptor or electron-pair donor), and an inert solvent such as a saturated hydrocarbon solvent (e.g., cyclohexane). These solvent mixtures have comparatively small excess free energies of mixing, and large increases in solubility on addition of this interactive component would be a result of specific solvation.

It is assumed that the following equilibria operate between the solute S and the complexing agent, ligand, or interacting cosolvent L:

$$S + L \rightleftharpoons SL \overset{+L}{\rightleftharpoons} SL_2 \overset{+L}{\rightleftharpoons} \cdots SL_{(n-1)} \overset{+L}{\rightleftharpoons} SL_n \qquad (4.4)$$

SL, SL_2, and SL_n are stoichiometric complexes, the formation of each of which is defined by an equilibrium constant, thus,

$$K_{1:1} = \frac{[SL]}{[S]_o [L]_f} \qquad (4.5)$$

$$K_{1:n} = \frac{[SL_n]}{[SL_{n-1}][L]_f} \qquad (4.6)$$

Since the thermodynamic activities are replaced by molar concentrations, the activity coefficients are all assumed to be unity and the standard state of each dissolved substance is a solution of unit molarity which behaves as an ideal dilute solution. Hence, $[S]_o$ is the solubility of the solute in the pure, noninteracting solvent, which is assumed to represent the free solute concentration in the cosolvent mixtures and $[L]_f$ is the concentration of free (uncomplexed) ligand. Although a single solute molecule is assumed to be present in each complex, the mathematical form of the resulting equation remains unaffected by allowing more than one solute molecule in each molecular complex. The total solubility of solute $[S]_t$ in any system is then given by

$$[S]_t = [S]_o + [SL] + [SL_2] + \cdots \qquad (4.7)$$

and the total concentration of complexing agent $[L]_t$ added to the system is expressed as

$$[L]_t = [L]_f + [SL] + 2[SL_2] + \cdots \qquad (4.8)$$

Introducing the expressions for $[SL]$, $[SL_2]$, and so on from Eqs. 4.5 and 4.6, Eq. 4.7 becomes

$$[S]_t = [S]_o + K_{1:1}[S]_o[L]_f + K_{1:1}K_{1:2}[S]_o[L]_f^2 + \cdots \qquad (4.9)$$

and Eq. 4.8 becomes

$$[L]_t = [L]_f + K_{1:1}[S]_o[L]_f + 2K_{1:1}K_{1:2}[S]_o[L]_f^2 + \cdots \qquad (4.10)$$

In the absence of solute, the total concentration of interactive cosolvent $[L]_t$ is equal to $[L]_f$, provided that the extent of self-association of cosolvent is negligible. This assumption does not hold for cosolvents such as alcohols, which self-associate as discussed in Chapter 6.

4.3.2 Experimental Studies with Cosolvent

Equation 4.9 indicates that the plots of solubility versus concentration of cosolvent are linear only if 1:1 complexation alone is significant, as in Figures 4.2–

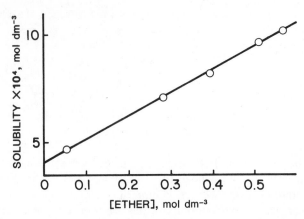

Figure 4.2 Molar solubility of picric, acid versus molar concentration of diethyl ether in ether–cyclohexane solvent mixtures at 25°C, demonstrating the linearity of the plot of solubility versus the concentration of cosolvent, in contrast to that in Figure 4.1. $1\,mol\,dm^{-3} = 1\,mol\,L^{-1} = 1\,M$. Reproduced with permission of the copyright owner, the American Pharmaceutical Association, from Fung, H. L. and Higuchi, T. (1971). *J. Pharm. Sci.*, **60**, 1782–1788.

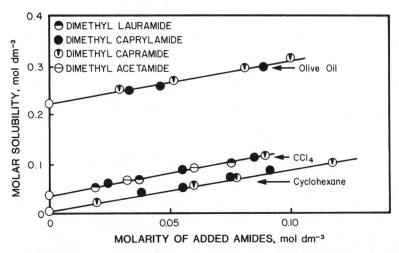

Figure 4.3 The apparent molar solubility of salicylic acid in the presence of various molarities of several disubstituted monoamides in carbon tetrachloride, cyclohexane, and olive oil at 30°C, demonstrating linear plots whose slopes are independent of the noninteractive solvent and of the hydrocarbon chain length of the amide cosolvent or ligand, but whose incercept reflects the solubility of the solute in the former solvent. $1\,mol\,dm^{-3} = 1\,mol\,L^{-1} = 1\,M$. Reproduced with permission of the copyright owner, American Association of Colleges of Pharmacy, from Higuchi, T., Hydrogen Bonded Complexes in Non-Polar Solutions—Influence of Structure and Solvent on Formation Tendency and Stoichiometry. In *Proceedings of the American Association of Colleges of Pharmacy Teachers' Seminar*, Vol. 13, Lemberger, A. P. (Pharmacy Ed.), July 9–15, 1961, School of Pharmacy, University of Wisconsin, Madison, WI.

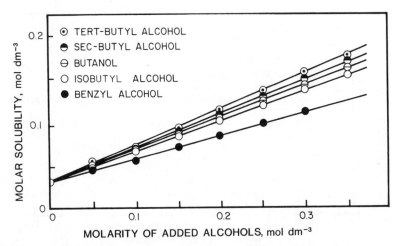

Figure 4.4 Shows the linear molar solubility plots of salicylic acid with various molar concentrations of added alcohols in carbon tetrachloride at 30°C. $1 \, mol \, dm^{-3} = 1 \, mol \, L^{-1} = 1 \, M$. Reproduced with permission of the copyright owner, American Association of Colleges of Pharmacy, from Higuchi, T., Hydrogen Bonded Complexes in Non-Polar Solutions—Influence of Structure and Solvent on Formation Tendency and Stoichiometry. In *Proceedings of the American Association of Colleges of Pharmacy Teachers' Seminar*, Vol. 13, Lemberger, A. P. (Pharmacy Ed.), July 9–15, 1961, School of Pharmacy, University of Wisconsin, Madison, WI.

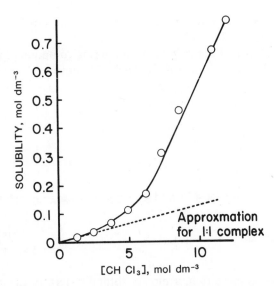

Figure 4.5 Molar solubility of phthalic anhydride versus molar concentration of chloroform in chloroform–cyclohexane solvent mixtures at 25°C, demonstrating the formation of complexes of higher order than 1:1 with respect to chloroform. $1 \, mol \, dm^{-3} = 1 \, mol \, L^{-1} = 1 \, M$. Reproduced with permission of the copyright owner, the American Pharmaceutical Association, from Fung, H. L. and Higuchi, T. (1971). *J. Pharm. Sci.*, **60**, 1782–1788.

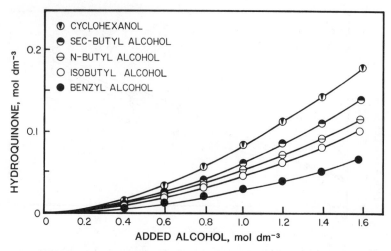

Figure 4.6 Nonlinear molar solubility plots of hydroquinone (i.e., quinol) with various molar concentrations of added alcohols in carbon tetrachloride at 30°C. $1 \, mol \, dm^{-3} = 1 \, mol \, L = 1 \, M$. Reproduced with permission of the copyright owner, American Association of Colleges of Pharmacy, from Higuchi, T., Hydrogen Bonded Complexes in Non-Polar Solutions—Influence of Structure and Solvent on Formation Tendency and Stoichiometry. In *Proceedings of the American Association of Colleges of Pharmacy Teacher's Seminar*, Vol. 13, Lemberger, A. P. (Pharmacy Ed.), July 9–15, 1961, School of Pharmacy, University of Wisconsin, Madison, WI.

4.4., and are concave upward if higher-order complexation is present, as in Figures 4.1, 4.5, and 4.6.

If only 1:1 complexes are important, the terms containing $[L]_f^2$ and higher powers in Eqs. 4.9 and 4.10 can be ignored. On subtracting

$$[S]_t - [S]_o = [L]_t - [L]_f \tag{4.11}$$

and, according to Eq. 4.10,

$$[L]_f = [L]_t/(1 + K_{1:1}[S]_o) \tag{4.12}$$

Eliminating $[L]_f$ from Eqs. 4.11 and 4.12 and rearranging yields

$$\text{increase in solubility} = [S]_t - [S]_o = [L]_t \frac{K_{1:1}[S]_0}{K_{1:1}[S]_0 + 1} \tag{4.13}$$

Thus, for 1:1 complexation, a plot of solubility $[S]_t$ in the presence of the interacting solvent, against the total concentration $[L]_t$ of the interacting solvent, gives a straight line. From the slope, $K_{1:1}$ can readily be calculated. If the increase in solubility, $[S]_t - [S]_o$, due to the presence of the interacting solvent is plotted instead of $[S]_t$, the straight line passes through the origin. These principles are

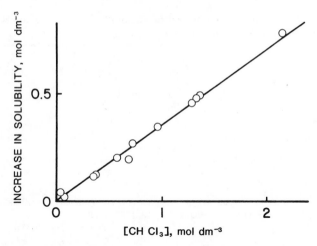

Figure 4.7 Increase in molar solubility of phenyl benzoate versus molar concentration of chloroform in chloroform–cyclohexane solvent mixtures at 25°C (solubility in cyclohexane = 0.393 mol dm^{-3}). 1 mol dm^{-3} = 1 mol L^{-1} = 1 M. Reproduced with permission of the copyright owner, the American Pharmaceutical Association, from Fung, H. L. and Higuchi, T. (1971). *J. Pharm. Sci.*, **60**, 1782–1788.

exemplified by Figure 4.7. Specimen calculations exemplifying Eq. 4.13 are presented in Section 5.9.

At low concentrations of L, only 1:1 complexes are usually significant, and therefore Eq. 4.13 applies in these situations. If $[S]_o$ is very low, the fraction of complexing agent in the complexes is small and thus $[L]_t \cong [L]_f$. Consequently, Eq. 4.12 indicates that $1 + K_{1:1}[S]_o \cong 1$. Equation 4.13 can then be approximated by

$$\text{fractional increase in solubility} = \frac{[S]_t - [S]_o}{[S]_o}$$

$$\cong K_{1:1}[L]_t \qquad (4.14)$$

The equation may be applied to the low-concentration region of Figures 4.1, 4.5, and 4.6.

Chloroform can act as a hydrogen-bond donor (Gutowsky and Saika, 1953), and its ability to increase solubility has been cited as evidence for chloroform-base association (for references see Fung and Higuchi, 1971 and Anderson et al., 1980). Stability constants of complexes are usually determined by PMR studies. Some 1:1 stability constants with chloroform have been obtained for several organophosphorus compounds and long-chain tertiary amines and for piperidine, benzene, aromatic hydrocarbons, acetonitrile and dimethyl sulfoxide, and amines (for references see Fung and Higuchi, 1971). Chloroform and base concentrations were sufficiently low to indicate 1:1 complexation.

Table 4.8 Solvate Equilibrium Constants of Various Solutes with Chloroform at 25°C Obtained from the Solubility Method Using Eq. 4.14[a]

Solute	$K_{1:1}$ (dm^3 mol^{-1})	K_2[b] (dm^3 mol^{-1})^2	K_3[c] (dm^3 mol^{-1})^3
Phenyl benzoate	0.89		
Benzoquinone	1.06		
Benzophenone	1.01		
Phthalic anhydride	0.61	0.214	0.043

[a] $1\, dm^3\, mol^{-1} = 1\, L\, mol^{-1} = 1\, M^{-1}$.

[b] $K_2 = K_{1:1}K_{1:2}$.

[c] $K_3 = K_{1:1}K_{1:2}K_{1:3}$.

Reproduced with permission of the copyright owner, the American Pharmaceutical Association, from Fung, H. L. and Higuchi, T. (1971). *J. Pharm. Sci.*, **60**, 1782–1788.

Fung and Higuchi (1971) attempted to explain the solubility in chloroform in terms of stoichiometric complexes by measuring the solubility of the solute in solvent mixtures consisting of various molar concentrations of chloroform in cyclohexane. They plotted the increase in solubility due to the chloroform against the molar concentration of chloroform as shown in Figure 4.7. The presence of a small amount of the interactive solvent increases the solubility far beyond that which could be attributed to athermal, that is, entropic corrections or regular solution theory. The plots for phenyl benzoate, *p*-benzoquinone, and benzophenone were linear, indicating 1:1 complexes. From the slopes, the stability constants $K_{1:1}$ were calculated from a linear plot expressed by Eq. 4.14 and are tabulated in Table 4.8. The solubility of phthalic anhydride gives a curve of increasing slope (Fig. 4.5), which indicates the formation of higher-order complexes (Eqs. 4.4, 4.6–4.10). The complexes formed between chloroform and the solutes presumably have the following hydrogen-bonded structures:

The 1:1 interaction constants shown in Table 4.8 are all about 1 dm^3 mol^{-1} and are similar to, or greater than, those for amines (slightly less than 1 dm^3 mol^{-1}). The interaction constant between chloroform and ether is 1.52 dm^3 mol^{-1} from PMR. Thus, in a nonpolar solvent such as cyclohexane, amines have a basic strength similar to that of ketones, esters, and anhydrides,

Table 4.9 Solvate Equilibrium Constants of Various Solutes with Ether at 25°C Obtained from the Solubility Method Using Eq. 4.14[a]

Solute	$K_{1:1}$ $(dm^3 mol^{-1})$	K_2[b] $(dm^3 mol^{-1})^2$
Carbazole	3.53	
Picric acid	2.73	
Phthalic anhydride	0.606	0.10
Succinic anhydride	0	0.392

[a] $1\,dm^3\,mol^{-1} = 1\,L\,mol^{-1} = 1\,M^{-1}$.
[b] $K_2 = K_{1:1}K_{1:2}$.

Reproduced with permission of the copyright owner, the American Pharmaceutical Association, from Fung, H. L. and Higuchi, T. (1971). *J. Pharm. Sci.,* **60**, 1782–1788.

whereas ether is a stronger base toward chloroform. These results substantiate established findings that pK_a values in water are not satisfactory measures of acidic or basic character in nonpolar solvents (Gurka and Taft, 1969).

The solubility of picric acid (Fig. 4.2) and carbazole in ether–cyclohexane mixtures shows linear plots against ether concentration from which the 1:1 interaction constants shown in Table 4.9 were calculated. At higher ether concentrations, positive deviations occurred which may be attributed to the formation of higher-order complexes. The interaction mechanisms between ether and picric acid and between ether and carbazole seem to involve hydrogen bonding, presumably through structures I and II, respectively, whereas strong charge-transfer interactions are suspected to occur between the carbonyl groups of phthalic or succinic anhydride and the ether-oxygen atoms. In cyclohexane, carbazole is more strongly bound to ether than is picric acid, whereas the latter is the stronger acid in aqueous solution. Reversal of acidity scales in nonaqueous solution presumably arises from the differential influences of the surrounding environment on the other parts of the molecule.

If both 1:1 and 1:2 complexes are formed, the stability constant of each complex can be calculated from the following version of Eq. 4.9:

$$\frac{[S]_t - [S]_o}{[L]_f} = K_{1:1}[S]_o + K_{1:1}K_{1:2}[S]_o[L]_f \tag{4.9a}$$

Provided that the cosolvent L predominates in the free form of concentration $[L]_f$, so that only a relatively small proportion exists in the two complexes, that is, $[L]_f = [L]_t$, then a plot of $([S]_t - [S]_o)/[L]_t$ against $[L]_t$ should yield a straight line. Since $[S]_o$ is the known solubility of S in the absence of the cosolvent, $K_{1:1}$ can be calculated from the intercept, while $K_{1:2}$ may be calculated from the slope of the plot. By means of this treatment, interactions between ether and phthalic anhydride and between ether and succinic anhydride were studied over the entire range of mole fractions. The increase in solubility/[ether] against [ether] gave linear plots as shown in Figure 4.8, which indicates that 1:2 as well as 1:1 complexes are formed. From the slopes, the stability constants $K_{1:1}$ and $K_{1:2}$ were calculated as described above and are tabulated in Table 4.9.

The stability constants of the higher-order complexes are designated K_2 and K_3 in Tables 4.8 and 4.9 and are defined by the following equilibrium:

$$S + nL \rightleftharpoons SL_n \tag{4.15}$$

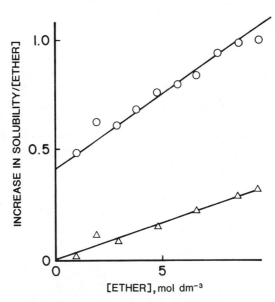

Figure 4.8 A plot of the increase in molar solubility per unit molarity of added cosolvent (ether). This plot facilitates the determination of solvate equilibrium constants between ether and phthalic anhydride (○) and ether and succinic anhydride (△) from solubility data of the anhydrides in ether–cyclohexane solvent mixtures at 25°C. Reproduced with permission of the copyright owner, the American Pharmaceutical Association, from Fung, H. L. and Higuchi, T. (1971). *J. Pharm. Sci.*, **60**, 1782–1788.

so that

$$K_n = \frac{[SL_n]}{[S]_o[L]_f^n}, \qquad \text{e.g.,} \qquad K_2 = \frac{[SL_2]}{[S]_o[L]_f^2} \qquad (4.16)$$

Comparison of Eqs. 4.16 and 4.6 shows that

$$K_2 = K_{1:1}K_{1:2} \quad \text{and} \quad K_3 = K_{1:1}K_{1:2}K_{1:3} \qquad (4.17)$$

We note that $K_{1:1}, K_{1:2}, K_{1:3} \cdots K_{1,n}$ have units $dm^3\,mol^{-1}$ (or M^{-1}), whereas K_2 has units $dm^6\,mol^{-2}$ (or M^{-2}) and K_3 has units $dm^9\,mol^{-3}$ (or M^{-3}). These new definitions of stability constants are required since the interaction between succinic anhydride and ether appears not to involve the formation of 1:1 complexes, so $K_{1:1}$ and $K_{1:2}$ as defined by Eqs. 4.5 and 4.6 have no meaning. However, K_2 as defined by Eq. 4.16 appears to be significant, as shown in Table 4.9. Although charge-transfer interactions may occur between the carbonyl groups of phthalic or succinic anhydrides and the ether oxygen atoms, the low values of $K_{1:1}$ and K_2 for these systems may also be explained in terms of nonspecific effects.

Higuchi and Connors (1965) have stated that a complex may be said to be formed when, in the immediate vicinity of the solute S, significantly more molecules of ligand L or cosolvent are found than would be predicted on a purely statistical basis. Anderson (1977) has calculated an apparent $K_{1:1}$ value from this rule. To illustrate this principle, let us consider a typical solute S, which has molar solubility of $10^{-3}\,mol\,dm^{-3}$ in a hydrocarbon solvent of molar volume $100\,cm^3\,mol^{-1}$. In a solution containing $0.1\,mol\,dm^{-3}$ of an interactive cosolvent of molar volume $100\,cm^3\,mol^{-1}$ (and therefore of mole fraction 0.01), one in every 100 solvent molecules is a cosolvent molecule. If a particular solute molecule is assumed to have 10 nearest neighbors typical of the liquid state, the probability that one neighbor is a cosolvent molecule is about 0.1. Therefore, roughly 10% of the solute molecules would have the cosolvent as a nearest neighbor, and the maximum increase in solubility expected would be 0.1 $\times 10^{-3}\,mol\,dm^{-3} = 10^{-4}\,mol\,dm^{-3}$. This would correspond to an apparent $K_{1:1}$ of the following value:

$$K_{1:1} = \frac{10^{-4}}{10^{-3} \times 0.1} = 1\,dm^3\,mol^{-1}$$

An apparent $K_{1:1}$ as much as $1\,dm^3\,mol^{-1}$ could be rationalized on a purely statistical basis and may not therefore reflect complex formation. Thus, when interpreting situations which give rise to low apparent values of $K_{1:1}$, it is advisable to bear in mind this statistical principle.

Anderson et al. (1980) investigated the solubility of p-nitrophenol, p-iodophenol, p-phenylphenol, 2, 4, 6-triiodophenol, carbazole, and anthraquinone in mixtures of isooctane with each of the cosolvents—dibutyl ether, dipentyl ether, butyl butyrate, and chloroform. These workers noted that graphical methods are quite cumbersome when an appreciable amount of the added

Table 4.10 Solvate Equilibrium Constants of Various Solutes with Polar Solvents in Isooctane at 25°C[a]

	Butyl Ether			Pentyl Ether			Butyl Butyrate			Chloroform			
Solute	$K_{1:1}$ (dm³ mol⁻¹)	$K_{1:2}$ (dm³ mol⁻¹)	$\sigma\%$[b]	$K_{1:1}$ (dm³ mol⁻¹)	$K_{1:2}$ (dm³ mol⁻¹)	$\sigma\%$	$K_{1:1}$ (dm³ mol⁻¹)	$K_{1:2}$ (dm³ mol⁻¹)	$\sigma\%$	$K_{1:1}$ (dm³ mol⁻¹)	$K_{1:2}$ (dm³ mol⁻¹)	$K_{1:3}$ (dm³ mol⁻¹)	$\sigma\%$
p-Nitrophenol	102.5	0.56	2.2[c]										
p-Iodophenol	26.0	—	0.5[d]										
p-Phenylphenol	8.9	0.42	2.7	8.3	0.27	2.1	21.2	0.78	3.3	0.73	0.25	0.15	2.8
Carbazole	2.5	0.25	1.1	1.9	0.23	1.9	6.5	0.44	3.3	0.68	0.30	0.06	4.3
2,4,6-Triiodophenol	1.1	0.33	1.4										
Anthraquinone										0.96	0.39	0.15	2.5

[a] 1 dm⁻³ mol⁻¹ = 1 L mol⁻¹ = 1 M^{-1}.
[b] The $\sigma\%$ value is the square root of the sum of the squares of the percent deviation between the calculated and observed solubilities divided by degrees of freedom.
[c] Fit of data below 0.1 mol dm⁻³
[d] Fit of data in 1:1 region only.

Reproduced with permission of the copyright owner, the American Pharmaceutical Association, from Anderson, B. D., Rytting, J. H., and Higuchi, T. (1980). J. Pharm. Sci., 676–680.

cosolvent exists in the complexed form, because $[L]_t$ then does not equal the free-ligand concentration. They therefore wrote a computer program using the simplex method of least squares (Deming and Morgan, 1973; Morgan and Deming, 1974) to solve simultaneously Eqs. 4.9 and 4.10 (which relate $[S]_t$ and $[L]_t$ to the concentrations of the various complexes) and to optimize the variables $K_{1:1}$, $K_{1:2}$, and $K_{1:3}$. The sum of squares of the percent deviation of the experimental solubilities was minimized to give equal weight to each point. The complexation constants, for those instances where specific interactions can reasonably be assumed, were calculated. These constants are listed in Table 4.10. Constants determined graphically, as in Figures 4.2–4.4, 4.7, and 4.8, were usually within 10% of the computer-optimized values.

The $K_{1:2}$ values in Table 4.10 are small ($< 1 \, dm^3 \, mol^{-1}$) and are relatively insensitive to the molecular structure of the solute or cosolvent and therefore cannot justifiably be termed specific interaction constants. The fact that each of the $K_{1:2}$ values is appreciably smaller than the corresponding $K_{1:1}$ value suggests that the solute–cosolvent interactions are effectively saturable at the 1:1 stoichiometry, a characteristic of specific interactions.

4.3.3   Free Energies and Stability Constants

If the $K_{1:1}$ values obtained in dilute solutions of the interactive cosolvents represent specific interactions, the contribution ΔG_{HB}^{θ} of this specific interaction to the overall standard free energy of transfer ΔG^{θ} from isooctane to pure cosolvent is given by

$$\Delta G_{HB}^{\theta} = - RT \ln (K_{1:1}/V_m) \qquad (4.18)$$

where V_m is the molar volume of the pure cosolvent and is expressed in the same units as $K_{1:1}$ (i.e., $L \, mol^{-1}$ in Table 4.10), so that $1/V_m$ represents $[L]_f$ in Eq. 4.6. Then $K_{1:1}/V_m$ is a dimensionless quantity, since ΔG^{θ} is expressed in terms of a dimensionless ratio in Eq. 3.55. This comparison, which is shown in Table 4.11, ignores the possibility that complexes larger than 1:1 may exist in the pure interactive cosolvent. For example, 1:2 complexes are expected to be important in solutions of anthraquinone with chloroform, since anthraquinone has two hydrogen-acceptor sites. Therefore, the contributions of specific interactions to solubility expressed in Table 4.11 are probably minimum estimates. However, nonspecific contributions to the $K_{1:1}$ values in dilute solution may also be significant, particularly when the $K_{1:1}$ values are small.

In most cases, as shown in Table 4.11, specific interactions make the major contribution to the overall transfer free energy. The differences are generally relatively small and may arise from higher-order molecular complexes as well as from nonspecific effects. The results show that the tendency toward specific interactions should receive the greatest emphasis in attempts to predict solubility of polar solutes in the presence of polar solvents.

We have seen that stability constants of higher-order complexes may be defined in two ways, namely, according to Eq. 4.6 or according to Eq. 4.16. (The

Table 4.11   Standard Free Energies of Transfer of Solute $(\Delta G^{\theta})^a$ from Isooctane to Various Interactive Solvents and the Calculated Hydrogen-Bonding Contribution to the Free Energy from $K_{1:1}$ Values $(\Delta G_{HB}^{\theta})^a$

Solute	*Butyl Ether*		*Pentyl Ether*		*Butyl Butyrate*		*Chloroform*	
	ΔG^{θ}	Δ_{HB}^{θ}	ΔG^{θ}	ΔG_{HB}^{θ}	ΔG^{θ}	ΔG_{HB}^{θ}	ΔG^{θ}	ΔG_{HB}^{θ}
p-Nitrophenol	−4.9	−3.6						
p-Iodophenol	−3.2	−2.5						
p-Phenylphenol	−3.0	−2.3	−2.7	−2.2	−3.7	−2.8	−2.6	−1.4
Carbazole	−2.1	−1.6	−1.9	−1.4	−2.9	−2.2	−2.4	−1.3
2,4,6-Triiodophenol	−1.8	−1.2						
Anthraquinone							−3.0	−1.5

^{a}Units are kcal mol^{-1}. 1 cal = 4.184 J.

Reproduced with permission of the copyright owner, the American Pharmaceutical Association, from Anderson, B. D., Rytting, J. H., and Higuchi, T. (1980). *J. Pharm. Soc.*, **69**, 676–680.

stability constants of 1:1 complexes, $K_{1:1}$, which are defined by Eq. 4.5, are the same according to either criteria.) Both of these types of stability constant are based on the assumption that the activity coefficients of all the species are unity. As stated earlier, the standard state of each species, with the exception of the noninteracting solvent, is a hypothetical 1 mol dm^{-3} solution which is assumed to behave as an ideal dilute solution. It is, of course, theoretically preferable to define equilibrium constants in terms of activities. However, since the values of the activity coefficients are frequently unknown, we have to be content with apparent or practical equilibrium constants based on a convenient concentration scale, usually the molar scale. The true thermodynamic stability constant of a 1:1 complex is defined by

$$K_{T,1:1} = \frac{\{SL\}}{\{S\}_o\{L\}_f} \tag{4.19}$$

where $\{SL\}$, $\{S\}_o$, and $\{L\}_f$ are the activities of the dissolved complex, the dissolved uncomplexed solute, and the dissolved uncomplexed ligand, respectively. Ignoring the activity coefficients has small effects on $K_{1:1}$ at low concentrations and solubilities ($< 0.1\,M$) in a given system of low polarity. Intermolecular interactions at higher concentrations and/or solubilities ($> 1\,M$) and/or for highly polar interactions may cause the apparent $K_{1:1}$ or other stability constant of a given complexation system to vary by a factor of 2 or more. Solvent effects will be discussed in Section 5.7, while concentration effects will be considered in Section 5.8.

4.4 INFLUENCE OF MOLECULAR STRUCTURE ON COSOLVENCY AND COMPLEXATION

4.4.1 1:1 Complexation

When comparing the complexation of solutes which have different solubilities $[S]_o$ in the pure, noninteracting solvent, the standard state of unit activity of the uncomplexed solute is best defined as the pure undissolved solute. Since the solution is saturated with the solute in the solubility experiments being considered in this chapter, the dissolved uncomplexed solute is in equilibrium with the pure undissolved solute, which is usually a liquid or a solid. Consequently, the activity of the solute, $\{S\}_o$, is equal to unity, as in the pure state. On the other hand, the activities of the other dissolved species, L_f, SL, and higher-order complexes are replaced by molar concentrations as before. Equations 4.19 and 4.5 then become

$$K_{DA} = \frac{[SL]}{[L]_f} = \frac{[DA]}{[A]_f} \qquad (4.20)$$

To indicate the change of standard state of the uncomplexed solute, the notations of the species and the stability constant are also changed. S is replaced by D, L by A, SL by DA, $K_{1:1}$ by K_{DA}, K_2 by K_{DA_2}, and so on. Similarly, Eq. 4.16 becomes

$$K_{DA_n} = \frac{[DA_n]}{[A]_f^n}, \quad \text{for example,} \quad K_{DA_2} = \frac{[DA_2]}{[A]_f^2} \qquad (4.21)$$

Frequently, the dissolved solute D is the hydrogen-bond donor and the ligand or cosolvent A is the hydrogen-bond acceptor.

Comparison of Eq. 4.15 with Eq. 4.20, while bearing in mind that $S \equiv D$ and $L \equiv A$, shows that

$$K_{1:1} = K_{DA}/[D]_o \qquad (4.22)$$

where $[D]_o \equiv [S]_o$ is the solubility of the solute in the pure noninteracting solvent (i.e., in the absence of cosolvent). Similarly, comparison of Eq. 4.15 with Eq. 4.20 indicates that

$$K_n = K_{DA_n}/[D]_o, \quad \text{for example,} \quad K_2 = K_{DA_2}/[D]_o \qquad (4.23)$$

Simple calculations exemplifying Eqs. 4.4–4.23 are given in Section 5.9.

By calculating K_{DA} and K_{DA_2} values, Higuchi and Chulkaratana (1961) analyzed the relationships between functional group and solubility. The solubility behavior of monofunctional compounds could be rationalized in terms of

1:1 complexes and that of bifunctional compounds in terms of 1:2 complexes, and will now be discussed.

Fully substituted amides, such as N, N-dimethylacetamide and its homologues, can act as hydrogen-bond acceptors or bases in the presence of proton donors such as salicylic acid. Accordingly, one would expect these amides to form hydrogen-bonded complexes with salicylic acid in organic solvents of low polarity, such as cyclohexane, carbon tetrachloride, or even olive oil, thereby increasing the solubility of salicylic acid. Figure 4.3 shows that this is the case. The linear relationships indicate the formation of 1:1 complexes, for which the following structure may be postulated:

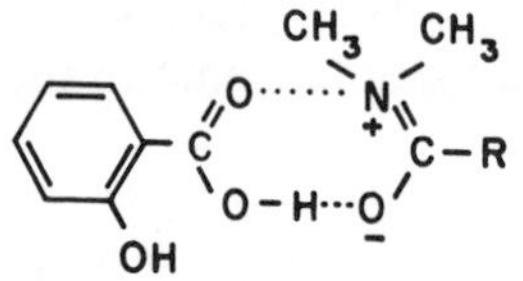

Since the slopes of the lines are identical, the formation constants of the complexes, K_{DA}, are also identical and independent of both (a) the length of the hydrocarbon chain R in the amide and (b) the nature of the low-polarity solvent. Although each pure solvent permits different solubilities of salicylic acid, the absolute increase in solubility brought about by a given concentration of tertiary amide is constant and unaffected by the nature of the solvent or the hydrocarbon chain length of the amide. The lack of dependence of K on the intrinsic solubility of the solute in the pure solvent occurs because the activity of the solute in all the saturated solutions is a constant. This constant is equal to the activity of the pure solid with which the saturated solutions are in equilibrium. For weak complexes the value of K, which would be small, depends greatly on the solvent (Section 5.7) and on the detailed structure of the complexing agent (or ligand).

The intrinsic solubility of the polar solute salicylic acid in the pure inert solvent cyclohexane is very small, as expected, but is greater in the more polarizable solvent carbon tetrachloride, owing to Debye interactions with the solute (Sections 2.11.4 and 2.11.5, Table 3.13, Section 3.11). The solubility is much larger in olive oil, which consists mainly of glyceryl trioleate, since the presence of polarizable double bonds and polar proton-accepting ester groups gives rise to Debye forces, and especially hydrogen-bonding between the ester and solute. Further discussion of the interactions involving amides is given in Section 5.5.6.

The solubility of salicylic acid in carbon tetrachloride is increased by the presence of alcohols owing to 1:1 complexation, as shown in Figure 4.4. The stability constant K_{DA} is virtually the same for the normal straight-chain alcohols above methanol, as shown in Table 4.12. For the other alcohols, K_{DA} follows the order expected on the basis of their proton-accepting abilities (which is the reverse of their proton-donating abilities), that is, phenol < benzyl alcohol < methanol < n-alkanols < cyclohexanol (sterically favored), and for the butanols, iso < normal < secondary < tertiary. The substituent effect is discussed

Table 4.12 Formation Constants $K_{DA} = [DA]/[A]$ for Salicylic Acid with Alcohols in Carbon Tetrachloride at 30°C

Methanol	0.584	sec-Butyl alcohol	0.718
Ethanol	0.685	tert-Butyl alcohol	0.836
Butanol	0.679	Cyclohexanol	0.785
Octanol	0.684	Benzyl alcohol	0.404
Dodecanol	0.662	Phenol	0.009
Isobutyl alcohol	0.616		

Table 4.13 Formation Constants $K_{DA} = [DA]/[A]$ for Salicylic Acid Complexes with Aldehydes, Ketones, and Esters in Carbon Tetrachloride at 30°C

n-Butyraldehyde	0.287	n-Propyl propionate	0.134
n-Heptaldehyde	0.258	n-Butyl n-butyrate	0.139
3-Pentanone	0.186	Isobutyl isobutyrate	0.115
4-Heptanone	0.163	β-Propiolactone	0.096
Cyclohexanone	0.342	γ-Butyrolactone	0.298
Ethyl acetate	0.168		

further in connection with linear free-energy relationships in Section 5.5.1. Alcohols can also act as proton donors and may, therefore, act bifunctionally. Further discussion of interactions involving alcohols is given in Section 5.5.2.

Carbonyl goups in aldehydes, ketones, esters, and lactones can also act as proton acceptors (bases) and these compounds increase the solubility of salicylic acid in nonpolar solvents as expected. Some K_{DA} values determined for these compounds are given in Table 4.13. Although ketones are more basic than aldehydes, steric factors presumably bring about the observed reduction in K_{DA}. Cyclization in cyclohexanone eliminates the steric factors and so increases K_{DA}. Esters are weaker proton acceptors than ketones or ethers. Therefore, K_{DA} is generally smaller for esters. Since γ-butyrolactone has no steric hindrance, the K_{DA} is higher than for the other esters. β-Propionolactone, although also a cyclic ester, has a strained ring system which probably inhibits charge separation and reduces K_{DA}. The hydrogen-bonding interactions of esters and diesters, which also increase the apparent solubility of organic acids, are discussed further in Section 5.5.4. Ethers and diethers act in the same way, as discussed in Section 5.5.5.

This discussion has shown how specific solvation ascribable to hydrogen bonding leads to increased solubility of salicylic acid. In every case where the association constant has been determined by solubility measuremnt, the result has been consistent with expectations based on physical organic chemistry.

Other phenolic hydrogen bond donors behave in a way similar to salicylic acid. Tables 4.14 and 4.15 compare the K_{DA} values arising from the increase in solubility of various phenols brought about by the presence of hydrogen-bond donors containing oxygen. The more strongly acidic the phenol, the stronger the complex with a given hydrogen-bond acceptor (base) will be. The $-\log K_{DA}$ values for a given hydrogen-bond acceptor with various phenolic hydrogen-bond donors is found to be a linear function of $pK_a(= -\log K_a)$ of the phenols, as might be expected. Thus, the enhancement of solubility brought about by an interacting solvent follows a linear free-energy relationship. The application of Table 4.14 in the prediction of the solubility of salicylic acid in the presence of an ester is presented in Section 5.9.

Table 4.15 also shows that salicylic acid with acetic acid gives a strong complex with a high value of K_{DA}. Presumably, both acids can act mutually as donor and acceptor, giving rise to a very specific lock-and-key fit similar to the relationship between an enzyme and a substrate or between an antigen and an antibody. This concept could have great value in solvent selection. The nature of the salicylic acid and acetic acid interaction will be further discussed on p. 197.

Table 4.14 Formation Constant K_{DA} of Complexes of Some Phenols with Some Esters and Lactones in Carbon Tetrachloride at 30°C

Ester	Salicylic Acid	p-Phenylphenol	p-Nitrophenol
Ethyl acetate	0.168	0.174	0.415
n-Propyl propionate	0.134	0.165	0.420
n-Butyl butyrate	—	—	0.416
Isobutyl isobutyrate	0.115	0.153	—
Diethyl succinate	0.213(0.107)[a]	0.297(0.149)[a]	—
β-Propiolactone	0.096	0.080	—
γ-Butyrolactone	0.298	0.320	—

[a]Value in parentheses refers to K_{DA} for each of the two ester groups.

Table 4.15 Formation Constant K_{DA} of Complexes of Some Phenols with Some Ethers and Carboxylic Acids in Carbon Tetrachloride at 30°C

Cosolvent	p-Phenylphenol	Salicylic Acid	Pyrocatechol
Ethyl ether	0.148	0.261	0.449
n-Propyl ether	0.107	0.186	0.284
n-Butyl ether	0.104	0.189	0.282
1-Propyl ether	0.157	0.244	0.504
1,4-Dioxane	0.176	0.330	0.427
Acetic acid	0.106	12.0	0.095
Monochloroacetic acid	0.059	5.7	0.075

4.4.2 Bifunctional Compounds

Hydroquinone (quinol) and other bifunctional hydrogen donors might be expected to interact with two hydrogen acceptors (bases) to form a DA_2-type complex according to the equation

$$D + 2A \rightleftharpoons DA_2 \qquad (4.24)$$

If the tendency to form a 1:1 complex is negligible, as in the case of the interaction between succinic anhydride with ether (Fung and Higuchi, 1971), then only 1:2 complexation need be considered, for which K_{DA_2} is given by Eq. 4.21. If most of the cosolvent or base A is in the free uncomplexed form $[A]_t \cong [A]_f$, then

$$K_{DA_2} = \frac{[DA_2]}{[A]_f^2} = \frac{[D]_t - [D]_o}{[A]_t} \qquad (4.25)$$

where $[D]_o$ is the concentration of free uncomplexes solute (e.g., hydroquinone).

$$[D]_t = K_{DA_2}[A]_t^2 - [D]_o \qquad (4.26)$$

According to this scheme the solubility of hydroquinone in carbon tetrachloride at 30°C is, in fact, increased by the presence of each member of a series of alcohols as hydrogen acceptors. When $[D]_t$ is plotted against $[A]_t$, the term $[A]^2$ in Eq. 4.26 leads to a quadratic-type plot, as shown in Figure 4.6 for each alcohol. According to Eq. 4.26, these plots can be linearized as shown in Figure 4.9 by

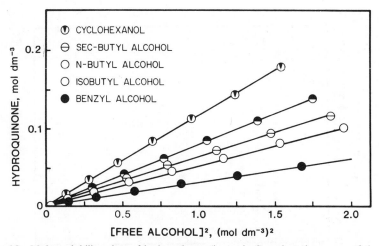

Figure 4.9 Molar solubility plots of hydroquinone (i.e., quinol) against the square of the molar concentration of added alcohols in carbon tetrachloride at 30°C. $1 \, mol \, dm^{-3} = 1 \, mol \, L^{-1} = 1 \, M$. Reproduced with permission of the copyright owner, American Association of Colleges of Pharmacy, from Higuchi, T., Hydrogen Bonded Complexes in Non-Polar Solutions—Influence of Structure and Solvent on Formation Tendency and Stoichiometry. In *Proceedings of the American Association of Colleges of Pharmacy Teacher's Seminar*, Vol. 13, Lemberger, A. P. (Pharmacy Ed.), July 9–15, 1961, School of Pharmacy, University of Wisconsin, Madison, WI.

plotting $[D]_t - [D]_o$ against $[A]^2$ (or by simply plotting $[D]_t$ against $[A]^2$, since $[D]_o$ is very small). This 1:2 stoichiometry emphasizes that the alcohols are undergoing specific interactions with hydroquinone. The slope of each plot yields K_{DA_2} directly. The various alcohols give K_{DA_2} values when interacting with hydroquinone which follow the same rank order as the K_{DA} values for interactions with salicylic acid. In fact, the logarithms of the K values are linearly related according to a linear free-energy relationship. Further discussion on this hydrogen-bonding interaction of hydroquinone is given in Section 5.5.6 (Fig. 5.16).

4.4.3   Indomethacin Solubility in Esters

The increases in solubility discussed above are brought about by the presence of an interactive cosolvent in an inert or noninteractive solvent. If the solubilizing effect is the result of specific solute–cosolvent interactions, the same complexation equilibrium should apply when the interactive and noninteractive species are present together in the same molecule in each member of an homologous series. The interactive species then represents a suitable functional group, such as an ester or amide group, and the noninteractive species represents the hydrocarbon or equivalent moiety. Figure 4.10 illustrates these considerations for the solubility of a given solute, namely, the drug indomethacin (Scheme 4.1), in the members of an homologous series of pure interactive solvents, namely, esters (Inagi et al., 1981). The solubility of the solute, $[D]_t$, is plotted against a suitable

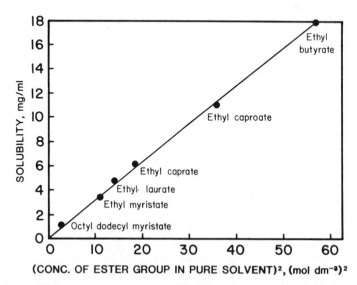

Figure 4.10   Solubility of indomethacin in several esters at 25°C plotted against the square of the molar concentration of the ester group in the pure solvent. $1\,\mathrm{mol\,dm^{-3}} = 1\ \mathrm{mol\,L^{-1}} = 1\,M$. Reproduced with permission of the copyright owner, the Pharmaceutical Society of Japan, from Inagi, T. (1981) *Chem. Pharm. Bull.* (*Tokyo*), **29**, 1708–1714.

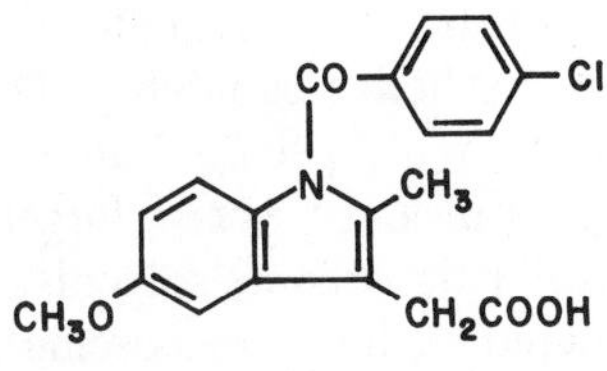

Scheme 4.1 The molecular structure of indomethacin.

function of the molar concentration of the functional group, $[A]_t$, in each pure liquid solvent. Thus, $[A]_t$ is given by

$$[A]_t = \frac{1000\rho n}{M_1} \tag{4.27}$$

where M_1 is the molecular weight of the solvent, n is the number of functional groups in each solvent molecule, and ρ is the density of the solvent in g cm^{-3}. The lower the ester in the homologous series, the smaller is M_1, so the larger is $[A]_t$ and the greater is the solubility. This type of calculation is possible because the esters do not self-associate to any significant extent.

To obtain the linear plot shown in Figure 4.10 for the solubility $[D]_t$ of indomethacin in various esters, $[D]_t$ is plotted against the *square* of $[A]_t$. Thus, the relevant equation turns out to be Eq. 4.26, representing a 1:2 complex, indomethacin :2 ester groups.

The slope of the plot gives the complexation constant K_{DA_2}. The line passes through the origin, indicating that $[D]_0 = 0$. Scheme 4.1 shows that indomethacin has only one obvious proton-donor group which is capable of hydrogen bonding to one ester group. Consequently, the polarizable aromatic ring probably interacts with the second ester group through a Debye interaction. Knowledge of the plot of solubility against a suitable function of $[A]_t$, representing the value of the complexation constant, makes possible the prediction of the solubility of the compound in any monofunctional aliphatic ester, either in the pure state or when mixed with a noninteractive cosolvent such as a hydrocarbon. It is, however, necessary to ascertain the value of $[D]_0$ in the latter solvents. Using the example of indomethacin, K_{DA_2} depends only on the solute and the interactive functional group and is the same in pure monofunctional aliphatic ester or in their mixtures with noninteractive solvents.

4.5 CONCLUSIONS

We have attempted to illustrate by means of simple examples how the solubility of polar substances is increased considerably by the presence of an interactive solvent which forms stoichiometric complexes with the dissolved solute. In the examples cited, a highly directional bond is formed in most instances between the solute molecules and the solvent. Regular solution and athermal treatments are demonstrably inadequate for rationalizing these observations. Moreover, they lack predictive capability. The authors, therefore, have elected to develop the concept of specific solute-solvent interaction of the hydrogen-bonding or charge-transfer type being largely responsible for solubility behavior of most organic solids in organic solvents.

The major evidence offered in favor of specific complexation is that the stoichiometry of the complexes accords with the number and nature of the interacting functional groups and that the complexation constants depend only on the electronic nature and the steric environment of functional groups and not on the length of the attached hydrocarbon chain.However, because the systems discussed in this chapter can fit complexation models does not prove conclusively that solubility changes are due totally to specific interactions. Nevertheless, as discussed in Chapter 5, values of the stability constants ($K_{1:1}$, $K_{1:2}$, K_{DA}, K_{DA_2}, etc.) obtained from solubility data often agree with those obtained by other methods and are consistent with the relative hydrogen-donating and hydrogen-accepting abilities of the respective molecules (Higuchi et al., 1969). Thus, the presence of specific interactions, although circumstantial, is probably the most satisfactory explanation for the solubility of polar solids in the presence of polar solvents. The next chapter explores in some depth the specific hydrogen-bonding interactions proposed above.

Chapter V

INFLUENCE OF COMPLEXATION, ESPECIALLY HYDROGEN BONDING, ON SOLUBILITY IN ORGANIC SOLVENTS: PREDICTION OF ASSOCIATION CONSTANTS

5.1 INTRODUCTION

In Chapter 4 evidence was provided to support the concept that most of the solubility differences between organic solids in various organic solvents can be ascribed to different extents of complex formation arising, for example, from hydrogen bonding between the solute and the solvent. In this chapter we shall discuss in detail the various methods used to study complexation between polar solutes and polar solvents and the possibility of predicting their association constants. Much of the data presented in this chapter are those obtained in the presence of an essentially nonpolar cosolvent to allow more effective analyses of these interactions. These specific complexations are primarily responsible for the solubility behavior of most organic solids of interest in organic solvents. However, as pointed out in Chapter 4, marked changes in solubility behavior can arise when complexation among organic solute molecules occurs in aqueous environments, a situation which will be discussed in Chapter 10.

Because of the importance of complexation to the subject matter of this book, we shall in this chapter review in detail the nature of these interactions and their quantitative measurements. The influence of stoichiometry and the effect of the magnitude of stability constants on the solubility relationships are described. The chapter also includes methods for determining the stoichiometry and stability constants and discusses the influence of chemical structure on the latter in terms of linear free-energy relationships. With this type of information, the solubility of a given solute of known structure in the presence of a given composition of solvent can then be usefully predicted in many situations.

The theoretical background and practical details have been reviewed by Kostenbauder and Higuchi (1956a), Higuchi et al. (1969), and Anderson et al. (1980). They are also exemplified by the work of Chulkaratana (1964) and Higuchi and Connors (1965), who studied hydrogen bond formation between salicylic acid, substituted phenols or hydroquinone (as proton donors), and ethers, ketones, esters, lactones, alcohols, or acids (as proton acceptors).

A nonpolar solvent, such as a paraffinic hydrocarbon or carbon tetrachloride, is maintained in a saturated condition with respect to the dissolving solid. This solid may be thought of as the proton donor D. Incremental amounts of an interactive cosolvent or ligand which may be thought of as the proton acceptor A are added to the noninteractive cosolvent to form a hydrogen-bonded complex or complexes. At equilibrium, the total amount of D in solution is determined spectrophotometrically. If the complex(es) formed are not highly soluble, a plot of solubility (i.e., $[D]_t$) versus the concentration of complexing agent present will increase until it reaches the saturation solubility of the complex itself, at which point the curve will level off.

This behavior is shown by the complex formed between p-hydroxybenzoic acid and ethyl theophylline (Bolton et al., 1957) in water (Figure 10.10). If a very soluble complex is formed, as in the case of the complex between oxytetracycline and nicotinamide (Higuchi and Bolton, 1959) (Figure 10.6), the curve will not plateau. The principles are essentially unchanged if D and A are interchanged. For this reason, the dissolving solid is sometimes called the substrate or solute S and the interactive cosolvent is called the ligand L (Chapter 4 and 10).

5.2   STABILITY CONSTANTS

5.2.1   Theoretical Background

If the noninteractive solvent contains the proton donor D as the only solute, D may exist as various associated species in equilibrium with each other. Thus,

$$D_1 \rightleftharpoons D_2 \rightleftharpoons D_3 \rightleftharpoons D_4 \cdots \rightleftharpoons D_n \qquad (5.1)$$

Each subscript refers to the stoichiometric aggregation number of D in the complex, D_1 being the monomer (self-association is discussed further in Chapter 6). The observed molar solubility $[D]_t$ of D in the solvent is given by

$$[D]_t = D_1 + 2[D]_2 + 3[D]_3 + \cdots + n[D]_n = \sum_{n=1} n[D]_n \qquad (5.2)$$

An example is shown by Scheme 5.1a.

If we introduce into this system a third component at a subsaturation level, namely, the cosolvent, the ligand, or the second solute A, which is capable of undergoing specific hydrogen-bonding interactions with D, we can write the solute species as

$$D_1 \rightleftharpoons D_2 \rightleftharpoons D_3 \rightleftharpoons \cdots \rightleftharpoons D_n$$
$$A_1 \rightleftharpoons A_2 \rightleftharpoons A_3 \rightleftharpoons \cdots \rightleftharpoons A_n$$
$$DA \rightleftharpoons DA_2 \rightleftharpoons DA_3 \rightleftharpoons \cdots \rightleftharpoons DA_n$$
$$D_2A \rightleftharpoons D_2A_2 \rightleftharpoons D_2A_3 \rightleftharpoons \cdots \rightleftharpoons D_2A_n$$
$$D_3A \rightleftharpoons D_3A_2 \rightleftharpoons D_3A_3 \rightleftharpoons \cdots \rightleftharpoons D_3A_n \qquad (5.3)$$

A simple example is shown in Scheme 5.1b in which the hydrogen acceptor is

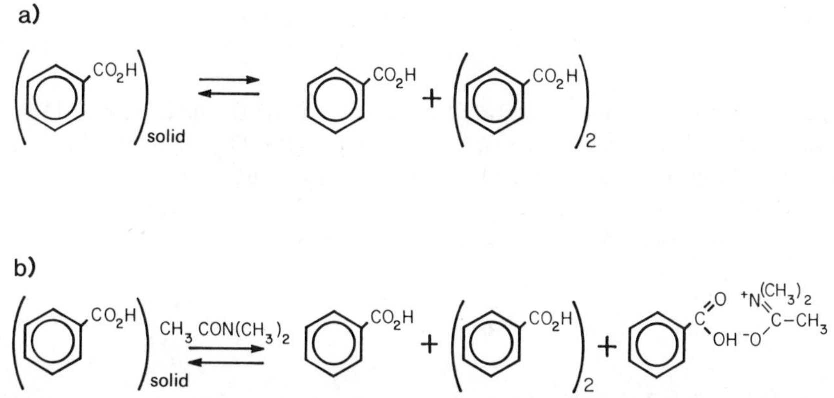

a)

b)

Scheme 5.1 The various species involved when solid benzoic acid is in equilibrium with its solution in a weakly interacting or noninteracting solvent, such as a paraffin or carbon tetrachloride, in the absence (upper equilibrium) or presence (lower equilibrium) of dimethylacetamide, a complexing ligand, or cosolvent. The dimers of benzoic acid are also present, as discussed in Section 6.7.

dimethylacetamide. If we assume that both components are present in relatively low concentrations, the apparent total solubility $[D]_t$ of D is given by

$$[D]_t = [D]_o + ([DA] + [DA_2] + \cdots)$$
$$+ 2([D_2A] + [D_2A_2] + \cdots)$$
$$+ n([D_nA] + [D_nA_2] + \cdots) \qquad (5.4)$$

The higher terms are not normally significant for most systems at low concentrations, so that

$$[D]_t = [D]_o + [DA] + [DA_2] + [DA_3] \qquad (5.5)$$

The equilibrium concentration of each solute species present in these mixtures follows the law of mass action. For complex species we can write, in general,

$$nD + mA \rightleftharpoons D_nA_m \qquad (5.6)$$

$$K^{eq}(D_nA_m) = a(D_nA_m)/a^n(D)a^m(A) \qquad (5.7)$$

Each bracket indicates the species to which the stability constant K^{eq} or the activity a refers. In the forthcoming discussion, we shall indicate the relevant species by subscripts. For particular complexes, Eq. 5.7 becomes

$$K^{eq}_{DA} a_D = a_{DA}/a_A \qquad (5.8)$$

$$K^{eq}_{DA_2} a_D = a_{DA_2}/a_A^2 \qquad (5.9)$$

and

$$K_{D_2A}^{eq} a_D^2 = a_{D_2A}/a_A \tag{5.10}$$

Since the solution is always saturated with component D, the dissolved D is in equilibrium with solid D. Therefore, the activity a_D of D in solution is always constant, and in the equations shown below it is taken into the constant K. If we define our standard state such that the activity coefficients are unity in infinitely dilute solution (see Section 3.3) Eq. 5.8–5.10 become

$$K_{DA} = \frac{[DA]}{[A]}; \qquad K_{DA_2} = \frac{[DA_2]}{[A]^2}; \qquad K_{D_2A} = \frac{[D_2A]}{[A]} \tag{5.11}$$

where $[DA]$, $[DA_2]$, and $[D_2A]$ are the concentrations of the respective complex species and $[A]$ is the concentration of free complexing agent, that is, proton acceptor in solution. K_{DA}, K_{DA_2}, and K_{D_2A} have been termed the interaction constants to distinguish them from the corresponding equilibrium constants K^{eq}. From Eq. 5.6 we see that for all complexes which are first order with respect to A, $m = 1$, and Eq. 5.11 can be written in the form $[D_nA] = K_{D_nA}[A]$.

5.2.2 First-Order (1:1) Complexes

If a 1:1 complex ($n = m = 1$) is assumed,

$$D + A \rightleftharpoons DA$$

$$K_{DA}[A] = [DA] \tag{5.12}$$

$$K_{DA}([A]_t - [DA]) = [DA] \tag{5.13}$$

where $[A]_t$ is the total concentration of complexing agent added. Eq. 5.13 can be written

$$K_{DA}[A]_t = [DA](1 + K_{DA}) \tag{5.14}$$

therefore

$$[DA] = \frac{K_{DA}}{1 + K_{DA}}[A]_t \tag{5.15}$$

For this system, Eq. 5.4 simplifies to

$$[D]_t = [D]_o + [DA] \tag{5.16}$$

and Eq. 5.15 can be written

$$[D]_t = [D]_o + \frac{K_{DA}}{1 + K_{DA}}[A]_t \tag{5.17}$$

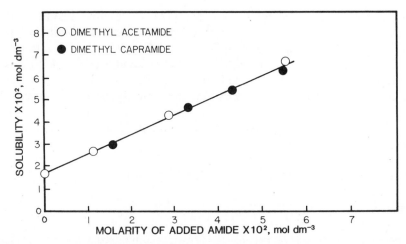

Figure 5.1 The apparent molar solubility of p-phenylphenol in the presence of various molarities of dimethylacetamide, and dimethylcapramide in carbon tetrachloride at 30°C, demonstrating a linear plot independent of the chain length of the amide cosolvent (i.e., ligand). $1 \, mol \, dm^{-3} = 1 \, mol \, L^{-1} = 1 \, M$. Reproduced with permission of the author. Shami, E. G. (1964). Ph.D. Thesis, Formation of Hydrogen Bonded Complexes by Amides in Nonaqueous Solvents, University of Wisconsin, Madison, WI.

where $[D]_t$ is the total solubility of D in the solution, which contains complexing agent A at a total concentration $[A]_t$. $[D]_o$ is the intrinsic solubility of D in the absence of A. A specimen calculation exemplifying Eq. 5.17 is presented in Section 5.9. Plots of $[D]_t$ against $[A]_t$ should give a straight line of slope $K_{DA}/(1 + K_{DA})$ from which K_{DA} can readily be calculated, that is, $K_{DA} = \text{slope}/(1 - \text{slope})$. An example of such a plot is shown in Figure 5.1 for the interaction between p-phenylphenol and two N,N-dimethyl acylamides. We note that K_{DA} is the same for both amides, which indicates that the acyl hydrocarbon chain has no influence on complexation, as found for the interaction with salicylic acid discussed in Section 4.4.1.

As mentioned in Chapters 1 and 4, the length of the hydrocarbon chain in the molecule of a cosolvent or ligand often exerts little or no influence on the stability constant of the complex formed with the solute or on the solubility of the solute in the presence of a given concentration of cosolvent. This point is important and indicates that the specific interactions play a dominant role between the solute molecules and the functional groups in the cosolvent or ligand molecules.

5.2.3 Second-Order (1:2) Complexes

If DA_2 is the only type of complex formed (of the second order with respect to the complexing agent), we have

$$D + 2A \rightleftharpoons DA_2 \tag{5.18}$$

and

$$K_{DA_2} = \frac{[DA_2]}{[A]^2} \tag{5.19}$$

This is a simple quadratic equation. When $[DA_2]$ is plotted against $[A]^2$, a straight line of slope K_{DA_2} passing through the origin is obtained, where $[DA_2] = [D]_t - [D]_o$. This simple quadratic is exemplified by the interaction of hydroquinone (quinol) with alcohols (see Fig. 4.6 and Section 4.4.2, Fig. 4.9) and arises because $K_{DA_2} \gg K_{DA}$, the latter being effectively zero.

Although several complex species can theoretically exist in solution, the majority of D will, in practice, usually be present in solution as DA or DA_2, or as a mixture of DA and DA_2. The total solubility $[D]_t$ of D will be given by Eq. 5.5:

$$[D]_t = [D]_o + [DA] + [DA_2] \tag{5.5a}$$

Therefore,

$$[D]_t - [D]_o = [DA] + [DA_2] = (\text{complexed D}) \tag{5.20}$$

Substitution of $[DA]$ by $[A]K_{DA}$ and $[DA_2]$ by $[A]^2 K_{DA_2}$ according to Eq. 5.11 leads to

$$[D]_t - [D]_o = [A]K_{DA} + [A]^2 K_{DA_2} \tag{5.21}$$

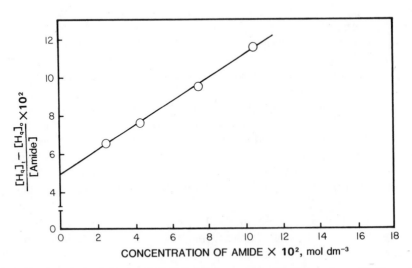

Figure 5.2 Complexing behavior of N,N-dimethylcapramide with hydroquinone (Hq) in carbon tetrachloride at 30°C, demonstrated by a plot based on Eq. 5.22.

$([Hq]_t - [Hq]_o)/[\text{amide}] = K_{DA} + K_{DA_2}[\text{amide}]$. $K_{DA} = 0.05$. $K_{DA_2} = 0.0061$ dm^3 mol^{-1}. 1 mol dm^{-3} = 1 mol L^{-1} = 1 M.

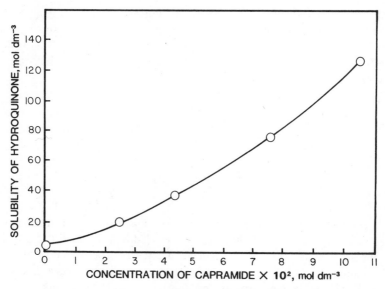

Figure 5.3   Complexing behavior of capramide [A] with hydroquinone [D] in carbon tetrachloride at 30°C, demonstrated by a simple plot of [D] versus [A]. The rising curve indicates the formation of complex(es) of higher order than DA (cf. Eq. 5.21). $1 \text{ mol dm}^{-3} = 1 \text{ mol L}^{-1} = 1 M$.

Therefore,

$$([D]_t - [D]_o)/[A] = (\text{complexed D})/[A] = K_{DA} + [A]K_{DA_2} \qquad (5.22)$$

A specimen calculation exemplifying Eq. 5.21 is presented in Section 5.9.

When $([D]_t - [D]_o)/[A]$ is plotted againt [A], which is the concentration of *free* complexing agent in the solution, a straight line is obtained for which the intercept is K_{DA} and the slope is K_{DA_2}, as shown in Figure 5.2 and in Figure 4.8 for complexation between diethyl ether and two anhydrides. Figure 5.3 is a plot similar to that of Figure 5.2 using axes suitable for evaluating only K_{DA} for 1:1 complexes. The slope of the tangent to the curve at low [A] of Figure 5.3 is equal to K_{AD} according to Eq. 5.21. This means that very low concentrations of A allow much smaller amounts of DA_2 to be formed than of DA. We note that hydroquinone interacts with N,N-dimethylacylamides to form both DA- and DA_2-type complexes, whereas when it interacts with alcohols, only DA_2-type complexes are formed (Figs. 4.6 and 4.9), as discussed in Section 4.4.2, K_{DA} being effectively zero. The interaction of hydroquinone with diamides (Fig. 5.16) is discussed in Section 5.5.6.

5.2.4   Third-Order (1:3) Complexes

If third-order complexes, DA_3, are formed in significant amounts in addition to first- and second-order complexes (i.e., DA, DA_2), K_{DA} can be estimated from the rate of increase in solubility at small concentration of the interaction cosolvent A. Under these circumstances, Eq. 5.16 and 5.17 are approximately obeyed, as illustrated in Figure 4.5. At higher concentrations of A, Eq. 5.5 forms the basis of subsequent calculations. Introducing the expressions for [DA], [DA$_2$], and [DA$_3$] using Eq. 5.11, we obain

$$[D]_t = [D]_o + [A]K_{DA} + [A]^2 K_{DA_2} + [A]^3 K_{DA_3} \qquad (5.23)$$

which on rearrangement affords

$$\left(\frac{[D]_t - [D]_o}{[A]} - K_{DA} \right) \bigg/ [A] = K_{DA_2} + K_{DA_3}[A] \qquad (5.24)$$

where [A] is the molar concentration of the free cosolvent. K_{DA} is determined at low [A] as described above. At higher [A], the left-hand side of Eq. (5.24), which is designated F in Figure 5.4, is calculated from the experimental data and is plotted against [A]. If only first-, second-, and third-order complexes are present, a straight line is obtained whose intercept gives K_{DA_2} and whose slope gives K_{DA_3}. Table 4.8 lists values of $K_{1:1} = K_{DA}/[D]_o$, $K_2 = K_{DA_2}/[D]_o$, and $K_3 = K_{DA_3}/[D]_o$ for the interaction between phthalic anhydride ($= D$) and chloroform ($= A$) in cyclohexane at 25°C, where $[D]_o$ is the molar solubility of this solute in cyclohexane in the absence of chloroform. The relationships between the equilibrium constants are given by Eqs. 4.22 and 4.23.

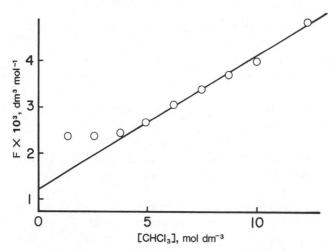

Figure 5.4 Plot for determining K_{DA_2} and K_{DA_3} for the formation of complexes between phthalic anhydride A and chloroform D in cyclohexane at 25°C. This plot demonstrates the use of Eq. 5.24, the left-hand side of which is

$$F = (\Delta_{solubility}/[CHCl_3] - K_{DA})/[CHCl_3].$$

$1 \, mol \, dm^{-3} = 1 \, mol \, L^{-1} = 1 \, M$. Reproduced with permission of the copyright owner, the American Pharmaceutical Association, from Fung, H. L. and Higuchi, T. (1971) *J. Pharm. Sci.*, **60**, 1782–1788.

Anderson et al. (1980) noted that graphical methods of analyzing the solubility data are quite cumbersome when an appreciable amount of added cosolvent exists in the complexed form. They accordingly determined complexation constants from a computer analysis of the data as described in Section 6.6.2 (see Figs 6.17–6.20) and on p. 155.

5.3 EXPERIMENTAL METHODS FOR DETERMINING ASSOCIATION CONSTANTS IN NONAQUEOUS SOLUTION

5.3.1 Introduction

The previous section has shown that the association or stability constants of solute–cosolvent (donor–acceptor) complexes are of vital importance for assessing the significance of specific interactions in solution and for the prediction of solubility in the presence of an interactive cosolvent or ligand. Although solubility measurements may be used to determine these quantities, other methods independent of solubility, such as spectroscopic techniques, may be preferable to dispel any doubts about the importance of specific interactions or complexation between polar solutes and cosolvents.

In this section various methods for determining stability constants of complexes are described, including the partition method and various spectroscopic techniques. A useful procedure for determining the stoichiometry of complexes is also considered. It is important to note that different experimental methods of determining the complexation constant between two given species under defined experimental conditions give very similar values (Table 5.1).

The solubility method may sometimes give values which are about 20% higher than those determined by other methods because solubility is influenced by nonspecific interactions in solution, as discussed in Chapter 2, in addition to specific solute–solvent interactions. Higuchi et al. (1969) have reviewed various experimental methods for determining association constants for hydrogen-bonding interactions in nonaqueous solutions using the literature pertaining to phenols and phenolic compounds, or hydrogen-bond donors, with the following compounds as hydrogen-bond acceptors: ketones, ethers, aromatic organophosphorus compounds, alcohols, vinyl compounds, amides, esters, and the drug griseofulvin.

5.3.2 Liquid–Liquid Partition Method

The partitioning of a hydrogen-bond donor D between a nonpolar solvent and a polar solvent such as water is influenced by the extent of any complex formation in the nonpolar phase when a hydrogen-bond acceptor A is present in the system. The complex formed (e.g., DA) is normally extracted into the organic phase. If the interaction is limited to a 1:1 complex and takes place only in the organic phase, Eq. 5.6 becomes

$$D + A \rightleftharpoons DA \qquad (5.25)$$

Table 5.1 Equilibrium Constants[a] for Various Hydrogen-Bonding Interactions at 25°C

Hydrogen Donor	σ^f	$pK_a{}^g$	Cyclohexane Solvent Hydrogen Acceptor					Carbon Tetrachloride Solvent Hydrogen Acceptor			
			TBP[e]		Sarin[e]		l-PTP	l-PTP[h]			Griseo-fulvin[c]
			LLP[b]	UV	LLP[b]	UV	ORD	ORD[e]	NMR[d]		ORD
Phenol	0.0	9.98	950	930	215		231	105	105		33
p-CH₃-phenol	−0.171	10.14	640	620	130						
p-C₂H₅-phenol	−0.151	10.00	755		140						
p-t-C₄H₉-phenol	−0.197	10.25	575		100		49				
p-C₆H₅-phenol	+0.009	9.55	1380	1270	210						
m-OCH₃-phenol	−0.115	9.65						125			35
p-OCH₃-phenol	−0.268	10.21	712		135		247	69			23
m-F-phenol	+0.337	9.28		3320				283			84
p-F-phenol	−0.062	9.95	1900	1820	260		459	216			55
m-Cl-phenol	+0.373	9.13						418			109
p-Cl-phenol	+0.227	9.38	2840	2900	470		1801	273			87
p-Br-phenol	+0.232	9.36	2985		520		1145	296			89
m-I-phenol	+0.352	9.06						344			125
p-I-phenol	+0.276	9.31	3227		540		1696	268			82

Compound								
2,3,5,6-tetra-F-phenol	5.6				990		608	129
Penta-F-phenol	5.53				1720		950	265
p-COCH$_3$-phenol		+0.874	6620					
p-COC$_2$H$_5$-phenol			4680		800			
p-CHO-phenol	7.62	+1.126	7782		1270			
m-CN-phenol	8.61	+0.678	15550	9020	2300		1506	263
p-CN-phenol	7.95	+1.00	20600	17100	2150		1245	423
p-NO$_2$-phenol	7.15	+1.270			3000	33088	1925	830
Catechol	9.13		4950	5250	820	2328	596	
4-i-C$_3$H$_7$-catechol			3440	3360	600			
4-NO$_2$-catechol			6210					
1-Naphthol	9.85			1290	235			
2-Naphthol	9.93			1130	240			

[a] $K_{1:1}$ (dm^3 mol^{-1}, L mol^{-1}, or M^{-1}).

[b] LLP = liquid–liquid partition.

[c] Nakano, N. I., Ph.D. thesis University of Wisconsin (1967).

[d] Higuchi, T., Nakano, N. I., and Richards, J. H., unpublished results.

[e] Hou, J. P., Ph.D. thesis, University of Wisconsin (1967). Sar in = isopropyl methylphosphonofluoridate. TBP = tri-n-butylphosphate.

[f] Hammett σ values from Nakano, N. I., Ph.D. thesis, University of Wisconsin (1967) and from Hou, J. P., Ph.D. thesis, University of Wisconsin (1967).

[g] Nakano, N. I., Ph.D. thesis, University of Wisconsin (1967) and Kostenbauder and Higuchi (1956a).

[h] l-PTP = 1-S-ethyl-O-ethylphenylphosphonothionate (see Scheme 5.2 in Section 5.3.4).

and Eq. 5.7 becomes

$$K' = \frac{a_{DA}}{a_D a_A} = \frac{[DA]}{[D][A]} \frac{\gamma_{DA}}{\gamma_D \gamma_A} = K \frac{\gamma_{DA}}{\gamma_D \gamma_A} \qquad (5.26)$$

where A and D represent the free species, DA the complex, a the activity, and γ the activity coefficient, and the square brackets signify molar concentrations. K' represents the true stability constant of the complex in terms of activities and K represents the apparent stability constant in terms of concentration.

If the calculation of K is based on D, its total concentration $[D]_t$ in the organic phase, assuming that some dimerization can occur, is given by

$$[D]_t = [D] + [DA] + 2[D_2] \qquad (5.27)$$

$$[D]_t = [D] + [DA] + 2K_d[D]^2 \qquad (5.28)$$

where D_2 refers to the dimer and K_d to the dimerization constant in the organic phase. Therefore,

$$[DA] = [D]_t - [D] - 2K_d[D]^2 \qquad (5.29)$$

The terms [A] in Eq. 5.26 and [D] in Eqs. 5.28 and 5.29 refer to the organic phase and can be determined accurately using the partition law, .once the partition coefficients and the concentrations of [D] and [A] in the aqueous phase are known. If A is not sufficiently soluble in the organic phase, [A] is obtained by subtracting DA in the organic phase from $[A]_t$.

For systems containing both DA and D_2A complexes, Eq. 5.28 contains a term for D_2A, thus,

$$[D]_t = [D] + [DA] + 2[D_2] + 2[D_2A] \qquad (5.30)$$

Therefore

$$[DA] + 2[D_2A] = [D]_t - [D] - 2K_d[D]^2 \qquad (5.31)$$

Comparison of the left-hand sides of Eqs. 5.29 and 5.31 indicates that the apparent stability constant K_{app} is given by

$$K_{app} = ([DA] + 2[D_2A])/[D][A] \qquad (5.32)$$

Now,

$$K_{1:1} = [DA]/[D][A] \qquad (5.33)$$

and if we can assume that

$$DA + D = D_2A \qquad (5.34)$$

then

$$K_{1:2} = [D_2A]/[DA][D] \qquad (5.35)$$

and Eq. 5.32 can be written

$$K_{app} = K_{1:1} + K_{1:1}2K_{1:2}[D] \tag{5.36}$$

If K_{app} is plotted against [D], then $K_{1:1}$ is given by the intercept and $K_{1:2}$ is calculated from the slope and intercept.

Some examples of systems examined by the liquid–liquid partition (LLP) method are presented in Table 5.1.

5.3.3 Infrared Spectroscopic Method

The vibration of molecules, including the stretching-compression of valency bonds and the bending oscillation of bond angles, can be studied in the infrared (IR) region. The fundamental stretching–compression transition of the bond A—H is found to be reduced when it undergoes hydrogen bonding with the proton acceptor B. The lowering of energy gives a measure of the strength of the A—H····B hydrogen bond. The O—H····O systems will serve to illustrate the method. The hydroxyl group when not hydrogen-bonded, that is, when "free," has a fundamental stretching vibration corresponding to a sharp peak around $3600 \, cm^{-1}$ ($= 110 \, THz \equiv 43 \, kJ \, mol^{-1}$). The peak is observed most simply when the substance concerned is in the gaseous state. However, the peak can be studied in a suitable inert solvent or even in the solid state, although the spectrum becomes increasingly difficult to interpret.

To observe the free OH peak in an inert solvent, the substance must be so dilute that hydrogen bonding is negligible. When the hydroxyl group becomes a donor in a hydrogen bond in a given state or concentration of the substance, the O—H stretching peak (a) moves to a lower frequency, (b) becomes broader (Bournay and Robertson, 1978), that is, less sharp, and (c) becomes more intense, in the sense that the area beneath the peak is increased. The frequency in the range $3600-2000 \, cm^{-1}$ correlates well with the O—O distance of between 3.4 and 2.5 Å. As mentioned above, the decrease in frequency correlates positively with hydrogen-bonding energy, as originally reported by Badger and Bauer (1937). This extrathermodynamic relation unfortunately has limited applicability (Ghersetti and Lusa, 1965; Arnett et al., 1970, 1974; Pimentel and McClellan, 1971; Kollman and Allen, 1972). For shorter, stronger bonds, the spectral changes becomes more complicated.

Similar considerations apply to other types of hydrogen bonds involving nitrogen, oxygen, or other atoms. Comparison of covalent-bond vibration frequencies ($\sim 10^{12} \, Hz$) with rates of hydrogen isotope exchange in solution ($\sim 10^8-10^9 \, s^{-1}$) indicates that a hydrogen bond between two molecules persists for about 10^3 or 10^4 molecular vibrations, which demonstrates that hydrogen-bonded complexes can exist as distinct species in solutions (Hofacker et al., 1976).

Infrared spectrophotometry of solutions of the donor A with the acceptor B in a suitable solvent can be used to determine (Arnett et al., 1970, 1974) the stability constant of the donor acceptor (or hydrogen-bonded) complex, which is the equilibrium constant K for the following reaction:

A—H + B $\rightleftharpoons$ A—H———B

Donor   Acceptor$\rightleftharpoons$Complex (C)

[A]$_o$ [B]$_o$ O   initial concentration   (5.37)

[A]   [B]   [C]   equilibrium concentration

$$K = [C]/[A][B] \qquad (5.38)$$

But since

$$[B] = [B_o] - [C] \quad \text{and} \quad [A] = [A_o] - [C] \qquad (5.39)$$

$$K = \frac{[A]_o - [A]}{[A]([B]_o - [A]_o + [A])} \qquad (5.40)$$

For the spectroscopic method, since [A]$_o$ and [B]$_o$ are known quantities, [A] can be determined from the IR A—H stretching peak (e.g., near $3600 \, \text{cm}^{-1}$ for "free" O—H) and [C] from the new, broad peak which has been shifted to a lower frequency due to hydrogen bonding. Insertion of the data into Eq. 5.40 enables K to be calculated. This treatment assumes that dimerization of A is negligible, an assumption for which tests must be run and allowances made. Linearity of the Beer–Lambert plot of absorbance of A against [A] over the relevant concentration range in the absence of B is usually taken to mean that A is not dimerized. Absence of dimerization of B is shown by the constancy of K at various [B]$_o$ values.

Infrared and other spectroscopic observations (Fenby and Hepler, 1976) usually indicate that self-associated molecules of the donor are stable even in dilute solutions of inert or nonpolar solvents. Phenol, for example, forms species larger than dimers, as shown by spectroscopic (Fenby and Hepler, 1976) and vapor-pressure evidence (Anderson et al., 1979), by distribution between water and carbon tetrachloride, and by enthalpies of dilution (Fenby and Hepler, 1976).

The standard free-energy change for complex formation can be calculated using

$$\Delta G^\theta = -RT \ln K \qquad (5.41)$$

After allowing for the effect of temperature on the relevant peaks, the standard enthalpy of complexation can be calculated from the temperature dependence (Spencer et al., 1975) of K as follows:

$$\Delta H^\theta = -R \left(\frac{\partial (\ln K)}{\partial (1/T)} \right)_p \qquad (5.42)$$

The standard entropy of complex formation can be calculated from the well-known equation

$$\Delta G^\theta = \Delta H^\theta - T\Delta S^\theta \quad \text{or} \quad \Delta S^\theta = (\Delta H^\theta - \Delta G^\theta)/T \qquad (5.43)$$

The solvent necessarily undergoes some kind of interaction with A, B, and C, and the nature and strength of each of these interactions will inevitably affect K, ΔG^{θ}, ΔH^{θ}, and ΔS^{θ}. Solvent effects are discussed later in this chapter (Section 5.7).

5.3.4 Spectropolarimetric (ORD) Method

Spectropolarimetry is frequently referred to as optical rotatory dispersion (ORD), the theory and application of which have been discussed fully by Djerassi (1960). The use of ORD in the study of hydrogen-bond formation has been discussed by Meier and Higuchi (1965) and by Nakano and Higuchi (1968), who succeeded in overcoming problems connected with lack of sensitivity and precision by the use of an equation of Rossotti and Rossotti (1961). Iterative procedures for calculating stability constants have been developed (Meier and Higuchi, 1965; Nakano and Higuchi, 1968), and the favored method will be discussed later. Examples include (a) the camphor–phenol complex in carbon tetrachloride in which open-chain phenol dimers are the donor species (Meier and Higuchi, 1965), and (b) the interactions between phenols and griseofulvin (Table 5.1) or *l*-S-ethyl-O-ethylphenylphosphonothionate (*l*-PTP).

The sensitivity of the ORD method using camphor as a quantitative indicator of hydrogen bonding is poor because the donor–acceptor site and the chromophore are relatively far removed from the chiral center. A much greater sensitivity can be achieved by (a) the use of an indicator structure in which the donor–acceptor site and the chromophore are directly attached to the chiral center and (b) the use of a phosphorus atom instead of a carbon atom as the chiral center. For these reasons, Higuchi et al. (1969) synthesized and used *l*-PTP (see Scheme 5.2) as an ORD indicator of hydrogen bonding. *l*-PTP has a large molecular rotation which increases by the order of $1000°$ on electron donation (proton acceptance in hydrogen bonding) because of a great increase in the polarization of the P—O bond. It is possible merely to titrate the proton donor D with the proton acceptor A in the presence of a trace amount of the *l*-PTP indicator. An excess of A produces a high specific rotation. The sensitivity of this method is several powers of ten greater than that of the coventional ORD method. Furthermore, the oxygen atom in $P \rightarrow O$ is a potent electron-pair donor with about the same degree of proton acceptance as the oxygen atoms in amides and amine oxides.

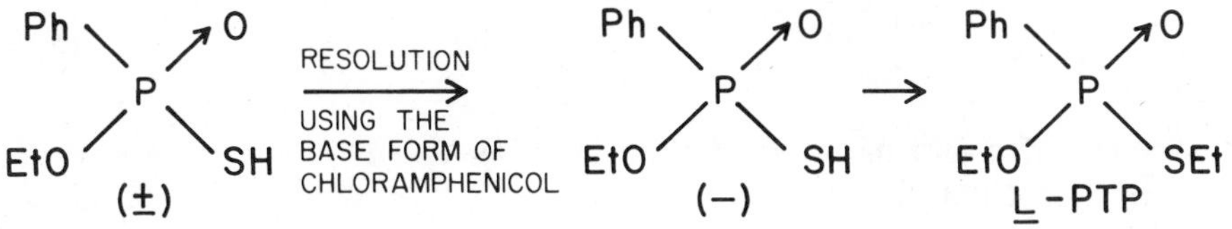

Scheme 5.2 The synthesis and structure of *l*-S-ethyl-O-ethylphenylphosphonothionate (*l*-PTP).

5.3.5 Ultraviolet Spectrophotometric Method

The ultraviolet (UV) absorption spectra of hydrogen-bonded complexes differ from those of the uncomplexed interactants in the wavelength, slope, and intensity of the absorption (Baba, 1958; Baba and Suzuki, 1961) such that the stability constant of the complex can be determined from suitable measurements. The technique is so sensitive that a very low concentration (10^{-5}–10^{-4} M) of the hydrogen donor is usually required. Consequently, self-association can often be neglected. Furthermore, reproducible data can easily be obtained, and the technique is simpler and quicker than the others discussed here. The main disadvantage of the UV method is that the stoichiometry of the complex can be determined less easily. The Job method of stoichiometry determination, to be discussed later, is applicable only when one type of complex is formed. Usually a simple stoichiometry (e.g., 1:1) is assumed in calculating stability constants using the UV method. On the whole, the advantages of the UV method outweigh the disadvantages.

Benesi and Hildebrand (1949) developed the original, fundamental procedure (B–H) for the UV determination of the stability constant K_c of a charge-transfer complex and of the molar absorption (or extinction) coefficient ε_C of the charge-transfer bond. The concentration of the complex C is related to the absorbance (or extinction) E_C of the charge-transfer peak according to the equation

$$E_C = [C]l\varepsilon_C \tag{5.44}$$

where l is the cell length. The acidic solute species will be represented here by the proton donor D and the base by the proton acceptor A. If $[D]_o \gg [A]_o$, very little D will be used to form the complex, so $[D]_o \gg [C]$ and Eqs. 5.38 and 5.39 become

$$K_c = [C]/[D]_o([A]_o - [C]) \tag{5.45}$$

Therefore,

$$[D]_o[A]_o - [D]_o[C] = [C]/K_c \tag{5.46}$$

or

$$[A]_o = \frac{[C]}{K_c}\frac{1}{[D]_o} + [C] \tag{5.47}$$

Introducing the expression for $[C]$ given by Eq. 5.44 and multiplying throughout by l/E_C, we obtain the Benesi–Hildebrand equation

$$\frac{[A]_o l}{E_C} = \frac{1}{K_c\varepsilon_C}\frac{1}{[D]_o} + \frac{1}{\varepsilon_C} \tag{5.48}$$

E_C is measured at some wavelength at which neither D nor A absorb, or else a correction has to be applied. For constant $[A]_o$, a plot of $[A]_o l/E_C$ against $1/[D]_o$ gives a straight line of slope $1/K_c\varepsilon_C$ and intercept $1/\varepsilon_C$. Therefore, K_c = intercept/slope. By means of their Eq. 5.48, Benesi and Hildebrand (1949) determined the stability constants $(K_c)_{B-H}$ of the charge-transfer complexes between iodine and various aromatic and polar molecules. When derived in this way, K_c is sometimes referred to as K_{B-H}. The equation seems to fail for very weak complexes due to solute–solvent interactions to be discussed in Section 5.7.

Iterative procedures for determining the stability constants of complexes from UV measurements have been described by Higuchi and co-workers (1969) and by Nakano and Higuchi (1968). The favored method is discussed later. Some data for the interactions between phenols and tri-n-butylphosphate (TBP) or isopropyl methylphosphonofluoridate (sarin) are presented in Table 5.1.

5.3.6 Determination of the Stoichiometry of Complexes

Job's method of continuous variations (Jones, 1964) is commonly employed for determining the stoichiometry of complexes. This method involves the measurement of some physical quantity, which is proportional to the concentration of the complex, such as UV-visible absorbance and optical rotation. The optical rotation method is illustrated here by the tryptophan–caffeine and the tryptophan–theophylline interaction in aqueous solution, as shown in Figure 5.5. Although these complexes are derived from stacking the molecules to form molecular clusters in aqueous solution, rather than simple hydrogen bonding, they do serve to illustrate the method. This technique requires the assumption that only a single complex species AB_n is formed from A and B according to the equation

$$A + nB = AB_n \tag{5.49}$$

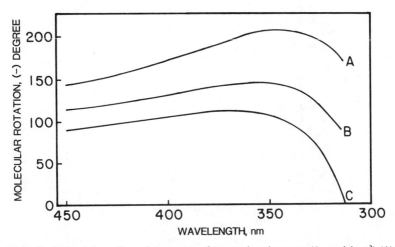

Figure 5.5 Optical rotatory dispersion spectra of tryptophan in water $(4\,\text{mmol}\,\text{dm}^{-3})$ (A) with $32\,\text{mmol}\,\text{dm}^{-3}$ caffeine, (B) with $32\,\text{mmol}\,\text{dm}^{-3}$ theophylline, and (C) without alkylxanthines. $1\,\text{mmol}\,\text{dm}^{-3} = 10^{-3}\,\text{mol}/\text{Liter} = 1\,\text{m}M$. Reproduced with permission of the copyright owner, the American Pharmaceutical Association, from Nakano, M. and Higuchi, T. (1968). *J. Pharm. Sci.*, **57**, 1865–1868.

By mixing a solution of A with one of B, various mixtures were prepared in which the sum of the molar concentrations of A and B had the constant value S (in mol dm^{-3}), thus,

$$[A]_o + [B]_o = S \tag{5.50}$$

If $[B]_o = xS$, then $[A]_o = (1 - x)S$ and the stoichiometry is given by

$$n = x/(1 - x) \tag{5.51}$$

The physical property was measured for the original solutions of A, Y_A, and B, Y_B, and for each mixture under constant conditions (e.g., constant cell length). Let Y_C represent the physical property of the complex C. In terms of Y_A, Y_B, and Y_C, the difference ΔY between the measured property and that expected if no interaction took place is given by

$$\Delta Y = Y_A[A] + Y_B[B] + Y_C[C] - Y_A[A]_o - Y_B[B]_o \tag{5.52}$$

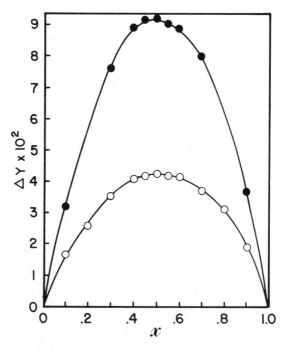

Figure 5.6 Plot of the difference ΔY between the measured optical rotation in degrees at 300 nm and that assuming no interaction, against the mole fraction of tryptophan x. Total concentration $S = 30\,\text{mmol dm}^{-3}$. Key: ●, tryptophan–caffeine; ○, tryptophan–theophylline. $1\,\text{mmol dm}^{-3} = 10^{-3}\,\text{mol L}^{-1} = 1\,mM$. Reproduced with permission of the copyright owner, the American Pharmaceutical Association, from Nakano, M. and Higuchi, T. (1968). *J. Pharm. Sci.*, **57**, 1865–1868.

After interaction,

$$[A] = [A]_o - [C] \quad \text{and} \quad [B] = [B]_o - n[C] \tag{5.53}$$

Therefore,

$$\Delta Y = (Y_C - Y_A - nY_B)[C] \tag{5.54}$$

$$\Delta Y = Y[C] \quad \text{where} \quad Y = Y_C - Y_A - nY_B \tag{5.55}$$

Figure 5.6 shows two examples of a plot of ΔY against x. The stoichiometry of the complex corresponds to that value of x for which $[C]$ and hence ΔY have maximum values, that is, $dC/dx = 0$ and $d(\Delta Y)/dx = 0$. Figure 5.6 shows that the maximum value occurs at $x = 0.5$ for the complexes tryptophan–caffeine and tryptophan–theophylline. According to Eqs. 5.50 and 5.51, $n = 1$ for these complexes.

5.3.7 Iterative Methods for the Determination of Stability Constants of Complexes

The iterative method enables stability constants of complexes to be calculated from the measurement of some physical property which is proportional to the concentration of the complex, as mentioned above. Such properties include absorbance from UV measurements and optical rotation from ORD data. In the case of 1:1 stoichiometry ($n = 1$), Eqs. 5.38 and 5.39 may be combined as follows:

$$K[A]_o[B]_o - K[C]\{[A]_o + [B]_o - [C]\} = [C] \tag{5.56}$$

and Eqs. 5.54 and 5.55 may be written

$$\Delta Y = (Y_C - Y_A - Y_B)[C] \tag{5.57}$$

$$= Y[C]$$

where

$$Y = Y_C - Y_A - Y_B \quad \text{and} \quad [C] = \Delta Y/Y \tag{5.58}$$

Equations 5.56 and 5.58 can be combined to give

$$\frac{[A]_o[B]_o}{\Delta Y} = \frac{1}{KY} + \frac{[A]_o + [B]_o - [C]}{Y} \tag{5.59}$$

K and Y_C were computed from the difference in a measured property, ΔY, and from the initial concentrations of A and B. This iterative procedure (Nakano and Higuchi, 1968) is exemplified by Figure 5.7 and is described below.

Step 1. $[A]_o[B]_o/Y$ is plotted against $([A]_o + [B]_o)$ and the slope of the line m_o is evaluated.

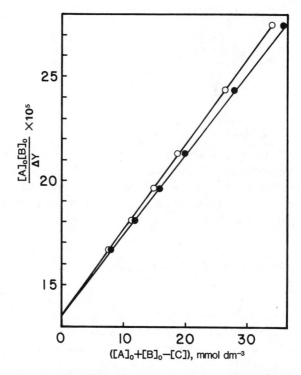

Figure 5.7 Plots based on Eq. 5.59 for the system tryptophan ($4\,\text{mmol dm}^{-3}$) plus caffeine (4–$32\,\text{mmol dm}^{-3}$). $[A]_o[B]_o/\Delta Y$ has the units $\text{mol}^2\,\text{dm}^{-6}\,\text{degree}^{-1}$, where ΔY is the difference between the measured optical rotation in degrees of 300 nm and that assuming no interaction. Key: ●, first approximation; ○, third and final approximation. $1\,\text{mmol dm} = 10^{-3}\,\text{mol L}^{-1} = 1\,\text{m}M$. Reproduced with permission of the copyright owner, the American Pharmaceutical Association, from Nakano, M. and Higuchi, T. (1968). *J. Pharm. Sci.*, **57**, 1865–1868.

Step 2. $[C]_1$, the first approximate value of $[C]$, is calculated by substituting m_o into Eq. 5.58, that is, $[C]_1 = \Delta Y m_o$.

Step 3. $[A]_o[B]_o/Y$ is plotted against $([A]_o + [B]_o - [C]_1)$ and the slope m_1 is obtained.

Step 4. $[C]_2$ is calculated as in Step 2, that is, $[C]_2 = \Delta Y m_1$.

Step 5. Steps 3 and 4 are repeated until the convergent slope is obtained.

Step 6. K is calculated from the slope and intercept of the last cycle and Y_C from the final slope, thus,

$$K = \frac{\text{slope}}{\text{intercept}}; \quad Y_C = Y_A + Y_B + \frac{1}{\text{slope}} \tag{5.60}$$

Application of this procedure to tryptophan solutions with varying concentrations of amides using ORD measurements at 330 nm, indicated by Figure 5.5, afforded the values shown in Table 5.2. The linear plots involved in

Table 5.2 Stability Constants $K_{1:1}$ of Tryptophan Complexes with Some Amides in Water at $25°C^a$

Compound	$K_{1:1}$ (dm^3 mol^{-1})
Theophylline	35
8-Methoxycaffeine	50
1, 3-Dimethyluracil	5.8
Sarcosine anhydride	$\cong 0$

^{a}1 dm^3 mol^{-1} = 1 L mol^{-1} = 1 M^{-1}.

Reproduced with permission of the copyright owner, the American Pharmaceutical Association, from Nakano, M. and Higuchi, T. (1968). J. *Pharm. Sci.*, **57**, 1865–1868.

steps 1–6 above converge quickly, as shown in Figure 5.7, when the stability constants are moderately high, $> 20\,\text{dm}^3\,\text{mol}^{-1}$.

Proton magnetic resonance (PMR) data have been widely used to determine stability constants of hydrogen-bonded and charge-transfer complexes in ternary systems. Some of the methods require that the concentration of one of the interactants, usually the base for hydrogen-bonding studies, be substantially in excess of the other so that the reduction in its concentration resulting from complex formation can be neglected. This procedure is hampered (a) by limited solubility of an interactant, (b) by insufficient sensitivity of the PMR method, (c) by modification of the properties of the solvent, (d) by formation of higher-order complexes, and (e) by spectroscopic anomalies at high concentrations. The interative procedure of Nakano et al. (1967) overcomes these problems and enables stability constants to be calculated from prior measurements on systems in which the two interactants have similar concentrations of the order of 2–100 mM. The method takes into account the amount of both reactants present in the complex by successive approximation.

The basic equilibrium situation is represented by Eq. 5.38 and usually involves very rapid reactions. Consequently, the observed chemical shift of the acidic protons on A, δ_{obs}, is the time-weighted average of the chemical shift of the protons in the complexed form δ_C, and that in the uncomplexed form δ_A. The relationship is

$$\delta_{obs} = \frac{[C]}{[A]}\delta_C + \frac{[A] - [C]}{[A]}\delta_A \tag{5.61}$$

Algebraic rearrangement gives

$$[C] = \frac{\delta_{obs} - \delta_A}{\delta_C - \delta_A}[A] \tag{5.62}$$

Substituting for (C) in Eq. 5.56, while leaving the term ($[A] + [B] - [C]$) intact, we obtain

$$\frac{[B]}{\Delta\delta_{obs}} = \frac{1}{\delta_C - \delta_A}\{[A] + [B] - [C]\} + \frac{1}{K(\delta_C - \delta_A)} \tag{5.63}$$

where

$$\Delta\delta_{obs} = \delta_{obs} - \delta_A$$

Equation 5.63 contains two unknowns, [C] and δ_C, which can be evaluated by a simple iterative procedure (Nakano et al., 1967) similar to that described above and illustrated by Figure 5.8. $[C]/\Delta\delta_{obs}$ is plotted against $([A] + [B])$ to yield a slope which Eq. 5.63 indicates is approximately equal to $1/(\delta_C - \delta_A)$. This quantity is substituted into Eq. 5.63 to give the first approximate value of [C]. The [C] values thus obtained is then inserted into Eq. 5.63 and $[C]/\Delta\delta_{obs}$ is plotted against $([A] + [B] - [C])$ to yield an improved value of the slope. These steps are repeated until two successive cycles afford essentially identical convergent values for the slope. The final stability constant K is then calculated from the limiting slope and intercept values. The chemical shift δ_C of the bonded proton in the complex C can also be obtained from the final slope.

If $[B] \gg [A]$, Eq. 5.63 reduces to equivalent relationships given by the following equations:

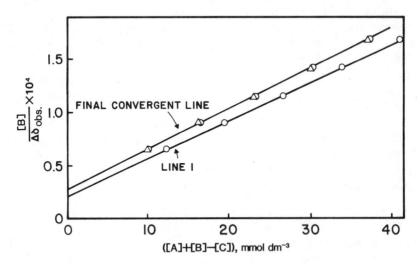

Figure 5.8 Plots showing the application of the iterative method to plots of $[B]/\Delta S_{obs}$ (mol dm^{-3} Hz^{-1}) versus $([A] + [B] - [C])$ for calculation of the stability constant from PMR data for the interaction of phenol with dimethylacetamide in carbon tetrachloride at 25°C. Line 1 corresponds to the first approximation with [C] set equal to zero. Points indicated by (○) on the final convergent line are for the second approximation and (△) for the final convergent value of [C]. Reprinted with permission from Nakano, M, Nakano, N. I., and Higuchi, T. (1967) *J. Phys. Chem.*, **71**, 3954–3959. Copyright © 1967 by the American Chemical Society.

(a) Hanna and Ashbaugh's (1964) form:

$$\frac{1}{\Delta\delta_{obs}} = \frac{1}{K(\delta_C - \delta_A)[B]} + \frac{1}{\delta_C - \delta_A} \tag{5.64}$$

(b) Foster and Fyfe's (1965) form:

$$(\Delta\delta_{obs}/[B]) + \Delta\delta_{obs}K = (\delta_C - \delta_A)K \tag{5.65}$$

It can be shown that Eq. 5.63 can also be directly developed from these equations if the higher-order terms are not neglected during their derivation. Equation 5.63 is generally applicable to PMR data for computation of both stability constants and δ_C values of 1:1 complexes in tertiary systems. The equation is applicable not only to hydrogen bonding but to other types of rapidly reversible intermolecular interactions in solution.

The number of iterations necessary is dependent on the concentration ratio of the two interactants as well as on the degree of interaction. Essentially one iteration cycle brought the values of [C] very close to the final convergent values for the relatively strong interaction between phenol, [A], and dimethyl-acetamide, [B], shown in Figure 5.8. The horizontal shifts of each point from line 1 to the final convergent line correspond to the [C] values at equilibrium, and the total added concentration of the two interactants is indicated by the points on line 1.

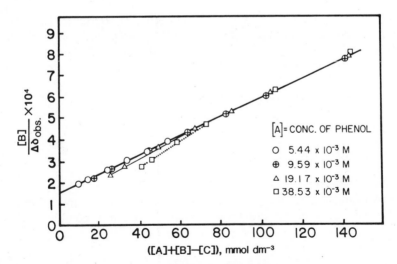

Figure 5.9 The effect of concentration of phenol on the plots of $[B]/\Delta\delta_{obs}$ ($mol\,dm^{-3}\,Hz^{-1}$) versus $([A] + [B] - [C])$ for interaction of phenol with isophorone in CCl_4 at 25°C. The straight line drawn is the regression line for $[A] = 5.44\,mmol\,dm^{-3}$ and $9.59\,mmol\,dm^{-3}$ only. $1\,mmol\,dm^{-3} = 10^{-3}\,mol\,L^{-1} = 10^{-3}\,M$. Reprinted with permission Nakano, M., Nakano, N. I., and Higuchi, T. (1967). *J. Phys. Chem.*, **71**, 3954–3959. Copyright 1967 by the American Chemical Society.

Since at the concentration of phenol employed for PMR there was some tendency for the hydroxyl signals to change on further dilution, the values of δ_A were estimated for each solvent by extrapolating the curve obtained at the concentration range of above 6 mM to infinite dilution (Nakano et al., 1967).

Higher concentrations of phenol, of the order of 10 mM and above, elicited significant deviations from Eq. 5.63, as shown in Figure 5.9. These deviations can be attributed to self-association, since associated forms of phenol appear to form hydrogen bonds more readily than the monomer (Baker and Harris, 1960; Bellamy and Pace, 1966). Theoretically, all points should converge into one straight line after iteration if the determination of stability constants is totally independent of the phenol concentrations.

5.3.8 Calorimetric Studies of Hydrogen Bonding and Polyfunctionality

Among various calorimetric methods for determining the enthalpy of donor–acceptor interaction (Fenby and Hepler, 1976), two stand out as being particularly valuable, namely, the high-dilution method and the pure-base method of Arnett and co-workers (1970).

In the high-dilution method, a small amount of a base B is injected into a dilute solution of the hydrogen-bond donor A—H to produce a 1:1 complex C, as indicated by reaction (Eq. 5.37). The enthalpy change Q associated with the formation of n moles of complex is related to the standard enthalpy of complex formation at infinite dilution ΔH_{fm} by

$$Q = n\Delta H_{fm} = [C]V\Delta H_{fm} \qquad (5.66)$$

where V represents the total volume of the solution. Q is determined experimentally by subtracting the enthalpy of solution for the given quantity of base B in the pure solvent, Q_I, from the enthalpy of solution of the base in the same amount of solvent containing the proton donor, Q_{II}. This is illustrated by the following cycle:

$$
\begin{array}{c}
A(S, \text{dilute}) + B(\text{pure}) \\
+ S \\
\end{array}
\quad
\begin{array}{c}
Q_{II} \\
\longrightarrow \\
\end{array}
\quad
AB(S, \text{dilute})
$$

$$
Q_I \downarrow
$$

$$
A(S, \text{dilute}) + B(S, \text{dilute}) \qquad Q = Q_{II} - Q_I
$$

Application of Hess's law indicates that $Q = Q_{II} - Q_I$. In some variations A and B are interchanged or the pure component, particularly if it is a solid, is replaced by a concentrated solution. Equations 5.38 and 5.39 can be combined to give

$$K = \frac{[C]}{([A]_o - [C])([B]_o - [C])} \qquad (5.67)$$

which can be rearranged into the quadratic form

$$[C]^2 - ([A]_o + [B]_o + K^{-1})[C] + [A]_o[B]_o = 0 \qquad (5.68)$$

This can be solved to give two roots, of which only one is acceptable:

$$2[C] = -([A]_o + [B]_o + K^{-1})$$
$$\pm \{([A]_o + [B]_o + K^{-1})^2 - 4[A]_o[B]_o\}^{1/2} \qquad (5.69)$$

Insertion of a K value determined independently (e.g., from spectroscopic data) into Eq. 5.69 affords a $[C]$ value for each pair of $[A]_o$ and $[B]_o$ values. Each of these $[C]$ values corresponds to an experimental Q value. Equation 5.66 indicates that a plot of Q against $[C]V$ should be linear with slope ΔH_{fm} and should pass through the origin (Arnett et al., 1970).

Various methods have been proposed to calculate both K and ΔH_{fm} simultaneously from Eq. 5.66 and 5.68 using trial-and-error methods. Fenby and Hepler (1976) have critically reviewed these and have proposed a preferable alternative, outlined below.

In the pure-base method (Arnett et al., 1970), a small quantity of hydrogen-bond donor A—H is injected into a much larger quantity of pure base B and the enthalpy change is measured. If the equilibrium constant K is large, all the added donor will become hydrogen-bonded to the acceptor. However, when this condition does not apply, alternative approaches may be used (Fenby and Hepler, 1976). Assuming complete conversion of added donor to the 1:1 complex, part of the measured enthalpy change ΔH_{fm} can be ascribed to hydrogen bonding, ΔH (A—H $\cdots$ B), and part to nonspecific interactions between A and B, ΔH (ns) involving van der Waals forces, thus,

$$\Delta H = \Delta H(\text{A}\text{—}\text{H} \cdots \text{B}) + \Delta H(\text{ns}) \qquad (5.70)$$

By proper choice of a model compound or homomorph (Brown et al., 1964) whose proton donor group has been replaced by a nonacidic group, the observed enthalpy change should correspond closely to ΔH(ns). In the case of a phenol, a suitable nonacid homomorph would be its methyl ether (i.e., an anisole) and n-butanol could be replaced by n-butyl chloride. When the homomorph is injected into the pure base as the solvent, the observed enthalpy of solution is a sum of the enthalpy which is expected in an "inert" solvent and that due to specific interactions. As possible "inert" solvents, various chloromethanes and liquid hydrocarbons were proposed. To determine ΔH(A—H $\cdots$ B), the measurement of the following partial molar enthalpies of solution at infinite dilution is necessary: that of the acid in the pure base, L^θ (A in B); that of the homomorph in the pure base, L^θ (M in B); and that of the acid in the "inert" solvent, L^θ (M in S). Then,

$$\Delta H(\text{A}\text{—}\text{H} \cdots \text{B}) = [L^\theta(\text{A in B}) - L^\theta(\text{M in B})]$$
$$- [L^\theta(\text{A in S}) - L^\theta(\text{M in S})] \qquad (5.71)$$

Although the ΔH values obtained (Arnett et al., 1970) agree with those from alternative methods, critical analysis (Duer and Bertrand, 1970) indicates that they depend greatly on the choice of "inert" solvent and "model" homomorph.

Duer and Bertrand (1970) recommended determining the standard enthalpies of formation of two complexes AHB and AHB′, both in the same solvent, thus,

$$\Delta H^{\theta}(A—H \cdots B) - \Delta H^{\theta}(A—H \cdots B') = [L^{\theta}(A \text{ in } B) - L^{\theta}(M \text{ in } B)]$$
$$- [L^{\theta}(A \text{ in } B') - L^{\theta}(M \text{ in } B')]$$
$$(5.72)$$

This procedure eliminates the effect of the inert solvent, and, furthermore, the ΔH^{θ} difference was found to be almost independent of the choice of model compound M.

Calorimetric measurements have also been used to investigate various self-association reactions of a given monomer A in a solvent S to give an associated species A_n, thus,

$$nA(S) \rightleftharpoons A_n(S) \qquad (5.73)$$

Self-association is discussed in greater detail in Chapter 6.

We have already seen that some hydrogen-bond donors and acceptors are capable of forming more than one hydrogen bond. Spencer et al. (1975) determined the enthalpy of interaction of each of the two protons of resorcinol (1,3-dihydroxybenzene) with four bases, including diethyl ether, in carbon tetrachloride by measuring the temperature depenedence of the O—H stretching frequency. It was found that the stepwise enthalpy changes for the addition of each proton acceptor are approximately the same ($\Delta H_1 = -6.1 \pm 0.4$, $\Delta H_2 = -5.9 \pm 0.2$ kcal mol^{-1} with diethyl ether), indicating that hydrogen bonding at one part of the molecule does not substantially alter the enthalpy of interaction at the other part of the molecule.

The negative entropy change, signifying that hydrogen bonding increases the order of the system, is greater for the second hydrogen bonding than for the first ($\Delta S_1 = -14$, $\Delta S_2 = -18$ J K^{-1} mol^{-1} with diethyl ether), since the first donor–acceptor interaction reduces the number of free sites available for the second. Thus $(-\Delta G_1) > (-\Delta G_2)$ and $K_1 > K_2$. In the case of catechol (1,2-dihydroxybenzene), the enthalpies of interactions of the two hydroxyl groups differ (as well as the entropies, free energies, and equilibrium constants) because the close proximity of the two hydroxyl groups causes the interaction at one to affect the interaction energy at the other.

The—NH$_2$ group is also theoretically capable of acting twice as a hydrogen-bond donor. Indeed, Spencer and co-workers (1978) found that aniline (PhNH$_2$) forms two hydrogen bonds with ethyl acetate, pyridine, or dimethylformamide using the pure-base calorimetric method. Since drugs are usually polyfunctional compounds, these considerations are important in understanding the interactions of drugs in solutions and are discussed further in this chapter and in Chapter 4.

5.4 CORRELATION AND PREDICTION OF STABILITY CONSTANTS FOR COMPLEXATION BETWEEN POLAR MOLECULES IN SOLUTION

In Chapter 4 and in other sections in the present chapter, the stability constants of solute–cosolvent (donor–acceptor) complexes were shown to determine the solubility of a solute in the presence of an interactive cosolvent or ligand. The total solubility of the solute in the presence of a given concentration of cosolvent or ligand is determined (1) by K_{DA} according to Eq. 5.17 for a single 1:1 complex; (2) by K_{DA_2} according to Eq. 5.19 for a single 1:2 complex; (3) by K_{DA} and K_{DA_2} according to Eq. 5.21 when both a 1:1 and 1:2 complex are formed; and (4) by K_{DA}, K_{DA_2}, and K_{DA_3} according to Eq. 5.23 when 1:1, 1:2, and 1:3 complexes are formed. If the relevant stability constants can be predicted, the solubility in the presence of a given concentration of cosolvent can also be predicted from the appropriate equation. In order to predict stability constants and other equilibrium constants from the properties or substituents of the interacting molecules, we must first examine the relevant linear free-energy relationships.

5.5 LINEAR FREE-ENERGY RELATIONSHIPS (LFERs)

5.5.1 Theory

The original linear free-energy relationships (LFERs) of Hammett (1940) have the general form

$$\log K = m \log K' + c \qquad (5.74)$$

where m and c are constants. K is the equilibrium (or rate) constant of one reaction and K' is the equilibrium (or rate) constant of the other reaction, with both reactions being subjected to the same variations in one condition (e.g., substituents in a reactant molecule), while all the other reaction conditions remain constant. According to this relationship, a plot of $\log K$ versus $\log K'$ should be linear with the slope m and the intercept c.

The term linear free-energy relationship can be understood as follows. The logarithm of the equilibrium constants (or rate constants) is proportional to the standard free-energy change associated with the reaction, ΔG^θ (or the standard free energy of activation) according to the well-known equation

$$\Delta G^\theta = -2.303 \, RT \log K \qquad (5.75)$$

Equation 5.74 therefore implies a linear relationship between the standard free-energy changes:

$$\Delta G^\theta = m\Delta G^{\theta'} + c_1 \qquad (5.76)$$

in which m has the same value as in Eq. 5.74, but the value of c is altered to $c_1 = -2.303 \, RTc$.

For a given reaction A, the standard free-energy change $\Delta G^{\theta A}$ may be considered to be a function of a number of independent variables x, y, and so forth. The variation of $\Delta G^{\theta A}$ at constant temperature and pressure can be expressed as the complete differential with respect to each of these variables as follows:

$$d\Delta G^{\theta} = \left(\frac{\partial \Delta G^{\theta A}}{\partial x}\right)_{T,P} dx + \left(\frac{\partial \Delta G^{\theta A}}{\partial y}\right)_{T,P} dy + \cdots \tag{5.77}$$

Provided the other variables remain unchanged and $(\partial \Delta G^{\theta A}/\partial x)_{T,P}$ remains constant, a variation of the variable x from a value x_0 to a value x_1 brings about the following change in $\Delta G^{\theta A}$:

$$\Delta G_1^{\theta A} - \Delta G_0^{\theta A} = \left(\frac{\partial G^{\theta A}}{\partial x}\right)_{T,P} (x_1 - x_0) \tag{5.78}$$

Application of Eq. 5.75 to Eq. 5.78 for the specific values $\Delta G_0^{\theta A}$, K_0 at $x = x_0$ and $\Delta G_1^{\theta A}$, K_1 at $x = x_1$, leads to

$$\log\left(\frac{K_1}{K_0}\right)^A = \frac{(\partial \Delta G^{\theta A}/\partial x)_{T,P}(x_0 - x_1)}{2.303\, RT} \tag{5.79}$$

For another reaction B, the effect of the same change in the variable x can be expressed by equations analogous to the previous two:

$$\log\left(\frac{K_1}{K_0}\right)^B = \frac{(\partial \Delta G^{\theta B}/\partial x)_{T,P}(x_0 - x_1)}{2.303\, RT} \tag{5.80}$$

Combining the last two equations affords

$$\log\left(\frac{K_1}{K_0}\right)^B = [\log(K_1/K_0)]^A \frac{(\partial \Delta G^{\theta B}/\partial x)_{T,P}}{(\partial \Delta G^{\theta A}/\partial x)_{T,P}} \tag{5.81}$$

This equation reflects the hypothesis that the effect of the change in the variable on reaction B may be considered to be composed of two parts. The first part expresses the effect on $\log(K_1/K_0)^A$, which depends solely on the change in the variable x. The second part represents the effect on $(\partial \Delta G^{\theta B}/\partial x)_{T,P} (\partial \Delta G^{\theta A}/\partial x)_{T,P}$, which expresses the relative sensitivities of reactions B and A to changes in x and which depends on the reaction conditions.

The relationship given by Eq. 5.81 has been applied to many types of chemical reaction and has been used to correlate much information on the effects of substituents on the reactants and changes in reaction conditions. For example, the well-known Hammett (1940) equation

$$\log(K/K_0) = \rho\sigma \tag{5.82}$$

where K and K_0 are the respective equilibrium constants (or rate constants) of the substituted and unsubstituted species, ρ is the reaction constant, and σ is a substituent constant.

5.5.2 LFERs for Phenol and Alcohol Interactions

The Hammett (1940) equation has been modified by Taft (1956) and has been applied by Higuchi et al. (1969) and by Hine (1975) to hydrogen-bonded systems. Figure 5.10 exemplifies the applicability of the Hammett equation to $K_{1:1}$ for two systems, namely, tributylphosphate and sarin, as proton acceptors, each with a series of phenols as proton donors. Similarly, Laurence and Wojtkowiak (1967, 1968) obtained linear correlations between substituent constants and the extent of hydrogen bonding, as measured by the shift in IR vibration frequency of OH stretching $\Delta\nu_{OH}$.

In hydrogen-bonding interactions, the change in free energy and hence $\log K_{DA}$ (or $\log K_{1:1}$) will depend on the hydrogen-donating and hydrogen-accepting abilities of the respective interacting species. Higuchi and co-workers (1969) obtained good (negative) linear correlations between $\log K_{1:1}$ and pK_a values of substituted phenols, the $K_{1:1}$ values referring to the interactions with

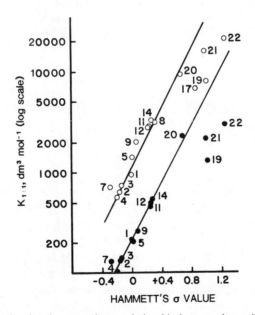

Figure 5.10 A plot showing the approximate relationship between the equilibrium constant $K_{1:1}$ (log scale) of TBP (○) and sarin (●) phenol complexes and the Hammett ρ values of the phenolic compounds, as listed in Table 5.1. $1\ dm^3\ mol^{-1} = 1\ L\ mol^{-1} = 1\ M^{-1}$. Reproduced with permission of the copyright owner, the American Pharmaceutical Association, from Higuchi, T., Richards, J. H., Davis, S. S., Kamada, A., Hou, J. P., Nakano, M., Nakano, N. I., and Pitman, I. H. (1969). *J. Pharm. Sci.*, **58**, 661–671.

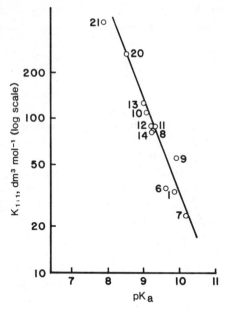

Figure 5.11 A plot of equilibrium constant $K_{1:1}$ (log scale) against pK_a for the interaction of griseofulvin with substituted phenols in carbon tetrachloride at 25°C. The numbers correspond to the phenols in Table 5.1. $1\,dm^3\,mol^{-1} = 1\,L\,mol^{-1} = 1\,M^{-1}$. Reproduced with permission of the copyright owner, the American Pharmaceutical Association, from Higuchi, T., Richards, J. H., Davis, S. S., Kamada, A., Hou, J. P., Nakano, M., Nakano, N. I., and Pitman, I. H. (1969). *J. Pharm. Sci.*, **58**, 661–671.

griseofulvin (Fig. 5.11 and Eq. 5.83) in carbon tetrachloride at 25°C.

$$\log K_{1:1}(\text{apparent}) = 8.01 - 0.651\,pK_a$$

$$r = 0.988, \qquad \text{s.d.} = 0.07 \tag{5.83}$$

The application of this equation to predict $K_{1:1}$ (apparent) is illustrated in Section 5.9. A similar correlation was found for the interaction of griseofulvin with organophosphorus compounds in carbon tetrachloride. These correlations indicate that the strength of the interaction increases with increasing hydrogen-donating ability of the phenols, the oxygen functions of griseofulvin and the phosphorus compounds presumably acting as hydrogen acceptors. Since the hydrogen-donating ability in these cases is expressed as $-pK_a$ in water, the correlation would presumably be improved if it were expressed in terms of an acidity function in the solvent actually used in the determination of the K_{DA} values.

The dipole–dipole interaction, which contributes greatly to hydrogen bonding, will be influenced by inductive effects of polar substituents on the proton donor and proton acceptor. Consequently, the stability constants of hydrogen-bonded complexes will be related to the polar substituent constant σ of the substituent groups. For example, $\log K_{1:1}$ for the interaction of griseofulvin with substituted phenols in carbon tetrachloride at 25°C gives the following correlations, which obey the Hammett equation (Eq. 5.82):

$$\log K_{1:1}(\text{apparent}) = 1.36\sigma + 1.57$$

$$r = 0.997, \qquad \text{s.d.} = 0.093 \tag{5.84}$$

$$\log K_{1:1}(\text{apparent}) = 1.43\sigma^* + 1.53$$

$$r = 0.998, \qquad \text{s.d.} = 0.070 \tag{5.85}$$

where σ and σ^* are the Hammett and Taft polar substituent constants, respectively.

Figure 5.12 shows another example in which $\log K_{1:1}$, derived from the influence of the hydrogen-accepting cosolvent butyl ether on the solubility of para-substituted phenols in isooctane, is related to σ_p according to the Hammett equation. It should be mentioned that the σ values were obtained from the effect of the substituents on the ionization of benzoic acids in water (Hine, 1975). Because the ionization constants in nonpolar media have not been determined, a direct correlation with acidity under the same solvent conditions, as suggested above, is impossible. Taft σ^- substituent constants give poor correlations with

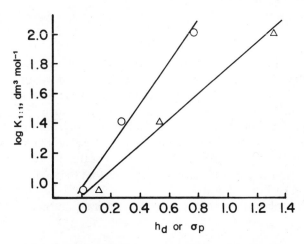

Figure 5.12   Plot of $\log K_{1:1}$ for the interaction of parasubstituted phenols (p-phenylphenol, p-iodophenol, and p-nitrophenol) with butyl ether in isooctane at 25°C versus the Hammett σ_p substitutent ($\triangle$) constant and versus Higuchi h_d values ($\bigcirc$). $1\,\text{dm}^3\,\text{mol}^{-1} = 1\,\text{L}\,\text{mol}^{-1} = 1\,M^{-1}$. Reproduced with permission of the copyright owner, the American Pharmaceutical Association, from Anderson, B. D., Rytting, J. H., and Higuchi, T. (1980). *J. Pharm. Sci.*, **69**, 676–680.

$\log K_{1:1}$ because resonance effects, which are included in the Taft values, influence hydrogen bonding much less than ionization (Barton, 1975).

Steric effects, as well as electronic effects, come into play when bulky groups are present close to the hydroxyl group of the phenol. Thus, the steric hindrance of the ortho-iodo groups accounts for the relatively low $K_{1:1}$ value and low solubility of 2,4,6-triiodophenol in the presence of dibutyl ether, a hydrogen-accepting solvent (see Tables 4.7 and 4.10). However, while solubility is quite sensitive to the functional groups in the solute and cosolvent, structural alterations which are not in the vicinity of the interacting group and which do not alter the acidity or basicity of the functional group exert little influence on the solubility. This is illustrated by the very small influence of hydrocarbon chain length of the cosolvent molecule on the stability constant of complexation and on solubility.

Similarly, the $K_{1:1}/\text{dm}^3 \text{ mol}^{-1}$ values for the interaction of carbazole with ethers varying in chain length are much the same, 3.53 for diethyl ether (Fung and Higuchi, 1971), 2.5 for dibutyl ether, and 1.9 for dipentyl ether, as shown in Table 4.10. This table also shows that $K_{1:1}/\text{dm}^3 \text{ mol}^{-1}$ for the interaction of p-phenylphenol with dibutyl ether and dipentyl ether are even closer, being 8.9 and 8.3, respectively. The solubilities in the pure ethers, however, differ by a much larger amount (Table 4.7), partly as a result of the greater concentration of acceptor oxygen atoms in the pure ethers of shorter chain length.

Other examples of the applicability of the Hammett equation (Eq. 5.82) to hydrogen bonding will now be considered. The solubility of salicylic acid in carbon tetrachloride is increased by the presence of various alcohols, and the 1:1 stability constants $K_{1:1}$, which describe the interaction, can be reasonably correlated with σ^* in the following equation:

$$\log (K_{1:1}/K_{1:1}^{n\text{-BuOH}}) = -0.4(\sigma^* + 0.13) \qquad (5.86)$$

where $K_{1:1}^{n\text{-BuOH}}$ is the stability constant for the n-butanol–salicylic acid system, σ^* is the Taft polar substituent constant for the R group in the alcohol ROH, and -0.4 is the value of the Taft reaction parameter ρ^*, describing the susceptibility of the reaction to polar effects (Taft, 1956). A preliminary discussion of the interaction between salicylic acid and alcohols was given on p. 158.

The question arises of whether salicylic acid or the alcohol is the proton donor in this interaction. The negative value of ρ^* indicates that the reaction is facilitated by electron release in the alcohol, which suggests that the alcohol is the hydrogen acceptor. The carboxylic proton of salicylic acid must be the hydrogen-bond donor, since the phenolic proton is involved in an intermolecular hydrogen bond. Thus, the $K_{1:1}$ suggests the following interaction (a):

(a)   (b)

Structure a is supported by the fact that $K_{1:1}$ for the salicylic acid–monochloroacetic acid interaction ($161\,dm^3\,mol^{-1}$) is less than that for the salicylic acid–acetic acid interaction ($340\,dm^3\,mol^{-1}$). If monochloroacetic and acetic acids were proton donors, the stronger one (monochloroacetic acid) would give the higher $K_{1:1}$, which is the opposite of the order observed. It is, however, possible that these acids may still be behaving as proton donors. The values of $K_{1:1}$, especially with acetic acid, suggest that these acids are capable of forming a second hydrogn bond in which they may be acting as proton donors to the phenolic hydroxyl of salicylic acid (b). This structure corresponds to a rather specific lock-and-key fit between the salicylic and acetic acid, which may be a prerequisite for extraordinarily high values of stability constants. This possibility has been mentioned in Chapter 4 (see Table 4.15).

The catechol–ROH systems in carbon tetrachloride are more complicated since they give rise to 1:1 and 1:2 complexes. The data can be correlated on the basis of the following three reasonable assumptions: (a) the 1:1 interaction is affected by polar properties but not by steric effects; (b) the 1:2 inteaction has the same susceptibility to polar effects as does the 1:1 complexation; (c) the 1:2 interaction is affected by the steric nature of the alcohol. The following linear free-energy relationships can be developed from the available data:

$$\log(K_{1:1}/K_{1:1}^{n\text{-BuOH}}) = -1.8(\sigma^* + 0.13) \tag{5.87}$$

$$\log(K_{1:2}/K_{1:2}^{n\text{-BuOH}}) = -1.8(\sigma^* + 0.13) + 0.7(E_s + 0.39) \tag{5.88}$$

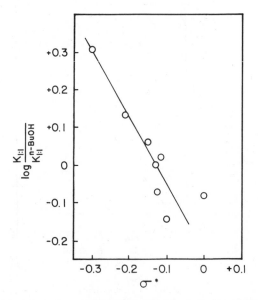

Figure 5.13 Correlation of $K_{1:1}$ for catechol–alcohol systems in carbon tetrachloride at 30°C with the Taft polar substituent constants. Reproduced with permission of the copyright owner, John Wiley and Sons, Inc. Higuchi, T. and Connors, K. A. (1965). *Advances in Analytical Chemistry and Instrumentation*, Vol. 4, John Wiley and Sons, Inc., New York.

where E_s is the steric parameter of Taft (1956) and 0.7 is the value of δ, which is a parameter that describes the sensitivity of the reaction to steric effects. The use of Eqs. 5.87 and 5.88 to predict $K_{1:1}$ and $K_{1:2}$ is illustrated in Section 5.9.

Figures 5.13 and 5.14 show plots of these equations and Table 5.3 gives the experimental values of the equilibrium constants and the estimates calculated from Eqs. 5.87 and 5.88. Agreement is fairly good for the $K_{1:1}$ values and satisfactory for the $K_{1:2}$ series, among which only less sterically hindered cyclohexanol gives a serious deviation. The higher susceptibility of the catechol–ROH reaction to polar ($\rho^* = -1.8$) effects compared with the susceptibility of the salicylic acid–ROH reaction ($\rho^* = -0.4$) may reflect the greater proton-donating ability (acidity) of salicylic acid, which probably swamps the susceptibility of the reaction to the polar properties of the alcohol.

The hydroquinone (or quinol)–alcohol interaction contrasts remarkably with the catechol–alcohol interaction. One might reasonably expect that both 1:1 and 1:2 interactions would be controlled largely by polar effects with little steric hindrance, but this is not the case. Instead, $K_{1:1}$ and $K_{1:2}$ are inversely proportional to each other, thus,

$$(K_{1:1})(K_{1:2}) = 200 \tag{5.89}$$

This equation implies a mutually exclusive relationship between first- and second-order complexation, such that the structural features conductive to one

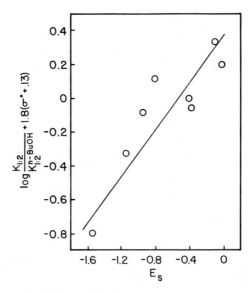

Figure 5.14 Correlation of $K_{1:2}$ for catechol–alcohol systems in carbon tetrachloride at 30°C with the Taft steric substituent constant E_s. Reproduced with permission of the copyright owner, John Wiley and Sons, Inc. Higuchi, T. and Connors, K. A. (1965). *Advances in Analytical Chemistry and Instrumentation*, Vol. 4, John Wiley and Sons, Inc., New York.

Table 5.3 Comparison of Observed and Calculated Stability Constants for the Salicylic Acid–Aliphatic Alcohol Interaction[a] in Carbon Tetrachloride at 30°C

R in ROH	$K_{1:1}(dm^3\,mol^{-1})^b$		$K_{1:2}(dm^3\,mol^{-1})$	
	Observed[a]	Calculated[c]	Observed[a]	Calculated[d]
Me	16.3	11.6	14.1	17.0
Et	14.2	17.5	29.3	22.8
n-Pr	20.8	19.3	12.7	15.4
n-Bu	19.8	19.8	15.4	15.4
i-Bu	16.8	16.1	10.3	5.3
s-Bu	26.9	27.6	10.2	6.6
tert-Bu	40.2	40.1	5.0	5.1
Cyclo-C_6H_{11}	22.8	21.5	22.0	8.9

[a] Higuchi, T. and Chulkaratana, S., unpublished observations.
[b] $1\,dm^3\,mol^{-1} = 1\,L\,mol^{-1} = 1\,M^{-1}$.
[c] With Eq. 5.87.
[d] With Eq. 5.88.

Reproduced with permission of the copyright owner, John Wiley and Sons, Inc. Higuchi, T. and Connors, K. A. (1965). *Advances in Analytical Chemistry and Instrumentation*, Vol. 4, John Wiley and Sons, Inc., New York, pp. 117–212.

are unfavorable to the other. A preliminary discussion of the hydroquinone–alcohol interaction was presented in Section 4.4.2.

5.5.3 LFERs for Carboxylic Acids and Their Anions

Carboxylic acids and their salts interact with many substances in aqueous solution (Higuchi and Connors, 1965). Some series (e.g., theophylline with carboxylates) give a rough correlation between the polar nature of the substituent and the extent of complex formation. None of these correlations can be expressed by Hammett and Taft relationships, probably because more than one mechanism is involved. The carboxylates are less amenable to such correlations than the corresponding free acids because of the incursion of resonance between the polar substituent (e.g., p-nitro) and the carboxyl group.

Carboxylates complex with many molecules, including theophylline, prednisolone, and phenacetin (Higuchi and Pisano, 1964). The $K_{1:1}$ values for a series of carboxylates with one of these compounds have similar proportions to the $K_{1:1}$ values for the same series of carboxylates with another of the compounds, which suggests that a specific lock-and-key fit is not a requirement for complexation with these systems.

Complexation is augmented by increasing the planar aromatic surface of the carboxylates, which suggests that hydrophobic, London, and Debye interactions contribute signficantly to the interaction. Complexation between theophylline and various halobenzoates gave rise to a rather broad UV absorption band at about 295 nm which can be attributed to charge-transfer interactions (Higuchi and Pisano, 1964). Hydrogen bonding probably does not contribute much to

Table 5.4 Comparison of Observed and Calculated Stability Constants $K_{1:1}$[a] for Dihydroxybenzoates[b,c] in Water at 30°C

Dihydroxy-benzoate	Theophylline		Hydrocortisone		Prednisolone		Acetophenetidin	
	Observed	Calculated	Observed	Calculated	Observed	Calculated	Observed	Calculated
2,4	56	45	13	10	12	10	2.4	2.3
2,5	43	42	10	12	12	12	2.4	2.1
2,6	300	75	13	18	11	14	3.3	3.6
3,4	14	25	6.8	6.9	6.4	8.4	1.5	1.3
3,5	20	23	5.6	8.3	6.5	10	2.0	1.2

[a] $1\,dm^3\,mol^{-1} = 1\,L\,mol^{-1} = 1\,M^{-1}$.
[b] From Higuchi and Drubulis, 1961.
[c] Calculated using Eq. 5.91.

Reproduced with permission of the copyright owner, John Wiley and Sons, Inc. Higuchi, T. and Connors, K. A. (1965). *Advances in Analytical Chemistry and Instrumentation*, Vol. 4, John Wiley and Sons, Inc., New York, pp. 117–212.

complexation in these aqueous systems, as indicated by the constancy of $K_{1:1}$ for the theophylline–trihydroxybenozate and theophylline–trimethoxybenzoate systems (Higuchi & Connors, 1965).

Comparison of dihydroxybenzoate complexation with monohydroxybenzoate complexation indicates that the additional hydroxyl group enhances the interaction (Higuchi and Drubulis, 1961). The addition allows two possibilities to arise. (a) If the added group increases the strength of the interaction by forming a complex with essentially the *same structure* as the monofunctional compounds, the *energies* of the interaction should be approximately additive. (b) If the new group leads to the formation of a new complex of *different structure*, the *extents* of interaction should be roughly additive. For the present case, let A and B represent different monohydroxybenzoates, D the corresponding dihydroxybenzoates, and O the unsubstituted benzoate. Then (a) can be represented by

$$\Delta G_{D}^{\theta} = \Delta G_{A}^{\theta} + \Delta G_{B}^{\theta} - \Delta G_{O}^{\theta} \tag{5.90}$$

where ΔG_{O}^{θ} is subtracted to ensure that each side of the equation has a contribution from just one aromatic ring and just one carboxyl group. Since $\Delta G^{\theta} = -RT \ln K$, Eq. 5.90 can be expressed as

$$K_{D} = K_{A} K_{B} / K_{O} \tag{5.91}$$

On the other hand, (b) can be expressed by

$$K_{D} = K_{A} + K_{B} - K_{O} \tag{5.92}$$

In the present systems, Table 5.4 shows that Eq. 5.91 predicts the data reasonably well, which supports (a).

5.5.4 LFERs for Ester Interactions

Esters may interact with many types of compound in both aqueous and nonaqueous environments. Thus, for example, ester–xanthine complexes are formed in water, thereby rendering the ester group resistant to hydrolysis (Higuchi and Lachman, 1955, 1957; Lachman et al., 1956, 1957). In nonpolar solvents, phenols, acids, and other proton donors interact with esters by hydrogen bonding with the ester acting as the acceptor. This has already been mentioned in Section 4.4.1 and in Tables 4.13 and 4.14.

It is interesting to consider the data of Chulkaratana (unpublished observations, 1961), in tabulated form (Higuchi and Connors, 1965) for the interaction of salicylic acid, catechol, or quinol (hydroquinone) with the diesters of straight-chain dicarboxylic acids or diacyl glycols in carbon tetrachloride at 30°C. The data show that within a homologous series of esters, $K_{1:1}$ increases with increasing chain length until it reaches a limiting value. This increase can be accounted for by assuming that the acid or phenol can interact with only a single ester group. Those diesters in which the ester groupings are not widely separated will interact no more strongly than the monoester, because interaction at one grouping will sterically hinder interaction at the other grouping. However, a diester with widely separated groups provides twice the probability of independent interaction with the acid or phenol, so the maximum limiting value of $K_{1:1}$ for diesters should be twice that for a similar monoester. For the dimethyl esters studied here, the best comparison compound might be methyl n-octanoate, but ethyl acetate is probably a satisfactory substitute. Table 5.5 shows the stability constants and the ratio of the limiting value to the constant for ethyl acetate; both values are reasonably close to 2. The same argument may be applied to the limiting value for diacetylglycols, for which the observed ratio is also about 2.

The values of $K_{1:2}$ for the interaction of catechol or hydroquinone with the esters are undoubtedly influenced by steric effects when the chain length is small. The limiting value of $K_{1:2}$, however, though affected by the sizes and shapes of the reactants, may be expected to be less susceptible to such effects than when the ester functions are close together. Statistical arguments suggest that the limiting values are related as

$$K_{1:2} = K_{1:1}/4 \quad \text{or} \quad K_{1:1} = 4K_{1:2} \tag{5.93}$$

$K_{1:1}$ is favored over $K_{1:2}$ by a factor of 2 by the availability of twice as many ester groups in forming the 1:1 complex and by a further factor of 2 because of the greater porbability of dissociation of the 1:2 complex. The total statistical effect gives a ratio of 4, and the observed ratios are close to this.

Table 5.5   Statistical Effect on the Apparent Stability Constants[a] of Difunctional Esters[b] in Carbon Tetrachloride at 30°C

Item	Salicylic Acid	Catechol	Hydro-quinone
A Ethyl acetate ($K_{1:1}$)	4.8	25	10.3
B Dimethyl Ester limiting value ($K_{1:1}$)	8.0	53	24
C Ratio B/A	1.8	2.1	2.3
D Diacetyl glycol limiting value ($K_{1:1}$)	11.0	51	—
E Ratio D/A	2.3	2.0	—
F Dimethyl ester limiting value ($K_{1:2}$)	—	12	6.8
G Ratio B/F	—	4.4	3.5

[a] $K_{1:1}$ or $K_{1:2}$ (1 dm^3 mol^{-1} = 1 L mol^{-1} = 1 M^{-1}).

[b] Higuchi, T. and Chulkaratana, S., unpublished data, Systems 460–511, from Table IV in Higuchi and Connors (1965).

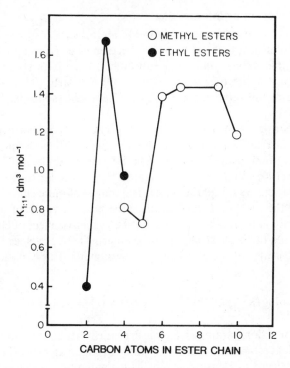

Figure 5.15 Interaction constants $K_{1:1}$ of catechol with straight-chain dicarboxylic acid esters in carbon tetrachloride at 30°C as functions of chain length. $1\,dm^{-3}\,mol^{-1} = 1\,L\,mol^{-1} = 1\,M^{-1}$. Reproduced with permission of the author. Chulkaratana, S. (1964). Ph.D. thesis, Hydrogen Bonding and Solubility in Nonaqueous Systems, University of Wisconsin, Madison, WI.

The strength of the interaction, as indicated by the values of $K_{1:1}$, between catechol and various dicarboxylic diesters, $ROOC\cdot(CH_2)_{n-2}\cdot COOR$, $R = Me$ or Et, in carbon tetrachloride at 25°C is plotted in Figure 5.15 against the length (n value) of the ester chain. $K_{1:1}$ peaks very sharply at $n = 3$, troughs at $n = 5$, and peaks broadly at $n = 6-9$. This behavior clearly indicates that a specific steric fit of the lock-and-key type occurs when $n = 3$ and when $n = 6-9$ in the structure depicted below. When $n = 1$, 4, or 5, $K_{1:1}$ is low, presumably because of severe steric strain.

5.5.5 LFERs for Ether Interactions

Although ethers possess relatively weak intermolecular forces and are rather unreactive chemically, they are powerful electron donors (and proton acceptors) and form complexes with numerous substances. They form Hassel-type charge-transfer complexes with the halogens (see p. 80). Similarly, polyethylene glycol, a polyether, forms insoluble complexes in iodine–iodide solutions (Guttman and Higuchi, 1955).

Stability constants in a nonpolar environment for ethers complexed with salicylic acid (Table 5.6), catechol, or p-phenylphenol have been determined by Chulkaratana (unpublished observations, 1961) and tabulated fully by Higuchi and Connors (1965). Plots of log $K_{1:1}$ against the Taft polar substituent constant σ^* gave similar curves for all the series. The shapes of these curves suggest steric hindrance to the reaction (which reduces the basic strength of the ether). The steric hindrance is partially overcome by the inductive effect of electron release (which increases the basic strength of the ether). The $K_{1:1}$ values for four ethers and three substrates fit the following four-parameter Taft equation reasonably well:

$$\log(K_{1:1}/K_{1:1}^{n\text{-Bu}_2\text{O}}) = -4.6(\sigma^* + 0.13) + (E_s + 0.39) \qquad (5.94)$$

The steric reaction parameter E_s was arbitrarily set at unity in the absence of any independent estimate. The negative value of ρ indicates that the interaction is facilitated by electron-releasing groups which enhance the basic strength of the ether. The converse is true for positive values of ρ. The large numerical value of ρ indicates that the interaction is extremely sensitive to these polar effects.

Examination of Table 5.6 gives an interesting qualitative picture of the interplay between electronic and steric effects in the interaction of ethers with salicylic acid in carbon tetrachloride at 30°C. ($K_{1:1}$ for the difunctional ethers should be divided by 2 to facilitate comparison with $K_{1:1}$ for the monofunctional compounds.) Dimethoxyethane ($CH_3OCH_2CH_2OCH_3$) is a weaker base electronically but a stronger base sterically than diethyl ether, but since $K/2 (= 0.238_5)$ is less than $K (= 0.261)$ for diethyl ether, the electronic factor predominates in the comparison. On the other hand, $CH_3O(CH_2)_8OCH_3$ ($K/2 = 0.318$) has two virtually independent $-OCH_3$ groups which confer steric advantages over $CH_3OCH_2CH_2OCH_3$ ($K/2 = 0.238_5$) and over the mono-functional ethers. 1,4-Dioxane suffers from electronic suppression in its

Table 5.6 Formation Constants $K_{DA} = [DA]/[A]$ for Salicylic Acid Complexes with Aliphatic Ethers in Carbon Tetrachloride at 30°C

Diethyl ether	0.261	Dimethoxyethane	0.477
Di-n-propyl ether	0.186	Dimethoxyoctane	0.636
Di-n-butyl ether	0.184	1,4-Dioxane	0.330
Diisopropyl ether	0.240		

Reproduced with permission of the author. Chulkaratana, S. (1964). Ph.D. thesis, Hydrogen Bonding and Solubility in Nonaqueous Systems, University of Wisconsin, Madison, WI.

interaction as a base, but has no steric hindrance; since, however, $K/2 (= 0.165)$ is less than for the monofunctional ethers, the former effect predominates.

5.5.6 LFERs for Amide Interactions

Amides are good hydrogen-bond acceptors (electron donors). A hydrogen atom attached to the nitrogen in a primary or secondary amide can form the hydrogen-bond donor. These features give strong amide–amide interactions which lead to high melting and boiling points. If the nitrogen is fully substituted, this particular self-association due to hydrogen bonding is eliminated and the amide can interact with other substances by acting as a hydrogen-bond acceptor. An example which has previously been discussed on p. 158 and 169 (see also Fig. 4.3) is the interaction between salicylic acid and N,N-dimethylacylamides (N,N-dimethylacetamide and its higher acyl homologues).

The interaction between tetramethyldiamides L and hydroquinone (quinol) S in carbon tetrachloride at 30°C, in general, proceeds by both 1:1 and 1:2 complexation. Figure 5.16 shows that 1:1 complexation is only significant when at least six methylene groups separate the two amide groups. $K_{1:1}$ rises steeply to a maximum of between 8 and 12 methylene groups and then decreases for greater numbers of methylene groups. This behavior suggests the 1:1 complexation structure

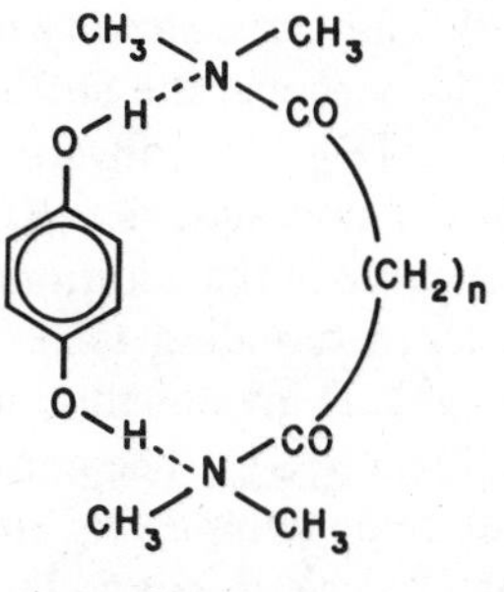

This structure enables one easily to imagine that n must be appreciable to maintain a good steric fit, but if n is too large, greater flexibility might discourage 1:1 complexation. In contrast, $K_{1:2}$ has generally significant values for virtually any number of methylene groups, presumably because SL_2-type complexation, according to the structure shown below, is sterically rather undemanding.

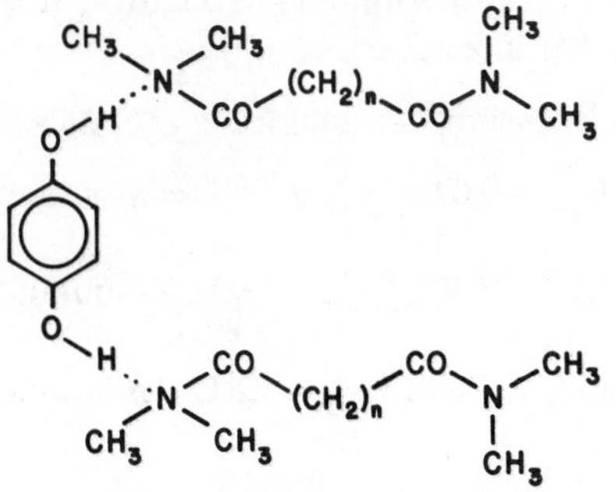

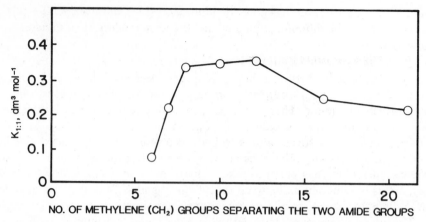

Figure 5.16 A plot showing the interaction constants $K_{1:1}$ of hydroquinone with N, N, N', N'-tetramethyldiamides in carbon tetrachloride at 30°C as a function of chain length. $1\,dm^3\,mol^{-1} = 1\,L\,mol^{-1} = M^{-1}$. Reproduced with permission of the author. Shami, E. G. (1964). Ph.D. thesis, Formation of Hydrogen Bonded Complexes by Amides in Nonaqueous Solvents, University of Wisconsin, Madison, WI.

5.5.7 Solubility and Type of Molecular Association

The solubility $[S]_o$ of a given solute S in a given solvent I is closely related to the propensity of S for molecular association with an added cosolvent L, as represented by $K_{1:1}$ or $K_{1:2}$. The stronger the S–S interactions, as compared with the S–I interactions, the lower is $[S]_o$. The converse is also true. (We note that $[S]_o$ depends on both S–S and S–I interactions.) When L is added, the apparent solubility of S is increased because of the additional S–L interactions. When the S–I interactions are weak (corresponding to a lower $[S]_o$), then $K_{1:1}$ or $K_{1:2}$ resulting from the addition of L is greater than when the S–I interactions are strong (corresponding to a higher $[S]_o$). The converse is also true.

Table 5.7 lists 8 ($= 2^3$) extreme complexing situations in which (a) the S–L interactions are rated strong or weak, (b) the S–I interactions are rated strong or weak, and (c) the solubility of S in I alone, that is, $[S]_o$, is rated high or low. Where possible, the extent of complex formation has been predicted. The data presented by Higuchi and Connors (1965) provide good examples of Cases $\alpha 1$ and $\alpha 2$ (Table 5.7).

The inverse relationships, suggested above, between the apparent stability constants and $[S]_o$, the intrinsic solubility of S in the solvent I in the absence of L are of the following form in case α.

For p-nitrophenol in carbon tetrachloride, cyclohexane, and heptane:

$$K_{1:1} = 0.046/[S]_o + 39\,(L = n\text{-butanol}) \tag{5.95}$$

$$K_{1:2} = 0.006/[S]_o + 19\,(L = n\text{-butanol}) \tag{5.96}$$

$$K_{1:2} = 0.032/[S]_o + 12(L = n\text{-butanol}) \tag{5.97}$$

Table 5.7 Predicted Effects of Relative Strengths of Intermolecular Interactions and Solute Solubility on the Strength of Complexation with an Added Ligand (Cosolvent) as Measured by $K_{1:1}$ or $K_{1:2}$

| Case | Relative Strengths of Inter-molecular Interactions | | $[S]_o{}^b$ | $K_{1:1}$ or $K_{1:2}{}^c$ |
	S–L[a]	S–I[a]		
$\alpha 1$	Strong	Weak	Low	High
$\alpha 2$	Strong	Weak	High	Low
$\beta 1$	Weak	Weak	Low	Low
$\beta 2$	Weak	Weak	High	High
$\gamma 1$	Strong	Strong	Low	?
$\gamma 2$	Strong	Strong	High	?
$\delta 1$	Weak	Strong	Low	?
$\delta 2$	Weak	Strong	High	Low

[a]S represents the solute. I represents the original solvent. L represents the added ligand (cosolvent).
[b]$[S]_o$ represents the solubility of S in I alone and depends on the relative strengths of the S–S and S–I interactions.
[c]$K_{1:1}$ and $K_{1:2}$ represents the extent of complexation between S and L according to Eqs. 4.5 and 4.6 or Eqs. 5.33 and 5.35.

Reproduced with permission of the copyright owner, John Wiley and Sons, Inc. Adapted from Higuchi, T. and Connors, K. A. (1965). *Advances in Analytical Chemistry and Instrumentation*, Vol. 4, John Wiley and Sons, Inc., New York.

$$K_{1:1} = 0.0065/[S]_o + 55$$

$$(L = \text{ethyl acetate, } n\text{-propyl propanoate, } n\text{-butyl butanoate}) \qquad (5.98)$$

For *p*-phenylphenol in carbon tetrachloride, cyclohexane, and heptane,

$$K_{1:1} = 0.013/[S]_o + 9.3 \, (L = n\text{-octanol}) \qquad (5.99)$$

$$K_{1:1} = 0.009/[S]_o + 9.3 \, (L = n\text{-butanol}) \qquad (5.100)$$

The intercept appears to represent the limiting value of the stability constant, below which the constant cannot be expected to fall as long as the system belongs to Case α.

Cases γ and δ are exemplified by the sarcosine anhydride systems. Detailed statements about these cases cannot yet be made since they depend intimately on the chemical nature of the S–L solvent system. It seems, however, that K is roughly an inverse function of $[S]_o$.

Case β systems are rather rare but may be exemplified by the complexation behavior of *p*-hydroxybenzoic acid with some tetraalkylamines in aqueous solution, for which the apparent stability constants increase with the solubility of the amide. Salicylic acid complexation, however, showed no correlation with amide solubility.

5.6 ASSIGNING DONOR AND ACCEPTOR CONTRIBUTIONS TO THE STABILITY CONSTANTS

5.6.1 Separation of the Donor and Acceptor Contributions to K_{DA}

The simplest possible equation for separating the donor and acceptor contributions to the stability constant is of the form

$$K_{DA} = k_D k_A \tag{5.101}$$

where K_{DA} is the stability constant of the complex DA and k_A and k_D are parameters reflecting the bonding capabilities of the hydrogen acceptor and donor (Higuchi et al., 1969). The existence of this relationship is supported by a proportionality between K_{DA} values for the interaction of a given acceptor molecule with various phenolic hydrogen donors and K_{DA} values for the

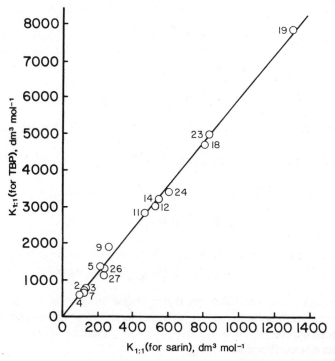

Figure 5.17 A plot showing the linear relationship between equilibrium constants $K_{1:1}$ of complexes formed by TBP and sarin with a series of phenolic compounds in cyclohexane at 25°C. The phenolic compounds are numbered as in Table 5.1. $1 \, dm^3 \, mol^{-1} = 1 \, L \, mol^{-1} = 1 \, M^{-1}$. Reproduced with permission of the copyright owner, the American Pharmaceutical Association, from Higuchi, T., Richards, J. H., Davis, S. S., Kamada, A., Hou, J. P., Nakano, M., Nakano, N. I., and Pitman, I. H. (1969). *J. Pharm. Sci.*, **58**, 661–671.

interaction of a different A molecule with the same donors. Equation 5.101 is an example of a free-energy relationship for which the general form may be represented by Eq. 5.81. Figure 5.17 shows such a relationship, which is clearly linear and passes through the origin. The data were taken from Table 5.1. If an arbitrary number is assigned to the k_D value of one donor (namely, $k_D = 10$ for phenol itself), it is possible to obtain k_A values for a few hydrogen acceptors, including TBP and sarin, and for the k_D values of several substituted phenols. Good agreement is obtained between the k_D values of these phenols derived from different series of interactions, which again supports Eq. 5.101.

When the equation was tested (Higuchi et al., 1969) with systems involving nonphenolic hydrogen donors, it was found that plots of K_{DA} values for the interaction of two given hydrogen donors with a similar series of acceptors, or for the interaction of two given hydrogen acceptors with a similar series of hydrogen donors, do not always give straight lines which pass through the origin. It therefore appears that Eq. 5.81 is not general but is limited to systems involving a series of chemically similar hydrogen donors or acceptors. This point will be further discussed later.

In hydrogen-bonding interactions, the change in free energy will depend on the hydrogen-donating and hydrogen-accepting abilities of the respective molecules, and these will in turn depend on the structures of the two molecules. It is therefore reasonable to expect that the interaction constant should obey an equation such as Eq. 5.81 in which the factor on the left, designated h_D, reflects the effect of change in structure of the hydrogen donor, and the ratio on the right, designated h_A^{-1}, gives a measure of the effect of structural change in the hydrogen acceptor. Equations 5.81 or 5.101 may therefore be written

$$\log (K_1/K_0)_A = h_D h_A \tag{5.102}$$

where K_1 and K_0 are the equilibrium constants for the separate interactions of two hydrogen donors with a common hydrogen acceptor A. The factor h_D is equal to $\log(K_1/K_0)_A$, where K_1 and K_0 are in this case the equilibrium constants for the interaction of the previous two donors with a given hydrogen acceptor A_o, which is chosen as an arbitrary standard ($h_A = 1$) [e.g., tri-n-butylphosphate (TBP) as the hydrogen acceptor for the series of phenols]. The factor h_A, which replaces the ratio of the partial differentials, reflects the effect of any acceptor A on the hydrogen-bonding equilibrium relative to that of A_o. Numerical examples exemplifying Eq. 5.102 are given in Section 5.9.

Reference to Eq. 5.102 and to the definition of h_A indicates that h_A is unity for interactions involving the arbitrarily chosen standard hydrogen acceptor A_o. A standard hydrogen donor must also be chosen in order that h_D values may be compared quantitatively, and from the definition of h_D, this quantity is zero for the arbitrarily chosen standard donor.

This concept has been generally confirmed by Taft and co-workers (Taft et al., 1969; Joris et al., 1972; Mitsky et al., 1972), who found that the stability constants K_0 for 1:1 complexes of a series of bases in carbon tetrachloride followed the relationship

$$\log K = m'' \log K_0 + c'' \tag{5.103}$$

where K_0 is the stability constant in carbon tetrachloride for the reference proton donor (in this case p-fluorophenol) with each base; m'' and c'' are empirical constants. (Taft designated $\log K_0$ as pK_{HB}). An excellent correlation was obtaned for several hydroxylic proton donors with a wide variety of bases.

5.6.2 Choice of Solvent and Temperature in Assigning Contributions to K_{DA}

The definitions of h_D and h_A given above assume that the solvent exerts a constant minimal effect on the interaction between any donor–acceptor combination. This means that Eqs. 5.81 and 5.102 will only be useful if solvent effects, especially those of a specific nature, can be ignored. Because of this limitation, the solvent used for the reference interaction should be inert, and saturated hydrocarbons, such as hexane, heptane, isooctane, and cyclohexane, have been chosen as being satisfactory. We shall return later to the effect of the solvent in Section 5.7.

Hydrogen-bonding interactions involve enthalpy and entropy changes and are therefore sensitive to changes in temperature. Consequently, it is necessary to specify a fixed temperature for the reference interaction, and a value of 25°C has been generally chosen because most studies have been made either at this temperature or over a range that includes this value.

5.6.3 Choice of the Reference Hydrogen-Bond Donor for $h_D = 0$

Phenol has been chosen by Higuchi et al. (1969) as the reference hydrogen donor (i.e., $h_D = 0$) because it is a strong donor, and because phenol and many substituted phenols have been involved in many of the more recent and more reliable investigations of hydrogen-bonding equilibria with a wide variety of acceptors. It should be pointed out, however, that the p-fluorophenol is the favored reference hydrogen-bond donor for calorimetric studies (Taft et al., 1969; Arnett et al., 1970; 1974; Spencer et al., 1979) and ^{19}F NMR studies (Gurka and Taft, 1969).

5.6.4 Choice of the Reference Hydrogen-Bond Acceptor for $h_A = 1$

The choice of reference hydrogen-bond acceptor was more difficult since relatively few investigations in an inert solvent had been involved with a single hydrogen acceptor. After careful consideration of the hydrogen acceptors, (i.e., bases) listed in Table 5.8, tri-n-butylphosphate (TBP) was selected by Higuchi et al. (1969) as the reference acceptor, for which $h_A = 1$. Table 5.8 shows that the h_A values (and the $\log K_0$ values) for sarin, l-PTP, trimethylamine, and methyl acetate in saturated hydrocarbon solution at 25°C increase with increasing strength of the base, as expected. These values were obtained from plots of $\log K_{DA}$ for each of these series (Table 5.1) against the h_D values derived from the

Table 5.8 Values of h_A and $\log K_{o(calc)}$ for Hydrogen-Bonding Interactions in Saturated Hydrocarbon Solution at 25°C

Hydrogen Acceptor	Solvent	h_A	s^a	r^b	n^c	$\log K_{o(calc)}{}^d$
TBP[e]	Cyclohexane	1.00	—	—	—	—
Sarin[e]	Cyclohexane	0.90	0.073	0.986	18	2.25
l-PTP[e]	Cyclohexane	1.63	0.179	0.978	9	2.30
$(CH_3)_3N^f$	Cyclohexane	0.74	0.022	0.996	4	1.94
$CH_3 \cdot COOCH_3{}^g$	n-Heptane	0.35	0	1.000	4	1.08

[a] s = SD of experimental measurements from the regression line.
[b] r = correlation coefficients.
[c] n = number of interactions involved in the calculation of h_A.
[d] $\log K_{o(calc)}$ = intercept of regression line with ordinate ($h_D = 0$).
[e] Calculations based on results given in Table 5.1. For chemical names see Table 5.1.
[f] Calculations based on results given by Denyer et al. (1955).
[g] Calculations based on results given by Nagakura (1954).

Reproduced with permission of the copyright owner, the American Pharmaceutical Association, from Higuchi, T., Richards, J. H., Davis, S. S., Kamada, A., Hou, J. P., Nakano, M., Nakano, N. I., and Pitman, I. H. (1969). J. Pharm. Sci., **58**, 661–671.

TBP series by a correlation method similar to that outlined by Jaffé (1953) (Fig. 5.18). The application of Table 5.8 for predicting $K_{1:1}$ values is given in Section 5.9.

The TBP series was the most suitable reference acceptor available to Higuchi et al. (1969), but it is subject to the disadvantage that the TBP molecule has more than one electrophilic negative center capable of associating with a hydrogen donor. Although the intermolecular association is likely to involve only the coordinated oxygen atom of the phosphate groups, the presence of the remaining oxygen atoms detracts somewhat from the ideality of TBP as the reference hydrogen acceptor.

5.6.5 Comparisons of h_D Values for Various Hydrogen-Bond Donors

The values of h_D shown in Table 5.9 were calculated from the results of each series by means of the following equation:

$$h_D = \bar{h}_D - (\bar{Y} - Y)r^2/h_A \qquad (5.104)$$

where $\bar{h}_D$ is the mean value of h_D, r is the correlation coefficient, $Y = \log K_{DA}$, and $\bar{Y}$ is the mean value of Y. Inspection of each horizontal row in Table 5.9 shows that the h_D values obtained for the same hydrogen donor from the different series agree quite well, supporting the general applicability of Eq. 1.102. A list of preferred h_D values is also shown in Table 5.9 and is based on the average or weighted average of available results. Inspection of the vertical columns in Table 5.9 shows that the h_D value increases with increasing proton-donating ability (i.e., acidity) of the proton donor. For example, h_D increases in the rank

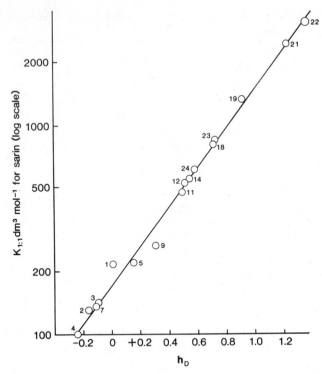

Figure 5.18 A plot of $K_{1:1}$ (log scale) for sarin complexes against the h_D values derived for this series from the TBP series using Eq. 5.102. $1 \, dm^3 \, mol^{-1} = 1 \, L \, mol^{-1} = 1 \, M^{-1}$. Reproduced with permission of the copyright owner, the American Pharmaceutical Association, from Higuchi, T., Richards, J. H., Davis, S. S., Kamada, A., Hou, J. P., Nakano, M., Nakano, N. I., and Pitman, I. H. (1969). *J. Pharm. Sci.*, **58**, 661–671.

order amines $\sim$ alcohols < alkylphenols < phenol < halophenols < *p*-nitrophenol. The application of Table 5.9 for predicting $K_{1:1}$ values is given in Section 5.9.

The extensive work of Gramstad and van Binst (1966) (see Higuchi et al., 1969) shows that Eq. 5.79 is not restricted to phenolic hydrogen donors. Using the interaction between triethylphosphine oxide and phenol in carbon tetrachloride at 20°C as the reference interaction (i.e., for triethylphosphine oxide $h_A^* = 1$ and for phenol $h_D^* = 0$), it is possible to calculate h_D^* values for 1-naphthol, methanol, indole, carbazole, and pyrrole. (The asterisks in h_A^* and h_D^* indicate that these series of values are based on a different reference interaction from that used in the calculation of h_A and h_D values.) The initial set of h_D^* values may then be used to determine the h_A^* values of the series of hydrogen acceptors shown in Table 5.10 in a similar manner to that described previously for Table 5.9. The calculated h_D^* values obtained from the equilibrium constants for each hydrogen donor with the various bases in the horizontal rows also show a reasonable agreement with only

Table 5.9 h_D Values for Hydrogen Donors in Saturated Hydrocarbon Solvents at 25°C

Solvent Hydrogen Donor	Cyclohexane Hydrogen Acceptor				n-Heptane	Preferred Values
	TBP[a]	Sarin[a]	$(CH_3)_3N$[b]	l-PTP[a]	CH_3-$COOCH_3$[c]	
Phenol	0	0	0	0	0	
p-CH_3-phenol	−0.17	−0.14	−0.17		−0.16	−0.16
p-C_2H_5-phenol	−0.10	−0.11				−0.11
p-t-C_4H_9-phenol	−0.21	−0.27		−0.34		−0.27
p-C_6H_5-phenol	0.15	0.09				
p-OCH_3-phenol	−0.12	−0.12		0.07		−0.12
p-F-phenol	0.30	0.19		0.23		0.24
p-Cl-phenol	0.49	0.47	0.49	0.58	0.48	0.48
p-Br-phenol	0.50	0.51		0.46		0.49
p-I-phenol	0.54	0.53		0.56		0.54
2,3,5,6-Tetra-F-phenol		0.03				
Penta-F-phenol		0.29				
p-$COCH_3$-phenol	0.85					
p-COC_2H_5-phenol	0.70	0.72				0.71
p-CHO-phenol	0.92	0.93				0.93
m-CN-phenol	0.98	1.21				
p-CN-phenol	1.24	1.18				1.21
p-NO_2-phenol	1.34	1.34		1.32		1.33
Catechol	0.74	0.73		0.64		0.74
4-i-C_3H_7-catechol	0.56	0.48				0.57
4-NO_2-catechol	0.82					
1-Naphthol	0.14	0.18	0.14		0.17	0.16
2-Naphthol	0.08	0.18	0.57		0.26	0.17
Triphenylmethanol			−1.15			
n-Heptanol			−1.87			
2,5-Dimethylpyrrole			−1.87			
Indole			−1.32			

[a]Calculations based on results given in Table 5.1.
[b]Calculations based on results given by Denyer et al. (1955).
[c]Calculations based on results given by Nagakura (1954).

one or two exceptions, emphasizing the general applicability of Eq. 5.102 to hydrogen-bonded systems. The h_D^* values also increase with increasing proton-donating ability of the H donors.

Figure 5.12 shows that log $K_{1:1}$, derived from the solubility data of para-substituted phenols in dibutyl ether, is a linear function of the h_D values which were derived from both thermodynamic and spectroscopic data. The correlation further emphasizes that the $K_{1:1}$ values reflect hydrogen-bonding interactions.

Table 5.10 h_A^* and h_D^* Values for Various Hydrogen Donors in Carbon Tetrachloride at 20°C

Hydrogen Acceptor	$(C_2H_5)_3PO$	$(C_6H_5)_3PO$	C_5H_5N	$N(C_2H_5)_3$	$(CH_3)_3PO_4$	$CCl_3P(O)(OC_2H_5)_2$	$HP(O)(OC_2H_5)_2$	$(C_2H_5)_2N-P(O)(OC_2H_5)_2$
h_A^* Value	1	0.88	0.83	0.65	0.71	0.60	0.69	0.71
H-Donor				h_D^* Values				
Phenol[a]	0	0	0	0	0	0	0	0
1-Naphthol[b]	0.18	0.06	0.17	0.12	0.21	0.08	0.09	0.06
Carbazole[b]	−1.24	−1.22	−1.53					
Indole[b]	−1.37	−1.40	−1.41		−1.39	−1.42	−1.42	−1.43
Pyrrole[b]	−1.64	−1.69	−1.60	−1.77				
Methanol[c]	−1.74	−1.85	−1.20		−1.69	−1.72	−1.72	−1.80

[a] Aksnes, G. and Gramstad, T., *Acta Chem. Scand.,* **14** (1960) 1485.

[b] Gramstad, T., *Acta Chem. Scand.,* **15** (1961) 1337.

[c] Gramstad, T. and Fuglevik, W. J., *Spectrochim. Acta,* **21** (1965) 503.

5.7   SOLVENT EFFECTS IN MOLECULAR COMPLEXATION

5.7.1   General Effects of Solute–Solvent Interaction

The values of stability constants of complexes formed by hydrogen bonding depend on the solvent used (Barriol and Weisbecker, 1967; Higuchi et al., 1969; Christian and Lane, 1976). It is convenient to divide the interactions between solute and solvent into two types, specific and nonspecific interactions. Specific interactions are relatively strong and are usually of the donor–acceptor type, resulting in molecular complexes with well-defined stoichiometries and structure. In order for this to occur, the solvent must usually be a proton donor (e.g., chloroform), a proton acceptor (e.g., ethers or amines), or an electron-pair donor or acceptor (i.e., a Lewis base or acid).

If the solvent is particularly polarizable (e.g., carbon disulphide), an aromatic compound, or even carbon tetrachloride, it will be capable of participating in significant Debye interactions of largely nonspecific nature with polar solute molecules (Sections 2.11.4–2.11.6, Section 3.11). Gramstad (1963) suggested that the interaction between triethylphosphine oxide and carbon tetrabromide is of this type. Non-specific interactions also involve London dispersion forces and, related to the bulk properties of the solvent, usually are much less influenced by steric requirements.

Even changes from one saturated hydrocarbon solvent to another will be accompanied by changes in the polarizability of the solvent and consequently by changes in the London and Debye interactions with the polar solute molecules. The effects, both enthalpic and entropic, of these changes will be small compared with those of the relatively strong specific hydrogen-bonding interactions, and can normally be regarded as second-order effects (Arnett et al., 1974). Nakano et al. (1967) found that the stability constant of the phenol–isophorone complex was quite similar in eight hydrocarbon solvents (Table 5.11).

Hydrogen-bonded complexes almost invariably become less stable in more strongly interactive solvents (Christian and Lane, 1976). An increase in the number of interactions is characterized by an increase in the stability constant, an increase in the $-\Delta U^{\theta}$ and $-\Delta H^{\theta}$ values of interaction, and an increase in specificity of the solvent–solute interaction. Organic π-donor–π-acceptor complexes, and especially the dimers of the carboxylic acids, have considerably smaller stability constants, $-\Delta U^{\theta}$ and $-\Delta H^{\theta}$ values, in more polarizable and more polar solvents. Furthermore, the extent of any deviation from Eq. 5.81, 5.101, and 5.102 will be much smaller for nonspecific, weakly interacting hydrocarbon solvents than for more specific, more strongly interacting polarizable or polar solvents. A number of examples have been reviewed by Christian and Lane (1976), and some particularly relevant cases of hydrogen-bond interactions will now be discussed.

Nakano et al. (1967) found that the stability constant of the phenol–isophorone complex, determined precisely by an interactive PMR method, depended significantly on the interactive solvent used, as shown in Table 5.11. K has clearly higher values in saturated hydrocarbon solvents than in carbon

Table 5.11 Proton Magnetic Resonance Data for Interaction of Phenol with Isophorone (3,5,5-Trimethyl-2-cyclohexene-1-one) in Various Solvents at 25°C

Solvent	$K_{1:1}$ (dm³ mol⁻¹)	Chemical Shift of Unbonded Phenolic Hydrogen[b]	Chemical Shift of Bonded Phenolic Hydrogen[b]
CCl₄	29.6	257.5	482.7
C₂Cl₄	39.7	256.5	479.7
CS₂	39.8	255.5	482.8
n-Pentane	67.9	240.0	481.3
n-Hexane	65.1	240.5	483.0
n-Heptane	67.2	241.5	483.4
Cyclopentane	63.2	241.5	483.0
Cyclohexane	69.8	242.5	484.6
Methyclocylo- hexane	71.0	243.3	485.4
Isooctane	81.0	240.6	482.7
Decalin	78.3	243.8	489.8

[a] $1 \, dm^3 \, mol^{-1} = 1 \, L \, mol^{-1} = 1 \, M^{-1}$.
[b] Units are in Hertz in reference to tetramethylsilane.

Isophorone

Reproduced with permission from Nakano, M., Nakano, N. I., and Higuchi, T. (1967). *J. Phys. Chem.*, **71**, 3954–3959. Copyright © 1967, American Chemical Society.

disulphide, carbon tetrachloride, and tetrachloroethylene. This effect may be due to specific interactions between the solutes and the latter solvents. From purely entropic consideration, a direct relationship between stability constant and molar volume of the solvents was expected. However, only a slight trend in this direction was apparent. There was no clear correlation between the stability constant and the polarizability (as measured by refractive index) or dielectric constant, although the nonhydrocarbon solvents, in general, possessed higher values.

The chemical shift of the bonded proton of the phenol–isophorone complex appeared to be relatively independent of the solvents used (Table 5.11), although that of free phenol was distinctly higher (and $K_{1:1}$ lower) in the nonhydrocarbon solvents. This may be attributed to the fact that a hydrogen atom involved in bonding is shielded from the solvent molecules, its chemical shift being relatively independent of the nature of the solvent shell and dependent largely on the properties of the hydrogen acceptor. On the other hand, the unbonded phenolic proton is free to interact electrically with the solvent molecules. The influence of solvents on spectroscopy has been reviewed by Jauquet and Laszlo (1975).

Solvent effects in the phenol–dimethylacetamide system have been studied by Takahashi et al. (1967), who found that the stability constant varied from $295 \, \text{dm}^3 \, \text{mol}^{-1}$ in cyclohexane to $130 \, \text{dm}^3 \, \text{mol}^{-1}$ in carbon tetrachloride to $16 \, \text{dm}^3 \, \text{mol}^{-1}$ in chloroform. In mixtures of chloroform and carbon tetrachloride, the equilibrium constant K was a linear function of $[\text{CHCl}_3]$ due to the formation of a 1:1 dimethylacetamide–chloroform complex.

5.7.2 Comparisons Between Carbon Tetrachloride and Paraffins as Solvents

A solvent which undergoes an interaction with either the hydrogen-donating or hydrogen-accepting solute will obviously influence values of h_D or h_A, respectively. Many solvents, including carbon tetrachloride, interact with both donor and acceptor and consequently affect both h_D and h_A (Higuchi et al., 1969). As carbon tetrachloride is commonly used as a solvent in studies involving IR spectroscopy and hydrogen bonding (Gramstad and van Binst, 1966; Higuchi et al., 1969; Arnett et al., 1970; Joris et al., 1972; Mitsky et al., 1972; Jauquet and Laszlo, 1975), the influence of this solvent on hydrogen bonding will be briefly discussed.

In spite of the dual type of interaction for carbon tetrachloride, an analysis of equilibrium constants involving the inclusion of all solvent effects in the h_A values provides results for h_D values which show reasonable agreement with those obtained from direct measurements in saturated hydrocarbon solvents. These results, which are shown in Table 5.12, were derived from plots of $\log K_{1:1}$ in carbon tetrachloride against the preferred h_D values shown in Table 5.9. As demonstrated in Table 5.10, h_A increases with increasing basic strength of the proton acceptor. As shown in Table 5.11, h_D increases with increasing acidic strength of the proton donor. The use of Table 5.12 is illustrated by an example in Section 5.9.

5.7.3 Comparison of K_{DA} (and $K_{1:1}$) in Carbon Tetrachloride and Paraffins

Chapter 4 contains values of complexation constants K_{DA} for donor–acceptor pairs in carbon tetrachloride at 30°C (Chulkaratana, 1964; Shami, 1964) and in saturated hydrocarbon solvents such as cyclohexane at 25°C (Higuchi et al., 1969; Fung and Higuchi, 1971) or isooctane at 25°C (Anderson et al., 1980). Since carbon tetrachloride is more polarizable and therefore more prone to Debye interactions than isooctane, the molecular environments in the two solvents will be different. Consequently, the K_{DA} value of a given system in one solvent will differ from that in the other. However, the discussion above indicates that the two

Table 5.12 h_A, h_D, and log $K_{o(calc)}$ Values for Hydrogen-Bonding Interactions in Carbon Tetrachloride Solutions at 25°C

Hydrogen Acceptor Solvent	l-PTP[a] CCl$_4$	Griseofulvin[b] CCl$_4$	Quinoline-1-Oxide[c] CCl$_4$	Pyridine[d] CCl$_4$	DMSO[e] C$_2$Cl$_4$	Average h_D Value
h_A	0.94	1.01	0.74	0.73	0.94	
s^f	0.053	0.071	0.063	0.033	0.029	
r^g	0.994	0.992	0.985	0.994	0.999	
n^h	9	8	3	6	5	
log K_{calc}[i]	2.00	1.48	2.24	1.75	1.85	

H Donor			h_D Values			
Phenol	0					
p-CH$_3$-phenol			−0.11	−0.18	−0.13	−0.14
p-t-C$_4$H$_9$-phenol				−0.22		−0.22
p-OCH$_3$-phenol	−0.25	−0.10		−0.18	−0.15	−0.17
p-F-phenol	0.33	0.27				0.30
p-Cl-phenol	0.46	0.46	0.50	0.47	0.45	0.47
p-Br-phenol	0.49	0.47				0.48
p-I-phenol	0.44	0.43		0.54		0.47
p-NO$_2$-phenol	1.44	1.42			1.34	1.39
p-CN-phenol	1.23	1.13				1.18

m-CN-phenol	1.32	0.93			0.83
Catechol	0.83			−0.16	−0.15
m-CH₃-phenol	0.03	0.07	−0.14		0.05
m-OCH₃-phenol	0.47	0.44			0.46
m-F-phenol	0.67	0.56	0.71	0.66	0.65
m-Cl-phenol	0.57	0.62		0.67	0.60
m-I-phenol	0.85	0.63			0.74
2, 3, 5, 6-Tetra-F-phenol	0.99	0.93			0.96
Penta-F-phenol				1.17	
m-NO₂-phenol					1.17

[a] Calculations based on results given in Table 5.1.

[b] Calculations based on results given by Nakano, M., Ph.D. thesis, University of Wisconsin, 1967.

[c] Calculations based on results given by Kubota, T., *J. Pharm. Soc. Japan*, **74** (1954) 831; **75** (1955) 1540.

[d] Calculations based on results given by Rubin, J., Senkowski, B. Z., and Panson, G. S., *J. Phys. Chem.*, **68** (1964) 1601.

[e] Calculations based on results given by Ghersetti, S. and Lusa, A., *Spectrochim. Acta*, **21** (1965) 1067.

[f] s = SD of experimental measurements from the regression line.

[g] r = correlation coefficient.

[h] n = number of interactions involved in the calculation of h_A.

[i] $\log K_{calc}$ = intercept of regression line with ordinate ($h_D = 0$).

Reproduced with permission of the copyright owner, the American Pharmaceutical Association, from Higuchi, T., Richards, J. H., Davis, S. S., Kamada, A., Hou, J. P., Nakano, M., Nakano, N. I., and Pitman, I. H. (1969). *J. Pharm. Sci.*, **58**, 661–671.

values will be related by multiplying factor. The value of this factor can be calculated by comparison of the K_{DA} as follows.

K_{DA} values in CCl_4 at 30°C:

for p-phenylphenol with di-n-butyl ether, 0.104 (Table 4.15);

for p-phenylphenol with n-propyl-n-propionate, 0.165 (Table 4.14).

$K_{DA} = K_{1:1} [D]_o$ (Eq. 4.22) in isooctane at 25°C:

for p-phenylphenol with di-n-butyl ether, $K_{DA} = 8.9 \times 12.4 \times 10^{-4} = 1.1036 \times 10^{-2}$ (Tables 4.7 and 4.10);

for p-phenylphenol with n-butyl-n-butyrate, $K_{DA} = 21.2 \times 12.4 \times 10^{-4} = 2.6288 \times 10^{-2}$ (Tables 4.7 and 4.10).

$$\frac{K_{DA} \text{ in isooctane at } 25°C}{K_{DA} \text{ in } CCl_4 \text{ at } 30°C} = 1.1036 \times 10^{-2}/0.104$$

$$= 0.106 \text{ with the ether as proton acceptor}$$

$$= 2.6288 \times 10^{-2}/0.165$$

$$= 0.159 \text{ with the esters as proton acceptors}$$

The ether value 0.16 is probably more reliable, since the proton acceptor is the same compound, di-n-butyl ether. A numerical example which employs the derived factor is given in Section 5.9.

5.7.4 Choice of Concentration Scale for Expressing Complexation Constants

The solvent dependency of stability constants of complexes is contingent upon the standard state or concentration scale in which these equilibrium constants are expressed. A volume-based state involving the calculation of stability constants K_c in terms of molarities is preferable when treating the equilibrium of associated solutes in condensed phases. This was the approach taken in the work discussed above. In fact, K_c values are much more constant among a group of similar solvents than are the K_x values based on mole fraction states, as shown by the data of Buchowski et al. (1966), presented in Table 5.13. K_x depends greatly on the molar volume of the solvent, even within a series of similar saturated hydrocarbon solvents. Stability constants based on molalities, K_m, also suffer from this disadvantage.

Kuntz et al. (1968) have developed equations for converting apparent equilibrium constants K_x, derived from the slopes and intercepts of the Benesi–Hildebrand (B–H) plots (Eq. 5.48) to the molarity-based equilibrium constant K_c. Kuntz et al. (1968) advanced the hypothesis that K_c is the universally

Table 5.13 Association Constants K_c and $(K_x)_{B-H}$ for the Reaction 1-Heptyne + Acetone $\rightleftharpoons$ Heptyne–Acetone at 25°C

Solvent	$K_c(\mathrm{dm^3\,mol^{-1}})^a$	$(K_x)_{B-H}{}^a$
Hexane	0.45	3.0
Decane	0.43	1.6
Tetradecane	0.43	0.9
Cyclohexane	0.44	3.7

[a] $1\,\mathrm{dm^3\,mol^{-1}} = 1\,\mathrm{L\,mol^{-1}} = 1\,M^{-1}$. $(K_x)_{B-H}$ is dimensionless. $(K_x)_{B-H}$ values have been calculated from experimental K_c values (Buchowski et al., 1966) using Eq. 5.105 and the molar volumes of solvent and acetone.

Reproduced with permission of the copyright owner, John Wiley and Sons, Inc. Christian, S. D. and Lane, E. H. (1976). Solvent Effects on Molecular Complex Equilibria. In *Solutions and Solubilities*, Part I. *Techniques in Chemistry*, Vol. VIII. M. R. Dack (Ed.), John Wiley and Sons, Inc., New York.

appropriate equilibrium expression for reacting systems and then demonstrated experimentally its veracity, as supported by Table 5.13. Since K_c is constant, both K_x and K_m must necessarily vary with the compositions of equilibrium mixtures. This point is emphasized by writing the last two quantities as $(K_x)_{B-H}$ and $(K_m)_{B-H}$ in Eqs. 5.105 and 5.106. These equations are not generally applicable for converting K_x and K_m values to K_c; they are valid only for the limited purpose of converting apparent values from B–H plots to K_c. For this purpose, Kuntz et al. (1968) derived the following equations, assuming that the concentrations of the acceptor and the complex are small compared with that of the donor, and that there is no volume change on mixing:

$$K_c = (K_x)_{B-H}V_s + (V_S - V_D) \tag{5.105}$$

$$K_c = (K_m)_{B-H}/\rho_S - V_D \tag{5.106}$$

where V_S and V_D are the partial molar volumes of the solvent and donor and ρ_S is the density of the solvent.

The general relationship between K_c and K_x, defined as $X_{DA}/X_D X_A$ and not derived from B–H plots, is stated below according to Kuntz et al. (1968). It is again assumed that the concentrations of A and DA are negligible compared with that of D. This general equation is

$$K_c = K_x[V_S - X_D(V_S - V_D)] \tag{5.107}$$

where V_S and V_D are the partial molar volumes of the solvent and donor respectively, and X_D is the mole fraction of the donor. This equation demonstrated that if K_c is independent of composition of the equilibrium mixtures, K_x is not, as shown experimentally.

5.7.5   Treatment of Solvent Effects as Competing Equilibria

Solvent effects in complexation equilibria have been reviewed by Christian and Lane (1976). Following the philosophies expressed in Section 2.11.8, some schools of thought tend, whenever possible, to propose specific interactions between the solvent and the donor and/or the acceptor and/or the complex, each corresponding to a new complex formed with the solute. This view resembles that of the Dolezalek school of chemical effects in solution phenomena. Other schools of thought tend, whenever possible, to propose nonspecific interactions between the solvent and each solute species, including the complex. This view resembles that of the van Laar school of physical effects in solution phenomena.

The models which postulate specific (chemical) solute–solvent interactions include (a) a single equilibrium between saturated species and (b) competing equilibria.

The single equilibrium between saturated species can be represented by the reaction

$$(A, S_n) + (D, S_m) = (DA, S_p) + qS \tag{5.108}$$

in which S represents a solvent molecule and n, m, and p are the number of solvent molecules in the saturation shells of the acceptor, donor, and complex, respectively, such that $q = n + m - p$. In the formation of a DA complex in solution, some of the solvent has to be "squeezed out" or excluded from contact with the reacting parts of the separated D and A molecules. (This effect tends to reduce α and α', which is discussed in Section 5.7.10.)

This model was developed for weak complexes (e.g., contact charge-transfer complexes) as well as for strong complexes and was presented by Carter, Murrell, and Rosch (1965) (CMR). The equilibrium constant is defined by

$$K_{CMR} = \frac{[DA, S_p]x_S^q}{[A, S_n][D, S_m]} \tag{5.109}$$

where x_S is the mole fraction of the solvent in the mixture, which is given by

$$x_S = \frac{[S]}{[S]_o + [D]_o + [A]_o} \tag{5.110}$$

The subscript o refers to the initial, uncomplexed value. K_{CMR} has, by definition, the same dimensions as K_{B-H} (i.e., K_c in Eqs. 5.45–5.48). Equations were derived in which the variation of x_S was allowed for in calculating K_{CMR}. Analysis of the methylbenzene–iodine, methylbiphenyl-1, 3, 5-trinitrobenzene, and methylbenzene–chloranil systems shows that K_{B-H} is less than K_{CMR} by about $3\,dm^3\,mol^{-1}$. However, the superiority of the CMR over the B–H approach is somewhat controversial.

The competing equilibria model was developed from the idea that one of the most important causes of the solvent dependency of stability constants was the formation of solvent–solute complexes ($S + A \rightleftharpoons SA$, molarity-based equilibrium constant K'_{SA}) which compete with the usual equilibrium ($D + A \rightleftharpoons DA$, molarity-based equilibrium constant K'_{DA}). Tamres (1973; Tamres and Yarwood, 1973) demonstrated that, assuming $[S]_o \gg [SA]$, $[A][D] \gg [DA]^2$, and $[D] \gg [A]$,

$$\frac{1}{K'_{DA}} = \frac{1}{(1 + K'_{SA}[S]_o)} \frac{[D]_o[A]_o \varepsilon l}{E} - [D]_o \qquad (5.111)$$

where l is the optical path length, E is the absorbance (or extinction) of DA and ε is the molar absorptivity of DA. Presentation in the form of the B–H Eq. 5.48 gives

$$\frac{[A]_o l}{E} = \frac{1 + K'_{SA}[S]_o}{K'_{DA}\varepsilon} \frac{1}{[D]_o} + \frac{1}{\varepsilon} \qquad (5.112)$$

Comparison with the B–H equation affords

$$K_{B-H} = \frac{K'_{DA}}{1 + K'_{SA}[S]_o} \qquad (5.113)$$

The same considerations apply if the solvent is an acceptor rather than a donor. The competing equilibrium is now $S + D \rightleftharpoons SD$, with the molarity-based equilibrium constant being K'_{SD}, which replaces K'_{SA}. If both SA and SD complexes are formed as well as DA, it can be shown (Bishop and Sutton, 1964) that

$$K'_{DA} = K_{B-H}(1 + K'_{SA}[S]_o)(1 + K'_{SD}[S]_o) \qquad (5.114)$$

where K_{B-H} is the molarity-based equilibrium constant obtained from the simple treatments (e.g., B–H analysis or solubility method). Consideration of the equilibrium $S + DA \rightleftharpoons SDA$, in which SDA represents a complex containing solvent, donor, and acceptor, may be advantageous. Extended treatments also include higher-order solvent–solute complexes (e.g., S_2A as well as SA) and solvent mixtures. The analysis of all these systems has been reviewed by Christian and Lane (1976).

Higuchi et al. (1969) used Eq. 5.114 to calculate the K'_{SD} for the interaction of hydrogen-donating solutes (phenols) with carbon tetrachloride as the solvent, assuming that this solvent did not form a significant complex with the hydrogen acceptors (i.e., $K'_{SA} = 0$). Equation 5.114 then simplifies to

$$K'_{SD} = \frac{1}{[S]_o}\left(\frac{K'_{DA}}{K''_{DA}} - 1\right) \qquad (5.115)$$

where $K''_{DA} (= K_{B-H})$ represents the apparent molarity-based equilibrium

Table 5.14 The Equilibrium Constants[a] for the Association of Carbon Tetrachloride with Phenol (K'_{SD}) and with Various Bases (K'_{SA}) at 25°C

Hydrogen Donor	Hydrogen Acceptor	K'_{DA} (Paraffin)	K''_{DA} (CCl$_4$)	K'_{SD}	K'_{SA}
Phenol	Benzene	0.48	0.31	0.05	0
Phenol	Dioxane	19	8.5	0.12	0.05
Phenol	Isophorone	70	30	0.13	0.05
Phenol	Pyridine	84	45	0.08	0.02
Phenol	$(C_2H_5)_3N$	85	58	0.05	0
Phenol	$CH_3COOC_2H_5$	10.1	9.2	0.01	—
Phenol	$(C_2H_5)_3PO$	9402	2512	0.26	0.14
Phenol	TBP[b]	940	276	0.23	0.12
Phenol	DMA[c]	295	136	0.12	0.04

[a]Units: $1 \, dm^3 mol^{-1} = 1 \, L \, mol^{-1} = 1 \, M^{-1}$. K'_{SD} was calculated from Eq. 5.115. K'_{SA} was estimated as indicated in the text.
[b]Tri-*n*-butylphosphate.
[c]Dimethylacetamide.

Reproduced with permission of the copyright owner, the American Pharmaceutical Association, from Higuchi, T., Richards, J. H., Davis, S. S., Kamada, A., Hou, J. P., Nakano, M., Nakano, N. I. and Pitman, I. H. (1969). *J. Pharm. Sci.*, **58**, 661–671.

constant determined experimentally in the presence of the interfering solvent (e.g., carbon tetrachloride) and K'_{DA} is the more fundamental molarity-based equilibrium constant for $D + A \rightleftharpoons DA$ obtained experimentally in the presence of an inert solvent such as a paraffin.

From the K'_{DA} and K''_{DA} data in Table 5.14 for the interaction between phenol and a series of bases, K'_{SD} values were calculated. The variability of the K'_{SD} values in Table 5.14 might indicate that the solvent, carbon tetrachloride, is also interacting with the acceptors. If a plausible value of K'_{SD} ($\cong 0.05 \, dm^3 \, mol^{-1}$) is assumed, it is possible to calculate K'_{SA} using the concept of competing equilibria (Table 5.14). Alternatively, the nonpolar but polarizable solvent may undergo nonspecific London and Debye interactions with the polar solute molecules which vary from solute to solute, depending on molecular size, shape, dipole moment, and polarizability of the solute molecules. The latter explanation is generally favored if the equilibrium constant is small (e.g., $< 1 \, dm^3 \, mol^{-1}$ for a 1:1 complex) (Christian and Lane, 1976); see the discussion on p. 153.

5.7.6 Treatment of Solvent Effects as Nonspecific Interactions

Some authors believe that solvent–solute complexation is not the only effect to consider in order to describe the variation of K with solvent. Others believe that to consider the solvent as a reactant and to introduce mass-action expressions involving the solvent will not prove worthwhile until the specific and non-specific effects can be unequivocally separated. Unfortunately, this separation is not yet possible (Christian and Lane, 1976).

Nonspecific interaction models have arisen from the finding that the energies and free energies of transfer of donor, acceptor, and complex molecules from the gas phase into relatively inert solvents, such as paraffinic hydrocarbons, are frequently as large as the energies and free energies of the complexation reactions themselves. Hence, even in the absence of strong specific interactions minor changes in solvent properties can induce relatively large variations in K and ΔU. The following approaches have been developed: (a) solvent-induced changes in activity coefficients, involving theories of solutions of nonelectrolytes; (b) change in bulk properties of the solvent; (c) use of homomorphs or analogues; and (d) transfer-energy and free-energy relationships.

Solute activity coefficients or chemical potentials can be calculated and treated using theories of solutions of nonelectrolytes, such as regular solution or solubility parameter theory, pair-potential calculations, and quasi-lattice theory of polymer solutions. The best known and most often applied of these theories is the regular solution theory of Scatchard (1931) and Hildebrand (Hildebrand and Scott, 1962), which has been particularly successful in treating solvent–solute interactions in iodine solutions.

5.7.7 Treatment of Solvent Effects by Regular Solution Theory

Assuming that the solubility parameter theory, modified by Flory–Huggins corrections to account for size differences between solvent and solute molecules, applies to D, A, and DA individually in solution, Buchowski et al. (1966) derived the relation

$$\log K_c = a + b\delta_S \qquad (5.116)$$

where K_c is the stability constant expressed in molarities; δ_S is the solubility parameter of the solvent and a and b are constants depending on the properties of D, A, and DA. The solubility parameters of these solutes were assumed to be independent of solvent, and the molecular volume of DA was assumed to be equal to the sum of the molecular volumes of D and A.

This model correctly predicts linear plots of $\log K_c$ against δ_S for the complexes 1-heptyne–acetone (Buchowski et al., 1966) and pyridine–iodine in several nonpolar media. Linear plots can also be predicted for the phenothiazine–chloranil complex (Barigand et al., 1973) in solvents of varying polarity using this model. The Flory–Huggins entropy corrections are necessary for satisfactory correlations. This model resembles the α method of Christian et al. (1970; Christian and Lane, 1976) and can also be related to specific solvent–solute interactions.

5.7.8 Influence of Dielectric Constant and Dipole Moment of the Solvent

Attempts have been made to relate changes in K and ΔH^θ to bulk properties of the solvent, such as the dielectric constant D. It should be stressed, however, that such simple relationships with solvent properties do not generally hold for other hydrogen-bonded or charge-transfer complex equilibria (e.g., Nakano et al., 1967) because of the complicating effects of solvent–solute interactions. Davies (1959) found that the K and $-\Delta H^\theta$ for the dimerization of acetic and propanoic acids *decreased* with increasing D of the medium in the order: vapor > hexane > benzene > carbon tetrachloride > carbon disulphide > nitrobenzene > chlorobenzene. The $-\Delta H^\theta$ values were found to be linear functions of $(1 + (\partial \ln D/\partial T)_p)/D$ for the solvent, as predicted from electrostatic theories and from the Gibbs–Helmholtz equations for an association which is purely electrostatic in origin.

Franzen and co-workers (1963, 1964) observed a linear relation between ΔG^θ of dimerization and D^{-1} of the solvent for N-methylacetamide and ε-caprolactam in mixtures of carbon tetrachloride and 1, 1, 1-trichloroethane. The ΔH^θ of dimerization of both amides in the *cis*- and *trans*-dichloroethane mixtures varied with D according to the Davies relationship mentioned above. Hirano and Kozima (1966) found that ΔU^θ for hydrogen-bond formation between methanol and triethylamine in carbon tetrachloride, chlorobenzene, or dichloromethane varied linearly with D^{-1} of the solvent. Similarly, Hou (1967) observed that for relatively nonpolar solvents, the stability constant for the phenol–tri-n-butylphosphate complex decreased with increasing dielectric constant.

The linear free-energy relationship of Taft (1956), represented by Eq. 5.103, which fitted stability constants in carbon tetrachloride, was further tested for complexes of p-fluorophenol in hexane and in several polar aprotic solvents. Results in cyclohexane followed the expected trend, but stability constants for amine complexes in chlorinated solvents such as chlorobenzene, p-dichlorobenzene, 1, 2-dichloroethane, and methylene chloride were significantly greater than the values predicted by the linear relationship. The enhancement of stability is attributed to increased proton transfer in the p-fluorophenol–amine complexes which is induced by the more polar solvents. This is similar to the trend observed by Nakano et al. (1967) and opposite to the relationships of Davies (1959), Franzen (1963, 1964), Hirano and Kozima (1966), and Hou (1967).

The explanation of these two opposite effects depends on the presence of a dipole moment. Hydrogen bonds are special cases of charge-transfer complexes involving dipole enhancement, which is defined as the magnitude of the dipole moment of the complex minus the magnitude of the vector sum of the dipole moments of the donor and acceptor oriented as in the complex. It may be postulated that when an increase in the dielectric constant of the solvent causes an increase in the stability constant of the complex, the dipole enhancement is positive and appreciable. On the other hand, when an increase in dielectric constant of the solvent reduces the stability constant, the dipole enhancement is probably negative.

5.7.9 Use of Homomorphs to Infer Nonspecific Contributions

Nonspecific contributions to solute–solute and solute–solvent interactions can be inferred by comparing the results involving specifically interacting polar molecules with those involving nonpolar molecules of nearly equivalent dimensions and structure, known as "model" compounds or as homomorphs (Brown et al., 1953). For aliphatic alcohols, ethers, and ketones, suitable homomorphs might be aliphatic hydrocarbons. Homomorphs have been used for the determination of stability constants, directly or chromatographically (Sheridan et al., 1972), and for measuring enthalpies of hydrogen-bond formation using the pure-base method of Arnett and co-workers (1970, 1974), discussed in Section 5.3.8.

5.7.10   Study of Solvent Effects Using Solvation Cycles

Christian and co-workers (1970, 1976) favor investigation of the thermodynamics of the transfer reactions involved in the solvation cycle:

$$D(V) + A(V) \rightleftharpoons DA(V) \quad (V = \text{vapor state})$$

$$D(S) + A(S) \rightleftharpoons DA(S) \quad (S = \text{solution state}) \tag{5.117}$$

The fraction of the standard internal energy $\Delta U^{\theta}_{V \to S}$ of transfer or standard free energy $\Delta G^{\theta}_{V \to S}$ of transfer of the donor plus acceptor, which is retained by the complex, is defined by

$$\alpha = \Delta U^{\theta}_{DA,V \to S} / (\Delta U^{\theta}_{D,V \to S} + \Delta U^{\theta}_{A,V \to S}) \tag{5.118}$$

$$\alpha' = \Delta G^{\theta}_{DA,V \to S} / (\Delta G^{\theta}_{D,V \to S} + \Delta G^{\theta}_{A,V \to S}) \tag{5.119}$$

The determination of the quantities involved and the value of this approach, which appears most promising, have been reviewed by Christian et al. (1970, 1976).

The constants α and α' can be used to infer something about the physical nature of the complexes and their interaction with solvents (Christian and Lane, 1976; Christian et al., 1970), and Table 5.15 presents some experimental values.

(a)   For a given complex, α and α' are usually approximately equal unless there is an unusual entropy effect, for example, diethylamine with methanol (or, to a lesser extent, with water), for which hydrogen bonding requires a specific molecular orientation.

(b)   Moderately weak complexes are usually characterized by values of α and α' less than unity, which do not change greatly with changes in the solvent, for example, the dimerization of trifluoroacetic acid. The squeezing out of the solvent molecules as the complex forms from the monomers is not compensated for by a correspondingly large excess energy of interaction between the dipole of the complex and the medium.

(c)   For hydrogen-bonded systems, α and α' are rarely greater than unity even for moderately strong complexes. In general, dipole enhancement of hydrogen-bonded complexes (e.g., between diethylamine and water or methanol) is smaller than those of charge-transfer complexes of comparable strength (e.g., between

Table 5.15 Experimental Values of α and α' for Typical Molecular Complexes[a] Calculated from Experimental Data Using Eqs. 5.118 and 5.119

Complex	Solvent	α	α'
$(CF_3COOH)_2$	Cyclohexane	0.72	0.55
	CCl_4	0.59	0.55
	Benzene	0.43	0.39
	1,2-Dichloroethane	0.48	0.39
Pyridine–H_2O[b]		—	0.71
Diethylamine–H_2O	Hexadecane	1.01	1.05
	Diphenylmethane	0.85	1.01
	Benzyl ether	0.80	0.90
Diethylamine-CH_3OH	Hexadecane	0.74	0.93
	Diphenylmethane	0.73	0.89
	Benzyl ether	0.70	0.85
m-F-phenol-dimethyl sulfoxide[c]		0.58	—
m-F-phenol-ethyl acetate[c]		0.45	—
m-F-phenol-pyridine[c]		0.61	—
m-F-phenol-triethylamine[c]		0.68	—
Benzene-I_2	CCl_4 or heptane	~1.0	0.8–0.9
Diethyl ether-I_2	Heptane	1.0	0.9
Pyridine-I_2[b]		—	0.93
Trimethylamine-SO_2	Heptane	1.30	1.21
	Chloroform	1.43	1.44
	Dichloromethane	1.62	1.51
TCNE–triphenylene[b]		—	0.65

[a] References to calculation of α' and α and to original data are in Christian and Lane (1976).
[b] Estimated from free-energy data for several nonpolar and slightly polar solvents; gas-phase data not available.
[c] Estimated from enthalpy data for several nonpolar and slightly polar solvents; gas-phase data not available.
Reproduced with permission of the copyright owner, John Wiley and Sons, Inc. Christian, S. D. and Lanes, E. H. 1976. Solvent Effects on Molecular Complex Equilibria. In *Solutions and Solubilities*, Part I. *Techniques in Chemistry*, Vol. VIII. M. R. Dack (Ed.), John Wiley and Sons, Inc., New York.

trimethylamine and sulfur dioxide). Evidently, the squeezing-out effect almost always predominates over the opposing effect of complex dipole–solvent interactions.

(d) Cyclic complexes should have abnormally small values of α and α'. For instance, dimers of carboxylic acids, such as trifluoroacetic acid, are much less polar than the monomers and must form complexes with the solvent which result in considerable loss of interfacial molecular area. These two effects should diminish the dimer solvation energy compared with the solvation energies of the unreacted monomers.

(e) Values of α and α' which are greatly in excess of unity indicate strong complexes with pronounced dipole enhancement. Examples include charge-transfer systems involving significant polar dative bonding, such as the complex between trimethylamine and sulfur dioxide. In general, α and α' tend to increase with increasing polarity and polarizability of the solvent (e.g., heptane $<$ chloroform $<$ dichloromethane).

5.8 THE PREDICTION OF STABILITY CONSTANTS AND OF SOLUBILITIES OF POLAR SOLUTES IN THE PRESENCE OF POLAR SOLVENTS

The stoichiometry of the hydrogen-bonding interaction between the polar solute and the interactive solvent or ligand can be deduced from the functionality of the solute and ligand according to the principles outlined above. For example, meta- and para-dihydric phenols (D) are expected to interact with alcohols (A) in nonpolar solvents to form predominantly DA_2-type complexes. Ortho-dihydric phenols, however, are expected to form DA- and DA_2-type complexes, owing to steric crowding around the hydroxyl groups. If in doubt, it seems reasonable to consider mainly 1:1-type complexes, since these usually predominate at low ligand concentrations.

At higher ligand concentrations, results may be more difficult to predict, owing to deviation of activity coefficients from unity as a result of nonspecific effects and to increased interaction due to higher-order complexes. However, since the former effect usually decreases solubility, whereas the latter effect increases solubility, the two effects will often partially cancel each other. This mutual cancellation presumably permits the solubility of the solute to increase linearly with the concentration of the ligand over quite a wide concentration range as, for example, in Figure 4.7. Such a diagram might lead one to suppose that only first-order complexes are formed at a higher concentration ($2 \, \text{mol dm}^{-3}$) of ligand. Although such an assumption may be used to explain and predict the results, it may merely represent a useful approximation.

The most generally applicable methods of predicting complexation constants are based on linear free-energy relationships. Equation 5.102

$$\log(K_1/K_o) = h_D h_A \tag{5.120}$$

which involves a contribution from the hydrogen donor, h_D, and one from the hydrogen acceptor, h_A, has given particularly striking and useful correlations (Higuchi et al., 1969). As more stability constants of hydrogen-bonded and charge-transfer complexes become available, the number of available h_D and h_A values are likely to increase and this will enable more values of K_{DA} to be predicted.

In the absence of h_D and h_A values, other linear free-energy relationships of physical organic chemistry are likely to be valuable for prediction, such as relationships based on Equation 5.74

$$\log K = m \log K' + c \tag{5.121}$$

or on the Hammett equation (Eq. 5.82):

$$\log(K/K_o) = \rho\sigma \tag{5.122}$$

or on related variations. Some of these variations, such as Eqs. 5.83–5.89 and 5.93–5.100, should be regarded as useful empirical correlations. It seems that the most common parameters for correlating with and for consequent prediction of stability constants are the polar substituent constants σ or σ^* in Eqs. 5.84–5.88 and 5.94 and the Taft steric parameter E_s in Eqs. 5.88 and 5.94. Other quantities also give useful correlation. One example is pK_a of proton donors, as in Eq. 5.83. Another example is the solubility $[S]_o$ of the solute in the noninteractive solvent, which gives a measure of the strength of solute–solute interactions, as in Eqs. 5.95–5.100.

5.9 EXAMPLES OF THE PREDICTION OF COMPLEXATION CONSTANTS FOR HYDROGEN BONDING AND THE RESULTANT SOLUBILITY BEHAVIOR

Example 5.1 What concentration of *p*-methoxyphenol will increase the solubility of griseofulvin from $2.04\,\text{mmol dm}^{-3}$ in pure carbon tetrachloride at 25°C to $10\,\text{mmol dm}^{-3}$?

For griseofulvin, Table 5.12 affords $h_A = 1.01$ and $\log(K_o/\text{mol dm}^{-3}) = \log K_{calc} = 1.48$, and for *p*-methoxyphenol $h_D = -0.17$.

Insertion of the hydrogen-bonded data into Eq. 5.102 affords $K_1 = 20.3\,\text{dm}^3\,\text{mol}^{-1}$. This is $K_{1:1}$(apparent) and compares favorably with the experimental value of $23\,\text{dm}^3\,\text{mol}^{-1}$ in Table 5.1.

From Equation 4.13

$$[L]_t = \{[S]_t - [S]_o\}\frac{K_{1:1}[S]_o + 1}{K_{1:1}[S]_o}$$

Inserting $[S]_t = 10\,\text{mmol dm}^{-3}$, $[S]_o = 2.04\,\text{mmol dm}^{-3}$, and $K_{1:1} = 20.3\,\text{dm}^3\,\text{mol}^{-1}$ affords $[L]_t = 0.20\,\text{mol dm}^{-3}$. (When the literature value $K_{1:1} = 23\,\text{dm}^3\,\text{mol}^{-1}$ is used, $[L]_t = 0.18\,\text{mol dm}^{-3}$.) We see that $0.2\,\text{mol dm}^{-3}$ of *p*-methoxyphenol is required.

Example 5.2 *p*-Bromophenol has $pK_a = 9.36$. Calculate $K_{1:1}$ for the interaction between this phenol and griseofulvin in carbon tetrachloride at 25°C.

Insertion of $pK_a = 9.36$ into Eq. 5.83 yields $K_{1:1} = 82.5\,\text{dm}^3\,\text{mol}^{-1}$. This is $K_{1:1}$ (apparent) and compares favorably with the experimental value of $89\,\text{dm}^3\,\text{mol}^{-1}$ in Table 5.1.

Example 5.3 Calculate the solubility of griseofulvin in the presence of $0.10 \, \text{mol dm}^{-3}$ p-bromophenol in carbon tetrachloride at 25°C. The solubility in pure carbon tetrachloride is $2.04 \, \text{mmol dm}^{-3}$ at 25°C.

Insertion of $[L]_t = 0.10 \, \text{mol dm}^{-3}$, $[S]_o = 2.04 \, \text{mmol dm}^{-3}$, and $K_{1:1} = 82.5 \, \text{dm}^3 \, \text{mol}^{-1}$ into Eq. 4.13 leads to $[S]_t = 16.5 \, \text{mmol dm}^{-3}$. (When the literature value of $K_{1:1} = 89 \, \text{dm}^3 \, \text{mol}^{-1}$ is used, $[S]_t = 17.4 \, \text{mmol dm}^{-3}$.) Therefore, the solubility obtained is $17 \, \text{mmol dm}^{-3}$.

Example 5.4 What concentration of methyl acetate will increase the solubility of p-nitrophenol from $3.3 \times 10^{-4} \, \text{mol dm}^{-3}$ in pure isooctane at 25°C to $10^{-3} \, \text{mol dm}^{-3}$?

For methyl acetate in n-heptane at 25°C, $h_A = 0.35$, $\log(K_o/\text{dm}^3 \, \text{mol}^{-1}) = 1.08$ from Table 5.8, and for p-nitrophenol in cyclohexane at 25°C, $h_D = 1.33$ from Table 5.9.

It is reasonable to assume that the interactions in isoocatane are quantitatively similar in n-heptane and cyclohexane. Consequently, Eq. 5.102 can be applied as before to give $K_{1:1} = 35.1 \, \text{dm}^3 \, \text{mol}^{-1}$. The $K_{1:1}$ value does not assume that the activity of the substance in the pure state (solid or liquid) is unity and so can be applied to the solubility of the donor as well as to that of the acceptor.

Equation 4.13 can be applied as before to calculate $[L]_t$ after inserting the following values: $[S]_o = 3.3 \times 10^{-4} \, \text{mol dm}^{-3}$ (Table 3.4) $[S]_t = 10^{-3} \, \text{mol dm}^{-3}$, $K_{1:1} = 35.1 \, \text{dm}^3 \, \text{mol}^{-1}$. We predict that about $0.058 \, \text{mol dm}^{-3}$ of ethyl acetate should be required.

Example 5.5 Catechol interacts with alcohols in nonpolar solvents to give 1:1 and 1:2 complexes. Calculate $K_{1:1}$ and $K_{1:2}$ for the interaction of *tert*-butanol at 30°C in carbon tetrachloride using Eqs. 5.87 and 5.88. These correlations require the following information: for the tertiary butyl group $\sigma^* = -0.30$ and $E_s = -1.51$, and for the reference compound n-butanol, $K_{1:1} = 19.8 \, \text{dm}^3 \, \text{mol}^{-1}$ and $K_{1:2} = 15.4 \, \text{dm}^3 \, \text{mol}^{-1}$.

Equation 5.87 leads to

$$\log \frac{K_{1:1} \, tert\text{-butanol}}{K_{1:1} \, n\text{-butanol}} = -1.8(-0.30 + 0.13) = 0.306$$

Hence $K_{1:1}$ (*tert*-butanol) $= 40.1 \, \text{dm}^3 \, \text{mol}^{-1}$, which is in excellent agreement with the experimental value (Higuchi and Connors, 1965) of $40.2 \, \text{dm}^3 \, \text{mol}^{-1}$.

Equation 5.88 leads to:

$$\log \frac{K_{1:2} \, tert\text{-butanol}}{K_{1:2} \, n\text{-butanol}} = -1.8(-0.30 + 0.13) + 0.7(-1.51 + 0.39) = -0.478$$

Hence, $K_{1:2}$ (*tert*-butanol) $= 5.12\,dm^3\,mol^{-1}$ which also agrees well with the experimental value (Higuchi and Connors, 1965) of $5.0\,dm^3\,mol^{-1}$.

Example 5.6 Calculate the solubility of catechol in carbon tetrachloride containing $0.05\,mol\,dm^{-3}$ *tert*-butanol at $30°C$ given that the solubility in pure carbon tetrachloride is $0.0197\,mol\,dm^{-3}$ and assuming the values of the complexation constants calculated above.

Equation 5.21 can be applied thus:

$$[D]_t = [D]_o + [A]K_{DA} + [A]^2 K_{DA_2}$$

where $[A] = 0.05\,mol\,dm^{-3}$, $[D]_o = 0.0197\,mol\,dm^{-3}$, $K_{DA} = [D]_o K_{1:1} = 0.0197 \times 40.1 = 0.790$ (Eq. 4.22), and $K_{DA_2} = [D]_o K_{1:2} = 0.0197 \times 5.12 = 0.1009$ (Eq. 4.23). Hence, $[D]_t$ is given by

$$[D]_t/mol\,dm^{-3} = 0.0197 + (0.05 \times 0.790) + (0.05^2 \times 0.1009)$$

$$= 0.0197 + 0.0395 + 0.00025$$

$$= 0.060$$

The predicted solubility is $0.060\,mol\,dm^{-3}$, a threefold increase in the presence of only $0.05\,mol\,dm^{-3}$ of *tert*-butanol. The calculation shows that the last term, which represents 1:2 complexation, is negligible compared with the term for the 1:1 complex. The latter term reflects the steric interactions of the bulky *tert*-butanol group when hydrogen-bonded to the neighboring hydroxyl groups of catechol. When the experimental $K_{1:1}$ and $K_{1:2}$ values are used, the same result to two significant figures is obtained. Higher concentration of *tert*-butanol will result in a greater contribution from the term for 1:2 complexation.

Example 5.7 Predict the approximate solubility of salicylic acid in triolein (glyceryl trioleate) at $25°C$.

Triolein $= C_{57}H_{104}O_6 = 885.40$, has three ester groups and $d_4^{15} = 0.915\,g\,cm^{-3}$ (*Merck Index*, Windholz et al., 1983). The molarity of the ester group in the liquid therefore approximates to $(3 \times 1000 \times 0.915/885.40)\,mol\,dm^{-3} = 3.1003\,mol\,dm^{-3}$. This represents the molar concentration of the ester group, a hydrogen-bond acceptor, in a hydrocarbon environment. The environment for the ester group may be approximated by isooctane, in which the concentration of the ester group $[A]_t$ is $3.10\,mol\,dm^{-3}$.

The 1:1 interaction constant K_{AD} between salicylic acid and the ester group of straight-chain esters (e.g., *n*-propyl propionate) in carbon tetrachloride at $30°C$ is about 0.134 (Table 4.14). To convert K_{AD} in carbon tetrachloride at $30°C$ to that in isooctane at $25°C$, we must multiply by 0.16 (see Section 5.7.3). Thus, K_{AD} (salicylic acid + ester in isooctane at $25°C$) $= 0.134 \times 0.16 = 0.02144$. We also need to know the solubility of salicylic acid in a hydrocarbon environment (e.g., isooctane) at $25°C = 2.3 \times 10^{-3}$ mol dm^{-3} (Table 4.4).

Let us assume that the solubility of salicylic acid in triolein is so low that negligible amounts of complexes of higher order than 1:1 are formed. The simple 1:1 solubility–complexation equation (Eq. 5.17) may then be applied; thus,

$$[D]_t = [K_{DA}/(1 + K_{DA})][A]_t + [D]_o$$
$$= [0.02144/(1 + 0.02144)]3.1003 \, \text{mol dm}^{-3} + 2.3$$
$$\times 10^{-3} \, \text{mol dm}^{-3}$$
$$= 0.067375 \, \text{mol dm}^{-3}$$

The predicted solubility of salicylic acid in triolein is 0.067 mol dm^{-3} at 25°C.

The measured solubility of salicylic acid in olive oil at 30°C is 0.225 mol dm^{-3} (Shami, 1964). Olive oil is a mixed triglyceride containing approximately 83.5% w/w glyceryl trioleate (*Merck Index*, Windholz et a., 1983) and various other substances which may increase the specific interactions and account for the greater solubility than predicted for triolein.

Chapter VI

STRUCTURE OF SOLVENTS AND EFFECTS OF SELF-ASSOCIATION ON SOLUBILITY

6.1 INTRODUCTION

In Chapters 4 and 5 we discussed in detail the role of specific interactions, particularly hydrogen bonding, in affecting solubility behavior of polar organic molecules. If these solute–solvent interactions were the primary determinants of these equilibria, it would be a relatively simple task to set forth mathematical equations which would predict solubility behavior of polar molecules. Unfortunately, such an attempt is greatly complicated by the tendencies toward self-association which exist in many of our common solvents, particularly in hydroxylic solvents. Because of the importance of self-association in predicting solubility behavior, the characteristics of common solvents are discussed in considerable detail in this chapter.

An understanding of the nature of solvents is most important in our effort to provide some rational basis for predicting solubility behavior. Most common solvents of low polarity remain in their liquid condensed phase as a direct consequence of London interactions among their constituent molecules. Thus, most hydrocarbons, fluorocarbons, carbon tetrachloride, and simple esters show limited specific preferences and therefore form solutions with each other which tend largely to obey Raoult's law. Solvents composed of highly dipolar molecules, such as ketones, amides, and even certain esters, energetically prefer more polarizable solute environments because of Debye interactions (Sections 2.11.4–2.11.6 and 3.11). Since these systems have already been discussed in some detail in Chapters 2 and 3, we shall not consider them extensively in this chapter.

A number of very important solvents differ significantly from those listed above and need special attention. These are the associated solvents. The most important of this class is, of course, water. The special characteristics and structural features of water are not treated in detail in this book because of their complexity, which arises from hydrogen bonding in three dimensions, and especially because there are excellent monographs on this subject (e.g., Franks, 1983). Instead, the solubility of a variety of solutes in water is treated according to the semiempirical group contribution approach in Chapter 8. Other self-associated solvents, whose structure is less complex than water owing to a more limited ability to form hydrogen bonds, are however considered in this chapter. These solvents are the alcohols, phenols, and carboxylic acids.

The self-association of organic cosolvents, such as alcohols, phenols, amines, and carboxylic acids, will modify the complexation with a solute and its solubility. Furthermore, the self-association of many common solutes, such as phenols, modifies their solubility behavior in virtually all solvents. The major exceptions include aqueous systems in which solute–solvent hydrogen bonds effectively overwhelm the solute–solute hydrogen bonds. The impact of such associative tendencies on the diffusive transport of these molecular species is treated at the end of the chapter, though transport processes are not directly related to solubility.

Self-association of an organic compound, in the pure liquid state or in solution, may arise when at least one group in the molecule is able to act as both a proton donor (electron-pair acceptor) and a proton acceptor (electron-pair donor). Such groups include the hydroxyl and amino groups; compounds containing them can form intermolecular hydrogen bonds in the appropriate environment. Many of these compounds are of biological and pharmaceutical interest.

The hydroxyl group is the most important donor–acceptor group and, since the saturated aliphatic alcohols are the simplest compounds which contain a hydroxyl group, they are considered first. The profound influence of self-association on the physical properties of alcohols has been studied extensively. This influence may be seen by comparing the boiling point of any alcohol with that of the ether of the same empirical formula (e.g., Table 2.11). The breaking of hydrogen bonds, which occurs on dilution of alcohols in nonpolar solvents, results in changes in enthalpy, IR and PMR spectra, dielectric properties, partial molar volumes, vapor pressures, and other physical properties, each of which has been used to investigate self-association.

The names for the various self-associated species depend on the number n of single molecules, termed the monomers ($n = 1$), of which the former are composed. The following types of self-associated species are recognized: dimers, $n = 2$; trimers, $n = 3$; tetramers, $n = 4$; pentamers, $n = 5$; hexamers, $n = 6$; octomers, $n = 8$, and so on; oligomers, n is small; polymers or multimers, n is large; n-mers, the size of n is not specified. In other branches of science the word "polymer" implies a macromolecule in which the monomer units are convalently linked together. In this chapter, therefore, the word "multimer" is preferred to prevent any possible confusion with macromolecular chemistry. Indeed, all the terms stated above are employed to indicate self-associated species containing monomer units that are linked one to another by hydrogen bonds.

Continuing controversy persists over the number and size of associated species of primary alcohols in solution. Much of the early work, summarized by Pimentel and McClellan (1960), has treated alcohol polymerization as a stepwise process resulting in a continuum of species, for example, the $1-2-\infty$ model or the Kretschmer–Wiebe (1954) model. In one example Van Ness et al.(1967) compared IR data with heat-of-mixing data for ethanol–heptane and ethanol–toluene mixtures, concluding that the results were best explained by a model consisting of monomers, cyclic dimers, and linear polymers having 20 or more monomer units in a chain. Many of these earlier studies suggested a predominance of the dimer. The more recent studies described below suggest that higher multimers predominate.

6.2 MODELS FOR SELF-ASSOCIATION

6.2.1 Monomer–Trimer Models

Tucker and Becker (1973) favored a trimer rather than a dimer on the basis of the vapor pressure, 220-MHz-PMR, and IR data for *tert*-butanol in hexadecane. They also proposed the existence of higher multimers. A monomer–trimer–octomer model was earlier proposed by Tucker et al. (1969) to explain the vapor-pressure data of solutions of methanol in *n*-hexadecane and pressure–volume–temperature measurements on methanol vapor. The vapor-pressure technique is discussed in detail in Section 6.3.2.

6.2.2 Monomer–Tetramer Models

A monomer–tetramer model was shown by Fletcher and Heller (1967) to be consistent with the IR data for 1-octanol and 1-butanol in *n*-decane. Fletcher (1972) later concluded that the extent of self-association of primary alcohols in saturated hydrocarbons was on the order of 1-octanol > 1-butanol > ethanol > methanol. Dixon (1970) also found that the monomer–tetramer model agreed well with his PMR data on the hydroxyl proton shift for methanol in cyclohexane. Most of the evidence for the existence of tetramer comes from IR, PMR and enthalpy of mixing data.

6.2.3 Monomer–Dimer–Tetramer Model

The monomer–dimer–tetramer model in which the tetramer is the dominant species in very dilute solution has been frequently proposed. Examples include the work of Aveyard et al. (1973) on the association of 1-dodecanol, 1-octanol, or *n*-octane using vapor-pressure osmometry up to a limited concentration of $0.13 \, \text{mol dm}^{-3}$, and the work of Anderson et al. (1975) on the enthalpy of dilution of *n*-alkanols in isooctane. This calorimetric approach will be considered in some detail in Section 6.3.1.

6.2.4 Monomer–Pentamer Model

A monomer–pentamer model was proposed by Anderson et al. (1978) from vapor-pressure studies of the *n*-alkanols in isooctane solution. The vapor pressure of the alcohol above the solution was measured by means of the gas chromatography head-space technique mentioned in Section 3.4 and will be considered in more detail in Section 6.3.2.

6.3 EXPERIMENTAL METHODS FOR STUDYING SELF-ASSOCIATION

6.3.1 Calorimetry

In the studies of Anderson et al. (1975), solutions of each *n*-alkanol in isooctane were prepared by weight and the molalities were converted to molarities by means of accurate density measurements. Initial concentrations of the alcohols were less than $1 \, \text{mol dm}^{-3}$. The measured change in enthalpy ΔH, which occurred when diluting an alcohol solution of initial total molarity c_i to a final total molarity c_f, was interpreted as resulting from the breaking of bonds as an associated n-mer dissociates to a monomer. Association of an alcohol by a single equilibrium can be represented by

$$(\text{monomer})n\text{A}_1 \rightleftharpoons \text{A}_n(n\text{-mer}) \tag{6.1}$$

with the equilibrium constant being expressed as

$$K_{1,n} = [\text{A}_n]/[\text{A}_1]^n \tag{6.2}$$

The enthalpy of dilution, ΔH, according to these interpretations, is equal to the negative value of the standard molar enthalpy of association, $\Delta H^{\theta}_{1,n}$, for Eq. 6.1, multiplied by the fraction of n-mer dissociating, thus,

$$\Delta H = -\Delta H^{\theta}_{1,n}\{[\text{A}_n]_i/c_i - [\text{A}_n]_f/c_f\} \tag{6.3}$$

Eliminating the term $[A_n]_i$ from Eq. 6.2 and 6.3 affords

$$\Delta H = -\Delta H^{\theta}_{1,n}K_{1,n}\{([\text{A}_1]_i^n/c_i - [\text{A}_1]_f^n/c_f)\} \tag{6.4}$$

The material mass balance equation of any concentration c is given by

$$c = [\text{A}_1] + nK_{1,n}[\text{A}_1]^n \tag{6.5}$$

By means of Eqs. 6.4 and 6.5 the two variables $K_{1,n}$ and $\Delta H^{\theta}_{1,n}$ can be optimized for several sets of experimental values of ΔH, c_i, and c_f using a computer program of least squares to fit the data. This process is repeated for different values of n (2–5), and the residual standard deviation σ between the experimental ΔH and the value calculated from the preferred values of $K_{1,n}$ and ΔH^{θ} is calculated for each n value. The preferred model (i.e., the preferred value of n) is that corresponding to the lowest value of σ.

In the case of 1-butanol association, Table 6.1 indicates that the monomer–tetramer (1–4) model is preferred to the 1–2, 1–3, and 1–5 models, since σ is minimal for $n = 4$. Models in which two or more competing equilibria are combined show dominance of $n = 4$ and give little improvement of fit. Of the various n-mer models in Table 6.1, the 1–2–4 model is statistically preferable. Similarly, Figure 6.1 shows that the monomer–dimer–tetramer (1–2–4) model provides only a marginally better fit than the monomer–tetramer model for 1-octanol self-association in isooctane. This result indicates that the assumption of additional competing equilibria, each with its own pair of adjustable parameters $K_{1,n}$ and $\Delta H^{\theta}_{1,n}$, is unnecessary.

Any small difference in temperature of the order 0.3°C will increasingly

Table 6.1 Optimization of Parameters $K_{1,n}$ and $\Delta H^\theta_{1,n}$ in Eqs. 6.4 and 6.5 Using Various Models to Fit Calorimetric Data for the Self-Association of 1-Butanol in Isooctane at 25°C[a]

Self-Association Model	$\Delta H^\theta_{1,2}$ (kcal mol⁻¹)	$K_{1,2}$ (dm³ mol⁻¹)	$\Delta H^\theta_{1,3}$ (kcal mol⁻¹)	$K_{1,3}$ (kcal mol⁻¹)²	$\Delta H^\theta_{1,4}$ (kcal mol⁻¹)	$K_{1,4}$ (dm³ mol⁻¹)³	$\Delta H^\theta_{1,5}$ (kcal mol⁻¹)	$K_{1,5}$ (dm³ mol⁻¹)⁴	σ^b
1-2	−15.2	3.98							0.20
1-3			−17.5	48.2					0.076
1-4					−21.4	587			0.030
1-5							−25.5	7442	0.061
1-2-4	−425	0.017			−21.3	395			0.020
1-2-4	−123	0.059			−21.3	411			0.020
1-2-4	−5.3c	0.38			−21.6	632			0.027
1-2-3	−100	0.0001	−17.5	48.2					0.076
1-2-5	−120	0.102					−25.8	3251	0.033
1-2-3-4	−15.8	0.337	−38.8	0.003	−21.6	530			0.024
1-2-3-4-5	−67.3	0.107	−19.9	0.093	−21.4	422	−192	0.071	0.021

[a] See Section 6.3.2 for further details of the procedure employed. 1 cal = 4.184 J. 1 dm³ mol⁻¹ = 1 L mol⁻¹ = 1 M^{-1}.

[b] σ = relative standard deviation between experimental and calculated ΔH when the given parameters are used in the regression equation.

[c] A constant value of −6.3 kcal mol⁻¹ was entered into the equation to determine a K for dimer formation based on a reasonable ΔH^θ value. As is shown, $K_{1,2}$ remains small in this treatment.

Reproduced with permission from Anderson, B. D., Rytting, J. H., Lindenbaum, S., and Higuchi, T. (1975). J. Phys. Chem., 79, 2340–2344. Copyright © 1975 by the American Chemical Society.

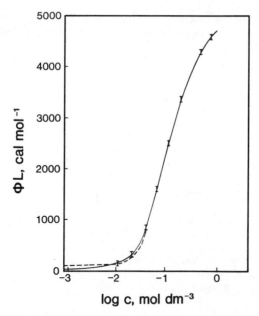

Figure 6.1 A comparison of the experimental calorimetric versus theoretical values of $-\phi_L$ in Eq. 6.8 for 1-octanol self-association in isooctane: theoretical curve using the 1–2–4 model (———); theoretical curve using the 1–4 model (– – –); I, range of experimental data. 1 cal = 4.184 J. 1 mol dm^{-3} = 1 mol L^{-1} = 1 M. Reprinted with permission from Anderson, B. D., Rytting, J. H., Lindenbaum, S., and Higuchi, T. (1975). *J. Phys. Chem.*, **79**, 2340–2344. Copyright © 1975 by the American Chemical Society.

affect $K_{1,n}$ as n increases. $K_{1,n}$ may be corrected for this effect by means of the following consecutive steps:

(1) the conversion of K based on molarity, K_c, to K based on molality, K_m, thus,

$$K_m = K_c(\rho - 0.001cM)^{n-1} \qquad (6.6)$$

where ρ is the measured density of the solution and M is the molecular weight of the solute;

(2) the correction of K_m from the experimental temperature to 25°C using the van't Hoff isochore (Eq. 1.8) in the following form:

$$\frac{d\ln K_m}{dT} = \frac{\Delta H_{1,4}^\theta}{RT^2} \qquad (6.7)$$

(3) the conversion of the corrected value of K_m back to K_c by means of Eq. 6.6

Table 6.2 shows that this correction has a significant but small effect on $K_{1,4}$, but exerts a negligible effect on $\Delta H_{1,4}^\theta$.

Table 6.2 Comparison of Temperature-Corrected versus Temperature-Uncorrected Values for $K_{1,4}$, and $\Delta H_{1,4}^{\theta}$ for the Self-Association of Various *n*-Alkanols in Isooctane at 25°C[a].

	Temperature-Uncorrected ($\leq 25°C$)		Temperature-Corrected ($= 25°C$)	
	$K_{1,4}$ $(dm^3 \, mol^{-1})^3$	$\Delta H_{1,4}^{\theta}$ $(kcal \, mol^{-1})$	$K_{1,4}$ $(dm^3 \, mol^{-1})^3$	$\Delta H_{1,4}^{\theta}$ $(kcal \, mol^{-1})$
Ethanol	674	−22.6	651	−22.6
1-Propanol	555	−21.4	533	−21.4
1-Butanol	587	−21.4	555	−21.4
1-Pentanol	657	−21.1	623	−21.2
1-Hexanol	673	−21.1	651	−21.2
1-Heptanol	633	−21.2	599	−21.2
1-Octanol	691	−21.2	662	−21.2

[a] 1 cal $= 4.184$ J. 1 dm^3 mol^{-1} = 1 L mol^{-1} = 1 M^{-1}.

Reproduced with permission from Anderson, B. D., Rytting, J. H., Lindenbaum, S., and Higuchi, T.(1975). *J. Phys. Chem.*, **26**, 705–721. Copyright © 1975 by the American Chemical Society.

The closeness of fit for a particular model may be examined by comparing the computer-generated value of relative apparent molar enthalpy, ϕ_L, from the model of best fit, with the experimental value of ϕ_L extrapolated to infinite dilution. Here ϕ_L is equal to the negative value of the enthalpy of dilution of 1 mole of the alcohol from a solution of given molarity to infinite dilution and is given by

$$\phi_L = \sum_{j=2}^{n} \Delta H_{1,j}^{\theta} K_{1,j}([A_1]_i^j / c_i) \tag{6.8}$$

The monomer–dimer–tetramer (1–2–4) model gives excellent agreement with the data, although it is only slightly better than the monomer–tetramer (1–4) model at high dilutions. Values of $-\phi_L$ at 1 mol dm^{-3} reflect hydrogen-bonding enthalpies, since each alcohol is largely associated at this concentration. For the 1–2–4 model, $-\phi_L$ kcal mol^{-1} has the following values: ethanol = 4.95, propanol = 4.65, butanol = 4.68, pentanol = 4.67, hexanol = 4.65, heptanol = 4.66, and octanol = 4.68. The enthalpy of formation of each hydrogen bond in the alcohols is equal to $\Delta H_{1,n}^{\theta}/4$ from Table 6.2. These values (5.65 kcal mol^{-1} for ethanol and 5.3 kcal mol^{-1} for the other alcohols) accord with those from other studies (for references see Anderson et al., 1975).

This treatment of the calorimetric data for alcohol association assumes (1) that the heat of dilution is due only to the dissociation of polymers and (2) that activity coefficients of all species are equal to unity. Neither of these assumptions is absolutely valid. Enthalpies of mixing of alcohols and alkanes result not only from the breaking of hydrogen bonds, but also from the disruption of Keesom interactions as the dipoles separate. This effect has been discussed by Smith and Brown (1973) and by Anderson et al. (1975), who calculated the nonhydrogen-bonding contribution to heats of dilution of alcohols, but concluded that it is small. There will also be changes in Debye interactions and possibly in London forces on dilution, but these effects will probably be negligible. The assumption that activity coefficients are unity in solution may, however, introduce appreciable errors over the concentration range used, that is, $1 \, \text{mol} \, \text{dm}^{-3}$ to zero.

6.3.2 Vapor-Pressure Method

The vapor-pressure technique is the most direct method for determining the thermodynamic activity of the solute in solution and is therefore potentially the most reliable procedure for obtaining a mathematical model for self-association.

Tucker and Becker (1973) studied the self-association of *tert*-butanol in hexadecane by measuring the total vapor pressure above each mixture and above the solvent. This measurement was made using a quartz pressure gauge, calculating the vapor pressure of the alcohol by subtracting the value from the solvent, and then analyzing the relationship between vapor pressure of the alcohol and the concentration according to the procedure to be described. To prevent the vapor pressure of the solvent from overwhelming the vapor pressure of the alcohol, this procedure requires a solvent of low volatility and alcohols of high volatility. Hence, this method is restricted to longer-chain alkanes such as hexadecane, and to alcohols of low molecular weight such as *tert*-butanol.

The chromatographic head-space technique, described immediately below, separates the vapors of the solvent and solute and gives a direct measure of the vapor pressure of the solute from the area of the solute peak. Although this procedure has a precision ($\pm 1\%$ at $> 1 \, \text{torr}$; $\pm 3\%$ at $< 0.1 \, \text{torr}$) less than that of the total pressure method of Tucker and Becker (1973) (± 0.001 to $0.003 \, \text{torr}$), it is far less restricted and has even been applied to solutes with saturated vapor pressures as low as $0.0557 \, \text{torr}$, for example, γ-butyrolactone in isooctane at $2°C$ (Grant et al., 1984).

A series of solutions of the solute, an alcohol, in isooctane or other solvent, with concentrations ranging from $0.005 \, \text{mol} \, \text{dm}^{-3}$ to pure solute was prepared by weight, in brown glass bottles of volume 20, 50 or $100 \, \text{cm}^3$. Molarities were calculated using the densities of the pure liquids to obtain the total volume of the solution. Changes in the partial molar volumes of the components on mixing were presumed to have a negligible effect on the total volume. Vapor samples can be withdrawn from the bottle through a small hole drilled in the screw cap which was made of a thermosetting plastic sealed to the bottle by means of a punctured polytetrafluoroethylene (PTFE) cap liner, as shown in Figure 6.2. The aluminum foil provides a temporary gas-tight seal.

The bottles are immersed up to the caps in a thermostatic water bath and equilibrated for at least 15 min. Vapor samples are withdrawn for gas chromatographic analysis by means of a PTFE sample loop, as shown in Figure 6.3. For this purpose, the loop is evacuated through the sampling needle (No. 5 in

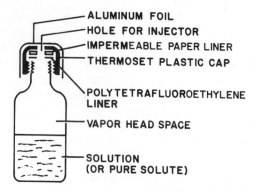

Figure 6.2 Bottles used for vapor sampling.

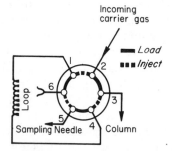

Figure 6.3 Valve system for sampling the vapor and injecting it onto the column.

Fig. 6.3) at the end of the equilibration process. Carrier gas then flows into the column through the bypass positions 2 and 3 in the valve by keeping the valve in the LOAD mode. The valve is then turned to the intermediate OFF mode, the PTFE sampling needle is pushed through both the aluminum foil and the hole in the cap into the vapor head space in the sample bottle; the valve is immediately returned to LOAD. The sample vapor is thereby drawn into the PTFE loop through positions 5 and 4 in the valve. The valve should not be kept in the OFF mode for longer than about 1 sec, as this deprives the column of carrier gas. The carrier gas is then diverted through the loop via positions 2–1–loop–4–3 and introduced onto the column by turning the valve to the INJECT mode.

The nature of the stationary phase in the gas chromatographic column is selected to give good separation and shape of the solute and solvent peaks. The vapor pressure p of the solute above a solution is normally found to be proportional to the peak area A, which is preferably determined by means of an electronic integrator. If the saturated vapor pressure of the pure solute is p^*, which corresponds to a measured peak area A^*, we can calculate p for any solution using the equation

$$p = Ap^*/A^* \tag{6.9}$$

Using the technique described, alcohol vapors having partial pressures as low as about 0.003 torr can be analyzed accurately.

The underlying assumptions of the vapor-pressure method of studying self-association in solution are (1) that the monomer does not associate in the vapor phase and that it behaves as an ideal gas, and (2) that the laws of dilute solution, and in particular Henry's law, apply to the dissolved monomer.

The first assumption is supported by the data of Cheam et al. (1970), who showed that self-association of methanol vapor at 25°C is negligible at vapor pressures lower than 30–40 torr. If the higher straight-chain alcohols behave similarly, self-association can be neglected since the highest vapor pressure encountered in the study was 20.85 torr for pure 1-propanol. At such low vapor pressures, it is a likely assumption that the monomer, which is represented by the subscript 1, obeys the ideal gas laws or, in other words, has a partial vapor pressure p_1 that is equal to its fugacity f_1. If we define the standard state, following previous suggestions discussed in Chapter 3, as a hypothetical 1 mol dm^{-3} solution of the monomer acting as if it were infinitely dilute in isooctane at 25°C, the activity a_1^c of the monomer is given by Eqs. 3.12 and 3.7 in the form

$$a_1^c = p_1/h_1^c = y_1^c c_1 \qquad (6.10)$$

The quantity h_1^c is the Henry's law constant based on molarity, while y_1^c is the activity coefficient of the monomer in solution. Equation 6.10 means that the vapor pressure of the solute above a dilute solution, p_1, is equal to the product of h_1^c, y_1^c, and c_1, the molar concentration of the monomer. The first assumption is equivalent to the statement that only the monomer is able to enter the vapor so the total pressure p of the alcohol is equal to p_1 (i.e., $p = p_1$). In the case of *tert*-butanol, which has a saturated vapor pressure of 41.98 torr at 25°C, Tucker and Becker (1973) have estimated that equating p with p_1 introduces an error of not more than 1% and have emphasized that correcting vapor pressures for nonideality in the vapor phase makes no difference in the relative fits using various models.

The second assumption is equivalent to assuming that y_1^c is unity at all concentrations of the monomer, which are usually low, so that

$$c_1 = p/h_1^c \qquad (6.11)$$

In fact, deviations of the total vapor pressure p of the alcohol from Henry's law are attributed to association, a process which will tend to maintain c_1 at the dilute levels necessary for this second assumption and Eq. 6.11 to be valid. The assumption is reasonable at low concentrations, but may become less valid at higher concentrations when the environment of the monomer consists of a large fraction of polymer molecules. For this reason, only data at concentrations below 1 mol dm^{-3} were accepted for testing the association model.

Henry's law constants h_1^c were determined from the slopes of plots of p/p^*

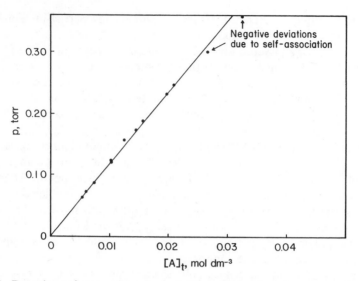

Figure 6.4 Dependence of vapor pressure (p, torr) of 1-pentanol on its molar concentration $[A]_t$ in isooctane at 25°C. Henry's law is obeyed up to $\sim 0.021 \text{ mol dm}^{-3}$. ($r = 0.9987$, slope $= h_1^c = 11.8 \text{ torr dm}^3 \text{ mol}^{-1}$. $1 \text{ mol dm}^{-3} = 1 \text{ mol L}^{-1} = 1 M$. 1 torr $= 133.3$ Pa. Reproduced with permission of the author. Anderson, B. D. (1977). Ph.D. thesis, Specific Interactions in Nonaqueous Systems. I. Self-Association of Alcohols and Phenol in Nonpolar Solvents. II. Solubilities of Organic Compounds in Organic Solvents and Cosolvent Mixtures, University of Kansas, Lawrence, KS.

against the total concentration, up to about 0.02 mol dm^{-3}, where p^* is the saturated vapor pressure of the pure alcohol. Figure 6.4 shows a typical plot. The linearity of such plots indicates the absence of significant association within the low concentration range. Thus,

$$h_1^c = \lim_{c \to 0} (p/c) \qquad (6.12)$$

From Eq. 6.9 we obtain

$$h_1^c = (p^*/A^*) \lim_{c \to 0} (A/c) \qquad (6.13)$$

The activity of the pure solute, a^{c*}, is given by Eq. 3.14, which on combining with Eq. 6.13 affords

$$a^{c*} = p^*/h_1^c = A^* \lim_{c \to 0} (c/A) \qquad (6.14)$$

Similarly, the concentration of the monomer, c_1, is given by introducing

Eqs. 6.9 and 6.13 into Eq. 6.11 to yield

$$c_1 = p/h^c = A \lim_{c \to 0} (c/A) \tag{6.15}$$

Knowledge of the saturated vapor pressure of the pure solute, p^*, is not required when using head-space or vapor analysis to determine (1) the activity of the pure solute, a^{c*}, or (2) the monomer concentration c_1. However, accurate instrumental peak areas for the condition of interest (A^* or A) and in the Henry's law region are required to give the limiting value of c/A.

For the various n-alkanols, plots of monomer concentration c_1 against the total alcohol concentration coincide below 1 mol dm^{-3}, as shown in Figure 6.5. Similarly, values of α, the ratio of the monomer molarity to the total molarity, when plotted against total molarity, also coincide for each alcohol, as shown in Figure 6.6. This coincidence suggests a similar association model for the various alcohols from 1-propanol to 1-octanol.

If only one n-mer species, A_n, is assumed to be in equilibrium with the monomer A_1, Eqs. 6.1 and 6.2 may be applied at concentrations above the Henry's law region. The total alcohol concentration $[A]_t$ is then given by

$$[A]_t = [A_1] + n[A_n] \tag{6.16}$$

$$[A]_t = [A_1] + nK_{1,n}[A_1]^n \tag{6.17}$$

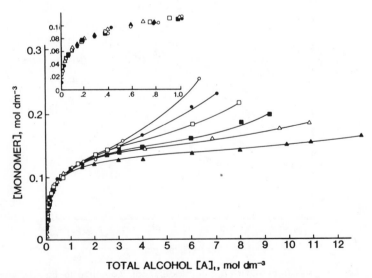

Figure 6.5 Monomer concentrations $[A_1]$ calculated from vapor pressure data versus total (i.e., formal) alcohol concentration $[A]_t$ for 1-propanol (▲), 1-butanol (△), 1-pentanol (■), 1-hexanol (□), 1-heptanol (●), and 1-octanol (○) in isooctane at 25°C. The region between 0 and 1 mol dm^{-3} is expanded as shown in the insert. 1 mol dm^{-3} = 1 mol L^{-1} = 1 M. Reproduced with permission of the copyright owner, Elsevier Science Publishing Co., Inc., from Anderson, B. D., Rytting, J. H., and Higuchi, T. (1978). *Int. J. Pharm.*, **1**, 15–31.

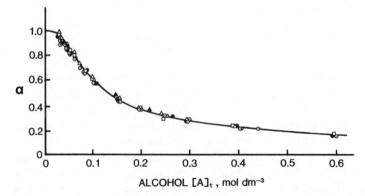

Figure 6.6 Plot of α, the ratio of monomer concentration to total (i.e., formal) alcohol concentration $[A_1]/[A]_t$, from vapor-pressure data versus $[A]_t$ for 1-propanol (△), 1-butanol (▲), 1-pentanol (□), 1-hexanol (●), 1-heptanol (⊙), and 1-octanol (○) in isooctane at 25°C. $1\,\text{mol}\,\text{dm}^{-3} = 1\,\text{mol}\,\text{L}^{-1} = 1\,M$. Reproduced with permission of the copyright owner, Elsevier Science Publishing Co., Inc., from Anderson, B. D., Rytting, J. H., and Higuchi, T. (1978). *Int. J. Pharm.*, **1**, 15–31.

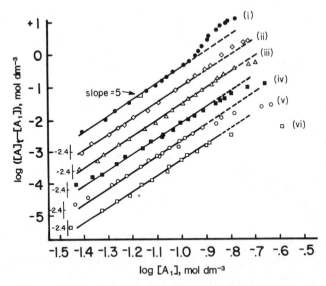

Figure 6.7 Log–log plots from vapor-pressure measurements according to Eq. 6.18 for (i) 1-propanol (●), (ii) 1-butanol (□), (iii) 1-pentanol (△), (iv) 1-hexanol (■), (v) 1-heptanol (○), and (vi) 1-octanol (□) in isooctane at 25°C. The ordinate is shifted downward 0.6 units for each alcohol after 1-propanol. The slope n indicates the stoichiometry of self-association. The straight lines having slopes of 5 represent the theoretical log–log plots based on the parameters in Table 6.4. $1\,\text{mol}\,\text{dm}^{-3} = 1\,\text{mol}\,\text{L}^{-1} = 1\,M$. Reproduced with permission of the copyright owner, Elsevier Science Publishing Co., Inc., from Anderson, B. D., Rytting, J. H., and Higuchi, T. (1978). *Int. J. Pharm.*, **1**, 15–31.

Therefore,

$$\log([A_t] - [A_1]) = \log(nK_{1,n}) + n\log[A_1] \qquad (6.18)$$

Hence, if only one associated species is present, a plot of $\log([A_t] - [A_1])$ against $\log[A_1]$ should give a straight line with a slope of n and intercept $(nK_{1,n})$. If more than one polymer is present, the line will curve upward. Figure 6.7 shows that all the n-alkanols studied give values of n very close to 5, corresponding to a monomer–pentamer (1–5) model. The $K_{1,n}$ values calculated from the intercept are also very similar for these alcohols, as shown in Table 6.3.

If more than one n-mer species is assumed to be in equilibrium with the monomer, it is possible to employ a computer program, such as the simplex method of least squares described by Deming and Morgan (1973) and Morgan and Deming (1974), to optimize values of $K_{1,n}$ for each self-associated species. The model which gives the smallest residual standard deviation fits the data best.

Table 6.4 shows an example in which several models are applied to 1-butanol. The addition of a smaller species to the 1–5 model did not significantly lower the standard deviation. The most noticeable improvement in fit occurred for the monomer–tetramer–hexamer (1–4–6) model, but even then the gain in fit over the 1–5 model based on a comparison of percent standard deviations was less than 1%. The same behavior was exhibited by the other alcohols. Figure 6.8 illustrates the fit to the 1–5 model in the case of 1-butanol.

The curve for the 1–4–6 model was nearly identical with that for the 1–5 model. This model fitting was also done for the other alcohols, but the magnitudes of $K_{1,4}$ and $K_{1,6}$ fluctuated considerably from alcohol to alcohol. On the other hand, the 1–5 model gives values of $K_{1,5}$ which vary only gradually from alcohol to alcohol, as shown in Table 6.3. Therefore, the advantage gained in describing the vapor-pressure data by including two adjustable parameters $(K_{1,4}$ and $K_{1,6})$ in the 1–4–6 model as against one $(K_{1,5})$ in the 1–5 model is

Table 6.3 Equilibrium Constants $K_{1,5}$ and Standard Free Energies of Pentamer Formation $\Delta G^{\theta a}$ from Vapor Pressure Data

Alcohol	$K_{1,5} \times 10^{-3}$ $(dm^3\ mol^{-1})^4$	$\Delta G^{\theta}_{1,5}$ $(kcal\ mol^{-1})$	*Emprical Slopes in Fig. 6.7*
1-Propanol	8.6 ± 0.2	-5.4	5.02
1-Butanol	8.7 ± 0.3	-5.4	4.97
1-Pentanol	9.4 ± 0.2	-5.4	4.99
1-Hexanol	10.0 ± 0.4	-5.5	4.69
1-Heptanol	10.9 ± 0.5	-5.5	4.78
1-Octanol	10.3 ± 0.6	-5.5	4.93

[a] $1\ cal = 4.184\ J.\ 1\ dm^3\ mol^{-1} = 1\ L\ mol^{-1} = 1\ M^{-1}$.

Table 6.4 Results of Computer Fits Using Various Self-Association Models for Vapor-Pressure Measurements on Solutions of 1-Butanol in Isooctane at 25°C[a]

Model	Standard Deviation ($\times 10^3$)	$K_{1,n}$ $(dm^3\,mol^{-1})^{n-1}$
1–3	8.1	$K_{1,3} = 129$
1–4	3.4	$K_{1,4} = 1{,}005$
1–5	0.75	$K_{1,5} = 8{,}700$
1–6	1.9	$K_{1,6} = 80{,}000$
1–2–5	0.75	$K_{1,2} \sim 0, K_{1,5} = 8{,}700$
1–3–5	0.75	$K_{1,3} \sim 0, K_{1,5} = 8{,}700$
1–4–5	0.75	$K_{1,4} \sim 0, K_{1,5} = 8{,}700$
1–4–6	0.51	$K_{1,4} = 358, K_{1,6} = 48{,}500$
1–3–8	1.3	$K_{1,3} = 41, K_{1,8} = 3.7 \times 10^6$

[a]Standard deviations and equilibrium constants are presented for various n-mer models. $1\,dm^3\,mol^{-1} = 1\,L\,mol^{-1} = 1\,M^{-1}$.

Reproduced with permission of the copyright owner, Elsevier Science Publishing Co., Inc., from Anderson, B. D., Rytting, J. H., and Higuchi, T. (1978). *Int. J. Pharm.*, **1**, 15–31.

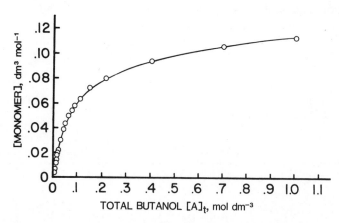

Figure 6.8 Monomer concentration $[A_1]$ from vapor-pressure measurements versus total (i.e., formal) 1-butanol molarity $[A]_t$ between 0 and $1\,mol\,dm^{-3}$ in isooctane at, 25°C. The solid line reresents the best fit using a monomer–pentamer model. The line based on a monomer–tetramer–hexamer model was nearly indistinguishable from the line drawn. $1\,mol\,dm^{-3}\,1\,mol\,L^{-1} = 1\,M$. Reproduced with permission of the copyright owner, Elsevier Science Publishing Co., Inc., from Anderson, B. D., Rytting, J. H., and Higuchi, T. (1978). *Int. J. Pharm.*, **1**, 15–31.

largely outweighed by the resulting loss in physical significance and by the difficulty in making comparisons between alcohols. Thus, the 1–4–6 model is prey to William of Occam's razor, leaving the 1–5 model as the most acceptable.

The assumption of distinct chemical "species" in addition to the monomer is justified (1) if association can be attributed to hydrogen bonding and (2) if the lifetime of the species is equivalent to a large number of molecular vibrations. The existence of hydrogen bonding between alcohol molecules has been established by six decades of research (Pimentel and McClellan, 1960) and with spectroscopic studies of alcohol solutions. The latter studies show correlations between O–H stretching frequency and solution behavior (Tucker et al., 1969). Since the interaction energy of the hydrogen bond (Pimentel and McClellan, 1960), is about $5 \, kcal \, mol^{-1}$ ($20 \, kJ \, mol^{-1}$), its average lifetime at 25°C will correspond to several thousand molecular vibrations. On these grounds, therefore, alcohol polymers may be postulated as distinct chemical species.

The wide concentration range of linearity in the plots of $\log([A]_t - [A_1])$ against $\log[A_1]$ shown in Figure 6.7 indicates that polymers, instead of ranging in size, exist at a specific optimum size. A minimum free energy with increasing polymer size is most often found in a cyclic species. Bellamy and Pace (1966) emphasized that the formation of a linear polymer will increase the acidity of the terminal proton, which will favor further hydrogen bonding. The formation of one additional bond in a cyclic species would then provide the decrease in free energy necessary for stability, the favorable enthalpic contribution being partly balanced by some decrease in entropy arising from the loss of accessible conformations. Dielectric studies on alcohols in alkane solvents, such as the work of Campbell et al. (1975) on octanols, show a decrease in apparent dipole movement as a function of concentration, which is well explained by the formation of cyclic polymers of diminished polarity.

6.4 RELATIVE MERITS OF PENTAMER AND TETRAMER MODELS FOR n-ALKANOL SELF-ASSOCIATIONS

The monomer–pentamer model (Anderson et al., 1978) indicated by vapor-pressure data is clearly at variance with tetramer models discussed in much of the literature (Fletcher and Heller, 1967; Dixon, 1970; Fletcher, 1972; Anderson et al., 1975). In fact, the planar cyclic pentamer has an energetically more favorable structure than has the planar cyclic tetramer for the following reasons. The cyclic pentamer can accommodate (1) O–H···O hydrogen bonds which are linear and therefore particularly stable (see Section 2.11.7) and (2) covalent hydrogen bond angles, H–O–H, of 108°, which is close to the bond angle of 109.1° found by Peterson and Levy (1957) for the equivalent angle in ice. The planar cyclic tetramer, on the other hand, would either (1) suffer some bending strain of its O–H···O hydrogen bonds or (2) have its H–O–H angles constricted to 90°. Both alternatives, according to Pauling (1960), are unfavorable energetically.

As mentioned earlier, most of the evidence for the existence of the tetramer comes from IR, PMR and enthalpy of mixing data, whereas the evidence for the pentamer arises from the vapor-pressure method. The relative merits of the techniques have been reviewed by Tucker and Becker (1973) and by Anderson et al. (1978) and will be briefly considered here.

Equation 6.17 indicates that the concentration of the monomer as a function of molarity must be known in order to test a particular self-association model. The vapor-pressure method is capable of determining this relationship within the limitations pointed out earlier. The measurable quantity in both PMR and calorimetry cannot, however, be converted directly to monomer concentration. Furthermore, additional parameters are required to define the system completely. For example, the chemical shift obtained from PMR measurements is a weighted average and includes contributions from monomers as well as from higher n-mers. Similarly, in calorimetric studies, the enthalpy change caused by mixing alcohols with hydrocarbons or by diluting the solutions is again not related directly to the monomer concentration. Moreover, two parameters must be determined for each polymer present, namely, $K_{1,n}$ and $\Delta H_{1,n}^{\theta}$, as discussed earlier. An additional parameter which corrects for enthalpy changes due to factors other than the breaking of hydrogen bonds is probably also necessary (Van Ness et al., 1967; Smith and Brown, 1973).

In the IR technique, it is often assumed that the monomer band accurately reflects monomer concentration over the entire range of interest. However, absorption by hydroxyl groups at the ends of polymer chains may overlap the monomer peak, thereby causing an error in the apparent monomer concentration, as found by Tucker and Becker (1973) for the *tert*-butanol–hexadecane system. Their results also indicated that the upper limit for direct use of monomer absorbance data was approximately $0.1 \, \text{mol} \, \text{dm}^{-3}$ *tert*-butanol.

On the other hand, the IR, PMR, or calorimetric methods each provide their own characteristic data which the vapor-pressure method cannot give. Furthermore, even the vapor-pressure data are at present not sufficiently precise to support one particular model to the total exclusion of all other. Thus, the final decision between the 1–4–6 and the 1–5 model is essentially a value judgment.

The existence and importance of the dimer in n-alkanol solutions is highly debatable (Fletcher, 1969; Aveyard et al., 1973; Tucker and Becker, 1973). Results of the vapor-pressure method shown in Table 6.4 do not completely exclude dimers or trimers, but the computer fits of the vapor pressures are not significantly improved by considering these smaller species, which suggests that their formation constants are very small and do not contribute significantly to the overall behavior. In fact, the experimental vapor pressures can be explained by the monomer–pentamer model to within $\pm 3\%$ over the whole range from 0 to 1 $\text{mol} \, \text{dm}^{-3}$ for all the n-alkanols studied, as exemplified by Figure 6.8.

6.5   EFFECTS OF CHAIN LENGTH AND CHAIN BRANCHING

6.5.1   Influence of Chain Length on Free Energies of Transfer

From the data for n-alkanols in Table 6.3, the standard free energies of transfer defined in Scheme 3.1 were calculated and are presented in Table 6.5.

One of the methods of correlating and predicting behavior in solution and on phase transfer is the semiempirical group-contribution approach reviewed by Davis et al. (1974). The basic premise of this method is that the free energy is additively composed of independent contributions from the constituent functional groups. This approach will be discussed further in Chapter 7. For the moment it is sufficient to point out that a plot of $\Delta_{iso}^{vap}G^\theta$ against carbon number shown in Figure 6.9 gives a straight line whose slope represents the contribution of a single methylene group to the free energy of transfer from the isooctane environment to the vapor state. The value obtained for $\Delta(\Delta_{iso}^{vap}G^\theta)CH_2$ is $+710\, cal\, mol^{-1}$, which is very close to the average value of $700\, cal\, mol^{-1}$ for the CH_2 group found by Davis et al. (1972) in a survey of the literature. This value refers to each alcohol in the monomeric state, since the reference state in isooctane is infinitely dilute and since the vapor state behaves ideally.

The vapor-pressure data (Anderson et al., 1978) summarized in Figure 6.6 and Table 6.3 show a remarkable similarity in the standard free energy of the higher alcohols in the pure state, resulting in a higher escaping tendency for monomer in the higher alcohols.

6.5.2   Influence of Chain Length on Self-Association

The vapor-pressure data (Anderson et al., 1978) summarized in Figure 6.6 and Table 6.3 show a remarkable similarity in the standard free energy of self-association among the n-alkanols from 1-propanol to 1-octanol. Calori-

Table 6.5   Standard Free Energies of Transfer of Alcohol from Isooctane to Vapor $\Delta_{iso}^{vap}G^\theta$, and from Alcohol to Isooctane $\Delta_{alc}^{iso}G^\theta$, and the Mole Fraction-Based Henry's Law Constant of Alcohol in Isooctane h^x, All at $25°C^a$

Alcohol	h^x (torr)	$\Delta_{iso}^{vap}G^\theta$ (kcal mol^{-1})	$\Delta_{alc}^{iso}G^\theta$ (kcal mol^{-1})
1-Propanol	765	-0.004	$+2.13$
1-Butanol	200	$+0.79$	$+2.06$
1-Pentanol	71	$+1.40$	$+2.02$
1-Hexanol	22.6	$+2.08$	$+1.96$
1-Heptanol	6.02	$+2.86$	$+1.93$
1-Octanol	1.76	$+3.59$	$+1.87$

^{a}1 cal $= 4.184$ J. 1 torr $= 133.3$ Pa.

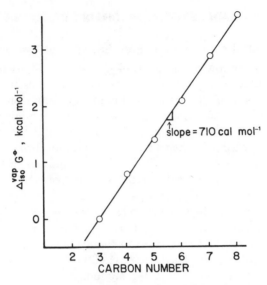

Figure 6.9 Standard molar free energy of transfer of alcohol from isooctane to the vapor phase $\Delta_{iso}^{vap}G^{\theta}$ at 25°C versus alcohol chain length. The slope gives the methylene group contribution to the transfer process. 1 cal = 4.184 J. Reproduced with permission of the copyright owner, Elsevier Science Publishing Co., Inc., from Anderson, B. D., Rytting, J. H., and Higuchi, T. (1978). *Int. J. Pharm.*, **1**, 15–31.

metric data of Anderson et al. (1975) indicate that the standard enthalpy of self-association, shown in Table 6.2, is also remarkably similar for the same *n*-alkanols. (The use of a different association model will not alter the observed trend.) This suggests a constancy in the entropy of self-association of the alcohols and implies that the loss of entropy on forming the polymer is not appreciably altered by the addition of methylene groups to the alkyl chain.

At higher alcohol concentrations, however, distinct differences in self-association appear, as shown in Figure 6.5, the extent of association in the pure state decreasing with increasing chain length. Figure 6.10 illustrates that 1-propanol (and to a small extent 1-butanol) shows negative deviations from the 1–5 model at higher concentration, whereas the longer-chain alcohols, such as 1-hexanol, exhibit positive deviations. The negative deviations for 1-propanol suggest further association to form higher polymers or networks. Networks could result from the formation of two hydrogen bonds to one atom, as suggested by Brink et al. (1977) and by Smith (1977b). For the longer-chain alcohols, the decreased association could arise from the greater bulk of the alkyl chains, which could either hamper the formation of these networks (Anderson et al., 1978) or exert a screening effect inhibiting short-range dipole–dipole interactions between neighboring polymers (Smith (1977b). Work with methanol (Tucker et al., 1969) and ethanol (Anderson et al., 1978) appears to continue the trend toward enhanced network and polymer formation at lower concentrations, with decreasing chain length shown in Figure 6.5.

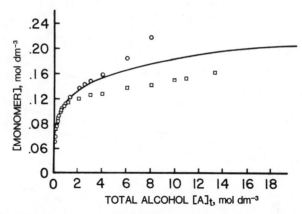

Figure 6.10 Plot of monomer concentration $[A_1]$ from vapor-pressure data versus total (i.e., formal) alcohol molarity $[A]_t$ for 1-propanol ($\square$) and 1-hexanol ($\bigcirc$) showing the deviation from a simple monomer–pentamer model. The solid line assumes $K_{1,5} = 10{,}000$ $(dm^3\,mol^{-1})^4$ at higher concentrations. $1\,mol\,dm^{-3} = 1\,mol\,L^{-1} = 1\,M$. Reproduced with permission of the copyright owner, Elsevier Science Publishing Co., Inc., from Anderson, B. D., Rytting, J. H., and Higuchi, T. (1978). *Int. J. Pharm.*, **1**, 15–31.

A very important nonspecific effect may account for the positive deviations of the vapor pressures (shown in Figure 6.10) from those predicted by the 1–5 model, namely an increase in the activity coefficient of the monomer as the environment changes progressively from dilute solution in almost pure isooctane to the pure alcohol state consisting of alcohol polymers. This effect corresponds to the breakdown of the assumption $y_1^c = 1$ in the derivation of Eq. 6.11 from Eq. 6.10. The activities of the pure alcohols, shown in Table 3.2, inversely reflect the degree of association in the pure state, which decreases with increasing length of the hydrocarbon chain.

6.5.3   Influence of Chain Branching on Self-Association

Rytting et al. (1978a) studied the vapor-pressure of the four isomeric butanols at various concentrations in isooctane to obtain information on the effects of chain branching. Figure 6.11 shows the effect of total alcohol concentration on the monomer concentration ($c_1 = p/h_1^c$; Eq. 6.11) for each isomer and indicates that self-association decreases with increased chain branching. At concentrations above the Henry's law region, application of Eq. 6.18 to the plots in Figure 6.12 at 25°C shows that the stoichiometric number n for self-association is 5 for *n*-butanol, corresponding to the pentamer, but 4 for *tert*-butanol, corresponding to a tetramer.

The lines for all four isomers diverge at increasing monomer concentration, corresponding to a gradual reduction of slope n from 5 to 4 with increasing chain branching. This behavior can be interpreted as a shift in the size of the dominant polymer as a result of chain branching. The curvature in Figure 6.12

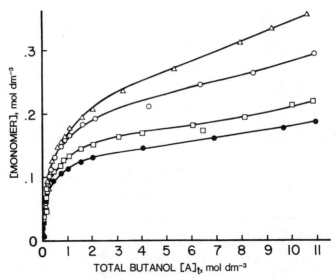

Figure 6.11 Plot of monomer concentration $[A_1]$ from vapor-pressure measurements versus total (i.e., formal) alcohol concentration for the four isomers of butyl alcohol in isooctane at 25°C; n-alcohol (●); isobutyl (□); sec-butyl alcohol (○); $tert$-butyl alcohol (△). $1 \, mol \, dm^{-3} = 1 \, mol \, L^{-1} = 1 \, M$. Reprinted with permission from Rytting, J. H., Anderson, B. D., and Higuchi, T. (1978). *J. Phys. Chem.*, **82**, 2240–2245. Copyright © 1978 by the American Chemical Society.

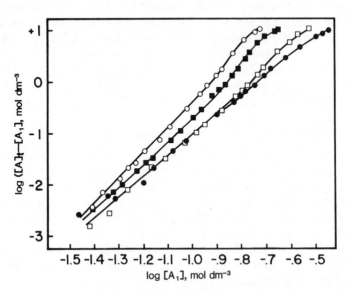

Figure 6.12 Log–log plots from vapor-pressure data according to Eq. 6.18 for n-butyl alcohol (○), isobutyl alcohol (■), sec-butyl alcohol (□), and $tert$-butyl alcohol (●) in isooctane at 25°C. The slope (i.e., mean value of n) is shifted from 5 to 4 between n-butyl and $tert$-butyl alcohol. $1 \, mol \, dm^{-3} = 1 \, mol \, L^{-1} = 1 \, M$. Reprinted with permission from Rytting, J. H., Anderson, B. D., and Higuchi, T. (1978). *J. Phys. Chem.*, **82**, 2240–2245. Copyright © 1978 by the American Chemical Society.

Table 6.6 Effect of Branching on the Relative Stabilities of Tetramers and Pentamers in the Butyl Alcohols from Vapor-Pressure Data in Isooctane at 25°C

Alcohol	Temperature (°C)	$K_{1,4}{}^{a}$ $(dm^3 \, mol^{-1})^3$	$K_{1,5}{}^{a}$ $(dm^3 \, mol^{-1})^4$	Model Giving Best Fit
n-Butyl	15	0	32500 ± 2700^b	1–4–6
	25	0	8700 ± 540	1–4–6
	35	71 ± 80^b	2740 ± 660	1–4–6
Isobutyl	25	72 ± 78	3690 ± 640	1–4–6
sec-Butyl	25	159 ± 32	610 ± 190	1–3–5
tert-Butyl	15	530 ± 110	1130 ± 770	1–4–5
	25	230 ± 9	0	1–3–4
	35	110 ± 6	0	1–3–5

[a]Computer-optimized values for $K_{1,4}$ and $K_{1,5}$ using a monomer–tetramer–pentamer model. $1 \, dm^3 \, mol^{-1} = 1 \, L \, mol^{-1} = 1 \, M^{-1}$.

[b]Error limits represent mean 90% confidence limits based on Fisher's F statistic. Note that the estimated error limits exceed the values for $K_{1,4}$ in isobutyl and n-butyl alcohol at 35°C, and are approximately $\pm 70\%$ of $K_{1,5}$ for the tert-butyl alcohol at 15°C.

Reprinted with permission from Rytting, J.H., Anderson, B.D., and Higuchi, T. (1978) *J. Phys. Chem.*, **82**, 2240–2245. Copyright © 1978 by the American Chemical Society.

at higher monomer concentrations occurs either from increasing contributions of other polymers, from changes in the activity coefficient of the monomer, or from both.

When both pentamer and tetramer were allowed to compete, a least-squares computer-fitting technique showed a gradual shift from dominance of pentamer in n-butanol toward dominance of tetramer in tert-butanol, as shown by the relative magnitudes of $K_{1,4}$ and $K_{1,5}$ in Table 6.6. Application of the computer-fitting procedure to a mixed model gave ambiguous results, as shown in Table 6.6, which illustrates the difficulty in choosing models more complex than a monomer–single n-mer model.

The effect of temperature was studied only for n-butanol and tert-butanol, as shown in Table 6.6. With increasing temperature (15–25–30°C), the lines in Figure 6.12 had the same slope, corresponding to the same value of n, but were displaced downward, corresponding to decreasing $K_{1,n}$. Plots of log $K_{1,n}$ (based on molarity) against $1/T$ are linear according to the van't Hoff isochore equation (Eq. 1.9), but yield ΔH^θ values which contain a small contribution from the thermal expansion of the solvent. The slopes were corrected for this small expansion according to the equation

$$\Delta H_{1,n}^\theta = -R[\partial \ln K/\partial(1/T)]_p + RT^2(1-n)\alpha \qquad (6.19)$$

where α is the thermal expansivity, that is, the coefficient of thermal expansion, of the solvent (Wooley et al., 1971).

The value of α was estimated to be roughly $1.2 \times 10^{-3} \, K^{-1}$ for isooctane using reported density data for isooctane at 20 and 25°C (Riddick and Bunger,

Table 6.7 Thermodynamic Quantities per Mole of Monomer from Vapor-Pressure Measurements for n-Mer Formation of Alcohols in Isooctane at 25°C[a]

Alcohol	$\Delta G_{1,n}^{\theta}/n$ (kcal mol^{-1})	$\Delta H_{1,n}^{\theta}/n$ (kcal mol^{-1})	$\Delta S_{1,n}^{\theta}/n$ (cal K^{-1} mol^{-1})
n-Butyl	-1.1 ± 0.7	-4.5 ± 0.6	-11.5 ± 4
tert-Butyl	-0.6 ± 0.3	-3.6 ± 0.6	-10.0 ± 3

[a] 1 cal = 4.184 J.

Reprinted with permission from Rytting, J.H., Anderson, B.D., and Higuchi, T. (1978). *J. Phys. Chem.*, **82**, 2240–2245. Copyright © 1978 by the American Chemical Society.

1970). The resulting quantity $\Delta H_{1,n}^{\theta}$ is the standard enthalpy for formation of the n-mer at the mean temperature of 25°C. The $\Delta H_{1,n}^{\theta}$ values obtained were -22.7 ± 2.9 kcal mol^{-1} for pentamer formation in n-butanol and -14.5 ± 2.4 kcal mol^{-1} for tetramer formation in tert-butanol. Division of these values by 5 or 4, respectively, gives the enthalpy of association per mole of monomer. Values of $\Delta G_{1,n}^{\theta}$ were calculated from the $K_{1,n}$ constants reported in Table 6.6 using $\Delta G_{1,n}^{\theta} = -RT \ln K_{1,n}$. Hence, values of $\Delta S_{1,n}^{\theta} = (\Delta H_{1,n}^{\theta} - \Delta G_{1,n}^{\theta})/T$ were calculated. Table 6.7 lists these thermodynamic quantities per mole of monomer associating. The values of $\Delta H_{1,n}^{\theta}$ are within the expected range for hydrogen bond formation (Joesten and Schaad, 1974; see also Table 2.12). The shift from the pentamer to the tetramer with increased chain branching among the butanols can be explained using space-filling models, assuming the species are cyclic. A cyclic pentamer of n-butanol has been shown to be stable on the basis of bond-angle considerations (Anderson et al., 1978), and molecular models of this species show the absence of steric interferences. However, a molecular model of the cyclic pentamer of tert-butanol shows restricted rotation for at least two of the alkyl groups in any conformation. The cyclic tetramer of tert-butanol allows free rotation of all alkyl groups simultaneously, so that steric effects could account for the observed shift in the association number n.

The decreased self-association in the highly branched alcohol shown in Figures 6.12 and 6.13 cannot wholly account for the fact that the boiling point of tert-butanol is about 35°C lower than that of n-butanol. For example, at 25°C the fractions of monomer to total alcohol concentration at 1 mol dm^{-3} are approximately 0.174 for tert-butanol and 0.113 for n-butanol. Thus, in both cases over 80% of the alcohol molecules are associated at 1 mol dm^{-3}.

The log–log plots and computer-fitting technique are both based on certain assumptions regarding a self-association model, whereas Tucker and Christian (1977b) have developed a model-independent mathematical treatment for estimating average polymer size $\bar{n}$ in solution. The total (i.e., formal) alcohol concentration is given by Eq. 6.17 with a series of consecutive values of n, thus,

$$[A]_t = [A_1] + 2K_{1,2}[A_1]^2 + 3K_{1,3}[A_1]^3 + \cdots + nK_{1,n}[A_1]^n \quad (6.20)$$

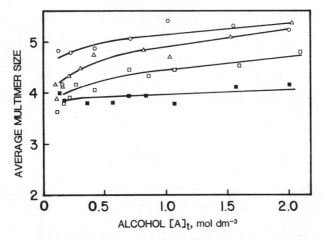

Figure 6.13 Plots of average multimer size (i.e., mean n) from vapor-pressure data versus total (i.e., formal) alcohol molarity for *tert*-butyl (■), *sec*-butyl (□), isobutyl (△), and *n*-butyl alcohol (○) in isooctane at 25°C. $1 \, mol \, dm^{-3} = 1 \, mol \, L^{-1} = 1M$. Reprinted with permission from Rytting, J. H., Anderson, B. D., and Higuchi, T. (1978). *J. Phys. Chem.*, **82**, 2240–2245. Copyrright © 1978 y the American Chemical Society.

where $[A_1]$ is the concentration of the monomer. The total concentration of all the species present is given by

$$[S] = [A_1] + K_{1,2}[A_1]^2 + K_{1,3}[A_1]^3 + \cdots + K_{1,n}[A_1]^n \qquad (6.21)$$

Differentiation of Eq. 6.21 shows that

$$d[S]/d[A_1] = [A]_t/[A_1] \qquad (6.22)$$

Integration leads to

$$[S] = \int_0^{[A]} ([A]_t/[A_1])d[A_1] \qquad (6.23)$$

From the calculated $[S]$, the average polymer size is given by

$$\bar{n} = ([A]_t - [A_1])/([S] - [A_1]) \qquad (6.24)$$

By trapezoidal inttegration of the $[A]_t/[A_1]$ versus $[A_1]$ curves above $0.1 \, mol \, dm^{-3}$, $[S]$ was evaluated. Hence, $\bar{n}$ was calculated and plotted against $[A]_t$ in Figure 6.13. This graph confirms that $\bar{n}$ decreases with increasing branching, being close to 5 for *n*-butanol and to 4 for *tert*-butanol, and increases only slightly with increasing concentration. Dimers are relatively unimportant for all the alcohols, even at low concentrations. Polymers larger than tetramers are not required to explain the *tert*-butanol data even up to $2 \, mol \, dm^{-3}$.

The boiling points of the four isomeric butanols decrease markedly with increased chain branching, thus: n-butanol, 117–118°C; isobutanol, 108°C; sec-butanol, 99.5°C; tert-butanol, 82°C. This effect may be attributed to either (1) a decrease in association and/or (2) an increase in the escaping tendency of the monomer reflected in the Henry's law constants (Table 6.1) as chain branching increases. Brink et al. (1977) postulated that the lower boiling points for the more hindered octanols was the consequence of networks being less readily formed in the branched alcohols. However, factors in addition to association must be considered, since chain branching also lowers the boiling point of hydrocarbons (e.g., for n-pentane, b.p. = 36.1°C; for neo-pentane, b.p. = 9.5°C).

The free energies of transfer for the isomeric butanols are listed in Tables 6.8 and 6.9 and have been calculated from the Henry's constants and vapor

Table 6.8 Thermodynamic Values from Vapor-Pressure Measurements for the Transfer of Alcohol from Isooctane to Vapor at 25°C and Mole Fraction-Based Henry's Law Constant of Alcohol in Isooctane h^{xa}

Alcohol	h^x (torr)	$\Delta_{iso}^{vap}G^\theta$ (kcal mol^{-1})	$\Delta_{iso}^{vap}H^\theta$ (kcal mol^{-1})	$\Delta_{iso}^{vap}S^\theta$ (cal K^{-1} mol^{-1})
n-Butyl	200 ± 3^b	0.79 ± 0.01	7.56 ± 0.13^c	22.7 ± 0.5
Isobutyl	282 ± 4	0.59 ± 0.01	—	—
sec-Butyl	376 ± 5	0.42 ± 0.01	—	—
tert-Butyl	716 ± 11	0.035 ± 0.01	6.53 ± 0.04	21.8 ± 0.2

[a] 1 cal = 4.184 J. 1 torr = 133.3 Pa

[b] Error limits in h^x represent standard error of the slope of vapor pressure mole fraction.

[c] Error limits in $\Delta_{iso}^{vap}H^\theta$ were obtained from the standard error of the slope of $\ln h^x$ versus $1/T$ multiplied by R.

Reprinted with permission from Rytting, J.H., Anderson, B.D., and Higuchi, T. (1978). *J. Phys. Chem.*, **82**, 2240–2245. Copyright © 1978 by the American Chemical Society.

Table 6.9 Free Energies and Enthalpies for the Transfer of Alcohol from the Pure Liquid to Vapor and to Isooctane at 25°C^a

Alcohol	$\Delta_{alc}^{vap}G^\theta$ (kcal mol^{-1})	$\Delta_{alc}^{vap}H^\theta$ (kcal mol^{-1})	$\Delta_{alc}^{iso}G^\theta$ (kcal mol^{-1})	$\Delta_{alc}^{iso}H^\theta$ (kcal mol^{-1})
n-Butyl	2.85^b	13.4 ± 0.14^c	2.06	5.84 ± 0.27^d
Isobutyl	2.55	—	1.96	—
sec-Butyl	2.21	—	1.79	—
tert-Butyl	1.71	11.5 ± 0.10	1.68	5.0 ± 0.14

[a] 1 cal = 4.184 J

[b] Estimate of errors unavailable from literature data.

[c] Error limits represent standard errors of slopes of $R \ln p_1^*$ versus $1/T$.

[d] Error limits represent sum of errors in $\Delta_{alc}^{vap}H^\theta$ and $\Delta_{iso}^{vap}H^\theta$.

Reprinted with permission from Rytting, J.H., Anderson, B.D., and Higuchi, T. (1978). *J. Phys. Chem.*, **82**, 2240–2245. Copyright © 1978 by the American Chemical Society.

pressures stated in Table 3.2. For n-butanol and $tert$-butanol, the standard enthalpy of transfer, $\Delta_{iso}^{vap}H^{\theta}$, of the monomer was calculated from the temperature dependence of $\Delta_{iso}^{vap}H^{\theta}$. Similarly, $\Delta_{alc}^{vap}H^{\theta}$ was calculated from linear plots of $\ln p^*$ versus $1/T$ according to Eq. 3.15.

$$\Delta_{alc}^{iso}H^{\theta} = \Delta_{alc}^{vap}H^{\theta} - \Delta_{iso}^{vap}H^{\theta} \qquad (6.25)$$

Figure 6.14 shows the relationship between $\Delta_{iso}^{vap}G^{\theta}$ and molecular surface area, which were estimated as described by Harris et al. (1973) and as outlined in Section 7.2.4. The decrease in surface area as a result of increased branching leads to an increase in Henry's law constants, presumably because of a decrease in the dispersion interaction energy between the monomer and the solvent. From the quantities in Table 6.9, we can estimate the relative contributions of dispersion interactions and hydrogen bonding to the vapor-pressure differences between n-butanol and $tert$-butanol at 25°C.

The difference between n-butanol and $tert$-butanol in the free energy of transfer from pure alcohol to vapor, $\Delta(\Delta_{alc}^{vap}G^{\theta})$, is 1.14 kcal mol^{-1}, while the contribution to this difference from variation in the Henry's law constants (which reflect the monomer's escaping tendency), $\Delta(\Delta_{iso}^{vap}G^{\theta})$, is 0.76 kcal mol^{-1}. The remainder is then attributed to differences in extent of association. Evidently, the largest factor contributing to the differences in vapor pressures between these two alcohols at room temperature is the volatility of the monomer, while self-association differences are significant but less important. These results cannot be extrapolated to the normal boiling points of the alcohols. However, if the relative contributions remain about the same, the observed differences in boiling point may arise mainly from differences in the escaping tendency of the monomer due to branching, presumably because their molecular surface areas differ.

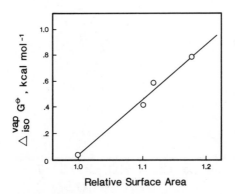

Figure 6.14 Plot of standard free energies of transfer of alcohol monomer from infinite dilution in isooctane to the vapor phase $\Delta_{iso}^{vap}G^{\theta}$ at 25°C versus relative surface area of the alkyl portion of the molecule. The relative molecular surface area of $tert$-butyl alcohol has been assigned the value of unity. 1 cal = 4.184 J. Reprinted with permission from Rytting, J. H., Anderson, B. D., and Higuchi, T. (1978). *J. Phys. Chem.*, **82**, 2240–2245. Copyright © 1978 by the American Chemical Society.

Although the preceding discussion has been based primarily on data drawn from experimental studies in isomeric butyl alcohols, it is apparent that much of it applies directly to other monohydric aliphatic alcohols. Other homologues of these 4-carbon alcohols would be expected to behave in nearly the same manner both qualitatively and quantitatively. Indeed, even phenols appear to form similar cyclic structures, vapor-pressure data indicating the formation of pentamers, as we shall see in Section 6.8.1.

Self-association behavior varies considerably in nonpolar solvents such as alkanes, carbon tetrachloride, and benzene (Fletcher, 1969; Woolley and Hepler, 1972; Campbell et al., 1975; Ben Naim, 1980), partly as a result of differing strengths of the solute–solvent interactions. These effects are partially accounted for by contrasts in polarizability, size, and shape of the solvent molecules, as discussed in Section 5.7 in connection with molecular complexation. Nevertheless, it is generally assumed and observed that in unlike saturated aliphatic hydrocarbon solvents, solute–solvent interactions are similar and molar activity coefficients $(y^c)^\infty$ vary only slightly because the choice of molarity apparently compensates partially for molar volume effects (Fung and Higuchi, 1971; see also Section 4.2). However, a comparison of solute activities of acetanilide, carbazole, phthalic anhydride, picric acid, and salicylic acid in isooctane and hexadecane (in Table 3.4 and discussed in Section 3.5.3) shows that only rough agreement should be expected for molar activity coefficients in hydrocarbon solvents of widely different molar volumes.

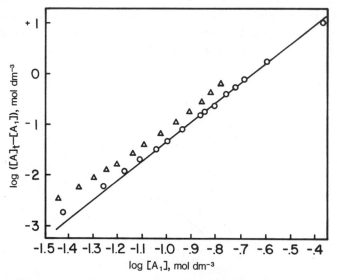

Figure 6.15 Comparison of the vapor-pressure data for the *tert*-butyl alcohol at 35°C obtained by Rytting et al. (1978a) in isooctane (○) with those of Tucker and Becker (1973) in hexadecane (△). The data are plotted according to Eq. 6.18. 1 mol dm^{-3} = 1 mol L^{-1} = 1 M. Reprinted with permission from Rytting, J. H., Anderson, B. D., and Higuchi, T. (1978). *J. Phys. Chem.*, **82**, 2240–2245. Copyright © 1978 by the American Chemical Society.

Figure 6.15 shows a direct comparison of data for *tert*-butanol at 35°C in isooctane (Rytting et al., 1978a) and in *n*-hexadecane (Tucker and Becker, 1973), the Henry's law constants being 173 and 288 torr dm^3 mol^{-1}, respectively. The similarity of the slopes in Figure 6.15 suggest similarities in the self-associated species in the two solvents. Yet Tucker and Becker (1973) favored a 1–3–6 or 1–3–∞ model, while Rytting et al. (1978a) favored a 1–4 model, as emphasized in Figure 6.12. Rytting et al. (1978a) did not feel justified in including additional parameters to gain only slight improvement in the fit to the data in accordance with Occam's razor.

Comparison of the intercepts in Figures 6.12 and 6.15 indicates that the equilibrium constant for association, when calculated using the same model, increases as the solvent is changed from isooctane to *n*-hexadecane. The observation that association constants and Henry's law constants increase with increases in the carbon number and in molar volume of the alkane solvent, even when molarity in the concentration unit, is supported by other work (Wóycicka and Recko, 1972; Smith and Brown, 1973; Smith, 1977a). Solvent effects of this and similar types are currently under investigation in a number of laboratories.

6.6 STUDIES WITH ALCOHOLS AS COSOLVENTS

6.6.1 Influence of Alcohols on the Solubility of Polar Solutes in Nonpolar Solvents

Alcohols are weak proton donors and quite strong acceptors. We have seen in certain sections of Chapters IV and V how hydrogen-bond donors and acceptors can increase the solubility of polar solutes in hydrocarbon solvents by complexation involving solute–alcohol hydrogen bonding. Self-association is expected to exert a pronounced influence on the solubility of polar solutes because of its effect on the structure of the solvent and because alcohol–alcohol hydrogen bonding is then competing with alcohol–solute hydrogen bonding. The influence of self-association on solubility is of particular importance to solubility in water, so the information gained in alcohol systems may be of considerable help in the development of theories for aqueous solubility. As we shall see in Chapter 8, solubility in water is so complex that predictions using semiempirical group-contribution methods are in common use at the present time. The influence of solvent structure among the alcohols, however, decreases with increasing length of the alkyl side chain, since even in totally associated alcohol the side chains constitute relatively unstructured regions.

As has already been mentioned, self-association is a cooperative phenomenon among the alcohols, since the hydrogen bonds linking polymers are stronger than those in dimers (Bellamy and Pace, 1966). For example, higher-order complexes between long-chain tertiary amines and methanol form more readily than the 1:1 complexes and are therefore stronger bases than the uncomplexed amines (Tucker and Christian, 1975a, b, 1977a). Anderson (1977) found that higher-order complexes in alcohols are significant with hydrogen-donor solutes, but are much less important for pure hydrogen acceptors.

The methods used to treat solubility data in unassociated solvents can be extended to self-associated alcohol solvents if the thermodynamic activity of alcohol as a function of concentration is known. The concentration of the free monomer is lowered by self-association, is related to the total molarity of the alcohol by Eq. 6.5, and is given approximately by the alcohol vapor pressure (Eq. 6.11). This relationship also applies to the alcohol monomer in saturated solutions of solutes which have a sufficiently low solubility. Thus, the total solubility of the solute $[S]_t$, can be expressed by Eq. 4.9, in which the alcohol A represents the ligand L and $[L]_f = [A_1]$, the concentration of the alcohol monomer, thus,

$$[S]_t = [S]_o + K_{1:1}[S]_o[A_1] + K_{1:2}K_{1:1}[S]_o[A_1]^2 + \cdots$$

$$(6.26)$$

Therefore,

$$([S]_t - [S]_o)/[S]_o = K_{1:1}[A_1] + K_{1:2}K_{1:1}[A_1]^2 + \cdots$$

$$(6.27)$$

$$= \text{fractional change in solubility}$$

In very dilute solutions of alcohols in isooctane, alcohol monomers predominate and only 1:1 solute–alcohol complexes are expected. Therefore, Eq. 6.27 is obeyed and plots of fractional change of solubility, $([S]_t - [S]_o)/[S]_o$, against the total concentration of alcohol, $[A]_t$, should be linear. The solubility of several substances shows this behavior in dilute solutions of alcohols in nonpolar solvents, including saturated hydrocarbons, as we have seen in Sections 4.3.2 and 4.4.1.

The hydrogen-donating ability of phenols, which is expressed by their $K_{1:1}$ values, with the hydrogen acceptor dibutyl ether (Table 4.10), is in the same order as the $K_{1:1}$ values with the alcohols (Table 6.10): 4-nitrophenol > 4-iodophenol > 4-phenylphenol. This trend suggests that the alcohol monomer functions primarily as a hydrogen acceptor in these interactions. For a given phenol, the $K_{1:1}$ value with the alcohols is greater than that with the ethers, which indicates that the alcohol monomer is a more effective hydrogen acceptor than the ether oxygen atom. This effect is partially attributed (Higuchi and Chulkaratana, 1961) to the steric effect of the additional alkyl chain outweighing its electronic effect on the ether oxygen atom, as discussed in Sections 4.4.1 and 5.5.2.

The hydrogen-donating ability of the alcohol monomer can be evaluated by examination of the $K_{1:1}$ values with anthraquinone, which can only function as a hydrogen acceptor. The $K_{1:1}$ values of anthraquinone with alcohol monomers (Table 6.10) are of the order $1 \, \text{dm}^3 \, \text{mol}^{-1}$, similar to that with chloroform (Table 4.10), suggesting that alcohol monomers are relatively weak hydrogen donors.

Since the concentration of monomer is relatively small even in pure alcohols ($\sim 0.22 \, \text{mol} \, \text{dm}^{-3}$ in n-butanol and $0.35 \, \text{mol} \, \text{dm}^{-3}$ in $tert$-butanol at 25°C), 1:1 solute monomer interactions are not very important in increasing the solubility in alcohols. For example, the $K_{1:1}$ value for the 4-phenylphenol–n-butanol complex

Table 6.10 Complexation Constants $K_{1:1}$ and K_n (Eq. 4.5 and 4.16) of Various Solutes with Alcohols at 25°C Assuming 1:1 and 1:n Complexes[a]

Solute	1-Butanol			1-Octanol			2-Methyl-2-propanol		
	$K_{1:1}$ ($dm^3\,mol^{-1}$)	$K_5 \times 10^{-4}$ ($dm^3\,mol^{-1}$)^5	$\sigma\%$	$K_{1:1}$ ($dm^3\,mol^{-1}$)	$K_5 \times 10^{-4}$ ($dm^3\,mol^{-1}$)^5	$\sigma\%$	$K_{1:1}$ ($dm^3\,mol^{-1}$)	$K_4 \times 10^{-4}$ ($dm^3\,mol^{-1}$)^4	$\sigma\%$
4-Nitrophenol	189	—	15[b]	204	—	10.4[b]	221	—	7.7[b]
4-Iodophenol	52	—	4.5[b]	—	—	—	58	—	6.4[b]
4-Phenylphenol	26	420	3.2	28.5	689	2.8	37	9.8	6.4
Carbazole	5.3	28	3.4	7.1	43	4.4[c]	7.1	0.37	2.5[c]
2,4,6-Triiodophenol	4.7	15	3.7	3.6	31	5.5	—	—	—
Anthraquinone	3.6	0.9	5.9	2.4	3	5.1	1.8	0.017	4.0
Anthracene	1.4	0.076	2.9	1.0	0.9	3.5	—	—	—

[a]% represents the percentage standard deviation. $1\,dm^{-3}\,mol^{-1} = 1\,L\,mol^{-1} = 1\,M^{-1}$. 2-Methyl-2-propanol is *tert*-butanol.
[b]Fit of data below $0.02\,mol\,dm^{-3}$ used to determine $K_{1:1}$.
[c]Fit based on data up to $1\,mol\,dm^{-3}$.

Reproduced with permission of the author. Anderson, B.D. (1977). Ph.D. thesis, Specific Interactions in Nonaqueous Systems. I. Self-Association of Alcohols and Phenol in Nonpolar Solvents. II. Solubilities of Organic Compounds in Organic Solvents and Cosolvent Mixtures, University of Kansas, Lawrence, KS.

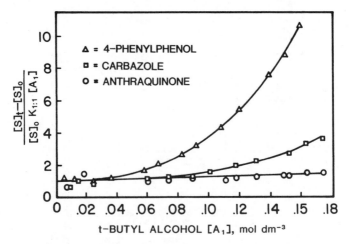

Figure 6.16 Plot of the ratio of higher-order complexes to 1:1 complexes (ordinate) for 4-phenylphenol, carbazole, and anthraquinone versus t-butyl alcohol monomer concentration $[A_1]$ in isooctane at 25°C. This plot is based on Eq. 6.27, the initial slope giving $K_{1:2}$, while upward curvature indicates the formation of complexes larger than 1:2. $1 \, \text{mol} \, \text{dm}^3 = 1 \, \text{mol} \, \text{L}^{-1} = 1 \, M$. Reproduced with permission of the author. Anderson, B. D. (1977). Ph.D. thesis, Specific Interactions in Nonaqueous Systems. I. Self-Association of Alcohols and Phenol in Nonpolar Solvents. II. Solubilities of Organic Compounds in Organic Solvents and Cosolvent Mixtures, University of Kansas, Lawrence, KS.

would itself account for a fivefold greater solubility of 4-phenylphenol in pure butanol than in pure isooctane. The actual increase is nearly 700-fold. Such large increases in solubility of polar organic substances in alcohols is largely dependent on the stability of higher-order complexes in alcohols.

By means of the plot shown in Figure 6.16, based on Eq. 6.27, the relative contributions of 1:1, 1:2, and 1:n ($n > 2$) for solute–alcohol complexes were assessed between various solutes and 0–1 mol dm^{-3} n-butanol. Anthraquinone gave a straight line of gentle slope, indicating that 1:1 and 1:2 complexes predominate. The increasing curvature with carbazole, and especially with 4-phenylphenol, indicates the increasing contribution of complexes larger than 1:2. This solubility behavior contrasts sharply with that in unassociated solvents discussed in Chapters 4 and 5. The initial slopes of all three systems, which reflect $K_{1:2}$, are comparatively small, suggesting that 1:2 complexes are relatively unimportant. Table 6.10 gives rough estimates of the amount of solute in 1:1 and higher complexes in alcohol solution. The concentration of solute in 1:1 complex is equal to $[S]_o K_{1:1}[A_1]$, where $[A_1] \cong [A]_t$ at low concentrations. The concentration of solute in higher-order complexes is equal to $[S]_t - [S]_o K_{1:1}[A_1] - [S]_o$.

Table 6.10 shows the overwhelming contribution of higher complexes of n-butyl alcohol to the solubility of three hydrogen-donor solutes and their smaller but significant contribution to that of anthraquinone, which is a hydrogen acceptor. Tucker and Christian (1975a, b, 1977a) found that higher-order complexes form to a greater extent in methanol and hydrogen acceptors, perhaps as a result of a difference in the nature of solvation in methanol. Whether the alcohol molecules form chain-like or cyclic species is debatable. Anthraquinone, which has no donatable hydrogen atoms, could not participate in a cyclic hydrogen-bonded complex, thereby explaining why it did not form complexes of higher order than 1:2. Hydrogen-donating solutes readily form higher-order complexes with alcohols, which is also consistent with the formation of cyclic hydrogen-bonded species since the alcohol molecules participating in a ring structure could only act as hydrogen acceptors. One can readily visualize, for example, a hydroxyl-containing solute molecule substituting for a solvent molecule in a pentameric alcohol structure.

Anthracene evidently forms weak complexes with alcohols in view of its low values of $K_{1:1}$ and K_5 in Table 6.10. The complexation may be attributed to weak donor–acceptor complexes involving the π-electrons of anthracene and to weak Debye forces between the polar alcohol molecules and the polarizable aromatic system; London dispersion forces may also be involved Sections 2.11.4–2.11.6.

6.6.2 Model For the Effect of Alcohol Self-Association on Solubility

Anderson (1977) found that a simple model involving SA and SA_n complexes could account for the observed solubility dependency of the solute S on the total concentration of alcohol $[A]_t$. In the remainder of Section 6.6, n is intended to refer to the stoichiometric number of alcohol molecules in the complex SA_n and not to the stoichiometry of alcohol self-association, A_n, as in previous and future sections. It turns out, however, that the experimental solubility data of various solutes in isooctane at 25°C agree well with a single model in which these two stoichiometric numbers are equal, for example $n = 5$ for the primary alcohols 1-butanol and 1-octanol, while $n = 4$ for the tertiary alcohol *tert*-butanol (Anderson, 1977).

According to this model, the total solubility $[S]_t$ is given by Eq. 6.26 in the following form:

$$[S]_t = [S]_o + K_{1:1}[S]_o[A_1] + K_n[S]_o[A_1]^n \qquad (6.28)$$

where

$$K_n = [SA_n]/[S]_o[A_1]^n = \prod_1^n (K_{1:1}K_{1:2}\cdots K_{1:n}) \qquad (6.29)$$

is the equilibrium constant of the reaction

$$S + nA_1 = SA_n \qquad (6.30)$$

and where A_1 refers to the alcohol monomer. In accordance with the self-association data, a single alcohol n-mer A_n, defined by Eqs. 6.28 and 6.29, is assumed to predominate. The total alcohol concentration $[A]_t$ is then

$$[A]_t = [A_1] + nK_{ass}[A_1]^n + K_{1:1}[S]_o[A_1] + nK_n[S]_o[A_1]^n \qquad (6.31)$$

where

$$K_{ass} = [A_n]/[A_1]^n \tag{6.32}$$

K_{ass} is used here instead of $K_{1,n}$ to denote the self-association constant so as to reduce possible confusion with the overall complexation constant K_n.

If $[S]_o$ is very small,

$$[A]_t = [A_1] + nK_{ass}[A_1]^n \tag{6.33}$$

Depending on the values of the constants $K_{1:1}$ and K_n, these equations gave good correlations with most of the curves of solubility versus total, that is, formal, alcohol molarity. Computer-simulated solubility profiles are shown in Figure 6.17. These curves are based on a process of solving Eqs. 6.28 and 6.31 simultaneously, while changing only the relative magnitudes of $K_{1:1}$ and K_n, assuming $n = 4$ and $K_{ass} = 200\,dm^9\,mol^{-3}$ for tetramer formation. Apparently linear plots are given by the hydrogen-donor solutes 4-phenylphenol and carbazole (e.g., Fig. 6.18). These curves are concave downward, as can be discerned in the anthraquinone–alcohol systems in Figure 6.19, and as predicted by this model.

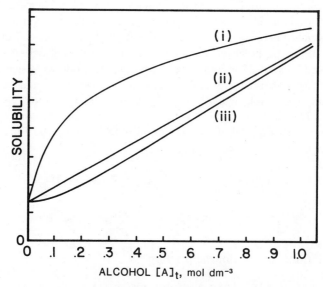

Figure 6.17   Computer-simulated profiles of solubility $[S]_t$ versus total (i.e., formal) alcohol molarity $[A]_t$ obtained by solving Eqs. 6.28 and 6.31 simultaneously while assuming only 1:1 and 1:n solute–alcohol complexes where n is the average stoichiometric number of the complex (i) Large $K_{1:1}$, K_n negligible; (ii) $K_n = nK_{1:1}K_{ass}$; (iii) large K_n, $K_{1:1}$ negligible. $1\,mol\,dm^{-3} = 1\,mol\,L^{-1} = 1\,M$. Reproduced with permission of the author. Anderson, B. D. (1977). Ph.D. thesis, Specific Interactions in Nonaqueous Systems. I. Self-Association of Alcohols and Phenol in Nonpolar Solvents. II. Solubilities of Organic Compounds in Organic Solvents and Cosolvent Mixtures, University of Kansas, Lawrence, KS.

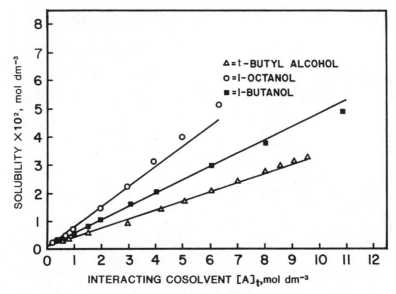

Figure 6.18 Molar solubility [S]$_t$ of carbazole versus total (i.e., formal) molarity [A]$_t$ of various interacting cosolvents in isooctane at 25°C. Lines represent the calculated solubilities based on the equilibrium constants in Table 6.10 using Eqs. 6.28 and 6.31. $1\,\mathrm{mol\,dm^{-3}} = 1\,\mathrm{mol\,L^{-1}} = 1\,M$. Reproduced with permission of the author. Anderson, B. D. (1977). Ph.D. thesis, Specific Interactions in Nonaqueous Systems. I. Self-Association of Alcohols and Phenol in Nonpolar Solvents. II. Solubilities of Organic Compounds in Organic Solvents and Cosolvent Mixtures, University of Kansas, Lawrence, KS.

Curves displaying concave downward curvature can be explained by the formation of only 1:1 solute–alcohol complexes (K_n negligible compared with $K_{1:1}$), and their shape should resemble those of plots of alcohol–monomer concentration versus formal alcohol molarity (e.g., Fig. 6.5). At low concentrations where self-association is absent, anthraquinone solubility is a linear function of [A]$_t$. With increasing concentration, self-association effectively competes with the solute for the monomer, so the slope of the curve decreases. At high alcohol concentrations, disregarding alcohol complexed with solute, the monomer concentration is a very small fraction of the total formal alcohol concentration, so Eq. 6.33 becomes

$$[A]_t \cong nK_{ass}[A_1]^n \qquad (6.34)$$

If solubility is effectively accounted for by 1:1 and 1:2 complexes, at high alcohol concentrations Eqs. 6.26 and 6.34 afford

$$[S]_t = [S]_o + K_{1:1}[S]_o([A]_t/nK_{ass})^{1/n} + K_{1:1}K_{1:2}[S]_o([A]_t/nK_{ass})^{2/n} \qquad (6.35)$$

which causes the downward curvature in the plot of [S]$_t$ versus [A]$_t$. Although

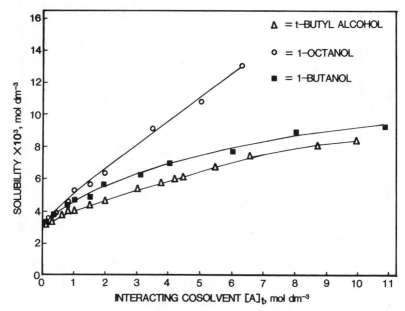

Figure 6.19 Molar solubility $[S]_t$ of anthraquinone versus total (i.e., formal) molarity $[A]_t$ of various interacting cosolvents in isooctane at 25°C. Lines represens the calculated solubilities based on the equilibrium constants in Table 6.10 using Eqs. 6.28 and 6.31. $1 \, mol \, dm^{-3} = 1 \, mol \, L^{-1} = 1 \, M$. Reproduced with permission of the author. Anderson, B. D. (1977). Ph.D. thesis, Specific Interactions in Nonaqueous Systems. I. Self-Association of Alcohols and Phenol in Nonpolar Solvents. II. Solubilities of Organic Compounds in Organic Solvents and Cosolvent Mixtures, University of Kansas, Lawrence, KS.

other models may also explain the data, the evidence suggests that higher-order anthraquinone–alcohol complexes are not important.

As discussed above, when 4-phenylphenol dissolves in alcohol solutions, complexes of order n greater than 1:2 are very important, indicating that K_n is large. At very low alcohol concentrations where self-association is negligible, a linear region corresponding to the 1:1 term in Eq. 6.28 is observed in accordance with the suggested model, and $K_{1:1}$ can be calculated from the initial slope, as shown in Figure 6.20. The computer-simulated curve based on the proposed model is in excellent agreement with the experimental points.

However, there is a break in the solubility curve of several hydrogen-donor solutes, including 4-phenylphenol, near 0.04–0.05 mol dm^{-3} 1-butanol. At high alcohol molarities, the solute solubility can be attributed almost entirely to the 1:n term in Eq. 6.28, which then becomes

$$[S]_t \cong K_n[S]_o[A_1]^n \tag{6.36}$$

whereas the total alcohol molarity can be similarly attributed almost entirely

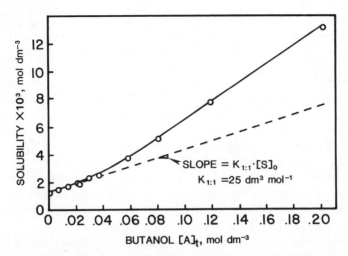

Figure 6.20 Molar solubility $[S]_t$ of 4-phenylphenol versus total (i.e., formal) 1-butanol molarity $[A]_t$ in isooctane at 25°C in the dilute region. The full curve represents the calculated solubilities based on the equilibrium constants in Tablg 6.10 using Eqs. 6.28 and 6.31. $1 \text{ mol dm}^{-3} = 1 \text{ mol L}^{-1} = 1 M$. Reproduced with permission of the author. Anderson, B. D. (1977). Ph.D. thesis, Specific Interactions in Nonaqueous Systems. I. Self-Association of Alcohols and Phenol in Nonpolar Solvents. II. Solubilities of Organic Compounds in Organic Solvents and Cosolvent Mixtures, University of Kansas, Lawrence, KS.

Table 6.11 Relative Contributions of 1:1 and Higher Complexes of Solutes to their Molar Solubilities in 1-Butanol at 25°C[a]

	Solubility due to			
Solute	"Free" Solute (mol dm^{-3})	1:1 Complex (mol dm^{-3})	Higher Complexes (mol dm^{-3})	Total Solubility (mol dm^{-3})
4-Phenylphenol	1.24×10^{-3}	6.0×10^{-3}	0.83	0.84
Carbazole	7.7×10^{-4}	7.8×10^{-4}	4.75×10^{-2}	4.9×10^{-2}
2,4,6-Triiodo-phenol	5.0×10^{-3}	4.4×10^{-3}	0.150	0.160
Anthraquinone	2.8×10^{-4}	1.9×10^{-4}	4.5×10^{-4}	9.2×10^{-4}

[a]The calculated contributions are based on the computer-optimized values of the equilibrium constants in Table 6.10 using Eqs. 6.28 and 6.31. $1 \text{ mol dm}^{-3} = 1 \text{ mol/Litre} = 1 M$.

Reproduced with permission of the author. Anderson, B.D. (1977). Ph.D. thesis, Specific Interactions in Nonaqueous Systems. I. Self-Association of Alcohols and Phenol in Nonpolar Solvents. II. Solubilities of Organic Compounds in Organic Solvents and Cosolvent Mixtures, University of Kansas, Lawrence, KS.

to the second and fourth terms in Eq. 6.31, thus,

$$[A]_t \cong nK_{ass}[A_1]^n + nK_n[S]_o[A_1]^n \tag{6.37}$$

Therefore,

$$[S]_t/[A]_t \cong K_n[S]_o/(nK_{ass} + nK_n[S]_o) \tag{6.38}$$

Since the right-hand side of this equation is constant for a given solute–solvent system, $[S]_t$ is a linear function of $[A]_t$ at higher alcohol concentrations, as is found for the solubility data of 4-nitrophenol, 4-phenylphenol, or carbazole in *tert*-butanol, 1-butanol, or 1-octanol (e.g., Fig. 6.18). Equation 6.38 enables the solubility of a solute in high alcohol concentrations to be predicted, provided that n, K_n, K_{ass}, and $[S]_o$ are known or can be estimated.

The break in the solubility curve would be absent and a straight line would be predicted over the whole range in alcohol concentration if $K_n = nK_{1:1}K_{ass}$, causing Eq. 6.28 to reduce to

$$[S]_t = [S]_o + K_{1:1}[S]_o\{[A_1] + nK_{ass}[A_1]^n\} \tag{6.39}$$

$$= [S]_o + K_{1:1}[S]_o[A]_t \tag{6.40}$$

The last equation incorporates Eq. 6.33, which assumes that the amount of alcohol involved in solute complexes is small.

A least-squares computer program was used to fit the solubility data to this model which assumes only 1:1 and 1:n solute–alcohol complexes. Taking K_{ass} values from Tables 6.3 and 6.6, Eqs. 6.28 and 6.31 were solved simultaneously, allowing the two parameters $K_{1:1}$ and K_n to vary. The computer-optimized values of $K_{1:1}$ and K_n are presented in Table 6.10, while the various contributions of each term in Eq. 6.28 to the total solubility of each solute in 1-butanol are stated in Table 6.11.

The computer plots of $[S]_t$ against $[A]_t$ embodying the optimized parameters show excellent agreement with the experimental points (e.g., Fig. 6.20). Furthermore, values of solubility calculated from this model agree closely with the experimental quantities as shown in Table 6.12. Although other models may give better fits, the precision of the data does not justify the use of additional parameters.

It should be emphasized that in the complex with each solute the number n of alcohol molecules which best fits the model turns out to be equal to the stoichiometric number for alcohol self-association in the absence of the solute, that is, $n = 4$ for *tert*-butanol and $n = 5$ for primary alcohols. Thus, the simple two-parameter model fits the data well. Still, we cannot conclude unequivocally that a unique 1:n complex predominates at high concentrations of alcohol without further evidence.

Table 6.12 Experimental and Calculated Solubilities[a] of Various Substances in Pure Organic Solvents at 25°C

Solvent	Anthracene		2,4,6-Triiodophenol		4-Phenylphenol		Carbazole		Anthraquinone	
	Obs.	Calc.[b]	Obs.	Calc.[b]	Obs.	Calc.[b]	Obs.	Calc.[b]	Obs.	Calc.[b]
1-Octanol	0.015	0.015	0.191	0.167	0.57	0.57	0.052	0.046	0.0013	0.0013
1-Butanol	0.0095	0.0098	0.160	0.179	0.84	0.80	0.049	0.052	0.00092	0.00104
tert-Butanol	—	—	—	—	0.86	0.78	0.033[c]	0.032	0.00084[b]	0.00089
Isooctane	6.6×10^{-3}	—	5×10^{-3}	—	1.24×10^{-3}	—	7.7×10^{-4}	—	2.8×10^{-4}	—

[a] $1\,\mathrm{mol\,dm^{-3}} = 1\,\mathrm{mol\,L^{-1}} = 1\,M$.

[b] Calculated solubilities using the computer-optimized values of the equilibrium constants in Table 6.10.

[c] tert-Butyl alcohol solidifies at 25°C. These values are solubilities in concentrated solutions of tert-butyl alcohol in isooctane.

Reproduced with permission of the author. Anderson, B. D. (1977). Ph.D. thesis, Specific Interactions in Nonaqueous Systems. I. Self-Association of Alcohols and Phenol in Nonpolar Solvents. II. Solubilities of Organic Compounds in Organic Solvents and Cosolvent Mixtures, University of Kansas, Lawrence, KS.

6.6.3 Influence of Molecular Structure on Solubility and Complexation in Alcohol Solutions

As discussed in Chapter 5, stability constants of hydrogen-bonded complexes often follow linear free-energy relationships. Figure 6.21 shows that $\log K_{1:1}$ values derived from solubility data of various phenols in the presence of various alcohols as cosolvents are linear functions of the Hammett polar substituent constant σ_p of phenol substituents (Anderson, 1977). In fact, $K_{1:1}$ values can readily be estimated from such linear free-energy relationships as shown in Section 5.5. The solubility of hydrogen-bonding solutes in the presence of low molar concentration of alcohols can then be predicted from the $K_{1:1}$ values, as described in Section 5.5.2, since alcohol monomers predominate in these solutions.

Chulkaratana (1964) and Higuchi and Chulkaratana, (1961) have shown that specific interaction constants do not greatly depend on the alkyl chain length of the cosolvent, as illustated by Figure 4.12 for alcohol cosolvents. Table 6.13 shows that the solubility of 4-phenylphenol in 1-mol dm^{-3} solutions of various n-alkanols in isooctane increases only marginally with chain length longer than C_3. Although the increases are small, they reflect differences in the formation of higher-order complexes with the solute, since 1:1 complexes are relatively unimportant for 4-phenylphenol.

The prediction of solubility of solutes in pure associated solvents or in strong solutions of associated solvents is difficult because the behavior of the constants K_n is not known even when K_{ass} values are available. Nevertheless, Table 6.13

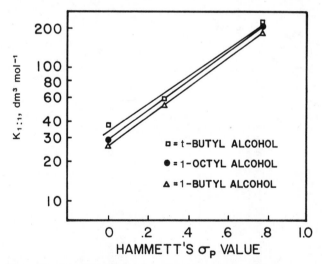

Figure 6.21 Plots of $K_{1:1}$ (log scale) for the interaction of *para*-substituted phenols (4-phenylphenol, 4-iodophenol, and 4-nitrophenol) with alcohol cosolvents in isooctane at 25°C versus the Hammett σ_p substituent constant. $1 \text{ dm}^3 \text{ mol}^{-1} = 1 \text{ L mol}^{-1} = 1 M$. Reproduced with permission of the author. Anderson, B. D. (1977). Ph.D. thesis, Specific Interactions in Nonaqueous Systems. I. Self-Association of Alcohols and Phenol in Nonpolar Solvents. II. Solubilities of Organic Compounds in Organic Solvents and Cosolvent Mixtures, University of Kansas, Lawrence, KS.

Table 6.13 Molar solubilities of 4-Phenylphenol in 1-mol dm^{-3} Solutions of Various Aliphatic Alcohols in Isooctane at $25°C^a$

Alcohol	Solubility $\times 10^2$ (mol dm^{-3})
Ethanol	6.9
1-Propanol	6.2
1-Butanol	6.8
1-Pentanol	6.9
1-Hexanol	7.4
1-Heptanol	7.9
1-Octanol	8.7

$^a 1\,\text{mol dm}^{-3} = 1\,\text{mol L}^{-1} = 1\,M.$

Reproduced with permission of the author. Anderson, B. D. (1977). Ph.D. thesis, Specific Interactions in Nonaqueous Systems. I. Self-Association of Alcohols and Phenol in Nonpolar Solvents. II. Solubilities of Organic Compounds in Organic Solvents and Cosolvent Mixtures, University of Kansas, Lawrence, KS.

can be used as a guide. In fact, the $K_{1:1}$ and K_n values for the proton donor solutes 4-phenylphenol, carbazole, and 2,4,6-triiodophenol with 1-butanol and 1-octanol give the following correlations:

$$\log K_n = 1.843 \log K_{1:1} + 4.022, \qquad n = 5$$

$$3 \text{ data points}, \quad r = 0.993 \text{ with 1-butanol} \qquad (6.41)$$

$$\log K_n = 1.572 \log K_{1:1} + 4.487, \qquad n = 5$$

$$3 \text{ data points}, \quad r = 0.973 \text{ with 1-octanol} \qquad (6.42)$$

Each of these relationships can be tested by using it to predict K_n for a proton donor, inserting K_n into Eq. 6.38 and comparing the solubility estimations from this equation with the experimental value. A somewhat detailed treatment is included here as an example of the general approach which can be used to predict solubility behavior in these systems.

For 4-nitrophenol with 1-butanol in isooctane at $25°C$, $K_{1:1} = 189\,\text{dm}^3\,\text{mol}^{-1}$ and $n = 5$ (Table 6.10, from which Eq. 6.41 predicts $K_n = 1.65 \times 10^8\,\text{dm}^{15}\,\text{mol}^{-5}$). Since for 4-nitrophenol in isooctane, $[S]_o = 3.3 \times 10^{-4}\,\text{mol dm}^{-3}$ at $25°C$ (Anderson et al., 1980), and since for 1-butanol $n = 5$ and $K_{ass} = 8700\,\text{dm}^{12}\,\text{mol}^{-4}$ in isooctane at $25°C$, Eq. 6.38 predicts $[S]_t/[A]_t = 0.1724$. Hence, in the presence of 0.110, 0.349, and $1.00\,\text{mol dm}^{-3}$ 1-butanol, the predicted solubility of 4-nitrophenol is 0.019, 0.060, and $0.172\,\text{mol dm}^{-3}$, respectively, whereas the experimental values (Anderson, 1977) are 0.018, 0.079, and $0.307\,\text{mol dm}^{-3}$. In view of the approximations made, the agreement is satisfactory, particularly in less than $0.5 \, \text{mol} \, \text{dm}^{-3}$ 1-butanol.

An Example of the Prediction of Solubility Behavior in The Presence of Alcohols Estimate the solubility of 3-nitrophenol at 25°C in isooctane containing 0.01, 0.1, and $1.0 \, \text{mol} \, \text{dm}^{-3}$ 1-butanol, given that the Hammett polar substituent constant of the 3-NO_2 group is 0.69.

$K_{1:1}$. Using the linear free-energy relationship for 1-butanol in Figure 6.21, interpolation at $\sigma = 0.69$ for 3-NO_2 leads to $K_{1:1} = 150 \, \text{dm}^3 \, \text{mol}^{-1}$ for 3-nitrophenol in isooctane at 25°C.

K_n. Using the linear free-energy relationship for 1-butanol in Eq. 6.41, the $K_{1:1}$ value given above corresponds with $K_n = 1.08 \times 10^8 \, \text{dm}^{15} \, \text{mol}^{-5}$ for $n = 5$.

$[S]_o$. The solubility of 3-nitrophenol in isooctane at 25°C is not known. However, the corresponding literature (Anderson et al., 1980) value for 4-nitrophenol, which is $3.3 \times 10^{-4} \, \text{mol} \, \text{dm}^{-3}$, can be used instead, since it is within the limitations of the required prediction. (The solubilities of the two compounds probably do not differ by a factor of more than 2. The solubility of the related compound 2,4,6-trinitrophenol in isooctane is $2.5 \times 10^{-4} \, \text{mol} \, \text{dm}^{-3}$ at 25°C. Fung and Higuchi, 1971.)

K_{ass} and n. These are given in Table 6.3 for 1-butanol. Thus, $n = 5$ and $K_{ass} = 8700 \, \text{dm}^{12} \, \text{mol}^{-4}$.

In the presence of $0.01 \, \text{mol} \, \text{dm}^{-3}$ 1-butanol ($< 0.05 \, \text{mol} \, \text{dm}^{-3}$), monomers predominate. Therefore, a total 1-butanol concentration $[A]_t = 0.01 \, \text{mol} \, \text{dm}^{-3}$ corresponds to $[A_1] = 0.01 \, \text{mol} \, \text{dm}^{-3}$. Under these conditions, the interaction between the dissolved solute and alcohol can be explained entirely by the 1:1 term in Eq. 6.28

Therefore,

$$[S]_t = [S]_o + K_{1:1}[S]_o[A_1]$$

Substitution into this equation gives $[S]_t = 8.25 \times 10^{-4} \, \text{mol} \, \text{dm}^{-3}$ or about $10^{-3} \, \text{mol} \, \text{dm}^{-3}$ as a reasonable estimate.

In the presence of 0.1 or $1.0 \, \text{mol} \, \text{dm}^{-3}$ 1-butanol ($> 0.05 \, \text{mol} \, \text{dm}^{-3}$), 1:1 and 1:$n$ complexes probably predominate and Eq. 6.38 can be used. This assumption then gives $[S]_t/[A]_t = 0.161$. Consequently, the solubilities of 3-nitrophenol in the presence of 0.1 and $1.0 \, \text{mol} \, \text{dm}^{-3}$ 1-butanol are about 0.02 and $0.2 \, \text{mol} \, \text{dm}^{-3}$, respectively.

6.7 INFLUENCE OF THE DIMERIZATION OF CARBOXYLIC ACIDS ON THEIR COMPLEXATION BEHAVIOR

Alkanoic acids (fatty acids) and other carboxylic acids (e.g., benzoic acid) have long been known to dimerize in nonpolar or low-polarity solvents (e.g.,

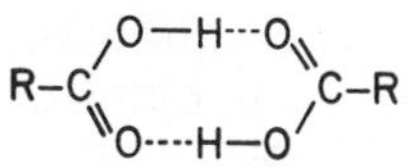

Scheme 6.1 The structure of dimers of carboxylic acid involving two O—H····O hydrogen bonds.

paraffins, benzene, carbon tetrachloride, and chloroform). Scheme 6.1 depicts the accepted hydrogen-bond interactions. Pimentel and McClellan (1960) have presented the classical evidence, which is based on (a) partition studies between these solvents and water (Goodman, 1958), (b) measurements of dielectric constants (Pohl et al., 1941), (c) PMR studies, (d) ebullioscopy and cryoscopy (Singleton, 1960), (e) IR spectroscopy (Hadzi and Sheppard, 1953), and (f) vapor-pressure measurements (Levy et al., 1975).

Carboxylic acids also dimerize in the vapor state, as shown by vapor-density measurements, while X-ray studies of the pure solid and pure liquid states indicate paired molecules (for references see Goodman, 1958). Carboxylic acids show insignificant dimerization in hydroxylic solvents such as water, because solute–solvent association effectively competes with solute–solute self-association. In the pure liquid state, however, the ability of fatty acids to dimerize and perhaps to form higher-order complexes (Levy et al., 1975) must influence the solubility behavior of solutes such as griseofulvin (Grant and Abougela, 1982).

Mehdizadeh and Grant (1984) studied the influence of dimerization on complex formation by means of phase solubility analysis with griseofulvin as the proton acceptor A, and each of various n-alkanoic acids as the proton donor D, in isooctane as the inert paraffinic solvent. Scheme 6.2 shows that griseofulvin contains two carbonyl groups, four ether-oxygen atoms, and an aromatic ring, each of which can accept protons to form a hydrogen bond. Since griseofulvin has no proton-donating groups, it acts only as a proton acceptor. Since each fatty acid molecule has only one donor site, whereas the griseofulvin molecule has about six acceptor sites, the total molar concentration of acid, $[D]_t$, and of griseofulvin, $[A]_t$, in solution can be expressed by the following mass balance equations:

$$[A]_t = [A]_o + [AD] + [AD_2] + \cdots + [AD_z] \tag{6.43}$$

$$[D]_t = [D] + 2[D_2] + [AD] + 2[AD_2] + \cdots + z[AD_z] \tag{6.44}$$

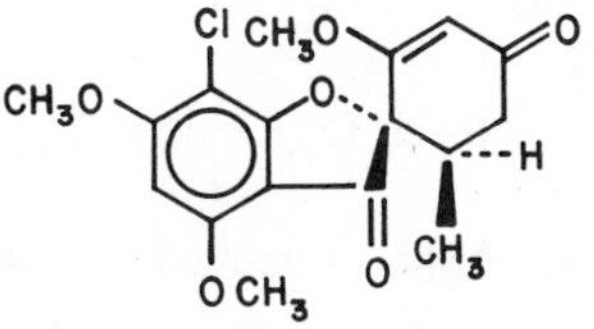

Scheme 6.2 The structure of the griseofulvin molecule showing two carbonyl groups, four ether-oxygen atoms, one aromatic ring, and conjugated unsaturation.

This notation is essentially analogous to that introduced in Section 5.2.

Here $[A]_0$ is the molar concentration of free, uncomplexed griseofulvin and is equal to the solubility in the inert solvent isooctane; $[AD_n]$ is the molar concentration of a typical complex of stoichiometric number n; $[D]$ is that of the fatty acid monomer; and $[D_2]$ is that of the fatty acid dimer, which usually predominates over the monomer in solvents of low polarity, such as hydrocarbons.

Taking as the standard state of unit activity a hypothetical 1-mol dm^{-3} solution which is assumed to behave as if it were infinitely dilute, the activity of X, $\{X\}$, is given by $\{X\} = \gamma_x[X]$, where γ_x is the molarity-based activity coefficient.

The formation of each complex can be written

$$A + nD \rightleftharpoons AD_n \quad \text{and} \quad [AD_n] = K'_n[D]^n \tag{6.45}$$

where

$$K'_n = K_{eq}\{A\}\gamma_D^n/\gamma_{AD_n} \tag{6.46}$$

such that K_{eq} is the thermodynamic stability constant (based on activity, see Section 5.2.1), whereas K'_n is the apparent stability constant, which is based on concentration and which will simply be referred to as the stability constant. K'_n is used instead of K_{AD_n} for simplicity (compare with Eqs. 4.21 and 5.11).

Dimerization of the fatty acid donor may be written

$$2D \rightleftharpoons D_2 \quad \text{and} \quad [D_2] = K_d[D]^2 \tag{6.47}$$

where

$$K_d = K_D\gamma_D^2/\gamma_{D_2} \tag{6.48}$$

such that K_D is the thermodynamic dimerization constant (based on activity) and K_d is the apparent dimerization constant (based on concentration).

Equations 6.45 and 6.47 give

$$[AD_n] = (K'_n/K_d^{-n/2})[D_2]^{n/2} \tag{6.49}$$

If two complexes AD_m and AD_n are present, Eqs. 6.43 and 6.44 become

$$[A]_t - [A]_0 = [AD_m] + [AD_n] \tag{6.50}$$

$$= (K'_m K_d^{-m/2})[D_2]^{m/2} + (K'_n K_d^{-n/2})[D_2]^{n/2} \tag{6.51}$$

$$[D]_t = [D] + 2[D_2] + m[AD_m] + n[AD_n] \tag{6.52}$$

Expressing Eq. 6.52 in terms of $[D_2]$ using Eqs. 6.47 and 6.49 affords

$$[D]_t = K_d^{-1/2}[D_2]^{1/2} + 2[D_2] + (mK'_m K_d^{-m/2})[D_2]^{m/2} + (nK'_n K_d^{-n/2})[D_2]^{n/2} \tag{6.53}$$

If D_2 is assumed to predominate over all other dissolved species containing D, Eq. 6.52 contracts to

$$[D]_t \cong 2[D_2] \tag{6.54}$$

This assumption is justified (a) by the high values of K_d of the order $10^4 \, dm^3 \, mol^{-1}$ (Table 6.14) in Eq. 6.47, such that $[D] \ll [D_2]$, and (b) by the relatively low solubility of griseofulvin, such that $[AD_m] \ll [D_2]$ and $[AD_n] \ll [D_2]$. Equation 6.51 then becomes

$$[A]_t - [A]_o = p[D]_t^{m/2} + q[D]_t^{n/2} \tag{6.55}$$

where

$$p = K'_m (2K_d)^{-m/2} \quad \text{and} \quad q = K'_n (2K_d)^{-n/2} \tag{6.56}$$

therefore,

$$([A]_t - [A]_o)/[D]_t^{m/2} \cong p + q[D]_t^{(n-m)/2} \tag{6.57}$$

This equation can be tested by linear regression analysis. If only one complex species AD_m is considered, Eqs. 6.43 and 6.55 become

$$[A]_t - [A]_o = [AD_m] \cong p[D]_t^{m/2} \tag{6.58}$$

Therefore,

$$\log\{[A]_t - [A]_o\} = \log\{K'_m (2K_d)^{-m/2}\} + (m/2)(\log[D]_t) \tag{6.59}$$

This equation can also be examined by linear regression analysis.

If only one complex AD_n is considered, but full account is taken of all species containing the fatty acid, including the monomer D, Eqs. 6.53 and 6.51 lead to

$$\frac{[D]_t - n([A]_t - [A]_o)}{([A]_t - [A]_o)^{1/n}} = c_1 + c_2 ([A]_t - [A]_o)^{1/n} \tag{6.60}$$

where

$$c_1 = K'^{-1/n}_n \quad \text{and} \quad c_2 = 2K_d^{1/2} K'^{-1/n}_n \tag{6.61}$$

If correct, this model enables K_d as well as K'_n to be calculated.

The apparent molar solubility of griseofulvin $[A]_t$ increases rapidly with increasing molar concentration of each fatty acid donor $[D]_t$ from the very small value in pure isooctane, $[A]_o = 9.358 \, \mu mol \, dm^{-3}$. The log–log plots (Figs. 6.22 and 6.23) according to Eq. 6.59 show two linear regions, each of which may be

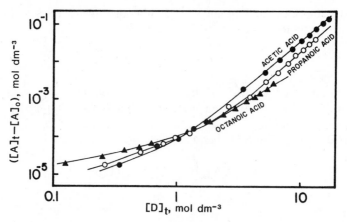

Figure 6.22 Plots of ($[A]_t - [A]_o$) (log scale) against $[D]_t$ (log scale), where ($[A]_t - [A]_o$) is the increase in the apparent molar solubility of griseofulvin brought about by the presence of acetic acid, 1-propanoic acid, and 1-octanoic acid with total molar concentration $[D]_t$ in isooctane at 25°C. $1 \, mol \, dm^{-3} = 1 \, mol \, L^{-1} = 1 \, M$. Reproduced with permission of the copyright owner, the American Pharmaceutical Association, from Mehdizadeh, M. and Grant, D. J. W. (1984). *J. Pharm. Sci.*, **73**, 1195–1202.

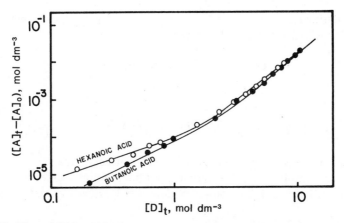

Figure 6.23 Plots of ($[A]_t - [A]_o$) (log scale) against $[D]_t$ (log scale), where ($[A]_t - [A]_o$) is the increase in the apparent molar solubility of griseofulvin brought about by the presence of 1-butanoic acid and 1-hexanoic acid with total molar concentration $[D]_t$ in isooctane at 25°C. $1 \, mol \, dm^{-3} = 1 \, mol \, L^{-1} = 1 \, M$. Reproduced with permission of the copyright owner, the American Pharmaceutical Association, from Mehdizadeh, M. and Grant, D. J. W. (1984). *J. Pharm. Sci.*, **73**, 1195–1202.

attributed to a complex AD_m or AD_n. This type of linearity is possible if m and n, and K'_m and K'_n, differ sufficiently, in which case their approximate values are given by the slopes and intercepts.

Accordingly, the two-complex model (Eq. 6.57) was tested with various pairs of (m, n) values using linear regression analysis (Figs. 6.24 and 6.25). The statistically most probable stoichiometries, corresponding to the linear re-

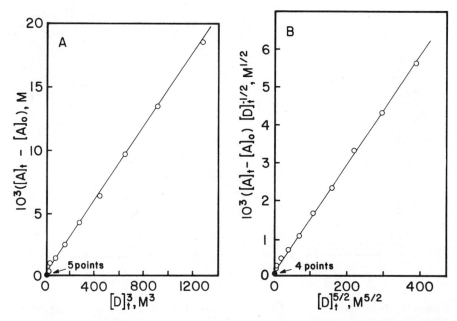

Figure 6.24 Plots of ($[A]_t - [A]_o$) and $[D]_t$ (Eq. 6.57) where ($[A]_t - [A]_o$) is the increase in the apparent molar solubility of griseofulvin brought about by the presence of butanoic acid of total molar concentration $[D]_t$ in isooctane at 25°C. These plots correspond to linear regressions assuming the formation of the following complexes: (A) AD_6 alone ($m = 0, n = 6$); (B) AD and AD_6 ($m = 1, n = 6$). For simplicity, the units of concentration are represented by M, where $1\,M = 1\,mol\,dm^{-3} = 1\,mol\,L^{-1}$. Reproduced with permission of the copyright owner, the American Pharmaceutical Association, from Mehdizadeh, M. and Grant, D. J. W. (1984). *J. Pharm. Sci.*, **73**, 1195–1202.

gression with the highest correlation coefficient r, and the lowest residual standard deviation s, are: (0, 5) for acetic acid; (0, 6) for propanoic and butanoic acid; (0, 6) or (2, 6) for hexanoic acid; (1, 5) for octanoic acid. The presence of lines of smaller slope in Fig. 6.22 and 6.23 clearly show that $m \neq 0$. If, however, too high a value of m is chosen (e.g., 2 or 3 for butanoic acid), Figures 6.24 and 6.25 show that the experimental points follow a sigmoid curve on either side of the regression line.

From these considerations the proposed complexation models are presented in Table 6.14 with their stability constants. Both K'_m and K'_n were calculated from the intercept p and slope q, using Eq. 6.56 and assuming literature values of K_d in n-heptane at 23–30°C (Pohl et al., 1941; Goodman, 1958; Pimental and McClellan, 1960; Jentschura and Lippert, 1971). Equilibrium constants for hydrogen bonding (e.g., K_d) are highest in paraffinic solvents and differ little within this homologous series (Pimentel and McClellan, 1960).

Since the K'_m values are also greatly influenced by n of the accompanying higher-order complex, K'_1 and K'_2 were calculated for the (1, 6) and (2, 6) models when assessing the influence of chain length of D on the stability constants

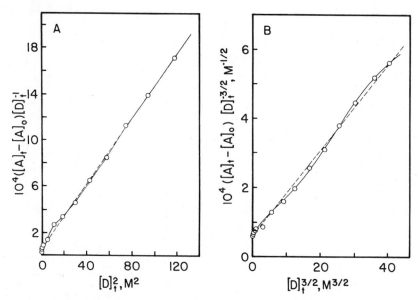

Figure 6.25 Pseudolinear plots of $([A]_t - [A]_o)$ and $[D]_t$ (Eq. 6.57) where $([A]_t - [A]_o)$ is the increase in the apparent molar solubility of griseofulvin brought about by the presence of butanoic acid of total molar concentration $[D]_t$ in isooctane at 25°C. These plots correspond to linear regressions assuming the formation of the following complexes: (A) AD_2 and AD_6 $(m = 2, n = 6)$; (B) AD_3 and AD_6 $(m = 3, n = 6)$. For simplicity, the units of concentration are represented by M, where $1\,M = 1\,mol\,dm^{-3} = 1\,mol\,L^{-1}$. Reproduced with permission of the copyright owner, the American Pharmaceutical Association, from Mehdizadeh, M. and Grant, D. J. W. (1984). *J. Pharm. Sci.*, **73**, 1195–1202.

Table 6.14 Dimerization Constants K_d, Complexation Stoichiometry m and n, and Complexation Constants K'_m and K'_n, which Best Fit Eqs. 6.55–6.57[a]

Straight-Chain Alkanoic Acid	K_d $(dm^3\,mol^{-1})$	AD_m	K'_m (Molarity Based)[b]	AD_n	K'_n (Molarity Based)[b]
Acetic	37,000[c]	AD	Very small[d]	AD_5	1.57×10^8
Propanoic	12,000[e]	AD	0.0167	AD_6	2.38×10^8
Butanoic	9,000[f]	AD	0.0106	AD_6	8.49×10^7
Hexanoic	6,000[f]	AD_2	0.959	AD_6	2.76×10^7
Octanoic	5,800	AD	0.0068	AD_5	3.58×10^5

[a] The units are: $K'_m (dm^3\,mol^{-1})^{m-1}$ and $K'_n (dm^3\,mol^{-1})^{n-1}$; $1\,dm^3\,mol^{-1} = 1\,L\,mol^{-1} = 1\,M^{-1}$.
[b] In isooctane containing various concentrations of several straight-chain alkanoic acids at 25°C.
[c] Dimerization constant in heptane at 23°C and 30°C (Goodman, 1958; Pohl et al., 1941) is assumed to represent the corresponding value in isooctane at 25°C.
[d] Cannot be estimated from the negative intercept.
[e] Dimerization constant in cyclohexane at 25°C (Jentschura and Lippert, 1971) is assumed to represent the corresponding value in isooctane at 25°C.
[f] Determined by interpolation of the K_d values (for references see Mehdizadeh and Grant, 1984).

Reproduced with permission of the copyright owner, the American Pharmaceutical Association from Mehdizadeh, M. and Grant, D. J. W. (1984). *J. Pharm. Sci.*, **73**, 1195–1202.

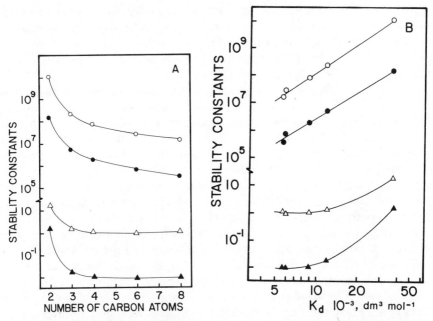

Figure 6.26   Influence of (A) the chain length of various straight-chain alkanoic acids and (B) the molarity-based dimerization constants K_d of various straight-chain alkanoic acids on the molarity-based stability constants K'_m or K'_n of their complexes AD_m or AD_n, with griseofulvin in isooctane at 25°C. Key: K'_1 of AD with AD_6 (▲); K'_2 of AD_2 with AD_6 (△); K'_5 of AD_5 with AD (●); K'_6 of AD_6 with AD (○). Units of stability constants are $(dm^3 \, mol^{-1})^{n-1}$, where $1 \, dm^3 \, mol^{-1} = 1 \, L \, mol^{-1} = 1 \, M^{-1}$. Reproduced with permission of the copyright owner, the American Pharmaceutical Mehdizadeh, M. and Grant, D. J. W. (1984). *J. Pharm. Sci.*, **73**, 1195–1202.

(Fig. 6.26A). $K'_1 (\cong 10^{-2}-10^{-1})$ and $K'_2 (\cong 1-10 \, dm^3 \, mol^{-1})$ are so small as to be accounted for by nonspecific interactions between D and A (Higuchi and Connors, 1965; Anderson et al., 1980) and are relatively insensitive to the chain length of the n-alkanoic acid. On the other hand, the K'_n values K'_5 and K'_6 are quite insensitive to m of the accompanying lower-order complex and are very large (Table 6.14 and Fig. 6.26),indicating powerful griseofulvin–fatty acid interactions. Since griseofulvin has 6 obvious acceptor sites (4 ethers + 2 carbonyls, Scheme 6.2), the favored stoichiometry is AD_6 (Table 6.14) unless steric crowding occurs, for example, AD_5 is favored for octanoic acid.

The nonlinear plots in Figure 6.26A may reflect (a) increasing disturbance to the D–A interactions by the longer hydrocarbon chains and (b) the partial replacement of these interactions by weaker dispersion forces (Section 2.11.4). Linear free-energy relationships are observed between $\log_{10}(K'_5$ or $K'_6)$ and $\log K_d$, but not between $\log_{10}(K'_1$ or $K'_2)$ and $\log_{10} K_d$ (Fig. 6.26B). These considerations accord with specific A–D interaction in AD_5 and AD_6 and with nonspecific interaction in AD or AD_2. Poor correlations are, however, found between $\log(K'_m$ or $K'_n)$ and pK_a of the acids, possibly reflecting the fact that proton transfer is more complicated than donor–acceptor bonding.

Although Eqs. 6.55–6.59 account well for the data, they are based on the following assumptions: (a) the activity coefficients (γ values) in Eqs. 6.46 and 6.48 are constant; (b) the activity of the solid phase, which is in equilibrium with each saturated solution (Eq. 6.46), is constant and equal to {A}, the activity of pure griseofulvin; (c) $[D]_t \cong 2[D_2]$ according to Eq. 6.54.

The activity coefficients, which refer to polar species, probably increase with increasing $[D]_t$ in isooctane. This effect will greatly influence K'_m and K'_n and may distort the apparent stoichiometries, especially n, thereby accounting for the unexpectedly low AD_5 stoichiometry for acetic acid (Table 6.14). The activity of each solid phase is effectively equal to {A} because after *standing* in contact with saturated solution the solid behaves analytically and physically as pure griseofulvin in its original form. On the other hand, when *crystallized* from any pure fatty acid except propanoic acid, griseofulvin forms a solid solvate or inclusion compound (Abougela et al., 1979; Grant and Abougela, 1981).

An alternative single-complex model in which all donor-containing species, including the monomer, are accounted for leads to Eq. 6.60, which fits the experimental data as well as do Eqs. 6.55–6.59. However, this model leads to unacceptable negative values of K'_n and unacceptably small values of K_d. Since fatty acid molecules in nonpolar solvents may form trimers and higher oligomers (Levy et al., 1975), and since lower-order complexes AD_m may be involved, the actual solutions may contain a range of species and complexes. Such a model would, however, be very complicated algebraically and would contain too many adjustable constants to be amenable to test.

To summarize, the solubility of griseofulvin in isooctane containing an n-alkanoic acid can be explained well by complex equilibria involving only AD_n ($n = 5$ or 6) and possibly also AD_m (where $m = 0$, 1, or 2), although other species, such as the monomeric acid, may also be involved.

6.8   STUDIES WITH PHENOLS AS SOLUTES

6.8.1   Phenol Self-Association: Solubility and Vapor-Pressure Studies

The influence of the nature of saturated aliphatic hydrocarbon solvents is particularly striking in the case of phenol solubility, which is only 0.30 mol dm^{-3} in isooctane, but as high as 4.0 mol dm^{-3} in cyclohexane (Anderson et al., 1979). The reason for this behavior was studied by Anderson et al. (1979) and can be summarized as follows. Application of Eq. 6.26 to a solution saturated with monomer leads to

$$a^{c*} = p_1^*/h_1^c = a_1^c = y_1^c c_2 \qquad (6.62)$$

$$= 0.089 \text{ in isooctane at } 25°C \text{ (Table 3.2)}$$

$$= 0.146 \text{ in cyclohexane at } 25°C \text{ (Table 3.2)}$$

Assuming that $y_1^c = 1$ for these dilute solutions, the concentration of monomer, c_1, in a saturated solution of cyclohexane (0.146 mol dm^{-3}) is a factor of about

1.64 greater than in a saturated solution of isooctane (0.089 mol dm^{-3}). As we shall see, phenol associates in both solvents probably forming pentamers. Apparently both polymer and monomer are stabilized to about the same extent in cyclohexane as in isooctane, so the equilibrium constant for association is of the same order in both solvents.

The small increase in the allowed monomer concentration of cyclohexane as compared with isooctane (factor of 1.64), when raised to the fifth power due to the association, results in a much higher concentration of polymer at saturation (factor $\cong 1.64^5 = 11.9$). The actual solubility is greater in cyclohexane than in isooctane by a factor of 13.3 (Anderson et al., 1979). No doubt other factors, such as differences of association constants and deviations of the individual species activity coefficients y_1^c from unity with increasing phenol concentrations, may also contribute to the enhanced solubility in cyclohexane. However, it is the fact that phenol self-associates which results in the present anomaly.

When there is little possibility of self-association, as in the case of the extremely low solubilities of carbazole, phthalic anhydride, picric acid, and salicylic acid in cyclohexane and isooctane (Fung and Higuchi, 1971), the solubility is about 1.6 times as great in cyclohexane than in isooctane, as shown in Table 3.4, and discussed in Section 3.5.3. This factor is similar to that calculated for the phenol monomer. (The acetanilide solubilities in Table 3.4 are an apparent exception.)

By means of the chromatographic head-space method, Anderson et al. (1979) obtained the data shown in Figures 6.27–6.29 for solutions of phenol in isooctane

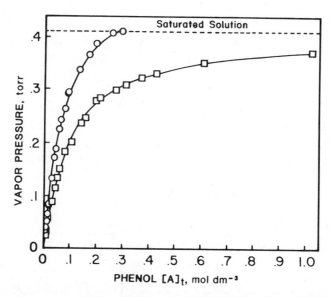

Figure 6.27 Vapor pressure of phenol versus total (i.e., formal) phenol molarity [A]$_t$ in isooctane (O) and cyclohexane (□) at 25°C. 1 mol dm^{-3} = 1 mol L^{-1} = 1 M. 1 torr = 133.3 Pa. Reprinted with permission from Anderson, B. D., Rytting, J. H. and Higuchi, T. (1979). *J. Am. Chem. Soc.*, **101**, 5194–5197. Copyright © 1979 by the American Chemical Society.

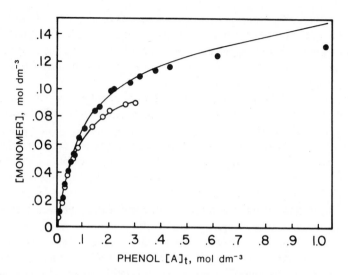

Figure 6.28 Monomer concentration versus total (i.e., formal) phenol molarity $[A]_t$ in isooctane (○) and cyclohexane (●) at 25°C. $1 \, mol \, dm^{-3} = 1 \, mol \, L^{-1} = 1 \, M$. Reprinted with permission from Anderson, B. D., Rytting, J. H., and Higuchi, T. (1979). *J. Am. Chem. Soc.*, **101**, 5194–5197. Copyright © 1979 by the American Chemical Society.

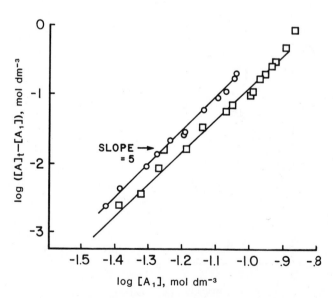

Figure 6.29 Log–log plots of phenol vapor-pressure data in isooctane (○) and cyclohexane (□) at 25°C (see Eq. 6.18). The slope n indicates the stoichiometry of self-association (cf. Fig. 6.7). $1 \, mol \, dm^{-3} = 1 \, mol \, L^{-1} = 1 \, M$. Reprinted with permission from Anderson, B. D., Rytting, J. H., and Higuchi, T. (1979). *J. Am. Chem. Soc.*, **101**, 5194–5197. Copyright © 1979 by the American Chemical Society.

and cyclohexane. The Henry's law constants, stated in Table 3.2 and discussed in Section 3.4, were obtained from the linear plots of vapor pressure against phenol concentrations below $0.02 \, \text{mol dm}^{-3}$, as shown in Figure 6.27.

Presentation of the data for higher concentrations according to Eq. 6.18 gave parallel straight lines for each solvent with slopes of 5 (Fig. 6.29). This behavior suggests that the associated species may consist mainly of pentamers, as was the case for the n-alkanols. This conclusion was confirmed using a least-squares computer-fitting technique, which gave a smaller standard deviation for the monomer–pentamer (1–5) model than for other monomer–single n-mer models. The monomer concentration calculated from the 1–5 model deviated from the experimental value by less than $\pm 3\%$, which is virtually within the experimental error of $\pm 2.6\%$ for estimating the monomer concentration from each data point. The equilibrium constants for pentamer formation in the two solvents are shown in Table 6.15.

Table 6.15 also lists the root-mean-square deviations in monomer concentration from computer fits of phenol vapor-pressure data employing one- and two-parameter models. As expected, the addition of more parameters improves the fit but greatly widens the confidence intervals for each parameter, making

Table 6.15 Root-Mean-Square Deviations for Various Computer Fits of Phenol Vapor Pressures in Cyclohexane and Isooctane at 25°C[a]

Self-Association Model	Cyclohexane[b] ($\times 10^3$)	Isooctane[c] ($\times 10^3$)
1–4	2.4	2.2
1–5	1.5[d]	0.94[e]
1–6	2.4	1.4
1–2–5	—	0.94
1–3–5	1.4	—
1–4–5	1.4	0.94
1–3–6	1.3	0.79
1–4–6	1.3	0.83
1–3–∞	2.2	1.5
1–4–∞	1.3	0.75
1–5–∞	1.5	0.90
1–6–∞	2.4	1.4

[a] $1 \, \text{mol dm}^{-3} = 1 \, \text{mol L}^{-1} = 1 \, M$.
[b] Determined from vapor-pressure data up to $0.43 \, \text{mol dm}^{-3}$.
[c] Determined from vapor-pressure data up to $0.3 \, \text{mol dm}^{-3}$.
[d] $K_{1,5} = 2660 \, \text{dm}^{12} \, \text{mol}^{-4}$.
[e] $K_{1,5} = 6260 \, \text{dm}^{12} \, \text{mol}^{-4}$.

Reprinted with permission from Anderson, B.D., Rytting, J.H., and Higuchi, T. (1979). *J. Am. Chem. Soc.*, **101**, 5194–5197. Copyright © 1979 by the American Chemical Society.

comparisons between models difficult. Table 6.15 also includes analysis of four 1–n–∞ models.

The 1–n–∞ models have been proposed over a number of years for the self association of phenol and alcohols with $n = 2$ or 3 (Coggeshall and Saier, 1951; Tucker and Becker, 1973). In applying this type of model, it is assumed that the formation of the first important species n is characterized by a unique equilibrium constant $K_{1,n}$. All the higher polymers are assumed to be formed in a stepwise manner with the same equilibrium constant K_∞ for the incorporation of an additional monomer unit. The general equation for the 1–n–∞ model, which is analogous to Eq. 6.17, is as follows:

$$[A]_t = [A_1] + \frac{K_{1,n}[A_1]^n(n - [n - 1]K_\infty[A_1])}{(1 + K_\infty[A_1])^2}$$ (6.63)

The relative goodness of fit of different models depends on whether the variation in monomer or the total (i.e., formal) molarity is minimized. Since most of the error is in the estimation of the monomer concentration, the former alternative was preferred by Anderson et al. (1979) in the work described above and analyzed in Table 6.15.

Lin et al. (1978) used a spectral method (IR and UV) to determine the vapor pressure of phenol above its solutions. Their data on the association of phenol in cyclohexane best fit a 1–3–∞ model when formal molarity is treated as the

Table 6.16 Summary of Literature Data on the Self-Association of Phenol in Various Solvents

Preferred Association Model[a] (1–n)	Association Constant[b] $(dm^3 mol^{-1})^{n-1}$	$-\Delta H$ (kcal mol^{-1})	Technique Used
	Solvent: Benzene		
1–2	$K_{1,2} = 0.575$ at 25°C		Partition[c]
1–2	$K_{1,2} = 0.570$ at 25°C	2.4	Vapor-pressure lowering[d]
1–2	$K_{1,2} = 61$ at 5°C		Cryoscopy[e]
1–2	$K_{1,2} = 125$ at 5°C		Cryoscopy[f]
1–3	$K_{1,3} = 450$ at 5°C		Cryoscopy[f]
1–2	$K_{1,2} = 0.128$ at 25°C	5.6	Calorimetry[g]
	Solvent: Carbon Tetrachloride		
1–2–∞	$K_{1,2} = 0.72$ at 25°C $K_\infty = 0.34$ at 25°C		IR[h]
1–3	$K_{1,3} = 4.78$ at 21°C		PMR[i]
1–3–6	$K_{1,3} = 4.1$ at 25°C $K_{1,6} = 432$ at 25°C		Partition between water and CCl$_4$[j]

(1–3, 1–4, 1–3–4, and 1–3–5 gave comparable standard deviations)

Table 6.16 (Continued)

Preferred Association Model[a] $(1-n)$	Association Constant[b] $(dm^3\,mol^{-1})^{n-1}$	$-\Delta H$ (kcal mol^{-1})	Technique Used
1–3–∞			Diffusion at 25°C[k]
1–3–∞	$K_{1,3} = 7.25$ at 20.7°C	9.3	IR[l]
	$K_\infty = 2.49$ at 20.7°C	5.2	
1–3	$K_{1,3} = 5.6$ at 25°C	8.54	Calorimetry and partition[m]
	$K_{1,3} = 14.9$ saturated with water		
1–3–∞	Discussion of data		Activity data[n] at 20–29°C
1–3–∞	$K_{1,3} = 4.38$ at 29.1°C	9.5	Activity from vapor
	$K_\infty = 2.18$ at 29.1°C	5.2	spectrum[o]

Solvent: Cyclohexane

1–3–∞	$K_{1,3} = 26.2$ at 22.2°C	11.3	IR[p]
	$K_\infty = 6.18$ at 22.2°C	5.5	
1–3	$K_{1,3} = 21.5$ at 25°C	10.98	Calorimetry[g]
1–3–∞	$K_{1,3} = 26.0$ at 22°C	13.4	Activity from vapor
	$K_\infty = 6.62$ at 22°C	6.0	spectrum[o]
1–5	$K_{1,5} = 2660$ at 25°C		Vapor pressure[p]

(1–3–6, 1–4–6, and 1–4–∞ gave comparable standard deviations)

Solvent: Isooctane

1–5	$K_{1,5} = 6260$ at 25°C		Vapor pressure[p]

(1–3–6 and 1–4–∞ gave comparable standard deviations)

1–5	$K_{1,5} = 6490$ at 25°C	22.5	PMR[q]
1–5 (or 1–6)	$K_{1,5} = 6000$ at 25°C	22.5	Calorimetry[q]

[a] Stoichiometric number, that is, number of monomer units in the polymer.
[b] Units: $K_{1,2}$, dm^3 mol^{-1}; $K_{1,3}$, dm^6 mol^{-2}; $K_{1,5}$, dm^{12} mol^{-4}; $K_{1,6}$, dm^{15} mol^{-5}; K_∞, dm^3 mol^{-1}.
[c] Philbrick (1934).
[d] Lasettre and Dickinson (1939).
[e] Bury and Jenkins (1934).
[f] Vandeborgh et al. (1970).
[g] Woolley and Hepler (1972).
[h] Coggeshall and Saier (1951).
[i] Saunders and Hyne (1958).
[j] Johnson et al. (1965).
[k] Longsworth (1966).
[l] Whetsel and Lady (1970).
[m] Woolley et al. (1971).
[n] Tucker et al. (1974).
[o] Lin et al. (1978).
[p] Anderson et al. (1979).
[q] Dressman (1981).

dependent variable, while a 1–5–∞ model gives the best fit when monomer molarity is treated as the dependent variable (Anderson et al., 1979). The change in goodness of fit probably arises from a difference in weighting of individual data points in different concentration regions by the two methods. Clearly, extreme caution must be exercised in interpreting the relative goodness of fit of various models to self-association data. Nevertheless, models are useful for summarizing the data and for comparing the associative behavior of a solute in different solvents.

Hence, it is not surprising that the choice of the most appropriate model to describe phenol and alcohol self-association has been the subject of considerable controversy. In the case of phenol self-association in benzene, carbon tetra-chloride, cyclohexane, or isooctane, the controversial and contradictory findings taken from the literature are highlighted in Table 6.16. The variation of the K values in benzene is particularly noteworthy.

In early studies of phenol self-association, dimers were considered to be important in the stepwise association process (Pimentel and McClellan, 1960). More recently, evidence against dimers has accumulated. Trimers were found to be the dominant phenol n-mer in several NMR studies (Saunders and Hyne, 1958; Bogachev et al., 1972; Dale and Gramstad, 1972) and calorimetric studies (Woolley et al., 1971; Woolley and Hepler, 1972). Other work has favored a 1–3–6 model (Johnson et al., 1965) or a 1–3–∞ model (Whetsel and Lady, 1970; Lin et al., 1978) for phenol self-association.

The 1–5 model, proposed by T. Higuchi's group for phenol self-association in isooctane, appears to be novel. This group reexamined the system by means of two complementary but different techniques, PMR and calorimetry, which will now be described.

6.8.2 Phenol Self-Association: An NMR Study

The advantage of the PMR method over other spectroscopic procedures for studying hydrogen bonding is the large chemical shift experienced by protons involved in hydrogen bonds and the accuracy to which these shifts can be measured. The rapid exchange between the monomer and the self-associated species in solution causes the hydroxyl protons to show only a single resonance. The observed shift for this resonance is the weighted average of the chemical shifts attributed to the individual species. Assuming the presence of only one polymeric species A_n of stoichiometric order n, and therefore possessing n exchangable protons, the observed chemical shift ν_{obs} with respect to a defined reference signal is given by

$$\nu_{obs}[A]_t = \nu_1[A_1] + \nu_n n[A_n] \tag{6.64}$$

where ν_1 is the chemical shift of the monomer and ν_n is that of the polymer with respect to the same reference signal. The other quantities $[A_1]$, n, $[A_n]$, and $[A]_t$ are defined by Eqs. 6.1, 6.2, 6.16, and 6.17. Equations 6.2, 6.16, and 6.64 can be combined with the elimination of $[A_1]$ and $[A_n]$ to give

$$(v_1 - v_{\text{obs}})/(v_1 - v_n) = [(v_{\text{obs}} - v_n)/(v_1 - v_n)]^n n K_{1,n} [A]_t^{n-1} \qquad (6.65)$$

This is the Lippert (1963) equation, which may be rearranged to the following form in which $(v_1 - v_{\text{obs}})$ is expressed as Δv:

$$(\Delta v/[A]_t^{n-1})^{1/n} = \{(v_1 - v_n) n K_{1,n}\}^{1/n} - \{(v_1 - v_n)^{1-n} n K_{1,n}\}^{1/n} \Delta v$$
$$= (\Delta v/f_A^{n-1})^{1/n} \qquad (6.66)$$

where $f_A = [A]_t$, the formal (i.e., total) molarity.

A plot of the left-hand side of this equation against the Δv will be linear only if the correct value of n is chosen, and $K_{1,n}$ can be calculated from the slope and intercept. If the curve approaches the ordinate exponentially for a given value of n, the corresponding polymer of order n is not present. If all the lines are curved, more than one polymeric species is present.

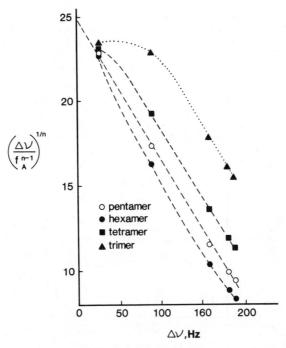

Figure 6.30 NMR data at 25°C for several concentration of phenol in isooctane ranging from 0.04 to 0.23mol dm^{-3}, plotted according to Lippert's method (Eq. 6.66) to determine the order n of the self-associated species formed: $n = 3$ ($\blacktriangle$); $n = 4$ ($\blacksquare$); $n = 5$ ($\circ$); $n = 6$ ($\bullet$). f_A is the formal (i.e., total) molarity of phenol. 1 mol dm^{-3} = 1 mol L^{-1} = 1 M. Reproduced with permission of the author. Dressman, J. (1981). Ph.D. thesis, The Effects of Association Equilibria on the Transfer Transport of Phenol, University of Kansas, Lawrence, KS.

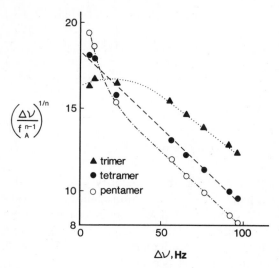

Figure 6.31 NMR data at 35°C for several concentrations of phenol in isooctane ranging from 0.04 to 0.23 mol dm^{-3}, plotted according to Lippert's method (Eq. 6.66) to determine the order n of the self-associated species formed at higher temperature: $n = 3$ (▲); $n = 4$ (●); $n = 5$ (○). f_A is the formal (i.e., total) molarity of phenol. 1 mol dm^{-3} = 1 mol L^{-1} = 1 M. Reproduced with permission of the author. Dressman, J. (1981). Ph.D. thesis, The Effects of Association Equilibria on the Transfer Transport of Phenol, University of Kansas, Lawrence, KS.

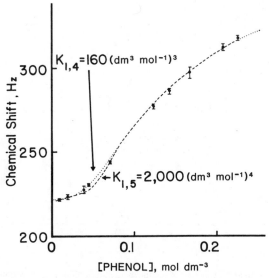

Figure 6.32 NMR data at 35°C plotted as chemical shifts Δv versus total molarity of phenol [phenol] in isooctane, with theoretical lines indicated for $K_{1,4}$ and $K_{1,5}$ and calculated from the Lippert equation (Eq. 6.66). 1 mol dm^{-3} = 1 mol L^{-1} = 1 M. Reproduced with permission of the author. Dressman, J. (1981). Ph.D. thesis, The Effects of Association Equilibria on the Transfer Transport of Phenol, University of Kansas, Lawrence, KS.

At 25°C, a straight line in the Lippert (1963) plot was obtained only in the case of $n = 5$, corresponding to the pentamer as the most probable species, as shown in Figure 6.30. It is unlikely that lower-order species such as the trimer are present since their plots are markedly nonlinear. The slope and intercept afford $K_{1,5} = 6490 \, dm^{12} \, mol^{-4}$ at 25°C in good agreement with $K_{1,5} = 6260 \, dm^{12} \, mol^{-4}$ from the vapor-pressure method.

At 35°C, the Lippert plot shown in Figure 6.31 was linear only for $n = 4$, corresponding to the tetramer as the dominant polymer with $K_{1,4} = 160 \, dm^9 \, mol^{-3}$. Thus, an increase in temperature caused a definite decrease in the size of the dominant polymer. Figure 6.32 also shows that the experimental data fit the tetramer model better than the pentamer model at 35°C.

6.8.3   Phenol Self-Association: A Calorimetric Study

Calorimetry provides a direct measure of $\Delta H^\theta/n$, the standard enthalpy of association per mole of monomer association by Eqs. 6.1 and 6.2. One way in which calorimetry may be applied for the measurement of standard enthalpy has already been described when we discussed the work of Anderson et al. (1975) on alcohol association. Woolley et al. (1971) developed an alternative method which was applied to phenol self-associating to form a single polymer to Eqs. 6.1 and 6.2.

From several measurements of the heat exchanged in diluting a solution of phenol, the quantity ϕ_L, defined by Equation 6.8 is evaluated. The value ϕ_L is the relative apparent molar enthalpy and is equal to the negative enthalpy of dilution of 1 mole of phenol from a solution of given molarity c to infinite dilution. Since at any finite concentration there will be both monomeric and associated species present, whereas at infinite dilution all species will be dissociated to monomers, then,

$$\phi_L = (\alpha/n)\Delta H^\theta_{1,n} \tag{6.67}$$

where α represents the fraction of monomers that is associated. Therefore,

$$[A_1] = c(1 - \alpha) \tag{6.68}$$

and

$$[A_n] = \alpha c/n \tag{6.69}$$

Substitution into Eq. 6.2 affords

$$K_{1,n} = \alpha/[n(c)^{n-1}(1 - \alpha)^n] \tag{6.70}$$

By definition of ϕ_L, or by combining Eqs. 6.67 and 6.69,

$$c\phi_L = \Delta H^\theta_{1,n}[A_n] \tag{6.71}$$

Combination with Eq. 6.2 yields

$$[A_1] = (c\phi_L/\Delta H^\theta_{1,n} K_{1,n})^{1/n} \tag{6.72}$$

Substitution into Eq. 6.5 gives

$$c = (c\phi_L/\Delta H^\theta_{1,n} K_{1,n})^{1/n} + nc\phi_L/\Delta H^\theta_{1,n} \tag{6.73}$$

Rearrangement affords the following linear equation which is suitable for graphical application:

$$\phi_L = \Delta H^\theta_{1,n}/n - (1/n)(\Delta H^\theta_{1,n})^{(n-1)/n}(1/K_{1,n})^{1/n}(\phi_L/c^{n-1})^{1/n} \tag{6.74}$$

A plot of ϕ_L against $(\phi_L/c^{n-1})^{1/n}$ will only be linear for that value of n corresponding to the predominant species in solution. It is necessary to try several plots, assuming a different value of n for each. If a single polymeric associated species predominates, only the appropriate value of n will give a linear plot. The slope and the intercept of this line are then used to calculate $\Delta H^\theta_{1,n}/n$ and $K_{1,n}$ for self-association.

Dressman et al. (1983) applied this procedure (Woolley et al., 1971) to the self-association of phenol in isooctane at 25°C. They obtained the plots shown in Figure 6.33, which indicate that trimers and tetramers cannot explain the data, whereas either pentamers or hexamers can. Unfortunately, solubility limitations

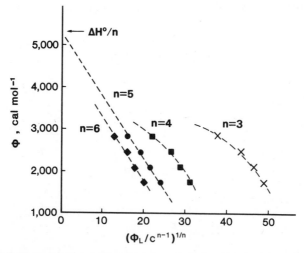

Figure 6.33 Calorimetric data at 25°C plotted according to Eq. 6.74 for various stoichiometries n of self-association. 1 cal = 4.184 J; units of c are mol dm^{-3}. Reproduced with permission of the author. Dressman, J. (1981). Ph.D. thesis, The Effects of Association Equilibria on the Transfer Transport of Phenol, University of Kansas, Lawrence, KS.

prevented data collection at smaller values of $(\phi_L/c^{n-1})^{1/n}$, so the method could not distinguish between hexamers or pentamers, although smaller species could be ruled out.

Since the vapor-pressure data of Anderson et al. (1979) and the PMR data of Dressman et al. (1982, 1983) clearly indicated that the pentamer was the dominant species, the data were replotted for $n = 5$, as shown in Figure 6.34. The abscissa is $n[A_n]$ and the ordinate is the heat absorbed when 1 L of the corresponding solution is taken to infinite dilution. The plot should be linear and pass through the origin, since it may reasonably be assumed that the quantity of heat absorbed is proportional to the amount of n-mer dissociated. Only when the correct value of K, as well as n, is chosen, will this be so. Several estimates of K were used for the plots in Figure 6.34. Only values of K close to $6000 \, \text{dm}^{12} \, \text{mol}^{-4}$ resulted in plots passing through the origin. The final results are $\Delta H^{\theta}_{1,5}/5 = 4.56 \, \text{kcal mol}^{-1} \pm 0.01$ and $K_{1,5} = 6000 \, \text{dm}^{12} \, \text{mol}^{-4}$, which agree well with the values obtained from PMR and the value of $K_{1,5}$ from vapor-pressure measurements.

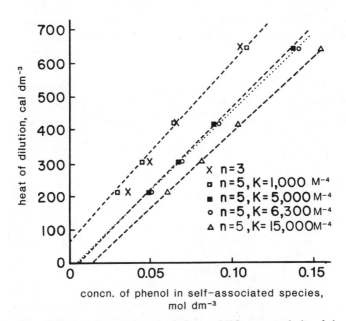

Figure 6.34 Heat of dilution of concentrated solutions at 25°C versus molarity of phenol present initially in the self-associated form, calculated using three different equilibrium constants for self-association and assuming that a pentamer is formed. The slope obtained from the $K_{1,5} = 6,300 \, \text{M}^{-4}$ plot gives the corresponding heat of dissociation per mole of phenol, that is, $\Delta H^{\circ}/n = 4.58 \, \text{kcal/mol}$ where $n = 5$. $1 \, \text{cal} = 4.184 \, \text{J}$. $1 \, \text{mol dm}^{-3} = 1 \, \text{mol L}^{-1} = 1 \, M$. Reproduced with permission of the author. Dressman, J. (1981). Ph.D. thesis, The Effects of Association Equilibria on the Transfer Transport of Phenol, University of Kansas, Lawrence, KS.

6.8.4 Phenol Self-Association: Discussion

The self-association results obtained by T. Higuchi's group (Anderson et al., 1979; Dressman, et al., 1983) using isooctane as the solvent may be compared with those obtained by other workers, such as Woolley et al. (1971, 1972), using other solvents at the same temperature of 25°C, by referring to Table 6.17. Dimers and trimers were favored in polarizable solvents such as benzene and carbon tetrachloride. Both solvents can interact with phenol by London and Debye forces. Benzene is a soft Lewis base and might interact in a PhO–H$\cdots\pi$-electron charge transfer, whereas carbon tetrachloride has a quadrupole which might interact with the phenol dipole. On the other hand, isooctane is an inert saturated hydrocarbon solvent which interacts only minimally with polar molecules and thereby allows phenol to form an associated species of higher order, probably a pentamer. The value of n in cyclohexane is controversial at present.

The data shown in Table 6.16 for phenol self-association in isooctane at 25°C illustrate the importance of obtaining results from several independent methods in order to dismiss association models which appear to fit data from just one or two techniques. For example, hexamer formation could not be ruled out by calorimetry, but is unlikely in the light of PMR and vapor-pressure data. Similarly, vapor-pressure data could be fitted to a 1–3–6 models, whereas calorimetry and PMR showed that trimer formation was insignificant. Overall, the results favor the monomer–pentamer model for the self-association of phenol in isooctane at 25°C.

Table 6.17 Linear Regression Analysis of Lippert Plots (Eq. 6.66) for Various Self-Association Models of Phenol in Isooctane at 25 and 35°C

Self-Association Model	Slope	Intercept	Correlation Coefficient	Standard Error of the Slope $\times 10^3$
		Temperature—25°C		
1–3	−0.0524	25.90	−0.961	1.214
1–4	−0.0739	25.36	−0.998	0.378
1–5	−0.0834	24.98	−0.999	0.184[a]
1–6	−0.0886	24.72	−0.996	0.451
		Temperature—35°C		
1–3	−0.0452	17.23	−0.963	0.495
1–4	−0.0946	18.53	−0.996	0.329[a]
	−0.0972	18.76	−0.999	0.197[b]
1–5	−0.1212	19.32	−0.986	0.788

[a] Favored model of lowest standard error.
[b] Excluding outlier.

Reproduced with permission of the author. Dressman, J. (1981). Ph.D. thesis, The Effects of Association Equilibria on the Transfer Transport of Phenol, University of Kansas, Lawrence, KS.

Phenol resembles the primary alcohols (Anderson et al., 1978) in its self-association behavior in saturated aliphatic hydrocarbon solvents at 25°C. Phenol is much more acidic than alcohols such as methanol, even in organic solvents, as indicated by its strong hydrogen-donating ability (Higuchi et al., 1969). Yet the equilibrium constant for the formation of phenol pentamer in isooctane ($K_{1,5} = 6300 \, dm^{12} \, mol^{-4}$) has a value similar to that for primary alcohols ($K_{1,5} = 8700 \, dm^{12} \, mol^{-4}$ for 1-butanol in isooctane). This similarity suggests that self-association interactions are not very sensitive to the acidity of the hydroxyl group, which is greater for phenols than for alcohols.

Fluoroalcohols, which are better proton donors than the corresponding alcohols (Kivinen et al., 1967), are apparently less associated than the latter (Kivinen et al., 1972). (The steric effect of the fluoro group on self-association is at present unknown.) Similarly, 3-and 4-fluorophenol are less associated than phenol itself, even though they are more acidic (Dale and Gramstad, 1972). In a study of the self-association of para-substituted acetophenone oximes, it was suggested that the charge density on the proton-acceptor atom is a more important factor than the acidity of the proton donor, for example, $CH_3O \cdot C_6H_4 \cdot C(CH_3) = NOH$ self-associates more strongly than $NO_2 \cdot C_6H_4 \cdot C(CH_3) = NOH$ (Reiser, 1959).

A phenyl group is sterically similar to a secondary alkyl (Charlton, 1975). Secondary alcohols are less strongly self-associated than primary alcohols, as indicated by the magnitudes of $K_{1,5}$ in Table 6.6, and are also self-associated to a lesser extent than phenol under the same conditions ($K_{1,5} = 6300 \, dm^{12} \, mol^{-4}$). Apparently the phenyl ring is less effective in hindering self-association than would be expected from steric considerations, possible because of its planarity.

Although the pentamer model is favored at 25°C, PMR evidence, summarized in Table 6.17, indicates that the predominant self-associated species of phenol at 35°C is a tetramer. The reduction in mean polymer size on increasing the temperature also occurs in liquid water and is a manifestation of the greater entropy and molecular motion at a higher temperature.

6.9   IMPLICATIONS OF SELF-ASSOCIATION STUDIES

Studies of self-association permit evaluation of the thermodynamic activity of the various species in solutions of alcohols, phenols, and related compounds and its relationship to solubility behavior. First, from this fundamental knowledge we can understand, for example, the order of magnitude of increase in solubility of phenol as the solvent is changed from isooctane to cyclohexane (see Fig. 6.27; Anderson et al., 1979). We have seen that the effect can be accounted for a 1.6-fold decrease in the activity coefficient of the monomer, which, when raised to the fifth power according to an experimentally derived association model, accounts well for the measured increase in solubility. This principle should facilitate an understanding of further unusual solubility differences.

Second, self-association of 2-naphthol and its complexation with dissolved solutes can explain the dissolution characteristic of this substance, which deviates from a modified Noyes–Whitney equation (Higuchi et al., 1972). This behavior will be discussed in Section 11.7, and the principles can be applied to other self-associating substances.

Third, self-association appears to exert a strong influence on mass transport across plastic film (Mikkelson et al., 1980). Mass-transfer processes are of fundamental pharmaceutical importance in drug delivery, including drug release from a dosage unit and drug transport across biological membranes. The rate and extent of these processes depend on the properties of the drug delivery unit, the film or absorbing membrane, and the physicochemical characteristics of the permeating drug.

6.10   INFLUENCE OF SELF-ASSOCIATION ON FREE DIFFUSION

6.10.1   Theoretical Considerations

Self-association of solutes in solution leads to the formation of larger molecules, which on average diffuse more slowly in solution. Since diffusional transport, like solubility, depends directly on the intermolecular interactions in solution, which are at the present time under study, and since it is an important factor accounting for the behavior of bioactive substances, it was thought appropriate to discuss the phenomenon at this point to complete this chapter.

A number of amphiprotic solutes which are capable of self-association exhibit a marked reduction in diffusion D with increasing concentration in nonpolar solvents. This follows from the Stokes–Einstein equation,

$$D = kT/6\pi\eta r \tag{6.75}$$

where D is the diffusivity of the solute, k is the Boltzmann constant (1.3806×10^{-23} J K^{-1}), T is the absolute temperature, η is the dynamic viscosity, and r is the radius of the diffusing particles. Other factors being equal, for rigid, almost spherical particles,

$$r = (3V/4\pi L)^{1/3} = (3M/4\pi\rho L)^{1/3} \tag{6.76}$$

where V is the molar volume, L is the Avogadro constant (6.022×10^{23} mol^{-1}), M is the molecular weight, and ρ is the density of the diffusing substance. Equations 6.75 and 6.76 indicate that:

$$D \propto 1/M^{1/3} \quad \text{and} \quad D \propto 1/V^{1/3} \tag{6.77}$$

Thus, an increase in apparent molecular weight (and apparent molar volume) due to self-association will reduce diffusivity. Owing to the cube-root dependency, diffusivity is somewhat insensitive to molecular size and is therefore not normally used for investigation self-association per se. However, because diffusion is of such importance in the pharmaceutical sciences, the influence of self-association on diffusion will be discussed.

Longsworth (1966) provided a number of examples of the dependence of D on self-association in his study of the free diffusion of various solutes at molalities from 0.01 to $0.5\,\text{mol}\,\text{kg}^{-1}$ in carbon tetrachloride at 25°C. Under these conditions, the Stokes–Einstein equation (Eq. 6.75) may be written

$$V^{1/3} = 32.99 \times 10^{-6}\,\text{cm}^3\,\text{s}^{-1}/D_o \qquad (6.78)$$

where V is the apparent molar volume in $\text{cm}^3\,\text{mol}^{-1}$ and D_o is the limiting diffusion coefficient (at low concentrations) in $\text{cm}^2\,\text{s}^{-1}$. Seven normal alkanes which do not associate show little concentration dependence on D (e.g., $C_{16}H_{34}$ in Fig. 6.35) and obey the empirical relationship

$$10^6 D_o/\text{cm}^2\,\text{s}^{-1} = 1.40 + 1639/m - 42900/m^2 \qquad (6.79)$$

where m (mol kg^{-1}) is the molality of the diffusing species.

The nature of this relationship arises from segmental diffusion, unlike those for homologous series of rigid molecules. The marked reduction in the diffusivity of $C_{15}H_{31}CH_2OH$ with increasing concentration in Figure 6.35 is due to self-association. The relatively low and constant diffusivity of $C_{15}H_{31}COOH$ is due to complete dimerization within the concentration range studied, as indicated in Figure 6.35. Other carboxylic acids behave similarly, which indicates that these molecules have high values of the dimerization constants $K_{1,2}$ (i.e., K_d).

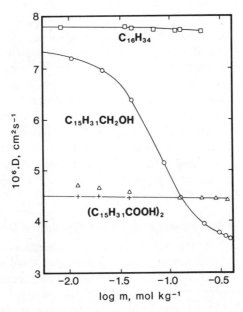

Figure 6.35 Diffusivities in carbon tetrachloride at 25°C as functions of molality: hexadecane, □; hexadecanol, ○; palmitic acid dimer, △ (obsd.); + (corrected). Reproduced with permission of the copyright owner, Academic Press, from Longsworth, L. G. (1966). *J. Colloid Interface Sci.*, **22**, 3–11.

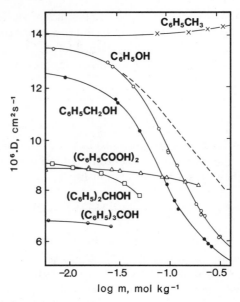

Figure 6.36 Diffusivities in carbon tetrachloride at 25°C as functions of molality: toluene, x; phenol, ○; benzyl alcohol, ●; benzoic acid dimer, △; diphenylmethanol, □; triphenylmethanol, ◑; computed (——) from association constants of Coggeshall and Saier (1951). Reproduced with permission of the copyright owner, Academic Press, from Longsworth, L. G. (1966). *J. Colloid Interface Sci.*, **22**, 3–11.

The alcohols and phenols shown in Figure 6.36 are present almost exclusively as monomers in dilute solution and associate with increasing concentration beyond the dimer, unless sterically hindered. In the case of alcohols, the dependence of D on concentration can be interpreted in terms of dimerization only in dilute solution and indicates that the association beyond the trimer is a more cooperative phenomenon than spectroscopic methods (Coggeshall and Saier, 1951) suggest. The decrease of D for N-methylacetamide (not shown) indicates that it associates to form a large multimer, not unlike a micelle, at concentrations approaching $0.2 \, mol \, kg^{-1}$, whereas the aprotic N,N-dimethylacetamide shows only a slight decrease in D at high concentrations, presumably as a result of Keesom forces.

6.10.2 Influence of Self-Association on the Membrane Diffusion of Phenol

Mikkelson et al. (1980) studied the diffusion of phenol across a high-density polyethylene film from a donor phase, consisting of a solution of phenol in isooctane in which phenol associates, to a receptor phase, consisting of aqueous sulfuric acid ($0.1 \, mol \, dm^{-3}$). They found that the steady-state flux of phenol across the film is directly proportional to its concentration in the donor phase only at low concentrations ($< 3 \times 10^{-4} \, mol \, cm^{-3}$), as shown in Figure 6.37. At higher concentrations in the donor phase, negative deviations in the steady-state

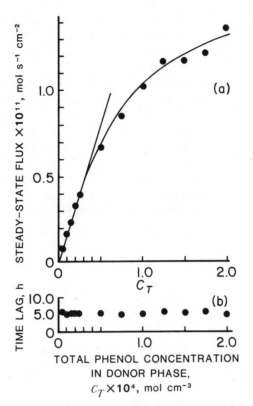

Figure 6.37 (a) Steady-state flux and (b) observed diffusion time lag as functions of the total phenol concentration in the donor phase C_T at 25°C. 1 mol cm^{-3} = 1000 mol L^{-1} = 1000 M. Reproduced with permission of the copyright owner, the American Pharmaceutical Association, from Mikkelson, T. J., Watanabe, S., Rytting, J. H., and Higuchi, T. (1980). *J. Pharm. Sci.*, **69**, 133–137.

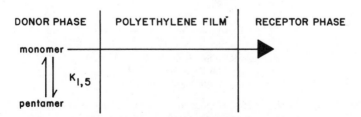

Scheme 6.3 Diffusion of phenol through a polyethylene film proceeds via the monomer, not the pentamer. Reproduced with permission of the copyright owner, the American Pharmaceutical Association, from Mikkelson, T. J., Watanabe, S., Rytting, J. H., and Higuchi, T. (1980). *J. Pharm. Sci.*, **69**, 133–137.

flux are observed which are the result of the reduction in thermodynamic activity of the phenol, arising from its self-association in the donor phase. This fact can be understood by reference to Scheme 6.3, in which only the monomeric species of phenol is assumed to be capable of permeation.

Assuming the monomer–pentamer model for phenol association, $n = 5$ and $K_{1,5} = 6260 \, dm^{12} \, mol^{-4}$, deduced by Anderson et al. (1979), the concentration of the monomer is calculated from Eq. 6.17. The steady state flux is found to be directly proportional to the concentration of the phenol monomer in the donor phase, as shown in Figure 6.38. Since the membrane was previously soaked in pure isooctane, the linear plot of phenol concentration in the receptor phase against time is preceded by a time lag, as shown in Figure 6.39. Since the diffusion time lag is independent of the phenol concentration, as shown in Figure 6.37, only one species is diffusing across the film, namely, the monomer. The lag reflects the intrinsic diffusivity of this species in the film according to the relationship

$$t_L = l^2/6D \qquad (6.80)$$

where t_L is the lag time, l is the film thickness, and D is the diffusivity of the diffusing species.

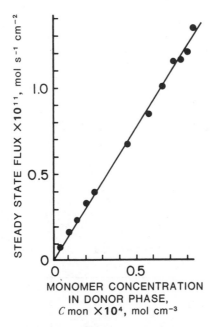

Figure 6.38   Steady-state flux at 25°C as a function of the monomer concentration of phenol in the donor phase. The slope of the line, which was obtained by regression analysis, equals 15.9 $\times 10^{-8} \, cm \, s^{-1}$ and represents the intrinsic permeability of the film to phenol contained in a donor phase of isooctane. $1 \, mol \, cm^{-3} = 1000 \, mol \, L^{-1} = 1000 \, M$. Reproduced with permission of the copyright owner, the American Pharmaceutical Association, from Mikkelson, T. J., Watanabe, S., Rytting, J. H., and Higuchi, T. (1980). *J. Pharm. Sci.*, **69**, 133–137.

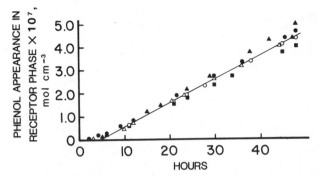

Figure 6.39 Phenol concentration appearing in the receptor phase as a function of time at 25°C. The phenol concentration in the donor phase was 2.00×10^{-4} mol cm^{-3}. The different symbols represent five individual experiments. The line is obtained by linear regression analysis of the data points at times equal to or greater than 15h. 1 mol cm^{-3} = 1000 mol L^{-1} = 1000 M. Reproduced with permission of the copyright owner, the American Pharmaceutical Association, from Mikkelson, T. J., Watanabe, S., Rytting, J. H., and Higuchi, T. (1980). *J. Pharm. Sci.*, **69**, 133–137.

Dressman et al. (1982) carried out similar diffusion experiments in which a silanized sintered glass disc replaced the plastic membrane, with the same receptor phases as before. However, this time the membrane was loaded with the donor-phase solution of phenol in isooctane. Consequently, a burst time was observed prior to the linear increase in the concentration of phenol in the donor compartment with time, as shown in Figure 6.40. The burst time increased with increasing concentration of phenol in the isooctane, suggesting that more than

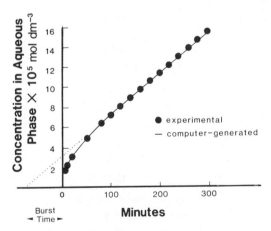

Figure 6.40 Cumulative concentration of phenol in the aqueous phase versus time for diffusion through a sintered glass filter at 25°C from a 0.1403-mol dm^{-3} solution of phenol in isooctane: experimental data (●); computer-generated results (——) obtained by numerical solution of Eq. 6.85. 1 mol dm^{-3} = 1 mol L^{-1} = 1 M. Reproduced with permission of the copyright owner, the American Pharmaceutical Association, from Dressman, J., Himmelstein, K. J., and Higuchi, T. (1982). *J. Pharm. Sci.*, **71**, 1226–1230.

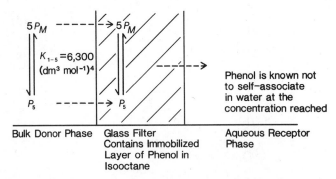

Scheme 6.4 Interactions of phenol in the silanized sintered-glass filter system. Reproduced with permission of the copyright owner, the American Pharmaceutical Association, from Dressman, J., Himmelstein, K. J., and Higuchi, T. (1982). *J. Pharm. Sci.*, **71**, 1226–1230.

one species was diffusing across the membrane. Scheme 6.4 shows a possible system in which both the monomer and the pentamer are the diffusing species.

The steady-state flux through the sintered glass disc, plotted in Figure 6.41, did not increase linearly with the total concentration of phenol in the donor phase. When the abscissa was converted to the initial concentration of phenol monomer in the donor phase, as before, the plot gave positive deviations from linearity at concentrations above $0.05 \, \text{mol dm}^{-3}$, as shown in Figure 6.42. This behavior indicated that the monomer was not the only species responsible for phenol

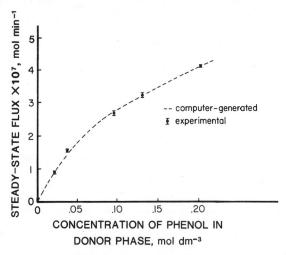

Figure 6.41 Steady-state flux of phenol at 25°C over the entire diffusion layer surface area versus total concentration of phenol in the donor phase: experimental mean with standard deviation (⬥); computer-generated results (---) obtained by numerical solution of Eq. 6.85. $1 \, \text{mol dm}^{-3} = 1 \, \text{mol L}^{-1} = 1 \, M$. Reproduced with permission of the copyright owner, the American Pharmaceutical Association, from Dressman, J., Himmelstein, K. J., and Higuchi, T. (1982). *J. Pharm. Sci.*, **71**, 1226–1230.

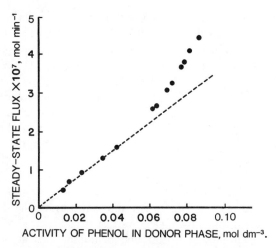

Figure 6.42 Steady-state flux of phenol at 25°C over the entire diffusion-layer surface area versus activity of phenol (concentration of monomeric phenol) in the donor phase: experimental data (●); linear plot (——) assuming contribution only from the monomer species. $1 \, mol \, dm^{-3} = 1 \, mol \, L^{-1} = 1 \, M$. Reproduced with permission of the copyright owner, the American Pharmaceutical Association, from Dressman, J., Himmelstein, K. J., and Higuchi, T. (1982). *J. Pharm. Sci.*, **71** 1226–1230.

transport. If it is supposed that both the monomer P and the pentamer P_5 are the diffusing species, each with their own characteristic diffusion coefficient, D_1 and D_5, respectively, the experimental data could be accurately predicted using Scheme 6.4. The following procedure was used. According to Fick's second law of diffusion,

$$\partial c / \partial t = D(\partial^2 c / \partial x^2) \tag{6.81}$$

where c is the concentration of a given species at a distance x into the disc, t is the time, and D is the diffusivity of that species. Applying the monomer–pentamer model,

$$c = P + 5P_5 \tag{6.82}$$

$$K_{1,5} = P_5 / P^5 \tag{6.83}$$

and

$$\partial c / \partial t = D_1(\partial^2 P / \partial x^2) + D_5 5(\partial^2 P_5 / \partial x^2) \tag{6.84}$$

By means of a series of applications of the chain rule, we can express all concentrations in terms of the concentration of the monomer, thus,

$$(\partial P / \partial t)(1 + 25K_{1,5}P^4) = (D_1 + 25K_{1,5}D_5P^4)(\partial^2 P / \partial x^2)$$
$$+ 100K_{1,5}D_5P^3(\partial P / \partial x)^2 \tag{6.85}$$

This model assumes that the association process is much faster than the diffusion process, so that association equilibrium is assured at all points and times within the membrane. The last differential equation can be solved by numerical computer methods. This model, when solved, enables (1) the receptor concentration versus time profile in Figure 6.40 to be reproduced accurately, (2) the steady-state flux as a function of total phenol concentration in Figure 6.41 to be reproduced accurately, (3) the burst time (Fig. 6.40) as a function of total phenol concentration to be reproduced within experimental error, (4) the concentration of monomer and total phenol as a function of distance across the membrane to be predicted at any time and at the steady state, and (5) the diffusivities of the monomer ($1.34 \times 10^{-5}\,\mathrm{cm^2\,s^{-1}}$) and of the pentamer ($2.42 \times 10^{-6}\,\mathrm{cm^2\,s^{-1}}$) to be calculated.

Since each pentamer unit carries five monomer units, the diffusivity of the pentamer may be multiplied by 5, yielding $1.21 \times 10^{-5}\,\mathrm{cm^2\,s^{-1}}$, for comparison purposes. This result indicates that one pentamer is a significantly slower transporter of phenol than five monomers, thus explaining the negative deviation from linearity in Figure 6.41 at higher concentrations. Finally, because the concentration of phenol in the aqueous receptor phase remained very low ($< 4 \times 10^{-4}\,\mathrm{mol\,dm^{-3}}$) compared with the solubility, self-association interactions within this phase could be ignored.

Thus, the model summarized in Scheme 6.4 explains well the diffusional behavior of phenol through a sintered glass disc. Since the pore diameter before silanization is about 4.5–5.5 μm, associated as well as monomeric species of

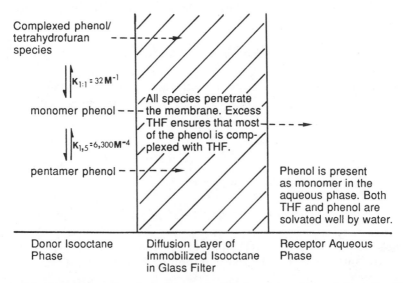

Scheme 6.5   The influence of tetrahydrofuran on the diffusion of phenol through an unstirred layer of isooctane. Reproduced with permission of the copyright owner, the American Pharmaceutical Association. Dressman, J., Himmelstein, K. J., and Higuchi, T. (1983). *J. Pharm. Sci.*, **72**, 12–17.

phenol can diffuse across the membrane, which consequently behaves as an unstirred layer of isooctane permitting self-association to occur within the diffusion layer as well as in the donor phase. The monomer–pentamer model for phenol association fits the data very well.

Incidentally, the diffusivity of the phenol monomer in isooctane at 25°C (1.34×10^{-5} cm^2 s^{-1}) (Harris et al., 1973) agrees well with that of phenol at 0.01 mol kg^{-1} in carbon tetrachloride at 25°C (1.37×10^{-5} cm^2 s^{-1}) (Longsworth, 1966). Phenol can exist only as the monomer under such conditions. Phenol at 0.5 mol kg^{-1} in carbon tetrachloride at 25°C has a diffusivity of 5.4×10^{-6} cm^2 s^{-1} (Longsworth, 1966) owing to self-association in this solvent.

The question which now arises is the possible influence of a phenol-complexing agent or cosolvent which could compete with self-association. Scheme 6.5 depicts the possible interactions resulting from the addition of tetrahydrofuran to the phenol–isooctane–sintered-glass disc system just considered. The interactions are expected to occur in the membrane as well as in the donor isooctane phase. The cyclic ether, tetrahydrofuran, binds more strongly to phenol ($K_{1:1} = 13.21$ dm^3 mol^{-1} in carbon tetrachloride at 25°C) than phenol binds to itself ($K_{1,2} = 1.45$ dm^3 mol^{-1} in carbon tetrachloride at 25°C). In isooctane (Scheme 6.5), the rank order is expected to be the same, so the concentration of pentamer present is expected to be much smaller than that of the phenol–tetrahydrofuran complex.

In view of the appreciable interaction constant of this complex, the concentration of monomer present is expected to be very small. Consequently, the appreciable diffusional flux of phenol across the sintered glass disc, shown in

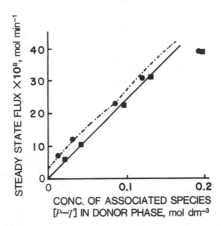

Figure 6.43 Steady-state flux of phenol at 25°C versus concentration of associated phenol–tetrahydrofuran [P–T] in the donor phase containing 4.76% (v/v) tetrahydrofuran in isooctane: (■) experimental data and (——) the computer generated results. For a 2:1 ratio of tetrahydrofuran to phenol: (●) the experimental data and (— —) computer-generated results. Standard deviations for experimental data are represented by the size of the symbols. 1 mol dm^{-3} = 1 mol L^{-1} = 1 M. Reproduced with permission of the copyright owner, the American Pharmaceutical Association, from Dressman, J., Himmelstein, K. J., and Higuchi, T. (1983). *J. Pharm. Sci.*, **72**, 12–17.

Figure 6.43, may probably be attributed almost entirely to the complex, in which case the diffusivity of the complex is a little lower than that of the phenol monomer. Since the burst time for the phenol–tetrahydrofuran system is independent of the concentration of phenol in the donor phase, within experimental error, only one species is responsible for virtually all the flux, and this species is presumably the phenol–tetrahydrofuran complex. The burst time t_B is a negative quantity which reflects the intrinsic diffusivity D of the diffusing species in the film thickness l, thus,

$$t_B = - l^2/3.1D \qquad (6.86)$$

The following conclusions may be drawn concerning the influence of associative processes on the rate of transport of diffusible species.

(a) Self-associations markedly decrease the rate of transport when the membrane material is too dense to allow passage of the associated species.

(b) When the associated species can move within the membrane, the overall rate of transport is governed by the relative diffusivity of the monomeric and associated species (multiplied by the association number).

(c) Even though a cosolvent or complexing agent may drastically lower the activity of the diffusing species, provided the complex can penetrate the membrane and has an appreciable diffusivity, the rate of transport may not be greatly affected.

(d) If the associated species or complex is large, is unable to penetrate the membrane, or both, the rate of transport is markedly retarded.

Thus, self-association not only affects equilibrium properties such as solubility and vapor pressure, but also kinetic properties such as rates of transport. When self-association is occurring with complexation phenomena, the situation may become more complicated. All these aspects must be considered when choosing excipients for pharmaceutical systems.

Chapter VII

GROUP CONTRIBUTIONS IN PREDICTION

7.1 INTRODUCTION

We have seen at the beginning of Chapter 2 that the solubility of a solute substance in a particular solvent system is an equilibrium property whose value is determined by the thermodynamic activity of the undissolved solute and the activity coefficient of the solute in the solution (Eq. 2.5). The definition and determination of the thermodynamic activity has been considered in Chapter 3. The activity coefficient of the dissolved solute may in many instances be predicted quite successfully from group contributions, which are the subject of this chapter.

Similarly, as indicated in Chapter 3 (Eqs. 3.33–3.36), the ratio of the solubilities of a given solute in two different solvents is often closely approximated by the corresponding partition coefficient of the solute and by the reciprocal of the ratio of its corresponding Henry's law constants. Partition coefficients and Henry's law constants may also be predicted from group contributions, thereby providing additional routes to solubility values.

In essence, the logarithm of the activity coefficients, partition coefficients, and Henry's law constants are given by the appropriate free-energy changes or chemical potentials, which can themselves be predicted using additivity concepts known as linear free-energy relationships (Wells, 1962; Higuchi et al., 1969; Johnson, 1975). Group contributions are the parameters which feature in the corresponding equations that make this prediction possible.

Ideally, it is desirable to be able to predict the chemical potential of a particular solute in a particular solvent and other thermodynamic properties simply from its molecular structure. However, rigorous methods of this type are limited almost entirely to mixtures of nonpolar species. The extension of statistical thermodynamics to binary systems does not yet provide good estimates of nonideal behavior and often the equations derived are based on mathematically convenient approximations that have little physical significance. Therefore, a semiempirical group-contribution approach appears to be an acceptable alternative. Group contributions have proved to be particularly useful and successful in predicting the solubility of a variety of solutes in water, in water–cosolvent mixtures, and in other aqueous systems; this is a major thrust of Chapter 8, while this chapter provides the essential background.

The group-contribution approach is essentially an additivity concept. Additive and constitutive concepts have been important in the development of classical physical chemistry, particularly in the early attempts to determine molecular structure from such physical quantities as molar volume, molar refractivity, and the parachor. Additive–constitutive properties of molecules have recently received renewed attention not only in the group-contribution approach (Davis et al., 1974), but also in their application to structure–activity relationships of drugs developed by Hansch, Leo, and co-workers in the 1960s and 1970s (Hansch and Fujita, 1964; Hansch, 1971; Leo et al., 1971; Leo, 1972; Hansch and Leo, 1979). In both of these concepts important roles are played by linear free-energy relationships, which are exemplified by Eqs. 7.1–7.14 and indicate that the free energy of a molecule is a linear function of certain properties of its constituent goups.

The concept of individual contributions of constituent groups to the thermodynamic properties of a molecule was first introduced by Langmuir (1925) in his principle of independent surface action, embodied in Eq. 8.3. The equation will be considered in Section 8.3 in connection with the molecular and group-surface area (MGSA) approach of Harris et al. (1973), Yalkowsky et al. (1972, 1975, 1976), Amidon et al. (1974, 1975), and Valvani et al. (1976).

Langmuir considered that the force field around a group, which is accessible to other molecules, is characteristic of that group and is independent of the rest of the molecule. This concept enabled Langmuir to make satisfactory solubility predictions for molecules consisting of one polar group bonded to one nonpolar group, that is, R–X.

A valuable contribution to the field of group contributions was made by Butler and co-workers in the 1930s (Butler et al., 1933; Butler and Ramchandani, 1935; Butler and Reid, 1936; Butler, 1937, 1962; Butler and Harrower, 1937). Butler recognized that in a homologous series (e.g., of aliphatic alcohols) the polar group and the methylene group each provide definite constant contributions to the molar Gibbs free energy of transfer and to the enthalpy of transfer (and hence to the entropy of transfer). This early work is notable for its elegance and scientific insight long before the phrase "group contributions" was in general use.

Deal and Derr (1968) extended the group-contribution concept to binary mixtures formed by members of two homologous series. Prausnitz and co-workers (Fredenslund et al., 1975, 1977a, b; Gmehling et al., 1978) have developed the concept into the UNIQUAC and UNIFAC approaches to help solve chemical engineering problems involving liquid–vapor and solid–liquid equilibria.

Group contributions have been reviewed by Davis et al. (1974) from the standpoint of the thermodynamics of functional groups in solutions of drug molecules. The group-contribution concept provides us with a powerful tool for relating the free energy of transfer of a chemical species from one phase to another in terms of the molecular structure of the species. This approach enables us to translate our understanding of solute behavior in a solvent system from observations we can make on separate phases in equilibrium with this solution.

One of the more useful applications of the concept is in understanding and applying the free energy of transfer from our idealized reference state to the gaseous state and to the aqueous phase. The approach appears to be strictly applicable only to rigid molecules, such as steroids, and less so to molecular structures which permit the adoption of various conformational structures, such as molecules containing a long alkyl chain. Furthermore, in cases where two polar or interactive groups are close to each other in a solute molecule, some modification of the additivity concept is necessary. However, whatever their shortcomings, additivity concepts, when properly applied, have shown remarkable powers of prediction and, by encouraging workers to seek the inevitable exceptions, have stimulated further work and increased understanding. In this chapter we shall attempt to explore and demonstrate the utility of this general concept.

In the group-contribution approach in its simplest (and possibly most useful) form, the molecule is considered to be composed of groups, each of which is assumed to act independently of the rest of the molecule and to make its own contribution to a specific thermodynamic property of a molecule. (The limitations of this concept have been summarized in the previous paragraph and will be further discussed later in the chapter.) Thus, since a given solute molecule is treated as follows,

$$\text{property of the solute molecule} = \sum \text{properties of its groups} \qquad (7.1)$$

Likewise, the standard molar free-energy of transfer ΔG_t^θ of the compound between two given phases (represented by Eq. 1.5) is assumed to be equal to the sum of the corresponding free-energy values, known as group contributions $\Delta(\Delta G)$ or $\Delta\Delta G$, for the various groups comprising the molecule, thus,

$$\Delta G_t^\theta \text{ of a compound} = \sum \Delta(\Delta G) \text{ of its groups} \qquad (7.2)$$

It is generally possible to treat the standard molar enthalpy of transfer, ΔH_t^θ, and the standard molar entropy of transfer, ΔS_t^θ, of the compound in an analogous fashion involving their respective group contributions, thus,

$$\Delta H_t^\theta = \sum \Delta(\Delta H) \quad \text{and} \quad \Delta S_t^\theta = \sum \Delta(\Delta S) \qquad (7.3)$$

The parentheses in Eqs. 7.2 and 7.3 are frequently omitted when referring to group contributions, that is, $\Delta\Delta G$, $\Delta\Delta H$, and $\Delta\Delta S$.

For an organic substance treated as a solute, Scheme 7.1 shows the various transfer processes to which group contributions have been applied and which are relevant to this book. The pure solute in its usual condensed phases, solid or liquid, has necessarily been omitted from this scheme because the intermolecular interactions in these phases are so specific, as explained in Sections 3.1 and 3.2 and emphasized when discussing Figure 7.1 later in this section, that group contributions to and from this phase cannot usefully be applied.

Each standard molar free energy of transfer ΔG_t^θ, in Scheme 7.1 is equal to $(-RT)$ times the natural logarithm of the corresponding equilibrium property Y, such as the Henry's constant in an organic solvent, h_o, or in water, h_w, or

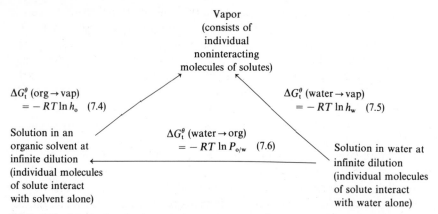

Scheme 7.1 Standard molar free energies of transfer, ΔG_t^θ, to which group contributions may be applied for solubility prediction, where h_o = Henry's law constant of the solute in the organic solvent, h_w = Henry's law constant of the solute in water, and $P_{o/w}$ = partition coefficient (organic solvent/water).

the partition coefficient between water and the organic solvent, $P_{o/w}$, according to Eqs. 7.4–7.6, which may be written in the general form (see Scheme 7.1)

$$\Delta G_t^\theta = - RT \ln Y(\text{molecule}) = - 2.303 RT \log Y(\text{molecule}) \qquad (7.7)$$

Therefore, $\log Y$ for an entire molecule is equal to the sum of each group contribution, $\Delta \log Y(\text{group})$, thus,

$$\log Y(\text{molecule}) = \sum \Delta \log Y(\text{groups}) \qquad (7.8)$$

where

$$\Delta \Delta G(\text{group}) = - 2.303 RT \Delta \log Y(\text{group}) \qquad (7.9)$$

Normally, the contribution of each group X to a given property is equal to the change in the property elicited by substitution of a hydrogen atom by the group X. If the rest of the molecule is designated R, the group contribution of X can be stated in either of the following ways:

$$\Delta \Delta G_t^\theta(X) = \Delta G_t^\theta(RX) - \Delta G_t^\theta(RH) \qquad (7.10)$$

$$\Delta \log Y(X) = \log Y(RX) - \log Y(RH) \qquad (7.11)$$

where

$$\Delta G_t^\theta(RX) = - 2.303 RT \log Y(RX) \qquad (7.12)$$

$$\Delta G_t^\theta(RH) = - 2.303 RT \log Y(RH) \qquad (7.13)$$

and

$$\Delta \Delta G_t^\theta(X) = -2.303 RT \Delta \log Y(X) \tag{7.14}$$

The alternative modes, $\Delta \Delta G$ and $\Delta \log Y$, of expressing group contributions can be readily interconverted by Eqs. 7.9 or 7.14. Table 7.1 shows values of group contributions to free energies of transfer (Eq. 7.2), whereas Table 7.2 shows some values of group contributions for enthalpies and entropies of transfer (Eq. 7.3).

Example 7.1. $\log P[(25\% \text{ v/v CHCl}_3 \text{ in CCl}_4)/\text{water}]$ at 30°C for the chloro group is 0.32, from which Eq. 7.9 affords $\Delta \Delta G^\theta$ (water $\rightarrow$ CHCl$_3$ + CCl$_4$) = -440 cal mol^{-1} = -1860 J mol^{-1} (Harris et al., 1973). We note that, by definition, $\Delta \Delta G(H) = 0$ and $\Delta \log Y(H) = 0$, where H refers to the unsubstituted hydrogen atom in a molecule. As we have seen in Chapter 3, paraffins are the most suitable solvents in routes leading to solubility prediction because the solute–solvent interactions reduce to Debye and London forces in these solvents.

Example 7.2. 1-Butanol is constituted as follows:

$$CH_3 \cdot CH_2 \cdot CH_2 \cdot CH_2 \cdot OH = CH_3 + 3CH_2 + OH$$

Consequently, the standard molar free energy of transfer of 1-butanol between an infinitely dilute solution in water and an infinitely dilute solution in a hydrocarbon solvent such as *n*-heptane is given by Eq. 7.2, the group contributions being taken from Table 7.1 thus,

$$\begin{aligned}
\Delta G_t^\theta(\text{1-butanol, water} \rightarrow \text{hc}) &= \Delta \Delta G_t^\theta(\text{CH}_3, \text{water} \rightarrow \text{hc}) \\
&\quad + 3\Delta \Delta G_t^\theta(\text{CH}_2, \text{water} \rightarrow \text{hc}) \\
&\quad + \Delta \Delta G_t^\theta(\text{OH}, \text{water} \rightarrow \text{hc}) \\
&= (-2000) + 3(-880) + (4350) \text{ cal mol}^{-1} \\
&= -290 \text{ cal mol}^{-1}
\end{aligned}$$

The standard molar free energies of transfer, ΔG_t^θ, in Scheme 7.1 are related as follows:

$$\Delta G_t^\theta(\text{water} \rightarrow \text{org}) = \Delta G_t^\theta(\text{water} \rightarrow \text{vap}) - \Delta G_t^\theta(\text{org} \rightarrow \text{vap}) \tag{7.15}$$

Comparison with Eqs. 7.4–7.6 shows that

$$P_{o/w} = h_w/h_o \tag{7.16}$$

as stated in Eq. 3.35 in an alternate form. The group-contribution concept can, of course, be applied to a given group X by using Eq. 7.15 in the form

$$\Delta \Delta G_t^\theta(\text{water} \rightarrow \text{org}) = \Delta \Delta G_t^\theta(\text{water} \rightarrow \text{vap}) - \Delta \Delta G_t^\theta(\text{org} \rightarrow \text{vap}) \tag{7.17}$$

Table 7.1 Group Contributions to the Standard Molar Free Energies of Transfer $\Delta\Delta G_t^{\theta a}$

Group	$\Delta\Delta G_t^{\theta}$ (cal mol^{-1})		
	$hc \to vapor$	$water \to vapor$	$water \to hc$
Substituents on Aliphatic Carbon Atoms			
—H	0	0	0
$>$CH$_2$	710	−170	−880
—CH$_3$	−1000	−3000	−2000
(N)CH$_3$			−1665
—C$_6$H$_5$		−440	
—C=CH$_2$		−2400	
—C≡CH		−790	
—OH	−390	3960	4350
—NH$_2$	−330	3420	3750
—COOH			5210
$>$CO		5750	
$>$O	510	2760	2250
—CHO	1650	3840	2190
—O·OC·H	1860	3310	1450
—CO·CH$_3$		2720	
—CO·O·CH$_3$		2000	
—CO·O·C$_2$H$_5$		1540	
—O·CO·CH$_3$		1840	
—O·CO·C$_2$H$_5$		1610	
—O·CO—			
—CO·NH$_2$			6400
F			−290
Cl	+435	−520	−955
Br	+970	−260	−1230
I	+1110	−330	−1440
Loss of H: Unsaturation and Ring Closure in Aliphatics			
Alkane–alkane			860
Alkane–alkyne			2155
Alkene–alkyne			1295
Alkyne–dialkyne			1665
Ring closure	0	950	950
Substituents on Benzenoid Carbon Atoms			
—H	0	0	0
$>$CH$_2$	710	−150	−860
—CH$_3$	710	−90	−800
(N)CH$_3$			−1665
—OH			2320
—COOH			955

Table 7.1   (Continued)

Group	$\Delta\Delta G_t^\theta$ (cal mol^{-1})		
	$hc \rightarrow vapor$	$water \rightarrow vapor$	$water \rightarrow hc$
3 or 4-F			-95
3 or 4-Cl			-765
3 or 4-Br			-1020
3 or 4-I			-1240
2-OCH$_3$ in anilines			-595
3-OCH$_3$ in anilines			305
4-OCH$_3$ in anilines			675
2-OCH$_3$ in benzaldehydes			0
4-OCH$_3$ in benzaldehydes			315
Substituents in Conjugated hetenoids			
2-OCH$_3$			410
3-OCH$_3$			615
3-OH			4365
4-OH			4775

[a] $\Delta\Delta G_t^\theta$ is in cal mol^{-1} at 25°C. The standard state of the solute in water or in a paraffinic hydrocarbon (hc) is an infinitely dilute solution based on the unit mole fraction scale. The standard state of the substance in the vapor state is an ideal gas at 1 atm pressure. 1 atm = 101,325 Pa. 1 cal = 4.184 J.

Data from Davis, S. S., Higuchi, T., and Rytting, J. H., in *Advances in Pharmaceutical Science*, Vol. 4, M. S. Bean, (Ed.), Academic Press, 1974, pp. 73–261, and from Abraham, M. H., *J. Chem. Soc., Faraday Trans., 1*, **80** (1984), 153–181.

To summarize, free energies of transfer, log (Henry's constant), and log(partition coefficient) of a molecule can each be calculated from the sum of certain values which are attributed to the different groups comprising the molecule and which are known collectively as group contributions. Table 7.1 lists values of the more important group contributions at 25°C (in practice, 20–30°C), the organic solvent being a nonpolar solvent, usually a paraffin, such as isooctane.

In contrast to more theoretical approaches, this method has so far been applied solely to the analysis and prediction of experimental data. However, the results often assume a form similar to those derived mathematically from statistical thermodynamics.

In Section 3.8 we saw that Butler (Butler et al., 1933; Butler and Ramachandani, 1935; Butler and Reid, 1936; Butler, 1937, 1962; Butler and Harrower, 1937) studied the quantities involved in the thermodynamic cycle represented by Scheme 3.1 for the homologous series of alcohols in the pure state, the gaseous state, and in an infinitely dilute aqueous solution. According to Eqs. 3.13 and 3.30

$$\gamma^\infty = h^x/p^*$$

Table 7.2 Group Contributions to the Standard Molar Enthalpies $\Delta\Delta H_t^\theta$ and Entropies $\Delta\Delta S_t^\theta$ of Transfer at 25°C[a]

Group	$hc \rightarrow vapor$		$water \rightarrow vapor$		$water \rightarrow hc$	
	$\Delta\Delta H_t^\theta$ (cal mol^{-1})	$\Delta\Delta S_t^\theta$ (cal K^{-1} mol^{-1})	$\Delta\Delta H_t^\theta$ (cal mol^{-1})	$\Delta\Delta S_t^{\theta}$ (cal K^{-1} mol^{-1})	$\Delta\Delta H_t^\theta$ (cal mol^{-1})	$\Delta\Delta S_t^\theta$ (cal K^{-1} mol^{-1})
Substituents on Aliphatic Carbon Atoms[b]						
—H	0	0	0	0	0	0
$>CH_2$	1400	2.32	780	3.20	−620	0.88
—CH_3			2360	18.1		
—C_6H_5			6320	22.7		
—OH			9740	19.4		
—NH_2			9640	20.9		
—COOH			10510			
—CO·CH_3			7640	16.4		
—CO·O·CH_3			7490	18.7		

Substituents on Benzenoid Carbon Atoms[c]

—CH$_3$	1160
1,4-(CH$_3$)$_2$	1060
1,3,5-(CH$_3$)$_3$	990
—CF$_3$	170
—OH	−70
—NH$_2$	1600
—CO·CH$_3$	2900
—CO·O·CH$_3$	5410
—NO$_2$	3190
—F	240
—Cl	2120
—Br	2650
—I	3390

[a] The standard state of the solute in water or in a paraffinic hydrocarbon (hc) is an infinitely dilute solution based on the unit mole fraction scale. The standard state of the substance in the vapor state is an ideal gas at 1 atm pressure. 1 atm = 101,325 Pa. 1 cal = 4.184 J.

[b] Abraham, M. H., J. Chem. Soc., Faraday Trans. I, **80** (1984), 153–181.

[c] The hydrocarbon is cyclohexane. Estimated accuracy ± 100–200 cal mol^{-1}. Fuchs, R., Young, T. M. and Rodewald, R. F., J. Am. Chem. Soc., **96** (1974) 4705–4706.

The standard state of the solute in water or in a paraffinic hydrocarbon (hc) is an infinitely dilute solution based on the mole fraction scale. The standard state of the substance in the vapor state is an ideal gas at 1 atm pressure in Scheme 3.1. The activity coefficients γ were calculated for shorter-chain compounds from the corrected Raoult's law equation (Eq. 3.23), that is,

$$\gamma = p/p^*x \qquad (3.23)$$

and the limiting activity coefficients γ^∞ at infinite dilution were obtained by extrapolation. For the longer-chain compounds, the solubility in water was so low that the saturated solution could be considered to be in equilibrium with the pure liquid solute so that

$$\gamma^\infty = 1/x^s$$

where x^s is the mole fraction solubilty (Eq. 2.104).

By means of Eq. 3.48 $[\Delta G_t^\theta(\text{pure solute} \rightarrow \text{water}) = \bar{G}^\theta - G^* = -RT \ln(p^*/h^x)]$ can be calculated. This, when plotted against carbon number, affords the following quantity as the slope:

$$\Delta\Delta G_t^\theta(CH_2, \text{ pure solute} \rightarrow \text{water}) = 800 \text{ cal mol}^{-1}$$
$$= 3350 \text{ J mol}^{-1}$$

Similarly, using Eq. 3.44, $\Delta G_t^\theta(\text{solute in water} \rightarrow \text{vapor}) = G_g^\theta - \bar{G}^\theta = -RT \ln h^x$ can be calculated and, when plotted against carbon number, gives the following quantity as the slope:

$$\Delta\Delta G_t^\theta(CH_2, \text{ solute in water} \rightarrow \text{vapor}) = 160 \text{ cal mol}^{-1}$$
$$= 670 \text{ J mol}^{-1}$$

Finally, by means of Eq. 3.40, $\Delta G_t^\theta(\text{pure solute} \rightarrow \text{vapor}) = G_g^\theta - G^* = -RT \ln p^*$ can be calculated, and this when plotted against carbon number yields the following quantity as the slope:

$$\Delta\Delta G_t^\theta(CH_2, \text{ pure solute} \rightarrow \text{vapor}) = 640 \text{ cal mol}^{-1}$$
$$= 2680 \text{ J mol}^{-1}$$

Comparison of the value of 800 cal mol^{-1} with the values of 160 and 640 cal mol^{-1} indicates that each additional methylene group causes an increase in $\Delta G_t^\theta(\text{pure solute} \rightarrow \text{water})$ and a decrease in aqueous solubility, of which $\frac{1}{5}$ is due to interaction of solute molecules in water and $\frac{4}{5}$ is due to the interaction of the solute molecules with each other. In this way, most of the differences in solubility of isomeric alkanols were attributed to differences in vapor pressure p^* rather than to the solute–solvent interactions represented by Henry's constant

$(h^x = p/x)$. In this connection, Hildebrand and Scott (1950) have pointed out that, when comparing the aqueous solubility of sparingly soluble liquids, allowance should be made for differences in vapor pressure by calculating solubilities at the same vapor pressure, assuming that Henry's law applies.

The group-contribution approach is particularly successful in correlating and predicting the free energy of transfer of a solute from the solution phase to the vapor, and therefore the Henry's law constant, and from one solution phase to another, and consequently the partition coefficient. Group contributions are often successful in these cases because the vapor pressures are usually so low that the vapor phase may be treated as an ideal gas which is devoid of intermolecular interaction, and because the solution phase is treated as being infinitely dilute, so that the only significant interactions of the solute are with the solvent. In the group-contribution approach, the solvents of interest are commonly either water (discussed in Chapter 8) or paraffinic hydrocarbons such as isooctane, heptane, or cyclohexane, in which the solute–solvent interactions are weak dipole–induced dipole (i.e., Debye) forces (Section 2.11.4).

The group-contribution approach is found to be less successful in predicting the free energy of transfer between the pure solute and the vapor state (and therefore the saturated vapor pressure) and between the pure solute and the solution phase and therefore the activity coefficient based on the Raoult's law mole fraction scale. The reason is that the molecular environment of each solute molecule in the pure state depends on the solute–solute interactions in that state, which are highly dependent on the nature of the solute. For this reason, as pointed out in Sections 3.1 and 3.2, the pure solute is less desirable as a reference state than an infinitely dilute solution or the vapor state. For the same reason, as noted in Sections 2.2.4 and 3.8, the prediction of the activity of a pure solute is extremely difficult.

It may, however, be possible to show correlation involving members of a homologous series of pure liquid solutes, such as the straight-chain aliphatic alcohols studied by Butler, simply because the molecular environment changes rather smoothly from one homologue to another. However, when applied to a homologous series of pure solid solutes, notable discontinuities arise even within the same homologous series, such as with the alkyl 4-aminobenzoate esters studied by Yalkowsky et al. and discussed in Section 8.3.

A discontinuity coincides with a marked change in the nature of the packing of the molecules in the crystal. Even with the homologous series of liquid straight-chain alkyl acetate esters $CH_3 \cdot (CH_2)_{n-3} \cdot O \cdot CO \cdot CH_3$, $\log p^*$ [$= -\Delta G_t^\theta$(pure liquid $\to$ vap)/$2.303RT$] is a curvilinear function of n, as shown in Figure 7.1, because the environment of CH_2 is changing continuously within the series of pure esters. However, the plot of $\log h^x$ [$= -\Delta G_t^\theta$ (isooctane $\to$ vap)/$2.303\,RT$] is virtually linear because the environment of CH_2 in isooctane is essentially constant.

Abraham (1984) has emphasized these points in a very informative and critical paper on the thermodynamics of the process gas $\to$ aqueous solution (the exact

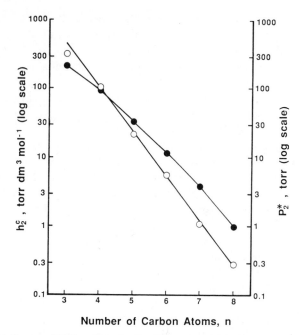

Figure 7.1 The influence of the number of carbon atoms n in the molecules of straight-chain alkyl acetate esters on their molarity-based Henry's law constants h_2^C in isooctane at 25°C ($\bigcirc$) (Warycha et al., 1980) and on the saturated vapor pressures p_2^* of the pure esters at 25°C ($\bullet$) (Boublik et al., 1973). (Note the degree of linearity of the different semilogarithmic plots.) 1 torr = 133.3 Pa. $1\,\text{mol}\,\text{dm}^{-3} = 1\,\text{mol}\,\text{L}^{-1} = 1\,M$.

negative of water → vap). He pointed out that only parameters for the process gas → water can be used to assess solute–water interactions and that the standard state of the pure liquid solute includes a different solute–solute interaction term for each standard state. Abraham (1984) also provided a useful table (Table 7.3) of correlation terms for converting values of the standard free-energy changes ΔG^θ (gas → solution) from one standard state to another. For consistency with this chapter and with most of this book, the exact negative of these values is presented as ΔG^θ (solvent → vap) in Table 7.3.

7.2 EVALUATION OF GROUP CONTRIBUTIONS

7.2.1 Henry's Law Constants

Rytting et al. (1978b) determined the hydroxyl, amino, and methylene group contributions to the free energy of transfer illustrated in Scheme 7.1. The ΔG_t^θ values for members of the aliphatic homologous series of hydrocarbons, alcohols, and amines were calculated from the Henry's law constants in dilute aqueous solution, h_w^x, and in heptane solution, h_{hc}^x. These Henry's law constants were

Table 7.3   Effect of Standard States on Values of ΔG^{θ} (solvent → gas) at 25°C[a]

Standard States of Solute		$\delta \Delta G^{\theta}$ at 298.15 K (cal mol^{-1})[b]		
Solvent	*Gas*	*Water*	*Methanol*	*Hexane*
Unit mol fraction, $x_2 \to 1$	1 atm	0	0	0
1 mol dm^{-3}	1 atm	−2378	−1896	−1201
1 mol kg^{-1}	1 atm	−2380	−2039	−1452
1 mol dm^{-3}	1 mol dm^{-3}	−4272	−3790	−3095
1 mol dm^{-3}	1 kPa	−3734	−3252	−2557
Solvent molecular weight (g mol^{-1})		18.015	32.04	86.18
Solvent density (g cm^{-3})		0.9971	0.7865	0.6548

[a] 1 cal = 4.184 J. 1 mol dm^{-3} = 1 mol L^{-1} = 1 M. 1 atm = 101, 325 Pa.
[b] Defined as the correction term: $\delta \Delta G^{\theta} = \Delta G^{\theta}$ (solute extrapolated to $x_2 = 0$ in solvent → solute gas, 1 atm) − ΔG^{θ} (solute in other standard state in solvent → solute gas in different standard state).

Reproduced with permission of the copyright owner, the Royal Society of Chemistry, adapted from Abraham, M. H., *J. Chem. Soc., Faraday Trans. I*, **80** (1984) 153–181.

determined by measuring the vapor pressure p by means of a vapor head-space GLC method similar to that described in Section 6.3.2, but using gas-tight syringes for injection. The Henry's law constants h^x were calculated in terms of mole fraction of the solute and had the dimension of pressure (torr = mm Hg = 133.3 Pa).

Figure 7.2 shows that $\log h_w^x$ is a linear function of carbon number n for members of the straight-chain alkanes, alcohols, and amines in water. The slope affords the methylene group contribution to the free energy of transfer from

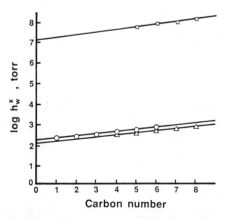

Figure 7.2   Logarithm of the Henry's law constants at 25°C versus carbon number for homologous series of straight-chain alkanes (□), alcohols (○), and amines (Δ) in water. 1 torr = 133.3 Pa. Reproduced with permission of the copyright owner, the American Pharmaceutical Association, from Rytting, J. H., Huston, L. P., and Higuchi, T. (1978). *J. Pharm. Sci.*, **67**, 615–618.

water to the vapor phase according to Eq. 7.9 when written in the form

$$\Delta\Delta G_t^\theta(CH_2, \text{water} \rightarrow \text{vap}) = -2.303 RT\Delta(\log h_2^x)/\Delta n$$
$$= -148 \text{ cal mol}^{-1} \; (-619 \text{ J mol}^{-1}) \tag{7.18}$$

An approximate value of the hydroxyl group contribution, $\Delta\Delta G_t^\theta(OH)$, assuming that the methyl group has the same group contribution as the methylene group, may be obtained from Figure 7.2 by subtracting the alkane in water value of $\log h^x$ from the alcohol in water value of $\log h^x$ for each corresponding carbon number n, thus,

$$\Delta\Delta G_t^\theta(OH) \approx \Delta\Delta G_t^\theta(H\cdots OH) = -2.303 RT[\log h^x(ROH) - \log h^x(RH)] \tag{7.19}$$

(An approximate value of the amino group contribution $\Delta\Delta G_t^\theta(NH_2)$, may be obtained in an analogous manner.) The symbol $(H\cdots OH)$ indicates the group contribution for the hydroxyl group together with that for the terminal hydrogen atom which is lost from the alkane RH when it is hypothetically converted to the alkanol ROH. The presence of this terminal hydrogen atom distinguishes the terminal methyl group from the methylene group. The importance of this distinction is discussed in Section 7.3.2, together with the so-called terminal methyl group correction. Table 7.1 shows the group-contribution values suitably corrected.

Plots of $\log h_{hc}^x$ against carbon number are also linear for the straight-chain alkanes, alcohols, and amines in heptane, as shown in Figure 7.3. The slope of each line gives the methylene group contribution to the free energy of transfer from heptane to the vapor phase according to an equation analogous to Eq. 7.18. In this case, $\Delta\Delta G_t^\theta(CH_2, \text{heptane} \rightarrow \text{vap}) = 730 \text{ cal mol}^{-1}$ (3050 J mol^{-1}). The approximate hydroxyl and amino group contributions were obtained by subtraction for each corresponding carbon number, as in Eq. 7.19. Table 7.1 lists the values obtained properly adjusted for the methyl group correction.

Application of the group-contribution concept to each group X according to Eq. 7.17 shows that

$$\Delta\Delta G_t^\theta(X, \text{water} \rightarrow \text{heptane}) = \Delta\Delta G_t^\theta(X, \text{water} \rightarrow \text{vap})$$
$$- \Delta\Delta G_t^\theta(X, \text{heptane} \rightarrow \text{vap}) \tag{7.20}$$

The values obtained for the group contributions to the free energy of transfer of the methylene group from water to heptane were $-907 \text{ cal mol}^{-1}$ (-3795 J mol^{-1}) for the alcohols and $-851 \text{ cal mol}^{-1}$ (-3560 J mol^{-1}) for the amines. These values agree well with those from other publications for the methylene group and the mean values are shown in Table 7.1. The corresponding quantities for the polar groups, namely, $\Delta\Delta G_t^\theta(H\cdots OH, \text{water} \rightarrow \text{heptane}) =$

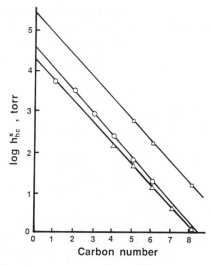

Figure 7.3 Logarithm of the Henry's law constants at 25°C versus carbon number for homologous series of straight-chain alkanes ($\square$), alcohols ($\bigcirc$), and amines ($\triangle$) in heptane. 1 torr = 133.3 Pa. Reproduced with permission of the copyright owner, the American Pharmaceutical Association, from Rytting, J. H., Huston, L. P., and Higuchi, T. (1978). *J. Pharm. Sci.*, **67**, 615–618.

5570 cal mol^{-1} (23.3 kJ mol^{-1}) and $\Delta\Delta G_t^\theta(\text{H}\cdots\text{NH}_2,\text{water}\rightarrow\text{heptane}) =$ 4980 cal mol^{-1} (20.8 kJ mol^{-1}), are greater than the mean values of $\Delta\Delta G_t^\theta(\text{OH})$ and $\Delta\Delta G_t^\theta(\text{NH}_2)$ given in Table 7.1 by about 1200 cal mol^{-1} (5020 J mol^{-1}), which corresponds to the methyl group correction mentioned above and discussed in Section 7.3.2.

The determination and the significance of group contributions from Henry's constants have been further discussed by Davis et al. (1974), Rytting et al. (1978), and Abraham (1984). The last author has provided a useful set of group contributions between water and the vapor state for a series of substituents on aliphatic carbon atoms. Tables 7.1 and 7.2 contain these values, that is, $\Delta\Delta G_t^\theta(\text{X}, \text{water}\rightarrow\text{vap})$, $\Delta\Delta H_t^\theta(\text{X}, \text{water}\rightarrow\text{vap})$ and $\Delta\Delta S_t^\theta(\text{X}, \text{water}\rightarrow\text{vap})$. Abraham (1984) has also critically reviewed the data from which he has calculated these group contributions.

Fuchs et al. (1974) have provided calorimetrically determined group contribution values, $\Delta\Delta H$, between various solvents (cyclohexane, benzene, methanol, and dimethylformamide) and the vapor state for a series of substituents on aromatic carbon atoms. Table 7.2 contains the particular values for transfer from the paraffinic hydrocarbon cyclohexane. These values, which were calculated as $\Delta\Delta H_t^\theta(\text{X}, \text{hc}\rightarrow\text{vap}) = \Delta H(\text{C}_6\text{H}_5\text{X}, \text{hc}\rightarrow\text{vap}) - \Delta H(\text{C}_6\text{H}_6, \text{hc}\rightarrow\text{vap})$, probably need no correction for the loss of a hydrogen atom attached to carbon atom, since the milieu is hydrocarbon. The individual values of ΔH (solvent $\rightarrow$ vap) for each solute were calculated by subtracting the calorimetric enthalpy of solution at infinite dilution from the enthalpy of vaporization of the solute. The latter were derived from ΔH_v values of Dreisbach (1955). The effects of the various solvents on the individual values of the enthalpic group contributions in Table 7.4 clearly show the influence of the various intermolecular interactions discussed in Section 2.11.

7.2.2 Partition Coefficients

At low concentrations, provided that effects due to dimerization and/or ionization are negligible, Eq. 7.16 and Eqs. 3.37 and 3.38 can be expressed as follows in terms of the Raoult's law mole fraction scale:

$$P_{o/w}^x = h_w^x/h_o^x = \gamma_w^\infty/\gamma_o^\infty \tag{7.21}$$

where the subscript w indicates aqueous solution, the subscript o indicates solution in an organic solvent, and the superscript x indicates that concentrations of solute are expressed on the Raoult's law mole fraction scale. Application of Eqs. 7.6 and 7.10 to a given compound RX and to a reference compound RH enables, in principle, the group contribution of X, namely, $\Delta\Delta G_t^\theta$ (X, water $\rightarrow$ org), to be calculated thus:

$$\Delta\Delta G_t^\theta(X) = \Delta G_t^\theta(RX) - \Delta G_t^\theta(RH) \tag{7.22}$$

$$= -2.303RT[\log P(RX) - \log P(RH)] \tag{7.23}$$

$$= -2.303RT\log[P(RX)/P(RH)]$$

$$= -2.303RT\log F(X) \tag{7.24}$$

where $P(RX)$ and $P(RH)$ are the partition coefficients of RX and RH and $F(X)$ is the ratio by which substituent X changes the partition coefficient, that is,

$$F(X) = P(RX)/P(RH) \tag{7.25}$$

The partition coefficient, when expressed in terms of molar concentrations, may be written $P_{o/w}^c$, and for dilute solutions is related to $P_{o/w}^x$ as follows:

$$P_{o/w}^x = P_{o/w}^c(V_o/V_w) \tag{7.26}$$

where V_o and V_w are the molar volumes of the organic and water phases.

To account for the partitioning behavior of drug molecules from water to lipid in living organisms, Hansch and Fujita (1964) proposed that the partition coefficients of drugs $P_{o/w}^c$, be calculated as the molar concentration in 1-octanol divided by the molar concentration in water. Organic solvents other than 1-octanol have also been proposed as models for biological lipids. By analogy with the Hammett equation and other linear free-energy relationships, Hansch and

Table 7.4 Aromatic and Group Enthalpies[a] of Transfer from Each of Several Solvents to the Vapor[b]

Compared with Substituent (X)	Group X Polarizability[c] ($\alpha \times 10^{24}$, cm^3)	Group X molar volume[d] (V_m, cm^3 mol^{-1})	Solvent $\Delta\Delta H$ (kcal mol^{-1})			
			Cyclohexane	Benzene	Methanol	Dimethyl-formamide
Benzene	—	—	(+7.15)	(+8.09)	(+7.73)	(+8.05)
Me	4.9	17.5	+1.16	+0.87	+0.90	+0.87
1,4-Me$_2$	4.9	17.5	+1.06	+0.90	+0.90	+0.87
1,3,5-Me$_3$	4.9	—	+0.99	+0.88	+0.86	+0.82
c-C$_6$H$_{12}$	—	—	(+7.91)	(+7.01)	(+6.70)	(+6.04)
c-C$_6$F$_{11}$CF$_3$	—	—	(+3.9)	(+2.2)	(+3.7)	(+1.7)
F	0	5.0	+0.24	+0.38	+0.73	+0.75
Cl	5.0	12.9	+2.12	+2.05	+2.27	+2.37
Br	7.2	16.2	+2.65	+2.73	+2.84	+2.97
I	13.1	23.0	+3.39	+3.56	+3.61	+4.16
CF$_3$	4.6	34.1	+0.17	+0.33	+1.30	+0.94
COCH$_3$	10.3	28.0	+2.90	+4.74	+4.26	+5.24
NO$_2$	6.7	13.5	+3.19	+4.54	+4.46	+5.29
COOEt	16.4	53.9	+5.41	+5.96	+5.16	+5.74
NH$_2$	4.4	23.0	+1.60	+3.63	+5.86	+7.60
OH	1.6[e]	—	-0.07	+2.72	+6.92	+8.69

[a] Estimated accuracy ±0.1 to 0.2 kcal mol^{-1}. Values in parentheses are for the entire molecule; others are those of individual groups.

[b] $\Delta\Delta H_t^\theta$ (X, solvent → vap), kcal mol^{-1}. 1 cal = 4.184 J.

[c] Calculated from Eqs. 2.132 and 2.133 followed by $\alpha(X) = \alpha(C_6H_5X) - \alpha(C_6H_6)$.

[d] Calculated as $V_m(X) = V_m(C_6H_5X) - V_m(C_6H_6)$.

[e] Calculated as $\alpha(OH) = \alpha(m\text{-cresol}) - \alpha(\text{toluene})$, where m-cresol is 3-methylphenol and toluene is methylbenzene.

Adapted and reprinted with permission from Fuchs, R., Young, T. M., and Rodewald, R. F. (1974). *J. Am. Chem. Soc.*, **96**, 4705–4706. Copyright 1974 by the American Chemical Society.

Fujita (1964) proposed the following equation for calculating the contributions $\pi_{(X)}$ of a substituent group X to the partition coefficient of the molecule RX:

$$\pi_{(X)} = \log\left(P^c_{(RX)}/P^c_{(RH)}\right) = \log P^c_{(RX)} - \log P^c_{(RH)} \tag{7.27}$$

where $P^c_{(RX)}$ and $P^c_{(RH)}$ are the molar concentration-based partition coefficients for compound RX and unsubstituted compound RH. $\pi_{(X)}$ is termed the hydrophobic group substituent constant of group X.

As defined above, $\pi_{(X)}$ is independent of the standard state of the concentration scale (x or c) used to define the partition coefficients, since the term (V_o/V_w) in Eq. 7.26 cancels by division in Eq. 7.27. However, in some cases, for example, when the intact molecule is taken as being equivalent to the π value, the choice of standard state or concentration scale determines the actual value of π, since Eq. 7.26 indicates that $P^x_{o/w}$ is not equal to $P^c_{o/w}$. Davis (1973b) drew attention to this point and suggested that π values, like the group-contribution values $\log F$, are additive only if, like $\log F$, they are referred to the mole fraction scale and not when referred to the molar concentration scale. Equation 7.26 indicates that

$$\log P^x_{o/w} = \log P^c_{o/w} + \log(V_o/V_w) \tag{7.28}$$

$$\pi \text{ (based on x)} = \pi \text{ (based on c)} + \log(V_o/V_e) \tag{7.29}$$

At 25°C for 1-octanol, $V_o = 158.42 \text{ cm}^3 \text{ mol}^{-1}$, whereas for water, $V_w = 18.07 \text{ cm}^3 \text{ mol}^{-1}$; therefore, the correction term $\log(V_o/V_w)$ is equal to 0.943, which is comparatively large. The rounded value 0.94 has been employed in Section 2.9.

Equation 7.23 can be written in the form

$$\log P_{(RX)} = \log P_{(RH)} + \pi_{(X)} \tag{7.30}$$

This equation refers to a molecule RX with only one functional group X. In general, for a molecule with more than one functional group, each with its own value of π, we may apply the group contribution concept and write

$$\log P(\text{substituted compound}) = \log P(\text{reference compound}) + \sum \pi \tag{7.31}$$

Taking pentachlorophenol C_6Cl_5OH as an example,

$$\log P_{(C_6Cl_5OH)} = \log P_{(C_6H_6)} + 5\pi_{(Cl)} + \pi_{(OH)}$$
$$= 2.13 + 5(0.71) - 0.67 = 5.01$$

Observed $\log P_{(C_6Cl_5OH)} = 5.12$.

When the substituents undergo strong electronic interaction, simple additivity implied by Eq. 7.31 fails. For example, in the case of 3,5-dinitrophenol

$3, 5\text{-}C_6H_3(NO_2)_2OH,$

$$\log P_{(3,5\text{-}C_6H_3(NO_2)_2OH)} = \log P_{(C_6H_6)} + 2\pi_{(NO_2)} + \pi_{(OH)}$$

$$= 2.13 + 2(-0.28) - 0.67 = 0.90$$

$$\text{observed } \log P_{(3,5\text{-}C_6H_3(NO_2)_2OH)} = 2.33$$

Intramolecular hydrogen bonding between neighboring groups also increases $\log P$, as in the case of the 2-hydroxy group and the carboxyl group of 2, 4-dihydroxybenzoic acid, $2, 4\text{-}C_6H_3(OH)_2COOH$, thus,

$$\log P_{(2,4\text{-}C_6H_3(OH)_2COOH)} = \log P_{(C_6H_6)} + 2\pi_{(OH)} + \pi_{(COOH)}$$

$$= 2.13 + 2(-0.67) - 0.32 = 0.47$$

$$\text{observed } \log P_{(2,4\text{-}C_6H_3(OH)_2COOH)} = 1.44$$

For the prediction of $\log P^c$(octanol/water) of aromatic compounds, π-substituent constants are widely used, whereas in the case of aliphatic compounds an alternative approach, known as the fragmentation method, is often preferred. As we have seen in the π system, groups are *substituted* for hydrogen atoms, whereas in the fragment system, the appropriate structural elements known as fragmentation constants f are *summed* over the entire molecule, thus,

$$\log P = \sum a_{(X)} f_{(X)} \tag{7.32}$$

where $a_{(X)}$ is the number of occurrences of the fragment X for which the fragmentation constant is $f_{(X)}$. This concept was introduced by Nys and Rekker (1973), who developed a set of fragmentation values from a statistical survey of partition data available up to 1973. Leo et al. (1975), Rekker (1977), and Chou and Jury (1980) have further elaborated this approach, while Acree et al. (1984) have provided an example of how the fragmentation constant of a group may be determined.

According to the definition of π from Eq. 7.23, the value for hydrogen, $\pi_{(H)}$, is zero. On the other hand, the fragmentation constant for hydrogen, $f_{(H)}$, which represents the intrinsic hydrophobicity of the hydrogen atom, cannot be zero because the hydrophobicity of a methyl group, $-CH_3$, is appreciably greater than that of a methylene group, $>CH_2$ (Davis, 1973a, b), as discussed in Section 7.3.2. Application of Eq. 7.32 to RH and RX shows that

$$\log P_{(RH)} = f_{(R)} + f_{(H)} \tag{7.33}$$

and

$$\log P_{(RX)} = f_{(R)} + f_X \tag{7.34}$$

Substitution into Eq. 7.30 shows that

$$f_{(X)} = \pi_{(X)} + f_{(H)} \tag{7.35}$$

This equation relates the π and f values of a given group and shows that $f_{(H)}$ is a fundamental quantity in the relationship. Nys and Rekker (1973) and Rekker (1977) obtained $f_{(H)} = 0.17$ in most aliphatic structures and 0.20 in most aromatic structures and, starting with log P values for a large number of molecules, used a *reductional* approach to calculate $f_{(CH_3)}, f_{(CH_2)}, f_{(CH)}$, and so on.

On the other hand, the Hansch school (Hansch and Leo, 1979) started with carefully measured log P values for a few simple molecules (e.g., H_2, CH_4, and C_2H_6), which presented no unsuspected interactions and used a *constructional* or *synthetic* approach to calculate f values of various groups. In fact, Hansch and Leo (1979) proposed the value $f_{(H)} = 0.23$ from $f_H = \frac{1}{2} \log P_{(H_2)} = \frac{1}{2} \times 0.45$. From log $P_{(CH_4)}$ or log $P_{(CH_3CH_3)}$, Hansch and Leo (1979) proposed the mean value $f_{(CH_3)} = 0.89$, which leads to $f_{(C)} = f_{(CH_3)} - 3f_{(H)} = 0.89 - (3 \times 0.23) = 0.20$. The Hansch school (1979) retains *constant* f values for the fundamental structural elements, for example, $f_{(H)}$ and $f_{(C)}$, and then seeks other factors F, which influence log P values of more complex solutes, thus,

$$\log P = \sum a_{(X)} f_{(X)} + \sum b_{(\xi)} F_{(\xi)} \tag{7.36}$$

where $b_{(\xi)}$ is the number of times the factor ξ occurs in the molecule. Examples of such factors include the type of bond attached to the atom or group, whether aromatic or aliphatic, chain branching, and proximity, which may be represented by one- or two-carbon separation between two interacting groups.

The treatment has remarkable powers of prediction, but is rather complex and may have to undergo further modification as more partition data become available and as more perturbing factors are recognized and understood. For example, Seiler et al. (1982) also found that partition coefficients of 5-(substituted benzyl)-2,4-diaminopyrimidines cannot be calculated with reasonable accuracy from fragmentation constants. Interactions between substituents can reduce P values by an order of magnitude below those obtained by applying Hansch's or Rekker's models.

Among the imperfectly understood anomalies is the fact that the observed value of P for $\Delta 9$-tetrahydrocannabinol is four orders of magnitude less than the value predicted by the fragmentation method (Hansch and Leo, 1979). This phenomenon, known as the *pot effect*, is beyond the scope of this book, as are further details of the fragmentation method. The complexities and interactions associated with the organic solvent 1-octanol may account for some of the anomalies.

The 1-octanol/water system chosen by Hansch and co-workers (Hansch and Fujita, 1964; Hansch and Leo, 1979) is by no means the only one used for determining $P_{o/w}$ and π values. Currie et al. (1966) employed cyclohexane/water, Flynn (1971) used ether/water, Kakeya et al. (1969) and Harris et al. (1973) used chloroform/water. The following equation of Collander (1951) enables the partition coefficient of any given solute, P_2, (based on organic solvent 2/water) to be calculated from the partition coefficient of the same solute, P_1, (based on organic solvent 1/water):

$$\log P_2 = a \log P_1 + b \tag{7.37}$$

where a and b are constants which depend on the given pair of solvent systems. This equation has received ample justification in the literature for a wide variety of solvent systems (e.g., Okada et al., 1984). At first sight, one might expect that the values of a and b would be independent of the nature of the solute. This is true for solutes in the same hydrogen bonding class, for example, donors in a standard solvent, acceptors in a standard solvent, and non-hydrogen-bonding solutes (Leo, 1972).

If solutes in different hydrogen bonding classes are included in the correlation, the correlation coefficient r decreases considerably from about 0.98 and the residual standard deviation s increases considerably from about 0.2 (Leo, 1972). If the solvents 1 and 2 possess similar hydrophobic characteristics as measured by the reciprocal of the solubility of water in the solvent (i.e., γ^{∞}), then the good correlation is restored for solutes of different hydrogen bonding class (Leo, 1972). This result can be rationalized by assuming that the standard free energy of transfer, ΔG_t^{θ}, in Eq. 1.5 and Eq. 7.6 in Scheme 7.1 has a contribution from polar or hydrogen bonding groups, ΔG_p^{θ}, and a contribution from hydrocarbon (or lipophilic) groups, ΔG_{hc}^{θ}, thus,

$$-2.303 \, RT \log P = \Delta G_t^{\theta} = \Delta G_p^{\theta} + \Delta G_{hc}^{\theta} \tag{7.38}$$

To obtain a good correlation, either of these two contributions can be varied proportionately or else only one may be varied while the other is held constant.

7.2.3 Chromatography

In order that partition data may be predicted without any experimental measurements, the data have been related to other molecular quantities and parameters. Since the most important factors controlling partitioning to or from water are the volume of the hole created by the solute in water (Leo et al., 1975) and the strength of the interaction between the solute and the polar water molecules, values of $\log P_{o/w}$, $\log F$, π, and f are often related to the following: Taft steric substituent constants, molar volume, molar refractivity, polarizability, electronic substituent constants, molar attraction constants, molecular orbital indices, and parachor (Dunn, 1977; Hansch and Leo, 1979). Since chromatography is frequently determined by partitioning behavior, partition data are often also related to chromatographic constants (Davis, 1973b).

Martin (1949) proposed that the group-contribution concept could be applied to the free energy of transfer, ΔG, in the partition process involved in paper chromatography, and Bate-Smith and Westhall (1950) showed that ΔG was related to the R_M value which is defined by

$$R_M = \log(R_F^{-1} - 1) \tag{7.39}$$

Since ΔG is related to R_M, the group contributions $\Delta\Delta G$ and π are also related to analogous group contributions of R_M, namely, ΔR_M. Tomlinson (1975) has reviewed the relationships between π and ΔR_M values for series of compounds. Tabulated substituent values ΔR_M and π enable both chromatographic constants and partition coefficients to be predicted (Davis et al., 1974).

In high-performance liquid chromatography (HPLC), retention data correlates closely with partition data, and the latter, such as the partition coefficients (n-octanol/water), can now be rapidly determined by means of this technique (Wright and Diamond, 1969a; Mirrless et al., 1976). Tomlinson et al. (1976) have determined by means of reversed phase HPLC the following values for methylene group contributions for transfer of monofunctional substituted benzoic acid esters from C_{18} hydrocarbon bonded to silicon to methanol + water (80:20 v/v) at 25°C: $\Delta\Delta G = -0.999$ kJ mol^{-1} (-0.239 kcal mol^{-1}); $\Delta\Delta H = -3.599$ kJ mol^{-1} (-0.860 kcal mol^{-1}); $\Delta\Delta S = -2.67$ JK^{-1} mol^{-1} (-0.638 cal K^{-1} mol^{-1}).

Comparison of these values with those in Table 7.5 suggests that the hydrocarbon-like bonded stationary phase behaves in a manner intermediate between

Table 7.5 Thermodynamics of Transfer of the Methylene Group from Aqueous Phase to Various Nonaqueous Phases at 25°C^a

System	Solutes	$\Delta\Delta G^{\theta}$ (kJ mol^{-1})	$\Delta\Delta H^{\theta}$ (kJ mol^{-1})	$\Delta\Delta S^{\theta}$ (J mol^{-1} K^{-1})
Transfer to pure liquid state	Various aliphatic solutes	-3.4 ± 0.3	-1.4 ± 0.3	8 ± 2
Transfer to nonpolar organic solvent	Various aliphatic solutes and ion-pair species	-3.5 ± 0.3	-1.6 ± 0.3	6.5 ± 1
Transfer to micelle	Various nonionic surfactants	-3.0 ± 0.3	-1.6 ± 0.8	5 ± 2
Transfer to pure solid state	Alkanoic acids	-3.5	-7.4	-12
Transfer to complex phase	Alkyl sulfates and BTPC^b	-3.2 ± 0.1	-2.0 ± 0.3	5 ± 1
Thermodynamics of fusion	Various aliphatic compounds		-5 ± 1	-5 ± 0.5
Transfer to C_{18} hydrocarbon column in HPLC	Substituted benzoate esters	-1.0	-3.6	-2.7
Transfer to liposome	Alkylphenols	-2.1	-5.7	-12

^{a}1 cal = 4.184 J.
^{b}Benzyltrimethylphosphonium chloride.

Reproduced with permission of the copyright owner, Academic Press, adapted from Mukhayer, G. I. and Davis, S. S. *J. Colloid Interface Sci.*, **59**, (1977) 350–359.

that of a solid and a nonpolar liquid, but more closely resembles a solid. Using a wide variety of model solutes, Hafkensheid and Tomlinson (1981) found that in reversed-phase HPLC x_2^w and $\log P_{o/w}$ were linearly related by κ, the HPLC analog of R_M, where x_2^w is the mole fraction solubility of the solute in water and $P_{o/w}$ is the partition coefficient of octanol/water. Here κ is defined by the following equation, which is analogous to Eq. 7.39:

$$\kappa = \log k = \log(t/t_o - 1) \tag{7.39a}$$

where k is the chromatographic capacity factor and t and t_o are corrected solute retention time and the retention time of the eluent slightly enriched with water.

For poorly water-soluble solutes k is obtained by extrapolating retention data in simple binary mixtures (water + organic modifier) to pure aqueous eluent. The linear free-energy relationship between ΔG and R_M, mentioned above, is exactly paralleled by that between ΔG and κ, so Hafkenshied and Tomlinson (1981) were able to deduce group contributions to κ. Like $\log x_2^w$ and $\log P$, κ was found to be a linear function of molecular surface area. These concepts have since been extended (Grünbauer and Tomlinson, 1983; Grünbauer et al., 1983).

Group values have been derived for the transfer of solutes from aqueous to nonaqueous biological phases (Wright and Diamond, 1969a; Vora et al., 1972; Davis et al., 1974), including liposomes (Diamond and Katz, 1974). Such values permit the prediction of the influence of changes in molecular structure on the passage of a drug molecule across a biological membrane, and also give some indication of the physical state of the membrane. For example, the methylene group contributions, particularly $\Delta\Delta H$ and $\Delta\Delta S$, for transfer from water to the interior of a liposome (shown in Table 7.5) suggest that the latter environment more closely resembles the solid state than the liquid state (Diamond and Katz, 1974). This suggestion is supported by the multilamellar structure of liposomes which imparts appreciable rigidity to the hydrocarbon chains of the packed phospholipid molecules.

A number of workers (Rogers and Davis, 1980; Anderson et al., 1983; Saket et al., 1984) have studied the partitioning of drug substances between aqueous solutions (subscript w, e.g., $0.15\,mol\,dm^{-3}$ or 0.9% w/w sodium chloride) and phospholipid vesicles such as liposomes (subscript p). Plots of $\log P_{p/w}$ against $1/T$ usually show a discontinuity (Fig. 7.4) at the endothermic phase transition temperature T_c of the phospholipid above which the fluidity of the phospholipid acyl chains greatly increase (Chapman et al., 1967).

For L-α-dimyristoyl phosphatidoyl choline (DMPC, lecithin), T_c is 23°C ($1/T_c = 3.377 \times 10^{-3}\,K^{-1}$), as shown in Figure 7.4. For the partitioning of hydrocortisone, cortisone, and their 21-alkanoate esters (Saket et al., 1984), $\Delta G_t^\theta(w \to p)$ is negative while $\Delta H_t^\theta(w \to p)$ and $\Delta S_t^\theta(w \to p)$ are positive. Thus, the transference process is entropy driven below and above T_c. As the hydrophobic, rigid steroid molecule transfers, the observed positive entropy change arises from

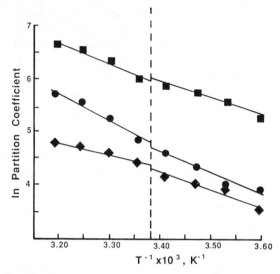

Figure 7.4 Temperature-dependent partitioning of hydrocortisone and its 21-derivatives between DMPC liposomes and 0.9% saline: hydrocortisone (◆), hydrocortisone acetate (●), and hydrocortisone propanoate (■). The discontinuity corresponds to the phase transition temperature T_c. Reproduced with permission of the copyright owner, Elsevier Science Publishing Co., Inc., from Saket, M. M., James, K. C., and Kellaway, I. W. (1984). *Int. J. Pharm.*, **21**, 155–166.

the loss of water structure surrounding the steroid molecules and disordering of the phospholipid bilayers.

The methylene group contribution to steroid partitioning into DMPC liposomes below and above T_c is also dominated by a similar entropy effect. For cortisone esters, the longer the acyl chain, the greater the structural change in the pohspholipid bilayers, as measured by the enthalpy of transition from the solid to the liquid crystalline state (Arrowsmith et al., 1983).

For the partitioning of 4-alkylphenols and 4-halophenols between $0.15 \, mol \, dm^{-3}$ sodium chloride and DMPC liposomes (Rogers and Davis, 1980), $P_{p/w}$ increases with increasing temperature below T_c, but decreases above T_c. The transfer process was found to be entropy-dominated below T_c [i.e., $\Delta S_t^\theta(w \rightarrow p)$ is positive] and enthalpy-dominated above T_c [i.e., $\Delta H_t^\theta(w \rightarrow p)$ is negative]. $P_{p/w}$ correlated well with the molecular size of the solute as did $P_{o/w}$ (from water to organic solvents). The interactions of the functional groups in the two environments, in addition to the nature of the hydrophobic moiety, undoubtedly influence the transfer process.

Evidently, a more complete understanding of the thermodynamics of partitioning requires knowledge of the enthalpies and entropies of transfer as well as the free energies of transfer as measured by the partition coefficients. As mentioned previously, calorimetry provides much more accurate values of ΔH_t^θ than do plots of $\ln P_{o/w}$ against $1/T$. The few calorimetric studies in the literature (e.g., Haberfield et al., 1984) generally reveal a wealth of information and usually show that ΔH_t^θ and ΔS_t^θ are more sensitive to solute structure and solvent properties than $\Delta G_t^\theta (= -RT \ln P_{o/w})$. This finding is another example of enthalpy–entropy compensation (Lumry and Rajender, 1970; Tomlinson, 1983).

7.2.4 Ion-Pair Partitioning

The limited solubility of many organic molecules in water often hinders the measurement of partition coefficients. To help solve this problem, Harris et al. (1973) developed a method using ion-pair extraction equilibria. Following the approach and procedure to be discussed in Section 9.3.1.2, the protonated form of dextromethorphan was chosen as the extracted cation to be paired with several inorganic anions and a series of alkyl sulphates whose structures were modified to enable the group-contribution approach to be used. The organic phase consisted of chloroform or chloroform–carbon tetrachloride mixtures in which chloroform acted as a solvating solvent to aid extraction. The composition of the aqueous phase was controlled with respect to pH, anion concentration, and ionic strength. A correlation was found between the corresponding free-energy differences and the effective surface areas of the alkyl substituents. The surface areas of the alkyl groups exposed to the aqueous phase exerted a much greater effect than the interactions of the same groups in the organic phase, as we shall see in Chapter 8.

Equation 9.6 can be written in the alternative form

$$D = K_o[X^-][M]^n + \text{constant} \tag{7.40}$$

where D is the observed partition coefficient defined by Eq. 9.5, $[X^-]$ is the concentration of anion in the aqueous phase, $[M]$ is the concentration of the solvating agent chloroform in the organic phase, and K_o is the equilibrium constant defined in terms of all the relevant concentrations by Eq. 9.4. The constant in this equation allows for competing equilibria such as (1) the extraction of the free base, (2) the extraction of buffer anions, and (3) the dissociation of the ion pair in the organic phase. Comparison of Eqs. 9.2 and 9.4 shows that

$$K_e = K_o[M]^n \tag{7.41}$$

that is

$$\log K_e = \log K_o + n \log [M] \tag{7.42}$$

where K_e is the ion-pair extraction constant defined by Eq. 9.2. Equation 7.40 may now be written as

$$D = K_e[X^-] + \text{constant} \tag{7.43}$$

D was plotted against $[X^-]$ to give a straight line whose slope gave K_e. The standard free energy of extraction of the ion pair species is given by

$$\Delta G_e^\theta = -RT \ln K_e \qquad (7.44)$$

The group contribution $\Delta\Delta G(R)$ of a given substituent R was given by the difference in the ΔG_e^θ value between a substituted compound and the unsubstituted parent compound, for example,

$$\Delta G_e^\theta(3,3\text{-dimethylbutyl sulfate}) - \Delta G_e^\theta(\text{ethyl sulfate}) = \Delta\Delta G(\textit{tert}\text{-butyl})$$

$$(7.45)$$

This method supplies true group contributions provided the hydrogen atom makes an insignificant contribution to the thermodynamics of transfer (Harris et al., 1973).

The enthalpy value ΔH_e^θ was calculated from the van't Hoff plot of the temperature dependence of ΔG_e^θ, while the entropy value ΔS_e^θ was derived using the same basic procedure as described previously (Eqs. 1.8 and 1.9 in Section 1.2.6). These data are inevitably less reliable than the ΔG_e^θ values. The group contributions $\Delta\Delta H$ and $\Delta\Delta S$ were calculated by a procedure entirely analogous to that for $\Delta\Delta S$ and are listed in Table 7.1 with the π and F values defined by Eqs. 7.24 (or 7.25) and 7.27 for group X.

The extent of interaction of the solvating agent M (i.e., chloroform in carbon tetrachloride) with polar groups (e.g., halogen substituents, phenyl groups, and ether oxygen atoms) was assessed from the solvation number n. This number was determined as the gradient of the plot of $\log D$ against $\log[M]$, as indicated by Eq. 9.7 and by analogy with the procedure discussed in Section 9.3.1.2. For most groups, n was about 3 and increased by 0.26 for halogen substituents and by 0.50 for ether linkages. Even though the solvation of these polar groups is small, it might still affect the thermodynamic values.

Free energies of transfer of alkyl groups to and from water are often linearly related to the surface area A and volume V, of the individual groups, as we shall see. Harris et al. (1973) found an excellent linear relationship (Fig. 7.5) between $\Delta\Delta G$ values (Table 7.6) and relative A values (Table 7.7) of various straight-chain, branched chain, and cyclic hydrocarbon groups, but rather scattered relationships (Fig. 7.6) between $\Delta\Delta G$ and V calculated from the van der Waals radii (Bondi, 1964) or from solute molar volumes (McAuliffe, 1966).

Harris et al. (1964) estimated group A values as the number of spheres (representing the hydrogen atoms of water) that could be placed around the model of the group or molecule. These values, determined using atomic models, were expressed relative to the *tert*-butyl group as an accessible standard whose relative surface area was set equal to unity (Table 7.7). Similarly, Hermann (1972), Yalkowsky et al. (1972a, b, 1975, 1976), Amidon et al. (1974, 1975), and Valvani et al. (1976) obtained excellent linear correlations between aqueous solubility of various molecules and molecular surface areas computed from bond lengths, atomic radii, and solvent molecule radii, as will be discussed in Section 8.3.

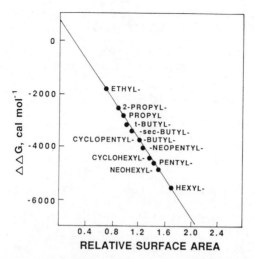

Figure 7.5 Correlation between the free energies of transfer $\Delta\Delta G$ from water to 25% v/v chloroform in carbon tetrachloride at 30°C and the relative surface areas of common organic groups. The relative surface area of the *tert*-butyl group is set arbitrarily at 1.0. 1 cal = 4.184 J. Reprinted with permission from Harris, M. J., Higuchi, T., and Rytting, J. H. (1973). *J. Phys. Chem.*, 77, 2694–2703. Copyright 1973 by the American Chemical Society.

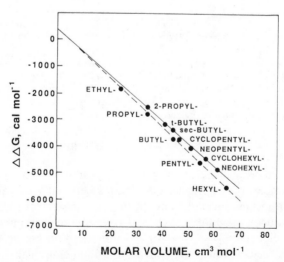

Figure 7.6 Correlations between the free energies of transfer $\Delta\Delta G$ from water to 25% v/v chloroform in carbon tetrachloride at 30°C and the van der Waals volumes of common organic groups: straight-chain alkyl groups (———); branched chain and cyclic alkyl groups (———). 1 cal = 4.184 J. Reprinted with permission from Harris, M. J., Higuchi, T., and Rytting, J. H. (1973). *J. Phys. Chem.*, 77, 2694–2703. Copyright 1973 by the American Chemical Society.

Table 7.6 Thermodynamic Values of Transfer for Various Organic Groups from Water to 25% v/v Chloroform in Carbon Tetrachloride at 30°C

Group	$\Delta\Delta G_{30}$ (cal mol^{-1})	$\Delta\Delta H$ (cal mol^{-1})	$\Delta\Delta S_{30}$ (cal K^{-1} mol^{-1})	π Value[a]	F Factor[a]
Neohexyl	−4831	−2860	6.53	3.49	3090
Cyclohexyl	−4405 (Et)[b]	−3661	2.45	3.18	1510
	−4414 (Pr)[b]	−3280	3.49	3.18	1510
Neopentyl	−4036	−1976	6.79	2.91	813
Cyclopentyl	−3699	−3126	1.89	2.67	468
sec-Butyl	−3376	−2723	2.15	2.43	269
tert-Butyl	−3136	−2702	1.43	2.26	182
Phenyl	−2872 (Bu)[b]	−2274	1.97	2.07	118
	−2828 (Pr)[b]	−3436	−2.01	2.04	110
	−2976 (Et)[b]	−3748	−2.55	2.14	139
Phenoxy	−2897	−3602	−2.33	2.08	120
Isopropyl	−2507	−1777	2.41	1.81	64.6
Butoxy	−2053	−92	6.47	1.48	30.2
Ethylene	−947	−1891	−3.09	0.68	4.79
Methylene	−916	−378	1.77	0.66	4.57
Bromo	−807	−1789	−3.24	0.58	3.80
Chloro	−439	−1203	−2.52	0.32	2.09
4-Fluoro (aromatic)[b]	−179	247	1.47	0.13	1.35
Fluoro (aliphatic)[b]	−38	−449	−1.36	0.03	1.06
Methoxy (—O—)[b]	303	−87	−1.50	−0.26	0.55
(MeOPr-)[b]	1214	1139	−0.25	−0.88	0.13
(BuOEt-)[b]	1541	3208	5.50	−1.11	0.08
(PhOEt-)[b]	79	146	0.22	−0.06	0.87

[a] F and π are related to $\Delta\Delta G$ by Eqs. 7.23–7.25 and 7.27, thus, $\Delta\Delta G = -2.303\,RT\log F = -2.303\,RT\pi$, where $RT = 602.4$ cal mol^{-1} and $2.303\,RT = 1387.1$ cal mol^{-1} at 30°C. 1 cal = 4.184J.
[b] When several derivatives were investigated to find the thermodynamic properties of a single group, the derivative from which the data were derived are shown in parentheses.

Reprinted with permission from Harris, M. J., Higuchi, T., and Rytting, J. H. (1973). *J. Phys. Chem.*, **77**, 2694–2703. Copyright 1973 by the American Chemical Society.

The major conclusions are that, although the volume of a molecule or group cannot be neglected, the interfacial interactions between water and the solute are of greater importance in explaining the thermodynamic properties of organic molecules in aqueous solution. Such a correlation further implies that it is possible to predict the aqueous solubility and partition coefficients for large hydrophobic molecules, which are difficult to obtain empirically from a simple determination of their surface areas. We shall return to these considerations in Chapter 8.

The plots of ΔH and $T\Delta S$ against carbon number n each show a discontinuity at $n = 5$ and are in mutual opposition, such that ΔG continues its linear course

***Table* 7.7** Relative Surface Area[a] of Several Alkyl Groups and the Statistics of their Estimation

Functional Group	Relative Surface[a] Area		Standard Deviation[b] (%)	Standard error of the mean (%)	n
	A	B			
Methyl	—	0.51	0	0	6
Ethyl	0.70	0.74	2.2	0.9	6
n-Propyl	0.93	0.96	2.2	1.0	5
Isopropyl	0.88	0.90	3.7	1.6	5
n-Butyl	1.18	1.18	3.8	1.2	10
sec-Butyl	1.08	1.15	1.6	0.6	8
tert-Butyl	1.00	1.00	2.2	1.0	5
n-Pentyl	1.42	1.42	0.9	0.3	10
Neopentyl	1.29	1.29	1.5	0.5	7
Cyclopentyl	1.18	1.20	3.5	1.1	10
Cyclohexyl	1.38	1.33	3.4	0.6	33
n-Hexyl	1.70	1.64	3.6	1.2	9
sec-Hexyl	—	1.40	1.5	0.7	5

[a]The relative surface areas of the *tert*-butyl group are set equal to one. The values in column A were determined by one individual and those in column B by another. The statistical data refer to the values in column B.
[b]The percent standard deviation in the n values reported in column B for that group.
[c]The percent standard error of the mean for the values in column B.

Reprinted with permission from Harris, M. J., Higuchi, T., and Rytting, J. H. (1973). *J. Phys. Chem.*, **77**, 2694–2703. Copyright 1973 by the American Chemical Society.

(Fig. 7.7). This behavior is an example of enthalpy–entropy compensation (see Tomlinson, 1983) and suggests that the group contributions are more applicable to free energies of transfer than to the corresponding enthalpies and entropies of transfer.

7.3 PARTICULAR GROUP CONTRIBUTIONS

7.3.1 Methylene Group

The methylene group has provided most group-contribution data (Davis et al., 1972, 1974), largely because of the ease with which the information may be calculated from equilibrium measurements on members of homologous series. Table 7.8 presents values of methylene group contributions involving the transfer of various solutes from water to solvents of low polarity. The values obtained are essentially independent of the nature of (a) the solute and (b) the latter solvent, emphasizing the value of the group-contribution concept and the constancy of the methylene–solvent interactions for relatively short hydrocarbon chains. These interactions consist essentially of London dispersion forces, while

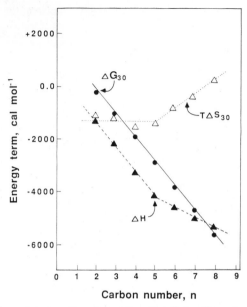

Figure 7.7   Thermodynamic properties of transfer of *n*-alkyl sulfates from water to 25% v/v chloroform in carbon tetrachloride, expressed in energy terms as function of carbon number for straight-chain alkyl groups at 30°C. 1 cal = 4.184 J. Reprinted from Harris, M. J., Higuchi, T., and Rytting, J. H. (1973). *J. Phys. Chem.*, **77**, 2694–2703. Copyright 1973 by the American Chemical Society.

short chains have sufficiently limited conformational arrangements to be considered relatively inflexible.

Table 7.9 provides a summary of the most acceptable values of $\Delta\Delta G_t^\theta(CH_2)$, $\Delta\Delta H_t^\theta(CH_2)$, and $\Delta\Delta S_t^\theta(CH_2)$ compiled (Davis et al., 1974) from various literature sources for various phase-transfer processes.

Group contributions calculated from Raoult's law mole-fraction-based activity coefficients γ^∞ correspond to the reciprocal of the mole-fraction solubility of the liquid solute (Eq. 2.103a), the transfer process being pure solute → solution. In pure alcohols, amines, and carboxylic acids, the polar groups participate so successfully in intermolecular hydrogen bonding that a methylene group mainly experiences London dispersion forces with the hydrocarbon chains in these solvents to about the same extent as in a paraffinic hydrocarbon solvent. Thus, for these solutes, the methylene group contribution (a) for pure solute → hydrocarbon is effectively zero (b) for pure solute → water closely approximates hydrocarbon → water, for example, $\Delta \log \gamma_w^\infty(CH_2) \cong \Delta \log P_{hc/w}(CH_2)$, and (c) for pure solute → vapor closely approximates hydrocarbon → vapor for example, $\Delta \log p^*(CH_2) \cong \Delta \log h_{hc}(CH_2)$.

$\Delta\Delta G_t^\theta(CH_2$, hydrocarbon → water) = 850 cal mol^{-1} = 3600 J mol^{-1}, and is essentially constant for lower members ($n \leqslant 10$) of the homologous series of solutes (e.g., alkanes, alkanols, and alkylamines) corresponding to linear plots of

Table 7.8 Methylene Group Contributions to the Thermodynamics of Phase Transfer of Various Species of Chain Length Greater than C_5 from Water to an Organic Solvent of Low Polarity[a]

Solute	Organic Solvent	$\Delta\Delta G_t^\theta(CH_2)$		$\Delta\Delta H_t^\theta(CH_2)$		$\Delta\Delta S_t^\theta(CH_2)$	
		(cal mol^{-1})	(J mol^{-1})	(cal mol^{-1})	(J mol^{-1})	(cal K^{-1} mol)	(J K^{-1} mol^{-1})
Case III ion pairs: dextromethorphan with alkyl sulfates[b]	25% CHCl$_3$ v/v in CCl$_4$ at 30°C	−917	−3840	−378	−1580	1.77	7.41
n-Alkyl 4-hydroxybenzoates[c]	Cycloexane at 25°C	−877	−3670	−345	−1440	1.78	7.47
n-Alkanols[d]	Inert hydrocarbons at 25°C	−850	−3560	−400	−1670	1.6	6.7

[a] 1 cal = 4.184 J.
[b] Harris et al. (1973).
[c] Crugman (1971).
[d] Davis et al. (1974).

Davis, S. S., Higuchi, T., and Rytting, J. H. (1974). *Adv. Pharm. Sci.*, **4**, 73–261.

337

Table 7.9 Methylene Group Contributions to the Thermodynamics of Phase Transfer of Alkanes and Alkanols at 25°C[a]

Transfer Process	$\Delta\Delta G_t^\theta(CH_2)$		$\Delta\Delta H_t^\theta(CH_2)$		$\Delta\Delta S_t^\theta(CH_2)$	
	$(cal\,mol^{-1})$	$(J\,mol^{-1})$	$(cal\,mol^{-1})$	$(J\,mol^{-1})$	$(cal\,K^{-1}\,mol^{-1})$	$(J\,K^{-1}\,mol^{-1})$
Solution (pure solute → water)	845	3540	350	1460	−2.0	−8.4
Vaporization (pure solute → vapor)	700	2930	1200	5020	1.6	6.7
Hydration (solute vapor → water)	137	573	−975	−4080	−3.8	−15.9

[a] 1 cal = 4.184 J.

Davis, S. S., Higuchi, T., and Rytting, J. H. (1974). *Adv. Pharm. Sci.*, **4**, 73–261.

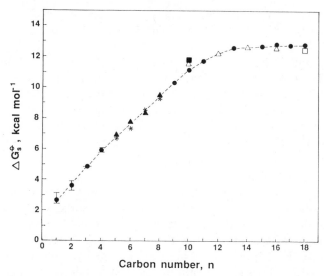

Figure 7.8 Influence of chain length (i.e., number of carbon atoms n) on the free energy of solution of pure n-alkanes, $\Delta G_S^\theta = \Delta G_t^\theta$ (pure solute → water, extrapolated to $x_2 \to 1$) = $-RT \ln x_2$, at 25°C: experimental data from different workers (I, $*$, $\triangle$, $\blacktriangle$, $\square$, $\blacksquare$, see Nelson and De Ligny, 1968; calculated ($\bullet$, unless obscured by experimental points) by Nelson and De Ligny (1968) from theory of Pierotti (1963). 1 cal = 4.184 J. Davis, S. S., Higuchi, T., and Rytting, J. H. (1974) *Adv. Pharm. Sci.*, **4**, 73–261.

$\log \gamma_w^\infty$ or $\log P_{hc/w}$ against carbon number n for these solutes. The slope of such plots, however, falls off for $n \geqslant 10$, as shown in Figure 7.8. This reduction in slope corresponds to a decrease in the methylene group contribution and reflects an increasing hydrophobic interaction to be discussed in Section 8.2. This behavior presumably results from curling of the hydrocarbon chains into a spherical droplet in water in order to minimize the area of contact between the hydrocarbon chain and water. The area of such a droplet and hence ΔG_t^θ(alkyl chain → water) then increase relatively slowly as a linear function of $n^{2/3}$ or approximately as (molecular weight)$^{2/3}$. Differentiation with respect to n affords the methylene group contribution $\Delta\Delta G_t^\theta$(CH_2, hydrocarbon → water), which is proportional to $n^{-1/3}$, corresponding to a weak dependence on n. Similarly, long alkyl chains also appear to curl up to form spherical molecules in the vapor phase under the influence of van der Waals forces (Huggins, 1939). In support of this suggestion, the enthalpy of vaporization, ΔH_v, of the n-alkanes is a linear function of $n^{2/3}$, the slope being 3.21 and the intercept being $(2.92 - 0.0193T)$, in kcal mol^{-1}, where T is the absolute temperature (Wall et al., 1970). Differentiation with respect to n, as before, affords the group contribution $\Delta\Delta H_t^\theta$(CH_2, alkane → vap) = $n^{-1/3}$ 2.14 kcal mol^{-1}.

If the polarity of the organic solvent (subscript o) is increased, the Debye interactions between the solvent molecules and the methylene groups of the solute also increase. Consequently, the following methylene group contributions decrease: $\Delta \log h_o(CH_2)$, $\log F(CH_2)$, $\Delta \log \gamma_o^\infty(CH_2)$, while the corresponding free energies increase, that is, become less favorable: $\Delta\Delta G^\theta(CH_2, \text{ org} \to \text{vap})$, $\Delta\Delta G^\theta(CH_2, \text{ org} \to \text{water})$, $\Delta\Delta G^\theta(CH_2, \text{ pure solute} \to \text{org})$. Intuitively, increasing polarity of the organic solvent will lead to increased mutual solubility of the solvent and water, which will tend to make the solute environment more alike in the two solvents and will therefore reduce all group contributions to partition (Treybal, 1963). Solubility parameter theory is unable to explain such trends quantitatively. However, $\log F(CH_2)$ decreases and $\Delta\Delta G^\theta(CH_2, \text{ org} \to \text{water})$ increases linearly with increasing logarithm of the solubility of water in the solvent (Davis et al., 1974), which may indicate the strength of the Debye forces.

7.3.2 Methyl Group

Since the methyl group has three hydrogen atoms, whereas the methylene group has only two, the two groups will have different surface areas, so we cannot assume that they will have the same group contributions. Furthermore, since the methyl group can take part in an inductive effect when attached to an aromatic ring, it will be slightly polar in this position. The methyl group contribution has been reviewed by Davis (1973a) and by Davis et al. (1974) and depends on the position of the group in the drug molecule. The methyl group may be (1) attached to an aromatic or saturated ring system, (2) in the terminal position in an alkyl chain, (3) in a more central position in a branched alkyl chain, or (4) attached to other atoms besides carbon (e.g., oxygen or nitrogen).

When attached to an aromatic ring, the methyl group gives generally slightly smaller $\log F$ values than the aliphatic methylene group for a range of different organic solvents, thus,

$$\log F(CH_3) = 1.12 \log F(CH_2) - 0.113 \tag{7.46}$$

If, however, it is assumed that the group contribution of the methyl group on an aromatic ring is, in general, equal to that of the aliphatic methylene group, the error will be small, provided that there is no interaction between functional groups (Davis, 1973a).

When the methyl group is in a terminal position in an alkyl chain, its contribution is more difficult to calculate. If $\Delta G_t^\theta(\text{hc} \to \text{water})$ of a homologous series of alkanes is plotted against carbon number, the line does not pass through the origin, which indicates that $\Delta\Delta G_t^\theta(CH_3, \text{ hc} \to \text{water}) \neq \Delta\Delta G_t^\theta(CH_2, \text{ hc} \to \text{water})$. The intercept is approximately $2000 \, \text{cal mol}^{-1}$ ($8370 \, \text{J mol}^{-1}$), which represents twice the so-called terminal methyl group correction (approximately $1000 \, \text{cal mol}^{-1}$ of $4184 \, \text{J mol}^{-1}$), since two of the carbon atoms of the alkanes are in methyl groups instead of methylene groups. We note that

$$\text{intercept} = 2 \times \text{terminal } CH_3 \text{ correction}$$

$$= 2(\Delta\Delta G_{CH_3} - \Delta\Delta G_{CH_2})$$

$$= 2[\Delta\Delta G_t^\theta(CH_3) - \Delta\Delta G_t^\theta(CH_2)] \tag{7.47}$$

If the free energy of a homologous series (of alkanols, alkylamines, or alkanoic acids represented by R–X) is plotted against carbon number the intercept is not simply the group contribution of X, $\Delta\Delta G_t^\theta(X)$, but in fact,

$$\text{intercept} = \text{terminal } CH_3 \text{ correction} + \Delta\Delta G_t^\theta(X) \qquad (7.48)$$

Comparison of the analogous free energy of transfer data for an alkane R–R with that for R–X generates two simultaneous equations which are analogous to Example 7.2 and which are special cases of Eq. 7.2. Since these equations also have two unknowns, $\Delta\Delta G_t^\theta(CH_3)$ and $\Delta\Delta G_t^\theta(X)$, they can be solved. Consequently, both the group contributions of methyl and of X can be calculated. This procedure has been carried out for a number of cases, for example: R–X = EtOH, R–R = butane = 2 × Et; R–X = PrOH, R–R = hexane = 2 × Pr. A similar procedure has also been applied to the side chains of proteins (Tanford, 1962). In general, for an aliphatic compound $CH_3(CH_2)_{n-1}X$,

$$\Delta G_t^\theta(CH_3(CH_2)_{n-1}X) = \Delta\Delta G_t^\theta(CH_3)$$
$$+ (n-1)\Delta\Delta G_t^\theta(CH_2) + \Delta\Delta G_t^\theta(X) \qquad (7.49)$$

$$= \Delta\Delta G_t^\theta(CH_3\text{correction})$$
$$+ n\Delta\Delta G_t^\theta(CH_2) + \Delta\Delta G_t^\theta(X) \qquad (7.50)$$

where

$$\Delta\Delta G_t^\theta(CH_3 \text{ correction}) = \Delta\Delta G_t^\theta(CH_3) - \Delta\Delta G_t^\theta(CH_2) \qquad (7.51)$$

Davis (1973a) has calculated and discussed the mean terminal methyl group contribution and correction.

Example 7.3. Calculate a value for $\Delta\Delta G_t^\theta(OH, w \to hc)$, where w = water and hc = paraffinic hydrocarbon cyclohexane, from the following data at 25°C: $\log P_{hc/w}^c = -2.10$ for ethanol and 3.62 for n-butane (see Hansch and Leo, 1979); density = 0.77389 g cm^{-3} for cyclohexane and 0.99705 g cm^{-3} for water.

$$V(\text{cyclohexane}) = 84.161/0.77389 = 108.75 \text{ cm}^3 \text{ mol}^{-1}$$
$$V(\text{water}) = 18.015/0.99705 = 18.068 \text{ cm}^3 \text{ mol}^{-1}$$
$$\log \frac{V(\text{cyclohexane})}{V(\text{water})} = \log \frac{108.75}{18.068} = 0.780$$

From Eq. 7.28

$$\log P_{hc/w}^x = \log P_{hc/w}^c + \log \frac{V(\text{cyclohexane})}{V(\text{water})}$$

$$= -2.10 + 0.78 = -1.32 \text{ for ethanol}$$

$$= 3.62 + 0.78 = 4.40 \text{ for } n\text{-butane}$$

From Eq. 7.6

$$\Delta G_t^\theta(\text{w} \to \text{hc}) = -RT \ln P_{\text{hc/w}}^{\text{x}} = -2.303 \, RT \log P_{\text{hc/w}}^{\text{x}}$$

$$= -1364 \, (\text{cal mol}^{-1}) \log P_{\text{hc/w}}^{\text{x}}$$

$$= 1801 \text{ cal mol}^{-1} \text{ for ethanol}$$

$$= -6003 \text{ cal mol}^{-1} \text{ for } n\text{-butane}$$

The Gibbs free energy of transfer, water → hydrocarbon, for each compound may be apportioned among the various group contributions as follows:

$$\Delta G_t^\theta(CH_3 \cdot CH_2 \cdot OH) = \Delta\Delta G_t^\theta(CH_3) + \Delta\Delta G_t^\theta(CH_2) + \Delta\Delta G_t^\theta(OH)$$

$$\Delta G_t^\theta(CH_3 \cdot CH_2 \cdot CH_2 \cdot CH_3) = 2\Delta\Delta G_t^\theta(CH_3) + 2\Delta\Delta G_t^\theta(CH_2)$$

Therefore

$$\Delta\Delta G_t^\theta(OH) = \Delta\Delta G_t^\theta(CH_3 \cdot CH_2 \cdot OH)$$

$$- \tfrac{1}{2}\Delta\Delta G_t^\theta(CH_3 \cdot CH_2 \cdot CH_2 \cdot CH_3)$$

$$= 1801 - \tfrac{1}{2}(-6003)$$

$$= 4802 \text{ cal mol}^{-1} = 20090 \text{ J mol}^{-1}$$

This value is 10% higher than the 4350 cal mol^{-1} tabulated in Table 7.1 since the latter is a mean value derived from data obtained by various workers from various alkanols, while the former is derived from isolated literature values.

Example 7.4. Calculate a value for (a) the methyl group contribution $\Delta\Delta G_t^\theta(CH_3, \text{w} \to \text{hc})$ and for (b) the corresponding terminal methyl group correction (Eq. 7.51) assuming $\Delta\Delta G_t^\theta (CH_2, \text{w} \to \text{hc}) = -880$ cal mol^{-1} at 25°C and using the data and notation of the previous example (7.3).

(a) $\Delta G_t^\theta(CH_3 \cdot CH_2 \cdot CH_2 \cdot CH_3) = 2\Delta\Delta G_t^\theta(CH_3) + 2\Delta\Delta G_t^\theta(CH_2)$

$$\Delta\Delta G_t^\theta(CH_3) = \tfrac{1}{2}\Delta G_t^\theta(CH_3 \cdot CH_2 \cdot CH_2 \cdot CH_3) - \Delta\Delta G_t^\theta(CH_2)$$

$$= \tfrac{1}{2}(-6003) - (-880) \text{ cal mol}^{-1}$$

$$= -2121 \text{ cal mol}^{-1} = -8875 \text{ J mol}^{-1}$$

This value compares favorably with -2000 cal mol^{-1} tabulated in Table 7.1.

(b) From Eq. 7.51

$$\Delta\Delta G_t^\theta(CH_3 \text{ correction}) = -2121 - (-880)\text{ cal mol}^{-1}$$
$$= -1241 \text{ cal mol}^{-1} = -5194 \text{ J mol}^{-1}$$

The value calculated from the methyl group contribution in Table 7.1 is $-1120 \text{ cal mol}^{-1}$ (-4686 J mol^{-1}). Agreement is satisfactory.

If it is difficult or impossible to obtain a terminal methyl group correction or to assign a group contribution value to the terminal methyl group, then the methyl group correction and the group contribution of X cannot be separated and are together included in the intercept obtained by extrapolating ΔG_t^θ or log Y values to zero carbon number (Davis, 1973d). The intercept then represents the uncorrected group contribution of X, thus,

$$\text{intercept (at } n=0) = \Delta\Delta G_t^\theta(H \cdots X) \tag{7.52}$$

Comparison with Eq. 7.48 shows that

$$\Delta\Delta G_t^\theta(H \cdots X) = \Delta\Delta G_t^\theta(X) + \text{terminal } CH_3 \text{ correction} \tag{7.53}$$

The effect of the methyl group in a nonterminal portion on an alkyl chain will be considered in the section on chain branching in Section 8.4.

The replacement of a hydrogen atom of an alcohol, phenol, or amine by a methyl group produces a profound change to the hydrogen bonding, corresponding to the loss of a proton. The partition coefficient (organic solvent/water) is increased and the water solubility is decreased. The hydroxyl group can form two hydrogen bonds, but the ether group can form only one. Therefore, two ether oxygens are about as effective as one hydroxyl group in reducing $P_{o/w}$ (Davis et al., 1974).

7.3.3 Cyclization, Double and Triple Bonds

Ring closure appears to increase $\Delta\Delta G_t^\theta$ by an amount comparable to that obtained by the removal of one methylene group from the equivalent straight-chain compound (McAuliffe, 1966; Tsonopoulos and Prausnitz 1971; Davis et al., 1974). This effect had been attributed (1) to the reduction in molar volume or surface area of the solute and (2) to the loss of one or two terminal methyl groups which have a larger group contribution than the methylene group (Davis, 1973a). The more compact structure of the cyclic compound as compared with the corresponding acyclic compound reduces the surface area of contact of the hydrocarbon moiety with water and so reduces $P_{o/w}$ (Hansch, 1971) and increases the solubility in water (Cabani et al., 1971a, b; Harris et al., 1973).

Phenyl group contributions are inevitably influenced by the London, Debye, and possibly charge-transfer interaction of the π electrons (Clint et al., 1968;

Harris et al., 1973). Comparison of the group-contribution data (Harris et al., 1973) in Table 7.6 for the phenyl and cyclohexyl groups, both from their propyl derivatives, shows that phenyl-group substitution increases the water solubility about 14 times that for cyclohexyl-group substitution. Extraction with a nonpolar organic solvent would increase this ratio. The effect may be attributed to the larger positive entropy of transfer associated with the cyclohexyl group. This property causes a greater increase of order in the water structure around the cyclohexyl group than around the phenyl group.

Comparison of $\Delta\Delta G_t^{\theta}(-CH=CH-) = -947$ cal mol^{-1} (Table 8.6) Harris et al., 1973) and other data (Lamb and Harris, 1960; Plakogiannis et al., 1970) with that for two methylene groups (2×-900 cal mol^{-1}; Table 7.8) indicates that the presence of each isolated double bond is approximately equivalent to the removal of one methylene group. The introduction of the double bond is known to reduce $P_{o/w}$ (Davis et al., 1974) and to increase the solubility in water, while the triple bond exerts an even greater effect. This behavior results (1) from a reduction in the molar volume and surface area of the solute and (2) from an increased interaction of the polarizable π-electron system with polar water molecules, presumably as a result of Debye forces (Section 2.11.4) and donor–acceptor interaction (Section 2.11.7).

The effect of chain branching will be discussed in Section 8.4 because of its ability to increase the solubility of hydrocarbons in water.

7.3.4 Polar Groups

The replacement of a hydrogen atom by a hydroxyl or carboxyl group will enable hydrogen bonds to be formed with water, thereby increasing ΔG_t^{θ} (water → vapor), ΔG_t^{θ}(water → hc), and the solubility in water and reducing h_w, $P_{o/w}$, and γ_w^{∞}. The somewhat stronger Debye interactions in hydrocarbons will reduce h_{hc}. The unfavorable $\Delta\Delta G_t^{\theta}$(OH, water → hc) results from an unfavorable enthalpy change, the entropy change being favorable (Némethy and Scheraga, 1962a, b; Laiken and Némethy, 1970). Hydroxyl and carboxyl group contributions have been widely discussed (Latimer and Rodebush, 1920; Diamond and Wright, 1969b; Davis, 1973d; Davis et al., 1974).

For partitioning into a series of organic solvents, $\log F(H \cdots OH)$ and $\log F(H \cdots COOH)$ are linearly correlated with $\log s$, where s is the solubility of water in the solvent (Leo et al., 1971; Davis, 1973a), while $\log F(H \cdots OH)$ is linearly correlated with the logarithm of the IR spectral shift of the hydroxyl group of methanol in that solvent, and $\log F(H \cdots COOH)$ is linearly correlated with the logarithm of the self-association constant of acetic acid in that solvent (Davis, 1973d; Davis et al., 1974).

The halogen group contribution $\Delta\Delta G_t^{\theta}$(X, water → hc) becomes more negative in the order F, Cl, Br, I, indicating increasingly favorable transfer from water to hydrocarbon (Currie et al., 1966; Hansch and Anderson, 1967; Davis, 1973c). This rank order corresponds to the order of increasing atomic volume, which disfavors entropically the interaction with water (Section 8.2), and to the order of decreasing electronegativity and, consequently, decreasing dipolar attractions between X and water (Hansch and Fujita, 1964; Hansch and Anderson, 1967; Davis, 1973c; Harris et al., 1973). The nature of the organic solvent significantly influences the group contribution (Currie et al., 1966; Davis, 1973c) because of X–solvent interactions.

Halogen atoms exert powerful mesomeric effects when attached to the 4 position of the aromatic ring. This disfavors $\Delta\Delta G_t^\theta(X, \text{water} \to hc)$, making it less negative than when X is attached to an aliphatic system. The position of X in the aromatic ring will significantly affect the group contribution, the mesomeric effect being abolished in the 3 position, while steric effects become important in the 2 position (McGowan, 1963; Davis, 1973c). Similarly, the group contributions of other polar groups, such as methoxy (OMe) and hydroxy (OH), will be significantly influenced by their electronic environment, particularly when attached to aromatic, heterocyclic, or conjugated systems (McGowan et al., 1986; Harris et al., 1973; Davis et al., 1974). Ether group contributions have a particularly broad literature (Butler, 1937, 1962; Hansch and Fujita, 1964; Deutsch and Hansch, 1966; Iwasa et al., 1965; Kertes, 1971; Davis et al., 1974), owing to the sensitivity of ether interactions to electronic and steric effects.

7.4 RATIONALE FOR SIMPLE VERSUS ELABORATE APPROACHES TO GROUP CONTRIBUTIONS

With increasing distinction between the functional groups, accuracy of correlation and prediction increases. When considering, for example, aliphatic alcohols, the first approximation makes no distinction between a primary and secondary hydroxyl group, the second approximation makes this distinction, while the third approximation considers the influence of groups adjacent to the —CH_2OH or $\diagdown$CHOH groups. In the limit, as more distinctions are made, we arrive at group contributions which are specific to each individual molecule. The advantage of the group-contribution approach is then lost. We can expect diminishing returns for each successive distinguishing step. For practical purposes, a compromise must be achieved. The number of groups must remain small, but must not be so small as to neglect significant effects of molecular structure on physical properties. This book emphasizes the simplest group-contribution approach, whereas chemical engineers may require a more refined method of prediction. The next paragraph presents the reasons for this refinement.

In the biological sciences (particularly the pharmaceutical, agricultural, and environmental sciences), the complexity of the molecules is perhaps greater than is usually found in general chemical engineering, whereas the conditions of temperature and pressure and the number of components in each phase are usually far less extreme. Furthermore, the degree of accuracy required of the prediction is much greater in chemical engineering than in the biological sciences. (In critical cases prediction must, of course, be followed by careful measurement.) At present, the most appropriate compromise for the biological sciences is to attain an accuracy of prediction within a factor of 2 or 3 by simple methods of calculation. This may be achieved by using simple group contributions shown in Tables 7.1 or 7.6 and a hand-held calculator. In chemical engineering an accuracy of prediction within a few percent is desirable. This greater accuracy requires the use of models with many parameters and, consequently, considerable computational effort, which can now be readily achieved. At this stage we shall attempt to indicate the general approaches taken by chemical engineers.

In chemical technology the number of pure compounds is already very large, while the number of different mixtures is larger still by many orders of magnitude. Thousands or perhaps millions of multicomponent liquid mixtures of interest in the chemical industry can be constituted from between 20 and 100 functional groups. Each group may be associated with one, two, or more parameters.

Following Langmuir (1925) and Butler (1937, 1962), the subject evolved as follows. Papadopoulos and Derr (1959) and Redlich et al. (1959) used group contributions to correlate heats of mixing. Wilson and Deal (1962) developed the solution-of-groups method for predicting activity coefficients. Derr and Deal (1969) expanded this approach by developing their analytical solutions of groups (ASOG) method, followed by Ronc and Ratcliff (1971). Subsequent developments are summarized in the next section.

7.5 PREDICTIONS OF PARTITION COEFFICIENTS AND SOLUBILITY USING THE UNIQUAC AND UNIFAC MODELS

7.5.1 The Principles of the UNIQUAC Model

Following Guggenheim's (1952) quasi-chemical theory of liquid mixtures, Abrams and Prausnitz (1975) proposed the UNIQUAC model. UNIQUAC is an acronym for universal quasi-chemical equation or model. This model is used to correlate and predict liquid–vapor and liquid–liquid equilibria, that is, solution properties such as partition coefficient, chromatographic retention capacity factor, Henry's law constant, or solubility of nonelectrolyte molecules (Anderson and Prausnitz, 1978a, b). Since the activity coefficient of the dissolved solute is related to each of these quantities, as we have seen in various parts of this book, its prediction is a major goal. The activity coefficient predicted or correlated by the UNIQUAC model is based on the Raoult's law mole fraction scale according to Eq. 3.21, the standard state being the pure liquid solute, supercooled, if necessary. This choice of standard state imposes on the model a requirement for prior knowledge of certain properties of the pure liquid components. In this way the interactions and properties of the molecules in the standard state are taken into account.

In the UNIQUAC model, the expression for the activity coefficient γ_h of a component substance h in a liquid mixture is composed of two parts (see Eq. 2.118), thus,

$$\ln \gamma_h = \ln \gamma_h^C + \ln \gamma_h^R \quad (\text{or } \gamma = \gamma_h^C \gamma_h^R) \tag{7.54}$$

γ_h^C represents the combinational part, which reflects the excess entropy of mixing due to differences in geometry (i.e., size and shape) of the molecules in the mixture. γ_h^R represents the residual contribution, which reflects the excess enthalpy of mixing due to the energies of mutual interaction between the molecules. This distinction is necessary because the geometric, (i.e., entropic) factors are not accounted for by the interactions (i.e., the enthalpic factors).

The combinational contribution is given by

$$\ln \gamma_h^C = \ln (\phi_h/x_h) + (z/2) q_h \ln (\theta_h/\phi_h) + l_h - (\phi_h/x_h) \sum^j x_j l_j \qquad (7.55)$$

where x_h is the mole fraction of h in the mixture,

$$l_h = (z/2)(r_h - q_h) - (r_h - 1) \qquad (7.56)$$

z is the number of nearest neighbors in the solution at the molecular level, almost invariably taken to be 10 (see Section 4.3.2)

θ_h is the molecular surface area fraction

$$\theta_h = q_h x_h \bigg/ \sum^j q_j x_j \qquad (7.57)$$

and ϕ_h is the molecular volume fraction

$$\phi_h = r_h x_h \bigg/ \sum^j r_j x_j \qquad (7.58)$$

The summations referring to j (and i overleaf) are over all the components, $1, 2, 3$, up to M, where M is the total number of components in the mixture. The quantities q_h and r_h represent the van der Waals surface and the van der Waals volume, respectively, of the molecules of h in the pure liquid h. Values of q_h and r_h may be calculated from the group contributions to the van der Waals surface area and van der Waals volume, respectively (Bondi, 1964, 1968; Fredenslund et al., 1977a, b; Nyburg and Faerman, 1985), or may be found in the literature (e.g., Abrams and Prausnitz, 1975).

The residual contribution to $\ln \gamma_h$ is given by

$$\ln \gamma_h^R = q_h \left[1 - \ln \left(\sum^j \theta_j \tau_{jh} \right) - \sum^j (\theta_j \tau_{hj} / \sum \theta_i \tau_{ij}) \right] \qquad (7.59)$$

where

$$\tau_{jh} \equiv \exp \left[-\frac{u_{jh} - u_{hh}}{RT} \right], \qquad \tau_{hj} \equiv \exp \left[-\frac{u_{hj} - u_{jj}}{RT} \right] \qquad (7.60)$$

Table 7.10 Parameters of the UNIQUAC Model

Type of Mixture	Designation of Molecules	Structural Parameters (Based on Geometry)	Interaction Parameters (Based on Energy)
Binary	1 and 2	r_1, r_2, q_1, q_2	τ_{12}, τ_{21}
Ternary	1, 2, and 3	r_1, r_2, r_3	τ_{12}, τ_{21}
		q_1, q_2, q_3	τ_{23}, τ_{32}
			τ_{31}, τ_{13}

which are referred to as interaction parameters. The terms u_{hj} and u_{jh} are pair-potential energies between j and h molecules and vice versa, such that $u_{jh} = u_{hj}$ (see Fig. 2.16). On the other hand, Eqs. 7.60 show that $\tau_{jh} \neq \tau_{hj}$. The summations that refer to jh are over all interacting pairs of h molecules with other molecules. The summations referring to ij are over all the possible interacting pairs of molecules.

Thus, each pair of neighboring molecules j and h interact with each other and the energies of mutual interaction u_{jh}, u_{jj}, and u_{hh} are converted to two interaction parameters τ_{jh} and τ_{hj}, which enter into the residual part of the UNIQUAC equation. Table 7.10 shows that binary mixtures of molecules 1 and 2 are characterized by one pair of complementary interaction parameters (τ_{12}, τ_{21}), whereas ternary mixtures are characterized by three pairs of complimentary interaction parameters, making a total of six parameters. Each pair of interaction parameters for a ternary system is derived from experimental information on the constituent binary systems and vice versa.

The UNIQUAC approach involves two successive correlation steps. In the first step, nonlinear regression analysis is employed to obtain those estimates of UNIQUAC parameters which give the best fit to the experimental quantity, such as the activity coefficient of a substance in a liquid mixture or solution. In the second step these parameters are used to estimate or predict a second experimental quantity such as the activity coefficient of the substance in another liquid mixture or solution. In many cases the UNIQUAC parameters have since been tabulated, thereby eliminating the first step.

7.5.2 The Principles of the UNIFAC Model

Fredensulund et al. (1975) extended the UNIQUAC model, which applies essentially to the compounds themselves (i.e., complete molecules), to provide a new group-contribution method for predicting the activity coefficients in liquid mixtures of nonelectrolytes. This new model is termed UNIFAC and combines the concept that a liquid mixture is composed of a solution of functional groups (Wilson and Deal, 1962) with an extension of the UNIQUAC theory of liquid mixtures of Abrams and Prausnitz (1975). The resulting model contains two adjustable parameters per pair of functional groups. The term UNIFAC is an acronym for UNIQUAC functional-group activity coefficients. This method also proceeds in two steps. In the first, the experimental activity coefficient data are reduced by regression to obtain UNIFAC parameters characterizing interactions between pairs of functional (or structural) groups in nonelectrolyte systems. Gas–liquid chromatographic data may also be used to derive UNIFAC parameters (Zarkarian et al., 1979). In the second step, the activity coefficients in other nonelectrolyte systems are predicted using the parameters appropriate to the various functional groups The UNIFAC interaction parameters have since been determined for a number of groups, thereby eliminating the first step.

In the UNIFAC system, as earlier in this chapter, a group is any convenient structural unit, such as $-CH_2-$, $-CH_3$, $-COCH_2-$, $-CH_2Cl$ (Fredenslund et al., 1975, 1977a, b). Following Wilson and Deal (1962), the fundamental assumptions of the solution-of-groups methods are stated below.

The first assumption is the underlying equation (Eq. 7.54) for the UNIQUAC method. The activity coefficient γ_i of a compound i in a liquid mixture is similarly considered to be composed of a combinational part γ_i^C and a residual part γ_i^R, each of which has the same significance as in Eq. 7.54.

The second assumption is that the residual part is equal to the sum of the individual contributions of each solute group in the solution minus the sum of the individual contributions in the pure component environment, thus,

$$\ln \gamma_i^R = \sum^k v_k^{(i)}[\ln \Gamma_k - \ln \Gamma_k^{(i)}] \qquad (7.61)$$

The summation is over all the groups of total number N in the molecule of i. As before, k represents a group of one kind, 1, 2, 3, up to N, while v_k is the number of groups of kind k in the molecule of i. Γ_k is the residual activity coefficient of group k in the solution, whereas $\Gamma_k^{(i)}$ is the residual activity coefficient of group k in a reference solution containing only molecules of type i. The term $\ln \Gamma_k^{(i)}$ in Eq. 7.61 is necessary for normalization to ensure that $\gamma_i \to 1$ as $x_i \to 1$. The activity coefficient of group k in molecule i depends on the molecule i in which k is situated. For example, $\Gamma_k^{(i)}$ for the CH_2OH group (represented as COH) in 1-butanol ($CH_3 \cdot CH_2 \cdot CH_2 \cdot CH_2OH$) refers to a solution containing 0.25 group mole fraction COH, 0.50 group mole fraction CH_2, and 0.25 group mole fraction CH_3 (Fredenslund et al., 1975).

In the UNIFAC model the second assumption of the solution-of-groups concept is extended. The residual activity coefficient Γ_k of any group of kind k is calculated from the following expression, which is similar to Eq. 7.59:

$$\ln \Gamma_k = Q_k \left[1 - \ln \left(\sum^n \Theta_m \Psi_{mk} \right) - \sum^m \left(\Theta_m \Psi_{km} \Big/ \sum^n \Theta_n \Psi_{nm} \right) \right] \qquad (7.62)$$

An analogous equation also holds for $\ln \Gamma_k^{(i)}$. In Eq. 7.62, the group-surface area fraction of group m in the mixture is

$$\theta_m = Q_m X_m \Big/ \sum^n Q_n X_n \qquad (7.63)$$

This equation is analogous to Eq. 7.57. In Eqs. 7.62 and 7.63 the mole fraction of group m in the mixture is

$$X_m = \sum^j v_m^{(j)} x_j \bigg/ \sum^j \sum^n v_n^{(j)} x_j \tag{7.64}$$

In Eqs. 7.62–7.64 the summations are over all the different groups in the mixture, this is, m and $n = 1, 2, 3$, up to N, which represents the total number of groups.

The quantity Ψ_{mn} in Eq. 7.62 allows for energetic interactions between the groups and is given by

$$\Psi_{mn} = \exp\left[-(U_{mn} - U_{nn})/RT\right] = \exp\left[-(a_{mn}/T)\right] \tag{7.65}$$

where, U_{mn} reflects the energy of interaction between groups m and n, while a_{mn}, which is termed the group-interaction parameter, is a measure of the *difference* in the energy of interaction between a group n with a group m and between two groups n. Equation 7.65 is analogous to Eq. 7.60 of UNIQUAC. We note that while $U_{mn} = U_{nm}$,

$$a_{mn}(= U_{mn} - U_{nn}) \neq a_{nm}(= U_{nm} - U_{mm}) \tag{7.66}$$

Thus, there are two interaction parameters a_{mn} and a_{nm} for each pair (i.e., binary mixture) of groups. Group-interaction parameters have the dimensions of absolute temperature, but are assumed to be independent of the temperature of the system. We have seen that Eqs. 7.62, 7.63, and 7.65 for each functional group of UNIFAC are analogous to Eqs. 7.59, 7.57, and 7.60, respectively, for the complete molecules of UNIQUAC. We also note that $\ln \Gamma_k$ in Eq. 7.62 (or $\ln \gamma_h^R$ in Eq. 7.59) depends on both the interactions between the groups (or molecules) and on the surface areas of the groups (or molecules).

The third assumption of the solution-of-groups models is that individual group contributions in any environment containing groups of kinds 1, 2, 3, up to N are only a function F of the mole fraction of each group (X_1, X_2, etc.) and the absolute temperature. The UNIFAC model fulfills this assumption through Θ_m in Eq. 7.63 and Ψ_{mn} in Eq. 7.65. According to this assumption, the residual activity coefficients for all mixtures of the same groups (e.g., ester + alkane mixtures), may be calculated from the same function F. For example, the same parameters are used to represent vapor–liquid equilibria in ethyl acetate + hexane mixtures as in decane + methyl tetradecanoate mixtures.

In the UNIFAC method, the combinatorial activity coefficient of each component is calculated in exactly the same way as in the UNIQUAC model. For this purpose, Eqs. 7.55–7.58 are used with the replacement of the subscript h, which is used to represent a molecule in UNIQUAC, by the subscript i, which is used to represent a molecule in UNIFAC. Furthermore, the molecular surface-area fraction θ_h is replaced by the group surface-area fraction Θ_m. Values of r_i

and q_i in the UNIFAC equivalent of Eqs. 7.57 and 7.58, respectively, are calculated as follows:

$$q_i = \sum^k v_k^{(i)} Q_k \qquad (7.67)$$

$$r_i = \sum^k v_k^{(i)} R_k \qquad (7.68)$$

where $v_k^{(i)}$ is the number of groups of type k in one molecule of i, and Q_k and R_k are the group surface-area parameter and the group volume parameter, respectively, which are calculated from Bondi's (1964, 1968) tabulated values of the van der Waals group surface-area parameter and group volume parameter A_{wk} and V_{wk}, respectively, using the following equations:

$$Q_k = A_{wk}/2.5 \times 10^9 \qquad (7.69)$$

$$R_k = V_{wk}/15.17 \qquad (7.70)$$

The constants 2.5×10^9 and 15.17 are normalization factors (Abrams and Prausnitz, 1975).

We note that γ_i^R depends on temperature, whereas γ_i^C does not. Furthermore, $\ln \gamma_i^C$ is often small numerically, but is far from negligible when the size and shape of the molecules differ considerably. Usually, but not invariably, $\ln \gamma_i^R$ is much larger than $\ln \gamma_i^C$. For long-chain molecules, q_i/r_i tends toward a constant limiting value and Eq. 7.55 reduces to a form similar to the Flory–Huggins equation (Eq. 2.116).

7.5.3 General Applications of the UNIFAC model

In the calculation of $\ln \gamma_i^R$, the group-interaction parameters are calculated from experimental data on binary liquid–vapor equilibria (e.g., Gmehling and Onken, 1977), including gas–liquid chromatographic data (Zarkarian et al., 1979). Each pair of interaction parameters a_{mn} and a_{nm} for a given pair of groups are constants that can be tabulated, along with the surface-area parameters Q_k and volume parameters R_k of the individual groups. Such values are, of course, necessary and sufficient with the UNIFAC equations to enable predictions to be made on the other binary systems and on multiple-component systems. For this purpose, tables of UNIFAC parameters for various functional groups are available (Fredenslund et al., 1977a, b; Skjold–Jørgensen et al., 1979; Magnussen et al., 1981; Gmehling et al., 1982) and have been revised and extended from time'to time. Tables 7.11 and 7.12 list some UNIFAC parameters for a few selected functional (or structural) groups. As described in Section 7.5.2, the appropriate UNIFAC parameters are inserted into the UNIFAC equations to predict activity coefficients and hence (a) solubilities (Eq. 2.5), (b) partition coefficients (Eq. 3.3), (c) chromatographic retention capacity factors (Eq. 7.39a),

Table 7.11 Some Examples of UNIFAC Group Surface-Area Parameters Q_k and Group Volume Parameters $R_k{}^a$

Main Group[b]	Subgroup[c]	Q_k	R_k	Sample Group Assignments[d]
1. CH_2	1. CH_3	0.848	0.9011	Butane: 2 CH_3, 2 CH_2
	2. CH_2	0.540	0.6744	Butane: 2 CH_3, 2 CH_2
	3. CH	0.228	0.4469	2-Methylpropane: 3 CH_3, 1 CH
	4. C	0.000	0.2195	2, 2-Dimethylpropane: 4 CH_3, 1 C
3. ACH[e]	9. ACH[e]	0.400	0.5313	Benzene: 6 ACH
	10. AC	0.120	0.3652	Styrene: 1 CH_2=CH, 5 ACH, 1 AC
4. $ACCH_2$[e]	11. $ACCH_3$[e]	0.968	1.2663	Toluene: 5 ACH, 1 $ACCH_3$
	12. $ACCH_2$	0.660	1.0396	Ethylbenzene: 1 CH_3, 5 ACH, 1 $ACCH_2$
	13. ACCH	0.348	0.8121	Cumene: 2 CH_3, 5 ACH, 1 ACCH
8. (Phenolic)	21. ACOH	0.680	0.8952	Phenol: 5 ACH, 1 ACOH
11. COOC	25. CH_3COO	1.728	1.9031	Butyl acetate: 1 CH_3, 3 CH_2, 1 CH_3COO
	26. CH_2COO	1.420	1.6764	Butyl propanoate: 2 CH_3, 3 CH_2, 1 CH_2COO

[a] From Fredenslund et al. (1977b).

[b] Because of differences in shape and size, various subgroups within each main group are distinguished.

[c] It is necessary to distinguish between the subgroups within each main group for the purposes of this table, but not for the group interaction parameters in Table 7.12.

[d] Other examples: 4-hexylresorcinol = 4-hexylbenzene-1, 3-diol: 1 CH_3, 4 CH_2, 1 $ACCH_2$, 2 ACOH, 3 ACH; hexane: 2 CH_3, 4 CH_2; ethyl accetate: 1 CH_3COO, 1 CH_2, 1 CH_3; ethyl myristate = ethyl tetradecanoate: 1 CH_2COO, 12 CH_2, 2 CH_3.

[e] AC indicates aromatic (benzenoid) carbon atom to which another group must be attached, for example, CH_2 in $ACCH_2$.

Table 7.12 Some Examples of UNIFAC Group Interaction Parameters[a] a_{mn} (K)

Main Groups m[b]	Main Groups n[b]				
	1 CH_2	3 ACH	4 $ACCH_2$	8 ACOH	11 COOC
1 CH_2	0	61.13	76.50	2789	232.1
3 ACH	−11.12	0	167.0	1397	5.994
4 $ACCH_2$	−69.70	−146.8	0	726.3	5688
8 ACOH	311.0	2043	6245	0	−713.2
11 COOC	114.8	85.84	−170	853.6	0

[a] Examples: $a_{ACOH, COOC}$ = −713.2 K; $a_{COOC, ACOH}$ = 853.6 K. Note that $a_{mn} \neq a_{nm}$. The rows refer to group m; the columns refer to group n; from Fredenslund et al., 1977b; Skjold-Jørgensen et al., 1979).

[b] Within each main group the various subgroups, some of which are shown in Table 7.11, possess the group-interaction parameters characteristic of the main group.

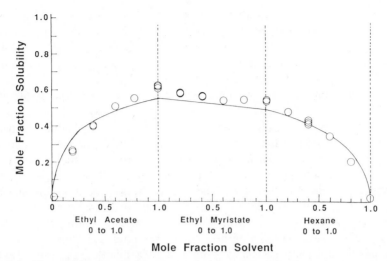

Figure 7.9 The mole fraction solubility of hexylresorcinol at 25.6°C in the three possible binary mixtures of ethyl acetate, ethyl myristate, and hexane. The open circles represent the experimentally determined solubilities. The full curve represents the predictions from UNIFAC. The dotted line represents the single-component solvents. Reproduced with permission of the copyright owner, The American Pharmaceutical Association, from Ochsner, A. B. and Sokoloski, T. D. (1985). *J. Pharm. Sci.*, **74**, 634–637.

(d) Henry's law constants (Eq. 3.35), and (e) other properties in which liquid–vapor and liquid–liquid phase equilibria are involved.

The UNIFAC method in its original or suitably modified form may also be employed directly to correlate or predict (a) Gibbs free energy and vapor pressure of pure components (Fredenslund and Rasmussen, 1979; Yair and Fredenslund, 1983), (b) liquid–vapor equilibria and Henry's law constants (Antunes and Tassios, 1983; Gupta and Daubert, 1986), and (c) liquid–liquid equilibria and extraction (e.g., of aromatics, Mukhopadhyay and Dongaonkar, 1983; Mukhopadhyay and Pathak, 1986).

7.5.4 Prediction of Solubilities Using the UNIFAC Model

For the prediction of the solubilities of solids, the activity (i.e., ideal mole fraction solubility in this case) of the solid may be calculated from the enthalpy (or entropy) of fusion and the melting point using Eq. 2.14, 2.15, or 2.16 (e.g., Ochsner and Sokoloski, 1985). If the melting point appreciably exceeds the temperature of prediction, the heat capacity difference should be introduced using Eq. 2.29.

Tables 7.11 and 7.12 list examples of UNIFAC parameters suitable for predicting the activity coefficients (and hence the solubilities) of alkylphenols in alkanes, in aliphatic esters, and in their binary mixtures. Using these values, Ochsner and Sokoloski (1985) applied UNIFAC to predict the actual (i.e., nonideal) solubility of 4-hexylresorcinol (4-hexyl-1, 3-benzenediol, of ideal mole fraction solubility 0.383 at 25.6°C) in ethyl acetate, ethyl myristate, hexane, and their binary liquid mixtures at 25.6°C. As shown in Figure 7.9, the predicted solubilities are within 10% of the experimental solubilities in 18 out of 21 of the solutions. This example shows that the UNIFAC method can account for positive and negative deviations from Raoult's law in hydrogen-bonding systems of molecules having different sizes.

For further information, we refer the reader to Fredenslund et al. (1975, 1977a, b) and Acree (1984), who have shown how the UNIQUAC and UNIFAC models are applied in practice and have provided some helpful numerical examples. Prediction of the activity coefficient of one of the components in a binary liquid mixture (e.g., acetone + pentane or ethyl acetate + hexane) of known composition (i.e., mole fraction) takes up about two pages of purely numerical calculation using a hand-held calculator; the accuracy of prediction is about 10-20%. Prediction of the activity coefficient of a component in a tertiary mixture (e.g., 4-hexylresorcinol in ethyl acetate + hexane) requires many pages of purely numerical calculation. For these and more complex systems, a computer is highly desirable, for which excellent programs are available (e.g., Prausnitz et al., 1980).

Chapter VIII

Solubility in and Partitioning into Water

8.1 INTRODUCTION

Water is the most common solvent. It is cheap and readily available and, because the human body and other living organisms contain an essentially aqueous environment, is of overriding importance in biological systems. Consequently, knowledge of aqueous solubilities is necessary in chemical engineering, chemistry, and the biological, pharmaceutical, and medical sciences, as for example in the design and development of a new drug (usually formulated in water for use in parenteral form). The environmental impact of many pollutants relates directly to their solubility in water. Many organic substances exhibit extremely high values of thermodynamic activity in water, that is, escaping tendency from water, a factor of major importance again from the biological and environmental standpoints.

In this chapter we discuss the factors that influence solubility of organic compounds in water and the methods by which the low aqueous solubilities of sparingly soluble substances may be determined. Procedures for predicting aqueous solubility will be described and illustrated with examples. The group-contribution approach described in the previous chapter will be utilized where possible, because it has proved to be one of the simplest methods for correlating and predicting the complex physical quantities to which this chapter is devoted.

This chapter, like previous chapters, considers only nonelectrolytes as the primary solutes. However, water is an ionizing solvent, as explained in Section 2.11.2, and can itself dissociate into H^+ and OH^- ions. Consequently, acidic and basic solutes can ionize in an aqueous environment to an extent which depends on the pK_a of the solute and the pH of the solution. Since the salt which is formed upon ionization often has a much higher aqueous solubility than the unionized solute, the apparent aqueous solubility of the solute will be governed by the intrinsic solubility of the unionized form and the fraction ionized, which is itself an exponential function of the difference between pH and pK_a (Eqs. 11.38 and 11.40). Detailed discussion of this effect is postponed until Chapter 11 (Section 11.3) since it plays a particularly important role in influencing the dissolution rates of weak acids and weak bases.

8.2 THE "HYDROPHOBIC" INTERACTION

8.2.1 Survey

The aqueous solubilities are very low for the hydrocarbons, for the higher members of homologous series of alcohols, amines, carboxylic acids and amides, and for the noble and diatomic gases. Many drug substances and other organic compounds also have low aqueous solubilities. The reciprocal relationship between activity coefficient and solubility represented by Eq. 2.5 indicates that the activity coefficient is very high. If the Raoult's law mole-fraction-based standard state is used, for which the activity is unity for the pure liquid solute, the molar excess free energy of mixing, which is given by

$$\Delta G^{E} = RT \ln \gamma_2^{\infty} \tag{8.1}$$

is strongly positive, indicating an extraordinary lack of affinity (defined in terms of free energy) of the compounds mentioned above for water.

Hartley (1936) commented on the lack of solubility of paraffin chains in water and noted "a strong attraction of water molecules for one another in comparison with which the paraffin–paraffin and paraffin–water attractions are slight." The tendency for hydrocarbon chains and other hydrophobic groups to come together so as to minimize contact with water molecules has been variously named "the hydrophobic interaction," "the hydrophobic bond" (a misnomer, since no chemical bond is formed (Hildebrand, 1968; Némethy et al., 1968)), "the hydrophobic effect" (Tanford, 1973), and "the entropic union" (Lauffer, 1966).

This effect includes not only (1) the lack of solubility of hydrocarbons and other nonpolar substances in water, but also such phenomena in aqueous media as (2) the adsorption of nonpolar molecules onto nonpolar surfaces, (3) micelle formation by surfactants, (4) the solubilization of drug molecules by micelles, (5) drug–protein binding, (6) drug–macromolecule interactions, and (7) drug–drug interactions.

The enthalpy of solution, ΔH_s^{θ}, of such nonpolar solutes in water is usually a small positive (or even negative) quantity at room temperature. Since the solubilities are very low, the free energy of solution, ΔG_s^{θ}, is positive when the standard state of the solute in solution is hypothetical ideal 1-mol dm^{-3} solution. Since $\Delta G_s^{\theta} = \Delta H_s^{\theta} - T \Delta S_s^{\theta}$, the standard entropy of solution ΔS_s^{θ} is negative. Since the solubility of nonpolar solutes in water is very low, the changes of thermodynamic quantities on forming the solution can be attributed almost entirely to the corresponding partial molar quantities of the solute. Thus the partial molar enthalpy of solution of the solute, $\Delta \bar{H}_2$, is a small positive (or negative) quantity, whereas the partial molar entropy of solution of the solute, $\Delta \bar{S}_2$, is large and negative.

Such systems cannot satisfactorily be treated by regular solution theory (or solubility parameter theory) since this treatment requires that the entropy of solution be ideal, and therefore positive. The enthalpy change also does not always accord with the theory and is considered in the next two paragraphs. The unfavorable entropy change, which is actually much more important, is discussed in Sections 8.2.2–8.2.4.

The surface tension σ of a liquid is a direct two-dimensional measure of the intermolecular forces (Shinoda, 1977). For saturated aliphatic hydrocarbons for which the intermolecular forces are purely London dispersion forces, $\sigma_2 = 20-$

$26 \, \text{mN} \, \text{m}^{-1}$ (or $\text{mJ} \, \text{m}^{-2}$). For water at $20°C$, $\sigma_1 = 72.8 \, \text{mN} \, \text{m}^{-2}$, which represents the sum of the contributions from the dispersion forces and from hydrogen bonding (and dipole) interactions. The interfacial tension σ_{12} between water and saturated hydrocarbons is about $50\text{--}52 \, \text{mN} \, \text{m}^{-1}$. The following equation, derived by Fowkes (1963),

$$\sigma_{12} = \sigma_1 + \sigma_2 - 2(\sigma_1^d \sigma_{12})^{1/2} \tag{8.2}$$

relates the σ values to σ_1^d, which represents the contribution of dispersion forces to the surface tension of water. Introducing the known numerical values yields $\sigma_1^d = 21.8 \pm 0.7 \, \text{mN} \, \text{m}^{-1}$ (or $\text{mJ} \, \text{m}^{-2}$), which is close to the value of σ_2 given above. Thus, the dispersion energy per unit surface area of water is close to that of hydrocarbon. The difference $(\sigma_1 - \sigma_1^d) = 51 \, \text{mN} \, \text{m}^{-2}$ is the contribution of hydrogen bonding (and dipole) interactions to the surface tension of water and is close to the value for the interfacial tension σ_{12}. Hence, the enthalpy of mixing of hydrocarbons with water is mostly attributable to changes in the hydrogen-bonding interaction in the water surrounding the solute.

We note that the interactions between hydrocarbon molecules (represented by σ_2) are weaker than those between hydrocarbon molecules and water (represented by σ_{12}), which are themselves weaker than those between water molecules (represented by σ_2). This suggests that, when applied to hydrocarbons, the terms lipophilic and hydrophobic are actually misnomers, because hydrocarbons interact more weakly with themselves than with water. The low solubility of hydrocarbons in water is evidently not a result of the weakness of the hydrocarbon–water interactions, but rather a result of the comparative strength of the water–water interactions which resist the intervention of hydrocarbon molecules.

8.2.2 Theories of the Hydrophobic Interaction

A negative value of ΔS_s^θ means that a nonpolar solute–water system becomes more ordered as the solute goes into solution. ΔS_s^θ in water is about $12\text{--}22 \, \text{kcal} \, \text{mol}^{-1}$ (or $50\text{--}90 \, \text{kJ} \, \text{mol}^{-1}$) less than in other solvents. This result has been interpreted according to various theories or points of view. A survey of these theories reveals two quite distinct concepts. The entropy decrease and the hydrophobic interaction could arise either (a) from the formation of a greater degree of structure among the water molecules in the immediate vicinity of the hydrocarbon chains or nonpolar groups, or (b) from the restriction of the internal degrees of freedom (i.e., motion) of the hydrocarbon chains or nonpolar groups by the surrounding network of hydrogen-bonded water molecules. Both points of view lead to the conclusion that the negative ΔS_s and the hydrophobic interaction will increase with increasing volume or surface area of the cavity created in the solvent by the hydrocarbon or nonpolar group. We shall return to this point later.

In most of the literature and at times in this book, hydrophobic interactions are attributed to an increase in the degree of structure of water. However, these interactions may well be explained by the restricted motion of the solute molecule in its cavity in water.

8.2.3 "More Structured Water" Theory

The first point of view envisages more ordered or more structured water around the solute molecules. Frank and Evans (1945) postulated that water molecules surrounding solute molecules form frozen patches or microscopic "icebergs." The concept of structured water was developed by Frank and Wen (1957), Klotz (1958), Kauzmann (1959), and Némethy and Scheraga (1962a–c). The underlying notion is that bulk water is believed to retain some of the properties of ice. Icebergs, or ice-like species, in which each water molecule is bonded tetrahedrally to four neighboring water molecules, are in equilibrium with monomeric water and with species of intermediate hydrogen-bonding states and of intermediate size. When a hydrocarbon or another nonpolar molecule or group is placed in this medium, the mean lifetime of the structured species or icebergs is increased. Thus, the degree of structuring of water is increased and the entropy of the system is decreased. Shinoda (1977) showed how the thermodynamic consequences of iceberg formation differ from those of regular solution theory.

Frank and Wen (1957) postulated flickering clusters of water molecules, but Stillinger (1976) considers that a bimodal distribution of species, "small" and "large," is more appropriate. The overall view is that liquid water consists of a random hydrogen-bonded network with a number of strained and broken bonds which can reform while other hydrogen bonds strain and break. It is, therefore, not surprising that the presence of nonpolar solutes will alter the statistical pattern of hydrogen-bonding in favor of more ordered structures. Goldammer and Hertz (1970), using nuclear magnetic relaxation measurements, have adduced evidence for a reduction in mobility of water molecules close to apolar residues.

Hermann (1971) has suggested that the negative ΔS_s results from the tendency of water dipoles in the layer of water adjacent to the hydrocarbon molecule to align themselves in an asymmetric electric field, which causes them to experience more restricted motion. The asymmetric electric field is similar to that responsible for the surface tension of water and is not found in the bulk liquid. The partition function in terms of the significant structure theory accounts well for the thermodynamic properties of aqueous solutions of hydrocarbons, except for the fact that the calculated heat capacity is too low.

8.2.4 "Restricted Solute Motion" Theory

The second point of view considers that the hydrophobic effect and the negative ΔS_s^θ arise from the restriction of the internal motion of the nonpolar groups by the surrounding network of water molecules. This suggestion was originally made by Aranow and Witten (1960). They recognized the special properties of water and that increased water structuring can occur upon insertion of an apolar group in this solvent. However, they attributed the negative ΔS_s^θ to the changes in the conformation of the hydrocarbon. The following evidence favors this theory over the previous one.

1. The entropy change due to the restriction of the internal conformational degrees of freedom of a liquid hydrocarbon is calculated to be $R \ln 3$ per methylene group on solidification, which is the same as the experimental value on transfer for a liquid hydrocarbon to water (Aranow and Witten, 1960).

2. The formation of ordinary ice from water is accompanied by an increase in molar volume of $1.6 \, cm^3 \, mol^{-1}$, whereas Masterton (1954) measured the following decreases in the partial molar volume $(\Delta V / cm^3 \, mol^{-1})$ on the transfer of the following hydrocarbons to water: -22.7 for methane (in hexane); -18.1 for ethane (in hexane); -6.2 for benzene (in the pure liquid). These decreases cannot be due to structuring in water, but reflect the restriction of the motion of the hydrocarbon chains. The finding by Florence (1966) accords with this, namely, that the partial molar volume of the monomer of nonionic surfactants increases with micelle formation in water due to the withdrawal of the hydrocarbon residues into the nonpolar interior of the micelle.

3. Howarth (1975) measured the ^{13}C spin-lattice relaxation time (T_1) of suitable groups in a number of solutes in various solvents, including water. The data indicate that water is unique in its ability to restrict the motion of hydrophobic parts of solute molecules. The motional restriction of the solute and solvent molecules was shown to account for all of the entropy change in solution which had previously been ascribed to structure-making, namely the $\Delta \bar{S}_2$ anomaly of 12–$22 \, kcal \, K^{-1} \, mol^{-1}$ or $(50$–$90 \, J \, K^{-1} \, mol^{-1})$ mentioned above. The contribution from the solute is the greater. No assumptions of structure are necessary other than that of reasonably close packing of solvent molecules around solute molecules.

The insertion of a solute molecule into any solvent requires the formation of a suitable cavity, which is then filled by the solute molecule. Water is unique in possessing zero enthalpy of cavity formation (Eley, 1939; Howarth, 1975), presumably because water can be regarded as an imperfect crystal with many defects that can be enlarged to form cavities with very little energy expenditure.

When any solute molecule is inserted into water, an increase in van der Waals interactions will make the process energetically favorable, but the additional interactions and cavity restrictions will reduce the molecular motion at or near the interface between solvent and solute, thereby reducing ΔS^{θ}. The tumbling and other motions of both solute and solvent will be reduced. This conclusion is supported by the evidence from ^{13}C relaxation times and from the correlation of the partial molar entropy of solution of solutes from the gas phase into water with the area of the solvent shell surrounding the solute (Miller and Hildebrand, 1968; Hermann, 1972).

If the enthalpy of the cavity formation is really zero, the partial molar enthalpy of solution, $\Delta \bar{H}_2$, of solutes from the gas phase into water should be proportional to the total van der Waals attractive potential of each solute molecule, which is directly proportional to the square root of the molar enthalpy of vaporization at

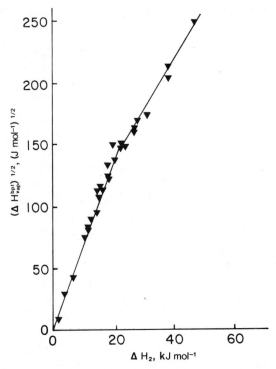

Figure 8.1   Plot of the partial molar enthalpy of solution in water ΔH_2 of gaseous solutes at 298 K against the enthalpy of vaporization at the boiling point ΔH_{vap}^{bpt}. Solutes in order of increasing ΔH_{vap}^{bpt} are He, H_2, Ne, N_2, Ar, O_2, CH_4, Kr, NF_3, CF_4, Xe, C_2H_4, C_2H_6, CO_2, CH_3F, C_3H_8, CH_3Cl, n-C_4H_{10}, SF_6, CH_3Br, n-C_5H_{12}, CH_3I, n-C_6H_{14}, benzene, I_2, naphthalene, and biphenyl. Reproduced with permission of the copyright owner, the Royal Society of Chemistry, from Howarth, O.W. (1975). *J. Chem. Soc. Faraday I*, **71**, 2303–2309.

the boiling point, ΔH_{vap}^{bpt} (Howarth, 1975). This is illustrated by Figure 8.1. Since the line passes through the origin, the enthalpy of cavity formation is either zero for all sizes of solute molecule, or it depends linearly on $\Delta \bar{H}_2$. The former is the simpler assumption. The change in gradient above $\Delta \bar{H}_2 = 20\,\text{kJ}\,\text{mol}^{-1}$ can be attributed to solute–solute self-association in water, corresponding to the hydrophobic effect, owing to the increasing size and polarizability of the solute molecules (Howarth, 1975). In general, the solute–solute interactions in these self-associated species are nonspecific van der Waals forces and contrast sharply with the hydrogen-bonding interactions in the self-associated species considered in Chapter 6.

The large partial molar heat capacity $\Delta(C_p)_2$, representing the large temperature dependency of $\Delta \bar{H}_2$ of nonpolar solutes in water, is then merely a result of temperature expansion of the cavities. This will cause a large change in r^{-6}, to which $\Delta \bar{H}_2$ is proportional, assuming that it arises from van der Waals forces

(Howarth, 1975). It should, however, be pointed out that the temperature dependence of hydrophobic interactions has also been explained by Ben-Naim and co-workers (Ben-Naim et al., 1973; Ben-Naim, 1974, 1980; Yaacobi and Ben-Naim, 1974) using statistical thermodynamics in relation to the "more structured water" theory.

Hydrophobic interactions have been reviewed by Ben-Naim (1980) and Franks (1983). Whichever view is adopted, it is apparent that the volume and the surface area of the cavity created in the solvent by the introduction of the solute molecule may be useful quantities for interpreting and correlating solubility behavior. The size of the cavity is not the only factor, however. The thermodynamic quantities (ΔG, ΔH, ΔS) associated with the solute–solvent interaction taking place at the surface of the cavity are also of crucial importance. These thermodynamic quantities will, of course, vary from one part of the molecule to another or from group to group within the molecule. The concept of individual group contributions is not new, but is one of the most effective in correlating, interpreting, and predicting aqueous solubility and related quantities.

8.3 THE MOLECULAR AND GROUP-SURFACE AREA (MGSA) APPROACH TO SOLUBILITY PREDICTION

In Section 2.8 this method of predicting the solubility of organic compounds in polar solvents was briefly introduced. The approach was developed in the 1970s by Harris et al. (1973), Yalkowsky et al. (1972b, 1975, 1976), Amidon et al. (1974, 1975), and Valvani et al. (1976). Yalkowsky, Amidon, Valvani, and co-workers viewed the concept as an extention of regular solution theory in which the surface area A_B of the solute molecule (or group) B in contact with the solvent A replaces the molar volume, while the interfacial tension σ_{AB} replaces the cohesive energy densities of interaction. As suggested by its name, this approach also utilizes the structural concept of "group contributions" discussed in Chapter 7.

Knowledge of how σ_{AB} and A_B vary with structure could extend the application of Eq. 2.100 to nonelectrolytes in polar solvents. The Langmuir (1925) principle of independent surface action (Section 7.1) proposes that the hydrocarbon (h) and polar groups (p) can be considered to make separate, additive contributions to the surface area A_B of the solute molecule, thus,

$$A_B = A_B^p + A_B^h \tag{8.3}$$

and similarly to the microscopic interfacial free energy, thus,

$$\sigma_{AB}A_B = \sigma_{AB}^p A_B^p + \sigma_{AB}^h A_B^h \tag{8.4}$$

where A_B^p and A_B^h are the surface areas of the polar and hydrocarbon groups, and σ_{AB}^p and σ_{AB}^h are the microscopic interfacial free energies of interaction (interfacial tensions) of the respective groups with the solvent. Equation 2.100 for the activity coefficient of the solute becomes

$$\ln \gamma_B = (\sigma_{AB}^p A_B^p + \sigma_{AB}^h A_B^h)/kT \tag{8.5}$$

or

$$\log \gamma_B = (\sigma_{AB}^p A_B^p + \sigma_{AB}^h A_B^h)/2.303\,kT \tag{8.5a}$$

Since we have replaced molar areas by molecular areas, the gas constant R must be replaced by the Boltzmann constant $k = R/L$, where L is the Avogadro constant. The polar term in the previous equation is often found to be small for homologous series of drugs in water and polar solvents (Yalkowsky et al., 1976). Furthermore, the shape of the cavity occupied by a solute molecule may be irregular, and portions of it may have a high curvature. The previous equation must then be modified by introducing a curvature correction factor c which is less than unity, thus,

$$\ln \gamma_B \cong c\sigma_{AB}^h A_B^h/kT \tag{8.6}$$

$$\log \gamma_B \cong c\sigma_{AB}^h A_B^h/2.303kT \tag{8.6a}$$

The previous two equations provide a means of predicting γ_B and hence solubilities. In practice, uncertainties in estimations of c, σ, and A values require that empirical correlations be established between γ_B and A.

Hermann (1972) developed a computer program for calculating the surface area of molecules and groups from the input information of bond lengths and angles, atomic radii, and the radius of a spherical solvent molecule (1.5 Å for H_2O). Using Hermann's procedure, Amidon et al. (1975) and Valvani et al. (1976) calculated the molecular areas A and group areas A^p and A^h for members of several homologous series, including hydrocarbons and alcohols by the three methods, all of which gave good linear correlations with $\log \gamma_B$ according to Eqs. 8.5a and 8.6a with an appropriate intercept (e.g., Fig. 8.2). This indicates that Langmuir's (1925) principle of independent surface action is applicable and that activity coefficients and solubilities can be predicted from computer calculations of molecular surface area.

The method in which the molecule was considered as a collection of spherical groups (e.g., CH_3, CH_2, OH), while omitting the solvent radius, was favored over the others by its marginally better correlation and, more importantly, by its independence of the nature of the solvent. The results of this method are probably close to true estimates of the surface area of molecules in the gas phase or in any solvent and are particularly useful in predicting the solubility and the partitioning of nonpolar solutes and the effect of molecular modification.

The prediction of solubility of solids, x^s, from molecular or group-surface areas requires that $\log \gamma$ first be predicted from its correlation with the area. Since the

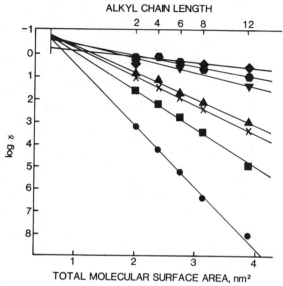

Figure 8.2 Plots of the logarithm of the mole-fraction activity coefficient γ against the total molecular surface area (and alkyl chain length) of alkyl 4-aminobenzoates in the following pure solvents: water ($\bullet$), glycerol ($\blacksquare$), formamide ($\times$), ethylene glycol ($\blacktriangle$), propylene glycol ($\blacktriangledown$), methanol ($\bullet$), and ethanol ($\blacklozenge$). Reproduced with permission of the copyright owner, the American Pharmaceutical Association, from Yalkowsky, S. H., Valvani, S. C., and Amidon, G. L. (1976). *J. Pharm. Sci.*, **65**, 1488–1494.

solubility is given by

$$\log x^s = \log a^s - \log \gamma \qquad (8.7)$$

the activity of the solid solute a^s is predicted as described for ideal and regular solutions, the standard state being the pure liquid state (Sections 2.2.1 and 2.2.2). A correlation between log (solubility) itself and molecular surface area will show a break whenever the crystalline structure changes, as occurs at C_4 (butyl) for the alkyl 4-aminobenzoate esters (Fig. 8.3).

For sparingly soluble liquid solutes the mole fraction solubility is given by $x^l = 1/\gamma$ (Eq. 2.104), and log (solubility) should correlate directly with molecular or group surface area. Amidon et al. (1975) and Valvani et al. (1976) confirmed that $-\log$ (solubility) (i.e., $\log \gamma$) of liquid aliphatic solutes in water gave good linear correlations with the surface area of the molecule or the hydrocarbon group, the gradient being close to $0.032 \, \text{Å}^2$, as shown in Figure 8.2.

The MGSA approach may be applied to polar solvents other than water. The log (solubility) and $-\log \gamma$ of solutes with a high proportion of hydrocarbon groups, such as the alkyl 4-aminobenzoate esters, tend to decrease not only

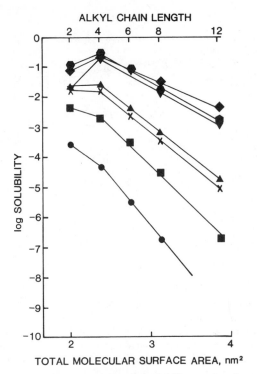

Figure 8.3 Plots of the logarithm of the mole-fraction solubility x against the total molecular surface area (and alkyl chain length) of alkyl 4-aminobenzoates in the following pure solvents: water ($\bullet$), glycerol ($\blacksquare$), formamide ($\times$), ethylene glycol ($\blacktriangle$), propylene glycol ($\blacktriangledown$), methanol ($\bullet$), and ethanol ($\blacklozenge$). Reproduced with permission of the copyright owner, the American Pharmaceutical Association, from Yalkowsky, S. H., Valvani, S. C., and Amidon, G. L. (1976). *J. Pharm. Sci.*, **65**, 1488–1494.

with increasing alkyl chain length and molecular or group-surface area of the solute, but also in the solvent rank order: ethanol, methanol, propylene glycol, formamide, glycerol, water (Figure 8.2, Yalkowsky et al., 1976). This solvent rank order is best depicted in Figure 8.4, which shows a negative correlation between $\Delta \log \gamma_B / \Delta n$ (i.e., the sensitivity of $\log \gamma_B$ to carbon number n, as discussed in Section 7.3.1) and the hydrocarbon–solvent interfacial tension σ_w^h, determined using tetradecane (Yalkowsky et al., 1975). This rank order also corresponds roughly to increasing solvent polarity as measured by P' of Snyder (1978, 1980), described in Section 2.7, by dipole moments (McClellan, 1963; Riddick and Bunger, 1970), discussed in Section 2.11.3, and by dielectric constants (Riddick and Bunger, 1970), considered in Section 2.11.2.

The slope of the line in Figure 8.4 indicates that the curvature correction c in Eq. 8.6 is about $\frac{1}{3}$, in good agreement with theoretical estimates and with the value obtained for hydrocarbons and alcohols in water. Figure 8.4 also has predictive value in relating the solubility of solutes in polar solvents to that in water.

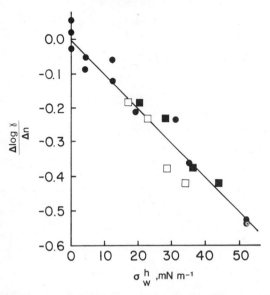

Figure 8.4   Dependence of methylene group contribution $\Delta \log \gamma / \Delta n$ to the logarithm of the activity coefficient γ in a solvent upon the solvent–tetradecane interfacial tension σ_w^h, thus: experimental interfacial tension of pure liquids ($\bullet$) and liquid mixtures ($\square$); calculated interfacial tension of liquid mixtures ($\blacksquare$). Reproduced with permission of the copyright owner, the American Pharmaceutical Association, from Yalkowsky, S. H., Amidon, G. L., Zografi, G., and Flynn, G. L. (1975). *J. Pharm. Sci.*, **64**, 48–52.

The MGSA approach also may be applied to mixed solvent systems (solvent + cosolvent), as shown in Figure 8.5. Since a good linear relationship exists between log solubility (or $\log \gamma_B$) and the volume fraction of the cosolvent, ϕ_c, as shown in Figures 8.6 and 8.7 (Yalkowsky et al., 1976), the MGSA approach suggests that the molecular surface energy parameter of a mixed solvent, σ_m, is given by

$$\sigma_m = \phi_w \sigma_w + \phi_c \sigma_c \qquad (8.8)$$

where ϕ_w and ϕ_c are the volume fractions of solvent and cosolvent, respectively, and where σ_w and σ_c are the respective molecular surface energy parameters. This equation is analogous to Eq. 2.91 for solubility parameters. This concept enables the solubility of solutes in many mixed-solvent systems to be predicted. Some systems, solute + cosolvent + solvent, however, show nonlinear dependence of the logarithm of solubility on the volume fraction of each solvent in a mixture of solvents, as discussed in Section 8.11.

As we have seen in Section 7.2.4 (Harris et al., 1973), and as we shall see later in this chapter, solubility, activity coefficients, and partition coefficients appear to correlate with molecular surface area slightly better than with molecular volume.

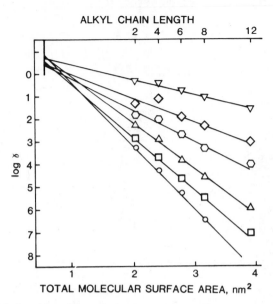

Figure 8.5 Plot of the logarithm of the mole-fraction activity coefficient γ of alkyl 4-aminobenzoates in propylene glycol–water mixtures at 37°C against the total molecular surface area (and alkyl chain length) of the solute. Mixtures: ∇, propylene glycol alone; $\diamond$, 80% propylene glycol in water; $\bigcirc$ 60% propylene glycol in water; $\triangle$, 40% propylene glycol in water; $\square$, 20% propylene glycol in water; and $\bigcirc$, water alone. Reproduced with permission of the copyright owner, the American Pharmaceutical Association, from Yalkowsky, S. H., Valvani, S. C., and Amidon, G. L. (1976). *J. Pharm. Sci.*, **65**, 1488–1494.

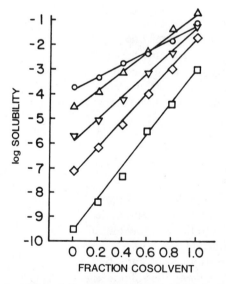

Figure 8.6 Plot of the logarithm of the mole-fraction solubility of alkyl 4-aminobenzoates in propylene glycol–water mixtures against the volume fraction of propylene glycol. Alkyl ester solutes: ethyl ($\bigcirc$), butyl ($\triangle$), hexyl (∇), octyl ($\diamond$), and dodecyl ($\square$). Reproduced with permission of the copyright owner, the American Pharmaceutical Association, from Yalkowsky, S. H., Valvani, S. C., and Amidon, G. L. (1976). *J. Pharm. Sci.*, **65**, 1488–1494.

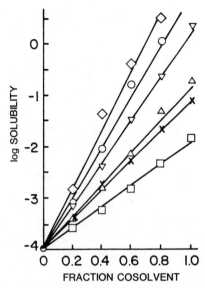

Figure 8.7 Plot of the logarithm of the mole-fraction solubility of hexyl 4-aminobenzoate in the following water–cosolvent systems: water alone (◕), glycerol–water (□), formamide–water (×), ethylene glycol–water (△), propylene glycol–water (▽), methanol–water (○), and ethanol–water (◇). Reproduced with permission of the copyright owner, the American Pharmaceutical Association, from Yalkowsky, S. H., Valvani, S. C., and Amidon, G. L. (1976). *J. Pharm. Sci.*, **65**, 1488–1494.

8.4 INFLUENCE OF CHAIN BRANCHING ON THE AQUEOUS SOLUBILITY AND PARTITION COEFFICIENT OF ALKANES

A branched alkane is usually more soluble in water than the straight-chain isomer (Hildebrand and Scott, 1950; Butler, 1962; McAuliffe, 1966). The MGSA approach (discussed above and in Section 7.2.4) predicts that the isomer with the greatest degree of branching exposes the smallest effective surface area to its aqueous surroundings and, as a result, exhibits the least unfavorable interaction energy and entropy, and is consequently the most water-soluble.

In accordance with the effect of branching on aqueous solubility, the partition coefficient (organic solvent–water) is generally found to decrease with increased branching (Kolarik and Pankova, 1966; Leo et al., 1971; Kertes, 1971). Group contributions of branching to log F are -0.20 in octanol–water (Hansch, 1971; Leo et al., 1971), -0.04 in ether–water (Flynn, 1971), and -0.14 for dialkyl-amines in various chlorinated aliphatic solvents, as calculated (Davis et al., 1974) from the data of Divatia and Biles (1961).

Detailed analysis of the data for branched-chain alkanoic acids shows that the influence of branching depends on the position of the side chain relative to the polar grouping, suggesting steric involvement (Davis et al., 1974). The decreased partition coefficient from water as a result of branching is exemplified by comparison of the $\Delta\Delta G_t^\theta$(water $\to$ 25% v/v CHCl$_3$ in CCl$_4$) values of *tert*- and *sec*-butyl in Table 7.6.

Comparison of the $\Delta\Delta G_t^\theta$(water $\to$ CCl$_4$ + CHCl$_3$) values of neopentyl and neohexyl (Table 7.6) indicates a methylene group contribution of -795 cal mol^{-1} (-3.33 kJ mol^{-1}). Since this value is quite similar to the mean value of $\Delta\Delta G_t^\theta$(CH$_2$, water $\to$ CCl$_4$ + CHCl$_3$) $= -916$ cal mol^{-1} (-3.83 kJ mol^{-1}), it is apparent that branching on nearby carbon atoms does not greatly influence the methylene group contribution. The large positive $\Delta\Delta S_t^\theta$ values for neopentyl and neohexyl are indicative of the water structuring brought about by these groups and suggest that these substituents increase ion-pair extraction primarily by an entropy-controlled process.

Again, $\Delta\Delta G_t^\theta$(water $\to$ organic solvent) of virtually any group seems to be independent of the compounds chosen, whereas $\Delta\Delta H_t^\theta$ and $\Delta\Delta S_t^\theta$ are quite sensitive to environment (Table 7.6). This is another example of enthalpy–entropy compensation (Lumry and Rajender, 1970; Tomlinson, 1983). Using 3, 3-dimethylbutyl sulphate and neooctyl sulphate to calculate the thermodynamic group contributions for the *tert*-butyl group affords $\Delta\Delta G_t^\theta$ values of -3136 and -3188 cal mol^{-1} (-13.1 and -13.3 kJ mol^{-1}), respectively, whereas the $\Delta\Delta H_t^\theta$ values are -2702 and -750 cal mol^{-1} (-11.3 and -3.14 kJ mol^{-1}), and the $\Delta\Delta S_t^\theta$ values are 1.43 and 8.04 cal K^{-1} mol^{-1} (5.98 and 33.6 J K^{-1} mol^{-1}), respectively. The bulky *tert*-butyl group may act (1) to shield the alkyl chain from water interactions and (2) to reduce the degrees of conformational freedom of the alkyl chain, as described earlier for the phenyl group (Section 7.3.3).

8.5   PREDICTION OF AQUEOUS SOLUBILITY VIA THE SUPERCOOLED LIQUID AND SOLUTION IN 1-OCTANOL

Amidon and Williams (1982) extended the treatment of Yalkowsky, Valvani, and co-workers, which is described in Section 2.9 and which requires knowledge of the partition coefficient (octanol–water) $P_{o/w}^c$ and the melting point (absolute value, T_m). By also allowing for variations in entropy of fusion, ΔS^f, molar volume, V_2, and solubility parameter, δ_2, of the solute (indicated by subscript 2), Amidon and Williams (1982) improved the predictive capability of the treatment. These workers considered the standard molar Gibbs free energy of transfer of pure solid solute to a saturated aqueous solution, $\Delta_s^{ssw} G_2^\theta$, to be made up of the following consecutive standard molar Gibbs free energy changes, based on the mole-fraction scale:

$$\text{solid} \xrightarrow{\Delta_s^{scl} G_2^\theta} \text{supercooled liquid (scl)} \tag{8.9}$$

$$\text{scl} \xrightarrow{\Delta_{scl}^{sso} G_2^\theta} \text{saturated solution in 1-octanol (sso)}, \ x_{2,o}^s \tag{8.10}$$

$$\text{sso} \xrightarrow{\Delta_{\text{sso}}^{\text{ssw}}G_2^\theta} \text{saturated solution in water (ssw), } x_{2,\text{w}}^{\text{s}} \tag{8.11}$$

Summation gives

$$\text{solid} \xrightarrow{\Delta_{\text{s}}^{\text{ssw}}G_2^\theta} \text{ssw, } x_{2,\text{w}}^{\text{s}} \tag{8.12}$$

that is

$$\Delta_{\text{s}}^{\text{ssw}}G_2^\theta = \Delta_{\text{s}}^{\text{scl}}G_2^\theta + \Delta_{\text{scl}}^{\text{sso}}G_2^\theta + \Delta_{\text{sso}}^{\text{ssw}}G_2^\theta \tag{8.13}$$

The standard molar Gibbs free-energy change in each step can now be calculated in sequence. From Eq. 2.16, which assumes that ΔC_p is constant,

$$\Delta_{\text{s}}^{\text{scl}}G_2^\theta = -RT \ln a_2 = \Delta S^{\text{f}}(T_m - T) \tag{8.14}$$

where a_2 is the thermodynamic activity of the solid 2 with respect to the pure supercooled liquid. From Eq. 2.89,

$$\Delta_{\text{scl}}^{\text{sso}}G_2^\theta = RT \ln \gamma_2 = v_2 \, \phi_1^2 (\delta_1 - \delta_2)^2 \tag{8.15}$$

where $RT \ln \gamma_2$ is the excess molar Gibbs free energy of a regular solution (with respect to an ideal solution), and the subscript 1 refers to the solvent which in this stage is 1-octanol. Now $\phi_1 \cong 1$ for dilute solutions and $\delta_1 = 10.3 \, \text{cal}^{1/2} \, \text{cm}^{-3/2}$, therefore,

$$\Delta_{\text{scl}}^{\text{sso}}G_2^\theta = V_2(10.3 - \delta_2)^2 \tag{8.16}$$

From the equation at the base of the triangle in Scheme 7.1, and expressing concentrations as mole fractions,

$$\Delta_{\text{sso}}^{\text{ssw}}G_2^\theta = -RT \ln(1/P_{\text{o/w}}^{\text{x}}) = RT \ln P_{\text{o/w}}^{\text{x}} \tag{8.17}$$

Substituting the free-energy changes of the individual steps (Eqs. 8.14, 8.16, and 8.17) in Eq. 8.13 for the total free-energy change affords

$$\Delta_{\text{s}}^{\text{ssw}}G_2^\theta = \Delta S^{\text{f}}(T_m - T) + V_2(10.3 - \delta_2)^2 + RT \ln P_{\text{o/w}}^{\text{x}} \tag{8.18}$$

According to Eq. 2.79,

$$\Delta_{\text{s}}^{\text{ssw}}G_2^\theta = -RT \ln x_{2,\text{w}}^{\text{s}} = -2.303 RT \log x_{2,\text{w}}^{\text{s}} \tag{8.19}$$

If we express ΔS^{f} in $\text{cal} \, \text{K}^{-1} \, \text{mol}^{-1}$ and δ_1 and δ_2 in $\text{cal}^{1/2} \, \text{cm}^{-3/2}$, then $R = 8.3143/4.184 = 1.9872 \, \text{cal} \, \text{K}^{-1} \, \text{mol}^{-1}$ and at $T = 298 \, \text{K}$, $1/2.303 RT = 7.33 \times 10^{-4} \, \text{mol} \, \text{cal}^{-1}$. Inserting these numerical values into Eq. 8.18 and eliminating the free-energy term from Eqs. 8.18 and 8.19 yields

$$\log x_{2,w}^s = -0.00073\Delta S^f(T_m - 298) - 0.00073V_2(10.3 - \delta_2)^2 - \log P_{o/w}^x \tag{8.20}$$

According to Eq. 2.103 for octanol–water partition coefficients,

$$\log P_{o/w}^x = \log P_{o/w}^c + 0.94 \tag{8.21}$$

According to Eq. 2.108 for solubilities in water,

$$\log c_{2,w}^s = \log x_{2,w}^s + 1.74 \tag{8.22}$$

Eliminating $P_{o/w}^x$ and $c_{2,w}^s$ from the last three equations gives the following equation of Amidon and Williams (1982):

$$\log c_{2,w}^s = 0.80 - 0.00073\Delta S^f(T_m - 298) - 0.00073V_2(10.3 - \delta_2)^2 - \log P_{o/w} \tag{8.23}$$

Numerical comparison shows that this equation with the predicted constants provides nearly as good a fit as a regression equation based on experimental data. Thus, Eq. 8.23 provides good estimates of the solubility of solids and liquids. Since the term for the solubility parameter (from Eqs. 8.15 and 8.16) makes only a small contribution, the assumption of ideal solubility of all compounds in 1-octanol ($\delta_2 = \delta_1 = 10.3\,\text{cal}^{1/2}\,\text{cm}^{-3/2}$) made by Yalkowsky and Valvani (1980) appears to be a reasonable approximation, especially when values of V_2 or δ_2 are not available.

A number of studies (e.g., Hansch et al., 1968; Yalkowski and Valvani, 1980) have led to a simple linear relationship between $\log c_{2,w}^s$ and $\log P_{o/w}$ (e.g., Eq. 2.114). Such relationships (e.g., Hansch et al., 1968) provide quite good predictions of the aqueous solubilities of organic liquids (Eq. 2.105). For solids, the semiempirical treatment of Yalkowski, Valvani, and co-workers discussed in Section 2.9 has led to Eqs. 2.114 and 2.115 in which the activity of the solid is also considered in a separate term. These equations provide good predictions of aqueous solubility. Less good predictions, within an order of magnitude, are given when the activity of the solid is omitted from the empirical equation (Eq. 2.110) linking $\log c_{2,w}^s$ and $\log P_{o/w}^c$ (e.g., Chiou et al., 1977). The resulting simplified empirical equation may, however, be satisfactory for environmental predictions of solubility and bioaccumulation of a set of similar compounds, such as polychlorinated biphenyls, organochlorines, or organophosphate pesticides.

One real advantage of this treatment of Amidon and Williams (1982), in addition to the inclusion of several properties of the solute, is that the contribution of each step of the thermodynamic treatment (Eqs. 8.9–8.12) can be assessed. This approach is useful for identifying the reasons for deviation of the predicted aqueous solubility from the experimental value. However, while the predictions hitherto accomplished appear to be very good, there is a inevitably some uncertainty concerning the intermolecular interactions in the hypothetical supercooled liquid (see Section 3.1) and in 1-octanol solutions (see Sections 2.7, 2.9, and 7.3) and concerning the reliability of solubility parameters in highly interactive systems (see Sections 2.6 and 4.2). Consequently, thermodynamic cycles via a solution in a paraffinic hydrocarbon or via the vapor phase may be preferable, as described in Sections 8.7–8.10. These alternative approaches to the estimation of aqueous solubility have the added advantage of being amenable to prediction using group contributions (Chapter 7).

Using alternative molecular properties, such as those that correlate with and are used to predict partition coefficients, other useful conditions for the estimation of aqueous solubility have been suggested. For example, Baker et al. (1984) have proposed the following equation for the mole-fraction aqueous solubility of polycyclic aromatic hydrocarbons:

$$\log x_{2,w}^{s} = a_o + a_1 \text{ (molecular size descriptor)}$$
$$+ a_2(\theta_m - 25) + a_3(\text{RMRI})r_d \tag{8.24}$$

where the molecular size descriptor may be molecular surface area, molecular volume, or molecular weight, θ_m is the Celsius melting point, RMRI is a newly defined molecular redundancy index which is related to the shape of the molecule, r_d is the summation of the interatomic distances of the molecule, and a_o–a_3 are regression coefficients.

8.6   DIRECT CORRELATION OF AQUEOUS SOLUBILITY USING SIMPLE EMPIRICAL GROUP-CONTRIBUTION APPROACHES

While examining homologous series of organic compounds, several workers have found that solubility in water or organic solvents follows an equation of the form

$$\log c_n^s = \log c_o^s - B'n \tag{8.25}$$

where c_n^s and c_o^s are the solubilities of the homologues containing n and zero methylene groups, respectively (for references, see Davis et al., 1974) and B' is the empirical methylene group contribution to $\log c_n^s$. The empirical group contribution $\Delta \log c_x^s$ of a given group X can be found by comparing $\log c_o^s$ with that of the corresponding compound without the group in question. In this way, a table of empirical group contribution $\Delta \log c_x^s$ for a series of functional groups X can be constructed (e.g., Saracco and Spaccamela Marchetti, 1958; Spaccamela Marchetti and Saracco, 1958; Korenman et al., 1971). The solubility c^s of a given compound may then be predicted using an equation analogous to Eq. 7.8, thus,

$$\log c^s = \sum \Delta \log c_x^s \tag{8.26}$$

The value of such correlations and tables for prediction of solubility is reduced because the environment of the molecules in the pure compound is changing from one compound to the next in a nonlinear fashion within a homologous series and between homologous series (see pp. 317, 318 and Fig. 7.1). Equations 8.25 and 8.26 may then not be sufficiently linear to be useful. This problem is particularly serious for solids, since crystal structures and the molecular environments within them are very difficult to predict (see Sections 2.2.4, 2.2.5, and 3.9).

The problem just stated is less serious for liquids, which are more disordered than solids and for which the molecular environments are more predictable. Thus, some succcess has been claimed in applying the group-contribution approach to activity coefficients at infinite dilution γ^∞, as exemplified by the following equations which are analogues of Eqs. 8.25 and 8.26 (for references, see Davis et al., 1974):

$$\log \gamma_n^\infty = \log \gamma_o^\infty - Bn \tag{8.27}$$

$$\log \gamma^\infty = \sum \Delta \log \gamma_x^\infty \tag{8.28}$$

If this approach is valid, then Eqs. 8.25 and 8.26 may be a good approximation for liquids, especially if their solubilities are sufficiently low that $x^s \cong 1/\gamma^\infty$, according to Eq. 2.104. These conditions are assumed in the simple empirical group-contribution approaches to the aqueous solubility of organic liquids (e.g., Koreman et al., 1971). These conditions have also been applied to the prediction of solid solubility by expressing the thermodynamic activity of the solid in terms of the pure hypothetical supercooled liquid as the standard state, assuming $\Delta C_p = 0$. For this assumption Eqs. 2.14–2.18 have been applied (see Sections 2.2.4 and 2.2.6). This approach to predicting the aqueous solubilities of aromatic solids has been taken by Tsonopoulos and Prausnitz (1971). Irmann (1965), utilizing Eq. 2.18, also developed empirical group contributions for estimating the aqueous solubilities of a variety of paraffins, olefins, acetylenes, and aromatic hydrocarbons and their halogen derivatives.

When using and comparing these essentially empirical group contributions, careful attention should be paid to the units of solubility, which define the standard-state concentration of the hypothetical supercooled liquid solute. For example, Tsonopoulos and Prausnitz (1971) chose the pure state of unit mole fraction, Korenman et al. (1971) chose unit molarity (following the literature data from which they obtained the aqueous solubilities), whereas Irmann (1965) chose a concentration of 1 g per gram of water. It should be emphasized that the liquid state of the solute (whether it be accessible or supercooled) will represent a varying molecular environment of the solute molecules, thereby limiting the value of these group-contribution approaches.

8.7 FOUR ALTERNATIVE ROUTES FOR THE PREDICTION OF SOLUBILITY IN WATER

8.7.1 Introduction

Scheme 8.1 represents four alternative thermodynamic cycles for predicting the standard molar free energy of solution of the solute in water, that is, ΔG_t^θ (pure $\rightarrow$ water), from other standard molar free-energy changes. ΔG_t^θ (pure $\rightarrow$ water) is directly related to the aqueous solubility by Eq. 8.29 and the other standard molar free-energy changes are related to their corresponding equilibrium properties by the analogous Eqs. 8.30–8.34. The group-contribution concept has been successfully applied to the quantities represented by Eqs. 8.31, 8.33, and 8.34 in the upper triangle, as discussed in Chapter 7.

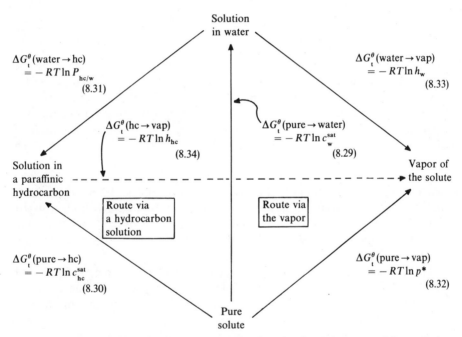

Scheme 8.1 Standard molar free energies of transfer of a solute for the purposes of the prediction of the solubility of the solute in water: p^* = saturated vapor pressure of the pure solute; $P_{hc/w}$ = partition coefficient (hydrocarbon/water); h_{hc} = Henry's law constant in a paraffinic hydrocarbon; h_w = Henry's law constant in water; c_{hc}^{sat} = solubility in the paraffinic solvent; and c_w^{sat} = solubility in water.

The four routes in Scheme 8.1 are (a) via the vapor (pure solute $\rightarrow$ vapor $\rightarrow$ solution in water), (b) via a hydrocarbon solution (pure solute $\rightarrow$ solution in a paraffinic hydrocarbon $\rightarrow$ solution in water), (c) first via a hydrocarbon solution and then via the vapor (pure solute $\rightarrow$ solution in a paraffinic hydrocarbon $\rightarrow$ vapor $\rightarrow$ solution in water), and (d) first via the vapor and then via a hydrocarbon solution (pure solute → vapor → solution in a paraffinic hydrocarbon → solution in water).

8.7.2   Route: Pure Solute → Vapor → Solution in Water

The thermodynamic cycle on the right-hand side of Scheme 8.1 indicates that

$$\Delta G_t^\theta(\text{pure} \to \text{water}) = \Delta G_t^\theta(\text{pure} \to \text{vapor}) - \Delta G_t^\theta(\text{water} \to \text{vapor}) \quad (8.35)$$

Introducing Eqs. 8.29, 8.32, and 8.33 gives

$$\ln c_w^{sat} = \ln p^* + \Delta G_t^\theta(\text{water} \to \text{vapor})/RT \quad (8.36)$$

or

$$c_w^{sat} = p^*/h_w \quad (8.37)$$

The solubility of the solute in water, c_w^{sat}, has units which are defined by the concentration scale employed for the Henry's constant of the solute in water, h_w, or for the standard state of $\Delta G_t^\theta(\text{water} \to \text{vapor})$. Solubility prediction requires knowledge of the saturated vapor pressure p^* of the pure solute and either the Henry's constant in water, h_w, (in which case Eq. 8.37 is used) or the value of $\Delta G_t^\theta(\text{water} \to \text{vapor})$ from group contributions (e.g., in Tables 7.1 or 7.6), in which case Eq. 8.36 is used.

Example 8.1   The saturated vapor pressure of 1-hexanol is 0.72 mm Hg at 25°C and its Henry's law constant in water, h_w^x, is 600 mm Hg (mole fraction scale) at 25°C (Rytting et al., 1978b). Substitution into Eq. 8.37 leads to $c_w^{sat} = 0.72/600 = 1.2 \times 10^{-3}$ (mole-fraction scale). Thus, the calculated mole-fraction a solubility of 1-hexanol in water is 1.2×10^{-3} at 25°C.

The experimental solubility of 1-hexanol in water is stated (Riddick and Bunger, 1970) to be 0.706% w/w at 25°C. The molecular weight of 1-hexanol and water are 102.2 and 18.02, respectively. The mole-fraction solubility in water at 25°C is therefore

$$\frac{0.706/102.2}{0.706/102.2 + (100 - 0.706)/18.02} = 1.25 \times 10^{-3}$$

at 25°C. An alternative experimental value for the solubility of 1-hexanol in water has been stated (Yalkowsky and Valvani, 1980) in the form $\log c_w^{sat} = -1.24$ ($c_w^{sat} = 0.0575 \text{ mol dm}^{-3}$) at 25°C. An analogous calculation affords $x_w^{sat} = 1.04 \times 10^{-3}$ at 25°C. The agreement between the predicted and each experimental value is excellent and is, in fact, rather better than expected in view of the assumptions inherent in the predictions and the slight mutual miscibility of 1-hexanol and water.

8.7.3 Route: Pure Solute → Solution in a Paraffinic Hydrocarbon → Solution in Water

The thermodynamic cycle on the left-hand side of Scheme 8.1 indicates that

$$\Delta G_t^\theta(\text{pure} \to \text{water}) = \Delta G_t^\theta(\text{pure} \to \text{hc}) - \Delta G_t^\theta(\text{water} \to \text{hc}) \qquad (8.38)$$

Introducing Eqs. 8.29–8.31 gives

$$\ln c_w^{sat} = \ln c_{hc}^{sat} + \Delta G_t^\theta(\text{water} \to \text{hc})/RT \qquad (8.39)$$

or

$$c_w^{sat} = c_{hc}^{sat}/P_{hc/w} \qquad (8.40)$$

This is essentially the same equation as that defining the partition coefficient (hydrocarbon → water) for dilute saturated phases (Eq. 3.33). Furthermore, since c_{hc}^{sat} is the thermodynamic activity of the solute as defined in Chapter 3 (Eq. 3.31), Eq. 8.40 is essentially the same as Eq. 3.36. The solubility of the solute in water c_w^{sat} has units which are defined by the concentration scale employed for c_{hc}^{sat}, the solubility in the paraffin solvent. The partition coefficient $P_{hc/w}$ should also be expressed in terms of the same concentration scale (see Chapter 7), otherwise an appreciable error could result. If group contributions to $\Delta G_t^\theta(\text{water} \to \text{hc})$ are available, Eq. 8.39 may be used with due regard to the units of concentration for the standard states in the two solvents.

Example 8.2 We wish to predict the mole-fraction solubility of n-octyl p-aminobenzoate in water at 25°C from data in the literature. $H_2N-C_6H_4-CO \cdot O \cdot (CH_2)_7 CH_3 = C_{15}H_{23}NO_2 = 249.3$; m.p. = 71°C; log x_{hc}^{sat} (hexane at 37°C) = -2.495 where x^{sat} is the mole-fraction solubility in hexane at 37°C (Yalkowsky et al., 1975); log $P_{hc/w}^c = 3.79$ (Hansch and Leo, 1979), where $P_{hc/w}$ is the partition coefficient based on molarity in hexane/molarity in water.

The calculation is based on the following reasoning and assumption:

(1) The temperature difference between 37 and 25°C will change the solubility according to the van't Hoff isochore (Eq. 1.9):

$$\log \frac{x^{sat} \text{ at } 25°C}{x^{sat} \text{ at } 37°C} = \frac{\Delta H^s}{2.303 R} \left(\frac{25 - 37}{298.15 \times 313.15} \right)$$

The enthalpy of solution is unknown, but as a rough estimate the enthalpy of fusion of the solute may be used. This assumes that a very dilute solution of n-octyl p-aminobenzoate in hexane obeys Raoult's law and ignores ΔC_p. We note that $\Delta H^s \cong \Delta H^f = T_m \Delta S^f = (273.15 + 71)K \times 26.0 \, cal \, K^{-1} \, mol^{-1} = 8948 \, cal \, mol^{-1}$. Substitution of the values in the equation given above affords log $x_{hc}^{sat} = -2.746$ or $x_{hc}^{sat} = 1.795 \times 10^{-3}$.

(2) It is more convenient to use the mole-fraction scale in Eq. 8.40, in which

case $P_{hc/w}^c$ must be converted to $P_{hc/w}^x$ by means of Eq. 7.26 in the form

$$P_{hc/w}^x = P_{hc/w}^c(V_{hc}/V_w) \tag{8.41}$$

where $P_{hc/w}^c = 10^{3.79}$, $V_{hc} = 131.6 \text{ cm}^3 \text{ mol}^{-1}$, the molar volume of hexane (Riddick and Bunger, 1970), and $V_w = 18.07 \text{ cm}^3 \text{ mol}^{-1}$, the molar volume of water (Riddick and Bunger, 1970), all at 25°C.

Therefore,

$$P_{hc/w}^x = 10^{3.79} \times 131.6/18.07 = 44905$$

Substituting this value and $x_{hc}^{sat} = 1.795 \times 10^{-3}$ into Eq. 8.40 affords $x_w^{sat} = 1.795 \times 10^{-3}/44905 = 3.99 \times 10^{-8}$. This is the predicted mole-fraction solubility of n-octyl p-aminobenzoate in water at 25°C. This may be compared with the following experimental values determined by Yalkowsky et al. (1975) and by Yalkowsky and Valvani (1980): $x_w^{sat} = 2.25 \times 10^{-8}$ (Table II in Yalkowsky et al., 1975), 7.21×10^{-8} (Table IV in Yalkowsky et al., 1975), and 7.17×10^{-8} (from Table VII in Yalkowsky and Valvani, 1980). Considering the approximations and assumptions inherent in the predictive process, agreement with the Table II value is remarkably close. The experimental differences in the measurement of such low solubilities emphasize the value of predictive procedures.

8.7.4   Route: Pure Solute → Solution in a Paraffinic Hydrocarbon → Vapor → Solution in Water

This thermodynamic cycle traces a zig-zag path through Scheme 8.1, thus,

$$\Delta G_t^\theta(\text{pure} \rightarrow \text{water}) = \Delta G_t^\theta(\text{pure} \rightarrow \text{hc}) + \Delta G_t^\theta(\text{hc} \rightarrow \text{vap}) - \Delta G_t^\theta(\text{water} \rightarrow \text{vap}) \tag{8.42}$$

Introducing Eqs. 8.29, 8.30, 8.33, and 8.34 gives

$$\ln c_w^{sat} = \ln c_{hc}^{sat} + [\Delta G_t^\theta(\text{water} \rightarrow \text{vap}) - \Delta G_t^\theta(\text{hc} \rightarrow \text{vap})]/RT \tag{8.43}$$

or

$$c_w^{sat} = c_{hc}^{sat} h_{hc}/h_w \tag{8.44}$$

Equation 8.44 is essentially a combination of Eqs. 3.35 and 3.36 with elimination of $P_{hc/w}$. As before, the concentration scale used to express the c and h values in Eq. 8.44 must be self-consistent. This route is less likely to be useful than the others, since single values of partition coefficients are generally more readily available than pairs of values of Henry's law constants. If group contributions to $\Delta G_t^\theta(\text{water} \rightarrow \text{vap})$ and to $\Delta G_t^\theta(\text{hc} \rightarrow \text{vap})$ are available, Eq. 8.43 may be used with due regard to the units of concentration and pressure for the various standard states. Again, such pairs of group-contribution values are usually less readily available than single values of $\Delta G_t^\theta(\text{water} \rightarrow \text{hc})$, as in Table 7.1, in which case Eq. 8.39 may be used.

8.7.5 Prediction of Solubility in Water via the Route: Pure Solute → Vapor → Solution in Paraffinic Hydrocarbon → Solution in Water

This thermodynamic cycle also traces a zig-zag path through Scheme 8.1, thus,

$$\Delta G_t^\theta(\text{pure} \rightarrow \text{water}) = \Delta G_t^\theta(\text{pure} \rightarrow \text{vap}) - \Delta G_t^\theta(\text{hc} \rightarrow \text{vap}) - \Delta G_t^\theta(\text{water} \rightarrow \text{hc})$$
(8.45)

Introducing Eqs. 8.29, 8.31, 8.32, and 8.34 gives

$$\ln c_w^{sat} = \ln p^* + [\Delta G_t^\theta(\text{hc} \rightarrow \text{vap}) + \Delta G_t^\theta(\text{water} \rightarrow \text{hc})]/RT \quad (8.46)$$

or

$$c_w^{sat} = p^*/h_{hc}P_{hc/w} \quad (8.47)$$

Equation 8.47 is essentially a combination of Eqs. 3.14, 3.31, and 3.36, eliminating $a_2^c{}^*$ and solubility in α, a paraffinic hydrocarbon. As before, self-consistent units must be used to express the various quantities in Eq. 8.47. If group contributions to $\Delta G_t^\theta(\text{hc} \rightarrow \text{vap})$ and to $\Delta G_t^\theta(\text{water} \rightarrow \text{hc})$ are available, Eq. 8.46 may be used with due regard to the units of concentration and pressure for the standard states in various phases.

Example 8.3 Calculate the solubility of 1-octanol in water at 25°C given that the saturated vapor pressure p^* of 1-octanol is 0.72 mm Hg; the mole fraction based Henry's law constant h^x is 17.1 mm Hg in isooctane (Rytting et al., 1978b), and log P^c(n-octane/water) is 0.45 (McGowan, 1952), all at 25°C. Hansch and Leo (1979) reported that log P^c(n-alkane/water) = 0.45 for all n-alkanes from pentane to decane.

As in Example 8.2, it is most convenient to use the mole-fraction scale, in which case log $P_{o/w}^c$ must be converted to log $P_{o/w}^x$ by means of Eq. 7.28 in the form

$$\log P_{hc/w}^x = \log P_{hc/w}^c + \log(V_{hc}/V_w) \quad (8.48)$$

The molar volumes at 25°C are $V_{hc} = 114.233/0.69848 = 163.5 \text{ cm}^3 \text{ mol}^{-1}$ (for n-octane) and $V_w = 18.016/0.99705 = 18.07 \text{ cm}^3 \text{ mol}^{-1}$ (for water). We then obtain log $P_{hc/w}^c = 1.406$, so $P_{hc/w}^c = 25.5$. (We note that a change in the concentration scale has a considerable effect on the magnitude of the partition coefficient.)

Substitution of the numerical values of p^*, h_{hc}^x, and $P_{hc/w}^x$ into Eq. 8.47 using the mole-fraction scale affords

$$x_w^{sat} = 0.72/17.1 \times 25.5 = 1.65 \times 10^{-3}$$

This predicted mole-fraction solubility, 1.65×10^{-3} for 1-octanol at 25°C, may be compared with the experimental values (Riddick and Bunger, 1970; Yalkowsky and Valvani, 1980) of 1.25×10^{-3} and 1.04×10^{-3}, respectively. As before, the agreement is good.

8.8   USE OF GROUP CONTRIBUTIONS TO ESTIMATE AQUEOUS SOLUBILITY

The determination of the aqueous solubility of very sparingly soluble organic solids poses a number of practical problems which are discussed in Sections 8.9.2–8.9.5 together with empirical techniques for overcoming them. Solids which are less soluble than a few $ng\,cm^{-3}$ ($\mu g\,L^{-1}$) present problems resulting from the low equilibration rate and from insufficient sensitivity in the measurement of dissolved concentration. In these cases, indirect procedures may be required to replace direct measurement.

One useful approach to these problems was proposed by Higuchi et al. (1979). It is based on the fact that the majority of hydrophobic compounds are sufficiently soluble in a water-immiscible organic solvent to allow direct measurement of the solubility s_o in this type of solvent. For this purpose, accurate measurement using the phase-solubility technique is preferred. From this value the solubility s_w in water can then be calculated using the partition coefficient (organic solvent/water) $P_{o/w}$ according to Eq. 8.40 or Eq. 3.36 in the form

$$s_w = s_o/P_{o/w} \qquad (8.49)$$

The partition coefficient may not be measurable directly if the solubility in water is very low, but it can be estimated by the usual group contribution approach based on experimental results obtained with the closest appropriate analogue. The direct prediction of the solubility of solids without prior knowledge of their experimental activities is at present difficult because it depends on the co-operative interactions between the molecules within the crystal structure (Section 2.2.4). However, changes in partition coefficients resulting from variations in molecular structure can be predicted with

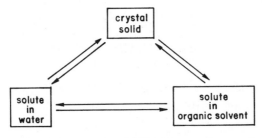

Scheme 8.2 Determination of the aqueous solubility of extremely water-insoluble hydrophobic solids by equilibration with an added organic solvent. Reproduced with permission of the copyright owner, the American Pharmaceutical Association, from Higuchi, T., Shih, F.-M. L., Kimura, T., and Rytting, J. H. (1979). *J. Pharm. Sci.*, **68**, 1267–1272.

reasonable confidence, since partitioning is essentially a noncooperative process. This approach is summarized by Scheme 8.2 and was tested with the steroidal compounds norethindrone and methyltestosterone and their acetate esters.

The measured solubilities and partition coefficients are listed in Table 8.1. The predicted partition coefficients were calculated by means of the group-contribution approach which, as we have seen, assumes that certain thermodynamic and other properties of a molecule can be calculated from the sum of the values for the various groups comprising the molecule (Harris et al., 1973; Davis et al., 1974). The chosen organic solvent was isooctane. The group contribution to the free energy of transfer between water and isooctane for the saturated hydrocarbon moieties was taken from the data of Harris et al. (1973), which are presented in Table 8.1, or was calculated by means of the relationship

$$\Delta\Delta G_t^\theta (\text{water} \rightarrow \text{isooctane})/\text{cal mol}^{-1} = -4070A + 1156 \qquad (8.50)$$

where A is the relative surface area with respect to the *tert*-butyl group and the correlation coefficient is 0.992 (Harris et al., 1973).

The partition coefficients $P_{o/w}^c$ for moieties having polar groups were determined from plots of concentration in isooctane against concentration in water and are reported in Table 8.2 together with the corresponding standard free energies of transfer, $\Delta G_t^\theta (\text{water} \rightarrow \text{isooctane})$, which were calculated using Eq. 8.11 in the form

$$\Delta G_t^\theta (\text{water} \rightarrow \text{isooctane}) = -RT \ln P_{o/w} \qquad (8.51)$$

The group-contribution approach should make possible the prediction of the ΔG_t^θ value and hence the partition coefficient of a molecule when the group contribution $\Delta\Delta G_t^\theta$ of each of the constituent goups is known, assuming that the group contributions are additive, thus,

$$\Delta G^\theta = \sum \Delta\Delta G \qquad (8.52)$$

This method should be valuable for very hydrophobic compounds for which the actual measurement is difficult. Scheme 8.3 illustrates such a stepwise calculation for methyltestosterone, starting from the 17-androstyl group. Table 8.3 summmarizes analogous calculations of the standard free energy of transfer, ΔG^θ, of the compounds studied. Table 8.3 also reports the calculated and measured partition coefficients which are seen to agree within a factor of 2, with the exception of methyltestosterone acetate. Calculation of the partition coefficient of this compound from the measured value for methyltestosterone plus the acetyl group contribution (calculated from the data for 1-methylcyclopentanol and methylcyclopentyl acetate) gave a value of about 5000 cal mol^{-1}, which is in much better agreement with the experimental value of 3300 cal mol^{-1}.

The group-contribution appproach of Higuchi et al. (1979) gives reasonable estimates of partition coefficients and is therefore particularly useful in those

Table 8.1 Measured Molar Solubilities and Molar Concentration-Based Isooctane–Water Partition Coefficients for Steroids[a]

Substance	Molar Solubility In Isooctane[b]	$(\text{mol dm}^{-3} = \text{mol L}^{-1} = M)$ In Water[b]	In Water in Contact with Isooctane	Isooctane–Water Partition Coefficient based on Molar Concentration
Norethindrone (I)	$6.90 \pm 0.13 \times 10^{-5}$ $(5.85 \times 10^{-5})^c$	$2.36 \pm 0.06 \times 10^{-5}$	$2.13 \pm 0.05 \times 10^{-5}$ $(1.67 \times 10^{-5})^c$	3.9 ± 0.17 $(4.4)^b$
Norethindrone acetate (II)	$2.52 \pm 0.04 \times 10^{-3}$ $(2.53 \times 10^{-3})^c$	$1.57 \pm 0.01 \times 10^{-5}$	$1.76 \pm 0.01 \times 10^{-5}$ $(1.65 \times 10^{-5})^c$	210 ± 5 $(183)^c$
Norethindrone enanthate	0.0232^b		$(7 \pm 3 \times 10^{-8})^c$	$(3.3 \times 10^{-5})^c$
Methylestosterone (III)	$1.3 \pm 0.03 \times 10^{-3}$	$1.12 \pm 0.02 \times 10^{-4}$ $(7.47 \times 10^{-5})^b$	$1.11 \pm 0.01 \times 10^{-4}$	16.0 ± 0.5
Methyltestosterone acetate (IV)	$5.3 \pm 0.05 \times 10^{-3}$	$1.43 \pm 0.03 \times 10^{-5}$ $(5.20 \times 10^{-6})^d$	$8.16 \pm 0.06 \times 10^{-7}$	3300 ± 173

[a] Uncertainties are expressed as standard errors of the mean.
[b] Obtained by conventional method.

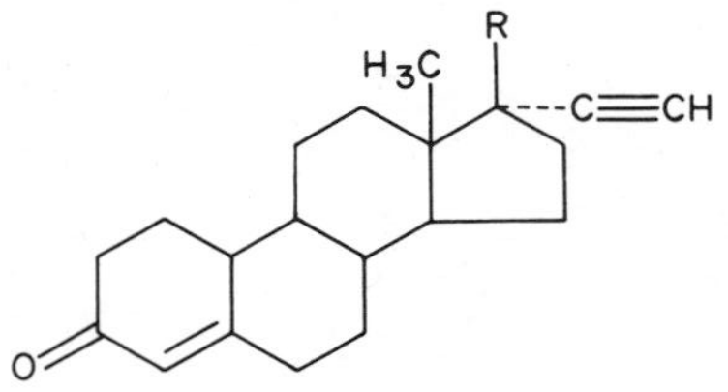

I : R=OH
II : R=OCOCH₃

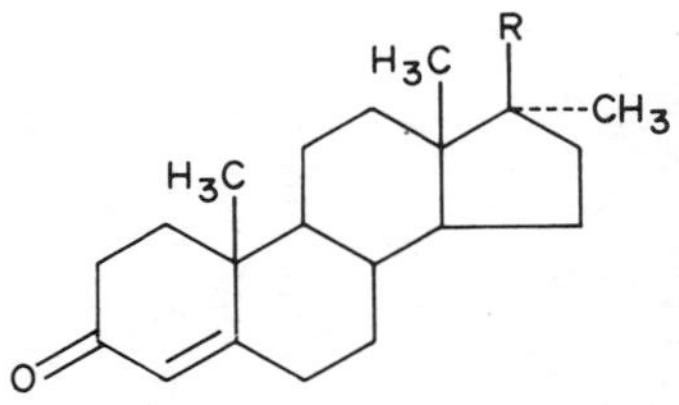

III : R=OH
IV : R=OCOCH₃

[c] R. E. Enever, University of London, London, England, personal communication.
[d] Literature value from Bowen et al. (1970).

Reproduced with permission of the copyright owner, the American Pharmaceutical Association, from Higuchi, T., Shih, F.-M. L, Kimura, T., and Rytting. J. H. (1979). *J. Pharm. Sci.*, **68**, 1267–1272.

Table 8.2 Molar Concentration-Based Partition Coefficients (Isooctane–Water) and the Corresponding Standard Molar Free Energy of Transfer for Organic Molecules[a]

Compounds	Partition Coefficient[b]	ΔG^{θ}(cal mol^{-1})
2-Cyclohexen-1-one	0.46 ± 0.01	460
1-Ethynylcyclopentanol	0.36 ± 0.01	580
1-Ethynylcyclopentyl acetate	36.5 ± 1.4	-2100
1-Methylcyclopentanol	0.46 ± 0.01	460
1-Methylcyclopentyl acetate	227 ± 11	-3200

[a] 1 cal $= 4.184$ J.

[b] Uncertainties are expressed as standard deviations.

Reproduced with permission of the copyright owner, the American Pharmaceutical Association, from Higuchi, T., Shih, F.-M. L., Kimura, T., and Rytting, J. H. (1979). *J. Pharm. Sci.*, **68**, 1267–1272.

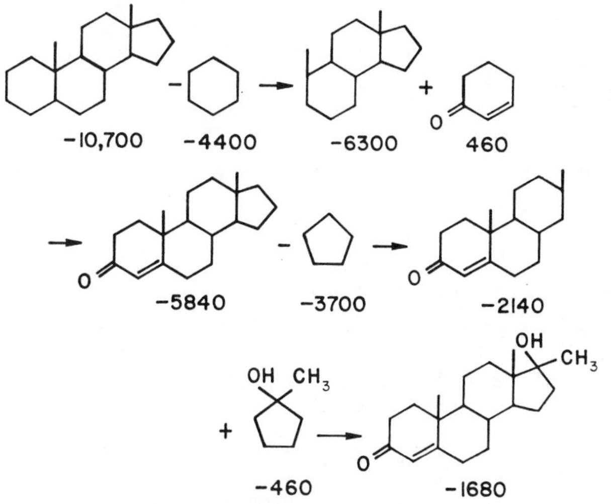

Scheme 8.3 Schematic diagram for calculating the standard molar free energy of transfer (cal mol^{-1}) for methyltestosterone using the 17-androstyl group as the beginning group. The numbers represent the values for each group and intermediate used in the calculations. 1 cal $= 4.184$ J. Reproduced with permission of the copyright owner, the American Pharmaceutical Association, from Higuchi, T., Shih, F.-M. L., Kimura, T., and Rytting, J. H. (1979). *J. Pharm. Sci.*, **68**, 1267–1272.

cases for which experimental measurements are difficult. Substitution of the measured solubility of a given compound in isooctane, s_o, and the calculated partition coefficient $P_{o/w}$ into Eq. 8.49 affords a calculated solubility in water, s_w, which agrees with the measured solubility, as shown in Table 8.1. Thus, the aqueous solubility of a compound which has a very low solubility in water can be estimated quite well by combining the known solubility in an organic solvent with the partition coefficient calculated from group contributions. In

Table 8.3 Calculation of the Standard Molar Free Energy of Transfer and Molar Concentration-Based Partition Coefficient (Isooctane–Water) for Norethindrone (I), Norethindrone Acetate (II), Methyltestosterone (III), and Methyltestosterone Acetate (IV)[a]

Functional Group	Structure	$\Delta(\Delta G^0)$ cal mol^{-1}	I	II	III	IV
17-Androstyl		−10,700[b]	+	+	+	+
Methyl	−CH$_3$	−900[b]	−	−		
Cyclohexyl		−4400[b]	−	−	−	−
2-Cyclohexen-1-one		460[c]	+	+	+	+
Cyclopentyl		−3700[b]	−	−	−	−
1-Ethynylcyclopentanol		580[c]	+			
1-Ethynylcyclopentyl acetate		−2100[c]		+		

1-Methylcyclopentanol

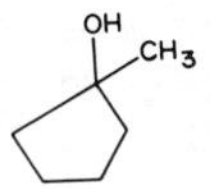

460[c]   +

1-Methylcyclopentyl acetate

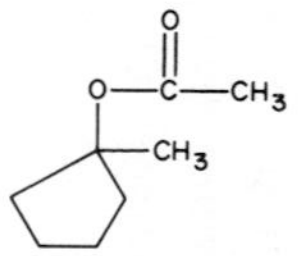

−3200[c]   +

	ΔG_t^θ (cal mol^{-1})		
−660	−3340	−1680	−5340
Partition coefficient (calculated)			
3.05	281	17.0	8200
Partition coefficient (measured)			
3.9	230	9.4	3300

[a] 1 cal = 4.184 J.
[b] Harris et al. (1973). The relative surface area used for the 17-androstyl group was 2.91; for the methyl group, it was 0.51, where the *tert*-butyl group has a value of 1.
[c] Table 8.2.

Reproduced with permission of the copyright owner, the American Pharmaceutical Association, from Higuchi, T., Shih, F.-M. L., Kimura, T., and Rytting, J. H. (1979). *J. Pharm. Sci.*, **68**, 1267–1272.

such cases, this procedure for estimating aqueous solubility may be the only feasible method.

8.9   DETERMINATION OF SOLUBILITY

8.9.1   Analytical and Synthetic Methods for Moderately Soluble Substances

Solubility determination on solids which are moderately soluble, that is, $> 1 \, mg \, cm^{-3}$ ($> 1 \, g \, L^{-1}$, $> 0.1\%$), normally poses no serious problems. Basically two methods may be used, the "analytical" or the "synthetic."

The analytical method involves first the preparation of a saturated solution at an accurately known temperature in the presence of an excess of the substance. This may be achieved by suitable agitation of the substance with the solvent in a thermostat until equilibrium is achieved. An aliquot of the saturated solution is then removed from the system by centrifugation, by filtration, or by straining through a plug of glass wool, and is analyzed by a suitable method. Losses of the dissolved solute by absorption onto the filter material may occur during filtration, but steps can be taken to avoid or minimize these effects (Florry, 1953; Prigogine, 1957; Kirkwood, 1968; Pierotti, 1976).

For the synthetic method, either a weighed amount of the solute substance (or a definite amount of solvent) is taken. While agitating at constant temperature, known amounts of solvent (or the solute substance) are added gradually until the solubility limit is reached. Alternatively, known amounts of the solvent and solute are agitated while gradually changing the temperature until the solubility limit is reached. Appropriate checks must be carried out to ensure that the system is very close to equilibrium when the content or temperature of the system is recorded. In the method of temperature variation, it is usual to focus attention on the appearance of the last small crystal. The equilibrium temperature is obtained by taking the mean of the two temperatures at which the crystal either slowly grows or slowly dissolves. This procedure may also be carried out under a hot-stage microscope.

For measuring the aqueous solubilities of adenine and guanine ($8.7 \, mmol \, dm^{-3}$ and $39 \, \mu mol \, dm^{-3}$, respectively, at $25°C$) De Voe and Wasik (1984) developed a column method, which should be applicable to other solutes of similar solubility. A column generator made of glass or stainless steel was packed with the solid. Water was forced downward through the column by applying pressure to the water vessel with helium gas. The saturated solution emerged from the bottom of the column and was collected and analyzed using HPLC. The authors applied a modified packing procedure for guanine because higher pressures were needed to force water through it. The authors claimed to be able to determine accurate values of the enthalpy of solution (and even its temperature derivative ΔC_p) from the temperature dependence of the solubilities so measured.

At various places in this book we have pointed out the experimental errors in the determination of apparent enthalpies of solution from the temperature variation of solubility using the van't Hoff isochore (Eq. 1.9). Enthalpies (i.e., heats) of solution should preferably be determined by direct calorimetic measurements, which, using modern calorimetry, can be exceedingly accurate, even for slightly soluble and/or weakly interacting solute–solvent systems such as hydrocarbons in water (Gill et al., 1976). For most purposes the differential enthalpy of solution (which is designated ΔH_s in this book) is more useful than the integral enthalpy of solution, since the former is related more directly to solubility by Eqs. 1.8 and 1.9 (note, however, Eq. 2.36). Differential enthalpies of solution are determined calorimetrically by extrapolating the experimental heat of solution to infinite dilution.

Calorimetric values of ΔH_s may be useful to establish the temperature dependence of solubilities (Gill et al., 1976), thereby enabling the solubility at an appropriate temperature to be predicted from the measured value at a different temperature. If the two temperatures are close, Eq. 1.9 may be used for this purpose. If the two temperatures are widely separated or if ΔH_s shows considerable temperature dependence, reflecting a high value of ΔC_p ($= b$ in Eq. 2.34), Eq. 2.33 should be used to predict the solubility at the temperature T of interest. If the temperature dependence of solubility is represented as follows:

$$\Delta H_s = a + b(T - 273.15 \text{ K}) + c(T - 273.15 \text{ K})^2 \qquad (8.53)$$

$$\Delta C_p = (\partial \Delta H_s / \partial T)_p = b + 2c(T - 273.15 \text{ K}) \qquad (8.54)$$

Gill et al. (1976) have found that c is insignificant for all cases studied. It is therefore usually sufficient to assume that ΔH_s is a linear function of $\Delta C_p (= b$ in Eq. 2.32) for the purposes of solubility prediction. It is, however, important to be aware of the possible errors involved in extrapolating any quantity, including solubility, beyond the range covered by the experimental data.

8.9.2 Special Problems with Very Sparingly Soluble Substances

The direct determination of solubilities much less than 1 mg cm^{-3}, corresponding to very sparingly soluble organic solids, is hampered by such problems as (1) slow equilibrium resulting from a low rate of dissolution during measurement, (2) the influence of impurities, and (3) the apparent heterogeneity in the energy content of the crystalline solid (Higuchi et al., 1979). These difficulties can lead to large discrepancies in reported values. The reported aqueous solubility of cholesterol, for example, ranges from 0.025 to 2600 mg cm^{-3} (g L^{-1}; Bagley and Scigliano, 1976).

8.9.3 Kinetics of Equilibration

The low equilibration rate during solubility measurement of slightly soluble species presents a serious problem which has been discussed by Higuchi et al. (1979). Solubility is an equilibrium property. Since it is impossible to attain a true equilibrium state, solubility values are in practice reported for systems that are reasonably close to equilibrium. Such systems are not particularly difficult to attain for moderately soluble substances. An aqueous suspension of benzoic acid crystals (of solubility 3.4 mg cm^{-3} at 25°C), for example, probably reaches a state close to equilibrium with moderate stirring at 25°C within a few hours. As we shall see in Chapter 11, the exact rate depends on the fineness of the crystals, the amount of solid added, and the degree of agitation.

Studies of the kinetics of dissolution discussed in Chapter 11 indicate that the classical Noyes–Whitney equation is an empirically useful relationship. For the present purpose, Eqs. 11.3, 11.11, and 11.20 may be combined in the form

$$dc/dt = (AD/Vh)(c_s - c) \tag{8.55}$$

where c is the concentration of solute dissolved at time t, dc/dt represents the rate of approach to equilibrium, A is the surface area available for dissolution, V is the volume of the solvent, D is the diffusivity of the solute which is relatively insensitive to molecuar weight and molecular structure, and c_s is the equilibrium solubility. The factor h is a constant which depends on the conditions of agitation, temperature, and other factors, and represents the apparent thickness of the diffusion layer immediately adjacent to the solid.

Since the rate of attainment of equilibrium increases with increasing solubility and with increasing surface area available for dissolution, the retarding influence of a low solubility can be partially reversed by increasing this area. Under experimental conditions, the dissolving crystals will become rounded, reducing the effective surface area and thereby delaying the attainment of equilibrium. This can be partially reversed by increasing the amount of solid—the larger the excess, the greater the dissolution rate. The final rate of approach to equilibrium is approximately proportional to the excess of solid present. For example, to obtain the same terminal rate of dissolution in water using a onefold excess of benzoic acid (solubility in water 3.4 mg cm^{-3} at 25°C), it would require, in the case of norethindrone acetate (solubility in water 5.34 μg cm^{-3} at 25°C), an approximate 600-fold excess (636 = 3400/5.34). The use of a large excess of a very sparingly soluble drug to reduce the equilibration time will greatly aggravate the effects of impurities and of energetic heterogeneity of the crystals and so may not be feasible.

The influence of impurities on the apparent solubility of crystalline solids will be discussed in Section 10.2. For the present purpose, however, Figure 8.8 shows a typical phase-solubility diagram and refers to a solid containing 10% of a soluble impurity. The amount in solution, which includes the dissolved impurity as well as the substance in question, increases with increasing amount of sample used until the solution becomes saturated with the soluble impurity as well as with the sparingly soluble substance. Although the true solubility indicated in Figure 8.8 corresponds to 4.5 mg, the apparent solubility continues to increase with sample size, the increase corresponding to onefold for every tenfold increase in the amount added. In an analogous way, for norethindrone acetate, for which a 600-fold excess was proposed, each 1% of total soluble

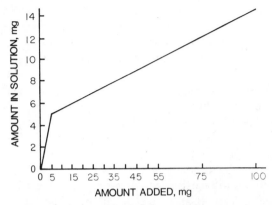

Figure 8.8 Phase-solubility diagram of a crystalline solid containing 10% soluble impurity. Reproduced with permission of the copyright owner, the American Pharmaceutical Association, from Higuchi, T., Shih, F.-M. L., Kimura, T., and Rytting, J. H. (1979). *J. Pharm. Sci.*, **68**, 1267–1272.

impurities will produce a sixfold increase in the apparent solubility of the material, with that going into solution being largely the impurities present rather than the main compound.

Thus, direct determination of the solubility of very sparingly soluble compounds presents a dilemma. If a large sample is employed to obtain a fairly short equilibration time (hours to weeks), the impurities may pose a problem. If a small sample is used to minimize the impurity effect, the equilibration time may extend to months or years. The use of only a onefold excess of norethindrone acetate, for example, will require approximately 600-fold greater equilibration time than for the benzoic acid system described above, with the corresponding time being of the order of months.

One answer to this dilemma is to use a highly specific method for determining the concentration of the main component. This would give a horizontal line in the phase-solubility diagram corresponding to a constant amount in solution (i.e., 4.5 mg in Fig. 8.8, independent of the amount of solid added). Frequently, the impurities are similar in chemical nature to the main component and/or may be present in the saturated solution at much higher concentrations, which can lead to analytical complications. This approach may therefore require highly selective and sensitive methods of analysis, especially for solubilities $< 1 \, \mathrm{ng \, cm^{-3}}$.

The solubility is also affected by the inherent heterogeneity of the energetic content of actual crystals, even though they may be chemicaly identical and pure. The origin of this effect is the different properties of individual crystals. In most samples of powders, individual crystals have different sizes and varying energetic defects, dislocations, cracks, and surface irregularities (Hüttenrauch, 1978). The more energetic particles or regions will dissolve faster and possess a greater intrinsic solubility. For example, Suryanarayanan and Mitchell (1985) observed that the solubility (and heat of solution) of dehydrated calcium gluceptate in water increased with increasing time of milling as a result of increasing lattice disruption (i.e., amorphization), corresponding to decreasing crystallinity. Interestingly, the increased energizing and disordering of this solid induced by milling exhibits enthalpy–entropy compensation (Grant and York, 1986; Vachon and Grant, 1987). If only a small amount of solid is added because it is moderately soluble, the effect of energy content will not be great, since the more energetic crystals and regions will dissolve, while material in a less energetic crystalline state will crystallize out to give an equilibrium value of the solubility.

If, because of low solubility, a very large sample is used, corresponding to a 10^5-fold or higher excess, only a very small fraction will dissolve and the solubility will correspond approximately to the thermodynamic activity of the most energetic component remaining out of solution. This will cause the equilibrium solubility to be exceeded. However, since the concentration is still very low, the rate of crystallization, which will lower the activity of the residual solid to the most stable state, will be extremely low. Consequently, the use of a large excess of a very sparingly soluble solid can give a supersaturated solution which can persist for an appreciable time.

These problems have been analyzed by Higuchi et al. (1979) and the following methods of overcoming them have been proposed: (a) prediction or calculation of solubility using the group-contribution method described in Section 8.8; (b) experimental measurement using a highly specific method of analysis, as discussed above; (c) experimental measurement of solubility using a facilitated dissolution method, which will now be discussed.

8.9.4 Facilitated Dissolution Method

In this procedure, the dissolution rate is enhanced by addition of an organic solvent which is immiscible with water. The organic layer is removed before filtration and analysis of the aqueous solution (Higuchi et al., 1979).

As already discussed, the dissolution rate of a very sparingly soluble solid becomes extremely low if only a small excess of the solid is used because of the limited surface area of the solid exposed to the solvent. Generally, the available surface area parallels the amount of solid used. Thus, for the benzoic acid systems cited above, a onefold excess would correspond to $6.8\ mg\ cm^{-3}$ or a surface area of $\sim 4\ cm^2$ per cubic centimeter of water at the beginning and a little more than $2\ cm^2$ per cubic centimeter at saturation, if the initial average diameter of the benzoic acid is $\sim 100\ \mu m$. Smaller calculations for norethindrone acetate would give $\sim 6 \times 10^{-3}\ cm^2$ per cubic centimeter and $4 \times 10^{-3}\ cm^2$ per cubic centimeter, much smaller values.

One way of increasing the interfacial area (A in Eq. 8.55) across which the solute is allowed to diffuse is to add a limited amount of a second immiscible solvent in which the solute is much more soluble. Thus, in isooctane the solubility of norethindrone acetate at 25°C is $860\ \mu g\ cm^{-3}$, nearly 200 times that in water. If the solvent system consists of $10\ cm^3$ of water and $0.2\ cm^3$ of isooctane, and if 0.5 mg of norethindrone acetate is added to it, the isooctane phase will contain at near equilibrium approximately 0.18 mg and the aqueous phase 0.05 mg.

Thus, about 40% of the added solid will go into solution. If we assume that all the hydrophobic steroid remains with the organic solvent, there will be 0.3 mg of the solid in contact with $0.2 \, cm^3$ of the isooctane solution or $1 \, cm^2$ per cubic centimeter of solution, assuming that the average particle diameter is $\sim 100 \, \mu m$ for the steroid. This surface-area–volume ratio can be expected to yield an adequate equilibration rate in the organic phase.

The equilibration rate from isooctane to water will be determined effectively by the interfacial area between the two and will be independent of the amount of solid present. Since any reasonable configuration will correspond to a fairly high interfacial-area–water-volume ratio, a relatively rapid equilibration rate would be expected under even mild agitation. The rate of equilibration in the organic phase will, of course, depend on the surface area of the solid available per cubic centimeter of solvent used.

This method assumes (1) that the solid phase effectively stays with the organic phase, (2) that the solubility in water is sufficient for measurement after equilibration, and (3) that the solubility of the organic solvent in water is sufficiently low so as not to affect significantly the aqueous solubility of the solid. Although some of these conditions may not be totally met in all instances, the method offers a solution to many situations in which equilibrium may be a problem. This approach may be inadvisable for solids containing impurities that are significantly more soluble in water than in the organic phase, as compared with the main component.

Figure 8.9 shows a phase-solubility diagram of norethindrone acetate. There was considerably greater scatter among the experimental points for the solubility determinations in pure water, particularly at the shorter equilibration times. The addition of an immiscible solvent definitely enhanced the equilibration

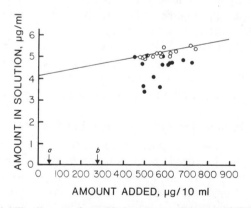

Figure 8.9 Phase-solubility diagram of norethindrone acetate in water at 25°C with an equilibration time of 51 h. Solvent medium: 10 mL of water and 0.25 mL of isooctane (○); 10 mL of water (●); a, amount of drug that saturated 10 mL of water; b, amount of drug that saturated 10 mL of water and 0.25 mL of isooctane. Reproduced with permission of the copyright owner, the American Pharmaceutical Association, from Higuchi, T., Shih, F.-M. L., Kimura, T., and Rytting, J. H. (1979). *J. Pharm. Sci.*, **68**, 1267–1272.

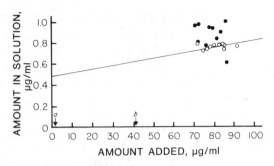

Figure 8.10   Phase-solubility diagram of methyltestosterone acetate in water at 25°C with an equilibration time of 34 h. Solvent medium: 10 mL of water and 0.2 mL of isooctane (○); 10 mL of water (●); a, amount of drug that saturated 10 mL of water; b, amount of drug that saturated 10 mL of water and 0.2 mL of isooctane. Reproduced with permission of the copyright owner, the American Pharmaceutical Association, from Higuchi, T., Shih, F.-M. L., Kimura, T., and Rytting, J. H. (1979). *J. Pharm. Sci.*, **68**, 1267–1272.

rate and yielded more precise results with at least comparable, and usually improved, accuracy.

The scatter in the data obtained with small excesses of solute in pure water may have resulted from particle size heterogeneities as well as from crystallinity differences, both of which are magnified greatly when small amounts are used (Hüttenrauch, 1978). The facilitated dissolution method prevented this problem even after 6 h, which represents an early stage before equilibration.

This technique also reduced the effects of impurities on the apparent solubility, if those impurities were more soluble in isooctane than water. Methyltestosterone acetate, for example, was prepared containing 14% soluble impurities as calculated from the phase diagram in isooctane. Extrapolation of the scattered data to obtain the aqueous solubilty directly from the phase-solubility diagram in water was unsuccessful. Figure 8.10 shows that the addition of 5% isooctane greatly reduced the scatter, thus facilitating extrapolation.

To summarize, the facilitated dissolution method offers an attractive alternative for aqueous solubility measurements where equilibration may be a problem. With radioactive labelling techniques, it should be feasible to apply this method to solubilities at the $ng\,cm^{-3}$ (i.e., $\mu g\,L^{-1}$) level.

8.9.5   Synthetic Methods for Very Sparingly Soluble Solids

Davis and Parke (1942) proposed a method for measuring the aqueous solubilities of various polycyclic hydrocarbons in the $ng\,cm^{-3}$ range. A fine dispersion is prepared by adding a solution of the substance (1 mg) in acetone (1 cm^{3}) to water (110 cm^{3}) with vigorous agitation. The solution of acetone plus 10 cm^{3} of water is evaporated under vacuum. The resulting stock solution is then diluted stepwise and the turbidity measured nephelometrically. The turbidity decreases as the concentration falls. The concentration at which the turbidity is the same as that of a liquid system prepared in the same way, but without the substance, is taken to represent the aqueous solubility.

Brooker and Ellison (1974) developed a similar method based on turbidimetric measurements. A solution ($20 \, \mathrm{mg \, cm^{-3}}$) containing the substance in a water-soluble solvent is released from a micrometer syringe at a rate of $0.1 \, \mathrm{cm^3 \, min^{-1}}$ into water ($45 \, \mathrm{cm^3}$) containing tragacanth ($90 \, \mu g$) as a protective colloid. As soon as a permanent turbidity appears, the absorbance is measured against that of a clear solution. The procedure is repeated at higher concentrations and extinctions are plotted against concentration. The aqueous solubility is the extrapolated value of the concentration which corresponds to the extinction of the clear solution. The method is applicable to substances with aqueous solubilities from 0.5 to $1000 \, \mu g \, \mathrm{cm^{-3}}$.

Fürer and Geiger (1977) prepared a microcrystalline suspension of the solid solute in water by means of a micro mill containing glass beads. The abraded glass particles can introduce additional turbidity and mild alkalinity, which must be subtracted and corrected, respectively. When the suspension is diluted in a stepwise fashion, there is a sudden disappearance of turbidity as soon as the aqueous concentration range is reached. The method is simple and rapid, and is suitable for determining aqueous solubilities in the range > 1000 to $\ll 1 \, \mu g \, \mathrm{cm^{-3}}$. Fürer and Geiger reported solubility values of pesticides at 20°C which, in most cases, agree to within a few percent of other literature values.

If the sample is contaminated with a less-soluble impurity, the synthetic methods give too low a value for the solubility of the main substance. They are inherently less accurate than the analytical methods, mainly because of the possibility that the system may not be close to equilibrium.

8.10   DETERMINATION OF PARTITION COEFFICIENTS

The usual method of measuring partition coefficients is to shake a solution of the solute in one (or both) solvents with the other solvent. When mass-transfer equilibrium and thermal equilibrium have been achieved, the two solutions are allowed to separate and are analyzed for the solute. A shake-flask or separatory funnel enclosed in a suitable jacket attached to a thermostatic system is often used for this purpose. Centrifugation may be used to facilitate and hasten the separation of the liquid layer. This type of batch method is laborious and frequently inaccurate and does not lend itself to studies at different temperatures. The partition coefficients so determined at different temperatures are not normally accurate enough for the calculation of enthalpies of transfer using the van't Hoff isochore (Eqs. 1.8 and 1.9). A complete determination of ΔH and ΔS is therefore not possible.

Reinhardt and Rydberg (1970) developed a much-improved partitioning apparatus, known as AKUFVE, which is available commercially and is manufactured in Sweden. AKUFVE consists of a mixer which provides rapid and efficient contact between two immiscible solvents, a unique centrifugal separator that ensures their rapid and complete separation, and connecting ports which allow sampling and flow facilities for measurement of the equilibrium concentrations of the solute in both solvents. Equilibrium normally takes a few minutes to achieve.

The experimental conditions can easily be changed so that a set of partition coefficients at different temperatures, pH values, or salt concentrations can be measured in a few hours. Davis and Elson (1974) have assessed the performance of AKUFVE using a range of solutes at different concentrations, a range of solvents with an aqueous phase of different pH values and ionic strengths, and a range of temperature. The derived partition coefficients agreed with literature values and were more accurate and reproducible than those determined by the classical method. Evidently, the AKUFVE apparatus rapidly provides accurate values of partition coefficient over a range of experimental conditions. It is an excellent method for studying the thermodynamics of phase transfer and for determining group contributions to partitioning (Section 7.3).

A rapid mix–filter probe system for rapid determination of partition coefficients was developed by Kinkel et al. (1981). The apparatus is a modification of that developed by Cantwell and Mohammed (1979) for two-phase acid–base titrations. It consists of a thermostated mixing chamber with a magnetic stirrer for vigorous mixing. The water and oil phases are sampled using filters of hydrophilic or hydrophobic material, respectively, linked to a high-performance pump. The phase under study is analyzed by pumping it through the flow cell of a UV spectrophotometer and returning it to the mixing chamber. The data, and their precision and accuracy, were found to be the same as those using the conventional shake-flask method. The mix–filter probe system is, however, more rapid (see Dunn et al., 1986).

If the organic solute which undergoes partitioning is unstable, decomposition will upset the distribution equilibrium and these methods of determining the partition coefficient P may not be applicable. Byron et al. (1980) used a stirred transfer cell containing equal volumes of the two solvents and determined P by examining the competing kinetics of partitioning and decomposition as shown in Scheme 8.4. The transfer cell is similar to that used by Higuchi and

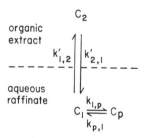

Scheme 8.4 The kinetics and equilibria of a solute which undergoes simultaneous partitioning and decomposition in the aqueous phase. C_1 is the concentration of solute (cyclohept-2-enone) in solvent 1 (water), C_2 is the concentration of solute in solvent 2 (light liquid paraffin), and C_p is the concentration of decomposition product (3-hydroxycycloheptanone) in water. The k values are the individual rate constants. The partition coefficient $P_{o/w} = k'_{1,2}/k'_{2,1}$. The equilibrium constant for pH-catalyzed decomposition in water $K = k_{1,p}/k_{p,1}$.

Michaelis (1968) in their studies of the influence of the solvation of ion pairs on their partitioning (Section 9.3.4).

As in the analogous case of solubility, a full thermodynamic description of the partitioning process requires knowledge of both ΔH_t^{θ} and ΔS_t^{θ}, as well as ΔG_t^{θ} ($= -RT \ln P$). Arguments similar to those presented in Section 8.9.1 indicate that ΔH_t^{θ} (and hence ΔS_t^{θ}) may be determined with greater accuracy using calorimetry than by applying the van't Hoff isochore (Eq. 1.9) to values of P measured at a range of temperatures. This has been achieved by Beezer et al. (1980, 1983) for m-alkoxyphenols partitioning between water and 1-octanol (mutually saturated). The enthalpy of solute transfer in partitioning from liquid α to liquid β may be readily calculated from the calorimetric enthalpies of solution of the solute in each of the two liquids, thus,

$$\Delta H_t^{\theta}(\alpha \to \beta) = \Delta H_s^{\theta}(\text{in } \beta) - \Delta H_s^{\theta}(\text{in } \alpha) \tag{8.56}$$

If the rate of solution of the solute in one of the solvents is too slow for accurate calorimetric measurements (as in the case of certain phenols dissolving in water), a two-phase titration method may be employed (Haberfield et al., 1984).

As with other processes of mass transfer, such as diffusion of solutes in solutions (Section 6.10.2) and dissolution of solids in liquids (Chapter 11), partitioning of a solute from one liquid across a phase boundary to another liquid proceeds at a finite rate. Rates of partitioning have been the subject of kinetic study (e.g., Guy and Honda, 1984; De Meere and Tomlinson, 1984). Detailed discussion of the kinetics of liquid–liquid phase transfer and partitioning is not considered to fall within the scope of this book.

8.11   INFLUENCE OF COSOLVENTS AND SALTS ON AQUEOUS SOLUBILITY

This section discusses the influence of organic solvents and salts on the aqueous solubility of organic compounds in the absence of specific complexation, which is considered in Chapter 10.

The volume fraction ϕ is usually found to be the most appropriate concentration scale for expressing the composition of a mixture of solvents, especially when the volumes of the solvent molecules differ by more than about 30%. Volume fraction expresses the relative influence of each solvent on the mean environment of the solute molecules in the absence of specific complexation. For a mixture of two solvents A and B, $\phi_A + \phi_B = 1$. In the following discussion water will be denoted by A and the organic cosolvent by B. The logarithm of the solubility c_2^s of an organic solute 2 is frequently found to be a linear function of ϕ_A and ϕ_B, thus,

$$\log c_2^s = \log (c_{2,B}^s) + k_{2,A} \phi_A \tag{8.57}$$

$$= \log (c_{2,A}^s) + k_{2,B} \phi_B \tag{8.58}$$

where $k_{2,A} = -k_{2,B}$ are constants which reflect the sensitivity of $\log c_2^s$ to ϕ_A and ϕ_B, respectively, $c_{2,B}^s$ is the solubility in pure $B(\phi_B = 1, \phi_A = 0)$, and $c_{2,A}^s$ is the solubility in pure A ($\phi_B = 0, \phi_A = 1$).

Equation 8.58 is illustrated by Figures 8.6 and 8.7. For a binary solvent mixture Eq. 8.58 is a necessary corollary of Eq. 8.57. Yalkowsky and Valvani (1977) have recommended that these equations be used in pharmaceutical formulation for expressing the solubilization of many drugs (e.g., steroids and diazepines) by a variety of water-miscible cosolvents [e.g., ethanol, propylene glycol, and polyethylene glycol (PEG) 400]. By means of Eqs. 8.57 and 8.58, Yalkowski and Valvani (1977) proposed a simple graphical technique for predicting whether a solution of a solubilized drug will become supersaturated and thus will have a potential for precipitation when it is diluted with water.

If, as is usually the case, the solid drug 2 exists in the same physical form in the presence of each mixture of water + cosolvent, its activity a_2 is constant with respect to any standard state. Therefore, according to Eq. 2.5, Eqs. 8.57 and 8.58 simply reflect the influence of solvent composition on the activity coefficient γ_2 of the solute, thus,

$$\log \gamma_2 = \log (\gamma_{2,B}) - k_{2,A} \phi_A \tag{8.59}$$

$$\log \gamma_2 = \log (\gamma_{2,A}) - k_{2,B} \phi_B \tag{8.60}$$

where $k_{2,A} = -k_{2,B}$ have the same significance as before and also reflect the sensitivity of $\log \gamma_2$ to ϕ_A and ϕ_B, respectively, while $\gamma_{2,B}$ is the activity coefficient of the solute in pure B, and $\gamma_{2,A}$ is that in pure A.

In the following discussion, which is based on Yalkowski and Valvani (1977), Gould et al. (1984), and Section 8.6, we shall take water as A and either alcohols, glycols, or PEG 400 as B. Relatively nonpolar solutes such as alkyl 4-aminobenzoate esters or tioconazole (Scheme 8.5) give linear plots (e.g., Figs. 8.6 and 8.7) according to Eqs. 8.57–8.60, while the $k_{2,B}$ values are positive since the environment of the dissolved solute is becoming more favorable as ϕ_B increases.

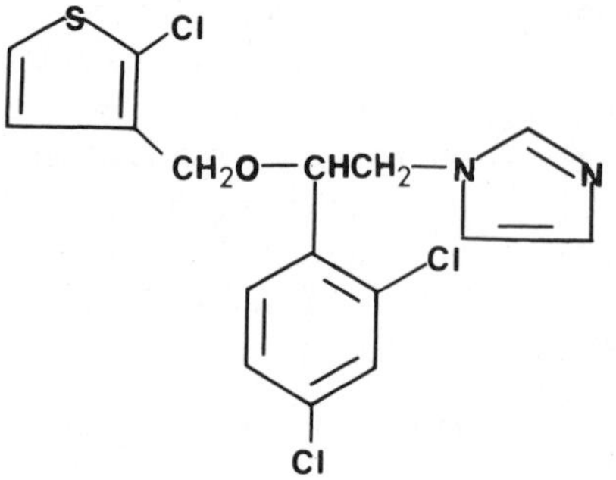

Scheme 8.5 The molecular structure of tioconazole.

A relatively polar solute such as oxfenicine [L-2-(4-hydroxyphenyl)glycine] also gives linear plots, but the $k_{2,B}$ values are negative (e.g., Fig. 8.11) since the environment of the dissolvent solute is becoming less favorable as ϕ_B increases.

A solute of intermediate polarity, such as caffeine, will often give plots with maxima (e.g., Fig. 8.12) or occasionally minima, indicating that values of $k_{2,B}$ are not constant. The origin of these maxima (and minima) have a complex

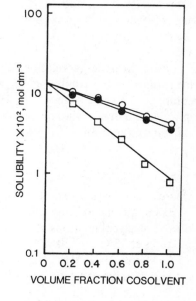

Figure 8.11 Plot of solubility (log scale) versus *vs.* cosolvent volume fraction (in water) for a polar drug oxfenicine [L-2-(4-hydroxyphenyl)glycine]. Cosolvent: polyethylene glycol 400 (○); propylene glycol (●); ethanol (□). $1 \, \text{mol dm}^{-3} = 1 \, \text{mol L}^{-1} = 1 \, M$. Reproduced with permission of the copyright owner, Elsevier Science Publishing Co., Inc., from Gould, P. L., Goodman, M., and Hanson, P. A. (1984). *Int. J. Pharm.*, **19**, 149–159.

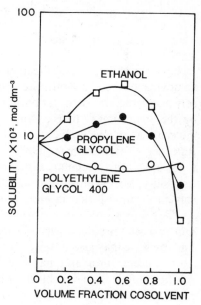

Figure 8.12 Plot of solubility (log scale) versus cosolvent volume fraction (in water) for a semipolar drug, caffeine. The cosolvents are ethanol, propylene glycol, and polyethylene glycol 400. $1 \, \text{mol dm}^{-3} = 1 \, \text{mol L}^{-1} = 1 \, M$. Reproduced with permission of the copyright owner, Elsevier Science Publishing Co., Inc., from Gould, P. L., Goodman, M., and Hanson, P. A. (1984). *Int. J. Pharm.*, **19**, 149–159.

thermodynamic basis, since they are a consequence of the influence of both enthalpy and entropy terms on the partial molar free energy (chemical potential) of the solute (Franks, 1983). These maxima (and minima) may be treated semiempirically by means of specific complexation models (Higuchi and Connors, 1965), regular solution theory (Gould et al., 1984), an extended solubility parameter approach (Martin et al., 1980), a NIBS approach (Acree and Bertrand, 1979; Acree and Rytting, 1982; Acree, 1984), an excess free-energy approach (Williams and Amidon, 1984a, b, c), UNIQUAC and UNIFAC treatments (see Section 7.10). Although the lists of semiempirical treatments and also theoretical treatments are extensive, definitive explanation has not yet been achieved.

Linear free-energy relationships have been shown to exist between the logarithms of the solubilities c_1^s and c_2^s of different but structurally related compounds 1 and 2 as they vary with changes in solvent or solvent composition in water and hydroxylic solvents (Manzo, 1982). The linear free-energy relationships have the form

$$\log c_2^s = e \log c_1^s + f \tag{8.61}$$

where e and f are constants which depend on the given pair of solutes.

Examples include any pair of alkyl 4-hydroxybenzoate esters in water + propylene glycol mixtures at 37°C, any pair of α-amino acids in water + ethanol mixtures at 25°C, and alkyl 4-hydroxybenzoate esters in a series of n-alkanols at 25°C (Manzo, 1982). If there is no change in physical form of each solid, so that its activity remains constant, Eq. 2.5 may be applied and Eq. 8.61 becomes

$$\log \gamma_2 = a \log \gamma_1 - b \tag{8.62}$$

where γ_1 and γ_2 are the activity coefficients of solutes 1 and 2 in a given solute medium.

Equations 8.61 and 8.62 are analogous to Eq. 7.37 and have an origin which is analogous to the equations discussed in Section 5.5. Furthermore, the solvent–solvent interaction energy for one type of solute molecule is presumably influenced by the environment to the same extent as that for the other type of solute molecule. Moreover, a change in entropy parallels a change in energy (Lumry and Rajender, 1970: Tomlinson, 1983), so that the free-energy changes, as reflected by $\log \gamma$, are linearly related. Equations 8.61 and 8.62 have obvious predictive capability, for example, in predicting the solubility of an amino acid in one solvent medium if that in another solvent medium is already known.

The composition of binary solvent mixtures may be expressed in mole-fraction units x when the component solvents have similar molecular sizes, and the mole-fraction scale may also have some advantages for certain theoretical treatments. This concentration unit has been used in thermodynamic investigations of solutions of alkali metal chlorides in water + methanol and in water + acetonitrile (De Valera et al., 1983) and of naphthalene in dilute aqueous solutions of ethanol (up to 0.07 mole fraction; Bennett and Canady, 1984). A plot of

ΔG_s^θ(naphthalene) against x(ethanol) is linear. ΔH_s^θ(naphthalene) is positive and increases as x(ethanol) increases, but is overwhelmed by the large, positive concomitant ΔS_s^θ(naphthalene) term. The net solubilization of naphthalene by ethanol is greater the higher the temperature.

Ethanol tends to solubilize the aromatic hydrocarbon probably by loss of water structure as hydrophobic interactions take place between the hydrocarbon and the alkyl group of ethanol. Dispersal of an iceberg or clathrate structure probably makes a significant contribution. This simple explanation may be applied to drug molecules in aqueous cosolvents, since most drugs have significant hydrophobic moieties (see Section 8.3). However, the functional groups in the drug and cosolvent molecules may exert significant and, occasionally, overwhelming effects.

Salts and other electrolytes can either increase or decrease the solubility of organic solutes according to equations analogous to Eqs. 8.58 and 8.60. However, the concentration of the salt B is not usually expressed as its volume fraction, but as its molar concentration c_B or molality m_B. Rarely is mole fraction x_B used. The "salt effect" was originally investigated by J. Setschenow (1889; sometimes stated as I. M. Sechenov), who proposed an empirical equation which may be stated in the form

$$\log c_2^s = \log(c_{2,o}^s) + k_2 c_B \tag{8.63}$$

where c_2^s is the molar solubility of the organic solute 2 in an aqueous solution of the salt B of molar concentration c_B; $c_{2,o}^s$ is the molar solubility of 2 in the absence of salt and k_2 is the salt-effect parameter, which here has the units $dm^3\, mol^{-1}$ (i.e., M^{-1}).

Clever (1983) drew attention to the various modes of expressing the salt-effect parameters, depending on the units used to express the solubility of the solute 2 and the concentration of the electrolyte B. Clever (1983) also stated the factors necessary for converting values of the salt-effect parameters from one pair of concentration scales to another. The choice of units depends on the application. We employ molarity for both c_2^s and c_B for practical reasons.

As we have seen for cosolvents, the origin of the salt effect is the influence of the electrolyte B of concentration c_B on the activity coefficient γ_2 of the nonelectrolyte 2. If the activity of the excess nonelectrolyte is constant, Eq. 8.63 leads to

$$\log \gamma_2 = \log \gamma_{2,o} - k_2 c_B \tag{8.64}$$

where $\gamma_{2,o}$ is the activity coefficient of the nonelectrolyte in the absence of electrolyte B. As before, k_2 is the salt-effect parameter, whose units are those of reciprocal concentration. Since Eq. 8.64 shows that electrolytes can change the activity coefficients of nonelectrolytes, the ionic strength of the solution and the nature of the electrolyte should be stated.

Ionic strength is defined as follows:

$$I = (1/2)\sum cz^2 \qquad \text{or} \qquad I = (1/2)\sum mz^2 \tag{8.65}$$

where c is the molar concentration, z is the valence of each type of ion, and the summation $\sum$ includes every type of ion in the solution. The molality m is frequently used to express concentration, especially in aqueous solution. Small differences in the concentrations of electrolytes from experiment to experiment are often eliminated by the addition of a relatively high ionic strength (often known as a swamping concentration) of a named electrolyte (e.g., sodium chloride or potassium chloride), as in the dissolution-rate studies discussed in Chapter 11.

If the added salt decreases the aqueous solubility of the nonelectrolyte, as is usually the case, the nonelectrolyte is said to be "salted-out" and k_2 in Eq. 8.63 is negative. In this case the electrolyte increases the activity coefficient of the solute (Eq. 8.64). A simple explanation is that the ions attract the water dipoles into their hydration sheaths, thereby reducing the effective concentration of free water or increasing the degree of structure within the solvent, making the aqueous environment enthalpically or entropically even less favorable for insertion of hydrocarbon groups.

If the added salt increases the aqueous solubility of the electrolyte, which is a less-common phenomenon, the nonelectrolyte is said to be "salted-in" and k_2 in Eq. 8.63 is positive. In this case the electrolyte decreases the activity coefficient of the solute (Eq. 8.64). Quaternary ammonium salts, guanidinium chloride, and certain nonelectrolytes such as urea salt-in organic compounds. The mechanism of this effect is complex. A simple explanation is that these compounds reduce the degree of structure within the solvent, water, making the aqueous environment entropically and/or enthalpically more favorable for hydrocarbon groups.

For a detailed discussion of salting-out, salting-in, structure-making, and structure-breaking in aqueous solutions, the reader is referred to a series of volumes by F. Franks (1983) entitled *Water, a Comprehensive Treatise*, and also Sarma and Ahluwalia (1973).

Since the presence of salts can alter the activity coefficients of nonelectrolytes in water (Eqs. 8.64, Long and McDevit, 1952), Eq. 3.34 indicates that the partition coefficient between aqueous solution and organic solvents will also be changed. Kojima and Davis (1984) studied the effect of salt concentration on the distribution of phenol between aqueous sodium chloride and carbon tetrachloride at various temperatures (10–30°C). Apparent $P(CCl_4/\text{aqueous NaCl})$ values increased with increasing salt concentration. This implies that γ_2 in water is increasing and that k_2 in Eqs. 8.64 and 8.63 is negative, indicating that phenol is being salted-out of water by sodium chloride. $\Delta H_t^\theta(\text{aqueous} \rightarrow CCl_4)$, calculated from the temperature dependence of the apparent partition coefficient, increases with increasing salt concentration, but $\Delta S_t^\theta(\text{aqueous} \rightarrow CCl_4)$ is constant. Evidently, the change in distribution is enthalpy-controlled. The positive enthalpy of transfer decreased with decreasing water activity, which was considered to reflect the concentration of free water.

The special property of certain salts to cease behaving as ordinary electrolytes in particular solvent environments is considered in some detail in the next chapter.

Chapter IX

ION PAIRS AND SOLUBILITY BEHAVIOR

9.1 INTRODUCTION

Oppositely charged ions are capable of acting as independent entities or of interacting strongly to form a new species known as an ion pair. An ion pair may behave as a neutral molecule or as an ion of lower resultant charge than its constituent ions.

The solubility behavior of an organic ion pair is, generally speaking, not very different from that of a polar covalent molecule, since the charges of the cation and anion partially cancel each other. Although chemists, biochemists, and pharmacists do not normally dissolve solid ion-pair materials, they may be frequently concerned with the transfer of ion pairs from one liquid phase to another and with the analysis of chemical species which are capable of forming ion pairs. The partitioning of ion pairs may be used to elucidate the characteristics of physicochemical processes such as thermodynamic changes and group contributions, as we have seen in Section 7.2.4.

Ion pairs that are familiar to organic chemists include organometallic compounds containing metals such as magnesium (Grignard reagents), lithium, and thallium; these compounds are soluble in diethyl ether, but are decomposed by water. Ion pairs known to pharmacists include certain salts, such as chlorpromazine hydrochloride and dextromethorphan hydrobromide, the latter being more soluble in chloroform than in water (Windholz et al., 1983). The ability of ion pairs to partition from water into organic solvents enables them to be absorbed by tissues more easily than the constituent ions. Ion pairs known to analytical chemists include those that form the basis of liquid–liquid extraction and of certain ion-selective electrodes.

This chapter is concerned with the solubility and partitioning behavior of ion pairs, the latter being related not only to the interaction between the ion pairs and the organic solvent, but also to the interactions between the individual ions and the aqueous phase. Before discussing solubility and partitioning of ion pairs, it is appropriate to consider the interactions between oppositely charged ions and the solvent, since these control solubility and partitioning behavior.

9.2 FORMATION OF ION PAIRS

9.2.1 Introduction

The pairing of oppositely charged ions is mainly a result of the coulombic, electrovalent, heteropolar, or ionic interaction. This force is discussed in

Section 2.11.2, together with the relevant equations for the potential energy (Eqs. 2.119 and 2.120). The potential energy is, of course, negative for the coulombic attraction between two oppositely charged ions. As we shall see, the formation of ion pairs may be assisted by van der Waals forces and by solvent structure. In the case of water, hydrophobic interactions may be involved.

The topic of ion pairs has been reviewed by Szwarc (1969) and has been the subject of a symposium which has particular relevance to the pharmaceutical sciences (Symposium on Ion-Pair Partition, 1972).

The lifetime of ion pairs is at least 10^{-5} s, which is equivalent to about 10^8 molecular vibrations, thereby clearly demonstrating that ion pairs may be regarded as independent species (Szwarc, 1972).

The formation of ion pairs may be studied by various methods, many of which have been discussed by Nancollas (1966). Methods of particular value include conductance studies, UV-visible spectrophotometry, IR spectrophotometry (Edgell et al., 1966), electron spin resonance (ESR) spectroscopy (Szwarc, 1969), and partition, distribution, or solvent extraction, with which we are particularly concerned.

9.2.2 Influence of the Solvent

The interacting ions may be organic (e.g., a cationic and an anionic surfactant in water) or inorganic (e.g., sodium chloride in liquid ammonia) or one may be organic and the other inorganic (e.g., dextromethorphan hydrobromide in chloroform). The importance of the solvent, and indeed the very existence of ion pairs, was first recognized by Bjerrum (1926). He observed that sodium chloride, which behaves as a strong electrolyte in water due to complete ionic dissociation, behaves as a much weaker electrolyte in liquid ammonia, suggesting incomplete ionic dissociation in this solvent. Bjerrum put forward a theory of ion–ion association which was based on electrostatic principles such as those embodied in Coulomb's law, namely, Eq. 2.119. On reducing the dielectric constant D of the solvent, the energy of attraction between oppositely charged ions is increased, permitting the ions to associate. Furthermore, the lower the dielectric constant of the solvent, the less polar it is and the weaker are the solvent–ion interactions. The reduction in competition of each ion for the solvent facilitates ionic association. The association of the oppositely charged ions reduces or neutralizes the electrostatic charges and consequently reduces the electrical conductivity.

Grunwald (1954) allowed for the existence of solvation and obtained the curve shown in Figure 9.1 for the variation of potential energy of interaction with separation r. This curve shows two relatively stable regions of minimum potential energy, a and s, which represent two different types of ion pairs. Point s represents a "solvent-separated ion pair," which is sometimes referred to as a "loose ion pair." To begin with, two oppositely charged ions in a solvent capable of solvating them are separated by a large distance. As the two oppositely charged ions approach each other, the coulombic attractive forces cause the potential energy to fall to a minimum at s. Smaller separations cause the solvent molecules which surround or solvate the ions to come into contact and therefore

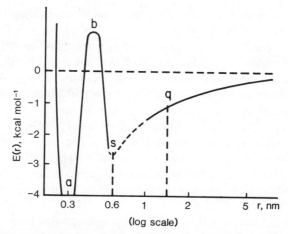

Figure 9.1 Dependence of the potential energy $E(r)$ of an ion pair on its distance of separation r (log scale): a is the well of the contact ion pairs; b is the maximum potential energy that creates a barrier between r and s; s is the position of the solvent-separated ion pairs, and q is the separation corresponding to the critical distance according to Bjerrum's theory (1:1 electrolyte in dielectric constant of 20 at 298 K). $1\,nm = 10\,Å$. $1\,cal = 4.184\,J$. Reprinted with permission from Grunwald, E. (1954). *Anal. Chem.*, **26**, 1696–1701. Copyright 1954 by the American Chemical Society.

to be squeezed out. Repulsive forces then cause the potential energy to rise to a maximum at b. At even shorter distances, when sufficient solvent has been squeezed out from between the ions, a strong ion attraction again occurs which allows the potential energy to fall to a new, lower minimum at a. Point a represents a "contact ion pair," which is also known as a "tight ion pair."

Further reduction of the separation causes the normal forces of repulsion to rise rapidly as the electron clouds of the ions interact. Point b represents the summit on an activation energy barrier between the two types of ion pair. The covalent species may not be stable and then represents a repulsive configuration arising from very close approach. Not all of the species shown in Scheme 9.1 may be formed by the two given ions. The solvent-separated ion pairs can only form when at least one of the ions is coordinated with solvent molecules (Grunwald, 1954; Evans and Gardam, 1969). The association of quaternary ammonium ions with anions in amphiprotic solvents, aqueous or nonaqueous, has been interpreted by Evans and Gardam (1969) in terms of the tight and loose ion pairs of Grunwald (1954). Their view rests on the belief that association

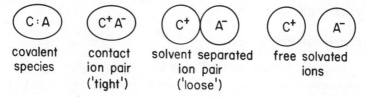

Scheme 9.1 Possible interactions between two oppositely charged ions in solution. The ellipses and circles represent the solvent sheaths.

of the salts is influenced by a specific cation–anion interaction which depends on the decrease in anion solvation.

Following Bjerrum's (1926) original observation and his theory for ion pairing of strong electrolytes, it was first believed that ion pairs could only exist in solvents of low dielectric constant. However, Diamond (1963) has shown that ion pairs can form in water if both the cation and anion are large. This is because the disturbance of the water structure by these large ions is reduced when the ions are forced into a single larger cavity by the associated water molecules. Thus, the Bjerrum view of electrostatic ion–ion interaction does not properly explain ion pairing in aqueous solutions and in other highly structured solvents. Rather, ion pairing in these structured phases is forced by the tendency of the water molecules to arrange themselves so as to minimize the disturbance to the solvent structure.

Ion pairing in water can be reinforced by hydrophobic interactions and by van der Waals forces between the ions. Indeed, owing to differences in solvent structure, the mechanism of ion pairing in water is often assumed to be different from that in other solvents, such as alcohols. This view, along with Diamond's (1963) suggestion that the difference could be explained by the active influence of water structure in enforcing ion pairing, has been confirmed by Accascina et al. (1972).

Ion pairs for which the stoichiometry indicates no resultant charge may nevertheless behave as dipolar molecules, because the positive and the negative centers can never coalesce completely. It is often possible to liken an ion pair to the zwitterions, such as those in the molecule of phosphatidyl choline, with close charge–separation. The creation of a hole in the solvent for the ion pair is associated with an increase in free energy which must be offset by a greater reduction in free energy by solvation of the ion-pair "dipole." This may be achieved by the usual attractive solute–solvent interactions previously discussed, namely: (1) a polarizable solvent, such as an aromatic molecule, which is capable of taking part in London, Debye, and charge-transfer interactions; (2) a polar solvent, such as acetonitrile, which is capable of taking part in Keesom, Debye, and London interactions; (3) a hydrogen-donor solvent, such as chloroform, a hydrogen-acceptor solvent, such as ether, or an amphiprotic solvent, such as ethanol, all of which are capable of hydrogen bonding. In order for ion pairs to be extracted from an aqueous phase by an organic phase, the high degree of solvation in water must be replaced by one or more of these solute–solvent interactions in the organic phase.

9.3   EXTRACTION OF ION PAIRS

9.3.1   Simple Ion-Pair Extraction

9.3.1.1   Influences of the Organic Phase and of the Relative Ionic Sizes

The formation of an ion pair causes a reduction in the overall electric charge of the original ions and a partial shielding of the charged centers. As a result, the aqueous solubility is reduced, but the solubility in organic solvents is increased. When an organic solvent, which is immiscible with water, is in contact with such an aqueous system, the disturbance to the water structure will be minimized by transference of the neutral ion pair across the phase boundary into the nonaqueous solvent. This can easily be explained in terms of free-energy differences. Organic ion pairs therefore tend to display high partition coefficients (organic solvent–water and oil–water). If the ions are unable to associate in water, perhaps because of strong hydration, they will tend to associate in the diffusion layer between the two phases where the dielectric constant falls sharply. Such transfer is often provoked by the presence of specific complexing agents or ligands in the nonaqueous organic phase (Hull and Biles, 1964; Higuchi et al., 1971).

The role of the organic phase in promoting ion-pair extraction has been explained in terms of the influence of the dielectric constant (Divatia and Biles, 1961) or solubility parameter (Freiser, 1969) of the solvent. However, Higuchi et al. (1971) have shown that the extraction data can be better explained in terms of specific solvation of the ion pairs in the organic phase. The evidence for this mechanism is given below.

As stated earlier, the presence of a solvating agent in the organic solvent frequently increases the distribution ratio of ion pairs in favor of the organic solvent. Higuchi et al. (1967) suggested that ion-pair solvation or, in other words, the masking of the ionic character of the ion–ion bond, would be a significant factor in the extractive equilibrium if the ion-pair structures were oriented in such a way as to expose surfaces of high charge density. Lipophilic ion pairs may, in a simplified view, be classified into three possible cases, as shown in Scheme 9.2.

Case I of Scheme 9.2 comprises a large cation, which is hydrophobic except for the positively charged center, paired with a small exposed anion, which would be expected to carry a relatively high negative charge per unit surface area. Such a system would be most effectively solvated by lipophilic molecules which have an exposed positively charged surface, such as dipolar molecules with acidic protons (e.g., chloroform, phenols, and alcohols). Since the solvating molecules would have their polar end buried adjacent to the anion, the appearance presented to the environment by the solvated ion pair is that of a relatively nonpolar aggregate.

Case II represents the situation opposite to that for Case I, such that the cationic charge is largely exposed. Lipophilic molecules which contain

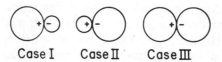

Case I   Case II   Case III

Scheme 9.2 Various types of ion pairs involving organic ions. Reprinted with permission from Higuchi, T., Michaelis, A., Tan, T., and Hurwitz, A. (1967). *Anal. Chem.*, **39**, 974–979. Copyright 1967 by the American Chemical Society.

nucleophilic sites would be particularly effective in solvating this type of ion pair (e.g., ethers, ketones, amides, and phosphate esters). Examples of Case II include certain organometallic compounds, particularly the Grignard reagents (e.g., $R^-Mg^{2+}Br^-$) and the organolithium compounds (Li^+R^-). These compounds are particularly soluble in ether, which stabilizes the ion pairs and provides an excellent anhydrous medium for reaction.

Case III comprises an ion pair with deeply buried charges. This may be termed a "buried ion pair." Having no exposed electrically unbalanced surface, it presents a nonpolar surface to the environment so that it would be readily extractable by nonpolar solvents. Solvation would then be minimal (Shinoda, 1954). Examples of Case III ion pairs include the 1:1 complexes between the dye Orange II (i.e., sodium 4-[2-hydroxy-1-naphthylazo]benzenesulfonate) and each of four long-chain quaternary ammonium salts (1-hexadecylpyridinium chloride, dodecylquinolinium bromide, [2-phenoxyethyl]dodecyldimethyl-ammonium bromide, and [diisobutylphenoxyethoxyethyl]dimethylbenzyl-ammonium chloride) (Zografi et al., 1964b). These ion-pair complexes are responsible for the enhanced distribution ratio of the dye, determined by visible spectrophotometry, from water (at a wide range of pH values and at various ionic strengths) to an organic solvent (chloroform).

On mixing solutions of the original salts in water containing sodium hydroxide, Zografi et al. (1964b) obtained the 1:1 ion-pair complexes as oils which, after extraction with a suitable organic solvent (ethyl acetate), could be crystallized as solids with melting points ranging from 90 to 142°C. The solubility products of the solid complexes were of the order 10^{-11} and $10^{-10} \, mol^2 \, dm^{-6}$ from 25 to 40°C and the enthalpies of solution ranged from 10 to 40 kcal mol^{-1} (42–170 kJ mol^{-1}). The corresponding entropies of solution were not favorable enough to compensate for the unfavorable enthalpy change; consequently, the aqueous solubilities of the complexes were small. This behavior emphasizes the fact that Case III ion pairs do not behave as electrolytes but as hydrophobic organic molecules, owing to the shielding of the ionic changes. In Section 9.5 we discuss further examples of Case III ion pairs.

These principles are well illustrated by the ion-pair extraction of amines such as dextromethorphan or chlorpheniramine from the aqueous phase into an organic solvent (Higuchi and Kato, 1966; Higuchi et al., 1967, 1971; Michaelis and Higuchi, 1969).

9.3.1.2   Extraction of Dextromethorphan

The pharmaceutical monoprotic tertiary amine dextromethorphan (d-3-methoxy-N-methylmorphinan) can, after protonation in water, be extracted as an ion pair with any one of a number of anions into cyclohexane containing some chloroform (Higuchi et al., 1967). The structure of dextromethorphan, given below, indicates that it has a large hydrophobic moiety. Since the hydrobromide salt is reported to be "soluble in chloroform" and "practically insoluble in ether" (Windholz et al., 1983), one may predict that this ion pair is an example of Case I.

Dextromethorphan

The extraction equilibrium can be written

$$BH^+_{aq} + X^-_{aq} \xrightarrow{K_e} BH^+X^-_{org} \qquad (9.1)$$

aqueous layer   organic layer

where BH^+_{aq} represents the protonated base in the aqueous phase, X^-_{aq} the anion in the aqueous phase, $BH^+X^-_{org}$ the ion pair in the organic phase, and K_e the extraction constant for the equilibrium; thus,

$$K_e = [BH^+X^-]_{org}/[BH^+]_{aq}[X^-]_{aq} \qquad (9.2)$$

A more rigorous equilibrium equation must include the contribution made by the dipolar solvating agent, as follows:

$$BH^+_{aq} + X^-_{aq} \underset{-nM_{org}}{\overset{+nM_{org}}{\rightleftharpoons}} [BH^+X^-, M_n]_{org} \qquad (9.3)$$

where M represents the solvating agent and n the effective molecularity of the solvating agent in the ion pair. This equilibrium will be represented by the equilibrium constant

$$K_0 = [BH^+X^-, M_n]_{org}/[BH^+]_{aq}[X^-]_{aq}[M]^n_{org} \qquad (9.4)$$

The experimentally determined quantity is the distribution ratio (or apparent partition coefficient)

$$D = [BH^+X^-, M_n]_{org}/[BH^+]_{aq} \qquad (9.5)$$

Equation 9.4 can therefore be written

$$K_0 = D/[X^-]_{aq}[M]^n_{org} \qquad (9.6)$$

Taking logarithms and rearranging, we obtain

$$\log D = \log K_0 + \log[X^-] + n\log[M] \qquad (9.7)$$

Although the subscripts have been omitted, $[X^-]$ is the concentration of free anion in the aqueous phase and $[M]$ is the concentration of free solvating agent in the organic phase. At constant $[X^-]$, Eq. 9.7 indicates that a plot of $\log D$ against $\log[M]$ should be a straight line of slope n and intercept

$(\log K_0 + \log[X^-])$. The apparent stoichiometry of the solvating agent with the ion pair is therefore given by the slope and the equilibrium constant by the intercept.

The experimental procedure is essentially the same as that used to determine partition coefficients. If the K_0 values are very high, a double-extraction technique may be used (Higuchi et al. 1967). Analysis of the aqueous and organic phases may preferably be carried out using UV-visible spectrophotometry. If the solvating agent also absorbs light, as in the case of phenols, a modified spectroscopic technique (Higuchi and Kato, 1966) or the acid-dye technique (Divatia and Biles, 1961) may be employed.

In all ion-pair extraction experiments, the solution should be buffered to a known pH and ionic strength and the temperature kept constant (e.g., at 25°C). For the ion-pair extraction of amines, it is usual to maintain the pH at a value at which the base is almost totally protonated (e.g., pH 3.0) so that negligible amounts of free base are extracted. The buffer must not itself contain types of ions which participate in ion pairing, such as phosphate or citrate. The same

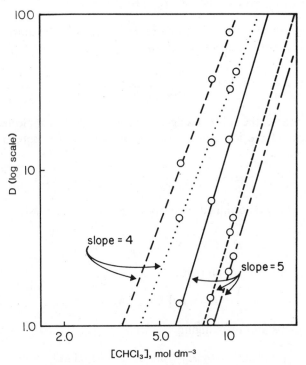

Figure 9.2 Effect of various anions on the distribution ratio D of dextromethorphan as a function of chloroform concentration in cyclohexane at pH 3.00, ionic strength $\mu = 0.11 \, \text{mol dm}^{-3}$, and 25°C. Concentrations: 0.5 mmol dm^{-3} dextromethorphan; 0.10 mol dm^{-3} anion; 0.10 mol dm^{-3} phosphate buffer; K$_2$SO$_4$ to adjust μ. 1 mol dm^{-3} = 1 mol L^{-1} = 1 M. The regression lines were fitted to the experimental data, thus: ---, iodide; ···, ethylsulfonate; ——, bromide; - - -, chloride; — — —, nitrate. Reprinted with permission from Higuchi, T., Michaelis, A., Tan, T., and Hurwitz, A. (1967). *Anal. Chem.*, **39**, 974–979. Copyright 1967 by the American Chemical Society.

consideration applies to the salt used to adjust the ionic strength; for this purpose potassium sulfate would normally be a good choice. It should be ascertained whether the distribution ratio D is proportional to the concentration of each counter ion which participates in ion pairing.

Figures 9.2 and 9.3 show plots of $\log D$ against $[M]$ according to Eq. 9.7 for various anions with M = chloroform. The linearity of the plots provides support for the model. The values and units of K_0 depend on the values of n, and so cannot be compared for different values of n. On the other hand, the value of K'_0 in the following equation, which has the dimensions of (concentration)$^{-1}$, does not depend on n and can be used for comparison purposes, as in Table 9.1.

$$K'_0 = K_0[M]^n_{org} = [BH^+X^-,M_n]_{org}/[BH^+]_{aq}[X^-]_{aq} = D/[X^-]_{aq} \quad (9.8)$$

K'_0 represents the value of K_0 at unit concentration of solvating agent, that is, $[M] = 1$ mol dm^{-3}.

Table 9.1 lists values of n and K'_0. For ions possessing highly charged surfaces, such as halides, nitrate, ethylsulfonate, and benzenesulfonate, about 4–5 chloroform molecules are bound to each ion pair. For trichloroacetate and picrate

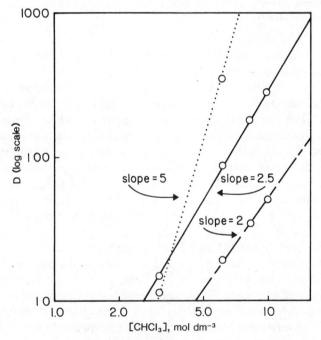

Figure 9.3 Effect of various anions on the distribution ratio D of dextromethorphan as a function of chloroform concentration in cyclohexane at pH 3.00, ionic strength $\mu = 0.11$ mol dm^{-3} at 25°C. Concentrations: 0.5 mmol dm^{-3} dextromethorphan; anion concentrations as shown; 0.10 mol dm^{-3} phosphate buffer; K_2SO_4 to adjust μ. 1 mol dm^{-3} = 1 mol L^{-1} = 1 M. The regression lines were fitted to the experimental data, thus: ———, 0.10 mol dm^{-3} trichloroacetate; ···, 1.0 mol dm^{-3} benzenesulfonate; ———, 1×10^{-3} mol dm^{-3} picrate. Reprinted with permission from Higuchi, T., Michaelis, A., Tan, T., and Hurwitz, A. (1967). *Anal. Chem.*, **39**, 974–979. Copyright 1967 by the American Chemical Society.

Table 9.1 Calculated Extraction Constants K_0' for Ion-Pair Extraction of Dextromethorphan with Various Anions and the Apparent Binding Numbers of Chloroform

Anion	K_0' $(dm^3\,mol^{-1})^a$	Average Number (n) of $CHCl_3$ Molecules Associated with Each Ion Pair
Nitrate	4.0×10^{-4}	4.98
Chloride	3.0×10^{-4}	4.94
Bromide	3.18×10^{-3}	5.02
Iodide	6.00×10^{-2}	4.20
Trichloroacetate	7	2.36
Picrate	5.6×10^2	2.08
Ethylsulfonate	3.6×10^{-2}	3.92
Benzenesulfonate	4.1×10^{-2}	4.82

$^a 1\,dm^3\,mol^{-1} = 1\,L\,mol^{-1} = 1\,M^{-1}$.

Reprinted with permission from Higuchi, T., Michaelis, A., Tan, T., and Hurwitz, A. (1967). *Anal. Chem.*, **39**, 974–979. Copyright 1967 by the American Chemical Society.

the resulting ion pair binds few solvating molecules (about 2), probably because of the shielding effect of the organic portion of the anion.

The values of K_0' depend greatly on the nature of the anion, as might be expected on the basis of their relative size and hydrophilic–hydrophobic tendencies. This dependency becomes apparent when comparing the extractability of identical ion pairs in the presence of different solvating agents.

When the primary aliphatic alcohol 1-pentanol was the solvating agent M, plots of log D against log[M] were linear, corresponding to $n = 1.96$ for the bromide and $n = 1.50$ for the trichloroacetate. Since 1-pentanol exists as an associated species (trimer, tetramer, pentamer, or hexamer, probably the pentamer) in cyclohexane as discussed in Chapter 6 (Sections 6.3.1–6.5.2), the actual molarity may be about five times the graphical values given above, namely, 9.8 for the bromide and 7.5 for the trichloroacetate.

p-tert-Butylphenol would be expected to associate like phenol, which exists largely as a pentamer (Anderson et al., 1979) (or trimer, Lin et al., 1978) in cyclohexane (see Section 6.8). To estimate the amount of monomer in the organic layer, D was determined at a series of concentrations of the phenol and extrapolated to zero concentration. From this extrapolated value, the partition coefficient of the monomer was obtained and hence the monomer concentration for any total phenol concentration was calculated. A plot of log D against log[monomer of *p-tert*-butylphenol] gave a straight line of slope 5.3 for the bromide ion pair.

The values of n in Table 9.2 for the bromide and trichloroacetate ion pairs represent a fairly high degree of solvation. According to Eq. 9.8, the extraction

Table 9.2 Effect of Various Solvating Agents on the Extraction Constant K'_0 for the Ion-Pair Extraction of Dextromethorphan with Several Anions and the Respective Molecularities[a] of the Solvating Agents

Anion	K'_0 (dm^3 mol^{-1})[b]	Molecularity[a]	Solvating Agent
Bromide	3.2×10^{-3}	5.0	Chloroform
	1.7	3.9	1-Pentanol
	5.1×10^7	5.3	*p-tert*-Butylphenol
Trichloroacetate	7.0	2.5	Chloroform
	51	3.0	1-Pentanol

[a] Apparent binding numbers n.
[b] $1\,\mathrm{dm^3\,mol^{-1}} = 1\,\mathrm{L\,mol^{-1}} = 1\,M^{-1}$.

Reprinted with permission from Higuchi, T., Michaelis, A., Tan, T., and Hurwitz, A. (1967). *Anal. Chem.*, **39**, 974–979. Copyright 1967 by the American Chemical Society.

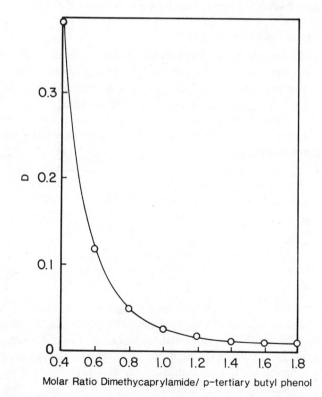

Figure 9.4 Influence of the interaction of dimethylcaprylamide with *p-tert*-butylphenol on the distribution ratio D of dextromethorphan hydrobromide at pH 2.1, ionic strength 0.26 mol dm^{-3}, and 25°C. $1\,\mathrm{mol\,dm^{-1}} = 1\,\mathrm{mol\,L^{-1}} = 1\,M$. Reprinted with permission from Higuchi, T., Michaelis, A., Tan, T., and Hurwitz, A. (1967). *Anal. Chem.*, **39**, 974–979. Copyright 1967 by the American Chemical Society.

constant K_0' in Table 9.2 corresponds to the value of D at which $[M] = 1 \, mol \, dm^{-3}$ and $[X^-] = 1 \, mol \, dm^{-3}$. The extraction constants increase with increasing acidity of the solvating agent, thus: chloroform < 1-pentanol < p-tert-butylphenol.

If dextromethorphan hydrobromide (and other dextromethorphan salts of small anions) form ion pairs belonging to Case I in Scheme 9.2, then addition of an electron-donor solvating agent such as dimethylcaprylamide to the organic phase should not enhance partitioning. When this hypothesis was tested using fixed concentrations of dextromethorphan, bromide ion, and p-tert-butylphenol, addition of dimethylcaprylamide did not, in fact, increase the distribution ratio D, but actually reduced it, as shown in Figure 9.4. This confirms that dextromethorphan hydrobromide belongs to Case I. The reduction in D was found to result from the removal of the solvating agent p-tert-butylphenol by the formation of a complex with the amide. The formation of strong hydrogen-bonded complexes between phenols and N,N-disubstituted amides has been discussed in Sections 4.4.1 and 5.5.6. These results emphasize the need for a proton donor as a solvating agent and the ineffectiveness of an electron donor.

9.3.1.3   Extraction of Chlorpheniramine

A similar set of experiments (Higuchi and Kato, 1966) was carried out with the pharmaceutical diprotic tertiary amine chlorpheniramine, whose structure, given below, also indicates the presence of a large hydrophobic moiety. Since the maleate salt is very soluble in chloroform ($240 \, mg \, mL^{-1}$ at 25°C), but is only "slightly soluble in ether" (Windholz et al., 1983), this ion pair may also be a representative of Case I. The major findings for ion pairs with various anions are described below.

Chlorpheniramine

Chlorpheniramine exists in aqueous solution as a mixture of uncharged, singly charged, and doubly charged species. The pH of the aqueous solution was adjusted to 3.5 and the experimental values of the distribution ratio D, which are given by Eq. 9.5, were made to apply to the singly charged species alone by multiplying by the correction factor $(1 + [H^+]/K_a)$. The ionization constant K_a of the doubly ionized form

$$BH^{2+} \rightleftharpoons BH^+ + H^+ \tag{9.9}$$

was determined spectrophotometrically to be $1 \times 10^{-4} \, dm^3 \, mol^{-1}$ at an ionic

Table 9.3   Calculated Values of the Extraction Constant $K'_0 = K_0[CHCl_3]^n$ for Chlorpheniramine Ion Pairs at Two Chloroform Concentrations and the Power Dependence on Chloroform Concentration[a]

	$(K'_0 = (D/[X^-])(1 + ([H^+]/K_a))$		*Mean Value*[b]
	[CHCl_3]		*of n in*
Anion	$1 \, mol \, dm^{-3}$	$10 \, mol \, dm^{-3}$	$[CHCl_3]^n$
Chloride	4.2×10^{-6}	1.9	5.4
Bromide	1.4×10^{-5}	5.8	5.2
Maleate	4.4×10^{-5}	15	5.4
Trichloroacetate	6.7×10^{-2}	550	3.9
Picrate	7.5×10^{-1}	2.2×10^4	5.2

[a]The extraction constant was corrected using Eq. 9.10 to correspond to a single protonation. $M = CHCl_3$ in Eqs. 9.1–9.11. $1 \, mol \, dm^{-3} = 1 \, mol \, L^{-1}$.

[b]Slope n of plots analogous to Figures 9.2 and 9.3 and represented by Eq. 9.7.

Reproduced with permission of the copyright owner, the American Pharmaceutical Association, from Higuchi, T. and Kato, K. (1966). *J. Pharm. Sci.*, **55**, 1080–1084.

strength of $0.55 \, mol \, dm^{-3}$. The correction given above eliminates the effect of the second protonation.

Plots of $\log D$ against $\log[M]$, where M is chloroform, give linear relationships whose slopes, stated in Table 9.3, indicate that n is close to 5 with a constant concentration of chloride, bromide, maleate, or picrate anions, and about 4 with a constant concentration of trichloroacetate.

Table 9.3 also gives values of K'_0 (defined by Eq. 9.8 and corrected for incomplete ionization) at two different concentrations of chloroform. The rank order of the K'_0 values for various ions with chlorpheniramine agrees with that for the same ions with dextromethorphan.

The buffer anion citrate, and the ionic strength-adjusting anion sulfate, did not form ion pairs even at varied pH values, because D was found to be independent of the concentration of these anions.

To ascertain the relative abilities of the singly protonated BH^+ and doubly protonated BH_2^{2+} forms of chlorpheniramine to form ion pairs, D was determined at various pH values at constant total concentration of chlorpheniramine, at constant concentration of anion (bromide or picrate), and at constant ionic strength and temperature. Figure 9.5 shows the plot of D against pH, with picrate as the anion, and suggests that the doubly protonated species BH_2^{2+} cannot form ion pairs, leaving only the singly ionized species BH^+ to do so. The plot for bromide followed a similar pattern.

When the correction factor $(1 + [H]^+/K_a)$ for double protonation is introduced into Eq. 9.6 we obtain

$$K_0 = \frac{D}{[X^-][M]^n}\left(1 + \frac{[H^+]}{K_a}\right) \tag{9.10}$$

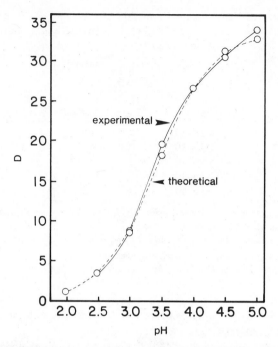

Figure 9.5   pH profile of the apparent partition coefficient D of chlorpheniramine–picrate ion pair. Chlorpheniramine, $0.1\,mmol\,dm^{-3}$; picrate, $2\,mmol\,dm^{-3}$; $0.05\,mol\,dm^{-3}$ citrate buffer; organic phase, 55% chloroform in cyclohexane. $1\,mol\,dm^{-3} = 1\,mol\,L^{-1} = 1\,M$. Reproduced with permission of the copyright owner, the American Pharmaceutical Association, from Higuchi, T. and Kato, K. (1966). *J. Pharm. Sci.*, **55**, 1080–1084.

in which $[M]^n$ refers to the organic phase, $[X^-]$, $[H^+]$, and K_a to the aqueous phase, and D to the molar concentration ratio of chloropheniramine (organic/aqueous). This equation can be arranged to give

$$\frac{1}{D} = \frac{1}{K_0[X^-][M]^n} + \frac{[H^+]}{K_0 K_a[X^-][M]^n} \tag{9.11}$$

Consequently, a plot of $1/D$ against $[H^+]$ should be linear with slope $1/K_0 K_a[X^-][M]^n$ and intercept $1/K_0[X^-][M]^n$.

Figure 9.6 shows that the data with picrate in Figure 9.5 conform to Eq. 9.11. The theoretical curve in Figure 9.4 was calculated from this equation after inserting the appropriate values of n, K_a, and K_0. The bromide data behaved in an analogous fashion. Equation 9.11 indicates that $K_a =$ intercept/slope in Figure 9.6. In fact, $K_a = 1 \times 10^{-4}\,dm^3\,mol^{-1}$, but the ratio (intercept/slope) was found to be $1.84 \times 10^{-4}\,dm^3\,mol^{-1}$ for the picrate system and $3.64 \times 10^{-4}\,dm^3\,mol^{-1}$ for the bromide system. These discrepancies could arise from ion-pair interactions in the aqueous layer.

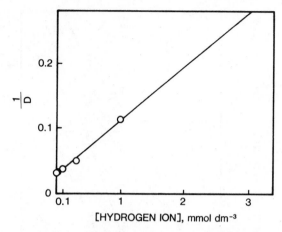

Figure 9.6 Reciprocal apparent partition coefficient D of chlorpheniramine as a function of hydrogen-ion concentration in the presence of $20\,\mu mol\,dm^{-3}$ picrate ion. $1\,mol\,dm^{-3} = 1\,mol\,L^{-1} = 1\,M$. Reproduced with permission of the copyright owner, the American Pharmaceutical Association, from Higuchi, T. and Kato, K. (1966). *J. Pharm. Sci.*, **55**, 1080–1084.

9.3.2 Selective Ion-Pair Extraction

Ion-pair extraction can often be made selective. If, for example, chlorpheniramine and dextromethorphan were present together in the same solution, they could be extracted as follows. At pH 1 and in the presence of bromide ions, dextromethorphan would form ion pairs and could be extracted by the organic phase (Highchi et al., 1967). Chlorpheniramine would exist almost totally in the doubly charged form and would not form ion pairs (Higuchi et al., 1971). After extraction of dextromethorphan, the pH could be increased to 3.5 and chlorpheniramine extracted.

9.3.3 Factors Influencing Ion-Pair Extraction

The distribution of the ion pair between water and an organic solvent is also altered (1) by side reactions (Modin et al., 1971), which can be allowed for by introducing a conditional extraction constant E^* into the left-hand side of Eqs. 9.2, 9.4, or 9.8; (2) by the polarizability of the ions and of the solvent, which influences the London and Debye interactions; and (3) by the ionic strength of the aqueous phase. For the extraction of hydrophobic ions, an increase in ionic strength decreases the extraction constant D, probably by shielding the charge of the interacting ions and consequently inhibiting the formation of the ion pair in the aqueous phase.

As mentioned earlier, ion-pair extraction has been related to factors other than solvation, such as molecular weight of the cosolvent, degree of branching on an aliphatic chain, and dielectric constant (Mukerjee, 1956; Divatia and Biles, 1961; Hull and Biles, 1964). Divatia and Biles (1961), working on the partitioning of amine salts, suggested that the choice of a suitable solvent for

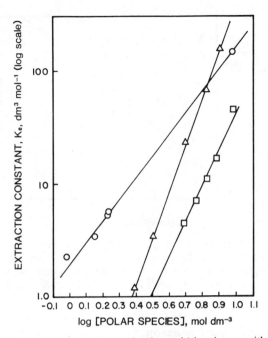

Figure 9.7 Effect of concentration of polar species (log scale) in mixtures with cyclohexane on the extraction constant K_e (Eq. 9.2, log scale) of dextromethorphan hydrobromide at pH 3.00 and 25°C. Concentrations: 0.5 mmol dm^{-3} dextromethorphan phosphate; 0.10 mol dm^{-3} NaBr; 0.10 mol dm^{-3} phosphate buffer. 1 mol dm^{-3} = 1 mol L^{-1} = 1 M. Polar species: chloroform, $\triangle$; cyclohexanone, $\bigcirc$; nitrobenzene, $\square$. Reprinted with permission from Higuchi, T., Michaelis, A., and Rytting, J. H. (1971). *Anal. Chem.*, **43**, 287–289. Copyright 1971 by the American Chemical Society.

extraction could be guided by the dielectric constant (relative permittivity, see Section 2.11.2). In a later study of the distribution of some amine salts of aromatic sulfonic acid dyes between water and a less polar solvent, Hull and Biles (1964) concluded that the formation of solvated species is more important in analyzing the distribution than are bulk properties, such as dielectric constant, of the solvent. The work of Higuchi et al. (1971b) supports the latter view. Figures 9.7 and 9.8 show that the extraction constants K_e (Eq. 9.2) for the chloroform–cyclohexane system often exceed those of either the cyclohexanone–cyclohexane or the nitrobenzene–cyclohexane systems, although the dielectric constants are smaller for the chloroform system.

Freiser (1969) has suggested that ion-pair extraction data can be better explained by regular solution theory than in terms of dielectric constants or specific solvation. However, since ion pairs are polar species, whereas regular solution theory applies to molecules which interact by van der Waals forces, such an interpretation may step outside the limits of the theory. As Freiser (1969) has pointed out, it is frequently possible to obtain good linear correlations between $\log K_e$ and the solubility parameter δ, as shown in Figure 9.9 (with the exception of one point). Such a linear relationship may not necessarily favor

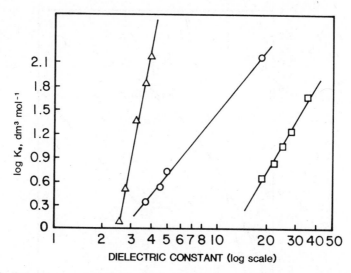

Figure 9.8 Extraction constant K_e (Eq. 9.2, log scale) of dextromethorphan hydrobromide as a function of dielectric constant (log scale) of polar solvent + cyclohexane mixtures at pH 3.00 and 25°C. Concentrations: 0.5 mmol dm^{-3} dextromethorphan phosphate; 0.10 mol dm^{-3} NaBr; 0.10 mol dm^{-3} phosphate buffer. 1 mol dm^{-3} = 1 mol L^{-1} = 1 M. Polar solvent: chloroform, △; cyclohexanone, ○; nitrobenzene, □. Reprinted with permission from Higuchi, T., Michaelis, A., and Rytting, J. H. (1971). *Anal. Chem.*, **43**, 287–289. Copyright 1971 by the American Chemical Society.

regular solution theory since, according to Eq. 2.87, the cohesive energy density $\delta^2 = \Delta U^V/V$ contains a ΔU term which may be constituent of the free-energy term and therefore related to $\log K_e$ and to concentration. One might therefore expect better correlations with δ^2 than with δ, but since δ only varies by about 1 in 9 cal$^{1/2}$ cm$^{-3/2}$, as in Figure 9.9, both δ and δ^2 should give approximately straight lines.

When the solubility parameter of the organic solvent mixture is closest to that of the solute (e.g., dextromethorphan hydrobromide), regular solution theory predicts minimal activity coefficient of the solute and hence maximum solubility and extraction. For such a very polar solute, ΔU^V and δ must be very high compared with those of the solvents. In other words, maximum $\log K_e$ values should occur when δ for the solvent mixture is as large as possible, which is contradicted by Figure 9.9. For example, the nitrobenzene–cyclohexane mixtures (of $\delta = 9.1$–10.0 cal$^{1/2}$ cm$^{-3/2}$) should give larger $\log K_e$ values than chloroform–cyclohexane mixtures (of $\delta = 8.5$–9.0 cal$^{1/2}$ cm$^{-3/2}$). However, the opposite occurs. Thus, at $\delta = 9.0$ cal$^{1/2}$ mol$^{-3/2}$, K_e in the chloroform–cyclohexane mixture is nearly two orders of magnitude greater than that in the nitrobenzene–cyclohexane mixture. Therefore, the ion-pair extraction data cannot be explained by the "like dissolves like" concept of regular solution theory.

The specific solvation theory readily explains why a high concentration of

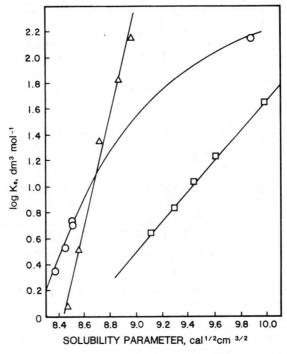

Figure 9.9 Extraction constant K_e (Eq. 9.2, log scale) of dextromethorphan hydrobromide as a function of solubility parameter δ of polar solvent + cyclohexane mixtures at pH 3.00 and 25°C. Concentrations: 0.5 mmol dm^{-3} dextromethorphan phosphate; 0.10 mol dm^{-3} NaBr; 0.10 mol dm^{-3} phosphate buffer. 1 mol dm^{-3} = 1 mol L^{-1} = 1 M. Polar solvent: chloroform, △; cyclohexanone, ○; nitrobenzene, □. Reprinted with permission from Higuchi, T., Michaelis, A., and Rytting, J. H. (1971). *Anal. Chem.*, **43**, 287–289. Copyright 1971 by the American Chemical Society.

chloroform gives higher extraction constants than equivalent concentrations of cyclohexanone or nitrobenzene, as shown in Figure 9.7. (The assumption that activity coefficients are unity in chloroform–cyclohexane mixtures should be quite sound, since Higuchi et al. (1971) have presented evidence that the excess free energy of mixing is very small.) As pointed out above, dextromethorphan hydrobromide belongs to Case I in Scheme 9.2 and is most readily solvated by polar proton donor molecules such as chloroform (and even more effectively by phenols and alcohols). Although nitrobenzene and cyclohexanone have higher solubility parameters than chloroform, they are electron-pair donors and are therefore equipped to solvate the large protonated dextromethorphan cation, though more weakly, since the charge there is less concentrated. Thus, solvation of Case I ion pairs by nitrobenzene or cyclohexanone would be less effective than in Case II, or when the small anion of Case I is solvated by chloroform. To summarize, specific solvation accounts for the data presented above, whereas explanations based on regular solution theory and dielectric constant fall short.

Specific solvation can also provide a useful basis for the selection of solvent media for extraction.

9.3.4 Mechanism and Kinetics of Ion-Pair Extraction

A postulated transfer mechanism (Doyle and Levine, 1967) responsible for extraction of ionic solutes from the aqueous to the organic phase through ion pairing can be represented by Scheme 9.3. The questions which arise are the relative rates of the initial ion pairings in Stage 1 and of the final phase transfer in Stage 2. For inorganic cations capable of forming inner coordination compounds, the rate-limiting step may be the formation of such compounds in the aqueous phase (McClellan and Freiser, 1964; Zolotov, 1965). For quaternary ammonium and several inorganic cations, Davies (1950) has suggested that the rate-limiting step is Stage 2, the slow process being the diffusion of water molecules away from the ions at the interface and resolution by molecules of the organic solvent. In these interphase transfer processes, the nature of the rate-limiting step may depend on the degree of agitation at or near the phase boundary.

Higuchi and Michaelis (1968) studied the rate of extraction of the dextromethorphanium ion paired with various inorganic ions from aqueous buffer to chloroform and obtained evidence which favors a modification of the classical two-film theory of Whitman (1923) discussed by Sherwood and Gordon (1955). The usual two-film theory concerns the diffusional transport of a neutral solute across two hypothetical film layers in contact with each other at the interface, the rate-limiting step(s) being the diffusional movements in the immediate vicinity of the interface and within the two contacting layers. The ion-pairing process differs from this situation in that the simultaneous transport of two ionic species, a cation and an anion, is involved. Figure 9.10 depicts such a model in which ions in a stirred aqueous layer are being extracted into a stirred organic layer with ion-pair formation occurring at the interfacial barrier. In the stirred aqueous organic phase the concentrations of the cation, $C_a M^+$, and of the anion, $C_a A^-$, are necessarily constant, but in the aqueous diffusion layer of thickness ΔX_a next to the interface, they decrease to the respective values $C_{ai} M^+$ and $C_{ai} A^-$ at the interface itself. Ion-pair formation, which occurs at the

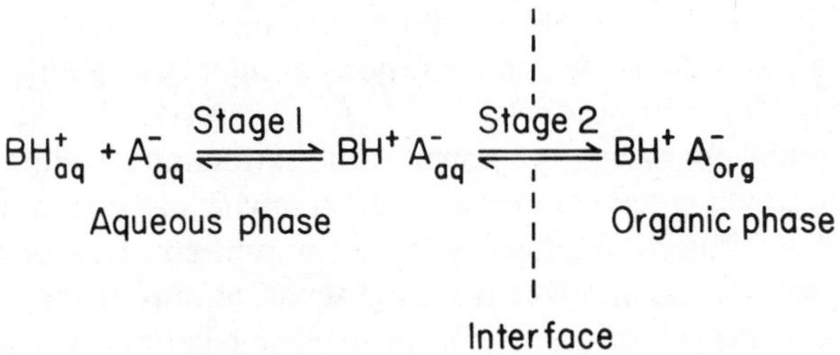

$$BH_{aq}^+ + A_{aq}^- \xrightleftharpoons{\text{Stage I}} BH^+ A_{aq}^- \xrightleftharpoons{\text{Stage 2}} BH^+ A_{org}^-$$

Aqueous phase   Organic phase

Interface

Scheme 9.3 Postulated transfer mechanism for extraction of an organic cation BH^+ by ion pairing with an anion A^-. Stage 2 is usually thought to be the rate-limiting step.

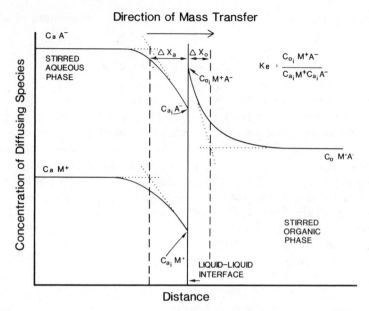

Figure 9.10 Schematic diagram of ion-pair transfer from aqueous to organic phase through diffusion layer ΔX_a in the former and ΔX_o in the latter. $C_a M^+$ and $C_a A^-$ are the concentrations of the M^+ and A^- ions in the bulk of the stirred aqueous phase. $C_{ai} M^+$ and $C_{ai} A^-$ are the concentrations of the M^+ and A^- ions immediately next to the interface in the unstirred aqueous diffusion layer. $C_o M^+ A^-$ is the concentration of the $M^+ A^-$ ion pair in the bulk of the stirred organic phase, whereas $C_{oi} M^+ A^-$ is the concentration of the ion pair immediately next to the interface in the unstirred organic diffusion layer. The extraction constant K_e represents the equilibrium constant for formation of the ion pair at the organic side of the interface from the separate ions at the aqueous side of the interface. K_e is given by an equation which is analogous to Eq. 9.2. Reprinted with permission from Higuchi, T. and Michaelis, A. F. (1968). *Anal. Chem.*, **40**, 1925–1931. Copyright 1968 by the American Chemical Society.

interfacial barrier, gives rise to $C_{oi} M^+ A^-$, which represents the concentration of the ion pair $M^+ A^-$ on the organic side of the interface. The concentration of the ion pair decreases across the organic diffusion layer, of thickness ΔX_0, until it reaches $C_0 M^+ A^-$, the necessarily constant value in the stirred organic phase. The outer limit of each diffusion layer is marked by the point of inter-section of the tangents to the concentration–distance curves in Figure 9.10.

At the interfacial barrier, the concentrations of the reacting ions, $C_{ai} M^+$ and $C_{ai} A^-$, and of the ion pair produced $C_{oi} M^+ A^-$, are related by the law of mass action. The equilibrium constant, known as the extraction constant K_e is given by the equation which is stated in Figure 9.10 and which is analogous to Eq. 9.2 in Section 9.3.1.2. Thus, K_e represents the formation constant of the ion pair at the organic side of the interface from the separate ions at the aqueous side of the interface, while the solute species in the immediate interfacial zones are assumed to be in equilibrium across the interface (Sherwood and Gordon, 1955; Higuchi and Michaelis, 1968). The rate-limiting steps involve diffusional movement effectively across a film in the aqueous layer of thickness ΔX_a, and a film in the organic layer of thickness ΔX_0, both directly adjacent to the interface.

The Whitman two-film theory leads to an equation relating the flux from one layer to that from the other in terms of (1) the concentrations of the cation and anion, (2) the diffusivities of the cation and anion in the aqueous phase and that of the ion pair in the organic phase, and (3) the effective thickness of the two diffusion layers. The experimental observations on the dextromethorphan systems mentioned above are in agreement with the predictions of the theory (Higuchi and Michaelis, 1968) and support the scheme in Figure 9.10.

9.4 THERMODYNAMICS OF ION PAIRING: INFLUENCE OF SOLUTE STRUCTURE

The standard free-energy change ΔG^θ, standard enthalpy change ΔH^θ, and standard entropy change ΔS^θ which accompany the formation of an ion pair are given by

$$\Delta G^\theta = -RT \ln K = \Delta H^\theta - T\Delta S^\theta \qquad (9.12)$$

where K is the equilibrium constant of the ion-pair association reaction

$$nC^{m+} + mA^{n-} \rightleftharpoons C_n A_m \qquad (9.13)$$

The enthalpy term (or more strictly the internal energy) can be considered to be made up of two contributions: (1) that resulting from electrostatic, that is, Coulomb, forces between the charged centers summarized by Eq. 2.119, and (2) that resulting from interactions other than Coulombic. The free-energy change can also be split up in an analogous fashion by allowing for the corresponding entropy differences. In the case of hydrophobic ions, such as ionic surfactants or tensides, the second contribution is essentially a hydrophobic interaction with some effect due to London forces, that is,

$$\Delta G = \Delta G(\text{electrostatic}) + \Delta G(\text{hydrophobic}) \qquad (9.14)$$

The electrostatic term is altered by changes in dielectric constant of the solvent. On the other hand, the hydrophobic term represents the free energy of formation of the cavity in the solvent to accommodate the hydrophobic solute and can be estimated from the isothermal compressibility $\beta = (\partial V/\partial p)_T$ of the solvent. The changes in enthalpy and entropy which accompany ion-pair formation are consistent with this concept. Any large positive entropy change which may be observed can be explained by the greater degree of freedom (lower resultant constriction of movement) of the ion pair in its new cavity as compared with the sum of the corresponding values for the two separate ions in their original cavities (Howarth, 1975).

The thermodynamics of transfer of a pharmaceutically relevant quaternary ammonium species paired with various anions from aqueous to organic phases have been studied by Michaelis and Higuchi (1969). The transfer of dextromethorphanium halides from water to chloroform involves a significant decrease in entropy (increase in order), whereas the opposite occurs for uncharged solute molecules. The evidence and significance will now be discussed.

The distribution ratios D defined by Eq. 9.5 were measured as described by Higuchi et al. (1967) at various temperatures. The enthalpy of transfer was calculated from the slope of the linear plot of $\ln D$ against $1/T$ according to the van't Hoff isochore (Eq. 1.9). The standard Gibbs free energy of transfer was calculated from $\Delta G^{\theta} = -RT \ln D$ and the standard entropy of transfer from $\Delta S^{\theta} = (\Delta H^{\theta} - \Delta G^{\theta})/T$, both at 25°C.

The influence of the composition of the organic phase, particularly chloroform content, on the transfer of dextromethorphan with three anions is shown in Table 9.4. It is reasonable to assume that the thermodynamic state of the aqueous phase is essentially unaffected by compositional changes in the organic phase. However, increasing chloroform content makes the enthalpy and entropy changes more negative. Since the influence of the enthalpy term is greater, the free-energy change becomes more negative, thereby increasing D. The entropic behavior indicates that the presence of the ion pair produces considerable reordering of chloroform molecules, which is to be expected since up to 5 molecules of chloroform may be incorporated into the structure of the ion pair (Higuchi et al., 1967), as shown in Table 9.1.

The general influence of the nature of the anion on the thermodynamics of

Table 9.4 Thermodynamic Values for Extraction Equilibria of Dextromethorphan Ion Pairs from Water to Organic Solvents at 25°C: Effect of Organic Solvent for Three Different Anions[a]

Anion	Organic Solvent (% v/v)	ΔH^{θ} (cal mol^{-1})	ΔG^{θ} (cal mol^{-1})	ΔS^{θ} (cal K^{-1} mol^{-1})
Bromide	100% CHCl$_3$	−8,780	−3,520	−17.6
	80% CHCl$_3$, 20% C$_6$H$_{12}$	−6,780	−2,960	−12.8
	40% CHCl$_3$, 60% C$_6$H$_{12}$	−4,000	−1,240	−9.3
	20% CHCl$_3$, 80% C$_6$H$_{12}$	−1,790	430	−7.4
Trichloro-acetate	80% CHCl$_3$, 20% C$_6$H$_{12}$	−7,670	−4,460	−10.8
	40% CHCl$_3$, 60% C$_6$H$_{12}$	−5,160	−3,470	−5.7
	20% CHCl$_3$, 80% C$_6$H$_{12}$	−2,840	−2,240	−2.0
Picrate	100% CHCl$_3$	−12,800	−9,300	−11.7
	40% CHCl$_3$, 60% CCl$_4$	−9,300	−7,630	−5.6
	100% CCl$_4$	−3,700	−5,800	+7.0

[a] 1 cal = 4.184 J.

Reproduced with permission of the copyright owner, the American Pharmaceutical Association, from Michaelis, A. F. and Higuchi, T. (1969). *J. Pharm. Sci.*, **58**, 201–204.

extraction is evident from Table 9.5, although it was not possible for experimental reasons to carry out all studies with a solvent of fixed composition. Since the anion is present in both phases, these thermodynamic values, unlike those brought about by changes in solvent composition, represent differences in the interaction of the ion pair with both phases.

On proceeding from chloride through bromide to iodide, the enthalpy and entropy changes become less negative. Since the influence of the entropy term is greater, the free-energy change becomes more negative, thereby increasing D. Thus, as the anion becomes larger and more polarizable, the comparatively greater ordering in the organic phase is reduced. Table 9.6 presents entropy data to help ascertain the relative contributions of the aqueous and organic phases to the entropy of transfer. The difference $\Delta\Delta S_{ext}$ between the observed values ΔS^{θ}_{ext} for extraction of the dextromethorphan halides is not approximated by the differences ΔS_{H_2O} between the published standard entropy values $S^{\theta}_{H_2O}$ of the anions in water. These comparisons reinforce the earlier suggestion that the structure of the organic solvent environment of the ion pair must contribute greatly to the energetics of extraction.

Perfluorocarboxylic acid anions are very hydrophobic and cause considerable ordering of water molecules in the aqueous phase. However, on transfer to the organic phase, it may be expected that the ordering will be reduced, even when the anions form part of an ion pair. This prediction is confirmed by the positive values ΔS^{θ} for trifluoroacetate and other perfluorocarboxylate anions, as shown in Table 9.5. With increasing length of the fluoroalkyl chain, the entropy change becomes more positive, reflecting the reduced entropy in the aqueous phase

Table 9.5 Thermodynamic Values for Extraction Equilibria of Dextromethorphan Ion Pairs from Water to Organic Solvents at 25°C: Effect of Nature of the Anion[a]

Anion	Organic Solvent (% v/v)	ΔH^{θ} (cal mol^{-1})	ΔG^{θ} (cal mol^{-1})	ΔS^{θ} (cal K^{-1} mol^{-1})
Chloride	100% CHCl$_3$	$-11,300$	$-2,450$	$-29,7$
Bromide	100% CHCl$_3$	$-8,780$	$-3,520$	-17.6
Iodide	100% CHCl$_3$	$-6,150$	$-4,300$	-6.2
Trifluoro-acetate	40% CHCl$_3$ in CCl$_4$	-835	$-2,620$	$+6.0$
Perfluoro-propionate	40% CHCl$_3$ in CCl$_4$	-25	$-3,600$	$+12.2$
Perfluoro-butyrate	40% CHCl$_3$ in CCl$_4$	$+760$	$-4,530$	$+17.8$
Perfluoro-pentanoate	40% CHCl$_3$ in CCl$_4$	$+1,570$	$-5,530$	$+23.8$
Picrate	100% CHCl$_3$	$-12,800$	$-9,300$	-11.7
	100% CCl$_4$	$-3,700$	$-5,800$	$+7.0$

[a] 1 cal $= 4.184$ J.

Table 9.6 Comparison of Differences of Entropies ΔS_{H_2O} for three Halide Ions in Water with the Corresponding $\Delta\Delta S_{ext}$ Values Calculated from Experimental Data for the Extraction of Dextromethorphan Salts from Aqueous to Organic Phases at 25°C[a]

Anion	$S^{\theta}_{H_2O}$ [b] $(cal\,K^{-1}\,mol^{-1})$	ΔS_{H_2O} $(cal\,K^{-1}\,mol^{-1})$	ΔS^{θ}_{ext} $(cal\,K^{-1}\,mol^{-1})$	$\Delta\Delta S_{ext}$ $(cal\,K^{-1}\,mol^{-1})$
Chloride	13.5		−29.7	
		+6.2		+12.1
Bromide	19.7		−17.6	
		+6.9		+11.4
Iodide	26.6		−6.2	

[a] 1 cal = 4.184 J.

[b] Weast, R. C., Astle, M. J., and Beyer, W. H., *CRC Handbook of Chemistry and Physics*, 68th ed., 1987–1988. CRC Press, Inc., Boca Raton, FL, 1987, pp. D-55, D-63, D-70.

Reproduced with permission of the copyright owner, the American Pharmaceutical Association, from Michaelis, A. F. and Higuchi, T. (1969). *J. Pharm. Sci.*, **58**, 201–204.

brought about by increased ordering of water molecules. This accords with Diamond's (1963) view of the role of solvent–solute interactions in directing ion-pair formation.

Table 9.5 also shows that increasing length of the fluorocarbon chain makes the enthalpy change less favorable (less negative or more positive). However, this is more than offset by the positive entropy change, with the result that the free energy of transfer becomes more negative and the extraction constant D increases. Figure 9.11 emphasizes this last point with three different pharmaceutical amines. Increasing the chain length by one CF_2 group increases D by a factor of about 5.3. This factor corresponds to a decrease in ΔG^{θ} of $990\,cal\,mol^{-1} = 4100\,J\,mol^{-1}$, which may be written $\Delta\Delta G_{CF_2} = -1.0\,kcal\,mol^{-1} = -4.1\,kJ\,mol^{-1}$, and represents the group contribution for the transfer of a CF_2 group in the defined ion-pair extraction process.

Banks (1937), working with a series of fatty acids, reported the corresponding value for the CH_2 group, which may be written $\Delta\Delta G_{CH_2} = -600\,cal\,mol^{-1} = -2500\,J\,mol^{-1}$. Similarly, Gibson and Weatherburn (1972), studying the extraction of quaternary ammonium cations paired with each member of a series of phosphonium cations, found that the distribution ratio D increased by a factor of about 2 for each additional CH_2 group in the alkyl chain. This indicates that $\Delta\Delta G_{CH_2} = -400\,cal\,mol^{-1} = -1670\,J\,mol^{-1}$. Evidently, the increase in ion-pair extractability due to each CF_2 increment is greater than that for each CH_2 increment, probably because of the greater tendency of a CF_2 group to organize water molecules in its vicinity. From Table 9.5 we note the following group contributions of CF_2 to the enthalpy and entropy changes in the defined ion-pair extraction process:

$$\Delta\Delta H_{CF_2} = +800\,cal\,mol^{-1} = +3300\,J\,mol^{-1}$$

$$\Delta\Delta S_{CF_2} = +6.0\,cal\,K^{-1}\,mol^{-1} = +25\,J\,K^{-1}\,mol^{-1}$$

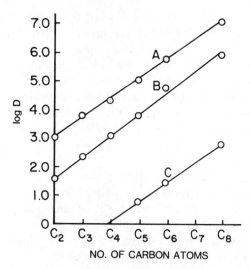

Figure 9.11   Effect of the number of carbon atoms of perfluoroalkanoic acid anions on the logarithm of the distribution ratio D from aqueous to organic phases for several protonated amines as cations. Temperature $= 25°C$. Concentrations: $0.5\,mmol\,dm^{-3}$ cation; $1.0\,mol\,dm^{-3}$ anion; $0.10\,mol\,dm^{-3}$ phosphate buffer. $1\,mol\,dm^{-3} = 1\,mol\,L^{-1} = 1\,M$. Organic phase $= 100\%$ $CHCl_3$. Higher values extrapolated from data for lower $CHCl_3$ concentrations. Amines at cations: dextromethorphan, A; chlorpheniramine, B; ephedrine, C. Reproduced with permission of the copyright owner, the American Pharmaceutical Association, from Michaelis, A. F. and Higuchi, T. (1969). *J. Pharm. Sci.*, **58**, 201–204.

Picrate was found to be a more effective anion than chloride for the ion-pair extraction of dextromethorphan into chloroform (Higuchi et al., 1967). This behavior is reflected in the much more negative ΔG^{θ} in Table 9.5. Since the ΔH^{θ} values are about equally exothermic, the difference in extractability is due to the more unfavorable (negative) entropy change for the chloride than for the picrate. This difference broadly parallels the number of chloroform molecules associated with the ion pair, which Table 9.1 shows to be 5 for the chloride but only 2 for the picrate. The greater solvation for the chloride is reflected in the greater degree of order.

The situation may, however, be complicated by the electron-rich nitro groups in the picrate anion whose oxygen atoms may form hydrogen bonds with chloroform. These effects can be seen by comparing the thermodynamic quantities shown in Table 9.5 for the extraction of the picrate ion pair into chloroform with that into carbon tetrachloride. The hydrogen bonding to chloroform releases more enthalpy but introduces more order, as compared with the nonspecific interactions in carbon tetrachloride.

Gibson and Weatherburn, studying the distribution of quaternary ammonium cations paired with an inorganic anion, found that the distribution ratio D generally increased with increasing size of the cation, the order being $ClO_4^- > SCN^- > I^- > ClO_3^- > NO_3^- > Br^- > BrO_3^- > Cl^-$.

The specific solvation of the type invoked for ion-pair extraction is not applicable to the extraction of nonionic solutes. Nevertheless, many nonionic pharmaceuticals, including the compounds represented in Figure 9.12, contain nitrogen or oxygen atoms with lone pairs which are capable of forming hydrogen bonds with the acidic proton of chloroform. Consequently, the observed distribution ratio D is expected to be related to the concentration of chloroform, [M], in the organic phase according to the equation

$$D = D_0 + D_0 K[M] \tag{9.15}$$

where D_0 is the distribution ratio in the presence of pure nonpolar solvent (e.g., carbon tetrachloride) and K is the stability constant for the interaction between solute and cosolvent (i.e., chloroform) in the organic phase. Equation 9.15 is analogous to Eq. 4.13, where D has replaced the solubility $[S]_t$ of the solute and [M] has replaced the total concentration $[L]_t$ of the ligand and when there is a 1:1 interaction between the solute and chloroform (i.e., $n = 1$, $m = 1$, so that $K = K_{1:1}$), while D_0 (i.e., $[S]_0$) is sufficiently small that $KD_0 \ll 1$.

In Figure 9.12 D is seen to increase linearly with increasing [M] in accordance with Eq. 9.15, D_0 being small. The linear relationship shown in Figure 9.12 between D and the concentration of chloroform for nonionic solutes contrasts sharply with the log–log relationship between these quantities for ion pairs, shown in Figure 9.2 and 9.3. The mechanistic explanation of the difference is that the number of chloroform molecules which solvate each ion pair is greater than unity; this behavior results in a power law which can be linearized by taking logarithms.

Nonionic solutes, when extracted by organic solvents from water under the same conditions as the ion pairs, undergo smaller negative free-energy changes than do ion pairs, and the enthalpy and energy changes are usually positive, as shown in Table 9.7. The enthalpy and entropy changes in Tables 9.4 and 9.5

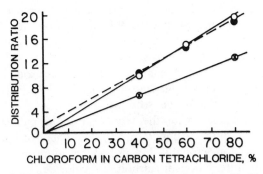

Figure 9.12 Effect of chloroform concentration (v/v) in carbon tetrachloride on the distribution ratio D of several organic solutes between water and chloroform at 25°C. Solutes at 0.5 mmol dm^{-3}: aniline (○); phenol × 10 (●); benzyl alcohol (⊗). 1 mol dm^{-3} = 1 mol L^{-1} = 1 M. Reproduced with permission of the copyright owner, the American Pharmaceutical Association, from Michaelis, A. F. and Higuchi, T. (1969). *J. Pharm. Sci.*, **58**, 201–204.

Table 9.7 Thermodynamic Values for the Extraction Equilibria of Some Nonionic Solutes from Water to Organic Solvents at 25°C[a]

Distributed Compound	Organic Solvent (% CHCl₃ in CCl₄)	ΔH^θ (cal mol⁻¹)	ΔG^θ (cal mol⁻¹)	ΔS^θ (cal K⁻¹ mol⁻¹)
Dextromethorphan	100	−1890	−5080	+10.7
Picric acid	100	−1370	−987	−1.3
Benzyl alcohol	80	+1270	−1500	+9.3
	60	+1300	−1340	+8.9
	40	+1495	−1110	+8.7
Aniline	80	+230	−1830	+6.9
	60	+150	−1570	+5.8
	40	+385	−1360	+5.9
Phenol	80	+2000	−407	+8.1
	60	+2500	−224	+9.1
	40	+2950	−177	+10.0

[a]Two ionic solutes capable of forming ion pairs are listed first for contrast and comparison. 1 cal = 4.184 J.

Reproduced with permission of the copyright owner, the American Pharmaceutical Association, from Michaelis, A. F. and Higuchi, T. (1969). *J. Pharm. Sci.*, **58**, 201–204.

show that the extraction of ion pairs (except with perfluorocarboxylates) is favored by the enthalpy and hindered by the entropy, whereas Table 9.7 illustrates that the extraction of most nonionic solutes is conversely hindered by the enthalpy change and favored by the entropy change. In other words, the extraction of most ion pairs is driven by a favorable enthalpy change, whereas the extraction of most nonionic solutes (and perfluorocarboxylate ion pairs) is driven by a favorable entropy change. Specific solvation in the organic phase frequently provides the simplest explanation of the observed effects, and the structuring within the organic phase is largely responsible for the negative entropy of extraction.

9.5 BEHAVIOR OF HYDROPHOBIC (CASE III) ION PAIRS IN WATER: CONDUCTIMETRIC STUDIES AND SOLUBILITY OF COMPLEX COACERVATES

Mukhayer and Davis (1975, 1976, 1977a, b, 1978a, b) carried out extensive studies on the influence of various factors on the formation and behavior of ion pairs between organic cations and organic anions in aqueous solution. An automated conductimetric titration procedure developed in collaboration with E. Tomlinson (Mukhayer et al., 1975) was employed. The interacting salts were benzyltriphenylphosphonium chloride (BTPC) or other quaternary phosphonium salts, and sodium dodecylsulfate (SDDS) or other sodium alkylsulfates. Scheme 9.4 shows that the interaction is an example of Case III in Scheme 9.1,

$$
\begin{array}{c}
\text{Ph} \\
| \\
\text{Ph—CH}_2\text{—P}^{\pm}\text{—Ph} \\
| \\
\text{Ph}
\end{array}
\qquad\qquad
\begin{array}{c}
\text{O} \\
\uparrow \\
\bar{\text{O}}\text{—S —O—C}_{12}\text{H}_{25} \\
\downarrow \\
\text{O}
\end{array}
$$

benzyltriphenylphosphonium

(BTP$^+$)

dodecylsulfate

(DDS$^-$)

Scheme 9.4 Two hydrophobic organic ions of opposite charge which can interact in water to form an ion pair of Case III (a "buried" ion pair).

the opposite charges being effectively "buried" in a predominantly hydrocarbon environment, which creates a region of low dielectric constant around the ionic charges.

The experimental procedure involves the continuous addition (e.g., $0.044\,\text{cm}^3\,\text{min}^{-1}$) of an aqueous solution of one of the salts (e.g., $0.04\,\text{mol}\,\text{dm}^{-3}$ SDDS) from a motorized automatic syringe to a stirred aqueous solution of the other salt (e.g., $50\,\text{cm}^3$ of $0.5\,\text{mol}\,\text{dm}^{-3}$ BTPC solution). Meanwhile, the conductivity G_t of the solution is plotted continuously against the volume of the added titrant, V_t. The addition causes the volume to increase continuously by the factor $F_t = [(V + V_t)/V]$, where V is the initial volume. At any given time, which is indicated by the subscript t, the conductivity (specific conductance) will be given by

$$
K = (G_i - G_t)F_t C \tag{9.16}
$$

where K is the conductivity, C is the cell constant, G_i is the initial conductance in the titration vessel at zero concentration of titrant, and the quantities G_t and F_t are as defined above.

Figure 9.13 shows a typical conductimetric titration curve of SDDS against butyltriphenylphosphonium chloride showing three distinct stages or regions. (The titration curve of SDDS against BTPC is analogous.) Region I corresponds to ion-pair associations, from which an association constant K_{IP} can be determined. Region II corresponds to phase separation of an ion-pair complex, from which a solubility product K_s can be obtained. Region III corresponds to micellar solubilization of the complex.

Region I occurs at low concentration of SDDS and is marked by a clear solution and a proportionality between the conductivity and the concentration of SDDS. The salts are almost completely ionized in solution; the degree of dissociation α is given by

$$
\alpha = \Lambda/\Lambda_0 \tag{9.17}
$$

where Λ is the molar conductivity of the salt in the solution and Λ_0 is the molar conductivity at infinite dilution. This equation will apply to BTP$^+$ and Cl$^-$ in equilibrium with its nondissociated, nonconducting ion pair BTPC.

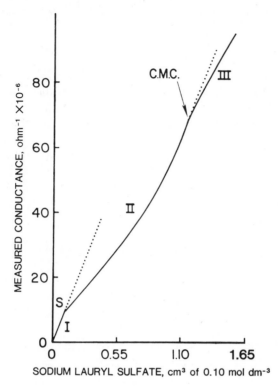

Figure 9.13 Conductimetric titration of sodium lauryl sulfate (i.e., sodium dodecylsulfate, $0.10 \, \text{mol} \, \text{dm}^{-3}$ solution) against butyltriphenylphosphonium bromide ($1.4 \, \text{mmol} \, \text{dm}^{-3}$ solution). $1 \, \text{mol} \, \text{dm}^{-3} = 1 \, \text{mol} \, \text{L}^{-1} = 1 \, M$. Region I corresponds to ion-pair associations, from which the association constants K_{IP} (Eq. 9.19) can be determined. Point S represents the "end point" of the titration and the beginning of Region II in which there occurs phase separation of a sparingly soluble ion-pair complex coacervate. From the concentrations at point S the stoichiometry and the solubility product K_s of the complex can be determined (Eq. 9.26). C.M.C. represents the critical micelle concentration of the system above which the complex is solubilized in the micelles of SDDS in Region III. Reproduced with permission of the copyright owner, the American Pharmaceutical Association, from Mukhayer, G. I., Davis, S. S., and Tomlinson, E. (1975). *J. Pharm. Sci.*, **64**, 147–151.

Addition of SDDS will give rise to dissociated monomers DDS^- and Na^+, dissociated dimers DDS_2^{2-} and $2Na^+$, and the undissociated nonconducting ion pairs SDDS and (BTP^+DDS^-). The hydrophobicity and size of both BTP^+ and DDS^- provide excellent conditions for the existence of water-structure-enforced ion pairs produced by the interaction

$$BTP^+ + DDS^- \rightleftharpoons BTP^+DDS^- \qquad (9.18)$$

The formation constant of the ion pairs is given by

$$K_{IP} = [BTP^+DDS^-]/[BTP^+][DDS^-] \qquad (9.19)$$

The conductivity K_{SDDS} (in ohm^{-1} cm^{-1}) of SDDS in water can be expressed by

$$K_{SDDS} \times 10^3 = \Lambda_{SDDS}[SDDS] \tag{9.20}$$

where Λ_{SDDS} is the molar conductivity (in cm^2 ohm^{-1} mol^{-1}) and [SDDS] is the molar concentration of SDDS (in mol dm^{-3}). A hypothetical value of K_{SDDS} can be calculated assuming that the degree of dimerization of SDDS does not change appreciably in the concentration range from 60 to 200 μmol dm^{-3} and that the BTP$^+$Cl$^-$ ion-pair association does not change with addition of SDDS.

The conductivity of the mixed system (SDDS + BTPC) is then considered to be a linear function of the various ionic species present in the solution at low concentrations. Therefore, the conductivity of the mixture K_m (in ohm^{-1} cm^{-1}) is given by

$$K_m \times 10^3 = \Lambda_{BTPC}[BTPC] + \Lambda_{SDDS}[SDDS] \tag{9.21}$$

where Λ_{BTPC} is the molar conductivity (in cm^2 ohm^{-1} mol^{-1}) and [BTPC] is the molar concentration of BTPC (in mol dm^{-3}). Since the total amount of BTPC in the titration vessel remains constant during the titration, the difference between the conductivity of the mixture and that for SDDS alone in water can be attributed (1) to the dilution of the BTPC by added SDDS in solution and (2) to the possible formation of the nonconducting ion pair (BTP$^+$DDS$^-$).

It is reasonable to assume that the inorganic ions Na$^+$ and Cl$^-$ do not form ion pairs in water at low concentrations. The conductivity K_m of an actual mixture is given by

$$K_m \times 10^3 = \Lambda_{BTPC}([BTPC] - [IP]) + \Lambda_{SDDS}([SDDS] - [IP])$$
$$+ [IP](\lambda_{Na^+} + \lambda_{Cl^-}) \tag{9.22}$$

where [IP] represents the concentration of the nonconducting ion pair [BTP$^+$DDS$^-$] and λ_{Na^+} ($= 50.10$ cm^2 ohm^{-1} mol^{-1}) and λ_{Cl^-} ($= 76.35$ cm^2 ohm^{-1} mol^{-1}) are the molar conductivities of the ions at infinite dilution at 25°C. Assuming that the observed deviation of Eq. 9.22 from Eq. 9.21 is a result of the formation of an ion pair, the change in conductivity ΔK_m due to removal of the species (BTP$^+$DDS$^-$) can be obtained by subtraction, thus,

$$\Delta K_m \times 10^3 = \Lambda_{BTPC}[IP] + \Lambda_{SDDS}[IP] - [IP](\lambda_{Na^+} + \lambda_{Cl^-}) \tag{9.23}$$

Rearrangement enables [IP] $\equiv$ [BTP$^+$DDS$^-$] to be calculated, thus,

$$[IP] = \frac{\Delta K_m \times 1000}{(\Lambda_{BTPC} + \Lambda_{SDDS}) - (\lambda_{Na^+} + \lambda_{Cl^-})} \tag{9.24}$$

Table 9.8 Values for the Ion-Pair (BTP^+DDS^-) Association Constant K_{IP} Formed by the Addition of SDDS ($40 \, mmol \, dm^{-3}$) to $50 \, cm^3$ BTPC ($0.5 \, mmol \, dm^{-3}$) in Aqueous Solution[a]

		$\Delta K_m \, (M\Omega^{-1} \, cm^{-1})$			
[SDDS] (mol dm^{-3})	Λ_{SDDS} ($\Omega^{-1} \, m^2 \, mol^{-1}$)	Dilution and ion-pair association	Ion-pair association	[IP] (mol dm^{-3})	K_{IP}^{b} (dm^3 mol^{-1})
6.0×10^{-5}	71.20	0.25	0.178	4.41×10^{-6}	159
8.0×10^{-5}	71.20	0.34	0.244	6.01×10^{-6}	162
1.0×10^{-4}	71.00	0.40	0.280	6.94×10^{-6}	152
1.2×10^{-4}	70.96	0.48	0.336	8.36×10^{-6}	153
1.4×10^{-4}	70.86	0.55	0.382	9.50×10^{-6}	149
1.6×10^{-4}	72.19	0.655	0.463	1.11×10^{-5}	154
1.8×10^{-4}	70.97	0.75	0.534	1.32×10^{-5}	163

[a] $1 \, mol \, dm^{-3} = 1 \, mol \, L^{-1} = 1 \, M; \Omega = ohm$.
[b] The mean $K_{IP} = 156 \, dm^3 \, mol^{-1}$.

Reproduced with permission of the copyright owner, Academic Press, from Mukhayer, G. I. and Davis, S. S. (1975). *J. Colloid Interface Sci.*, **53**, 224–234.

where ΔK_m is equal to the difference between the conductivity of the mixed system and the conductivity of SDDS solution alone at any given concentration of added SDDS. Values of K_{IP} were calculated from Eq. 9.19 and 9.24 and are listed in Table 9.8 for various concentrations of SDDS together with the measured quantities. Λ_{BTPC} was found experimentally to be $95.8 \, cm^2 \, ohm^{-1} \, mol^{-1}$ at $0.5 \, mmol \, dm^{-3}$. The mean value of K_{IP} at $25°C$ for the BTP–DDS ion pair was found to be $156 \, dm^3 \, mol^{-1}$, which is of the same order as other K_{IP} values for anions and cations of similar size.

Region II in Figure 9.13 commences at the break S in the conductivity curve. Beyond S the solution becomes turbid due to the separation of an ion-pair complex between the BTP and DDS ions, thus,

$$mBTP^+ + nDDS^- \rightleftharpoons BTP_mDDS_n \tag{9.25}$$

Electrical neutrality of the precipitated complex is maintained by uptake of inorganic ions of the opposite charge. Setting to unity the activity of the insoluble complex, its solubility product K_s is expressed in terms of activities by the equation

$$K_s = \{BTP^+\}^m \{DDS^-\}^n \tag{9.26}$$

Expressing the activity as the product of molar concentration and activity coefficient y we obtain

$$K_s = [BTP^+]^m[DDS^-]^n y_{BTP^+}^m + y_{DDS^-}^n \qquad (9.27)$$

The apparent (i.e., stoichiometric) solubility product is given by

$$K_s' = [BTP^+]^m[DDS^-]^n \qquad (9.28)$$

At low ionic strengths the activity coefficients are unity, so K_s' is a constant equal to K_s. Taking logarithms in Eq. 9.28 we obtain

$$\log[DDS^-] = (1/n)\log K_s' - (m/n)\log[BTP^+] \qquad (9.29)$$

The exact stoichiometry m/n was found by determining the complexation end point at S in Figure 9.13 for different concentrations of BTPC in the titration vessel. Equation 9.29 indicates that a plot of $\log[DDS^-]$ against $\log[BTPC^+]$ should be linear with slope m/n and with intercept $(1/n)\log K_s'$. Figure 9.14 shows that $m/n = 1$, indicating that the separating complex is an aggregate of 1:1 ion pairs. In confirmation of this, a plot of $[BTPC^+]$ against $[DDS^-]^{-1}$ is linear passing through the origin. Setting both m and n equal to unity, the derived mean value of $K_s = 9.97 \times 10^{-8}$ mol^2 dm^{-6} for the BTP–DDS ion pair.

In the complexation Region II in Figure 9.13 further addition of SDDS causes the conductivity to change in a curvilinear fashion as a result of the separation

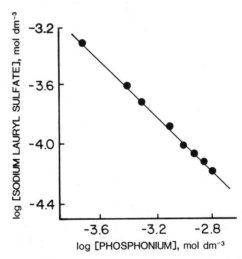

Figure 9.14 Titration of sodium lauryl sulfate (i.e., sodium dodecylsulfate) solutions against different concentrations of phosphonium salt BTPC (double log plot). The end point is incipient precipitation corresponding to point S in Figure 9.13. 1 mol dm^{-3} = 1 mol L^{-1} = 1 M. Reproduced with permission of the copyright owner, the American Pharmaceutical Association, from Mukhayer, G. I., Davis, S. S., and Tomlinson, E. (1975). *J. Pharm. Sci.*, **64**, 147–151.

of the 1:1 (BTP^+DDS^-) complex, which is in equilibrium with a solution containing free ions and ion pairs. Mukhayer and Davis (1975) derived a quadratic relationship for predicting free and complexed BTP^+ and DDS^- ions in solution and obtained values which agreed qualitatively with those calculated from conductivity measurements. The experimental values were corrected for the presence of ion pairs and for the "obstruction effect," which represents the diminution of the measured conductivity as a result of the insulating properties of the particles of the complex.

Mukhayer and Davis (1978b) showed that the BTP–DDS complex is a liquid-like substance known as a "coacervate" rather than a solid precipitate. Microscopy shows that the product has the typical vacuolated appearance of a complex coacervate. Surface tension measurements, which can also be used to study the interaction between the ions, show that the coacervate is much more efficient in reducing the surface tension of water than either of the two interacting salts alone. The electrophoretic mobility of the coacervate droplets is negative and indicates that absorbed SSD^- anions predominate on the surface.

Region III in Figure 9.13 begins when the critical micelle concentration (CMC) of the excess of SDDS has been reached in the presence of various BTPC species (BTP^+ in solution, BTP^+DDS^- in solution, complex coacervate). Beyond the CMC, the complex is progressively solubilized in the micelles on account of its hydrophobic surface and the solution progressively loses its turbidity. The change in slope of the CMC is less distinct than at the complexation points. The CMC of SDDS is influenced by the presence of the large organic cation BTP^+, by other ions, and by complex formation.

Using their conductimetric procedure (Mukhayer et al., 1975), Tomlinson and Davis (1978) studied the influence of structure and other factors on the apparent or stoichiometric solubility product K_s of complexes formed between monovalent cations from alkylbenzyldimethylammonium chlorides (ABDAC) and divalent anions from bischromones (Scheme 9.5). The bischromones used included sodium cromoglycate (SCG or cromolyn sodium), which is 5,5'-[(2-hydroxytrimethylene)dioxy]bis(4-oxo-4H-1-benzopyran-2-carboxylic acid disodium salt), and its 7,7-isomer.

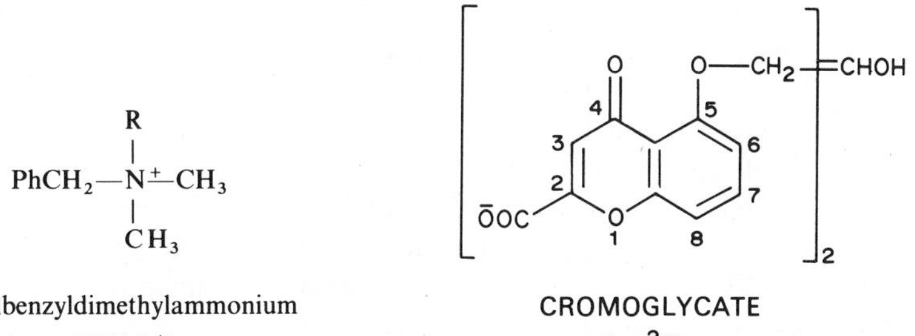

R
|
$PhCH_2$—$N^\pm$—CH_3
|
CH_3

alkylbenzyldimethylammonium
($ABDA^+$)

CROMOGLYCATE
(CG^{2-})

Scheme 9.5 Example of a divalent anion which is capable of forming an ion pair and a complex with two mole proportions of a monovalent cation.

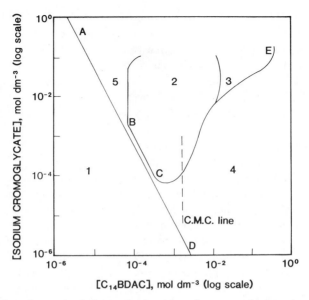

Figure 9.15 Phase diagram for the interaction between sodium cromoglycate and tetradecylbenzyl-dimethylammonium chloride (C_{14}BDAC) at 25°C (see text for explanation of areas). $1 \, mol \, dm^{-3} = 1 \, mol \, L^{-1} = 1 \, M$. Reproduced with permission of the copyright owner, Academic Press, from Tomlinson, E. and Davis, S. S. (1978). *J. Colloid Interface Sci.*, **66**, 335–344.

The phase diagram (Figure 9.15) of the system, SCG (Na_2CG) + tetradecyl-benzyldimethylammonium chloride (C_{14}BDAC), shows five distinctly observ-able regions. In Region 1 no insoluble complex is formed and its boundary is the solubility product line, for which $[Na_2CG] = K'_s/[C_{14}BDAC]^2$, that is, $\log[Na_2CG] = \log K'_s - 2\log[C_{14}BDAC]$, (cf. Eq. 9.29). In Regions 2 and 3 there is visual evidence of an insoluble complex, which appears as a somewhat milky grey-white emulsion in Region 2 and as a brown viscous oil in Region 3. Although Regions 4 and 5 are above the theoretical solubility product line, there was no visible evidence of turbidity. In Region 4 the concentration of C_{14}BDAC, the surface-active agent, exceed its CMC (Mukhayer et al., 1975) and a limited amount of complex, represented by the line CE, is solubilized within the micelles. Furthermore, above the CMC region, insufficient monomeric surfactant is available to produce the insoluble complex. In fact, the presence of free CG^{2-} ion will tend to suppress the measured CMC of the surfactant. At high concentrations of Na_2CG and at low concentrations of surfactant, corresponding to Region 5, no turbidity could be observed. Here the Na_2CG behaves as a stacked pseudomicelle at these concentrations (Champion and Meeten, 1973), thereby reducing the thermodynamic activity of the anion and, consequently, the tendency for complexation.

As the alkyl chain length of the $ABDA^+$ surfactant increases, the observed profile of the phase diagram exemplified by Figure 9.15 shifts to lower concentra-tions of Na_2CG and $ABDA^+$. Thus, an increase in chain length increases the tendency for coacervation to take place, probably because of an increase in the hydrophobic interaction discussed in Section 8.2.

Plakogiannis et al. (1970) obtained data which emphasized the influence of the hydrophobic nature of the ions on the extraction constant K_e of Case III ion pairs. K_e, defined by Eq. 9.2, is presented as the apparent partition coefficient between chloroform and water (Plakogiannis et al., 1970). For 48 ion pairs between alkylsulfates and substituted quinolinium ions, the following correlation was found with K_e expressed in the units of $mol\, dm^{-3}$:

$$\log K_e = 0.496(A_w 10^9)_X + 0.345(A_w 10^9)_R - 2.341$$

$$r = 0.924; \qquad s.d. = 0.272 \tag{9.30}$$

and for 19 ion pairs between alkylsulfates and substituted pyridinium ions, the correlation was

$$\log K_e = 0.169(A_w 10^9)_X + 0.253(A_w 10^9)_R - 1.366$$

$$r = 0.981; \qquad s.d. = 0.130 \tag{9.31}$$

where $(A_w 10^9)_R$ and $(A_w 10^9)_X$ are Bondi's (1964, 1968) group contributions to the surface area of the alkyl group R of the anion and of the substituent group X of the cation, respectively (cf. Eq. 7.69 in Section 7.5.2).

Little improvement in correlation was obtained on introducing the Hammett polar substituent constant σ. When Bondi's parameters were replaced by a linear combination of Hansch's π parameters and Hammett's σ parameters, the correlations were poorer. These results indicate that, for a given solvent system, the surface area of the alkyl groups and of the ring substituents are the most important factors in governing the extraction constants of Case III ion pairs. This suggestion is entirely consistent with the supposition that Case III ion pairs present a hydrophobic surface to the surrounding water molecules in aqueous solution, as we have seen in Section 9.4.

9.6 APPLICATIONS OF ION-PAIR FORMATION IN THE CHEMICAL AND PHARMACEUTICAL SCIENCES

These applications often result from the fact that the individual ions are highly water-soluble species, whereas the ion pairs are highly soluble in nonpolar media (including organic solvents, lipids, and cell membranes) and can sometimes be selectively solvated.

Interesting examples of ion pairs in the pharmaceutical sciences include the phenothiazines, which are psychoactive drugs (i.e., tranquilizers) whose molecular structures are shown in Scheme 9.6 (Murthy and Zografi, 1970). The planar aromatic moiety of these molecules confers hydrophobic properties and surface activity on the phenothiazines (Zografi et al., 1964a; Patel and Zografi, 1966; Zografi and Munshi, 1970). The phenothiazines concentrate at the air–

Scheme 9.6   The molecular structures of the phenothiazine drugs

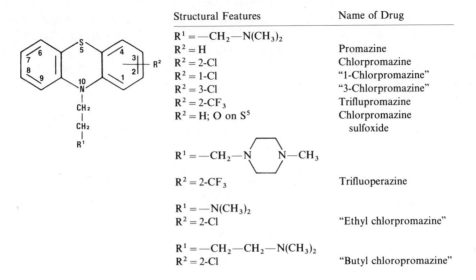

Structural Features	Name of Drug
$R^1 = -CH_2-N(CH_3)_2$	
$R^2 = H$	Promazine
$R^2 = 2$-Cl	Chlorpromazine
$R^2 = 1$-Cl	"1-Chlorpromazine"
$R^2 = 3$-Cl	"3-Chlorpromazine"
$R^2 = 2$-CF$_3$	Triflupromazine
$R^2 = H$; O on S^5	Chlorpromazine sulfoxide

$$R^1 = -CH_2-N\!\!\!\bigcirc\!\!\!N-CH_3$$

$R^2 = 2$-CF$_3$	Trifluoperazine

$R^1 = -N(CH_3)_2$	
$R^2 = 2$-Cl	"Ethyl chlorpromazine"

$R^1 = -CH_2-CH_2-N(CH_3)_2$	
$R^2 = 2$-Cl	"Butyl chloropromazine"

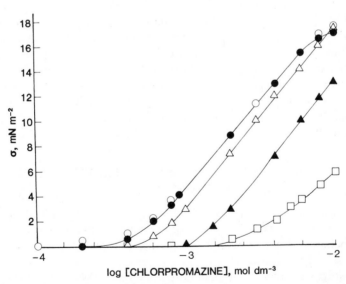

Figure 9.16   Plot of the water–air interfacial tension σ against log-molar concentration for chlorpromazine in water at pH 5.0 and 25°C in the presence of $0.1 \, \text{mol dm}^{-3}$ NH$_4$Cl or a tetraalkylammonium chloride: NH$_4$Cl ($\bigcirc$); (CH$_3$)$_4$N$^+$Cl$^-$ ($\bullet$); (C$_2$H$_5$)$_4$N$^+$Cl$^-$ ($\triangle$); (C$_3$H$_7$)N$^+$Cl$^-$ ($\blacktriangle$); (C$_4$H$_9$)$_4$N$^+$Cl$^-$ ($\square$). $1 \, \text{mol dm}^{-3} = 1 \, \text{mol L}^{-1} = 1 \, M$. Reprinted with permission of the copyright owner, the American Pharmaceutical Association, from Patel, R. M. and Zografi, G. (1966). *J. Pharm. Sci.*, **55**, 1345–1349.

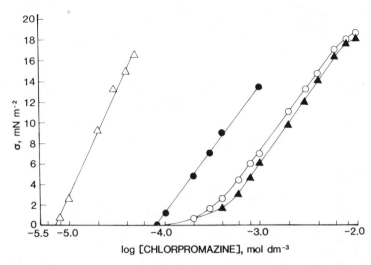

Figure 9.17 Plot of the water–air surface tension σ against log(molar concentration) for chlorpromazine in water at pH 5.0 and 25°C in the presence of 0.01 mol dm^{-3} sodium salt of an organic sulfonic acid: 2-naphthalenesulfonate ($\triangle$); benzenesulfonate ($\bullet$); propanesulfonate ($\circ$); methanesulfonate ($\blacktriangle$). 1 mol dm^{-3} = 1 mol L^{-1} = 1 M. Reprinted with permission of the copyright owner, the American Pharmaceutical Association, from Patel, R. M. and Zografi, G. (1966). *J. Pharm. Sci.*, **55**, 1345–1349.

water interface, particularly in the presence of insoluble monomolecular films (Zografi and Auslander, 1965). Thus, phenothiazines accumulate in biological membranes and are involved in metabolic processes controlled by the presence of biological interfaces, which may be related to their effect on the central nervous system (Zografi and Zarender, 1966).

Figures 9.16 and 9.17 illustrate the tendency of a typical phenothiazine, chlorpromazine, to concentrate at the air–water interface. This effect is shown by the approximately linear relationship between the surface tension σ at the interface, measured by the drop-volume method, and the logarithm of the molar concentration c of chlorpromazine in water at pH 5.0 according to the Gibbs adsorption isotherm. The latter may be stated in the following form in which activity coefficient of the solute is assumed to be constant:

$$\Gamma = -(1/RT)(d\sigma/d\ln c) \tag{9.32}$$

where Γ is the surface excess concentration (mol per unit area of solution surface), while R and T have their usual significance. The slope of these plots at a given concentration reflects the surface excess concentration. Leveling of the curve occurs at the CMC of the solute.

The tetraalkylammonium chlorides (Figure 9.16) and other organic cations reduced the surface tension and the surface excess concentration, which indicates an inhibition of the adsorption of chloropromazine at the air–water interface.

Table 9.9 The pK_a Values and Partitioning Behavior of Various Phenothiazine Drugs from Water to Dodecane and from Water to 1-Octanol at 30°C[a]

Phenothiazine Drug	pK_a	Dodecane–Water		1-Octanol–Water	
		$P^c_{o/w}$ (apparent) at pH 7.0	$P^c_{o/w}$ of Free Base	D with KCl (0.125 mol dm^{-3}) at pH 3.9	K'_0 (dm^3 mol^{-1}) with Cl$^-$
Promazine	9.4	42	10,400	8.1	58
Chlorpromazine	9.3	366	73,100	32.4	197
Chlorpromazine sulfoxide	9.0	0.0075	0.75	0.22	1.5
"1-Chlorpromazine"	9.4	245	61,200	17.0	114
"3-Chlorpromazine"	9.2	292	46,300	61.3	430
Trifluopromazine	9.2	863	137,000	60.0	384
Trifluoperazine	3.9	97	12,900	49.3	
	8.1				
"Ethyl chlorpromazine"	8.7	553	28,200		
"Butyl chlorpromazine"	9.7	232	116,000		

[a] 1 mol dm^{-3} = 1 mol L^{-1} = 1 M.

Reproduced with permission of the copyright owner, the American Pharmaceutical Association, from Murthy, K. S. and Zograf, G. (1970). J. Pharm. Sci., 59, 1281–1285.

This salting-in effect increased with increasing alkyl chain length in the cation and may be attributable to mutual nonspecific association on nonpolar groups in the aqueous solution corresponding to the hydrophobic interaction (Section 8.2).

On the other hand, increasing concentrations of propanesulfonate, benzenesulfonate, and naphthalenesulfonate ions (and also of bromide and iodide ions), each added as the sodium salt, progressively increased the surface tension (Figure 9.17) and surface concentration, indicating increased surface activity of chlorpromazine. This effect increased with increasing size (reflecting polarizability and hydrophobicity) and is probably caused by the formation of ion pairs (chlorpromazine cation + added anion). At high concentrations of chlorpromazine the more hydrophobic ion pairs may separate as an oil which can sometimes be induced to crystallize.

The partition coefficients of several phenothiazines from water to dodecane and from water to 1-octanol were measured at 30°C by Murthy and Zografi (1970) and are expressed in various ways in Table 9.9. The partition coefficients of the free bases were calculated by multiplying the measured (i.e., apparent) values at pH 7.0 by the factor $(1 + [H^+]/K_a)$. Since the pK_a values of the phenothiazines were mostly about 9, this correction factor is of the order 10^2. The partition coefficients of the free bases from water to dodecane were essentially independent of buffer concentration, ionic strength, and type and concentration of counter ions. These partition coefficients ranged from 10^4 to 10^5, indicating that the bases are hydrophobic compounds; the exception is the highly polar compound chlorpromazine sulfoxide.

The distribution ratios D of the phenothiazine cations from water (at pH 3.9 and at a defined ionic strength) to 1-octanol are also shown in Table 9.9. These ratios increased with increasing concentration of various anions, as exemplified by Figure 9.18. The data suggest that the phenothiazines form 1:1 Case II ion pairs in 1-octanol, which is presumably acting as a proton-donating electrophilic solvating agent, as do 1-pentanol and p-tert-butylphenol with dextromethorphan in cyclohexane (Table 9.2). The extraction constants K'_0, calculated using Eq. 9.8, of chlorpromazine with various anions (shown in Table 9.10), increased with increasing size (reflecting polarizability and hydrophobicity) and correspondingly with increasing alkyl chain length of the sulfonate anion. In the presence of tetraalkylammonium ions, however, partitioning into 1-octanol was reduced (Murthy and Zografi, 1970), indicating salting-in of the phenothiazine in aqueous solution as a result of the hydrophobic interaction (Section 8.2).

For different phenothiazines the K'_0 values into 1-octanol from water containing potassium chloride (Table 9.9) roughly parallel the partition coefficients of the free bases into dodecane. Both quantities reflect the influence of structural modification among the phenothiazines (Scheme 9.6), and might be amenable to the group-contribution approach if more compounds were studied, since the phenothiazines are essentially rigid molecules (Chapter 7). Overall, the results with the phenothiazines clearly demonstrate that their hydrophobic behavior and the formation of ion pairs with various anions

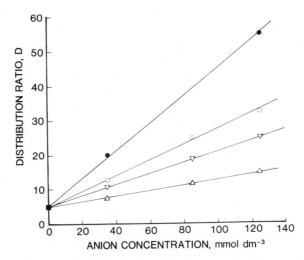

Figure 9.18 Plots of the distribution ratio D of chlorpromazine from aqueous buffers at pH 3.9 and 30°C to 1-octanol against the concentration of the following anions present as sodium salts: chloride (○); propanesulfonate (●); ethanesulfonate (▽); methanesulfonate (△); none (■). $1 \text{ mol dm}^{-3} = 1 \text{ mol L}^{-1} = 1 M$. Reproduced with permission of the copyright owner, the American Pharmaceutical Association, from Murthy, K. S. and Zografi, G. (1970). *J. Pharm. Sci.*, **59**, 1281–1285.

Table 9.10   Extraction Constants K'_0 of Chlorpromazine into 1-Octanol from Water Containing Various Anions at 30°C[a]

Anion	K'_0 $(\text{dm}^3 \text{ mol}^{-1})$	Anion	K'_0 $(\text{dm}^3 \text{ mol}^{-1})$
Chloride	197	Methanesulfonate	68
Bromide	383	Ethanesulfonate	163
Nitrate	362	Propanesulfonate	374
Acetate	213		

[a] $1 \text{ dm}^3 \text{ mol}^{-1} = 1 \text{ L mol}^{-1} = 1 M^{-1}$.

Reproduced with permission of the copyright owner, the American Pharmaceutical Association. Murthy, K. S. and Zografi, G. (1970). *J. Pharm. Sci.*, **59**, 1281–1285.

markedly influence the partitioning of these drugs and should consequently influence their uptake, interactions, and transport in lipids and biological membranes.

The following applications of ion-pair formation, illustrated by Scheme 9.3, are recognized as being particularly valuable:

(a) Extraction and analysis (Lagerström et al., 1972; Westerlund et al., 1972; Jonkman, 1975).

(b) Chromatography, especially HPLC.

(c) Ion-selective electrodes (Cattrall and Freiser, 1971; Baum, 1972; Muratsugu et al., 1977; Davis and Olejnik, 1979), which usually contain either a liquid membrane or a plastic membrane.

(d) Phase-transfer catalysis and organic chemical reactions, particularly esterifications, alkylations, inhibitions, and studies of steric effects.

(e) Biopharmaceutics and drug formulation (Schanker, 1960; Irwin et al., 1969; Gaginella et al., 1973; Tomlinson and Davis, 1976, 1978; Davis et al., 1978), particularly in the areas of drug absorption, membrane transport, the development and alteration of membrane potentials, drug–protein interactions, and sorption and desorption processes with polyelectrolytes.

Chapter X

PHASE-SOLUBILITY ANALYSIS AND COMPLEXATION IN AQUEOUS SOLUTION

10.1 INTRODUCTION

As mentioned in Chapter 1, phase-solubility analysis provides a number of analytical procedures for determining the purity of a substance and certain properties of the components of mixtures of solids, namely, solution properties, complexation, and interactions. Since the equilibrium solubility of a substance in a given solvent at a given temperature and pressure is a quantity characteristic of that substance, it may be used as a criterion of identity and purity, as discussed in the *U.S. Pharmacopeia XXI* (1985) to be described in Section 10.3.

The formation of a complex in the presence of a cosolvent or ligand in a given solvent is a natural extension of phase-solubility analysis. In Chapters 4 and 5 we considered the formation of complexes in more or less inert solvents, such as paraffinic hydrocarbons and carbon tetrachloride. In this chapter emphasis will be placed on the formation of complexes in aqueous solutions. In Chapter 8 we discussed the special role of water as a solvent for organic compounds. Section 10.6 in the present chapter discusses the nature and mechanism of the stabilization of certain complexes in aqueous solutions. In many of these complexes the molecules appear to be stacked one on top of the other, as in sandwich. The hydrophobic interaction, considered in Section 8.2, is believed to play an important role in this process.

10.2 GENERAL APPROACH AND INTERPRETATION OF PHASE-SOLUBILITY ANALYSIS

Phase-solubility analysis comprises two processes. The first is the analytical determination of the equilibrium weight of total dissolved solute, ρ, per unit quantity of solvent or solution for various weights of solid samples, σ, presented to unit quantity of solvent. The second process is the interpretation of the resultant equilibrium plots (e.g., Fig. 10.1 and 10.2) according to the phase rule to be discussed a little further on. Phase-solubility analysis was introduced by Northrop and Kunitz (1929) and has been reviewed by Higuchi and Connors (1965). The experimental procedure consists of adding a series of different weights of sample to a series of constant weights or volumes of solvent in which it is sparingly soluble, for example, 0.1–1% for the major component. As described in Section 10.3, the mixture is equilibrated by prolonged agitation at constant temperature and atmospheric pressure. The supernatant solution is removed

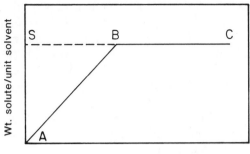

Figure 10.1 Phase-solubility diagram of a pure compound. Reproduced with permission of the copyright owner, John Wiley and Sons, Inc., from Higuchi, T. and Connors, K. A. (1965). Phase-Solubility Techniques, in *Advances in Analytical Chemistry and Instrumentation*, Vol. 4, John Wiley and Sons, Inc., New York.

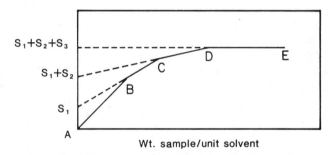

Figure 10.2 Phase-solubility diagram of a three-component mixture. The ordinate is the weight of dissolved solute per unit weight of solvent, as in Figure 10.1. Reproduced with permission of the copyright owner, John Wiley and Sons., Inc., from Higuchi, T. and Connors, K. A. (1965). Phase-Solubility Techniques, in *Advances in Analytical Chemistry and Instrumentation*, Vol. 4, John Wiley and Sons, Inc., New York.

and analyzed for *total* solute content, for example, by weighing the residue after evaporation of a given weight or volume of solution.

If the solid consists of a single substance, the phase-solubility diagram has the simple form shown in Figure 10.1, in which AB represents dissolution of all the added solid. The slope of AB is very close to unity if the solubility is low, with deviations from unity resulting from different concentration scales of the two axes. At point B the solution is saturated with the compound. Further addition of solid cannot increase the concentration of that already dissolved, so the slope of BC is zero. The solubility of the compound is given by extrapolating BC to the vertical axis and is equal to S in Figure 10.1.

If the solid sample consists of three substances, 1, 2, and 3, which do not interact in solution and whose solubilities are in the order $S_1 < S_2 < S_3$, the phase-solubility diagram has the form shown in Figure 10.2. As before, AB

represents the total dissolution of the solid, that is, of all three compounds. At point B, the solution becomes saturated with the least soluble compound 1, and therefore along BC only the remaining two components 2 and 3 continue to dissolve. At point C the solution becomes saturated with compound 2 as well as with 1, so from C to D only the final compound 3 continues to dissolve. At point D, the solution is saturated with all three components and the slope of DE is zero. The order in which the components saturate the solvent depends on their solubilities and their relative quantities in the solid sample.

From B to C only components 2 and 3 continue to dissolve, so the slope of BC is equal to the mass fraction of these components, $w_2 + w_3$, in the solid sample. Similarly, the slope of CD is equal to the fraction of component 3, w_3, since only this component is now continuing to dissolve. The fractional composition of the solid sample can therefore be calculated from the slopes. The intercepts of the extrapolated line segments BC, CD, and DE to the solution concentration axis (vertical axis) give the solubilities of the components, S_1, $(S_1 + S_2)$, and $(S_1 + S_2 + S_3)$, respectively, as shown in Figure 10.2. Between B and C, solid 1 is the only undissolved component and can therefore be isolated in a small quantity in the pure state. The behavior of the system, as described in this and in the previous paragraph and as summarized in Figure 10.2, can be interpreted by means of the phase rule.

The phase rule, which is due to Gibbs (see Lewis and Randall, 1961, Bowden, 1938, and Findlay et al., 1951), states that

$$P + F = C + 2 \tag{10.1}$$

P is the number of phases in the system. C is the number of components in the system, that is, the number of chemical substances minus the number of independent chemical equilibrium equations connecting them, if any. F is the number of degrees of freedom, that is, the number of independently variable factors (i.e., temperature, pressure, and the concentrations of components) which must be fixed in order for the composition and state of the system at equilibrium to be completely defined. The number 2 in Eq. 10.1 recognizes that there are two variables, temperature and pressure, in addition to the concentration values.

The behavior of the mixture of three solids summarized in Figure 10.2 is interpreted in Table 10.1 in terms of the phase rule. This treatment serves as a model for interpreting other systems consisting of solids and/or liquids. Such systems are known as condensed systems, since the vapor phase, which contains only a minute amount of material, is ignored.

The system in Figure 10.2 consists of three solid substances and one liquid substance, the solvent, and therefore comprises four components, as indicated in Table 10.1. Along line AB all three solids are in solution, so there is only one phase. Thus, the number of degrees of freedom is five; these are enumerated in Table 10.1. Any five of these variables may be altered within limits without changing the number of phases (i.e., one). Along line BC only solid 1 separates out and is in equilibrium with the solution, corresponding to two phases.

Table 10.1 Application of the Phase Rule to Figure 10.2[a]

Segment	C	P	F	Enumeration of Degrees of Freedom
AB	4	1	5	T, P, C_1, C_2, C_3
BC	4	2	4	T, P, C_2, C_3
CD	4	3	3	T, P, C_3
DE	4	4	2	T, P

[a] The interpretation of the relationship to Figure 10.2 is presented in the text.

Reproduced with permission of the copyright owner, John Wiley and Sons, Inc., from Higuchi, T. and Connors, K. A. (1965). Phase-Solubility Techniques, in *Advances in Analytical Chemistry and Instrumentation*, Vol. 4, John Wiley and Sons, Inc., New York.

Consequently, there are now only four degrees of freedom, that which was lost being C_1, the concentration of solid 1, which is fixed at the solubility value S_1. The remaining variables T, P, C_2 (i.e., concentration of solid 2), and C_3 (i.e., concentration of solid 3) can each be varied within limits without changing the number of phases (i.e., two). Along line CD both solid 1 and solid 2 separate out, giving rise to three phases. Now C_1 and C_2 are fixed at the corresponding solubility values S_1 and S_2, while three degrees of freedom, T, P, and C_3, remain. Finally, along line DE all three solids separate out and are in equilibrium with the solution. Since there are now four phases, there are only two degrees of freedom, namely, temperature and pressure, while C_1, C_2, and C_3 are each fixed at the corresponding solubility values S_1, S_2, and S_3. Thus, under defined conditions of temperature and pressure, the composition of the solution along line DE is fixed and the system appears to be invariant.

The limitations of phase-solubility analysis as summarized in Figure 10.2 and described in this section are as follows: (a) separation of the components, when present as the ratio of their solubilities, will be impossible; (b) the formation of a solid solution will greatly complicate the system; (c) the solubility of each component must not be affected by the presence of any of the others, that is, the components must not interact. By and large, only (c) is of relatively common occurrence. The treatment of interacting components will be discussed in Section 10.4.

10.3 EXPERIMENTAL PROCEDURE FOR PURITY DETERMINATION

The *U.S. Pharmacopeia XXI* (1985) provides a detailed description of the method normally employed for purity determination. Constancy of solubility, like constancy of melting point and other physical properties, indicates that the solid sample is a pure substance (i.e., an element or a compound) and is free from foreign admixture. The unique exception to this rule arises when the solid contains two or more component substances which are present in amounts that are in the ratio of their solubilities. Variability of solubility usually indicates the presence of one or more impurities or some other heterogeneity, for example, in energetic composition (see Section 8.9). Thus, phase-solubility analysis is applicable to all substances that are crystalline solids, provided that they are capable of forming a stable solution in a suitable solvent. The method cannot easily be applied to substances that form solid solutions with the impurities.

The standard method of phase-solubility analysis consists of the following six stages: (1) mixing, in a series of separate vessels, increasing quantities of solid sample with a constant measured amount of a defined solvent; (2) establishment of equilibrium in each vessel (i.e., each system) at the same temperature and pressure; (3) separation of the solid phase from each solution; (4) determination of the concentration of the material dissolved in each solution; (5) plotting the weight of the dissolved solute per unit of solvent (y axis and solution composition) against the weight of solid sample added per unit of solvent (x axis and system composition); and (6) extrapolation and calculation. Figures 10.1 and 10.2 illustrate the relevant plots.

A suitable solvent for phase-solubility analysis has the following properties. (a) The solvent is sufficiently volatile that it can be evaporated under vacuum but not so volatile that difficulty is experienced in transferring and weighing the solvent and its solutions. Solvents of boiling points between 60 and 150°C are usually suitable. (b) The solvent does not cause decomposition of the sample, react with the sample, or form solvates or salts with the sample. (c) The solvent is of known purity and composition. Accurately mixed solvents may be used, but trace impurities may strongly affect the solubility. (d) A solubility of $10-20 \, mg \, g^{-1}$ is optimal, but a wider working range can be employed.

The U.S. Pharmacopeia XXI (1985) recommends the use of 15-cm³ ampuls (ampoules) or solubility flasks as shown in Figure 10.3a. Other suitable containers may also be employed. The method of the U.S. Pharmacopeia XXI (1985) will now be outlined. Increasing amounts of solid S are weighed into a series of clean ampuls. The first ampul contains slightly less solid than will dissolve in 5 cm³ of the selected solvent. The second ampul contains slightly more material and each subsequent ampul contains increasingly more material than corresponds to the indicated solubility. Solvent in 5.0 cm³ aliquots is added to each ampul, which is then cooled in dry ice + acetone, sealed, allowed to attain room temperature, and weighed. The system composition is calculated as the weight of added solid per unit weight of the solvent.

The temperature of equilibration is usually maintained within ± 0.1°C or better by means of a water bath. Equilibration is assisted by rotating a horizontal shaft, equipped with clamps to hold the ampuls, at 25 rpm. Alternatively, the bath may contain a vibrator which is capable of agitating the ampuls at about 100–120 vibrations per second and which is fitted with a shaft and clamps to hold the ampuls. The time required for equilibration depends on the solid, the solvent, the method of agitation, and the temperature of equilibration. Equilibrium is normally attained more rapidly (1–7 days) by the vibration method than by the rotation method (7–14 days). To test whether equilibrium has been attained, one ampul, such as the next to last in the series, is warmed to a temperature about 15°C above the equilibrium temperature to produce a supersaturated solution. Equilibrium is assured if the analytical dissolved

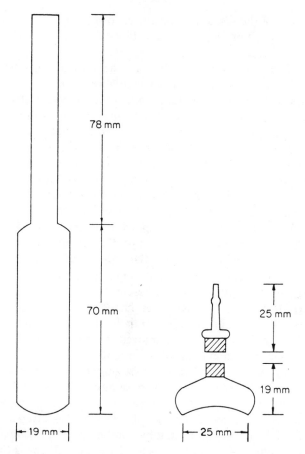

Figure 10.3a Ampul (left) and solublity flask (right) used in phase-solubility analysis. Reproduced with permission of the copyright owner, the United States Pharmacopeial Convention, Inc., Rockville, MD. *The U.S. Pharmacopeia XXI* (1985), p. 1343.

concentration obtained from the supersaturated solution falls in line with those obtained from the test specimens that approached equilibrium from an unsaturated solution.

The composition of the solution after equilibrium is determined by a suitable analytical method. First, the excess solid is allowed to settle in the ampuls while still in the water bath. Each ampul is opened and more than $2 \, cm^3$ of the solution is removed by means of a pipette fitted with a suitable filter. A 2-cm^3 aliquot of the clear solution from each ampul is transferred to a labeled, tared flask which is then weighed to determine the weight of the solution. For the classical method of phase-solubility analysis under present consideration (*U.S. Pharmacopeia XXI*, 1985), the weight of total solid dissolved is measured after cooling each flask in dry ice + acetone and after then evaporating the solvent under vacuum. The temperature is gradually increased to a value consistent with the stability of the test solid and the residue is dried to constant weight. The solution composition is calculated as the weight of the dissolved solid per unit weight of solvent.

The solution composition is plotted as the ordinate against the system composition for each ampul, as shown in Figures 10.1 and 10.2. The points for those ampuls, often only one, that contain an undersaturated solution fall on a straight line AB with a slope of unity passing through the origin. The points for those ampuls that contain a saturated solution plus excess of solid fall on another straight line BC, the slope m of which represents the weight fraction of impurity or impurities present in the solid sample under test. Any scatter of the points suggests that equilibrium may not have been achieved. A curve may indicate that the test sample is a solid solution. The percentage of the main component in the test sample, sometimes loosely termed the percentage purity, is given by $100(1 - m)$. As stated previously, the solubility of the defined main component in the solvent at the temperature of equilibration is given by the intercept of the straight line BC on the ordinate and is a constant.

The solution phase, from those ampuls represented by points on the straight line BC in figure 10.1 or 10.2, contains essentially all the impurities originally present in the test solid. On the other hand, the solid phase which is in equilibrium with this solution is essentially free from impurities. Hence, phase-solubility analysis can be employed to prepare (1) pure reference specimens of solid compounds and (2) concentrates of impurities from solid samples. A simple method of achieving these objectives is described in the U.S. Pharmacopeia XXI (1985) and will now be summarized. A weighed amount of solid test material is suspended in a suitable composition and amount of a nonreactive solvent such that about 10% of the solid is dissolved at equilibrium. The suspension is shaken in a suitable bottle at room temperature until equilibrium is attained, normally for 24 h. The mother liquor is then removed and evaporated to dryness near room temperature. The resulting solid residue contains the impurities, each of which is concentrated roughly in proportion to the ratio of the weight of sample taken to the weight of total solids dissolved in the volume of solvent used. The undissolved crystals remaining after withdrawal of the mother liquor and after rinsing and drying are often pure enough to be employed as a reference standard in pharmaceutical analysis (U.S. Pharmacopeia XXI, 1985).

10.4 COMPLEXATION SOLUBILITY ANALYSIS

The technique of complexation solubility analysis exploits the interactions in solution which limit the applicability of classical phase-solubility analysis. The experimental operation for this analysis is depicted in Figure 10.3b. The technique has been reviewed by Higuchi and Connors (1965). Equal masses of the compound of interest, known as the substrate S, and equal volumes of the solvent are added to each of several vials or ampuls. Increasing amounts of a relatively soluble complexing agent, known as the ligand L, are then added to

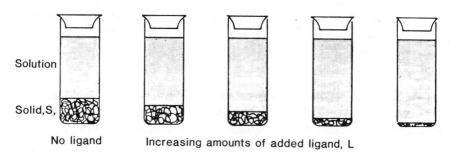

Solution

Solid, S,

No ligand   Increasing amounts of added ligand, L

Figure 10.3*b*   Simplified scheme illustrating the technique of complexation solubility analysis.

the vessels which are closed and equilibrated at constant temperature. S must be only slightly soluble in the solvent and must be present in considerable excess of its normal solubility in the pure solvent. The solution phase in each vessel is analyzed for the total molar concentration of compound S no matter what its molecular state may be, and this is plotted against the molar concentration of L that was added to the system. The phase-equilibrium diagrams obtained fall into two main classes, A (Fig. 10.4) and B (Fig. 10.9), each of which can be further subdivided.

In the subsequent discussion, the symbols have the following significance: $[S]_0$ represents the equilibrium solubility of S in the absence of L; $[S]_t$ represents the total molar concentration of dissolved S, both complexed and uncomplexed; $[L]_t$ represents the total molar concentration of uncomplexed and complexed L. In the forthcoming text and in several of the phase diagrams (e.g., Fig. 10.4–10.6), $[S]_t$ is termed the "apparent" solubility of S, since it includes not only free, uncomplexed S, but also complexes containing S, for example, SL, S_2L, and SL_2.

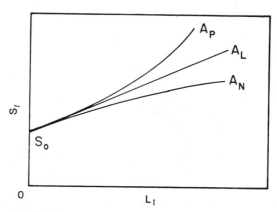

Figure 10.4   Phase-solubility diagram of Type A systems showing apparent increase in solubility of S caused by component L. The symbols S_t, S_0 and L_t here represent $[S]_t$, $[S]_0$, and $[L]_t$ in the text. Reproduced with permission of the copyright owner, John Wiley and Sons, Inc., from Higuchi, T. and Connors, K. A. (1965). Phase-Solubility Techniques, in *Advances in Analytical Chemistry and Instrumentation*, Vol. 4, John Wiley and Sons, Inc., New York.

Figure 10.4 illustrates type A phase diagrams, which are characterized by a continual increase in $[S]_t$ with increasing $[L]_t$, presumably because of the formation of soluble complexes between S and L. A solid phase consisting of pure S is in equilibrium with the solution so the thermodynamic activity of dissolved S remains constant.

A strictly linear rise in $[S]_t$ with increasing $[L]_t$, designated A_L in Figure 10.4, occurs when the molecules of the complexes formed contain just one molecule of L, that is, the complexes are of the first order in L, namely, $SL, S_2L, \ldots, S_mL$. Type A_L behavior is exemplified by salicylic acid as S with various water-soluble amides as L in water as the solvent (Fig. 10.5). This effect appears to begin below point Z in Figure 10.8.

Positive curvature, indicated by A_P in figure 10.4, arises when the complex(es) contain L at higher orders than one (e.g., $SL_2, SL_3, \ldots, SL_n$). Type A_P behavior is shown by oxytetracycline as S with nicotinamide as L in water at pH 5 (Fig. 10.6).

Negative curvature, indicated by A_N in Figure 10.4, may arise when the presence of L at high concentrations changes the stability constant of the complex or induces self-association of L. This behavior is rare.

Occasionally, a type A plot may show a plateau at which $[S]_t$ remains constant while $[L]_t$ is increasing. The break marking the beginning of the plateau may merely represent complete dissolution of solid S, which is normally avoided by adding more solid S; an example is shown in Figure 10.7. The break may also represent saturation of the solution with L, owing to its insufficient

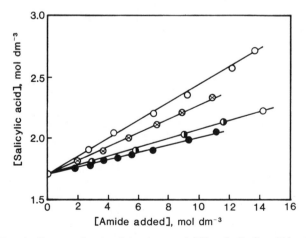

Figure 10.5 Type A_L diagrams showing the apparent solubility of salicylic acid in aqueous solution at 30°C in the presence of: tetramethylfumaramide ($\bigcirc$); tetramethylisophthalamide ($\otimes$); tetramethylsuccinamide ($\circleddash$); tetramethylphthalamide ($\bullet$) (Kostenbauder and Higuchi, 1956). $1 \, mol \, dm^{-3} = 1 \, mol \, L^{-1} = 1 \, M$. Reproduced with permission of the copyright owner, John Wiley and Sons, Inc., from Higuchi, T. and Connors, K. A. (1965). Phase-Solubility Techniques, in *Advances in Analytical Chemistry and Instrumentation*, Vol. 4, John Wiley and Sons, Inc., New York.

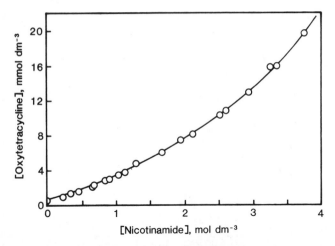

Figure 10.6 Type A_P diagrams showing the effect of nicotinamide on the apparent solubility of oxytetracycline at 25°C and pH 5 (Higuchi and Bolton, 1959). $1 \, mol \, dm^{-3} = 1 \, mol \, L^{-1} = 1 \, M$. Reproduced with permission of the copyright owner, John Wiley and Sons, Inc., from Higuchi, T. and Connors, K. A. (1965). Phase-Solubility Techniques, in *Advances in Analytical Chemistry and Instrumentation*, Vol. 4, John Wiley and Sons, Inc., New York.

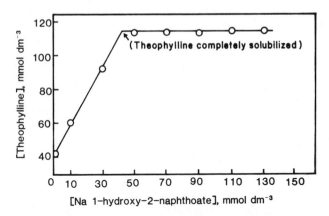

Figure 10.7 Type A diagram illustrating plateau formation due to disappearance of solid S (Higuchi and Drubulis, 1961). The system consists of theophylline, S, and sodium 1-hydroxy-2-naphthoate, L, in water at 30°C. $1 \, mol \, dm^{-3} = 1 \, mol \, L^{-1} = 1 \, M$. Reproduced with permission of the copyright owner, John Wiley and Sons, Inc., from Higuchi, T. and Connors, K. A. (1965). Phase-Solubility Techniques, in *Advances in Analytical Chemistry and Instrumentation*, Vol. 4, John Wiley and Sons, Inc., New York.

solubility, as illustrated in Figure 10.8. Water is also the solvent in these examples.

Type B phase-solubility diagrams, illustrated in Figure 10.9, arise when complexes of limited solubility are formed between S and L.

Type B_S in Figure 10.9 indicates that the solubility of the complex is not

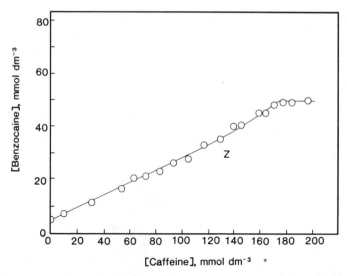

Figure 10.8 Type A diagram showing invariance due to appearance of solid L [Higuchi, T. and Lach, J. L. (1954), *J. Am. Pharm. Assoc., Sci. Ed.*, **43**, 527]. The system consists of benzocaine, S, and caffeine, L, in water at 30°C. $1 \, mol \, dm^{-3} = 1 \, mol \, L = 1 \, M$. Reproduced with permission of the copyright owner, John Wiley and Sons, Inc., from Higuchi, T. and Connors, K. A. (1965). Phase-Solubility Techniques, in *Advances in Analytical Chemistry and Instrumentation*, Vol. 4, John Wiley and Sons, Inc., New York.

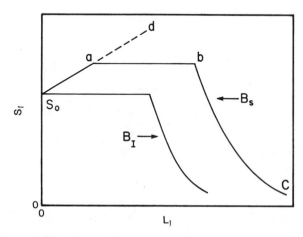

Figure 10.9 Phase-solubility diagrams of type B systems. The symbols S_t, S_0, and L_t here represent $[S]_t$, $[S]_0$, and $[L]_t$, in the text. Reproduced with permission of the copyright owner, John Wiley and Sons, Inc., from Higuchi, T. and Connors, K. A. (1965). Phase-Solubility Techniques, in *Advances in Analytical Chemistry and Instrumentation*, Vol. 4, John Wiley and Sons, Inc., New York.

negligible. The rise from $[S]_0$ to a, which is not necessarily linear, is interpreted as the formation of a complex which dissolves as in any of the type A diagrams. At point a, however, the solubility limit of the complex is reached and further addition of L results in the formation of more complex in a solid form. Since S is then also being precipitated with L, more solid S dissolves to keep constant the thermodynamic activity and the concentration of dissolved S, as required by the phase rule (Table 10.2). At point b, no more solid S remains and further addition of L results in the depletion of S from the solution by formation and precipitation of the complex. In certain experiments the solution may become supersaturated with small amounts of complex, as indicated by the dotted segment a–d.

Curve B_l is interpreted in the same way as B_S, except that the complex formed is so insoluble that the initial rise ($[S]_0$–a) in the concentration of S is phase rule (E1. 10.1).

If, in the B_S diagram, the nature of the complex responsible for $[S]_0$–a is the same as that finally precipitated along b–c, the increase in $[S]_t$ from $[S]_0$ to a must equal the final value of $[S]_t$ at c. When this is not the case, as is commonly found, the system must form two or more distinct complexes, one being responsible for the initial rise in the solubility of S and another being precipitated between b and c. Sometimes an increase in solubility is observed beyond point c, which apparently results from the formation of a more soluble complex than that responsible for the fall from b to c.

Figure 10.10 shows type B curves, the upper two curves being type B_S and the

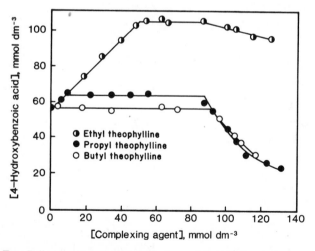

Figure 10.10 Type B diagrams showing the effect of several theophylline derivatives, as complexing agents L on the apparent solubility of 4-hydroxybenzoic acid in water at 30°C (Bolton et al., 1957). Complexing agents: ethyl theophylline (◑); propyl theophylline (●); butyl theophylline (○). $1 \, \text{mol} \, \text{dm}^{-3} = 1 \, \text{mol} \, \text{L} = 1M$. Reproduced with permission of the copyright owner, John Wiley and Sons, Inc. from Higuhi, T. and Connors, K.A. (1965). Phase-Solubility Techniques, in *Advances in Analytical Chemistry and Instrumentation*, Vol. 4, John Wiley and Sons, Inc., New York.

Table 10.2 Application of the Phase Rule to the Type B_S Diagram Showing the Effect of Several Theophylline Derivatives on the Apparent Solubility of 4-Hydroxybenzoic Acid in Water at 30°C[a]

Segment[b]	C	P	F	Enumeration of Degrees of Freedom
S_0-a	3	2	3	T, P, C_L
$a-b$	3	3	2	T, P
$b-c$	3	2	3	T, P, C_S

[a]The interpretation of the relationship to Figure 10.9 is presented in the text.
[b]From Figure 10.9.

Reproduced with permission of the copyright owner, John Wiley and Sons, Inc., from Higuchi, T. and Connors, K. A. (1965). Phase-Solubility Techniques, in *Advances in Analytical Chemistry and Instrumentation*, Vol. 4, John Wiley and Sons, Inc., New York.

lowest being type B_I. Table 10.2 shows a phase rule analysis (Eq. 10.1) of the type B_S plots illustrated in Figure 10.10. These B_S system consist of S, L, and the solvent and therefore comprise three components. In the plateau region of Figure 10.10 $P = 3$, corresponding to the saturated solution, excess of solid S, and an insoluble solid complex phase that is being precipitated. Therefore, $F = 2$, corresponding to complete definition of the system by P and T and to fixed values of C_S and C_L. In the ascending region S_0-a there are two phases, the saturated solution and excess of solid S, so $F = 3$. Consequently, T, P, and the concentration of S, C_S, may each be varied within limits without changing the number of phases and, when fixed, will define the system completely. Thus, the concentration of L, C_L, is fixed by the system. Similarly, in the descending region $b-c$, $F = 3$ (T, P, and C_S), so $P = 2$ here too, corresponding to the saturated

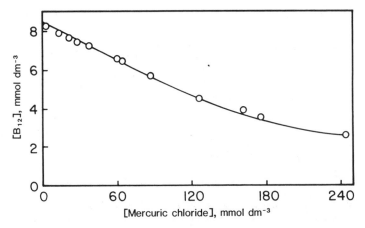

Figure 10.11 Phase diagram showing the effect of mercuric chloride on the apparent solubility of cyanocobalamin (vitamin B_{12}) in water at 25°C. $1 \, mol \, dm^{-3} = 1 \, mol \, L^{-1} = 1 \, M$. Reproduced with permission of the copyright owner, the American Pharmaceutical Association, from Havemeyer, R.N. and Higuchi, T. (1960). *J. Am. Pharm. Assoc.*, **49** (1960) 356–360.

solution with one solid phase only. Thus, there cannot be two solid phases; this means that only one complex can precipitate out at any given point along b–c, although at different points on b–c different complexes may appear. Alternatively, if two or more complexes are precipitated at a given point, they must form a single solid solution, thus ensuring that $P = 2$, as required by the phase rule (Eq. 10.1).

The unusual phase diagram in Figure 10.11 fits neither type A nor type B, but resembles the descending portion of type B just described. Application of the phase rule (Eq. 10.1) again indicates that $C = 3$, $F = 3$, and $P = 2$, corresponding to the saturated solution with one solid phase. It is possible that solid S may form a solid solution with the insoluble solid complex. Other explanations for this type of phase diagram have also been proposed by Havemeyer and Higuchi (1960).

10.5   DETERMINATION OF THE STOICHIOMETRY OF COMPLEXES

From a study of the phase-solubility diagram obtained in complex formation, it is sometimes possible to obtain information about the stoichiometry of the complex(es) formed, that is, the values of m and n in $S_m L_n$. If, for example, the initial slope in a type A or type B_S diagram exceeds unity, then at least one species must be present in which m exceeds unity. The value of n cannot normally be deduced, since the excess of solid S maintains the activity of dissolved S constant.

If a type A diagram exhibits a plateau due to the appearance of a solid L phase, a stoichiometric ratio m/n can be calculated. Such a system is shown in Figure 10.8, in which benzocaine is S and caffeine is L and the solvent is water. The increase in the molar solubility of each component above its normal value in the solvent alone is proportional to the amount entering into the complex. Since for caffeine, Z represents the normal solubility limit, which is $133 \, mmol \, dm^{-3}$, and the concentration at the break point is $175 \, mmol \, dm^{-3}$, then the increase in caffeine solubility is $175 - 133 \, mmol \, dm^{-3} = 42 \, mmol \, dm^{-3}$. For benzocaine, the solubility in water alone is $5 \, mmol \, dm^{-3}$, and the concentration at the break point is $49 \, mmol \, dm^{-3}$, so the increase in benzocaine solubility is $49 - 5 \, mmol \, dm^{-3} = 44 \, mmol \, dm^{-3}$. Thus, the stoichiometric ratio is given by

$$\frac{m}{n} = \frac{\text{benzocaine (S) content}}{\text{caffeine (L) content}} = \frac{44}{42} = 1.05 \cong 1$$

so the formula of the complex can be written SL. Since the diagram is really of the A_p type, a second complex is present to a minor extent.

A stoichiometric ratio can always be evaluated from a type B diagram and is that of the complex being precipitated under zero degrees of freedom. It is very unlikely that two or more complexes would be spontaneously precipitated, because this possibility would require them to have exactly the same solubilities when calculated as the concentration of S.

Figure 10.12 shows a typical type B_S diagram in which 1,3-dimethyl-2,4-dioxoquinoxaline (dimethylbenzoyleneurea) is S and catechol is L and the solvent is carbon tetrachloride. Knowing the total amount of S added to the system ($79.7 \, mmol \, dm^{-3}$), the stoichiometric ratio m/n of the components in the complex which is formed in the plateau region AB can be calculated from the phase diagram. This is possible because the excess of free S at point A is equal to the amount being converted into the complex between A and B which is ($79.9 \, mmol \, dm^{-3}$ initially $- 52.2 \, mmol \, dm^{-3}$ in solution at A to B) $27.7 \, mmol \, dm^{-3}$. The corresponding amount of L being converted to the complex is equal to that taken up between A and B which is ($86.2 \, mmol \, dm^{-3}$ at $B - 14.5 \, mmol \, dm^{-3}$ at A) $71.7 \, mmol \, dm^{-3}$. The stoichiometric ratio is then given by

$$\frac{m}{n} = \frac{\text{dimethylbenzoyleneurea (S) content}}{\text{catechol (L) content}} = \frac{27.5}{71.7} = \frac{1}{2.61}$$

indicating that the formula of the complex can probably be written as S_2L_5.

Figure 10.13 is a type B_I diagram. The total concentration of 2,5-dihydroxybenzoic acid (S) present in the system is $390 \, mmol \, dm^{-3}$. Calculation of the stoichiometry is based on the extent of the plateau region AB as in the case of Figure 10.12.

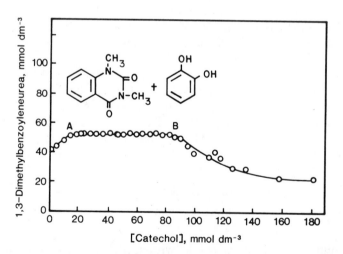

Figure 10.12   Type B_S phase diagram showing the interaction between 1,3-dimethylbenzoyleneurea and catechol in carbon tetrachloride at 30°C (Haddad et al., 1959). $1 \, mol \, dm^{-3} = 1 \, mol \, L = 1M$. Reproduced with permission of the copyright owner, John Wiley and Sons, Inc., from Higuchi, T. and Connors, K.A. (1965). Phase-Solubility Techniques, in *Advances in Analytical Chemistry and Instrumentation*, Vol. 4, John Wiley and Sons, Inc., New York.

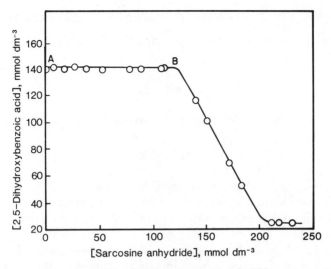

Figure 10.13 Type B_l diagram showing the effect of sarcosine anhydride on apparent solubility of 2, 5-dihydroxybenzoic acid in water at 25°C (Poole and Higuchi, 1959). $1 \, mol \, dm^{-3} = 1 \, mol \, L = 1 \, M$. Reproduced with permission of the copyright owner, John Wiley and Sons, Inc., from Higuchi, T. and Connors, K.A. (1965). Phase-Solubility Techniques, in *Advances in Analytical Chemistry and Instrumentation*, Vol. 4, John Wiley and Sons, Inc., New York.

$$S \text{ content} = (390 - 140) \, mmol \, dm^{-3} = 250 \, mmol \, dm^{-3}$$
$$L \text{ content} = (126 - 0) \, mmol \, dm^{-3} = 126 \, mmol \, dm^{-3}$$
$$m/n = S \text{ content}/L \text{ content} = 250/126 = 1.98$$

so the formula of the complex can be written as S_2L.

For those diagrams in which the descending portion is a straight line extending closely to the $[L]_t$ axis, m/n for the complex can be determined as follows. First, the down curve is extrapolated to $[S]_t = 0$. The difference between $[L]_t$ at the start of the down curve ($126 \, mmol \, dm^{-3}$) and $[L]_t$ at the extrapolated point ($202 \, mmol \, dm^{-3}$) minus $[L]_t$ present as the soluble complex (zero) is equal to the concentration of L needed to react with the free S in solution at AB ($140 \, mmol \, dm^{-3}$).

Concentration of L entering into the complex along the down curve $=$
   $(202 - 126 - 0) \, mmol \, dm^{-3} = 76 \, mmol \, dm^{-3}$

Concentration of S in solution $= 140 \, mmol \, dm^{-3}$

Therefore, $m/n = S \text{ content}/L \text{ content} = 140/76 = 1.84$

Thus, the formula of the complex can be written as S_2L.

These graphical methods for evaluation of the stoichiometry may be supplemented by isolation and chemical analysis of insoluble complexes.

10.6   ESTIMATION OF STABILITY CONSTANTS OF COMPLEXES

The general stoichiometric chemical equation for complex formation may be written

$$mS + nL \rightleftharpoons S_m L_n \tag{10.2}$$

The stability constant of the complex is the equilibrium constant, thus,

$$K = [S_m L_n]/[S]^m[L]^n \tag{10.3}$$

Obviously, the stoichiometry, represented by m and n, must be known before K can be calculated. In some cases more than one complex may be present, so that the equilibrium constant may have to be an apparent value based on a reasonable mean stoichiometric ratio. As we shall see, an approximate value of K can often be calculated from the data used in the phase equilibrium diagram, in which it will be expressed in the same units as the axis of the diagram, for example, on a molarity scale.

For type A systems and the initial portion of type B_S systems,

$$[S] = [S]_o; \quad [S_m L_n] = \frac{[S]_t - [S_o]}{m} \tag{10.4}$$

$$[L] = [L]_t - n[S_m L_n] = [L]_t - (n/m)\{[S]_t - [S]_o\} \tag{10.5}$$

Insertion into Eq. 10.3 gives us an expression for K in terms of m, n, and the known concentrations. From this expression, the following equation for the important special case of $n = 1$ can be derived:

$$[S]_t = \frac{mK[S]_o^m[L]_t}{1 + K[S]_o^m} + [S]_o \tag{10.6}$$

A plot of $[S]_t$ against $[L]_t$ for the formation of the soluble complex $S_m L$ should yield a straight line, corresponding to the type A_L diagram, the intercept being $[S]_o$ and the slope being given by

$$\text{slope} = \frac{mK[S]_o^m}{1 + K[S]_o^m} \tag{10.7}$$

If m is known, this equation enables K to be calculated from the slope of a type A_L diagram. For the special case of $m = 1$, we obtain

$$K_{1:1} = \text{slope}/[S]_o(1 - \text{slope}) \tag{10.8}$$

For the special case $m = 1$ and $n = 1$, combination of Eqs. 10.3–10.5 affords

$$K_{1:1} = \frac{[S]_t - [S]_o}{[S]_o([L]_t - [S]_t + [S]_o)} \tag{10.9}$$

If more complexes than SL were present (e.g., $S_2L, S_3L, \ldots, S_mL$), a type A_L diagram would still be observed, in which case Eq. 10.8 and 10.9 would yield apparent constants. If the slope of a type A_L diagram is greater than unity, the assumption of a 1:1 complex alone is untenable, so Eq. 10.7 should be used if a value of m can be stated. Type A_P and A_N diagrams are often approximated by a linear tangent to the curve at $[L]_t = 0$, the apparent stability constants being calculated with Eq. 10.8.

A more quantitative description of the type A_P diagram may be achieved by assuming that it results from the formation of the two complexes SL and SL_2, characterized by the constants

$$K_{1:1} = [SL]/[S][L] \tag{10.10}$$

$$K_{1:2} = [SL_2]/[SL][L] \tag{10.11}$$

The equations for the material balance are

$$[S]_t = [S] + [SL] + [SL_2] \tag{10.12}$$

$$[S] = [S]_o \tag{10.13}$$

$$[L]_t = [L] + [SL] + 2[SL_2] \tag{10.14}$$

Combining Eqs. 10.10–10.14 gives

$$[S]_t = \frac{[L]_t(K_{1:1}[S]_o + K_{1:1}K_{1:2}[S]_o[L])}{1 + K_{1:1}[S]_o + 2K_{1:1}K_{1:2}[S]_o[L]} + [S]_o \tag{10.15}$$

which shows that a plot of $[S]_t$ against $[L]_t$ will have an intercept $[S]_o$ and a slope which increases as $[L]_t$ increases, thus giving a type A_P diagram. Combining Eqs. 10.10–10.13 affords a more tractable expression:

$$[S]_t = [S]_o + K_{1:1}[S]_o[L] + K_{1:1}K_{1:2}[S]_o[L]^2 + \cdots \tag{10.16}$$

If the extent of complexation is fairly small, it may be permissible to assume that $[L] \cong [L]_t$, which gives

$$([S]_t - [S]_o)/[L]_t = K_{1:1}[S]_o + K_{1:1}K_{1:2}[S]_o[L]_t \tag{10.17}$$

A plot of $([S]_t - [S]_o)/[L]_t$ against $[L]_t$ should be linear. From the slope and intercept, values of $K_{1:1}$ and $K_{1:2}$ can be calculated. $([S]_t - [S]_o)/[L]_t$ is the slope of the chord from $[S]_o$ to $[S]_t$ on the type A_P diagram. A better approximation for $[L]$ is achieved by combining Eqs. 10.12–10.14 and by assuming that all the complex is either SL or SL_2. Higuchi and Bolton (1959) have analyzed a system containing SL, SL_2, and SL_3 complexes.

This discussion applies to the rising segments of type B_S as well as to type A diagrams. It is also possible to estimate stability constants from the descending portion of type B curves (Fig. 10.14) if the stoichiometry of the complex is known. The equilibrium constant is given by Eq. 10.3. At point X in Figure 10.14, the following equations hold:

$$S_X = [S] + m[S_m L_n] \qquad (10.18)$$

$$L_X = [L] + n[S_m L_n] \qquad (10.19)$$

Throughout the down curve, the solution is in equilibrium with the solid complex, so the dissolved concentration of the complex is constant. $[S]_B$, the concentration of S at point B where all free S has been removed from solution, may therefore be attributed to complex in solution, thus,

$$[S]_B = m[S_m L_n] \qquad (10.20)$$

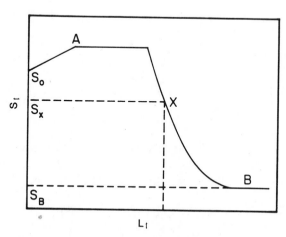

Figure 10.14 Type B diagram showing calculation of the stability constant from the descending portion of the curve. The symbols S and L here represent the concentrations of the solute and ligand, and are placed in square brackets in the text. Reproduced with permission of the copyright owner, John Wiley and Sons, Inc., from Higuchi, T. and Connors K.A. (1965). Phase-Solubility Techniques, in *Advances in Analytical Chemistry and Instrumentation*, Vol. 4, John Wiley and Sons, Inc., New York.

Insertion of Eqs. 10.18–10.20 for [S], [L], and $[S_mL_n]$ into Eq. 10.3 yields

$$K_{m:n} = [S]_B/m([S]_X - [S]_B)^m([L]_X - n[S]_B/m)^n \qquad (10.21)$$

If the same complex is considered to be responsible for the behavior over the entire curve, the $[S]_B$ may be replaced by the increase in solubility from $[S]_o$ to point A.

10.7   NATURE AND MECHANISM OF COMPLEXATION

The extensive phase-solubility studies of Higuchi and co-workers from 1952 to 1965 on the complexation of drugs and other pharmaceutically important compounds in solution have been reviewed and the data tabulated by Higuchi and Connors (1965). The data demonstrate the ubiquity of mutual solubility effects and allow some correlations between molecular structure and the extent of interaction to be made. The authors point out that the stoichiometry and equilibrium constants may be ambiguous quantities since more than one complex species is probably formed in most of the systems. In order to facilitate the comparison of complexing tendencies, it is common practice to calculate $K_{1:1}$ for type A and type B_S diagrams, even when no evidence is available to indicate that a 1:1 complex is formed. The equilibrium constants should be regarded as empirical parameters which describe approximately the increase in apparent solubility of S in the presence of L. The approximate solubility of some of the complexes formed in type B systems may be found in the original papers referenced by Higuchi and Connors (1965).

A complex may be considered to have been formed between S and L when, in the immediate vicinity of S, more molecules of L are found than would be expected statistically. The following example illustrates that an added agent L (e.g., sarcosine anhydride) can increase the solubility of a substrate S on a purely statistical basis without postulating a complex. The absence of complex formation means the S has no greater tendency to associate with L than with the solvent, and so any S–L association results only from a chance contact of S with L. Density measurements indicate that sarcosine anhydride 1% (w/v) in water has a volume fraction of 0.008, and so the probability that a sarcosine anhydride molecule is the nearest neighbor of an S molecule is 8/1000. Assuming that each S molecule has about 10 near neighbors, the probability is $10 \times 8/1000 = 0.08$. If such contacting S–L pairs are assumed to be separate species, an increase in the original solubility of about 8% may be expected in a 1% (w/v) sarcosine anhydride solution, even in the absence of specific complex formation. Any substantially greater increase in solubility may be ascribed to specific interactions which form discrete molecular species. A similar argument has been presented in Section 4.3.

The intermolecular interactions responsible for complex formation between S and L are not covalent but are the weaker intermolecular forces discussed in Section 2.11 namely, electrostatic (ion–ion, ion–dipole, ion–quadrupole, dipole–dipole, dipole–quadrupole, quadrupole–quadrupole, etc.), induction (ion–induced dipole, dipole–induced dipole), London dispersion forces (induced dipole–induced dipole), hydrogen bonding and other charge-transfer interactions, and solvophobic interactions (especially hydrophobic interactions in water, which are discussed in Section 8.2). Solvophobic interactions arise when relatively nonpolar molecules are present in a highly polar solvent. The relative contributions of these various interactions depend on the nature of molecules which are interacting.

Complex formation between a substrate S and a ligand L results from the combined effect of several of the intermolecular interactions quoted above and is the result of competition between several reactants. The S–L interaction is modified by (a) S–solvent and L–solvent interactions, which are in turn influenced by solvent–solvent interactions, and (b) intermolecular interactions and associations of the type S–S and L–L which form dimers, trimers, tetramers, and pentamers. All of these possibilities combine to produce the observed effect. The effects of individual contributions can be exposed by studying a series of related systems for which many experimental factors are held constant to minimize the alterations in some of the interactions. Specific interactions, such as hydrogen bonding and the charge-transfer complexation, are particularly amenable to structure–activity relationships, which have been discussed in Section. 5.5.

Higuchi and Kristiansen (1970) made some generalized deductions by comparing the relative magnitudes of the stability constants of various $S + L$ pairs in aqueous solution. The values of $K_{1:1}$ and $K_{1:2}$ were determined by phase-solubility analysis and by partition studies in the case of caffeine complexes. The binding between organic species in aqueous solution appears to take place most effectively between members of the large distinct classes of

Table 10.3 Classes of Organic Compounds which Form Complexes in Aqueous Solution[a]

Class A	Class B
Caffeine	Benzoates and salicylates
Theophylline	Cinnamates
Prednisolone	Cinnamamides
Tetramethylpyrimidopteridinetetrone (TMPPT)	Naphthoic acids
	Phenols and naphthols
	Aromatic amino acids
	Phenacetin
	Menadione
	Tryptophan

[a]The stability constants of complexes formed between members of different classes (i.e., A + B) are generally stronger than those formed between members within the same class (i.e., A + A or B + B).

Reproduced with permission of the copyright owner, the American Pharmaceutical Association, from Higuchi, T. and Kristiansen, H. (1970). *J. Pharm. Sci.*, **59**, 1601–1608.

structures classified arbitrarily as A and B (Table 10.3) which should not be confused with the types of phase diagram. Typical examples of Class A are the uncharged alkyl xanthines and the structurally related compound 1, 3, 7, 9-tetramethylpyrimido(5, 4-g)pteridine-2, 4, 6, 8(1H, 3H, 7H, 9H)-tetrone (TMPPT). Typical examples of Class B compounds are benzenoid compounds, phenols, and aromatic anions. Scheme 10.1 illustrates the relative stabilities of complexes between members of different classes and between members of the same class. $K_{1:1}$ values for A and B are greater than those for A with A or for B with B.

This principle is further illustrated by reference to Table 10.4 and to Figure 10.15, which shows two distinct linear free energy relationships (Section 5.5). Points near the steeper line in Figure 10.15 correspond to A class compounds, which form more stable complexes with cinnamamide (Class B) than with TMPPT (Class A). Points near the lower line correspond to B class compounds, which form more stable complexes with TMPPT (Class A) than with cinnamamide (Class B). From the slopes of such linear plots, Higuchi and Kristiansen (1970) calculated that the binding of Class A compounds to cinnamamide (Class B) is favored by about $600\,\text{cal}\,\text{mol}^{-1}$ $(2.5\,\text{kJ}\,\text{mol}^{-1})$ compared with the binding of the same class to TMPPT (Class A). For the binding of Class B compounds, TMPPT (Class A) is favored over the cinnamamides (Class B) by $1.0–1.5\,\text{kcal}\,\text{mol}^{-1}$ $(4.2–6.3\,\text{kJ}\,\text{mol}^{-1})$.

The strongest binding is generally given by polycyclic molecules in which there is extreme conjugation (Tables 10.4 and 10.5). This suggests that the area of planar overlap is important and that London dispersion forces (Section 2.9)

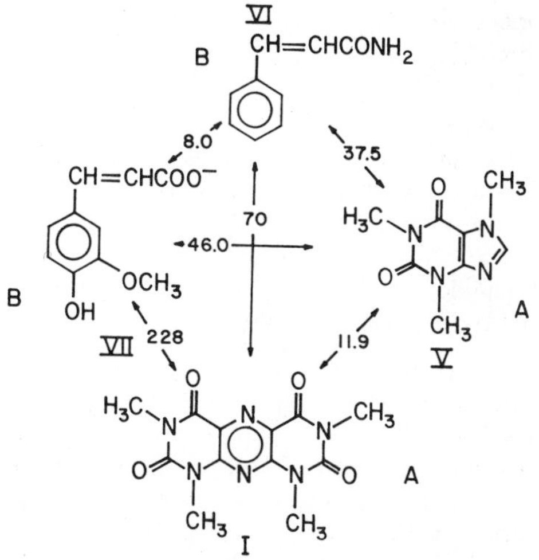

Scheme 10.1 A–B system interactions in water at 25°C. The numbers on the arrows represent the 1:1 stability constants ($\text{dm}^3\,\text{mol}^{-1}$, i.e., $\text{L}\,\text{mol}^{-1}$ or M^{-1}) for the interacting species. I, TMPPT (Class A); V, caffeine (Class A); VI, cinnamamide (Class B); and VII, ferulic acid anion (Class B). Reproduced with permission of the copyright owner, the American Pharmaceutical Association, from Higuchi, T. and Kristiansen, H. (1970). *J. Pharm. Sci.*, **59**, 1601–1608.

Table 10.4 Stability Constants[a] for the Interaction of Various Compounds with TMPPT, Cinnamamide, and N,N-Dimethylcinnamamide in Water at 25°C

Identification Number	Ligand	TMPPT	Cinnamamide	N,N-Dimethyl-cinnamamide
1	Caffeine	11.9	37.5	37.6
2	Theophylline	12.8	27.3	27.5
3	8-Methoxycaffeine	19.1	48.5	54.0
4	8-Chlorotheophylline	18.5	39.3	
5	Theophylline-7-acetic acid	5.0	22.7	
6	Theophylline-7-acetate	5.9	20.0	17.3
7	Phenylbutazone, sodium salt	6.0	5.0	
8	Antipyrine	1.9	3.6	
9	Imidazole	2.3	0.8	
10	Theophyllinate	56.2 (7.1)[b]	14.3	
11	8-Chlorotheophyllinate	154.1 (11.6)	25.8	
12	Phenacetin	43.3	8.0	6.8
13	Phenol	13.1 (5.7)	1.7	
14	4-Chlorophenol	29.2	2.7	
15	Sodium sorbate	4.2	1.4	
16	p-Anisidine	14.9[c]	2.4	
17	10-(2-Dimethyl-aminopropyl) pheno-thiazine hydrochloride	65.5 (2.7)	11.5	
18	7-(2-Dimethyl-aminoethyl) theophylline	10.0		31.0
19	β-Hydroxyethyl-theophylline	7.8		27.3
20	Sodium salicylate	44.3 (4.7)		1.9
21	2,6-Dihydroxybenzoate	102.1 (37.1)		5.9
22	2,6-Dihydroxybenzoic acid	96.6[c]		6.1
23	Nicotinamide	5.5		3.9
24	4-Hydroxycoumarin	99.0[c]		10.8

[a] $1\,dm^3\,mol^{-1} = 1\,L\,mol^{-1} = 1\,M^{-1}$.
[b] The numbers in parentheses are the 1:2 sequential stability constants ($K_{1:2}$, Eq. 10.11) for TMPPT as the ligand, also expressed in $dm^3\,mol^{-1}$.
[c] Calculated from initial increase in solubility of TMPPT. An insoluble complex is also formed.

Reproduced with permission of the copyright owner, the American Pharmaceutical Association, from Higuchi, T. and Kristiansen, H. (1970). *J. Pharm. Sci.*, **59**, 1601–1608.

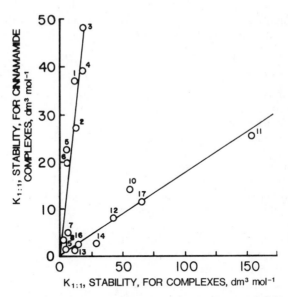

Figure 10.15 A plot of the 1:1 stability constants ($K_{1:1}$, Eq. 10.10) for cinnamamide and TMPPT with each of the indicated compounds. The abscissa refers to TMPPT complexes. The identification numbers refer to the compounds listed in Table 10.4. $1 \, dm^3 \, mol^{-1} = 1 \, L \, mol^{-1} = 1 \, M^{-1}$. Reproduced with permission of the copyright owner, the American Pharmaceutical Association, from Higuchi, T. and Kristiansen, H. (1970). *J. Pharm. Sci.*, **59**, 1601–1608.

or hydrophobic interactions are involved. The interacting organic molecules are presumed to associate by a stacking interaction which minimizes the planar area in contact with water molecules, as has been postulated for other systems which will be discussed.

Among the B class, the cinnamate ions form the strongest complexes with Class A compounds (Tables 10.4 and 10.5). The A class compound TMPPT undergoes a shift in the 362-nm absorption band toward longer wavelengths in the presence of cinnamates, benzoates, and theophyllinates, and the phase-solubility diagrams undergo upward curvature (as in Fig. 10.6), indicating the formation of higher-order complexes. The UV spectral shift suggests a type of charge–transfer interaction. If this is so, one can make the following speculations.

(a) The aromatic rings of Class A compounds appear generally to be subject to electronic effects which slightly deplete the π-electron system, enabling the surrounding nuclei to impart an overall small positive charge to the ring.

(b) Typical Class B compounds, on the other hand, appear to possess a concentration of negative charge which may reside on an anionic group or in the π-electron system of an aromatic ring. The latter effect may result from conjugation of the π-electrons with lone pairs of an electron-donating group, such as a phenolic hydroxyl group.

Table 10.5 Stability Constants Defined by Eqs. 10.10 and 10.11 for the Interaction of Some Organic Molecules in Water at 25°C[a]

Interacting Compounds	$K_{1:1}$ (dm^3 mol^{-1})	$K_{1:2}$ (dm^3 mol^{-1})
TMPPT–sodium benzoate	9.2	
TMPPT–*trans*-cinnamic acid anion	40.7	10.4
TMPPT–*d*,*l*-mandelic acid anion	4.2	
TMPPT–mandelamide	2.8	
TMPPT–5-phenyl-2,4-pentadienoic acid anion	128	25.9
TMPPT–*p*-aminohippuric acid anion	29.2	
TMPPT–*p*-coumaric acid anion	106	26.1
TMPPT–3-methoxy-4-hydroxybenzoic acid anion	50.5	14.6
TMPPT–3-methoxy-4-hydroxymandelic acid anion	139[b]	
TMPPT–caffeic acid anion	202	75
TMPPT–ferulic acid anion	228	85
TMPPT–cinnamamide	70.0	
TMPPT–*N*,*N*-dimethylcinnamamide	57.4	
TMPPT–*β*-naphthaleneacetic acid anion	141	16.0
TMPPT–4-methoxyphenylacetic acid anion	10.9	1.7
TMPPT–3-(*p*-methoxyphenyl)propionic acid anion	14.1	3.6
TMPPT–*N*,*N*-dimethyl-*p*-anisidine	8.2	
Caffeine–3-methoxy-4-hydroxybenzoic acid anion	17.8	
Caffeine–ferulic acid anion	46.0	
Phenacetin–caffeine	17.0	
Theophylline–caffeic acid anion	39.5	
Theophylline–2,6-dihydroxybenzoic acid anion	300	
Cinnamamide–ferulic acid anion	8.0	
N,*N*-Dimethylcinnamamide–caffeic acid anion	7.2	

[a]TMPPT is 1,3,7,9-tetramethylpyrimido(5,4–g)pteridine-2,4,6,8(1H,3H,7H,9H)-tetrone. 1 dm^3 mol^{-1} = 1 L mol^{-1} = 1 M^{-1}.
[b]See footnote *c* in Table 10.4.

(c) Intermolecular charge–transfer interactions will be stronger between members of different classes (i.e., A + B) than between members of the same class (i.e., A + A or B + B). The speculative charge–transfer interactions have not yet been proven.

In order to obtain additional information about the factors determining the stability of complexes which are stable in aqueous solution, Kristiansen et al. (1970) studied the influence of various organic solvents on the stability constants of the complexes. The procedures employed included spectrophotometry (Section 5.3.2), spectropolarimetry (Section 5.3.3), and phase-solubility analysis (the present chapter). Kristiansen et al. (1970) found that the presence of organic solvents invariably reduced the stability constants of all the complexes studied,

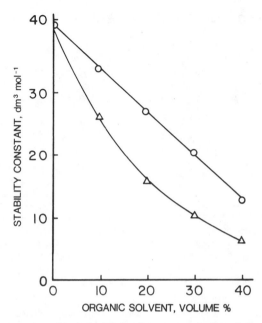

Figure 10.16 Stability constants ($K_{1:1}$, $dm^3\,mol^{-1}$, i.e., $L\,mol^{-1}$ or M^{-1}, from solubility at 25°C) of TMPPT–cinnamate ion complexes plotted against solvent composition: methanol–water mixtures (○); dioxane–water mixtures (△). Reproduced with permission of the copyright owner, the American Pharmaceutical Association, from Kristiansen, H., Nakano, N.I., and Higuchi, T. (1970). *J. Pharm. Sci.*, **59**, 1103–1106.

presumably by influencing the stacking interactions unfavorably. As the composition of the solvent medium is progressively altered from pure water toward about 50% v/v water by addition of organic solvent, the stability constants of all the complexes studied decreased monotonically, as illustrated by Figure 10.16.

Table 10.6 shows that the influence of a defined volume fraction of organic solvent (10% v/v) in reducing the stability constants of defined complexes generally correlates with the influence of solvent in reducing the surface tension of water. The hydroxy compounds exerted a much smaller destabilizing effect. Kristiansen et al. (1970) have extensively discussed these and related phenomena. A reasonable explanation is that water, and to a lesser extent other hydroxy compounds, forces the cyclic aromatic organic molecules to undergo stacking interactions by virtue of the hydrophobic effect (Sections 8.2.–8.5), thus stabilizing the complex. The addition of an organic solvent weakens the strong interactions between the water molecules, and thereby reduces the hydrophobic effect and the energetic necessity for the stacking interaction.

The destabilization of the complex parallels the weakening of the water–water interactions which, in Table 10.6, are measured by the surface tension. Although an aqueous environment is necessary for significant complexation, there may be some associative tendency even in pure organic solvents. For example, at

Table 10.6 Stability Constants $K_{1:1}$[a] of Complexes in 10% v/v Organic Solvent in Water at 20°C

Organic Solvent	TMPPT– DMCA[b] Complex	Menadione– Caffeine Complex	Surface Tension[c] (mN m^{-2})
None	60.0	48	72.0
Sucrose[d]	56.7	—	72.4
Glycerin	51.0	38	72.8
Methanol	48.7	34	58.2
Acetonitrile	39.9	25	53.8
Acetone	41.5	22	50.9
Dioxane	36.1	—	47.5[e]

[a] $1\,dm^3\,mol^{-1} = 1\,L\,mol^{-1} = 1\,M^{-1}$.
[b] N,N-Dimethylcinnamamide.
[c] *Handbook of Chemistry and Physics*, 49th ed., R. C. Weast (Ed.), The Chemical Rubber Co., Cleveland, OH, 1968, pp. 28–30.
[d] 10% w/v.
[e] Timmermans, J. *The Physical–Chemical Constants of Binary Systems in Concentrated Solutions*, Vol. IV, Interscience, New York, 1960, p. 14.

Reproduced with permission of the copyright owner, the American Pharmaceutical Association, from Kristiansen, H., Nakano, N. I., and Higuchi, T. (1970). *J. Pharm. Sci.*, **59**, 1103–1106.

25°C $K_{1:1}$ of the TMPPT + N,N-dimethylcinnamide complex is $60.0\,dm^3\,mol^{-1}$ in water, whereas it ranges from 2.5 to $4.6\,dm^3\,mol^{-1}$ in pure organic solvents such as methanol, acetone, and dioxane.

Simple, single-ring benzenoid compounds can form only very weak complexes with each other, if they form complexes at all, in aqueous solution. Using vapor pressure–solubility measurements Stellner et al. (1983) studied the thermodynamic properties of the benzene–phenol 1:1 complex (or "dimer") in aqueous solution. At 25°C, $K_{1:1} = 0.63\,dm^3\,mol^{-1}$, $\Delta H = 1.4\,kcal\,mol^{-1}$ ($5.9\,kJ\,mol^{-1}$), and $\Delta C_P = -140\,cal\,K^{-1}\,mol^{-1}$ ($-590\,J\,K^{-1}\,mol^{-1}$). $K_{1:1}$ has a maximum value of $0.68\,dm^3\,mol^{-1}$ at 35°C. The stabilization here appears to be entropic, primarily by hydrophobic interactions (Section 8.2).

The formation of soluble complexes between organic molecules in aqueous solution shares certain features with the phenomena (1–3) described in the remainder of this section.

1. The association of purine and pyrimidine bases in aqueous solution evidently occurs by plane-to-plane stacking, as in a multilayer sandwich. This process also occurs with nucleoside and nucleotide polymers and may be partially responsible for the structural stability of nucleic acids in aqueous solution. As before, the solvent water plays a crucial role in the stacking process, since addition of organic solvents reduces the stacking. The tendency to associate is stronger with purines than with pyrimidines, a phenomenon which may be related to the larger planar areas of purines.

The changes in enthalpy, entropy, and volume observed for self-association are all negative. This contrasts with the classical concept of hydrophobic interactions, which are marked by small negative or even positive enthalpy changes with the usual negative changes in entropy and volume (Section 8.2). Alkyl substituents enhance the association of the bases and their derivatives. Both dipole–induced dipole interactions (Section 2.9) and hydrophobic interactions have been implicated in the stacking process (Lönnberg et al., 1984). These phenomena have been studied by a variety of techniques and the topic has been reviewed by Eagland (1975) and by Lönnberg et al. (1984).

2. The self-association of polynuclear aromatic compounds has frequently been shown to occur in aqueous solution. Examples include caffeine and its alkyl homologues, and 7-ethyl- and 7-propyltheophylline, whose self-association has been studied by Guttman and Higuchi (1957) using partition measurements (Section 5.3.2). Other examples include the dimerization of methylene blue in aqueous solution (Giles and Duff, 1975) and in aqueous alcohols by means of spectroscopic techniques (Fornili et al., 1983) and the self-association of disodium cromoglycate (Scheme 9.5) in aqueous solution using light-scattering techniques (Attwood and Agarwal, 1984).

3. The formation of micelles and mixed micelles of surface-active agents in aqueous solution is a classical example of the hydrophobic interaction, although it is now known that other intermolecular interactions may also be involved (Attwood and Florence, 1983; Roux et al., 1984). The longer the hydrocarbon chain of the surface-active agent, the lower is the critical micelle concentration (CMC).

4. The ability of certain salts, such as those containing rings (e.g., sodium benzoate, phenylacetate, aromatic sulfonates, and their derivatives), to increase the aqueous solubility of sparingly soluble organic compounds has been termed "hydrotropy" by Neuberg (1916). Such salts are termed "hydrotropic salts" and their action cannot be attributed merely to the increased pH they impart to the aqueous solution. The solubilizing action of these salts is still apparent when the pH is held constant and is found to increase the solubility of neutral compounds, such as the higher alcohols, as described in the next section.

10.8 HYDROTROPY

Lindau (1932) and von Hahn (1933) related the increase in solubility to the reduction of surface tension of water by the hydrotropic salt, thus establishing a link with the solubilizing action of surface-active agents. Extending this concept, Winsor (1948a, b; 1950) considered hydrotropy to be similar to solubilization by amphiphiles and to emulsification. Winsor suggested that the hydrotropic salts and the amphiphiles may exert their effects either by forming micelles, by forming separate hydrophilic and lipophilic interactions, or by both effects. The hydrophilic affinity is simply hydrogen bonding of the ionic or polar groups with water, while the so-called lipophilic affinity is simply nonspecific van der Waals forces with any molecule, including the solute being solubilized. In support of Winsor's theory, Palit and Venkateswarlu (1954) found that a mixture containing suitable proportions of a hydrophilic and a lipophilic detergent gave a much higher solubilizing action than either type of detergent alone or mixed with another detergent of the same type.

The formation of molecular aggregates by hydrotropic salts is supported by discontinuities of the plots of the electrical conductivity of aqueous solutions against the concentration of the salt (e.g., sodium benzoate, salicylate, or phthalate; Badwan et al., 1980). Similar plots have long been known for amphiphiles (Winsor, 1950). The plot of aggregate concentration (analogous to the CMC of surfactants) of sodium salicylate when plotted against temperature is similar to that of the ionic surfactants (Badwan et al., 1980). The mechanism of this aggregation is, of course, the hydrophobic interaction which forces together the aliphatic chains, the aromatic rings, and/or other hydrophobic groups and molecules in aqueous solution. The chemical potential (and the activity coefficient) of the dissolved hydrophobic molecules is thereby reduced and the aqueous solubility increased.

The interaction or affinity concept of Winsor (1948a, b) involving hydrogen bonds has been extended by Ueda (1966), who analyzed the UV-visible spectral shifts by means of the Benesi–Hildebrand plot (Section 5.3.5). This technique has demonstrated complexes between the solute (e.g., pyrazinamide, riboflavin, and other cyclic amides) and the hydrotropic salt (substituted sodium benzoates) in aqueous solution at pH 7.0 at 25 or 60°C. From the values of the complexation constants at these temperatures, the thermodynamic functions were calculated. The values of ΔH (-1.0 to $-7.2\,\text{kcal mol} = -4.2$ to $-30.1\,\text{kJ mol}^{-1}$) and of ΔS (-4.0 to $-22.4\,\text{cal K}^{-1}\,\text{mol}^{-1} = -16.7$ to $-93.7\,\text{J K}^{-1}\,\text{mol}^{-1}$) suggested an enthalpically driven process corresponding to hydrogen bonding or charge-transfer interactions. This mechanism of hydrotropy is probably uncommon in aqueous solution because of preferential interaction of the salt with water rather than with the hydrophobic solute.

As an interesting corollary to the original definition of hydrotropy, Saleh et al. (1983, 1986) investigated the solubilization of water in aliphatic alcohols by sodium benzoate and its 2-hydroxy and 2, 5-dihydroxy derivatives. Figure 10.17 shows that increasing concentration of each of these hydrotropic salts in water increases the solubility of water in 1-butanol. The rank order of solubilizing power of the anion is salicylate (2-hydroxybenzoate) > benzoate > gentisate (2, 5-dihydroxybenzoate). The rank order is unchanged when the abscissa in Figure 10.17 is expressed as molar concentration, when the salt is dissolved in 1-butanol instead of water, or when 1-butanol is replaced by 1-hexanol. This rank order reflects the hydrophobic nature of the anion; salicylate is more hydrophobic because of intermolecular hydrogen bonding, as discussed in Section 3.9 in connection with Table 3.10. In contrast, sodium chloride and sodium acetate actually reduce the solubility of water in 1-butanol, whereas urea exerts a small solubilizing effect which levels off at high concentrations (Figure 10.17).

Saleh et al. (1983) measured the effect of these salts on the water O–H and benzenoid C–H stretching bands in the near IR spectrum. Increasing

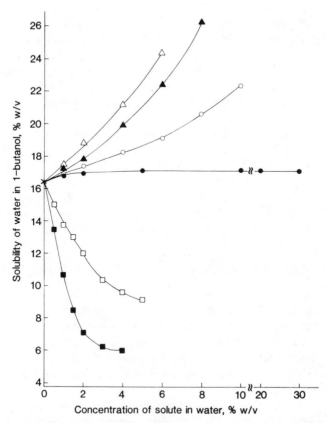

Figure 10.17 Effect of the concentration of added solute in water on the solubility of water in 1-butanol at 25°C. Solute: sodium salicylate (△); sodium benzoate (▲); sodium gentisate (○); urea (●); sodium acetate (□); sodium chloride (■); none, that is, pure water (×). Reproduced with permission of the copyright owner, The American Pharmaceutical Association, from Saleh, A.M., Ebian, A. R., and Etman, M.A. (1986). *J. Pharm. Sci.,* **75,** 644–647.

concentration of each salt increased the wavelength of maximum absorption λ_{max} of each band and increased the bandwidth at midheight of the O–H stretching band. Following Frank (1965) and Klotz (1965), the effect of O–H stretching was attributed to increased ordering of the water structure, suggesting that the hydrotropic salts are structure makers. Structured water should be more soluble in an organic medium. On the other hand, sodium chloride, a structure breaker, had the reverse effect on the IR spectrum and decreased the solubility of water in the alcohols. These results suggest that hydrotropy is essentially a direct manifestation of the hydrophobic interaction. With its cation, the carboxyl anion of the hydrotropic salt acts as a solubilizing group which brings the hydrophobic moiety (e.g., the benzene ring of benzoate) into aqueous solution. In explanation, the microscopic interfacial free energy between the hydrophobic group(s) of the sparingly soluble solute (e.g., 1-hexanol) and water is reduced by the van der Waals attraction to the hydrophobic moiety (e.g., benzene ring) of the salt so that the activity coefficient of the solute is reduced (see Sections 2.8 and 8.3). This effect is entropically driven corresponding to the hydrophobic interaction, as described in Section 8.2. The reduced activity coefficient allows a greater solubility of the solute of defined thermodynamic activity according to Eq. 2.5.

This increased aqueous solubility of a sparingly soluble solute caused by the dissolved hydrotropic salt is an example of the salting-in effect (Section 8.11), as emphasized by McKee (1946). Since hydrotropy brings solutes into solution, it can be employed to facilitate a number of industrial processes, such as electrochemical reactions, organic reactions, pulp processes, and even inorganic reactions (McKee, 1946), while its pharmaceutical uses are apparent in more recent literature (e.g., Ueda, 1966; Badwan et al., 1980).

10.9   ANALYTICAL APPLICATIONS OF SOLUBILITY COMPLEXATION

Complexation solubility analysis can be usefully applied to solve specific and difficult analytical problems. The technique is not destructive.

The initial rising segment of type A and B_S phase diagrams (Fig. 10.5 and 10.9) can be used for the analysis of component L by determining analytically the amount of S brought into solution by the presence of L. The concentration of L can be read from the diagram or calculated from $[S]_0$ and K. This procedure has the major advantage of being a one-point assay.

The type of diagram, solubility, stoichiometry, equilibrium constant, and equivalent weight could be criteria for identification. Some very closely related compounds might be differentiated in this way. The possibility of altering the solubility and reactivity of a substance by incorporating a suitable agent which will complex with it could be used in solvent extraction and chromatography.

When an insoluble complex is formed leading to a type B phase diagram, dissolved S could be determined by adding an excess of L, leading to its quantitative precipitation in an insoluble complex of known stoichiometry, followed by analytical determination of the excess of L. One example is the determination of polyethylene glycols by the formation of an insoluble iodine–polyethylene glycol complex, the excess of iodine then being titrated. Another example is the analysis of vitamin B_{12} (cyanocobalamin) by precipitation with phosphomolybdate. These analytical methods are one-point determinations, while considerable specificity may sometimes be possible.

Another procedure utilizing type B systems also requires L to be the analyte which is coupled with a suitable S component in stages until an excess has been added. The entire phase-solubility diagram is plotted, L being expressed in arbitrary units. Using the same arguments that were expressed for the calculation of the stoichiometry, the reverse procedure is adopted to calculate the true molar concentrations of L from the stoichiometry. As before, the main focus of interest is on the plateau region. The amount of L consumed in the plateau can be related directly to the number of moles of S consumed, if the stoichiometry is known.

10.10   PHASE TITRATIONS

Phase titrations are very general methods of analysis, reviewed by Higuchi and Connors (1965), defined as titrations whose end points are marked by the appearance or disappearance of a phase. This process may involve a chemical interaction or may merely reflect solubility limits of the titrant–titrand mixture. If the titration process appears to result from simple solubility equilibria without reaction between components, the system is classified as a solubility titration. If the phase appearance seems to be due to micelle formation, the titration is a micellar titration. If a true molecular species is formed and is responsible for the phase separation, the titration is a heterometric titration.

Solubility titrations can be illustrated by considering two miscible organic liquids A and B, A being miscible with water (e.g., ethanol) and B being immiscible with water (e.g., toluene). If the binary solution of A and B is titrated

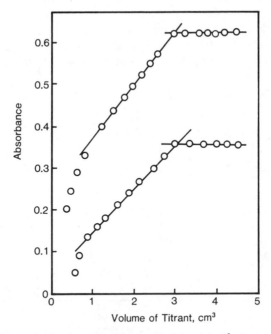

Figure 10.18 Heterometric titration of calcium with $0.1 \, mol \, dm^{-3}$ dipotassium phthalate in 90%(v/v) alcohol. Upper curve: titration of $3 \, cm^3$ of $0.1 \, mol \, dm^{-3}$ $Ca(NO_3)_2$ plus $7 \, cm^3$ of solvent. Lower curve: titration of $3 \, cm^3$ of $0.1 \, mol \, dm^{-3}$ $Ca(NO_3)_2$ plus $3 \, cm^3$ of $1 \, mol \, dm^{-3}$ $MgCl_2$ plus $4 \, cm^3$ of solvent (Bobtelsky, 1960). $1 \, mol \, dm^{-3} = 1 \, mol \, L^{-} = 1 \, M$. Reproduced with permission of the copyright owner, John Wiley and Sons, Inc., from Higuchi, T. and Connors, K.A. (1965). Phase-Solubility Techniques, in *Advances in Analytical Chemistry and Instrumentation*, Vol. 4, John Wiley and Sons, Inc., New York.

with water, at some volume of titrant, a phase separation will occur. The volume of water required to produce phase separation depends on the temperature, the nature of the two liquids A and B, and the composition of the binary mixture. Thus, the technique can be used to analyze such binary solutions by comparing the unknown titration result with a calibration curve prepared by titrating several A–B mixtures of known composition. By conducting the known and unknown titrations under similar conditions, many errors are cancelled. The technique is rapid, simple, and quite accurate, the mean error being about 0.1% absolute.

Rogers and co-workers have considerably developed the theoretical treatment and practical applications of this technique (Rogers et al., 1964; Rogers and Özsogomonyan, 1964). Examples of investigated mixtures include chloroform with alcohols, benzene with dioxane, and cyclohexane with acetone. It is possible to titrate mixtures of water and liquid A with liquid B until the mixture becomes turbid, for example, A = pyridine, B = chloroform (Roger and Özsogomonyan, 1964). The calibration curves can be plotted on triangular coordinate graph paper often used for other three-component systems. Binary solutions of similar liquids, such as cyclohexane and benzene, can be titrated with water after adding a substance such as ethanol, which is soluble in both the sample and titrant (Rogers et al., 1964).

Micellar titrations are often employed to determine the phase diagrams of water–organic substance–surfactant systems. W.I. Higuchi and Misra (1962) titrated a hydrocarbon–surfactant solution with water until a second phase appeared. Hall (1963) titrated solutions of polysorbate 80 and salicyclic acid with water until turbidity occurred. The appearance of certain micellar titrations

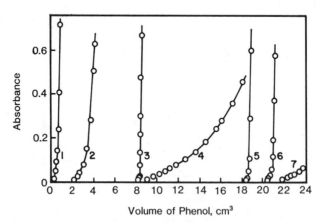

Figure 10.19 Heterometric titration of some polymers in aqueous solution with $0.55 \, mol \, dm^{-3}$ phenol: 1, polypropylene glycol (PPG) 1200; 2, PPG 750; 3, polyvinylpyrrolidone; 4, PPG 400; 5, polyethylene glycol (PEG) 6000; 6, PEG 4000; 7, PEG 1500 (Guttman and Higuchi, 1956). $1 \, mol \, dm^{-3} = 1 \, mol \, L^{-1} = M$. Reproduced with permission of the copyright owner, John Wiley and Sons, Inc., from Higuchi, T. and Connors, K.A. (1965). Phase-Solubility Techniques, in *Advances in Analytical Chemistry and Instrumentation*, Vol. 4, John Wiley and Sons, Inc., New York.

may change from transparent to gel-like to liquid crystalline (Lumb and Winsor, 1952).

Heterometric titrations have been largely developed by Bobtelsky (1960) and co-workers for the quantitative study of suspensions produced by chemical reactions. The photoelectric absorbance is plotted against volume of titrant added. The relevant sharp discontinuity, shown in Figures 10.18 and 10.19, corresponds to the end point. Figure 10.18 illustrates a method for titrating calcium ions in the presence of magnesium ions. In this method, Bobtelsky (1960) emphasized the transient nature of the suspensions formed and the state of physical nonequilibrium which distinguishes heterometry from turbidimetry. He noted that the absolute absorbance values are not highly reproducible, but the titres are. Although the original technique led to the appearance of solid phases, the method can also be applied when a liquid organic phase separates.

Figure 10.19 shows heterometric curves for the titration of some polymers with phenol in aqueous solution (Guttman and Higuchi, 1956). The increase in absorbance shown in this figure arises from the appearance of an oily liquid phase which seems to be a complex of the polymer with phenol. The technique is very sensitive and can be used as a qualitative test for the polymers in solution. The inverse relationship between chain length and concentration of phenol necessary to initiate the separation of a second phase indicates that the technique may be useful for differentiating polymers of the same type but of different molecular weight.

Chapter XI

DISSOLUTION RATES OF SOLIDS

11.1 INTRODUCTION

The rates at which solids dissolve are of primary importance in pharmaceutical development and quality control and in those aspects of chemical engineering that involve reactions of solid materials. The dissolution rates of solids may be determined experimentally, accounted for theoretically, and, in certain cases, predicted from the physicochemical properties of the system.

The many methods that have been developed for measuring the dissolution rates of solids may be broadly classified either as batch or as continuous-flow methods, as shown in Figure 11.1. Batch-type dissolution methods are more common and include various carefully standardized pharmaceutical methods, among which those defined by the *United States Pharmacopeia XXI* (1985) are the most used (Fig. 11.1*A*). One of the most popular batch methods to be used in the pharmaceutical sciences is the beaker–stirrer method of Levy and Hayes (1960). If *m* is the mass dissolved at time *t*, the dissolution rate is given by dm/dt.

If the solution in a batch-type dissolution method is sufficiently well stirred that the analyzed concentration c_b is representative of that in the entire volume V of the dissolution medium, then

$$m = Vc_b \tag{11.1}$$

and

$$dm/dt = V(dc_b/dt) \tag{11.2}$$

Figure 11.2*A* shows a typical batch-type dissolution curve. If the dissolved solute is allowed to accumulate in the dissolution medium and if sufficient solid is present, a saturated solution will eventually be formed.

The advantages of batch methods are: (1) they are simple to set up and operate, and (2) they are widely used and certain types have been carefully standardized, especially in the pharmaceutical sciences. The disadvantages of batch methods are: (1) the hydrodynamics of most batch methods are not well characterized with the notable exception of the rotating-disc method (Levich, 1962; (2) a small change in dissolution rate will often introduce such a small perturbation in the dissolution–time profile that it will not be measurable; and (3) the concentration c_b may not be uniform throughout the volume V.

Using a batch-type apparatus in which sticks of the sparingly soluble

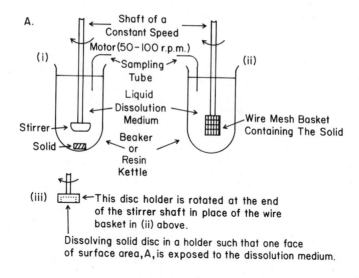

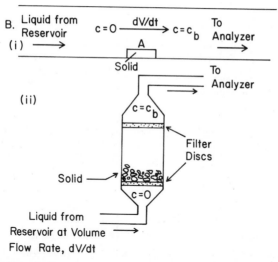

Figure 11.1 Schematic diagrams of various types of apparatus for measuring the dissolution rates of solids. (*A*) Batch-type dissolution apparatus: (i) USP XXI (1985) Dissolution Apparatus I; (ii) USP XXI (1985) Dissolution Apparatus II; (iii) rotating disc method (Levich, 1962; Wood et al., 1965). (*B*) Continuous-flow dissolution apparatus: (i) Schematic representation; (ii) column-type flow-through dissolution cell (Langenbucher, 1969). Each filter disc may be replaced by a stainless-steel sieve (40 US) on top of which glass beads (1 mm diameter) are placed.

compounds benzoic acid or lead chloride were rotated in water, Noyes and Whitney (1897) discovered that the dissolved concentration c_b of each solute increased with time t according to a first-order process, thus,

$$dc_b/dt = k_{NW}(c_s - c_b) \qquad (11.3)$$

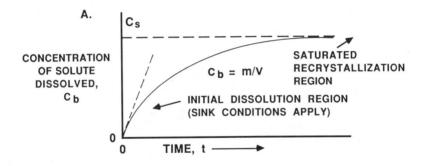

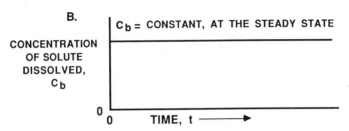

Figure 11.2 Dissolution–time profiles for (A) a batch-type dissolution apparatus and (B) a continuous-flow dissolution apparatus.

where k_{NW} is the first-order rate constant and c_s is the equilibrium solubility. This relationship is depicted in Figure 11.2A. Equations 11.2 and 11.3 may be combined thus:

$$dm/dt = k_{NW}V(c_s - c_b) \qquad (11.4)$$

From the beginning of dissolution ($t = 0$) until c_b reaches about 5–10% of c_s, it may often be assumed that $c_b \ll c_s$. In this case "sink conditions" are said to apply and Eqs. 11.3 and 11.4 lead to the following expressions for the initial dissolution rate:

$$(dm/dt)_{t \to 0} = (dc_b/dt)_{t \to 0}V = k_{NW}Vc_s \qquad (11.5)$$

Continuous-flow or column dissolution methods, shown schematically in Figure 11.1B (see also Groves and Alkan, 1975; Nelson and Shah, 1975; Shah and Nelson, 1975), are not yet in widespread use, but their applications are growing. After a short time of dissolution with a constant volume flow rate dV/dt of dissolution medium, a steady state is achieved, as shown in Figure 11.2B, such that the dissolution rate is given by

$$dm/dt = (dV/dt)c_b \qquad (11.6)$$

The advantages of continuous-flow methods are as follows: (1) sink conditions are readily achieved by adjustment of dV/dt, such that $c_b \ll c_s$; and (2) a change in dm/dt is readily measured by change in c_b. The disadvantages of continuous-flow methods are: (1) a large volume of dissolution medium may be required; and (2) for a solid with a low dissolution rate and solubility, c_b may be so low that a highly sensitive method of analysis is required.

Reliable measurements of dissolution rates are most important in the pharmaceutical sciences. This is because the rate of absorption of many solid drugs from the gastrointestinal tract is controlled by their dissolution rate. In other words, solid dissolution is frequently the rate-limiting step in drug absorption. There is a very extensive pharmaceutical literature on the measurement and interpretation of the rates of dissolution of solids from the pure drug forms and from the formulated drug products, for example, tablets, capsules, and suppositories (see Wagner, 1971; Swarbrick, 1970; Lesson and Carstensen, 1974; Carstensen, 1980; Hanson, 1982). Furthermore, considerable debate, scientific study, and technical development has been directed toward the design, operation, and interpretation of appropriate dissolution rate tests for pharmaceutical solids.

In general, the dissolution rate of a given solid material is found to be strictly proportional to the wetted surface area A of the dissolving solid (indicated in Fig. 11.1), that is,

$$dm/dt \propto A \tag{11.7}$$

The dissolution rate per unit surface area is often referred to as the "intrinsic dissolution rate" or "mass flux" J, and is given by

$$J = (dm/dt)(1/A) \tag{11.8}$$

Under sink conditions, that is, $c_b \ll c_s$, Eqs. 11.5 and 11.7 lead to the following expression for the initial dissolution rate:

$$(dm/dt)_{t \to 0} \propto Ac_s \tag{11.9}$$

Without making any assumptions about the mechanism of dissolution, the dissolution rate under constant defined conditions is given by

$$J = (dm/dt)(1/A) = k_1(c_s - c_b) \tag{11.10}$$

where k_1 is the first-order rate constant which has the dimensions of length/time. By comparing Eqs. 11.10 and 11.4, k_1 is related to k_{NW} as

$$k_{NW} = k_1(A/V) \tag{11.11}$$

11.2   THEORIES AND MECHANISMS FOR THE DISSOLUTION PROCESS

11.2.1   Introduction to the Theories

Various mechanisms for the dissolution of solids have been proposed (Bircumshaw and Riddiford, 1952; Higuchi, 1967). Two of the simplest theories which account for the dissolution rates of solids are illustrated in Figure 11.3. Both theories share the following features: (1) The solid–liquid interface creates an infinitesimally thin film of saturated solution which is in immediate contact with the solid and for which the concentration of the dissolving solid is the solubility c_s. (2) In the bulk of the solution, which is assumed to be well mixed, the concentration c of the dissolving solid at any given time is c_b.

If the transport and reaction processes both contribute to the observed dissolution rate, the observed rate constant k_1 in Eq. 11.10 is given by

$$1/k_1 = 1/k_T + 1/k_R \qquad (11.12)$$

or

$$k_1 = k_T k_R/(k_T + k_R) \qquad (11.13)$$

where k_T is the rate constant for transport and k_R is that for reaction.

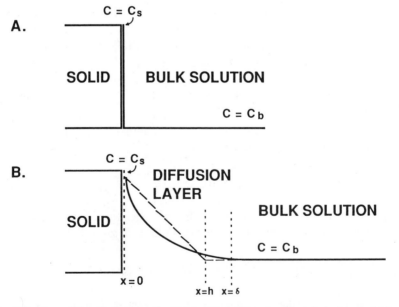

Figure 11.3   Two of the simplest theories for the dissolution of solids: (A) the interfacial barrier model and (B) the diffusion layer model, in the simple form of Nernst (1904) and Brunner (1904) (———); and in the more exact form of Levich (1962) (——). c is the concentration of the dissolving solid, c_s is the solubility, c_b is the concentration in the bulk solution, and x is the distance from the solid–liquid interface.

When transport is rate-limiting, $k_T \ll k_R$ and Eqs. 11.12 and 11.13 reduce to

$$k_1 \cong k_T \qquad (11.14)$$

When the surface reaction is rate-limiting, $k_R \ll k_T$ and Eqs. 11.12 and 11.13 reduce to

$$k_1 \cong k_R \qquad (11.15)$$

An additional theory according to Danckwerts (1950) for the dissolution of gases in reactive liquids may also apply to the dissolution of solids under certain conditions (Higuchi, 1967), but it has not often been used for the latter purpose.

11.2.2 Interfacial Barrier Theory

According to the interfacial barrier theory which is illustrated in Figure 11.3A, the intrinsic dissolution rate J is controlled by the surface reaction between the component(s) of the dissolution medium and the solid surface. If this reaction is of the first order,

$$J = (dm/dt)(1/A) = k_R(c_s - c_b) \qquad (11.16)$$

where k_R is the first-order rate constant for the interfacial reaction. Since in this case transport is not rate-limiting, the solute concentration c falls precipitously from the surface value c_s to the bulk value c_b over an infinitesimal distance. The interfacial barrier model probably applies when the rate of mass transport is limited by an adsorbed condensed film at the solid–liquid interface. The penetration energy in such a system is large and subject to a cooperative effect, so that there is a high activation energy barrier to the surface reaction and $k_R \ll k_T$.

Cases for which the interfacial barrier model controls the dissolution of organic compounds are rare. The dissolution of gallstones, which are composed largely of cholesterol, in bile acid media is such an example. Reaction-controlled dissolution may also occur when solid carbon acids, such as deuterated phenylbutazone, are dissolving in an ionizing medium, such as buffered water, under conditions of intense agitation, as described in Section 11.7. Under such conditions the rate-limiting step is the slow ionization of the carbon acid, such that $k_R < k_T$ (Mooney et al., 1981c).

11.2.3 Diffusion-Layer Theory

The best known and most useful model for transport-controlled dissolution is the diffusion-layer theory which is illustrated in Figure 11.3B. According to this theory, the dissolution rate dm/dt is controlled by the rate of diffusion of the solute molecules across a diffusion layer of thickness h. The observed dissolution rate constant k_1 is given by the transport rate constant k_T. With increasing distance x from the surface of the solid, c decreases from c_s at $x = 0$

to c_b at $x = h$. In practice, c will be a nonlinear function of x, the concentration gradient dc/dx decreasing from a high value at $x = 0$ to zero at $x = h$. In a stirred solution, the flow velocity, that is, the rate of shear of the liquid dissolution medium, will increase from zero at $x = 0$ to the bulk value at $x = h$.

The hydrodynamic aspects of the theory are fully discussed by Levich (1962). In the original and simplest form of the theory, Nernst (1904) and Brunner (1904) assumed:

(1) that the mass flux (i.e., intrinsic dissolution rate), is given by Fick's first law of diffusion, thus,

$$J = (dm/dt)(1/A) = -D(dc/dx) \qquad (11.17)$$

where dc/dx is the concentration gradient and D is the diffusivity (diffusion coefficient), and

(2) that the concentration gradient within the diffusion layer is constant, as indicated by the dashed line in Figure 11.3B, thus,

$$dc/dx = (c_b - c_s)/h \qquad (11.18)$$

at any given time t. Inserting this expression for the concentration gradient into Eq. 11.17 we obtain

$$J = (dm/dt)(1/A) = (D/h)(c_s - c_b) \qquad (11.19)$$

This is the dissolution rate equation according to the diffusion-layer theory in its simplest form. The observed dissolution rate constant k_1, which is here equal to k_T, is given by

$$k_1 = k_T = D/h \qquad (11.20)$$

The diffusion-layer model is generally found to be an acceptable mechanism for describing the dissolution rates of pharmaceutical solids in aqueous and nonaqueous systems. (The simple diffusion-layer model may be said to be 80% correct and to apply in more than 90% of the cases.) Attempts have therefore been made to predict the dissolution rates of pharmaceutical solids, including powders, by applying Eq. 11.19 (Hussain, 1972; Hersey, 1973).

For the prediction of dm/dt, the surface area A of the solid must be estimated, and this may not be an easy task for powders. As the particles dissolve, the change (a decrease) in the surface area is proportional to the $\frac{2}{3}$ power of the volume or mass of the powder. From this relationship, Hixon and Crowell (1931) derived a cube-root law which has proved useful in pharmaceutics (see Carstensen, 1980). Powders represent a special class of material whose dimensionally related properties are beyond the scope of this book.

11.2.4 Hydrodynamics of the Rotating Disc

The hydrodynamics of the rotating disc have been examined theoretically by Levich (1962). The Levich (1962) theory considers only forced convection due to rotation and ignores other mechanisms, such as natural convection, which may occur at low rotation speeds. Figure 11.4 shows the solvent flow field near the surface of the rotating disc under these circumstances. The apparent thickness h of the diffusion layer in immediate contact with the surface is given by

$$h = 1.612 \, (\text{cm}) D^{1/3} v^{1/6} \omega^{-1/2} \tag{11.21}$$

where D is the diffusivity of the dissolved solute, v is the kinematic viscosity of the fluid medium, and ω is rate of angular rotation (i.e., the angular velocity) in radians per second.

We note that

$$v = \eta/\rho \tag{11.22}$$

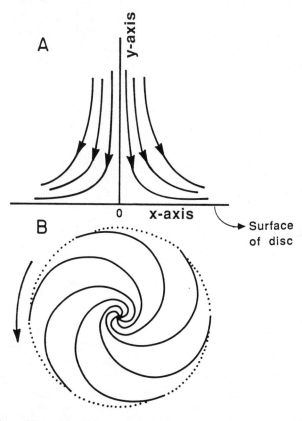

Figure 11.4 Lines of flow of liquid in the Levich rotating-disc method of determining dissolution rates. There is a transition from (A) flow essentially normal to the surface to (B) flow parallel with the surface, pointing to the existence of a viscous boundary layer. Reproduced with permission of the copyright owner, the Royal Society of Chemistry, from Bircumshaw, L.L. and Riddiford, A.C. (1952). *Quart. Rev.*, **6**, 157–185.

where η and ρ are the dynamic viscosity and density of the fluid, respectively. Since

$$\omega = 2\pi W \tag{11.23}$$

where W is the rotation speed in revolutions per second (Hz), then,

$$h = 0.643 \, (\text{cm}) D^{1/3} v^{1/6} W^{-1/2} \tag{11.24}$$

The actual thickness δ of the diffusion layer, corresponding to a nonlinear decrease in c with increasing x, as shown in Figure 11.3B, is related to the apparent thickness h as follows:

$$h = 0.893\delta \tag{11.25}$$

Because of its controlled hydrodynamics, the rotating-disc method, with or without modifications, continues to provide a useful method for studying intrinsic dissolution rates of solids (see Nicklasson et al., 1985).

Elimination of h from Eqs. 11.21 and 11.19 provides an accurate expression

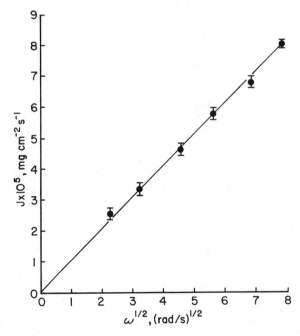

Figure 11.5 Plot of the flux J of 2-naphthoic acid as a function of the square root of the rotation speed ω in 0.01 mol dm^{-3} HCl, ionic strength $\mu = 0.5$ mol dm^{-3} (potassium chloride) at 25°C. The error bars represent the standard deviation for each point. 1 mol dm^{-3} = 1 mol L^{-1} = 1M. Reproduced with permission of the copyright owner, the American Pharmaceutical Association, from Mooney, K.G., Mintun, M.A., Himmelstein, K.J., and Stella, V.J. (1981). *J. Pharm. Sci.*, **70**, 13–22.

Table 11.1 Solubilities $c_0^a = [HA]_0$, Ionization Constants K_a, pK_a Values, and Experimental Diffusivities D_{HA} of Weak Acids from Dissolution Rates in Sodium Hydroxide Solutions at 25°C and at Ionic Strength $0.5\,\mathrm{mol\,dm^{-3}}$ [a]

Weak Acid	M^b	$c_0^s = [HA]_0^c$ $(\mathrm{mol\,dm^{-3}})$	$K_a \times 10^{5\,c}$ $(\mathrm{mol\,dm^{-3}})$	pK_a^c	$D_{HA} \times 10^6\,(\mathrm{cm^2\,s^{-1}})$			
					Experi-mental[d]	Theory[e] from $M^{1/3}$	Theory[f] from $M^{1/2}$	Literature[g]
Benzoic acid	122.1	2.16×10^{-2}	9.25	4.03	9.6	6.63	8.90	11.0 or 12.1
2-Naphthoic acid	172.2	1.3×10^{-4}	9.64	4.02	6.1	6.11	8.08	—
Indomethacin	357.8	2.62×10^{-6}	6.70	4.17	5.6	4.80	5.60	—

[a] Theoretical and literature values of D_{HA} are also included. $1\,\mathrm{mol\,dm^{-3}} = 1\,\mathrm{mol\,L^{-1}} = 1\,M$.
[b] Molecular weight.
[c] From Eq. 11.36 and Figure 11.6 (Mooney et al., 1981a).
[d] From Eq. 11.26 and Figure 11.5 (Mooney et al., 1981a).
[e] From the Stokes–Einstein equation (Eq. 11.29) (Mooney et al., 1981a).
[f] From Higuchi et al. (1972).
[g] In water at 25°C, from Higuchi et al. (1958) and Pakula et al. (1977).

483

for the intrinsic dissolution rate, thus,

$$J = 0.62D^{2/3}v^{-1/6}(c_s - c_b)\omega^{1/2} \tag{11.26}$$

Figure 11.5 illustrates the dependence of J on $\omega^{1/2}$ for 2-naphthoic acid dissolving in an aqueous solution of ionic strength $\mu = 0.5\,\mathrm{mol\,dm^{-3}}$ at 25°C. Equation 11.26 provides a means of determining the diffusivity of a solute by dissolution from a compressed disc. The slopes of the plot of J versus $\omega^{1/2}$ under sink conditions is $0.62D^{2/3}v^{-1/6}c_s$. From the slope of the line in Figure 11.5, the diffusivity of 2-naphthoic acid is calculated to be $6.1 \times 10^{-6}\,\mathrm{cm^2\,s^{-1}}$ in an aqueous solution of ionic strength $0.5\,\mathrm{mol\,dm^{-3}}$ at 25°C (Table 11.1). Tsuji et al. (1978) found good agreement between diffusivities derived from the Levich equation (eq. 11.26) and those determined by the diffusion-cell method of Goldberg and Higuchi (1968). In the rotating-disc method the dissolved solute may not be uniformly distributed throughout the volume V of the dissolution medium, an effect which may complicate the analytical procedure.

Equations 11.21 and 11.24–11.26 are based on a single diffusing species. If there are several diffusing species with different diffusivities, h will differ among the species, causing significant errors. However, many organic molecules have similar diffusivities of the order $10^{-5}\,\mathrm{cm^2\,s^{-1}}$ in water at 25°C. Furthermore, h is relatively insensitive to D in view of the $\frac{1}{3}$ power dependence in Eqs. 11.21 or 11.24. Consequently, the errors involved with several diffusing species only become significant if one (or more) species possesses very high diffusivity (i.e., the hydroxide ion), or very low diffusivity (i.e., a macromolecule).

11.2.5   Effect of Diffusivity and Agitation Rate

The various models for dissolution predict certain power dependence of the intrinsic dissolution rate J on the diffusivity D of the solute and on the rate R of agitation, stirring, or flow of solvent, thus,

$$J \propto D^a; \qquad J \propto R^b \tag{11.27}$$

Table 11.2 summarizes the values of the exponents a and b for various dissolution models. Reaction-controlled processes embodied in the interfacial barrier model are evidently not influenced by molecular motion within the solution, represented by D and R. In general, R may be represented by a hydrodynamic quantity known as Reynolds number Re (Levich, 1962), which is an important dimensionless index that characterizes the conditions of flow, thus,

$$\mathrm{Re} = \rho d(u/\eta) = d(u/v)[= r^2(\omega/v)] \tag{11.28}$$

where d is the mean hydraulic diameter, u is the linear flow rate of the flow stream, and η, ρ, and v have been defined above (r and ω refer to the radius and angular velocity, respectively, of the Levich rotating disc).

Table 11.2   Values of the Exponents a and b
Representing the Dependence of Intrinsic Dissolution
Rate J on Diffusivity D of the Solute and on the Rate R
of Agitation, Stirring, or Flow of Solvent[a]

Model or Theory	a	b
Diffusion layer (see Eq. 11.19)	1	0.4–1.0
Levich rotating disc (see Eq. 11.26)	$\frac{2}{3}$	$\frac{1}{2}$
Interfacial barrier (see Eq. 11.16)	0	0

[a]According to the proportionalities $J \propto D^a$ and $J \propto R^b$ (Eq. 11.27) (see Bircumshaw and Riddiford, 1952).

If flow is laminar, Re < 1000, but for turbulent flow Re > 2000, while the transition region between these two conditions is characterized by 1000 < Re < 2000. The condition of laminar flow is usually preferred in studies of dissolution rates, since the measured rates display greater selectivity, reproducibility, and predictability than under conditions of turbulence. Moreover, the hydrodynamic equations, such as Eqs. 11.21 and 11.24–11.26, for the rotating disc usually do not apply when flow is turbulent. Various studies have demonstrated the role of Reynolds number as a useful quantity for defining the influence of the flow rate of the solvent on the dissolution rate of solids, especially in continuous flow systems (e.g., Fee et al., 1976).

Diffusivity is evidently an important physical quantity for characterizing the diffusion-controlled dissolution of a solid. According to the Stokes–Einstein equation (Eq. 11.29) for small particles such as molecules, the diffusivity is inversely proportional to the dynamic viscosity of the liquid medium, thus,

$$D = kT/6\pi\eta r \tag{11.29}$$

where k is the Boltzmann constant, T is the absolute temperature, and r is the mean radius of the particles.

For approximately spherical molecules, $r \propto v^{1/3} \propto \nabla^{1/3} \propto M^{1/3}$, where v is the molecular volume, ∇ is the partial molar volume, and M is the molecular weight; therefore, $D \propto \nabla^{-1/3} \propto M^{-1/3}$ (Flynn et al., 1974). For approximately planar molecules, however, the apparent mean value of $r \propto a^{1/2} \propto v^{1/2} \propto M^{1/2}$, where a is the molecular area; therefore, $D \propto M^{-1/2}$ (Higuchi et al., 1972). Table 11.1 compares diffusivities determined using the Levich (1962) rotating disc, those calculated theoretically assuming $D \propto M^{-1/3}$ and $D \propto M^{-1/2}$, and literature values (see Mooney et al., 1981). In most instances the theoretical predictions agree well with the experimental values.

Since $J \propto D^a$ (Eq. 11.27), according to Eq. 11.29, $J \propto \eta^{-a}$. If other factors are equal, the latter proportionality may be used to evaluate the index a in

Table 11.2. However, if additional solutes such as macromolecules are added to increase η, they may also change the density ρ. Now the ratio $\eta/\rho = v$ (Eq. 11.22) directly influences h and J (Eqs. 11.21 and 11.26) in the Levich rotating-disc method and directly influences Re (Eq. 11.28), and hence J (Eq. 11.27), in methods involving other types of agitation.

11.3   INFLUENCE OF pH ON SOLUBILITY

Many organic compounds, particularly drugs, are weak acids or weak bases or are amphoteric in nature. The pH of the environment, if aqueous, exerts a pronounced effect on the solubility of these compounds and will therefore influence the dissolution rate. This has been considered by Higuchi et al. (1953). In the case of a weak acid HA, ionizing thus,

$$HA \rightleftharpoons H^+ + A^- \tag{11.30}$$

$$a_{H^+} a_{A^-} = a_{HA} K \tag{11.31}$$

where a_S indicates the thermodynamic activity of species S and K is the thermodynamic ionization constant of the weak acid. If molar concentrations, represented by square brackets, are used, then,

$$[H^+][A^-] = [HA](\gamma_{HA}/\gamma_{H^+}\gamma_{A^-})K \tag{11.32}$$

where γ_S indicates the activity coefficient of species S. If the solution is dilute or contains a swamping concentration (ionic strength) of a strong electrolyte, each activity coefficient may be assumed constant, so that an apparent ionization constant K_a can be defined, thus,

$$K_a = (\gamma_{HA}/\gamma_{H^+}\gamma_{A^-})K \tag{11.33}$$

Equation 11.32 may then be written in the following practical form:

$$[H^+][A^-] = K_a[HA] \tag{11.34}$$

The total molar solubility c_t^s of the weak acid is equal to the sum of the molar solubility c_o^s of the unionized species HA plus the molar concentration of the ionized species A^-, with which HA is in equilibrium, thus,

$$c_t^s = c_o^s + [A^-] \tag{11.35}$$

Now $[A^-]$ is given by Eq. 11.34 (or Eq. 11.32), while $[HA] = c_o^s$, so that

$$c_t^s = c_o^s + c_o^s K_a/[H^+] \tag{11.36}$$

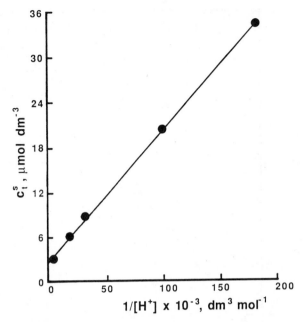

Figure 11.6 Plot of total solubility of indomethacin at 25°C versus $1/[H^+]$ for equilibrated $0.1 \, mol \, dm^{-3}$ acetate buffers, $\mu = 0.5 \, mol \, dm^{-3}$ (potassium chloride); c_o^s is obtained from the intercept and K_a is obtained from the slope. $1 \, mol \, dm^{-3} = 1 \, mol \, L^{-1} = 1 \, M$. Reproduced with permission of the copyright owner, the American Pharmaceutical Association, from Mooney, K.G, Mintun, M.A., Himmelstein, K.J., and Stella, V.J. (1981). *J. Pharm. Sci.*, **70**, 13–22.

and

$$c_t^s = c_o^s + c_o^s(\gamma_{HA}/\gamma_{H^+}\gamma_{A^-})K/[H^+] \tag{11.37}$$

Equation 11.36 predicts a linear plot of c_t^s against $1/[H^+]$ for a weakly acidic drug, as illustrated in Figure 11.6 for indomethacin. The intercept yields the value of c_o^s and the slope equals $K_a c_o^s$. Dividing the slope by the intercept provides the value of K_a. Values of c_o^s and K_a so determined for three weak acids are presented in Table 11.1.

Equation 11.36 may conveniently be written in the following logarithmic form:

$$\log(c_t^s - c_o^s) = \log c_o^s - pK_a + pH \tag{11.38}$$

This equation predicts that the logarithm of the increase in solubility of a weakly acidic drug is a linear function of pH with a slope of $+1$. Higuchi et al. (1953) measured the solubility of several weak acids (phenobarbital, barbital, and sulfathiazole) over a range of pH values at constant ionic strength and found that Eq. 11.38 was closely obeyed (with slopes 1.01, 1.05, and 1.03, respectively). Figure 11.7 presents an example. Deviations from the prediction

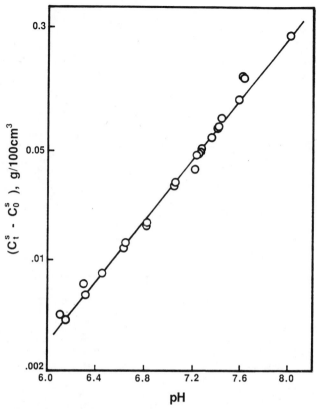

Figure 11.7   Influence of pH on the solubility of sulfathiazole in water at 25°C. The experimental data conform (log scale) to Eq. 11.38. Reproduced with permission of the copyright owner, the American Pharmaceutical Association, from Higuchi, T., Gupta, M., and Busse, L.W. (1953). *J. Am. Pharm. Assoc.*, **42**, 157–161.

may be attributed to small changes in the activity coefficients in Eq. 11.37 or, in the case of barbital at pH > 6.8, to the formation of some insoluble complex with the constituents of the buffer.

For a weak base B ionizing thus,

$$B + H^+ \rightleftharpoons BH^+ \tag{11.39}$$

an argument analogous to that of the previous two paragraphs leads to

$$\log(c_t^s - c_o^s) = \log c_o^s + pK_a - pH \tag{11.40}$$

where c_t^s is the total molar solubility of the weak base, which is given by the sum of the molar solubility c_o^s of the unionized species B plus the molar concentration of the ionized species BH^+. This equation is obeyed quite closely by oxytetracycline (terramycin), the coefficient of the pH (i.e., the slope) being -0.91 (Higuchi et al., 1953).

11.4 INFLUENCE OF pH, BASES, AND BUFFERS ON THE AQUEOUS DISSOLUTION RATES OF ACIDIC SOLIDS

11.4.1 Development of a Theory Based on a Single Equilibrium

Since pH greatly influences the equilibrium solubilities of weakly acidic (or weakly basic) solids, Eq. 11.9 suggests that their dissolution rate will also be affected. Since in most instances the dissolution of solids is found to be controlled by diffusion, the diffusion-layer model has provided a useful starting point in developing an understanding of the influence of pH, bases, and buffers on the rates of dissolution of weakly acidic solids.

In their studies of the dissolution rate of benzoic acid in dilute aqueous sodium hydroxide or potassium hydroxide, King and Brodie (1937) and Hixon and Baum (1944) satisfactorily explained their data according to a two-film model, shown in Figure 11.8. This model is a simple extension of the Nernst–Brunner diffusion-layer concept, illustrated by the dashed line in Figure 11.3B. In immediate contact with the surface of the solid in Figure 11.8 is an infinitesimally thin film of saturated solution represented by Eq. 11.36 and 11.37. Linear concentration gradients of all species are assumed to exist in the film of thickness $(h_1 + h_2)$ in Figure 11.8, and the dissolution rate is then proportional to the slope of the [HA] line, as suggested by Fick's first law of diffusion (Eq. 11.17). This model neglects the effects of base strength of the medium and the diffusivities of the reaction products. The model also assumes that the reaction takes place only at the plane AB, where the reactants come into contact, and disregards the possibility of neutralization elsewhere. Although the model describes quite well those dissolution processes for which the reaction equilibrium lies very far on the side of the products, it is expected to fail in other cases.

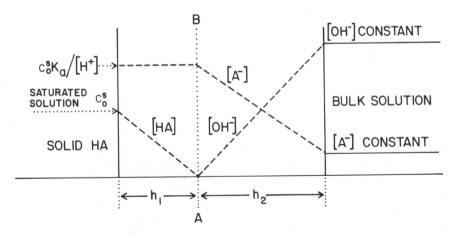

Figure 11.8 Linear concentration profiles in the Nernst–Brunner two-film model for the diffusion-controlled kinetics of dissolution of benzoic acid in aqueous sodium hydroxide or potassium hydroxide. The film thickness is $h_1 + h_2$. Reproduced with permission of the copyright owner, the American Pharmaceutical Association. Adapted from Higuchi, W.I., Parrott, E.L., Wurster, D.E., and Higuchi, T. (1958). *J. Am. Pharm. Assoc., Sci. Ed.,* **47**, 376–383.

Higuchi et al. (1958) extended the Nernst–Brunner model for diffusion-controlled dissolution kinetics to make it applicable to the rate of dissolution of a weak acid HA in a solution of an ionic base B^-. The relevant chemical equilibrium is essentially reversible, thus,

$$HA + B^- \overset{K}{\rightleftharpoons} A^- + HB \tag{11.41}$$

where

$$K = K_{HA}/K_{HB} = [A^-][HB]/[HA][B^-] \tag{11.42}$$

K is the apparent equilibrium constant, defined in terms of molar concentrations, and K_{HA} and K_{HB} are the apparent ionization constants of HA and HB.

In the steady-state dissolution of solid HA, represented by Figure 11.9, the boundary conditions are as follows:

Boundary conditions at $x = 0$:

$$[HA] = [HA]_0 = c_0^s, \qquad [HB] = [HB]_0$$
$$[A^-] = [A^-]_0, \qquad [B^-] = [B^-]_0 \tag{11.43}$$

Boundary conditions at $x = h$:

$$[HA] = [HA]_h, \qquad [HB] = [HB]_h$$
$$[A^-] = [A^-]_h, \qquad [B^-] = [B^-]_h \tag{11.44}$$

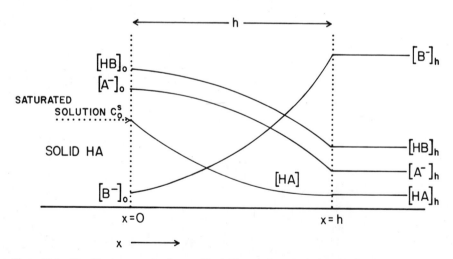

Figure 11.9 Curvilinear concentration profiles in the steady state during the dissolution of a weak acid HA in an aqueous solution of an ionic base B^-. The film thickness is h. Reproduced with permission of the copyright owner, the American Pharmaceutical Association. Adapted from Higuchi, W.I., Parrott, E.L., Wurster, D.E., and Higuchi, T. (1958). *J. Am. Pharm. Assoc., Sci. Ed.*, **47**, 376–383.

At steady state Fick's second law of diffusion in one dimension defined by x (Jost, 1960) leads to an equation of the following form for each of the four species involved (i.e., HA, HB, A^-, B^-):

$$d[S]/dt = D_S(d^2[S]/dx^2) + \phi_S = 0 \qquad (11.45)$$

where S represents a typical species and D_S is its diffusivity. Because equilibrium is assumed to take place instantaneously, the term ϕ_S for the rate of chemical reaction is zero in each case. The influence of rates of ionization, for which ϕ_S terms are required, is discussed in Section 11.6. The charge on species S will generally be omitted when S is a subscript (e.g., D_A and D_B).

For proper material balance across the diffusion layer:

$$D_{HA}\frac{d^2[HA]}{dx^2} = -D_{HB}\frac{d^2[HB]}{dx^2} = -D_A\frac{d^2[A^-]}{dx^2} = D_B\frac{d^2[B^-]}{dx^2} \qquad (11.46)$$

and

$$\frac{d[A^-]}{dt} = \frac{d[HB]}{dt} = -\frac{d[HA]}{dt} = -\frac{d[B^-]}{dt} \qquad (11.47)$$

Equations 11.46 and 11.47 express the stoichiometry inherent in Eq. 11.41. Any change in A^- flux in one direction must be accompanied by an equal flux of HB in the same direction and by equal fluxes of HA and B^- in the opposite direction. Hence, $\phi_{HA} = \phi_{HB} = \phi_A = \phi_B = 0$. Equation 11.46 means that a given amount of reaction of HA with B^- in any volume produces equivalent amounts of HB and A^-. The ϕ_S value in Eq. 11.45 is then the rate of neutralization of species S per unit volume.

To obtain an expression for the intrinsic dissolution rate, Eq. 11.46 is integrated twice. After each integration, the integration constants are defined by means of the boundary conditions at $x = 0$ and $x = h$, according to Eq. 11.43 and 11.44. The detailed procedure employed is described in Section 11.5.1. Under sink conditions, $D_A[A^-]_h = D_{HB}[HB]_h = D_{HA}[HA]_h = 0$, so the final expression for the intrinsic dissolution rate (J in Eq. 11.8) simplifies to

$$J = \frac{D_{HA}[HA]_0}{h} - \frac{D_{HB}D_A K[HA]_0}{2D_B h}$$
$$+ \frac{D_{HB}D_A K[HA]_0}{2D_B h}\left(1 + \frac{4D_B^2[B^-]_h}{D_{HB}D_A K[HA]_0}\right)^{1/2} \qquad (11.48)$$

When $K \to 0$, the complex equation is approximated by

$$J = \frac{dm}{dt}\frac{1}{A} = \frac{D_{HA}[HA]_0 - D_{HA}[HA]_h}{h} \qquad (11.49)$$

When K is very large, the complex equation reduces to

$$J = \frac{dm}{dt}\frac{1}{A} = \frac{D_{HA}[HA]_0 - D_{HA}[HA]_h + D_B[B^-]_h}{h} \qquad (11.50)$$

Thus, for a reaction very far over to the side of the products, Eqs. 11.48 and 11.50 predict linear concentration gradients of the acid and base in the diffusion layer, as assumed in the two-film model depicted in Figure 11.8 and applied by King and Brodie (1937) and Hixon and Baum (1944). Thus, Figure 11.8 describes a special case of the general model depicted by Figure 11.9.

Analogous equations apply to a dissolution process involving the nonionic equilibrium:

$$HA + B \rightleftharpoons HAB \qquad (11.51)$$

where

$$K = [HAB]/[HA][B] \qquad (11.52)$$

When the base reacts with two or more molar proportions of the acid or when more than one base is present, the algebra becomes more unwieldy, but the treatment is essentially similar to that given above. Higuchi et al. (1958) also discuss the situation in which solid HA is dissolving in a solution of a dibasic acid.

11.4.2 Experimental Test of the Single-Equilibrium Theory

Higuchi et al. (1958) tested the validity of the theory represented by Figure 11.9 using spherical compressed tablets of benzoic acid. Each tablet was dissolved under defined conditions of agitation of 25°C (Parrott et al., 1954). The aqueous medium contained a base at a defined concentration, with or without 0.75 mol dm^{-3} sodium chloride, to investigate the influence of neutral

Table 11.3 Acid–Base Equilibrium Constants K (Eqs. 11.42 or 11.82) for Reactions of Various Bases with Benzoic Acid and Experimental Diffusivities $D_B \times 10^5$, cm^2 s^{-1} of the Bases from Dissolution Rates of Benzoic Acid Compacts[a]

Substance	Base Type	K	D_B No Added Electrolyte	With 0.75 mol dm^{-3} NaCl	D_B From Literature[b]
Sodium hydroxide	OH$^-$	6×10^9	2.42	2.92	2.42
Sodium tetraborate	BO$_2$$^-$	1×10^5	1.03	0.93	—
Disodium phosphate	HPO$_4$$^{2-}$	1×10^3	0.97	0.87	1.06
Sodium bicarbonate	HCO$_3$$^-$	200	0.50	0.64	1.25
Sodium acetate	CH$_3$COO$^-$	3.7	—	—	—
Ethanolamine	Un-ionized	2×10^5	1.22	1.20	—

[a] 1 mol dm^{-3} = 1 mol L^{-1} = 1 M.
[b] For references, see Higuchi et al. (1958).

salt effects. After various times the extent of dissolution and the surface area were measured by the change in weight and diameter of the tablet (Parrott et al., 1954). The calculated dissolution rate was checked by titration of the solution. The diffusion layer thickness h was determined by substituting into Eq. 11.19 the literature values $c_s = 0.0285\,\text{mol}\,\text{dm}^{-3}$ and $D_{HA} = 1.11 \times 10^{-5}\,\text{cm}^2\,\text{s}^{-1}$ at 25°C, and the experimental value of J of benzoic acid into pure water ($c_b \ll c_s$). Under the defined condition of agitation, h was known (32.8 μm) and was assumed to be independent of the nature and concentration of the base. To test the validity of Eqs. 11.48–11.50, J was measured in the presence of various concentrations, $[B^-]_h$ or $[B]_h$, of the bases listed in Table 11.3. The plots of J against $[B^-]$ or $[B]$ were linear in the presence of sodium hydroxide (Fig. 11.10), ethanolamine (Fig. 11.11), sodium tetraborate, and disodium phosphate. This is because the acid–base equilibrium constants, defined in Eqs. 11.42 and 11.52

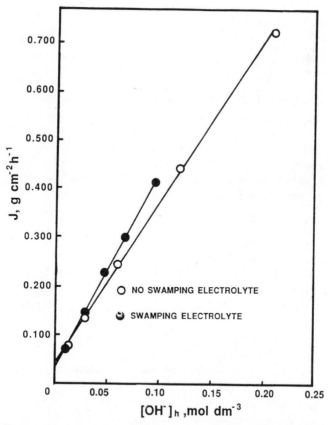

Figure 11.10   Intrinsic dissolution rate J of benzoic acid compacts in sodium hydroxide solutions. $1\,\text{mol}\,\text{dm}^{-3} = 1\,\text{mol}\,\text{L}^{-1} = 1\,M$. Reproduced with permission of the copyright owner, the American Pharmaceutical Association, from Higuchi, W.I., Parrott, E.L., Wurster, D.E., and Higuchi, T. (1958). *J. Am. Pharm. Assoc., Sci. Ed.*, **47**, 376–383.

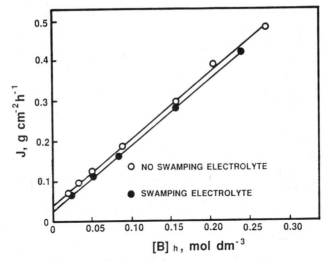

Figure 11.11 Intrinsic dissolution rate J of benzoic acid compacts in ethanolamine solutions. $[B]_h$ is the ethanolamine concentration. $1 \, mol \, dm^{-3} = 1 \, mol \, L^{-1} = 1M$. Reproduced with permission of the copyright owner, the American Pharmaceutical Association, from Higuchi, W.I., Parrott, E.L., Wurster, D.E., and Higuchi, T. (1958). *J. Am. Pharm. Assoc., Sci. Ed.*, **47**, 376–383.

and evaluated in Table 11.3, are sufficiently large to cause Eqs. 11.49 and 11.50 to be very good approximations. For these bases the basic strengths are high enough so as not to influence the dissolution rate of benzoic acid.

In the presence of sodium acetate (Fig. 11.12) or sodium bicarbonate, the plots of J against $[B^-]_h$ were concave upward. In the case of sodium bicarbonate this nonlinear behavior was attributed to the resistance imposed by pockets of liberated carbon dioxide that were observed on the surface of the solid. The nonlinear behavior in the presence of sodium acetate (Fig. 11.12) may be explained by the value of $K = 3.7$, which is too small for Eq. 11.50 to apply but too large for the effects of the base to be neglected. This situation provides a crucial test of the theory presented above.

Substitution of the following values into Eq. 11.48 leads to Eq. 11.53: $K = 3.67$; $[HA]_0 = 0.0285 \, mol \, dm^{-3}$; $h = 33 \, \mu m$; $D_{HB} = 1.28 \times 10^{-5} \, cm^2 \, s^{-1}$; $D_B = 1.19 \times 10^{-5} \, cm^2 \, s^{-1}$; $D_{HA} = 1.11 \times 10^{-5} \, cm^2 \, s^{-1}$; $D_A = 1.04 \times 10^{-3}$.

$$J \times 10^7 \, mol \, cm^{-2} \, s^{-1} = 1.78(1 + 4.13 \times 10^4 [B^-]_h)^{1/2} - 0.81 \quad (11.53)$$

This theoretical relationship, plotted in Figure 11.12, predicts intrinsic dissolution rates which are about 10% higher than the experimental values. Agreement can be considered to be satisfactory in view of the following: (a) the diffusivities may vary with concentration of each solute species and hence with distance across the film; (b) the thickness of the diffusion layer is not entirely independent of the diffusivities; (c) the salting-in effect of the benzoate ion will slightly affect [HA]; (d) the activity coefficients may vary and therefore the

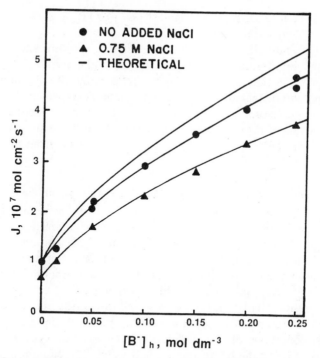

Figure 11.12   Intrinsic dissolution rate J of benzoic acid compacts in aqueous sodium acetate. $[B^-]_h$ is the sodium acetate concentration. $1 \, mol \, dm^{-3} = 1 \, mol \, L^{-1} = 1M$. Reproduced with permission of the copyright owner, the American Pharmaceutical Association. Higuchi, W.I., Parrott, E.L., Wurster, D.E., and Higuchi, T. (1958). *J. Am. Pharm. Assoc., Sci. Ed.,* **47**, 376–383.

equilibrium constant K may vary with the concentration of each solute species and hence with distance across the diffusion layer (Eqs. 11.42 and 11.52); and (e) the hydroxide ion may exert its own catalytic effect on dissolution, as considered later in this section and in Section 11.5.

In the absence of base, the slope of the apparently linear plot of J against c_s in the presence of 0–4.6 mol dm^{-3} electrolyte (NaCl) affords the value $D_{HA} = 1.16 \times 10^{-5} \, cm^2 \, s^{-1}$ using Eq. 11.49. In the absence of electrolyte, $D_{HA} = 1.11 \times 10^{-5} \, cm^2 \, s^{-1}$. In view of the small effect on D_{HA}, the influence of sodium chloride on the dissolution rate of benzoic acid is almost wholly explained by its salting-out effect in reducing c_s. We note that $c_s \equiv c_o^s$.

The presence of 0.75 mol dm^{-3} sodium chloride as a swamping electrolyte decreases the dissolution rate of benzoic acid in the presence of all but one base studied (e.g., Figs. 11.11 and 11.12); the exception is sodium hydroxide, with which sodium chloride increased the dissolution rate (Fig. 11.10). These influences correlate qualitatively with the effect of the electrolyte on the diffusivities of each base. D_B may be calculated from the experimental results using Eq. 11.50 and compared with values in the literature (Table 11.3). From the slope of each linear plot of J against $[B^-]$ or $[B]$ (e.g., Figs. 11.10–11.12), D_B may be calculated as $D = \text{slope} \times h$.

Electrolytes normally depress the diffusivity of ionic substances, as observed with sodium borate and disodium phosphate. However, sodium chloride at $0.75 \, \text{mol dm}^{-3}$ increases D_{OH} by about 12%, primarily by suppressing the diffusion potential created by the sodium ion, an effect which overwhelms the usual behavior. Except with bases involving abnormally rapid or slow ions, sodium chloride at $0.75 \, \text{mol dm}^{-3}$ may generally decrease the base-controlled dissolution rate of benzoic acid by about 10–20%, aside from that due to the salting-out effect.

11.4.3   Development of a Theory Based on Three Equilibria

Sections 11.4.1 and 11.4.2 considered the effects of bases on the dissolution rate of a solid acid assuming a *single* acid–base equilibrium represented by Eq. 11.41 or 11.51 within the diffusion layer. More recent studies of dissolution rates have shown that several acid–base equilibria within the diffusion layer may each influence the dissolution rates of acidic (or basic) solids. We shall see that the additional equilibria introduce further pairs of concentration terms into the equation for the intrinsic dissolution rate J (see Eqs. 11.49 and 11.50).

As the first example, Mooney et al. (1981a) studied the influence of pH under unbuffered conditions on the dissolution rates of carboxylic acids, the only reactive bases being the hydroxide ion and water. The equilibria involved are

$$H_2O + HA \overset{K'_a}{\rightleftharpoons} H_3O^+ + A^- \tag{11.54}$$

$$H_3O^+ + OH^- \overset{K'_w}{\rightleftharpoons} 2H_2O \tag{11.55}$$

These equilibria when combined lead to the following equilibrium, which is a special case of Eq. 11.41:

$$HA + OH^- \overset{K'_1}{\rightleftharpoons} H_2O + A^- \tag{11.56}$$

The pH was maintained at a constant, defined value by means of a pH-stat and the ionic strength was adjusted to $0.75 \, \text{mol dm}^{-3}$ by the presence of potassium chloride.

For simplicity, H_3O^+ may be replaced by the hydrogen ion H^+. In a dilute aqueous solution the activity of water and the activity coefficient may be assumed to be essentially constant. The apparent equilibrium constants of Eqs. 11.54–11.56 are then given by

$$K_a = [H^+][A^-]/[HA] \tag{11.57}$$

$$K_w = [H^+][OH^-] \tag{11.58}$$

$$K_1 = [A^-]/[HA][OH^-] \tag{11.59}$$

Fick's second law, Eq. 11.45, is applied to each of the four species involved in Eqs. 11.57–11.59 (i.e., HA, OH$^-$, A$^-$, H$^+$) with $\phi_S = 0$, corresponding to instantaneous chemical reaction, in each case.

For mass balance across the diffusion layer,

$$D_{HA}\frac{d^2[HA]}{dx^2} = -D_A\frac{d^2[A^-]}{dx^2} \tag{11.60}$$

and

$$D_{OH}\frac{d^2[OH^-]}{dx^2} = D_{HA}\frac{d^2[HA]}{dx^2} + D_H\frac{d^2[H^+]}{dx^2} \tag{11.61}$$

Equation 11.60 is given by the stoichiometry of both Eqs. 11.54 and 11.56. Equation 11.61 means that OH$^-$, while diffusing into a given volume element, will react with both HA, according to Eq. 11.56, and H$^+$, according to Eq. 11.55.

In the steady-state dissolution of solid HA (e.g., Fig. 11.9), the boundary conditions are as follows:

Boundary conditions at $x = 0$:

$$[HA]_0 = \text{solubility of HA}$$
$$[OH^-]_0, [H^+]_0, [A^-]_0 \text{ are unknown} \tag{11.62}$$

Boundary conditions at $x = h$:

$$[HA]_h = 0, [A^-] = 0, \text{ under sink conditions}$$
$$[OH^-]_h, [H^+]_h \text{ are defined by the bulk pH} \tag{11.63}$$

Integration of the second-order differential equations twice, assuming $\phi_s = 0$ and the boundary conditions given in Eqs. 11.62 and 11.63, leads to the following equations for the total flux J_t of the dissolving acid HA from the solid surface at $x = 0$ as a function of the changing conditions in the bulk solution (i.e., at $x = h$):

$$J_t = D_{HA}[HA]_0/h + D_H([H^+]_0 - [H^+]_h)/h$$
$$+ D_{OH}([OH^-]_h - [OH^-]_0)/h \tag{11.64}$$

J_t is equal to the intrinsic dissolution rate, thus,

$$J_t = J_{HA} + J_H + J_{OH} \tag{11.65}$$

where J_S is the flux of species S from $x = 0$ to $x = h$ across the diffusion layer and may be positive or negative depending on the direction of the concentration gradient. There is no expression for J_A because this is automatically accounted for by J_H and J_{OH}. Equation 11.64 differs from the equations derived by Higuchi et al. (1958) (e.g., Eqs. 11.49 and 11.50) since it contains terms for $[H^+]$ and $[OH^-]$, but it is similar to the equations of Hamlin and Higuchi (1966) and of Tsuji et al. (1978).

When the bulk pH is considerably lower than the pK_a of the dissolving acid (i.e., $[H^+] \gg K_a$), the ionization of the acid is suppressed within the diffusion layer. Then $[H^+]_0 = [H^+]_h$ and $[OH^-]_0 = [OH]_h$, so Eq. 11.64 reduces to

$$J_t = D_{HA}[HA]_0/h \qquad (11.66)$$

under sink conditions (i.e., $[HA]_0 \gg [HA]_h$).

When the pH of the bulk solution is increased so that it no longer supresses the ionization of the acid in the film, HA is able to dissociate by reaction with water or hydroxide ions according to Eq. 11.54 or 11.56. At pH values around the pK_a of the acid and up to neutrality in the bulk solution, the dominant form of Eq. 11.64 is

$$J_t = D_{HA}[HA]_0/h + D_H([H^+]_0 - [H^+]_h)/h \qquad (11.67)$$

In this region J_{OH} in Eq. 11.65 is not zero, but is negligible compared with J_{HA} and J_H.

When the pH in the bulk solution is increased such that $[OH^-]_h$ approaches $[HA]_0$, the controlling factor in dissolution is now the diffusion of hydroxide ion into the diffusion layer. Equations 11.56 and 11.59 now play a significant role. As HA is diffusing across the diffusion layer, the reaction with water and hydroxide ion occurs simultaneously. Hence J_t is now given by Eq. 11.64 in its complete form. The calculation of J_t therefore requires knowledge of the intrinsic solubility of the acid $[HA]_0$, the individual diffusivities D_{HA}, D_H, and D_{OH}, and the dissociation constant K_a.

11.4.4 Experimental Test of the Three-Equilibria Theory

Mooney et al. (1981a) tested the theoretical treatment discussed above using benzoic acid, 2-naphthoic acid, and indomethacin in the form of solid compressed discs. The initial dissolution rates of the three acids were followed using a procedure similar to that of Tsuji et al. (1978) and Underwood and Cadwallader (1978). The dissolution rates were determined using the rotating-disc method (Levich, 1962) adapted by Wood et al. (1965) and Nogami et al. (1966). Figure 11.13 shows the apparatus employed. The dissolution medium was an aqueous solution which was maintained at constant pH by addition of sodium hydroxide using an automatic recording pH-stat. A constant ionic strength of $0.5\,\mathrm{mol\,dm^{-3}}$ was maintained using potassium chloride. As the weak acid dissolved, its concentration was determined as a continuous function of time by circulating the solution through a flow cell in a UV spectrophotometer using

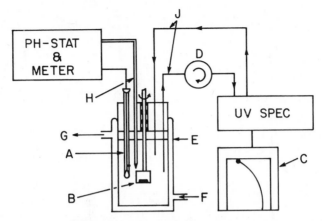

Figure 11.13 Diagram of the rotating disc and cell, the pH-stat and concentration-monitoring devices, the combination electrode (A), the Plexiglas disc holder with disc (B), the recorder for concentration versus time readout (C), the peristaltic pump (D), the thermostated–jacketed dissolution cell (E), the water jacket inlet (F), the water jacket outlet (G), the delivery tubing from the pH-stat (H), and the circulating tubing for the UV flow cell (J). Reproduced with permission of the copyright owner, the American Pharmaceutical Association, from Mooney, K.G., Mintun, M.A., Himmelstein, K.J., and Stella, V.J. (1981). *J. Pharm. Sci.*, **70**, 13–22.

a peristaltic pump. When steady-state conditions had been reached, the titrant was being delivered at a constant rate which depended on the release rate of the weak acid from the disc.

For the Levich (1962) rotating disc, Eq. 11.21 involving one diffusing species is assumed. Although several diffusing species are involved, Eq. 11.21 and 11.26 are expected to hold quite well for them all, except for hydroxide ions, as discussed in Section 11.4.3. Errors in the calculated fluxes are expected to occur only at high pH, when the hydroxide ion is a dominant term in Eq. 11.64. Table 11.1 presents the diffusivities D_{HA} of the unionized molecules, determined from the slope of the plot illustrated by Figure 11.5. For this purpose the Levich (1962) equation (eq. 11.26) was used in the following form appropriate to the sink conditions employed in the rotating-disc experiments:

$$J_t = 0.62 D_{HA}^{2/3} v^{-1/6} c_t^s \omega^{1/2} \qquad (11.68)$$

Table 11.1 also shows values of the solubilities $[HA]_0$ of the unionized weak acids and the ionization constants K_a determined by the plot shown in Figure 11.6 using Eq. 11.36.

Figure 11.14 shows the experimental initial dissolution rates of benzoic acid, 2-naphthoic acid, and indomethacin as a function of pH of the bulk solution and the theoretically predicted curves calculated from Eq. 11.64. For the theoretical predictions, values of diffusivities, equilibrium constants, and solubilities $[HA]_0$ were required, some of which are given in Table 11.1, while $[H^+]_h$ and $[OH^-]_h$ were calculated directly from the bulk pH. For comparative

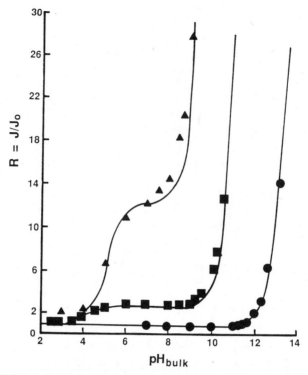

Figure 11.14   Relative dissolution rate R versus pH_{bulk} for several carboxylic acids at 25°C, $\mu = 0.5\,mol\,dm^{-3}$ (potassium chloride) using a pH-state to maintain constant pH_{bulk}. The solid lines are those predicted by Eq. 11.64 and the data points are those experimentally determined; J_0 refers to the dissolution on rate at pH 2.00, $\mu = 0.5\,mol\,dm^{-3}$. Key: indomethacin($\blacktriangle$); 2-naphthoic acid ($\blacksquare$); and benzoic acid ($\bullet$). Reproduced with permission of the copyright owner, the American Pharmaceutical Association, from Mooney, K.G., Mintun, M.A., Himmelstein, K.J., and Stella, V.J. (1981). *J. Pharm. Sci.*, **70**, 13–22.

purposes the relative dissolution rate ratio R was defined as

$$R = \frac{J}{J_0} = \frac{D_{HA}[HA]_0 - D_H([H^+]_0 - [H^+]_h) + D_{OH}([OH^-]_h - [OH^-]_0)}{D_{HA}[HA]_0}$$

$$(11.69)$$

This function normalizes the dissolution rates of each acid to the rate at pH = 2.00, at which the undissociated acid is the only diffusant. This procedure has the important effect of eliminating h, the thickness of the diffusion layer. The agreement between theory and experiment is evidently good. The discrepancies for indomethacin may be attributed to experimental error (by -30%) in the determination of D_{HA} or $[HA]_0$.

Figure 11.15 shows a plot of the pH at the solid–liquid interface pH_0, calculated from the theoretical model as a function of the pH of the bulk solution

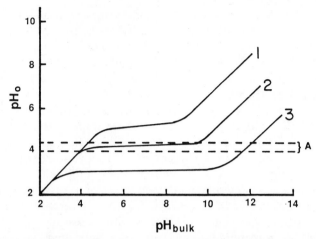

Figure 11.15 Plot showing the relationship between the calculated pH at $x = 0$, pH_0, and pH_{bulk} for the dissolution of several carboxylic acids, where water and hydroxide ion are the only bases in the dissolution medium. Key: 1, indomethacin; 2, 2-naphthoic acid; and 3, benzoic acid. A indicates the plateau region for 2-naphthoic acid. Reproduced with permission of the copyright owner, the American Pharmaceutical Association, from Mooney, K.G., Mintun, M.A., Himmelstein, K.J., and Stella, V.J. (1981). *J. Pharm. Sci.*, **70**, 13–22.

pH_{bulk} for each dissolving acid. The three clearly defined regions of each plot can be explained according to Eq. 11.64 and arise primarily from differences in the intrinsic solubility $[HA]_0$, and possibly in the diffusivity (D_{HA} and D_A) of each acid since the K_a values are all of the same order (Table 11.1). At low bulk pH, $pH_0 = pH_{bulk}$, because $[H^+]_{bulk}$ is sufficient to suppress dissociation of the dissolving acid within the diffusion layer. Under these circumstances Nernst–Brunner (1904) conditions exist, as defined by Eqs. 11.18 and 11.19.

At higher bulk pH, a plateau region A is evident in Figure 11.15, because the acid is dissociating into its conjugate base and hydrogen ions. This provides a larger concentration of total acid species at $x = 0$, but hydrogen ions will limit the degree and amount of dissociation. As the bulk pH increases, the acid will self-buffer the pH microenvironment of the diffusion layer to the approximate pH of a saturated solution of the acid, thus,

$$K_a[HA]_0 = [H^+]_0[A^-]_0 = [H^+]_0^2 \qquad (11.70)$$

where $[HA]_0$ is the solubility of the acid (in $0.5\,\text{mol dm}^{-3}$ potassium chloride at 25°C). In this region Eq. 11.64 approximates to

$$J = (D_{HA}[HA]_0 + D_H[H^+]_0)/h \qquad (11.71)$$

At the plateau region, a substantial increase in bulk pH increases the total concentration ($[HA]_0 + [A^-]_0$) by a relatively small amount, so J (and pH_0) increase by very little. 2-Naphthoic acid and indomethacin, which are less soluble (as measured by $[HA]_0$) than benzoic acid by factors of about 10^2 and 10^4, respectively, are less able to control or self-buffer the pH of the diffusion layer at $x = 0$ than is benzoic acid. The plateau region therefore commences at higher values of pH_{bulk} (benzoic acid $\cong 3.0$; 2-naphthoic acid $\cong 4.3$; indomethacin $\cong 5.2$), which correspond to the pH of the saturated solutions (in $0.5 \, mol \, dm^{-3}$ potassium chloride at $25°C$).

At the plateau pH_0 values of benzoic acid, 2-naphthoic acid, and indomethacin, the total acid species at $x = 0$ are respectively about 1.10, 2.65, and 11.35 times those at $pH_{bulk} = 2.00$. Hence, the dissolution rates of 2-naphthoic acid and indomethacin are more sensitive to changes in bulk pH than is that of benzoic acid. This underlines the importance of the intrinsic solubility of the acid in determining the sensitivity of its dissolution rate to pH. The pK_a values of the acids, together with $[HA]_0$, determine when deviation from Nernst–Brunner behavior commences with increasing bulk pH.

With a further increase in bulk pH, pH_0 rises, indicating that control of the dissolution rate will be affected by hydroxide ions diffusing into the diffusion layer from the bulk of the solution. The increasing hydroxide ions react with the acid according to Eqs. 11.56 and 11.59. Thus, the hydroxide term $D_{OH}([OH^-]_h - [OH^-]_0)$ now contributes to J_t in Eq. 11.64. The pH_0 rises at

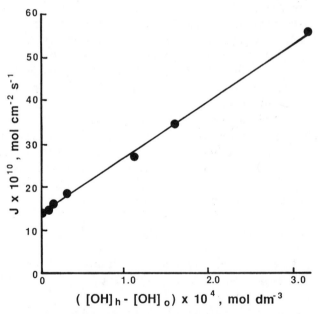

Figure 11.16 Plot of the total dissolution rate J of 2-naphthoic acid as a function of the hydroxide-ion concentration difference across the diffusion layer of a rotating disc at 450 rpm. Reproduced with permission of the copyright owner, the American Pharmaceutical Association, from Mooney, K.G., Mintun, M.A., Himmelstein, K.J., and Stella, V.J. (1981). *J. Pharm. Sci.*, **70**, 13–22.

Table 11.4 Hydroxide-Ion Diffusivity Calculated from the Slopes of the J_{obs} versus $[OH^-]_h - [OH^-]_0$ Plots for Benzoic Acid, 2-Naphthoic Acid, and Indomethacin

Compound	$Slope^a \times 10^2$ (cm s^{-1})	$h \times 10^{3\,b}$ (cm)	$D_{OH} \times 10^{5\,c}$ (cm^2 s^{-1})	r^d
Benzoic acid	1.174	2.29	2.69	0.9964
2-Naphthoic acid	1.334	1.97	2.63	0.9987
Indomethacin	1.826	1.67	3.04	0.9898

[a] Calculated from plots exemplified by Figure 11.16 with $[OH^-]$ expressed in the units mol cm^{-3}.
[b] Calculated from Eq. 11.21.
[c] Calculated from $D_{OH} = h \times$ slope of plots exemplified by Figure 11.16, according to Eq. 11.72.
[d] Correlation coefficient for each plot.

Reproduced with permission of the copyright owner, the American Pharmaceutical Association, from Mooney, K. G., Mintun, M. A., Himmelstein, K. J., and Stella, V. J. (1981). *J. Pharm. Sci.*, **70**, 13–22.

a rate which depends directly on the amount of hydroxide ion diffusing in. The bulk pH, at which the rise in pH_0 begins, decreases with decreasing buffering capacity, as defined by the solubility of the acid (i.e., at pH_{bulk} 11.0 for benzoic acid, 9.5 for 2-naphthoic acid, and 7.5 for indomethacin). At high bulk pH the terms $[H^+]_0$ and $[H^+]_h$ become negligible in comparison with the terms $[OH^-]_0$ and $[OH^-]_h$. Equation 11.64 then approximates to

$$J = \text{constant} + D_{OH}([OH^-]_h - [OH^-]_0)/h \qquad (11.72)$$

Figure 11.16 shows a plot which corresponds to this linear equation and Table 11.4 provides the regression data for all three acids. The slope of this plot is D_{OH}/h. If h is evaluated from Eq. 11.21, bearing in mind the assumption that h is the same for each species, D_{OH}, the diffusivity of the hydroxide ion, may be evaluated. Values of D_{OH} in Table 11.4 agree quite well among themselves and with those presented in Table 11.3. It is interesting to note that $[OH^-]_0$ is almost negligible in comparison to $[OH^-]_h$ for all three acids in this high pH range. At very high bulk pH Eq. 11.72 predicts that the fluxes of all the acids will be identical and independent of the properties of the acid.

Perhaps the most important observation and theoretical prediction of Mooney et al. (1981a) is that the increase in the dissolution rate is not a linear function of the increase in the apparent solubility (c_t^s in Section 11.3) on changing the pH. This conclusion appears contrary to the expectations of the Noyes–Whitney and Nernst–Brunner equations (Eqs. 11.10 and 11.19) because not one but several species, each participating in acid–base processes within the diffusion layer, are involved.

11.4.5 Concentration Profiles Across the Diffusion Layer Calculated from the Three-Equilibria Theory

From complete analytical solutions of the second-order differential equations (Eqs. 11.60 and 11.61), the concentration of each species may be calculated at any point x in the diffusion layer and the concentration profiles plotted. The detailed procedure employed is described in Section 11.5.2. In the present case, however, $[H^+]$ is calculated as a root of a cubic equation at each value of x.

Thus, Figure 11.17 shows idealized sections of the diffusion layer for benzoic acid, 2-naphthoic acid, and indomethacin at various bulk pH values. The fractional distance across the diffusion layer is calculated as x/h. Fractional concentrations at x are obtained for HA, A^-, and H^+ by dividing the concentration of the species at x by that at $x=0$. Since OH^- diffuses in the

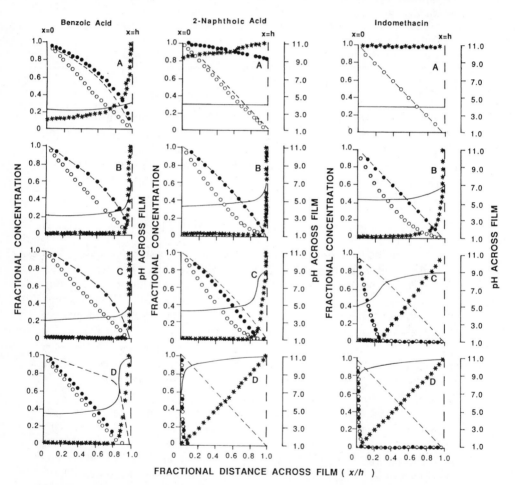

Figure 11.17 Idealized diffusion layer cross sections for benzoic acid, 2-naphthoic acid, and indomethacin, showing the fractional concentration profiles with the fractional distance across the film as calculated from Eqs. 11.21 and 11.60–11.61. Calculated values of pH across the film in each case also are included. Key: O, [HA]; ———, $[A^-]$; ●, $[H^+]$; *, $[OH^-]$; ———, pH; A, pH_{bulk} 4.00; B, pH_{bulk} 7.00; C, pH_{bulk} 9.00; and D, pH_{bulk} 11.00. Reproduced with permission of the copyright owner, the American Pharmaceutical Association, from Mooney, K.G., Mintun, M.A., Himmelstein, K.J., and Stella, V.J. (1981). *J. Pharm. Sci.*, **70**, 13–22.

opposite direction, its fractional concentration at x is evaluated as $[OH^-]_x/[OH^-]_h$. These fractional functions normalize the data across the diffusion layer and facilitate comparison of the concentration profiles for different compounds at the same pH value or for the same compound at different bulk solution pH values.

The left-hand column in Figure 11.17 clearly shows that benzoic acid gives virtually the same concentration profiles at bulk pH values of 4.0, 7.0, and 9.0. This similarity demonstrates the ability of benzoic acid to buffer the diffusion layer. Only at the high bulk pH of 11.0 does the hydroxide ion significantly affect pH_0 and consequently increase the dissolution rate. In the center column in Figure 11.17, 2-naphthoic acid demonstrates diminished self-buffering action by showing similar concentration profiles only at pH 7.0 and 9.0. The concentration profiles of indomethacin, shown in the right-hand column of Figure 11.17, are very sensitive to bulk pH. In other words, indomethacin is even less able to control its own microenvironment in the diffusion layer. Thus, bulk pH largely determines the concentration profiles, pH_0, and therefore the dissolution rate of this weak acid.

11.5 INFLUENCE OF BUFFERS ON THE RATES OF DISSOLUTION OF CARBOXYLIC ACIDS BASED ON SIX EQUILIBRIA

11.5.1 Development of the Theory

In their second paper Mooney et al. (1981b) extended their previous model (Mooney et al., 1981a) to predict the effect of buffers on the dissolution rate of a weak acid. The results emphasize the importance of the concentration of the buffer base and the pK_a value of the conjugate acid and play down the importance and direct involvment of the bulk pH. As before, dissolution is assumed to be controlled by diffusion. Figure 11.18 represents the postulated diffusion layer for the model used. Hydroxide ion, OH^-, and buffer base, B, may diffuse into the diffusion layer from the bulk solution and react with the carboxylic acid, HA, which is diffusing out. The products are the conjugate acid of the buffer, BH^+, the acid anion, A^-, and the hydrogen ion, H^+, all of which diffuse out.

The following six chemical equilibria are assumed to take place instantaneously within the diffusion layer so that no terms for rates of chemical reactions are required. In Section 11.6 the influence of rates of ionization are discussed.

$$H^+ + OH^- \rightleftharpoons H_2O \tag{11.73}$$

$$HA \overset{K_a^A}{\rightleftharpoons} H^+ + A^- \tag{11.74}$$

$$HA + OH^- \overset{K_1'}{\rightleftharpoons} H_2O + A^- \tag{11.75}$$

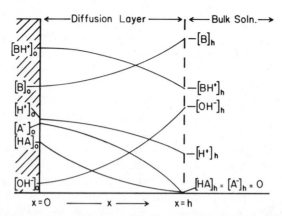

Figure 11.18 Diagrammatic representation of a solid carboxylic acid HA dissolving into a reactive medium containing hydroxide ion and buffer components B and BH^+ with a Nernst diffusion layer existing between the solid and the bulk solution. Sink conditions exist in the bulk solution, and the products BH^+, A^-, and H^+ diffuse out of the diffusion layer at a rate determined by their chemical reactivity and diffusivity. Reproduced with permission of the copyright owner, the American Pharmaceutical Association, from Mooney, K.G., Mintun, M.A., Himmelstein, K.J., and Stella, V.J. (1981). *J. Pharm. Sci.*, **70**, 22–32.

$$HA + B \underset{}{\overset{K_2}{\rightleftharpoons}} BH^+ + A^- \tag{11.76}$$

$$BH^+ \underset{}{\overset{K_a^B}{\rightleftharpoons}} H^+ + B \tag{11.77}$$

$$BH^+ + OH^- \underset{}{\overset{K_3'}{\rightleftharpoons}} H_2O + B \tag{11.78}$$

The apparent equilibrium constants, which are based on molar concentrations as before (Eq. 11.43 and Eqs. 11.85–11.87), are as follows:

$$K_w = [H^+][OH^-] \tag{11.79}$$

$$K_a^A = \frac{[H^+][A^-]}{[HA]} \tag{11.80}$$

$$\frac{K_1'}{[H_2O]} = K_1 = \frac{[A^-]}{[HA][OH^-]} = \frac{K_a^A}{K_w} \tag{11.81}$$

$$K_2 = \frac{[BH^+][A^-]}{[HA][B]} = \frac{K_a^A}{K_a^B} \tag{11.82}$$

$$K_a^B = \frac{[H^+][B]}{[BH^+]} \tag{11.83}$$

$$\frac{K'_3}{[H_2O]} = K_3 = \frac{[B]}{[BH^+][OH^-]} \tag{11.84}$$

The following six chemical species may be identified: HA, A^-, H^+, OH^-, B, and BH^+, each of which has its own second-order equation for diffusion according to Fick's second law (Eq. 11.45).

Equations 11.73–11.78 indicate the following mass balance within the diffusion layer:

$$\Delta(\text{flux } OH^- \text{ in}) + \Delta(\text{flux } B \text{ in}) = \Delta(\text{flux } HA \text{ out}) + \Delta(\text{flux } H^+ \text{ out}) \tag{11.85}$$

Therefore,

$$D_{OH}\frac{d[OH^-]}{dx} + D_B\frac{d^2[B]}{dx^2} = D_{HA}\frac{d^2[HA]}{dx^2} + D_H\frac{d[H^+]}{dx} \tag{11.86}$$

Any change in HA (and B) due to reaction automatically represents an opposite change in its conjugate species A^- (and BH^+), thus,

$$D_{HA}\frac{d^2[HA]}{dx^2} = -D_A\frac{d^2[A^-]}{dx^2} \tag{11.87}$$

$$D_B\frac{d^2[B]}{dx^2} = -D_{BH}\frac{d^2[BH^+]}{dx^2} \tag{11.88}$$

Integration of Eqs. 11.86–11.88 once with respect to x yields

$$D_{HA}\frac{d[HA]}{dx} = D_{OH}\frac{d[OH^-]}{dx} + D_B\frac{d[B]}{dx} - D_H\frac{d[H^+]}{dx} + I_1 \tag{11.89}$$

$$D_{HA}\frac{d[HA]}{dx} = -D_A\frac{d[A^-]}{dx} + I_2 \tag{11.90}$$

$$D_B\frac{d[B]}{dx} = -D_{BH}\frac{d[BH^+]}{dx} + I_3 \tag{11.91}$$

where I_1, I_2, and I_3 are constants of integration.

We now consider the mass balance into and out of the diffusion layer (Fig. 11.18). A^- is a product of the reaction between HA and either B, OH^-, or H_2O. However, B and OH^- also react with H^+ in the diffusion layer. Therefore,

$$\sum(\text{flux of reactants in}) = (\text{flux } A^- \text{ out}) - (\text{flux } H^+ \text{ out}) \tag{11.92}$$

or

$$\sum(\text{flux of reactants in}) = D_{\text{OH}}\frac{d[\text{OH}^-]}{dx} + D_{\text{B}}\frac{d[\text{B}]}{dx} \tag{11.93}$$

Fluxes into the diffusion layer have positive slopes. Now

$$(\text{flux A}^- \text{ out}) = -D_{\text{A}}\,(d[\text{A}^-]/dx) \tag{11.94}$$

and

$$(\text{flux H}^+ \text{ out}) = -D_{\text{H}}(d[\text{H}^+]/dx) \tag{11.95}$$

The previous four equations give

$$D_{\text{OH}}\frac{d[\text{OH}^-]}{dx} + D_{\text{B}}\frac{d[\text{B}]}{dx} = D_{\text{H}}\frac{d[\text{H}^+]}{dx} - D_{\text{A}}\frac{d[\text{A}^-]}{dx} \tag{11.96}$$

Comparison with Eq. 11.89 and 11.90 shows that $I_2 = I_1$. Equation 11.90 now becomes

$$D_{\text{HA}}\frac{d[\text{HA}]}{dx} = -D_{\text{A}}\frac{d[\text{A}^-]}{dx} + I_1 \tag{11.97}$$

Elimination of the terms in B and HA from Eqs. 11.89, 11.94, and 11.97 yields

$$-D_{\text{A}}\frac{d[\text{A}^-]}{dx} = D_{\text{OH}}\frac{d[\text{OH}^-]}{dx} - D_{\text{H}}\frac{d[\text{H}^+]}{dx} - D_{\text{BH}}\frac{d[\text{BH}^+]}{dx} + I_3 \tag{11.98}$$

The flux of A^- *out* of the diffusion layer is equal to the sum of the following three fluxes: the flux of H^+ *out* (Eq. 11.74), the flux of OH^- *in* (Eq. 11.75), and the flux of BH^+ *out* (Eq. 11.76). This statement may be expressed as follows, remembering that fluxes *in* are positive while fluxes *out* are negative:

$$-D_{\text{A}}\frac{d[\text{A}^-]}{dx} = -D_{\text{H}}\frac{d[\text{H}^+]}{dx} + D_{\text{OH}}\frac{d[\text{OH}^-]}{dx} - D_{\text{BH}}\frac{d[\text{BH}^+]}{dx} \tag{11.99}$$

Comparison of Eqs. 11.98 and 11.99 shows that $I_3 = 0$. Equation 11.91 now becomes

$$D_{\text{B}}\frac{d[\text{B}]}{dx} = -D_{\text{BH}}\frac{d[\text{BH}^+]}{dx} \tag{11.100}$$

Integration of Eqs. 11.89, 11.97, and 11.100 affords the following linear equations in x:

$$D_{HA}[HA] = D_{OH}[OH^-] + D_B[B] - D_H[H^+] + I_1 x + L_1 \qquad (11.101)$$

$$D_{HA}[HA] = -D_A[A^-] + I_1 x + L_2 \qquad (11.102)$$

$$D_B[B] = -D_{BH}[BH^+] + L_3 \qquad (11.103)$$

where L_1, L_2, and L_3 are additional constants of integration. Evaluation of I_1, L_1, L_2, and L_3 requires application of the following boundary conditions, illustrated in Figure 11.18:

Boundary conditions at $x = 0$:

$$[HA]_o = \text{solubility of HA} \qquad (11.104)$$

$[A^-]_o, [H^+]_o, [OH^-]_o, [B]_o, [BH^+]_o$ are all unknown

Boundary conditions at $x = h$:

$$[HA]_h = 0, [A^-]_h = 0, \text{ under sink conditions} \qquad (11.105)$$

$[H^+]_h, [OH^-]_h$ are defined by the bulk pH
$[B]_h, [BH^+]_h$ are defined by the buffer composition

Applying the boundary conditions at $x = 0$ yields:

$$D_{HA}[HA]_o = D_{OH}[OH^-]_o + D_B[B]_o - D_H[H^+]_o + L_1 \qquad (11.106)$$

$$D_{HA}[HA]_o = -D_A[A^-]_o + L_2 \qquad (11.107)$$

$$D_B[B]_o = -D_{BH}[BH^+]_o + L_3 \qquad (11.108)$$

Applying the boundary conditions at $x = h$ affords:

$$D_H[H^+]_h = D_{OH}[OH^-]_h + D_B[B]_h + I_1 h + L_1 \qquad (11.109)$$

$$I_1 h = -L_2 \qquad (11.110)$$

$$D_B[B]_h = -D_{BH}[BH^+]_h + L_3 \qquad (11.111)$$

The last six equations contain nine unknowns. The calculation of these unknowns requires three more equations which are provided by the expressions for the equilibrium constants (Eqs. 11.79–11.81). Since there are nine independent equations, each of the nine unknowns may be evaluated. Successive substitutions may be used to reduce these equations to one equation which may be solved to evaluate a particular unknown. For example, to express the unknown quantity $[H^+]_o$ in terms of the known quantities, the following cubic equation may be derived by successive substitution using the nine equations mentioned above:

$$p[H^+]_o^3 + q[H^+]_o^2 + r[H^+]_o + s = 0 \tag{11.112}$$

where

$$p = D_H D_{BH} \tag{11.113}$$

$$q = D_H D_B p K_a^B + D_{BH} u \tag{11.114}$$

$$r = D_B K_a^B (u - v) - D_A D_{BH} K_a^A [HA]_o - D_{OH} D_{BH} K_w \tag{11.115}$$

$$s = -D_A D_B K_a^A K_a^B [HA]_o - D_{OH} K_w K_a^B D_B \tag{11.116}$$

in which

$$u = D_{OH} [OH^-]_h + D_B [B]_h - D_H [H^+]_h \tag{11.117}$$

$$v = D_B [B]_h + D_{BH} [BH^+]_h \; (= L_3 \text{ also}) \tag{11.118}$$

Equation 11.112 may be solved as described by the method of Newton (see Carnahan et al., 1969), using a digital computer. Thus, for any given conditions of $[H^+]_h$, $[OH^-]_h$, $[B]_h$, and $[BH^+]_h$ in the bulk solution, $[H^+]_o$ may be calculated. From this value of $[H^+]_o$, the other unknowns may be successively calculated from the preceding equations, preferably using a digital computer.

Equation 11.89 may be written in the form

$$-I_1 = -D_{HA} \frac{d[HA]}{dx} - D_H \frac{d[H^+]}{dx} + D_{OH} \frac{d[OH^-]}{dx} + D_B \frac{d[B]}{dx} \tag{11.119}$$

Thus, I_1 represents the negative sum of the individual fluxes J_S across the diffusion layer:

$$-I_1 = J_{HA} + J_H + J_{OH} + J_B = J_t \tag{11.120}$$

I_1 is therefore the negative of the total acid flux J_t at $x = 0$ and is thus the dissolution rate of the solid acid. I_1 may be expressed in terms of the known quantities by subtracting Eq. 11.109 from Eq. 11.106 and rearranging, thus,

$$-I_1 = J_t = D_{HA} \frac{[HA]_o}{h} + D_H \frac{[H^+]_o - [H^+]_h}{h}$$
$$+ D_{OH} \frac{[OH^-]_h - [OH^-]_o}{h} + D_B \frac{[B]_h - [B]_o}{h} \tag{11.121}$$

Equation 11.121 differs from Eq. 11.64 (Mooney et al., 1981a) for an unbuffered medium simply by the inclusion of the term $D_B([B]_h - [B]_o)/h$ for the buffer in Eq. 11.121. By using the values of $[H^+]_o$, $[OH^-]_o$, and $[B]_o$ calculated from Eqs. 11.106–11.111, the intrinsic dissolution rate J_t of the solid weak acid may be calculated at any bulk pH or buffer concentration in the dissolution medium.

11.5.2   Experimental Test of the Six-Equilibria Theory

Mooney et al. (1981b) tested the theoretical treatment described above using the dissolution of rotating discs (Levich, 1962) of 2-naphthoic acid in aqueous solutions of buffers at an ionic strength of $0.5\,mol\,dm^{-3}$ (potassium chloride). The experimental system (Fig. 11.13) employed a pH-stat and UV spectrophotometer as before (Mooney et al., 1981a). The diffusivities and pK_a values of the buffer systems employed are presented in Table 11.5 and were used for the various calculations.

Figures 11.19–11.21 show plots of the intrinsic dissolution rate J_t against the total buffer concentration in the bulk solution. Both the observed J_t (J_{obs}) and the theoretical value (J_{theor}) calculated from Eq. 11.121 are shown together with the value of pH_o. The latter was calculated from Eqs. 11.112–11.118 and is indicated by the number close to each experimental point. The agreement between J_{obs} and J_{theor} was generally good.

In the absence of added buffer, that is, when pH_{bulk} is maintained only by the pH-stat, $[HA]_o$ and K_a of the acid determine J; the degree of dissociation of the acid is determined by reaction with water and hydroxide ion, as is J_t (Mooney et al., 1981a). Equation 11.121 for J_t in a buffered medium differs from Eq. 11.64 for J_t in an unbuffered medium simply by inclusion of the term $D_B([B]_h - [B]_o)/h$ for the buffer in Eq. 11.121. This term accounts for any buffer base species, B used up in reaction with HA, H^+, and H_2O. As $[B]_h$ increases, the corresponding term for its flux in the diffusion layer becomes more significant

Table 11.5 Diffusivities and pK_a Values for Imidazole, Morpholine, and Acetate for Aqueous Solutions at an Ionic Strength of $0.5\,mol\,dm^{-3}$ at $25°C^a$

Compound	M^b	pK_a		$D \times 10^6\,(cm^2\,s^{-1})$	
		Exp.^{c}	Thermo.^{d}	Theor.^{e}	Exp.^{f}
Imidazole	68.1	7.17	6.91	8.20	8.11
Morpholine	87.1	8.75	8.70	7.00	—
Acetate	59.0	4.60	4.75	8.84	8.61

^{a} $1\,mol\,dm^{-3} = 1\,mol\,L^{-1} = 1\,M$.
^{b} Molecular weight.
^{c} Determined experimentally (Mooney et al., 1981b).
^{d} Thermodynamic value from Butler (1964).
^{e} Theoretical value from the Stokes–Einstein equation (Eq. 11.29) (Mooney et al., 1981b).
^{f} Experimental value from the slope in Figure 11.6 using Eq. 11.36 (Mooney et al., 1981b).

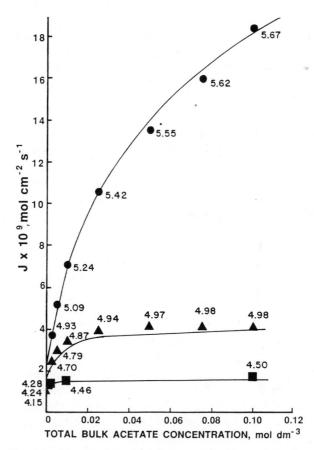

Figure 11.19   Plot of the dissolution rate of 2-naphthoic acid from a solid disc rotating at 450 rpm into media of varying acetate buffer concentrations and pH_{bulk} [$\mu = 0.5\,mol\,dm^{-3}$ (potassium chloride) at 25°C]. The pH_{bulk} was maintained constant by a pH-stat using $0.01\,mol\,dm^{-3}$ NaOH. The solid continuous lines represent J_{theor}, calculated using Eq. 11.121, and the symbols represent J_{obs}. The number with each point is the calculated pH_0 value calculated from Eqs. 11.168–11.174. Key: ■ pH_{bulk} 4.5; ▲, pH_{bulk} 5.0; and ●, pH_{bulk} 6.00. $1\,mol\,dm^{-3} = 1\,mol\,L^{-1} = 1\,M$. Reproduced with permission of the copyright owner, the American Pharmaceutical Association. Adapted from Mooney, K. G., Mintun, M. A., Himmelstein, K. J., and Stella, V. J. (1981). *J. Pharm. Sci.*, **70**, 22–32.

than the other terms. Eventually the pH within the diffusion layer and pH_0 are controlled by the swamping effect of the buffer as its concentration increases. This emphasizes the importance of the total buffer concentration in the bulk solution and the intrinsic solubility of the dissolving acid, $[HA]_0$, in determining the initial dissolution rate of that acid.

The pK_a for the acetate buffer (4.60) is similar to that for 2-naphthoic acid (4.02), and therefore the equilibrium constant for the reaction between them is relatively low ($K_2 = 3.8$). Thus the degree of catalysis of the dissolution of 2-naphthoic acid by acetate ion is relatively small compared with that by

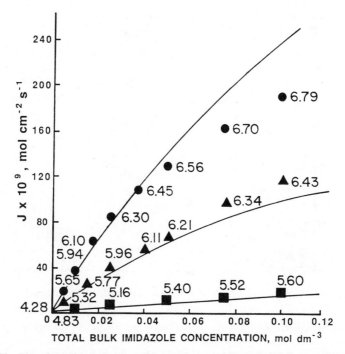

Figure 11.20 Plot of the dissolution rate of 2-naphthoic acid from a solid disc rotating at 450 rpm into media of varying imidazole buffer concentration and pH_{bulk} [$\mu = 0.5\ mol\ dm^{-3}$ (potassium chloride) at 25°C]. The pH_{bulk} was maintained constant by a pH-stat using $0.1\ mol\ dm^{-3}$ NaOH. The solid continuous lines represent J_{theor} calculated using Eq. 11.121. The number with each point is the calculated pH_o value. Key: ■, pH_{bulk} 6.00; ▲, pH_{bulk} 7.00; and ●, pH_{bulk} 8.00. $1\ mol\ dm^{-3} = 1\ mol\ L^{-1} = 1\ M$. Reproduced with permission of the copyright owner, the American Pharmaceutical Association. Adapted from Mooney, K.G., Mintun, M.A., Himmelstein, K.J., and Stella, V.J. (1981). *J. Pharm. Sci.*, **70**, 22–32.

comparable concentrations of stronger bases. At constant pH_{bulk} when the total buffer concentration in the bulk solution increases, the maximum possible value of pH_o is pH_{bulk} (Fig. 11.19). The asymptotic fluxes were observed experimentally only with the acetate system at pH_{bulk} 4.50 and 5.00, because the corresponding pH_o values in the absence of buffer do not differ greatly from the asymptotic value they must approach, that is, pH 4.50 and 5.00 with increasing bulk sodium acetate concentration. At pH_{bulk} 6.00 the total acetate concentration in the bulk solution must be much higher than that used experimentally for pH_o to approach pH_{bulk}, so the asymptotic flux of 2-naphthoic acid is not observed at this pH_{bulk} (Fig. 11.19).

In view of the higher pK_a values (i.e., higher basic strength) of imidazole and morpholine as compared with acetate (Table 11.5), considerably greater concentrations of the former buffers at pH 6.0–8.5 are required to make pH_o equal pH_{bulk}. Figures 11.20 and 11.21 show that the asymptotic dissolution rate of 2-naphthoic acid was not seen with these buffers at the concentrations and bulk

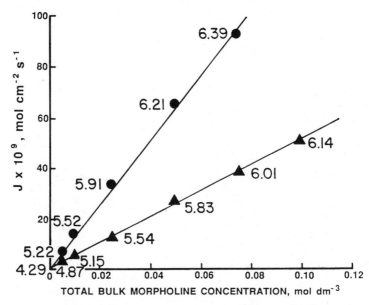

Figure 11.21 Plot of the dissolution rate of 2-naphthoic acid from a solid disc rotating a 450 rpm into media of varying morpholine buffer concentration and pH_{bulk} [$\mu = 0.5\,mol\,dm^{-3}$ (potassium chloride) at 25°C]. The pH_{bulk} was maintained constant by a pH-stat using $0.1\,mol\,dm^{-3}$ NaOH. The solid continuous lines represent J_{theor} calculated using Eq. 11.121, and the symbols represent J_{obs}. The number with each point is the calculated pH_o value calculated from Eq. 11.168–11.174. Key: ▲, pH_{bulk} 8.00; and ●, pH_{bulk} 8.50. $1\,mol\,dm^{-3} = 1\,mol\,L^{-1} = 1\,M$. Reproduced with permission of the copyright owner, the American Pharmaceutical Association. Adapted from Mooney, K.G., Mintun, M.A., Himmelstein, K.J., and Stella, V.J. (1981). *J. Pharm. Sci.*, **70**, 22–32.

pH values studied. The asymptotic dissolution rate at pH_{bulk} 8.00 is greater than that in the absence of buffer (i.e., at pH_{bulk} 4.29) by a factor of $10^{8.00-4.29} = 3400$. This corresponds to $J_t = 4.8 \times 10^{-6}\,mol\,cm^{-2}\,s^{-1}$, which is well beyond the range of measurement. (Figs. 11.20 and 11.21), showing that considerable bulk concentrations of imidazole or morpholine would be required. The solubility and pK_a of the dissolving acid also play important roles in defining the pH_{bulk} and buffer concentration for asymptotic dissolution rates.

The curvature seen in Figures 11.19 and 11.20 indicates that care should be taken in extrapolating buffer dissolution data. Such extrapolation is frequently employed (Higuchi et al., 1971; Lee et al., 1979) to obtain buffer-independent dissolution rates when no pH-stat is used and the buffer capacity is assumed to be large enough to resist bulk pH changes during dissolution. For example, extrapolation of least-squares linear regression of the four highest experimental individual concentrations at bulk pH 8.00 (Fig. 11.20) gives an extrapolated J_o value of $\cong 6.4 \times 10^{-8}\,mol\,cm^{-2}\,s^{-1}$, which is almost 50 times the true J_o value, $1.4 \times 10^{-9}\,mol\,cm^{-2}\,s^{-1}$, determined by the pH-stat without buffer. On the other hand, dissolution rates measured at the highest imidazole buffer concentration, $0.1\,mol\,dm^{-3}$, at pH_{bulk} 7.0, were not significantly different whether or not the pH-stat was operating. This finding is expected for a high buffer capacity and adds confidence to an early study (Higuchi et al., 1958) in which a pH-stat was not employed. However, without a pH-stat, dissolution of the acid causes pH_{bulk} to decrease as the buffer capacity is exceeded. Sink conditions will then no longer exist and the dissolution rate could decrease (Underwood and Cadwallader, 1978).

The basicity K_a^B of the buffer base B can greatly influence the dissolution rate (Fig. 11.22). K_a^B for imidazole is about 400 times that of acetate. Imidazole therefore reacts more extensively with the dissolving 2-naphthoic acid and has a much greater catalytic effect on its dissolution for the same concentration of B in the bulk solution. However, $[HA]_o$ and K_a^A of the dissolving acid are also important in describing the conditions at the solid–liquid interface (Eqs. 11.112, 11.115, and 11.116). In Figure 11.23 the ordinate data in Figure 11.22 are plotted against the difference in concentration of B across the diffusion layer. The linear relationships in Figure 11.23 are predicted by the following manipulation of Eq. 11.121:

$$J_t = \text{constant} + D_B([B]_h - [B]_o)/h \qquad (11.122)$$

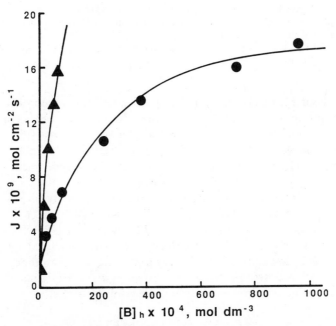

Figure 11.22 Plot of 2-naphthoic acid dissolution rate J versus buffer base concentration in the bulk solution for acetate (●) and imidazole (▲) at pH_{bulk} 6.00. The symbols represent J_{obs} and the continuous line represents J_{theor}. $1\,mol\,dm^{-3} = 1\,mol\,L^{-1} = 1\,M$. Reproduced with permission of the copyright owner, the American Pharmaceutical Association, from Mooney, K.G.. Mintun, M.A., Himmelstein, K.J., and Stella, V.J. (1981). *J. Pharm. Sci.*, **70**, 22–32.

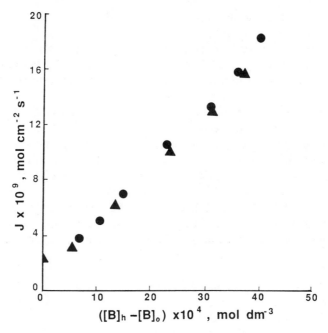

Figure 11.23 Plot of 2-naphthoic acid dissolution rate J versus buffer base concentration difference $([B]_k - [B]_o)$ across the diffusion layer at steady state for acetate ($\bullet$) and imidazole ($\blacktriangle$) at pH_{bulk} 6.00. The symbols represent observed rates (J_{obs}). $1\ mol\ dm^{-3} = 1\ mol\ L^{-1} = 1\ M$. Reproduced with permission of the copyright owner, the American Pharmaceutical Association, from Mooney, K.G., Mintun, M.A., Himmelstein, K.J., and Stella, V.J. (1981). *J. Pharm. Sci.*, **70**, 22–32.

The terms in [HA], [H$^+$], and [OH$^-$] in Eq. 11.121 here represent negligible contributions to the overall flux and are grouped together as a constant. According to Eq. 11.122 the slope of the plot of J_t against $([B]_h - [B]_o)$ in Figure 11.23 is D_B/h, which allows D_B to be evaluated if h is known. These D_B values (Table 11.5) are not obtained entirely from experiment, since the calculation of $[B]_o$ requires some initial estimate of D_B.

Table 11.6 shows a comparison between the experimental dissolution rates of benzoic acid in sodium acetate without a pH-stat (Higuchi et al., 1958) and those calculated using the present model (Mooney et al., 1981b). Agreement is good, even though pH_{bulk} in the earlier study was controlled by the buffer capacity of the added bases instead of by a pH-stat, and even though there are differences in ionic strength between the two systems. Equation 11.50 of the earlier study can be written as follows under sink conditions:

$$J_t = (D_{HA}[HA]_o + D_B[B]_h)/h \tag{11.123}$$

This equation assumes that the total concentration of buffer in the bulk solution is equivalent to that of the base species $[B]_h$. This assumption holds less well for weaker bases, such as acetate, than for stronger bases, such as

Table 11.6 Recalculated Data for Benzoic Acid Dissolving into Aqueous Solutions of Sodium Acetate at 25°C When No pH-Stat was Used[a]

Total Bulk Solution Concentration of Acetate (mol dm^{-3})	Calculated[b] from $[H^+]_h = \sqrt{K_w K_a^B/m}$ $\times 10^9$ (mol dm^{-3})	Calculated pH$_{bulk}$[b]	pH$_0$[c]	J_{obs}[d] $\times 10^7$ (mol cm^{-2} s^{-1})	J_{theor}[e] $\times 10^7$ (mol cm^{-2} s^{-1})
0.015	4.09	8.39	3.76	1.02	1.13
0.05	2.24	8.65	4.19	1.75	1.79
0.10	1.59	8.80	4.41	2.34	2.52
0.15	1.29	8.88	4.54	2.82	3.11
0.20	1.12	8.95	4.62	3.44	3.63
0.25	1.00	8.99	4.69	3.82	4.09

[a] From Higuchi et al. (1958). Solutions contained 0.75 mol dm^{-3} NaCl.
[b] Approximate hydrogen-ion concentration for the salt of a weak acid; K_w and K_a^B are defined by Eqs. 11.79 and 11.77; m is the total molar concentration of the salt; pH$_{bulk}$ is calculated from $[H^+]_h$.
[c] Calculated pH$_0$ assuming that pH-stat conditions are maintained in the bulk solution [Eq. 11.112 was used with parameters determined for benzoic acid from Mooney et al., 1981a (in Table 11.1) and acetate from Table 11.5].
[d] Calculated from data given by Higuchi et al. (1958).
[e] Theoretical value using an equivalent diffusion layer thickness $h = 28.2\ \mu$m (Mooney et al., 1981a).
Reproduced with permission of the copyright owner, the American Pharmaceutical Association, from Mooney, K. G., Mintun, M. A., Himmelstein, K. J., and Stella, V. J. (1981). *J. Pharm. Sci.*, 70, 22–32.

HPO$_4^{2-}$ and borate, with benzoic acid as the dissolving solid. Equation 11.122 is preferable for weaker bases and is more general than Eq. 11.123. For benzoic acid Eq. 11.123 is nevertheless a good approximation because [HA]$_0$ is sufficiently high to maintain its microenvironment within the diffusion layer by the self-buffering mechanism (Mooney et al., 1981a) discussed in Section 11.4.4. So, for strong bases [B]$_h$ will be significantly greater than [B]$_0$, thus validating Eq. 11.123.

The importance of the pH of the bulk solution has often been overemphasized in studies of dissolution rate. For the dissolution of weak acids in aqueous media, pH$_0$ will only equal pH$_{bulk}$ when acid dissolution is suppressed in the diffusion layer, when the acid is highly insoluble, or when a high bulk concentration of base swamps and controls completely the pH in the diffusion layer. On the other hand, the importance of the bulk concentration of the base and the basicity of the base on the dissolution rate of weak acids cannot be overemphasized. It is extremely important that the concentration and composition of the buffer medium be stated when recording the dissolution rates of weak acids. The pH of the medium is only one of the variables, and possibly a minor variable, controlling the dissolution rate (Mooney et al. 1981b).

Let us consider the consequences of (a) ignoring the various concentration gradients in the diffusion layer, with the exception of MA and A$^-$ (See Fig. 11.8), (b) calculating J from c_s using Eq. 11.19 and (c) calculating c_s from pH using Eq. 11.36. These assumptions are equivalent to the suppositions that the ionization of MA takes place at $x = 0$ next to the surface of the solid, and that $[B]_o = [B]_h$ and $[H^+]_o = [H^+]_h$. This situation is quite well approximated when 2-naphthoic acid (p$K_a = 4.02$) is dissolving in the presence of a relatively strong base such as sodium acetate buffer (p$K_a = 4.75$), as discussed in Section 11.5.2 and illustrated in Figs. 11.19 and 11.22 (steep curve). This state of affairs is not applicable to weaker bases such as imidazole (p$K_a = 6.91$) and morpholine (p$K_a = 8.70$), as discussed in Section 11.5.2 and illustrated in Figs. 11.20–11.22 (asymptotic curve). In these cases there are appreciable changes in the concentrations of all species within the diffusion layer shown in Fig. 11.18. Thus, the simple calculation of J from the c_s values calculated from pH is valid only under the special circumstances when the pK_a of the dissolved base is not significantly greater than the pK_a of the dissolving solid acid. This situation is somewhat analogous to the influence of a surface-active agent on the aqueous dissolution rate of a hydrophobic solid, when c_s can be simply calculated from the extent of micellar solubilization of the solid (Elworthy et al., 1968., Attwood and Florence, 1983).

11.5.3   Concentration Profiles Across the Diffusion Layer Calculated from the Six-Equilibria Theory

It is instructive to calculate the concentration of the various species at any point x in the diffusion layer, as has been done for the three-equilibria model (Fig. 11.17). In the case of $[H^+]$, Eqs. 11.101–11.103 may be solved simultaneously with Eqs. 11.79, 11.80, and 11.83. In this case a quartic equation in $[H^+]$ is obtained, thus,

$$a[H^+]^4 + b[H^+]^3 + c[H^+]^2 + e[H^+] + f = 0 \qquad (11.124)$$

where

$$a = D_H w \qquad (11.125)$$

$$b = w(L_2 - L_1) + D_H y \qquad (11.126)$$

$$c = D_B D_{HA} K_a^B (I_1 x + L_2 - L_3) - y(I_1 x + L_1) - D_{OH} K_w w + D_H z \qquad (11.127)$$

$$e = - D_{OH} K_w y - z(I_1 x + L_1 + L_3) \qquad (11.128)$$

$$f = - z K_w D_{OH} \qquad$$

in which

$$w = D_{HA} D_{BH} \qquad (11.129)$$

$$y = D_{HA}D_B K_a^B + D_A D_{BH} K_a^A \qquad (11.130)$$

$$z = D_A D_B K_a^A K_a^B \qquad (11.131)$$

Equation 11.124 may be solved for $[H^+]$ at given values of x using Newton's method (see Carnahan et al., 1969). Using the calculated value of $[H^+]$ at each value of x, the concentrations of the other species at any point x in the diffusion layer may be derived by solving Eqs. 11.101–11.103 using Eqs. 11.79, 11.80, and 11.83. The fractional concentrations are plotted against x/h, the fractional distance across the diffusion layer, as shown in Figure 11.24.

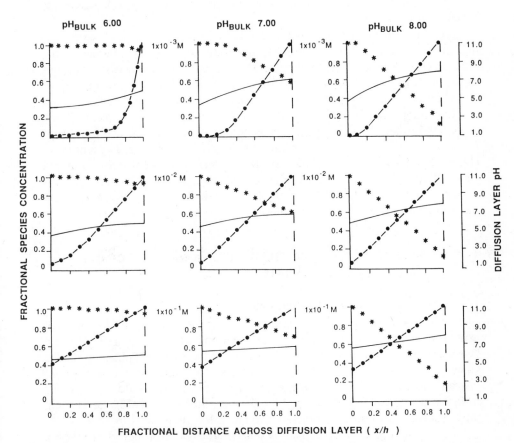

Figure 11.24 Idealized sections across the postulated diffusion layer adjacent to the surface of dissolving 2-naphthoic acid. They depict the fractional concentration profiles of the imidazole buffer species and diffusion layer pH as a function of the changing pH_{bulk} and total buffer concentration in the bulk solution. Key: *, imidazole conjugate acid BH^+; ●, imidazole base B; and (——), pH within the diffusion layer. $1\,M = 1\,mol\,L^{-1} = 1\,mol\,dm^{-3}$. Reproduced with permission of the copyright owner, the American Pharmaceutical Association, from Mooney, K.G., Mintun, M.A., Himmelstein, K.J., and Stella, V.J. (1981). *J. Pharm. Sci.*, **70**, 22–32.

Figure 11.24 shows idealized sections of the diffusion layer for the dissolution of 2-naphthoic acid in imidazole solutions of varying concentration and pH. The fractional distance x/h and concentrations of HA, A^-, H^+, and OH^- are as defined in Figure 11.17. The fractional concentration of BH^+ is evaluated as $[BH^+]/[BH^+]_o$ and that of B is evaluated as $[B]/[B]_h$. Only the B and BH^+ species and the pH are represented in Figure 11.24. As the total buffer concentration in the bulk solution is increased at all bulk pH values, the curvature of the pH profile becomes less pronounced as pH_o approaches pH_{bulk}.

Let us now compare the profiles obtained at pH_{bulk} 7.00 *with* imidazole buffer (Fig. 11.24, center column) with those obtained with 2-naphthoic acid at the same pH_{bulk} *without* a buffer (Fig. 11.17, center column, second row). We see that the pH within the diffusion layer shows a much less abrupt change in the buffered case.

Some investigations have suggested that a "plane of reaction" within the diffusion layer moves toward (or away from) the solid–liquid interface as the reactant concentration in the bulk solution is increased (or decreased) to account for an increase (or decrease) in the dissolution rate (King and Brodie, 1937; Hixon and Baum, 1944). When the hydroxide ion is the only base diffusing and reacting, the plane-of-reaction concept may be applied. For example, Mooney et al. (1981a) deduced the changing position of the point at which an abrupt pH shift occurs when $[OH^-]_h$ is varied (Fig. 11.17). However, Higuchi et al. (1958) and Mooney et al. (1981b) have found that the concept does not apply well to buffered solutions, since no abrupt plane of reaction is observed. This is because K_2 in Eq. 11.76 is generally lower than K'_1 in Eq. 11.75, since most bases are weaker than hydroxide ions.

11.6 POSSIBLE INFLUENCE OF THE RATE OF IONIZATION OF CARBON ACIDS ON DISSOLUTION RATE

11.6.1 Ionization Rates of Carbon Acids

The dissolution of carboxylic acids into aqueous basic media can be well explained by assuming that all reversible ionization reactions within the diffusion layer are instantaneous. However, carbon acids such as phenylbutazone (Scheme 11.1), in which the dissociating proton is bound to a carbon atom instead of an oxygen, nitrogen, or other heteroatom, do not ionize instantaneously (Stella and Pipkin, 1976; Jencks, 1969; Jones, 1972), that is, their rate constants for proton abstraction are much less than $10^{10} \, dm^3 \, mol^{-1} \, s^{-1}$.

The carbon acid considered here is phenylbutazone (Mooney et al., 1981c). The dissolution kinetics of this antiinflammatory drug appear to be central to its absorption in the body (see *J. Am. Pharm. Assoc.*, 1976, NS16, p. 365). As well as being of pharmaceutical interest, this drug displays transport kinetics which appeared not to be entirely explicable by models based on instantaneous ionization (Lovering and Black, 1974a, b; Stella, 1975).

Stella and Pipkin (1976) showed that in an aqueous solution of ionic strength

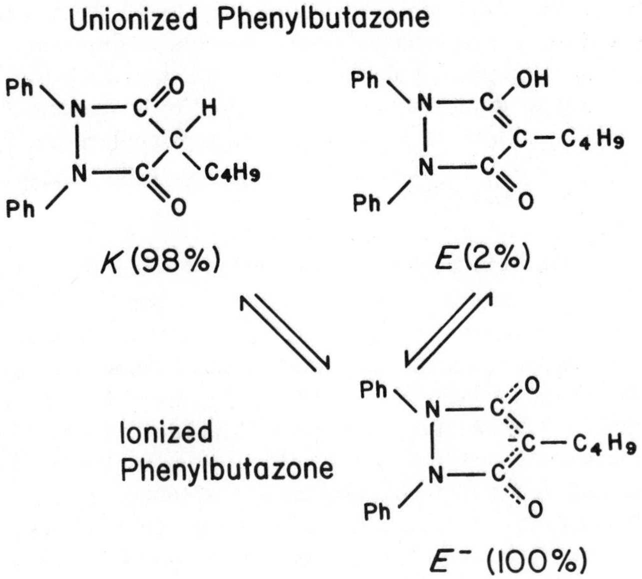

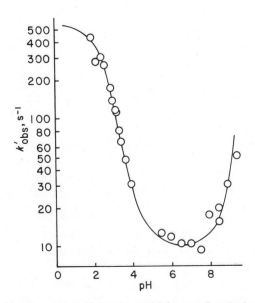

Unionized Phenylbutazone

K (98%) E (2%)

Ionized Phenylbutazone

E^- (100%)

Scheme 11.1 The ionization of phenylbutazone, a carbon acid, is not instantaneous.

Figure 11.25 Plot of $\log k'_{obs}$ versus pH for the establishment of an ionization equilibrium for phenylbutazone. Reproduced with permission of the copyright owner, the American Pharmaceutical Association, from Stella, V.J. and Pipkin, J.D. (1976). *J. Pharm. Sci.*, **65**, 1161–1165.

$0.1 \, \text{mol dm}^{-3}$ at 25°C, the enol form (E) of phenylbutazone contributes only about 1.8% of the total undissociated drug. Therefore, as shown in Scheme 11.1, it may be assumed that phenylbutazone exists predominantly in the diketo form (K), which undergoes noninstantaneous ionization to the enolate anion E^- as a result of reaction with bases. Figure 11.25 shows the influence of $[OH^-]$ and $[H^+]$ on the observed first-order rate constant of ionization k'_{obs} in the absence of other acids, bases, and buffers.

11.6.2   Dissolution Rate of Phenylbutazone in Aqueous Buffers

Mooney et al. (1981c) investigated the possible effects of such noninstantaneous reversible ionization on the dissolution rate of phenyl-butazone. They determined the dissolution rate of the solid drug using the rotating-disc method (Levich, 1962) as before. The dissolution medium was an aqueous solution of constant ionic strength $0.5 \, \text{mol dm}^{-3}$ (potassium chloride) and its pH was controlled by a pH-stat (Mooney et al., 1981a). Buffers were, however, omitted since they would prevent observation of any consequences of noninstantaneous ionization. Thus, the experimental conditions followed closely those described in Section 11.5. These conditions together with Scheme 11.1 indicate that the dissolution process involves the following equilibria:

$$K \underset{k_2}{\overset{k_1}{\rightleftharpoons}} H^+ + E^- \qquad (11.132)$$

diketo   enolate

$$K + OH^- \underset{k_4}{\overset{k_3}{\rightleftharpoons}} H_2O + E^- \qquad (11.133)$$

$$H^+ + OH^- \overset{K'_w}{\rightleftharpoons} H_2O \qquad (11.134)$$

At the steady state during the dissolution of phenylbutazone, a mass-balance equation for the rate processes of each diffusing species may be written. Each equation has the form of Eq. 11.45 and must account for both chemical ionization reactions and diffusion within the diffusion layer according to Fick's second law. Hence, the reaction rate function ϕ_S in the general equation (Eq. 11.45) for each species consists of mass-action terms for the formation and disappearance of the species in its various chemical reactions shown in Eqs. 11.132 and 11.133.

$$\frac{d[K]}{dt} = D_K \frac{d^2[K]}{dx^2} - \{[K](k_3[OH^-] + k_1) - [E^-](k_2[H^+] + k_4)\} = 0 \qquad (11.135)$$

$$\frac{d[E^-]}{dt} = D_E \frac{d^2[E^-]}{dx^2} - \{[E^-](k_2[H^+] + k_4) - [K](k_3[OH^-] + k_1)\} = 0 \qquad (11.136)$$

$$\frac{d[OH^-]}{dt} = D_{OH}\frac{d^2[OH^-]}{dx^2} - \{k_3[K][OH^-] - k_4[E^-]\} = 0 \qquad (11.137)$$

$$\frac{d[H^+]}{dt} = D_H\frac{d^2[H^+]}{dx^2} - \{k_2[E^-][H^+] - k_1[K]\} = 0 \qquad (11.138)$$

The variables k_1–k_4 are rate constants described in Eqs. 11.132 and 11.133 and the other symbols have their usual meanings. The equilibrium between $[H^+]$ and $[OH^-]$ is described by the usual ionic product K_w (Eqs. 11.58 or 11.79). However, the relationships between $[K]$ and $[E^-]$ and $[H^+]$ and $[OH^-]$ cannot be expressed by simple equilibria because the ionization reactions involved are not instantaneous.

Since the rates of interconversion between K and E^- at any point x in the film (i.e., diffusion layer) depend on their respective concentrations and $[H^+]$ and $[OH^-]$ at x, Eqs. 11.135–11.138 have no simple analytical solution. Therefore, to predict whether the noninstantaneous nature of the phenylbutazone ionization would affect the dissolution rate of the acid, numerical integration methods were used with Eqs. 11.135–11.138 to provide approximate film profiles for $[K]$ and pH_o at various bulk pH values. These methods have been outlined by Mooney et al. (1981c) and are beyond the scope of this book.

11.6.3 Effect of Deuteration of the Acidic Hydrogen Atom on Dissolution Rate

If noninstantaneous ionization of phenylbutazone influences the dissolution rate, substitution of the ionizable proton by deuterium should slow down the rates of ionization and the dissolution rate by the primary isotope effect. Accordingly, the dissolution behavior of phenylbutazone was compared with that of d-phenylbutazone. Two forms of 2-naphthoic acid, protonated and deuterated at the carboxyl group, were included as examples of instantaneously ionizing acids. Table 11.7 shows the relevant physical properties of these compounds.

All of these solid acids gave linear Levich (1962) plots of the observed values of J_t (J_{obs}) against $\omega^{1/2}$ in accordance with Eq. 11.68 and Figure 11.5, as indicated in Table 11.7. These linear plots show that the dissolution of both phenylbutazone and d-phenylbutazone is controlled by diffusion and, where possible, diffusivities were calculated from the slopes. The fact that the plots do not pass through the origin suggests that an additional mechanism, independent of rotation speed, such as natural convection (Section 11.2.4) is operating, as for indomethacin (Mooney et al., 1981a).

Table 11.7 shows no significant difference between the initial dissolution rates of 2-naphthoic acid and its deuterated analogue. Since the observed dissolution rates agree well with the theoretical predictions, the deuterated and nondeuterated forms of 2-naphthoic acid diffuse and react spontaneously with the base and to the same extent, as expected for a carboxylic acid. On the other hand, d-phenylbutzone dissolves appreciably more slowly than phenylbutazone

Table 11.7 Solubilities $[HA]_o$, pK_a Values, Dissolution Data (J_{obs} versus $\omega^{1/2}$ plots), and Derived Diffusivities D of Phenylbutazone, 2-Naphthoic Acid, and its Deuterated Analogues in Water at Ionic Strength 0.5 mol dm^{-3} (Potassium Chloride) at 25°C[a]

| | | | | J_{obs} versus $\omega^{1/2}$ Plots[e] | | | | |
| | | | | At pH$_{bulk}$ = 2.00 | | | At pH$_{bulk}$ = 6.50 | |
Dissolving Acid	M[b]	$[HA]_o$ (μmol dm^{-3})	pK_a	Slope × 10^6 (mg cm^{-2} s$^{-1/2}$ rad$^{-1/2}$)	Intercept × 10^5 (mg cm^{-2} s^{-1})	D × 10^6 (cm^2 s^{-1})	Slope × 10^6 (mg cm^{-2} s$^{-1/2}$ rad$^{-1/2}$)	Intercept × 10^5 (mg cm^{-2} s^{-1})
2-Naphthoic acid	172.2	130[c]	4.02[c]	10.44	0.01	6.52		
d-2-Naphthoic acid	173.2			10.30	0.01	6.39		
Phenylbutazone	308.4	26.1[c]	4.61[c], 4.54[d]	3.07	0.10	6.48	7.59	1.34
d-Phenylbutazone	309.4						5.46	1.23

[a] 1 mol dm^{-3} = 1 mol L^{-1} = 1M.

[b] Molecular weight.

[c] Determined by the solubility technique using plots (Figure 11.6) according to Eq. 11.36 (Mooney et al., 1981c).

[d] Determined by the UV spectrophotometric method.

[e] Plotted according to Eq. 11.68 ($J_{obs} = J_t$) as shown in Figure 11.5 (Mooney et al., 1981c).

Reproduced with permission of the copyright owner, the American Pharmaceutical Association from Mooney, K.G., Rodriguez-Gaxiola, M., Mintun, M., Himmelstein, K.J., and Stella, V. (1981c). J. Pharm. Sci., 70, 1358–1365 and therefore displays the primary isotope effect (Table 11.7), as expected for the ionization of a carbon acid.

The observed initial dissolution rate of phenylbutazone at various bulk pH values is compared in Figure 11.26 with the dissolution rate predicted assuming that this acid ionizes spontaneously in the diffusion layer according to Eq. 11.64. This assumption is equivalent to the supposition that phenylbutazone is behaving as a carboxylic acid. Agreement is good in the case of phenylbutazone, suggesting that its rate of ionization is not slow enough compared with the rate of diffusion to influence the dissolution rate. However, the experimental dissolution rate of d-phenylbutazone is significantly slower than the prediction (Fig. 11.26), suggesting that the primary isotope effect has slowed down

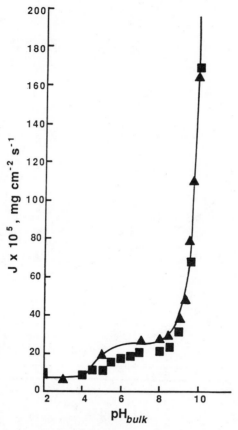

Figure 11.26 Dissolution rate J versus pH_{bulk} profile for phenylbutazone (▲) and d-phenylbutazone (■) from a rotating disc (600 rpm) dissolving in an aqueous medium (25°C at $\mu = 0.5\,mol\,dm^{-3}$ with potassium chloride). The pH_{bulk} was maintained with a pH-stat. The continuous line represents the profile generated when phenylbutazone is treated as a carboxylic acid, which ionizes rapidly (Eq. 11.64). $1\,mol\,dm^{-3} = 1\,mol\,L^{-1} = 1M$. Reproduced with permission of the copyright owner, the American Pharmaceutical Association, from Mooney, K.G., Rodriguez-Gaxiola, M., Mintun, M., Himmelstein, K.J., and Stella, V.J. (1981). *J. Pharm. Sci.*, **70**, 1358–1365.

ionization sufficiently in comparison with the rate of diffusion to influence the dissolution rate.

To understand these effects more clearly, it is useful to compare the average lifetime of a diffusing molecule in the diffusion layer, known as residence time t_D, with the mean life of a molecule in a first-order reaction, known as the

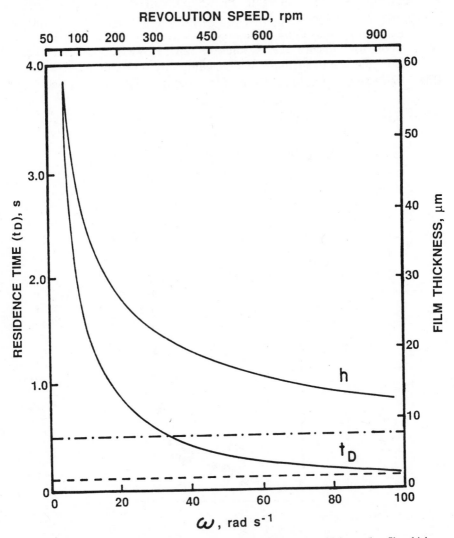

Figure 11.27 Plot of residence time t_D and apparent diffusion-layer thickness (i.e., film thickness h) for phenylbutazone (as defined by Eqs. 11.21 and 11.140) versus ω (angular velocity) for dissolution from a rotating disc. The dashed lines correspond to three half-lives ($3t_R \ln 2 = 2.08t_R$) for the deprotonation reaction (at pH_{bulk} 6.5) of phenylbutazone (———) and d-phenylbutazone ·—·—). Reproduced with permission of the copyright owner, the American Pharmaceutical Association, from Mooney, K.G., Rodriguez-Gaxiola, M., Mintun, M., Himmelstein, K.J., and Stella, V.J. (1981). *J. Pharm. Sci.*, **70**, 1358–1365.

reaction time t_R. Using the Nernst–Brunner diffusion-layer model, Higuchi et al. (1971) derived the following expression for t_D:

$$t_D = h^2/2D_K \qquad (11.139)$$

For the Levich (1962) rotating disc, h is given by Eq. 11.21, therefore,

$$t_D = 1.3v^{1/3}/D_K^{1/3}\omega \qquad (11.140)$$

Figure 11.27 shows plots of both t_D and h against the rotation speed of the disc, expressed as ω and in rpm. High rotation speeds cause marked reductions in t_D and h. The reaction time t_R is defined by Astarita (1967) as the reciprocal of the first-order rate constant k, thus,

$$t_R = 1/k \qquad (11.141)$$

Now t_R is equal to (half-life/ln 2) and is the time taken to complete $(1 - 1/e) \times 100\% = 63\%$ of the reaction by a first-order process. The rate constant k for the ionization of phenylbutazone was calculated from the data of Stella and Pipkin (1976) at the pH of interest ($pH_{bulk} = 6.5$ in Fig. 11.27) and t_R was obtained.

If t_R is much larger than t_D, under defined hydrodynamic conditions, the ionization reaction will only take place in the bulk solution. This might apply to d-phenylbutazone (Fig. 11.27, upper dashed line) at very high rotational speeds ($> 100\,\mathrm{rad\,s}^{-1}$ or $> 1000\,\mathrm{rpm}$). In fact, a rotational speed of 600 rpm ($62.83\,\mathrm{rad\,s}^{-1}$) was used to measure the dissolution rates of phenylbutazone and d-phenylbutazone. If t_R and t_D have similar orders of magnitude, the ionization reaction would influence the flux according to Eqs. 11.135–11.138. This situation evidently applies to d-phenylbutazone at angular velocities of 40–100 rad s^{-1} (rotational speeds of 400–1000 rpm) at bulk pH 5–8 (e.g., Figs. 11.26 and 11.27, upper dashed line). This situation might also apply to phenylbutazone itself at high rotational speeds ($> 60\,\mathrm{rad\,s}^{-1}$ or $> 600\,\mathrm{rpm}$; Fig. 11.27, lower dashed line).

When t_R is much shorter than t_D, dissolution will exhibit classical diffusion control, as for 2-naphthoic acid (Mooney et al., 1981a), which is discussed in Sections 11.4.3 and 11.4.4. Phenylbutazone dissolution under the conditions described by Mooney et al. (1981c) appears to just fall into this category (Figs. 11.26 and 11.27, lower dashed line). At a bulk pH greater than 9, however, even d-phenylbutazone dissolves by the classical diffusion-controlled mechanism (Mooney et al., 1981c) because the high concentration of hydroxide ion increases the rate of ionization of the deuterated carbon acid (Fig 11.25).

To summarize, it is unlikely that the noninstantaneous ionization kinetics demonstrated for phenylbutazone play a major role in determining the dissolution rate. This is because the average residence time in a typical aqueous diffusion layer for phenylbutazone dissolution is longer than the reaction time for its ionization. However, increasing the reaction time by the primary isotope effect using deuterium substitution for the ionizable proton caused a significant deviation of *d*-phenylbutazone from the classical behavior shown by phenylbutazone.

11.7   INFLUENCE OF MOLECULAR COMPLEXATION ON DISSOLUTION RATE

11.7.1   Development of a Theory

In Chapters 4 and 5 we have seen that molecular complexation can greatly influence solubility. Therefore, according to Eq. 11.10, complexation is expected also to affect the rates of dissolution of solids. Olander (1960) applied complexation equilibria (e.g. Eq. 4.5) to the Nernst—Brunner (1904) model for diffusion-controlled dissolution (Eq. 11.19).

If the dissolved solute is represented by S, the rate of dissolution in the absence of the complexing agent, $(dm/dt)_o$, can be expressed by Eq. 11.19 in the form

$$(dm/dt)_o = (D_S A/h)([S]_o - [S]_h) \tag{11.142}$$

where $[S]_o$ represents the solubility of the solid in the absence of complexing agent, $[S]_h$ is the concentration of the dissolved solute in the bulk solution and the other symbols have their usual meanings. The complexation equilibrium may be written

$$S + L \underset{}{\overset{K}{\rightleftharpoons}} SL \tag{11.143}$$

where

$$[SL] = K[S][L] \tag{11.144}$$

and K is the stability constant of the complex SL. The treatment of Olander (1960) makes the following assumptions. (a) The diffusion of species S, L, and SL can be represented by Fick's first law (Eq. 11.17). (b) The chemical equilibria are rapid compared with the transport process and are diffusion-controlled. (c) For given hydrodynamic conditions of dissolution, the thickness of the diffusion layer, h, is independent of diffusivities, D, which are themselves constant, and independent of concentrations. (d) A saturated ·concentration $[S]_o$ is maintained at the solid–liquid interface. (e) No concentration gradients exist in the bulk of the solution, that is, agitation is adequate.

From these assumptions Olander (1960) derived the following equation for the dissolution rate in the presence of complexing agent L:

$$\frac{dm}{dt} = \frac{A}{h} \{ D_S([S]_o - [S]_h) + D_{SL}([SL]_o - [SL]_h) \} \tag{11.145}$$

$$= \frac{A}{h} \{ D_S([S]_o - [S]_h) + D_L([L]_h - [L]_o) \} \tag{11.146}$$

where $[SL]_o$, $[L]_o$, $[SL]_h$, and $[L]_h$ are the concentrations of the complex and ligand at the solid–liquid interface and in the bulk solution, respectively. Expressions for $[SL]_o$ and $[L]_o$ are now required. From Eqs. 11.145 and 11.146 we see that

$$D_{SL}([SL]_o - [SL]_h) = D_L([L]_h - [L]_o) \qquad (11.147)$$

Applying Eq. 11.144 to the solid–liquid interface,

$$[SL]_o = K[S]_o[L]_o \qquad (11.148)$$

Elimination of $[L]_o$ from the last two equations gives

$$[SL]_o - [SL]_h = \frac{D_L}{D_{SL}}\left([L]_h - \frac{[SL]_o}{K[S]_o}\right) \qquad (11.149)$$

Therefore,

$$[SL]_o = \frac{K[S]_o(D_L[L]_h + D_{SL}[SL]_h)}{K[S]_oD_{SL} + D_L} \qquad (11.150)$$

Eliminating $[SL]_o$ from Eqs. 11.150 and 11.148,

$$[L]_o = \frac{D_L[L]_h + D_{SL}[SL]_h}{K[S_o]D_{SL} + D_L} \qquad (11.151)$$

In the presence of the complexing agent L, the total concentration of S in the bulk solution is given by

$$[S]_t = [S]_h + [SL]_h \qquad (11.152)$$

Eliminating $[S]_h$ and $[SL]_h$ from Eqs. 11.145, 11.150, and 11.152 we obtain

$$\frac{dm}{dt} = \frac{A}{h}\left\{D_S[S]_o + D_{SL}[SL]_o - [S]_t\left(\frac{D_S + D_{SL}K[L]_h}{1 + K[L]_h}\right)\right\} \qquad (11.153)$$

In the absence of the complexing agent L, $[L]_h = 0$, $[SL]_o = 0$, and $[S]_t = [S]_h$. Equation 11.143 may be written

$$\left(\frac{dm}{dt}\right)_o = \frac{D_SA}{h}([S]_o - [S]_t) \qquad (11.154)$$

Dividing Eq. 11.153 by Eq. 11.154,

$$\frac{dm}{dt}\bigg/\left(\frac{dm}{dt}\right)_o = [S]_o + \frac{D_{SL}}{D_S}[SL]_o - [S]_t\left(\frac{1 + K[L]_hD_{SL}/D_S}{1 + K[L]_h}\right)\bigg/([S]_o - [S]_t) \qquad (11.155)$$

11.7.2 Experimental Test of the Theory Using a Concentration-Jump Technique

This model was tested by Higuchi et al. (1972) using a batch method employing a round-bottom flask, a stirrer, and a compressed tablet of 2-naphthol. The initial dissolution rate was calculated at $t = 0$ by treating t as an approximately quadratic (i.e., parabolic function of c_b (Fig. 11.2A). For this purpose c_b was measured sepctrophotometrically at a series of times. Higuchi et al. (1972) derived Eq. 11.155 to calculate the ratio of the dissolution rate of 2-naphthol S, in cyclohexane plus ligand L, to that in cyclohexane alone. As discussed in Chapters 4 and 5, cyclohexane is a paraffin and is therefore an inert solvent, while L is an interactive cosolvent (i.e., a complexing agent). The calculated ratios were compared with the experimental values determined under constant conditions of (1) constant area A of the solid–liquid interface, (2) constant hydrodynamics of agitation (i.e., constant h), and (3) constant concentration of dissolved solute ($4 \times 10^{-4} \, \text{mol dm}^{-3}$).

To achieve these objectives the dissolution rate dm/dt, was determined by a "concentration-jump technique" (Higuchi et al., 1972). This method involved making a series of measurements of intrinsic dissolution rate immediately before

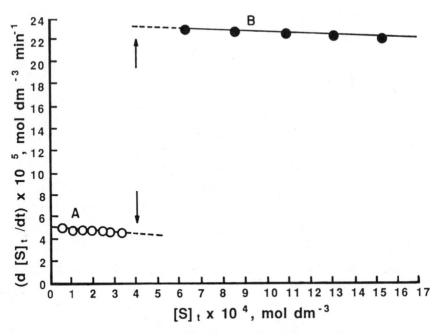

Figure 11.28 Plots of concentration of dissolved 2-naphthol [S]$_t$ against the rate of dissolution of a tablet of 2-naphthol ($332 \pm 5 \, \text{mg}$) in (A) cyclohexane and (B) cyclohexane containing $5 \times 10^{-2} \, \text{mol dm}^{-3} \, N, N$-dimethyldodecamide. The results were obtained from the same experiment at 25°C with a stirring rate of $100 \pm 2 \, \text{rpm}$. $1 \, \text{mol dm}^{-3} = 1 \, \text{mol L}^{-1} = 1 \, M$. Reproduced with permission of the copyright owner, the American Pharmaceutical Association, from Higuchi, T., Dayal, S., and Pitman, I.H. (1972). *J. Pharm. Sci.*, **61**, 695–700.

and immediately after a known amount of ligand L was rapidly injected into a flask in which a compressed tablet of the solid, in this case 2-naphthol, was dissolving at a steady rate. Figure 11.28 shows a typical plot of dissolution rate against the concentration of dissolved 2-naphthol. A short forward extrapolation of the curve for dissolution in pure cyclohexane and a short back extrapolation of that for dissolution in the mixed-solvent system are both made to a constant defined value of $[S]_t$.

As mentioned above, Higuchi et al. (1972) chose $[S]_t = 4 \times 10^{-4}\,\text{mol dm}^{-3}$ for 2-naphthol. In this way the dissolution rate in the presence and absence of the complexing agent L can be accurately estimated from Figure 11.28, when the same concentration of solid $[S]_t$ had dissolved in the same experimental run. The ratio of these rates is therefore determined under conditions of constant interfacial area A and constant solution hydrodynamics (i.e., constant h), both of which cancel in Eq. 11.155. Consequently, the ratio of dissolution rates will be independent of mechanical factors. Furthermore, the viscosities of the various solutions were found to be constant.

Equation 11.155 for the theoretical prediction of the dissolution rate contains various accessible physical quantities. Values of K, shown in Table 11.8, can be determined as described in Section 5.3. Values of diffusivities D appear as ratios which were calculated from the molecular weights, as shown in Table 11.9, assuming that $D \propto M^{-1/2}$, thus,

$$D_X/D_S = (M_S/M_X)^{1/2} \tag{11.156}$$

where D_X is the diffusivity of molecule X of molecular weight M_X and D_S is the diffusivity of the reference compound 2-naphthol of molecular weight $M_S = 144.2$. $[S]_t$, the total concentration of 2-naphthol, is $4 \times 10^{-4}\,\text{mol dm}^{-3}$. $[S]_0$ is the solubility of 2-naphthol in the absence of complexing agent $(1.099 \times 10^{-2}\,\text{mol dm}^{-3}$ in cyclohexane at 25°C). $[L]_h$ was taken to be the concentration of the added complexing agent, since it was always well in excess of $[S]_t$. $[SL]_0$ was calculated from Eq. 11.150.

The dissolution rates predicted theoretically from Eq. 11.155 are compared

Table 11.8   Stability Constants at 25°C for 1:1 Complexes of 2-Naphthol with Various Ligands in Cyclohexane[a]

Ligand (L)	Stability Constant K (dm⁻³ mol⁻¹)	Method
1-Propanol	33.8	Solubility
Undecanol	44.4	Solubility
N, N-Dimethylpropionamide	404.9	UV spectrophotometry
N, N-Dimethyldodecamide	448.8	UV spectrophotometry

[a] $1\,\text{dm}^3\,\text{mol}^{-1} = 1\,\text{L mol}^{-1} = 1\,M^{-1}$.

Reproduced with permission of the copyright owner, the American Pharmaceutical Association, from Higuchi, T. Dayal, S., and Pitman, I.H. (1972). *J. Pharm. Sci.*, **61**, 695–700.

Table 11.9 Ratio of the Diffusivities of Various Molecules to the Diffusivity of 2-Naphthol

Molecule (X)	Molecular Weight	D_X/D_A
2-Naphthol	144.2	1.00
1-Propanol	60.1	1.55
Undecanol	172.0	0.92
N,N-Dimethylpropionamide	101.2	1.19
N,N-Dimethyldodecamide	227.4	0.80
1:1 Complex of 1-propanol with 2-naphthol	204.3	0.84
1:1 Complex of undecanol with 2-naphthol	316.2	0.68
1:1 Complex of N,N-dimethylpropionamide with 2-naphthol	245.3	0.77
1:1 Complex of N,N-dimethyldodecamide with 2-naphthol	371.6	0.62

Reproduced with permission of the copyright owner, the American Pharmaceutical Association, from Higuchi, T., Dayal, S., and Pitman, I.H. (1972). *J. Pharm. Sci.*, **61**, 695–700.

Table 11.10 Ratios of Rate of Dissolution of 2-Naphthol in Mixed Solvents/Rate of Dissolution of 2-Naphthol in Cyclohexane when 4×10^{-4} mol dm^{-3} of 2-Naphthol has Dissolved[a]

Cosolvent in Cyclohexane	Ratio	
	Experimental	Theory
7.5×10^{-2} mol dm^{-3} 1-propanol	2.9	3.0
7.5×10^{-2} mol dm^{-3} undecanol	2.7	2.7
1×10^{-2} mol dm^{-3} N,N-dimethylpropionamide	1.9	2.0
1×10^{-2} mol dm^{-3} N,N-dimethyldodecamide	1.2	1.1
5×10^{-2} mol dm^{-3} N,N-dimethylpropionamide	5.2	7.1
5×10^{-2} mol dm^{-3} N,N-dimethyldodecamide	4.0	5.2

[a] 1 mol dm$^{-3} = 1$ mol L$^{-1} = 1\,M$.

Reproduced with permission of the copyright owner, the American Pharmaceutical Association, from Higuchi, T., Dayal, S., and Pitman, I.H. (1972). *J. Pharm. Sci.*, **61**, 695–700.

with the experimental values in Table 11.10. In general, there is good agreement between the predicted and observed dissolution rates, which supports the proposed complexation model. However, in the presence of higher concentrations of N,N-dimethylpropionamide or N,N-dimethyldodecamide (5×10^{-2} mol dm^{-3}), the discrepancies suggest that complexes of higher order than 1:1 are formed. The above theoretical treatment was based on the assumption that only 1:1 complexes are formed, which is valid only at low concentrations of substrate and ligand, as discussed in Sections 4.3 and 5.2.

The importance of the diffusivity of L in determining dissolution rates is emphasized in Table 11.10. 2-Naphthol dissolves faster in the presence of 1-propanol than in undecanol, even though it forms more stable complexes with the latter additive (Table 11.8). The dissolution rate is enhanced by the higher diffusivity of the smaller ligand (Table 11.9). A similar conclusion can be drawn by comparing the dissolution rates in solutions containing the two amides (Table 11.10). Thus, while the stability constants are similar (Table 11.8), dissolution is faster in the presence of the smaller additive because of its larger diffusivity (Table 11.9). The diffusivity of solutes is also influenced by the viscosity of the solution, as indicated by Eq. 11.29. However, in the system discussed the viscosities were found to be constant.

11.8 INFLUENCE OF AN IRREVERSIBLE REACTION ON DISSOLUTION RATE

11.8.1 Choice of a Model System and Its Experimental Study

In Section 11.6 the influence of a reversible reaction (ionization) on the dissolution rate of a solid was considered. For reactive solids which undergo a rapid irreversible reaction, the dissolution rate may also be affected. This is illustrated by the dissolution of solid 7-acetyltheophylline, which is rapidly and irreversibly hydrolyzed in aqueous solutions according to Scheme 11.2. The hydrolysis of this amide, A, is greatly facilitated by the quasi-aromatic character

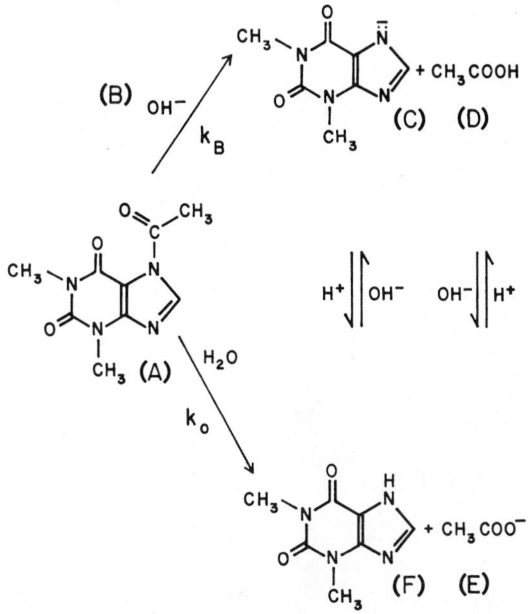

Scheme 11.2 The hydrolysis of 7-acetyltheophylline (A) is catalyzed by hydroxyl ions (B) leading to theophylline (F) and acetate ions (E). The theophylline anion (C) and acetic acid (D) are postulated intermediates. Reproduced with permission of the copyright owner. Higuchi, T., Lee, H.K., and Pitman, I.H. (1971). *Farm. Aikak.*, **80**, 55–90.

of the xanthine ring which causes the theophylline anion C to be an excellent leaving group.

The hydrolysis rate of dissolved 7-acetyltheophylline was determined spectrophotometrically by Higuchi et al. (1971a) in the presence of various buffer solutions at various pH values. The apparent first-order hydrolytic rate constants k_{app}, at various concentrations of buffer and hydroxyl ion, show that H^+ and OH^- per se exert negligible catalytic effect at pH $\leqslant$ 5.0, while at higher pH values the following relationship is obeyed:

$$k_{app} = k_0 + k_{OH^-}[OH^-] \qquad (11.157)$$

where $k_0 = 1.16 \times 10^{-2}\,s^{-1}$ and $k_{OH^-} = 1.13 \times 10^5\,dm^3\,mol^{-1}\,s^{-1}$ in aqueous

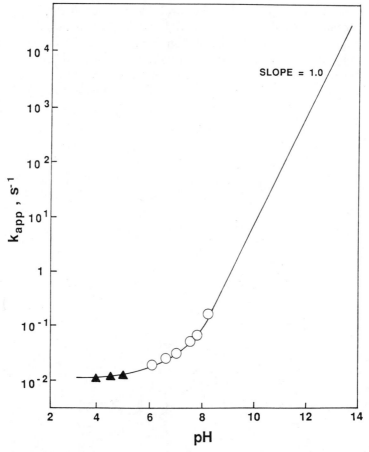

Figure 11.29 Influence of pH on the apparent first-order rate constant k_{app} (log scale); values for zero buffer concentration ($\blacktriangle$); values at ionic strength 0.1 mol dm^{-3} ($\bullet$); the line shown is based on the value predicted from Eq. 11.157. 1 mol dm^{-3} = 1 mol L^{-1} = 1M. Reproduced with permission of the copyright owner, from Higuchi, T., Lee, H.K., and Pitman, I.H. (1971). *Farm. Aikak.*, **80**, 55–90.

solution of sodium chloride at ionic strength $0.1 \, \text{mol dm}^{-3}$ and at 25°C. Thus, the rate of hydrolysis increases rapidly with pH above 8.0, as shown in Figure 11.29.

We would expect that this reaction will increase the dissolution rate of a solid only when the dissolved solute is undergoing a chemical reaction in the diffusion layer. Furthermore, an irreversible reaction will be more favorable to dissolution than a reversible reaction, because the accumulated solute in the diffusion layer is regenerated by the back reaction only in the reversible case (see Section 11.6).

Higuchi et al. (1971a) studied the dissolution kinetics of solid 7-acetyltheophylline in buffer solutions at various pH values both experimentally and theoretically. The experimental dissolution rates of single compressed tablets of 7-acetyltheophylline (200 mg, $2.64 \, \text{cm}^2$) into $100 \, \text{cm}^3$ of buffer solution in a 250-cm^3 two-necked flask at 25°C were determined under sink conditions by a batch method using a semicircular stirrer blade rotating at 76 rpm (see Fig. 11.1A).

The dissolved solute was determined as a function of time by UV spectrophotometry. The theoretical analysis was partially guided by earlier work on the dissolution of gases in reactive liquids (Van Krevelen and Van Hoftijzer, 1948; Van Krevelen and Van Hooren, 1948) and by the developments of Danckwerts (1950), Sherwood and Pigford (1952), and Higuchi et al. (1958).

11.8.2 Theory of the Influence of an Irreversible Reaction on Dissolution Rate

Scheme 11.2 may be represented by the general equations

$$A + B \xrightarrow{\text{slow step}} C + D \tag{11.158}$$

$$D + (n-1)B \rightleftharpoons E \tag{11.159}$$

Since the pK_a of the reactant, a xanthine, is approximately 8, $n = 1$ below $pH = 8$ and $n = 2$ above $pH = 8$. Let A be the concentration of 7-acetyltheophylline, the reactant, at a distance x within the diffusion layer of thickness h (see Fig. 11.3B), t the time, k_B the second-order rate constant for reaction between solute A and component B, and D_A the diffusivity of the reactant in the medium. The initial concentration of A in the bulk solution is taken to be uniformly equal to zero. The rate at which A crosses unit area of any plane of constant x within the diffusion layer in the direction of increasing x is

$$(dQ/dt)_x = -D_A(\partial A/\partial x) \tag{11.160}$$

and the rate at which the chemical reaction removes the solute (per unit volume) is $(k_0 + k_B B)A$.

If the chemical reaction is fast enough, it takes place in a thin zone within the diffusion layer. The amounts of A diffusing in, diffusing out, and reacting in a volume element of unit cross-section area between planes x and $x + dx$ (see Fig. 11.3B) in a time interval dt are as follows:

$$\text{amount diffusing in} = -D_A(\partial A/\partial x)\,dt \tag{11.161}$$

$$\text{amount diffusing out} = -\left[-D_A\left(\frac{\partial A}{\partial x} + \frac{\partial^2 A}{\partial x^2}\,dx\right)dt\right] \tag{11.162}$$

$$\text{amount reacting} = -(k_0 A + k_B AB)\,dx\,dt \tag{11.163}$$

If the resulting net change in concentration is dA, the net increase resulting from the last three equations can be equated with $dA\,dx$ giving

$$dA/dt = D_A(d^2 A/dx^2) - (k_0 + k_B B)A \tag{11.164}$$

For component B, the analogous differential equation in the diffusion layer, $0 < x < h$, will be

$$\frac{dB}{dt} = D_B(d^2 B/dx^2) - n(k_0 + k_B B)A \tag{11.165}$$

where D_B is the diffusivity of B. In the steady state,

$$dA/dt = (1/n)(dB/dt) = 0 \tag{11.166}$$

By combining the last three equations we obtain

$$D_A(d^2 A/dx^2) = (D_B/n)(d^2 B/dx^2) = (k_0 + k_B B)A \tag{11.167}$$

A general solution of the last equation by an exact mathematical method is not possible. However, solutions can be found for specific cases by numerical integration (Higuchi et al., 1971a). Furthermore, the left-hand side may be integrated thus:

$$D_A(dA/dx) + C_1 = (1/n)D_B(dB/dx) \tag{11.168}$$

where C_1 is the integration constant. Mass balance shows that

$$C_1 = -D_A\left(\frac{dA}{dx}\right)_{x=h} + \frac{1}{n}D_B\left(\frac{dB}{dx}\right)_{x=h} = -D_A\left(\frac{dA}{dx}\right)_{x=h} \tag{11.169}$$

Thus C_1 is equal to J_A, the rate of dissolution of A per unit surface area.

Integration of Eq. 11.168 affords

$$D_A A + C_1 x + C_2 = (1/n) D_B B \qquad (11.170)$$

where C_2 is the integration constant which is evaluated from the following boundary conditions across the diffusion layer.

At the interface, where $x = 0$, let $A = A_0$ and $B = B_0$; then

$$C_2 = (1/n) D_B B_0 - D_A A_0 \qquad (11.171)$$

At the bulk solution, where $x = h$, let $A = A_h$ and $B = B_h$; then

$$C_2 = (1/n) D_B B_h - D_A A_h - C_1 h \qquad (11.172)$$

Substituting J_A for C_1, the last two equations yield

$$J_A = -D_A \left(\frac{dA}{dx} \right)_{x=0} = \frac{D_A}{h}(A_0 - A_h) + \frac{1}{nh} D_B(B_h - B_0) \qquad (11.173)$$

Unfortunately, B_0, the concentration of hydroxide ion at the solid–liquid interface, is unknown, so J_A cannot be determined from the last equation.

For the special case in which the concentration of the liquid-phase reactant B remains constant within the diffusion layer (i.e., $B = B_h$), Eq. 11.167 may be readily solved, leading to

$$J_A = -D_A \left(\frac{dA}{dx} \right)_{x=0} = \frac{D_A}{h} \left[A_0 \frac{X}{\tanh X} - A_h \frac{X}{\sinh X} \right] \qquad (11.174)$$

where $X = [(k_0 + k_B B_h)/D_A]^{1/2}$.

When the reaction is so slow that it proceeds entirely within the bulk solution, diffusion through the liquid film is rate-determining; then $\sinh X \to X$ and $\tanh X \to X$ and the last equation reduces to

$$J_A = (D_A/h)(A_0 - A_h) \qquad (11.175)$$

which is analogous to Eq. 11.19 of the simple diffusion-layer theory of Nernst (1904) and Brunner (1904).

When the reaction is so fast that the reaction proceeds completely within the diffusion layer, $A_h = 0$, $\sinh X \to 1$, and $\tanh X \to 1$, so Eq. 11.174 reduces to

$$J_A = A_0 [(k_0 + k_B B_h) D_A]^{1/2} \qquad (11.176)$$

11.8.3   Test of the Theory Using Numerical Integration

For the purposes of numerical integration of Eq. 11.167, the following values of the various relevant physicochemical quantities were derived:

$D_A = 4.91 \times 10^{-6}\, \text{cm}^2\, \text{s}^{-1}$ (from the Stokes–Einstein Eq. 11.29)

$D_B = 2.92 \times 10^{-5}\, \text{cm}^2\, \text{s}^{-1}$ (from Higuchi et al., 1958)

$k_0 = 1.16 \times 10^{-2}\, \text{s}^{-1}$ (in Eq. 11.157)

$k_B = k_{OH^-} = 1.13 \times 10^{-8}\, \text{cm}^3\, \text{mol}^{-1}\, \text{s}^{-1}$ (in Eq. 11.157)

$h = 5.5 \times 10^{-3}\, \text{cm}$ (from the conditions of dissolution)

$A_0 = 1.27 \times 10^{-5}\, \text{mol cm}^{-3}$ (solubility, c_s, of A, calculated from the measured J_A at pH $= 4$ using Eq. 11.19)

For diffusion control under sink conditions, Eq. 11.175 gives $J_A = D_A A_0/h = 1.14 \times 10^{-8}\, \text{mol cm}^{-2}\, \text{s}^{-1}$; this situation will apply when the $[OH^-]$ is so low that the second term on the right in Eq. 11.173 can be ignored (e.g., at pH $= 4$).

In order to achieve numerical integration of Eq. 11.167, the variables are transformed to the following dimensionless forms:

$$\alpha = A/A_0, \qquad \beta = B/B_h, \qquad \xi = x/h \qquad (11.177)$$

α and β represent the dimensionless relative concentrations of A and B at a given point x in the diffusion layer with respect to that at the solid–liquid

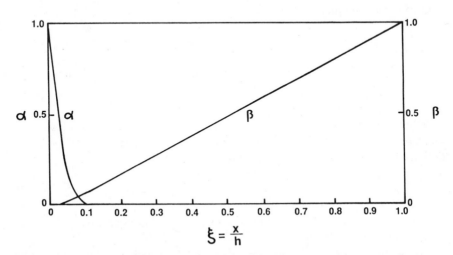

Figure 11.30 Dimensionless concentration profiles of 7-acetyltheophylline, α, and hydroxyl ions, β, plotted as functions of the dimensionless distance ξ throughout the diffusion layer at pH 13.0 and at 25°C. The plots were predicted by numerical integration of Eq. 11.167. Reproduced with permission of the copyright owner, from Higuchi, T., Lee, H.K., and Pitman, I.H. (1971). *Farm. Aikak.*, **80**, 55–90.

interface and the bulk solution, respectively. ξ represents the dimensionless distance within the diffusion layer with respect to the total thickness.

The numerical technique employed by Higuchi et al. (1971a) was a MIMIC program for solving ordinary differential equations on an IBM 7090 (7094) computer with a Fortran IV IBSYS Monitor. Figure 11.30 shows the calculated values of α and β throughout the diffusion layer at pH = 13.0. The relative

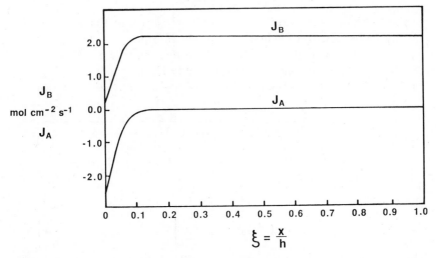

Figure 11.31 The flux of 7-acetyltheophylline J_A and of hydroxyl ions J_B plotted as functions of the dimensionless distance ξ throughout the diffusion layer at pH = 13.0 and at 25°C. The plots were predicted by numerical integration of Eq. 11.167. Reproduced with permission of the copyright owner, from Higuchi, T., Lee, H.K., and Pitman, I.H. (1971). *Farm. Aikak.*, **80**, 55–90.

Table 11.11   Reaction Times $t_R = 1/k_{app} = 1/k_o + k_{OH^-}[OH^-])^a$ for Hydrolysis of 7-Acetyltheophylline Under Various pH Conditions

pH	$[OH^-](mol\,cm^{-3})$	$t_R(s)^b$	$t_D(s)^c$
4.0	1.0×10^{-13}	8.6×10^1	3
6.0	1.0×10^{-11}	7.9×10^1	3
8.0	1.0×10^{-9}	8.0×10^0	3
10.0	1.0×10^{-7}	8.8×10^{-2}	3
12.0	1.0×10^{-5}	8.8×10^{-4}	3
13.0	1.0×10^{-4}	8.8×10^{-5}	3
13.5	3.2×10^{-4}	2.8×10^{-5}	3

a According to Eqs. 11.141 and 11.157.
b Reaction times were estimated under the conditions stated in the text.
c Diffusion time $t_D = h^2/2D_A = 3s$, independent of pH (Eq. 11.139).

Reproduced with permission of the copyright owner, from Higuchi T., Lee, H.K., and Pitman, I.H. (1971). *Farm. Aikak.*, **80**, 55–90.

concentration of A falls off rapidly within the first $\frac{1}{10}$th of the diffusion layer; this decrease becomes even steeper (not shown) at a higher pH (e.g., 13.5). The relative concentration of B (i.e., β) increases almost linearly across the diffusion layer and intersects the ordinate above the origin at pH = 13.5. The fluxes of A and B, respectively, that is,

$$J_A = -D_A(dA/dx)_x; \qquad J_B = (D_B/2)(dB/dx)_x \qquad (11.178)$$

throughout the diffusion layer at pH = 13.0 are shown in Figure 11.31. Both fluxes change rapidly over the first $\frac{1}{10}$th of the diffusion layer and then become constant over the remainder of the layer. At a higher pH (e.g., 13.5), the initial increases are steeper (not shown) than in Figure 11.31.

In Figure 11.32 the intrinsic dissolution rates J_A, predicted theoretically by

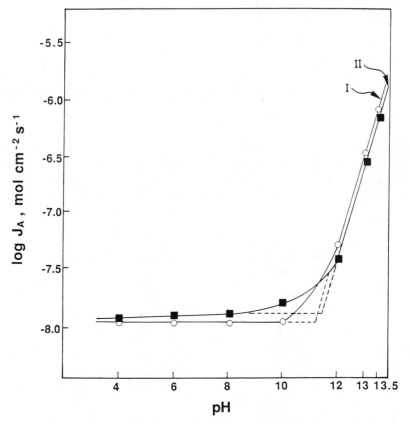

Figure 11.32 Logarithmic plots of the intrinsic dissolution rate J_A of 7-acetyltheophylline as a function of pH at 25°C: I (○) the theoretical line based on the values calculated from Eq. 11.173 assuming $B_o = 0$ at pH 4 ~ 12 and by numerical integration at pH 13 ~ 13.5; II (■) the experimental line from dissolution data manipulated by Eqs. 11.2 and 11.8. Reproduced with permission of the copyright owner. Higuchi, T., Lee H.K., and Pitman, I.H. (1971). *Farm. Aikak.*, **80**, 55–90.

numerical integration of Eq. 11.173, are compared at various pH values with those determined from the experimental measurements (manipulated by Eqs. 11.2 and 11.8). Agreement is seen to be good. Table 11.11 compares the reaction time t_R (calculated using Eqs. 11.141 and 11.157) with the diffusion time t_D (calculated from Eq. 11.139) for 7-acetyltheophylline at various pH values.

At pH $\leqslant$ 8.0, $t_R \gg t_D$, indicating that dissolution is diffusion-controlled according to Eq. 11.175, so that hydrolysis occurs essentially in the bulk of the solution. The dissolution rate should then be independent of pH, as shown theoretically and experimentally by Figure 11.32. At pH $\geqslant$ 10, $t_R \ll t_D$, indicating that dissolution is reaction-controlled, so that hydrolysis occurs essentially within the diffusion layer. The dissolution rate should then increase with increasing pH, as shown theoretically and experimentally in Figure 11.32. At pH $\gg$ 13, Figures 11.30 and 11.31 show that the reaction takes place within the first $\frac{1}{10}$th of the diffusion layer next to the solid–liquid interface. The influence of $[OH^-] = B_h$ on J_A can also be predicted qualitatively by considering the relative magnitudes of the terms on the right-hand side of Eq. 11.173.

Chapter XII

AFTERWORD

12.1 INTRODUCTION

In this book we have attempted to draw together various approaches for predicting solubilities, primarily of solid organic compounds, including their adducts, but excluding macromolecular species. The purpose of this chapter is to provide the reader with a summary of the methods for prediction of solubility and a few ideas for areas of future research.

12.2 SUMMARY OF METHODS OF SOLUBILITY PREDICTION

12.2.1 Introduction

Scheme 12.1 presents a very general thermodynamic cycle which summarizes various methods of predicting the solubility of a solid solute or a sparingly soluble liquid solute in a given solvent. Each apex of the triangle represents a defined chemical potential (partial molar free energy) of the solute of interest. The upper apex represents the pure solute and saturated solution which are in equilibrium and in which the solute therefore has the same chemical potential. The lower right-hand apex represents the solution standard state of the solute in the solvent, such as water, in which the solubility is required. The solubility of the solute in this solvent is calculated using the usual standard free-energy equation (Eq. 12.1) from which $\Delta G_{\text{pure}}^{\text{solution}}$, which is the difference in chemical potential of the solute between the upper and lower right-hand apices. To facilitate this calculation, Eq. 12.2 is applied, such that $\Delta G_{\text{pure}}^{\text{solution}}$ is calculated as the sum of two free-energy terms, one representing transfer of the solute from the pure state to a defined reference state $\Delta G_{\text{pure}}^{\text{reference}}$ (Eq. 12.3 in Scheme 12.1) and the other representing that from the reference state to the solution in the solvent of interest $\Delta G_{\text{reference}}^{\text{solution}}$ (Eq. 12.4 in Scheme 12.1). The reference state is represented by the lower left-hand apex in Scheme 12.1.

12.2.2 Three Types of Reference State

Three common reference states for the solute have been employed in this book and are designated in Scheme 12.1 as (1) the hypothetical pure supercooled liquid solute, (2) the vapor of the solute, and (3) an infinitely dilute solution of the solute in a paraffinic hydrocarbon solvent (e.g., isooctane) for which preference has been expressed in Chapter 3. The same designations (1), (2), and (3) are applied to the consequences of the respective reference states in Scheme 12.1. The choice of reference state of the solute may depend on the

Pure Solute⇌Saturated Solution

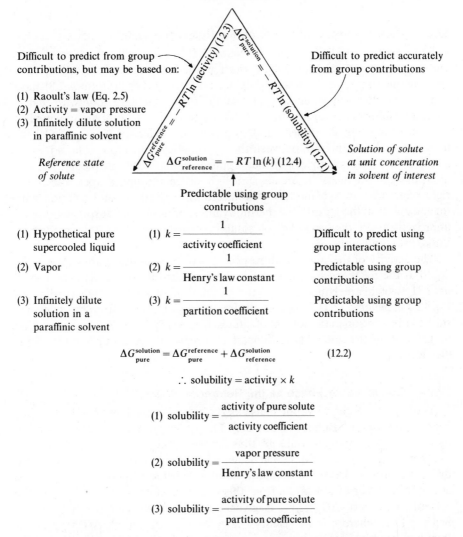

Difficult to predict from group contributions, but may be based on:

(1) Raoult's law (Eq. 2.5)
(2) Activity = vapor pressure
(3) Infinitely dilute solution in paraffinic solvent

Reference state of solute

$\Delta G^{\text{solution}}_{\text{reference}} = -RT\ln(k)$ (12.4)

Difficult to predict accurately from group contributions

Solution of solute at unit concentration in solvent of interest

Predictable using group contributions

(1) Hypothetical pure supercooled liquid

(2) Vapor

(3) Infinitely dilute solution in a paraffinic solvent

(1) $k = \dfrac{1}{\text{activity coefficient}}$

(2) $k = \dfrac{1}{\text{Henry's law constant}}$

(3) $k = \dfrac{1}{\text{partition coefficient}}$

Difficult to predict using group interactions

Predictable using group contributions

Predictable using group contributions

$$\Delta G^{\text{solution}}_{\text{pure}} = \Delta G^{\text{reference}}_{\text{pure}} + \Delta G^{\text{solution}}_{\text{reference}} \qquad (12.2)$$

$$\therefore \text{ solubility} = \text{activity} \times k$$

$$(1)\ \text{solubility} = \frac{\text{activity of pure solute}}{\text{activity coefficient}}$$

$$(2)\ \text{solubility} = \frac{\text{vapor pressure}}{\text{Henry's law constant}}$$

$$(3)\ \text{solubility} = \frac{\text{activity of pure solute}}{\text{partition coefficient}}$$

Scheme 12.1 The thermodynamic cycle which may be used for solubility prediction. Three methods are described depending on the chosen reference state: (1) hypothetical supercooled liquid, (2) vapor, and (3) infinitely dilute solution in paraffinic solvent. The scheme incorporates the standard molar free energies of transfer of solute between the pure state, reference state, and the solution.

available data or on one's philosophical viewpoint of intermolecular interactions in solution.

12.2.3   Pure Supercooled Liquid as the Reference State

The concepts of ideal solutions (Section 2.2), regular solutions (Section 2.6), and their extensions (Section 2.7) use the hypothetical pure supercooled liquid as the reference state. Yalkowsky, Valvani, and co-workers (Sections 2.8) have equated the chemical potential of the solute in this reference state with that in 1-octanol solution, since mixtures of the solute with 1-octanol are assumed to behave ideally. On the other hand, Amidon and Williams (Section 8.5) have maintained the distinction between the hypothetical supercooled liquid and a solution in 1-octanol, to which they applied regular solution theory, thereby interposing both these "reference states" between the upper apex and lower right-hand apex in Scheme 12.1. The insertion of the 1-octanol solution with regular solution theory into the thermodynamic cycle elicits a surprisingly small improvement in the accuracy of solubility prediction over the method of Yalkowsky and Valvani.

The activity of the solute with respect to the hypothetical pure supercooled liquid as the reference state can readily be evaluated using Eq. 2.14–2.16, 2.29, or 2.31 (which are equivalent to the evaualtion of $\Delta G_{\text{pure}}^{\text{reference}}$ according to Eq. 12.3 in Scheme 12.1). However, the free energy of transfer from this reference state to the solution in the solvent of interest (i.e., $\Delta G_{\text{reference}}^{\text{solution}} = RT \ln \gamma$, according to Eq. 12.4 in Scheme 12.1) is difficult to predict using group contributions (Section 7.1).

12.2.4   Gas or Vapor Phase as the Reference State

The use of the gas or vapor phase as the reference state in Scheme 12.1 has been emphasized by Butler (Section 3.8). The activity of the solute is then given by its saturated vapor pressure p^*, thus enabling $\Delta G_{\text{pure}}^{\text{reference}}$ to be evaluated as $-RT \ln p^*$ according to Eq. 12.3 in Scheme 12.1. This reference state is, however, difficult to apply directly to involatile solutes. Scheme 12.1 shows that the free energy of transfer of the solute from this reference state to the solution in the solvent of interest, $\Delta G_{\text{reference}}^{\text{solution}}$, is given by $RT \ln h$ according to Eq. 12.4 in Scheme 12.1, where h is the Henry's law constant. Although $\Delta G_{\text{reference}}^{\text{solution}}$ is in this case readily predictable using group contributions, complications arise whenever the solute molecule undergoes a conformational change upon entering or leaving the vapor state.

12.2.5   Dilute Solution in a Paraffinic Solvent as the Reference State

As discussed in Chapter 3 and summarized above, the reference state recommended in this book is an infinitely dilute solution of the solute in a paraffinic solvent such as isooctane. This reference state eliminates solute–solute interactions and reduces solute–solvent interactions to Debye and London forces. Chapter 3 describes procedures for evaluating the activity of the pure solute (based on the reference state), and hence $\Delta G_{\text{pure}}^{\text{reference}}$ (according to Eq.12.3 in Scheme 12.1). The free energy of transfer from this reference state to the desired solution phase (i.e., $\Delta G_{\text{reference}}^{\text{solution}}$) is given by $RT \ln P$ according to Eq. 12.4 in Scheme 12.1, where P is the partition coefficient (hydrocarbon/solvent of interest) and is often readily predictable using group contributions, as described in Sections 7.1 and 7.2.2.

12.3   USEFUL DIRECTIONS FOR FUTURE RESEARCH

While this book has shown that free energies of transfer, and hence solubilities and partition coefficients, of organic compounds in many solvent media can be predicted with reasonable accuracy, there are a number of gaps in our knowledge whose elucidation will improve our predictive capability and will provide interesting and useful directions for future research.

Flexible molecules can exist in a number of conformation states (e.g., ethyl acetate), as noted in Section 3.11. The relative proportions of the different conformations tend to differ from one solvent medium to another. Statistical mechanics is still unable to provide a priori predictions of the free energies of transfer of such molecules to a given solvent medium, partially because of the difficulty of assigning proper weighting of the contribution from each conformational state.

The group-contribution approach in a simple form is not applicable to all situations, particularly to flexible molecules. This limitation again arises from changes in the relative proportions of the different conformations, depending on the environment of the molecules.

Although the donor–acceptor complexation constants K_{DA}, K_{DA_2}, and so on are known or can be predicted for many donor–acceptor pairs in certain solvents, we would like to be able to predict the complexation constants for all species in all environments. An initial simplistic method has been described in Section 5.6. This approach could be broadened to examine the influence of a variety of common solvents other than carbon tetrachloride and paraffinic hydrocarbons. For this purpose, known values of complexation constants in the literature could be collected and analyzed using a computer. This method will reveal gaps in our knowledge of complexation constants which can be filled by recourse to experiment. This approach will facilitate the a priori prediction of complexation constants and hence solubilities in interactive solvent media.

Self-association of solutes possessing functional groups that can act as both donors and acceptors is normally limited to small stoichiometric numbers (e.g., 2, 4, and 5) at moderate concentrations in noninteractive solvents, as has been discussed in Chapter 6. There is experimental evidence that the stoichiometric numbers are higher in concentrated solutions and in the pure state, but the actual values are not usually known. Similarly, the nature and extent of self-association in interactive solvents and the competing roles of self-association and strong intermolecular interaction within the solvent phase are also usually unknown. Consequently, further investigation of these areas of solution chemistry will provide insight and predictive capability in self-associating systems.

Most organic molecules are polyfunctional and many of these are of biological interest. Polyfunctional molecules can undergo a variety of specific interactions which may compete to extents that depend on the environment and which are therefore in need of more detailed investigation. These specific interactions include complexation with other polar compounds or interactive solvent, intermolecular self-association, and intramolecular interactions among the polar functions. Examples of such compounds include glycols (1, 2-, 1, 3-, and 1, 4-diols). These compounds probably form cyclic structures in the vapor phase and in solution in a paraffinic solvent, due to internal hydrogen bonding, as suggested in Section 3.7. Because of the low fugacities of polyfunctional molecules such as diols, their vapor pressures and Henry's law constants are small and are therefore difficult to measure, thus introducing certain technical constraints in the study of self-association (Section 6.3.2). Nevertheless, powerful spectroscopic methods may help to elucidate the interactions involving polyfunctional compounds.

REFERENCES

Abboud, J.-L. M., Sraidi, K., Guiheneuf, G., Negro, A., Kamlet, M. J. and Taft, R. W. Studies on amphiprotic compounds. 2. Experimental determination of the hydrogen bond acceptor basicities of "monomeric" alcohols. *J. Org. Chem.*, **50** (1985), 2870–2873.

Abougela, I. K. A. and Grant, D. J. W. Fatty acid solvates of griseofulvin. *J. Pharm. Pharmacol.*, **31** (suppl.), (1979), 49P.

Abraham, M. H. Thermodynamics of solution of homologous series of solutes in water. *J. Chem. Soc. Faraday Trans. I*, **80** (1984), 153–181.

Abrams, D. S. and Prausnitz, J. M. Statistical thermodynamics of liquid mixtures; a new expression for the excess Gibbs energy of partly and completely miscible systems. *Am. Inst. Chem. Eng. J.*, **21** (1975), 116–128.

Accascina, F., Goffredi, M., and Triolo, R. Ion-pair formation of quaternary ammonium salts in hydrogen-bonded solvents: Electrical conductivities of some tetraalkylammonium perchlorates in water–dioxane mixtures at 25 degrees C. *Z. Phys. Chem. Neue Folge*, **81** (1972), 148–157.

Acree, W. E., Jr., *Thermodynamic Properties of Nonelectrolyte Solutions*. Academic, New York, 1984.

Acree, W. E., Jr. and Bertrand, G. L. Personal communication (1980).

Acree, W. E. and Bertrand, G. L. Thermochemical investigations of nearly ideal solvents. 4. Gas–liquid partition coefficients in complexing and non-complexing systems. *J. Phys. Chem.*, **83** (1979), 2355–2358.

Acree, W. E., Jr. and Bertrand, G. L. Thermochemical investigations of nearly ideal binary solvents. VII: Monomer and dimer models for solubility of benzoic acid in simple binary and ternary solvents. *J. Pharm. Sci.*, **70** (1981), 1033–1036.

Acree, W. E. and Rytting, J. H. Solubilities in binary solvent systems. II. The importance of non-specific interactions. *Int. J. Pharm.*, **10** (1982), 231–238.

Acree, W. E., Jr., Bacon, W. E., and Leo, A. J. Octanol/water partition coefficients of 4-substituted benzylidene *t*-butylamine *N*-oxides. *Int. J. Pharm.*, **20**, (1984), 209–211.

Adjei, A., Newburger, J., and Martin, A. Extended Hildebrand approach: Solubility of caffeine in dioxane–water mixtures. *J. Pharm. Sci.*, **69** (1980), 659–661.

Alexander, K. S., Laprade, B., Mauger, J. W., and Paruta, A. N. Thermodynamics of aqueous solutions of parabens. *J. Pharm. Sci.*, **67** (1978), 624–627.

Amidon, G. L. and Williams, N. A. A solubility equation for nonelectrolytes in water. *Int. J. Pharm.*, **11** (1982), 249–256.

Amidon, G. L., Yalkowsky, S. H., and Leung, S. Solubility of nonelectrolytes in polar solvents II: Solubility of aliphatic alcohols in water. *J. Pharm. Sci.*, **63** (1974), 1858–1866.

Amidon, G. L., Yalkowsky, S. H., Anik, S. T., and Valvani, S. C. Solubility of nonelectrolytes in polar solvents. V. Estimation of the solubility of aliphatic monofunctional compounds in water using a molecular surface area approach. *J. Phys. Chem.*, **79** (1975), 2239–2246.

Anderson, B. D. Specific Interactions in Nonaqueous Systems. I. Self-association of Alcohols and Phenol in Nonpolar Solvents. II. Solubilities of Organic Compounds in Organic Solvents and Cosolvent Mixtures. Ph.D. Thesis, University of Kansas, Lawrence, 1977.

Anderson, B. D., Rytting, J. H., Lindenbaum, S., and Higuchi, T. A calorimetric study of the self-association of primary alcohols in isooctane. *J. Phys. Chem.*, **79** (1975), 2340–2344.

Anderson, B. D., Rytting, J. H., and Higuchi, T. Vapor pressure studies of self-association of alcohols in isooctane I. The effect of chain length. *Int. J. Pharm.*, **1** (1978), 15–31.

Anderson, B. D., Rytting, J. H., and Higuchi, T. Influence of self-association on the solubility of phenol in isooctane and cyclohexane. *J. Am. Chem. Soc.*, **101** (1979), 5194–5197.

Anderson, B. D., Rytting J. H., and Higuchi, T. Solubility of polar organic solutes in nonaqueous systems: Role of specific interactions. *J. Pharm. Sci.*, **69** (1980), 676–680.

Anderson, N. H., Davis, S. S., James, M., and Kojima, I. Thermodynamics of distribution of *p*-substituted phenols between aqueous solution and organic solvents and phospholipid vesicles. *J. Pharm. Sci.*, **72** (1983), 443–448.

Anderson, T. F. and Prausnitz, J. M. Application of the UNIQUAC equation to calculation of multicomponent phase equilibria. 1. Vapor–liquid equilibria. *Ind. Eng. Chem., Process Des. Dev.*, **17** (1978a), 552–561.

Anderson, T. F. and Prausnitz, J. M. Application of the UNIQUAC equation to calculation of multicomponent phase equilibria. 2. Liquid–liquid equilibria. *Ind. Eng. Chem., Process Des. Dev.*, **17** (1978b), 561–567.

Antunes, C. and Tassios, D. Modified UNIFAC model for the prediction of Henry's constants. *Ind. Eng. Chem., Process Des. Dev.*, **22**, (1983), 457–462.

Aranow, R. H. and Witten, L. The environmental influence on the behavior of long chain molecules. *J. Phys. Chem.*, **64** (1960), 1643–1648.

Arnett, E. M., Joris, L., Mitchell, E., Murty, T. S. S. R., Gorrie, T. M., and Schleyer, P. v. R. Studies of hydrogen-bonded complex formation. III. Thermodynamics of complexing by infrared spectroscopy and calorimetry. *J. Am. Chem. Soc.*, **92** (1970), 2365–2377.

Arnett, E. M., Mitchell, E. J., and Murty, T. S. S. R., "Basicity." A comparison of hydrogen bonding and proton transfer to some Lewis bases. *J. Am. Chem. Soc.*, **96** (1974), 3875–3891.

Arrowsmith, M., Hadgraft, J., and Kellaway, I. W. Thermodynamics of steroid partitioning in dimyristoylphosphatidylcholine liposomes. *Biochim. Biophys. Acta*, **750** (1983), 149–156.

Ashworth, A. J. and Everett, D. H. The solubility of low-molecular-weight hydrocarbons in non-voletile liquids. *Trans. Faraday Soc.*, **56** (1960), 1609–1618.

Astarita, G., *Mass Transfer with Chemical Reaction*. Elsevier, New York, 1967, p. 9.

Attwood, D. and Agarwal, S. P. Self-association of disodium cromoglycate in dilute aqueous solution. *Int. J. Pharm.*, **22** (1984), 25–30.

Attwood, D. and Florence, A. T. *Surfactant Systems: Their Chemistry, Pharmacy and Biology*. Chapman and Hall, London and New York, 1983.

Atwood, J. L. and Davies, J. E. D., Eds. *Inclusion Phenomena in Inorganic, Organic, and Organometallic Hosts* (Proceedings of the Fourth International Symposium on Inclusion Phenomena and the Third International Symposium on Cyclodextrins, Lancaster, U.K., 20–25 July 1986). D. Reidel, Dordrecht, Holland, 1987.

Atwood, J. L., Davies, J. E. D., and MacNicol, D. D., Eds. *Inclusion Compounds*; Vol. 1, Structural Aspects of Inclusion Compounds Formed by Inorganic and Organometallic Host Lattices; Vol. 2, Structural Aspects of Inclusion Compounds Formed by Organic Host Lattices; Vol. 3, Physical Properties and Applications. Academic Press, London, 1984.

Aveyard, R., Briscoe, B. J., and Chapman, J. Activity coefficients and association of *n*-alkanols in *n*-octane. *J. Chem. Soc., Faraday Trans. I*, **69** (1973), 1772–1778.

Baba, H. Hydrogen bonding effect on the electronic absorption spectrum of *p*-hydroxyazobenzene. *Bull. Chem. Soc. Japan*, **31** (1958), 169–172.

Baba, H. and Suzuki, S. Electronic spectra and hydrogen bonding. I. Phenol and naphthols. *J. Chem. Phys.*, **35** (1961), 1118–1127.

Badger, R. M. and Bauer, S. H. Spectroscopic studies of the hydrogen bond. II. The shift of the O–H vibrational frequency in the formation of the hydrogen bond. *J. Chem. Phys.*, **5** (1937), 839–851.

Badwan, A. A., El-Khordagui, L. K., Saleh, A. M., and Khalil, S. A. The solubility of benzodiazepines in sodium salicylate solution and a proposed mechanism for hydrotropic solubilization. *J. Pharm. Pharmacol.*, **32** (Suppl.) (1980), 74P.

Bagley, E. B. and Scigliano, J. M. Polymer solutions. In *Solutions and Solubilities*, Part II, *Techniques of Chemistry*, Vol. VIII, Dack, M. R. J., Ed. Wiley, New York, 1976, pp. 437–485.

Baker, A. W. and Harris, G. H. Physical and chemical effects of substituent groups on multiple bonds. II. Thiolesters. *J. Am. Chem. Soc.*, **82** (1960), 1923–1928.

Baker, K. M. and Wilson, R. G. Benzene-induced solvent shifts and the stoichiometry of benzene–solute complexes. *J. Chem. Soc. B*, **1970** (1970), 236–239.

Baker, R. J., Acree, W. E., Jr., and Tsai, C-C. Correlation and estimation of aqueous solubilities of polycyclic aromatic hydrocarbons. *Quant. Struct. Act. Relat.*, **3** (1984), 10–16.

Banks, W. H. Dipole solvation. *Trans. Faraday Soc.*, **33** (1937), 215–224.

Barigand, M., Orszagh, J., and Tondeur, J. J. Complexes moléculaires. IV. La constante d'équilibre de la complexation de la phénothiazine avec le *p*-chloranile dans quelques solvants. *Bull. Soc. Chim. Fr.*, **1973** (1973), 48–50.

Barriol, J. and Weisbecker, A. Contribution à l'étude de l'effet de solvant sur la liaison hydrogène. *Compt. Rend. Acad. Sci. (Paris)*, **265C** (1967), 1372–1375.

Barry, B. W., Harrison, S. M., and Dugard, P. H. Vapour and liquid diffusion of model penetrants through human skin; correlation with thermodynamic activity. *J. Pharm. Pharmacol.*, **37** (1985), 226–236.

Barton, A. F. M. Solubility parameters. *Chem. Rev.*, **75** (1975), 731–753.

Barton, A. F. M., *CRC Handbook of Solubility Parameters and Other Cohesion Parameters*, CRC, Boca Raton, FL, 1983.

Bate-Smith, E. C. and Westhall, R. G. Chromatographic behaviour and chemical structure. I. Some naturally occurring phenolic substances. *Biochim. Biophys. Acta*, **4** (1950), 427–440.

Baum, G. The influence of hydrophobic interactions on the electrochemical selectivity ratios of liquid membranes responsive to organic ions. *J. Phys. Chem.*, **76** (1972), 1872–1875.

Beckett, A. H. and Moffat, A. C. Correlation of partition coefficients in *n*-heptane–aqueous systems with buccal absorption data for a series of amines and acids. *J. Pharm. Pharmacol.*, **21** (1969), 144S–150S.

Beezer, A. E., Hunter, W. H., and Storey, D. E. Quantitative structure–activity relationships: The Van't Hoff heats of transfer of resorcinol monoethers from water to *n*-octanol. *J. Pharm. Pharmacol.*, **32** (1980), 815–819.

Beezer, A. E., Hunter, W. H., and Storey, D. E. Enthalpies of solution of a series of *m*-alkoxy phenols in water, *n*-octanol and water–*n*-octanol mutually saturated: Derivation of the thermodynamic parameters for solute transfer between these solvents. *J. Pharm. Pharmacol.*, **35** (1983), 350–357.

Bellamy, L. J. and Pace, R. J. Hydrogen bonding by alcohols and phenols. I. The nature of the hydrogen bond in alcohol dimers and polymers. *Spectrochim. Acta*, **22** (1966), 525–533.

Benesi, H. A. and Hildebrand, J. H. A spectrophotometric investigation of the interaction of iodine with aromatic hydrocarbons. *J. Am. Chem. Soc.*, **71** (1949), 2703–2707.

Ben-Naim, A. *Water and Aqueous Solutions*, Plenum, New York, 1974.

Ben-Naim, A. *Hydrophobic Interactions*, Plenum, New York, 1980.

Ben-Naim, A., Wilf, J., and Yaacobi, M. Hydrophobic interaction in light and heavy water. *J. Phys. Chem.*, **77** (1973), 95–102.

Bennett, D. and Canady, W. J. Thermodynamics of solution of naphthalene in various water–ethanol mixtures. *J. Am. Chem. Soc.*, **106** (1984), 910–915.

Bent, H. A. The nature of intermolecular donor–acceptor bonds. In *Solutions and Solubilities*, Part II, *Techniques of Chemistry*, Vol. VIII, Dack, M. R. J., Ed. Wiley, New York, 1976, pp. 65–93.

Bentley, J. B., Everard, K. B., Marsden, R. J. B., and Sutton, L. E. Electric dipole moment studies on the conjugation and stereochemistry of some unsaturated ketones and aldehydes. Parts I and II. *J. Chem. Soc.*, **1949** (1949), 2957–2970.

Bircumshaw, L. L. and Riddiford, A. C. Transport control in heterogeneous reactions. *Q. Rev. Chem. Soc.*, **6** (1952), 157–185.

Bishop, R. J. and Sutton, L. E. The study of the equilibrium constant of complex formation between phenol and pyridine in solution by a dielectric constant method. *J. Chem. Soc.*, **1964** (1964), 6100–6106.

Bjerrum, N. Untersuchungen über ionenassoziation. 1. Der einfluss der ionenassoziation auf die aktivität der ionen bei mittleren Assoziationsgraden. *Kgl. Danske Videnskab. Selskab., Math.-fys. Meddelel.*, **7** (9) (1926), 1–48.

Black, C., Joris, G. G., and Taylor, H. S. The solubility of water in hydrocarbons. *J. Chem. Phys.*, **16** (1948), 537–543.

Blackadder, D. A. *Some Aspects of Basic Polymer Science*. Monograph for Teachers, No. 28, The Chemical Society, London, 1975.

Bobtelsky, M. *Heterometry*, Elsevier, Amsterdam, 1960.

Bogachev, Yu. S., Vasianina, L. K., Shapet'ko, N. N., and Alexeeva, T. L. An NMR study of phenol self-association. *Org. Magn. Resonance*, **4** (1972), 453–462.

Bogoslovskii, V. E., Mikhalyuk, G. I., and Shamolin, A. I. Vapor–liquid equilibrium in the systems water–dimethylacetamide, water–dimethylformamide, and water–methylpyrrolidone. *J. Appl. Chem. USSR*, **45** (1972), 1197–1199. Original article in Russian, *Z. Prikl. Khim.*, **45** (1972), 1154–1156.

Bolton, S., Guttman, D., and Higuchi, T. Complexes formed in solution by homologs of caffeine. Interactions between *p*-hydroxybenzoic acid and ethyl, propyl, and butyl derivatives of theobromine and theophylline. *J. Am. Pharm. Assoc., Sci. Ed.*, **46** (1957), 38–41.

Bondi, A. Van der Waals volumes and radii. *J. Phys. Chem.*, **68** (1964), 441–451.

Bondi, A. *Physical Properties of Molecular Crystals, Liquids and Glasses*, Wiley, New York, 1968.

Boublík, T., Fried, V., and Hála, E. *The Vapor Pressures of Pure Substances*. Elsevier, Amsterdam, 1973.

Boublík, T., Fried, V., and Hála, E. *The Vapour Pressures of Pure Substances: Selected Values of the Temperature Dependence of the Vapour Pressures of Some Pure Substances in the Normal and Low Pressure Region*, 2nd rev. ed., Physical Sciences Data 17, Elsevier, Amsterdam, 1984.

Bournay, J. and Robertson, G. N. Hydrogen bond dynamics in solution. *Nature (London)*, **275** (1978), 46–48.

Bowden, S. T. *The Phase Rule and Phase Reactions, Theoretical and Practical*, Macmillan, London, 1938.

Bowen, D. B. and James, K. C. The effect of temperature on the solubilities of testosterone propionate in low polarity solvents. *J. Pharm. Pharmacol.*, **22S** (1970), 104S–108S.

Bowen, D. B., James, K. C., and Roberts, M. An investigation of the distribution coefficients of some androgen esters using paper chromatography. *J. Pharm. Pharmacol.*, **22** (1970), 518–522.

Brink, G., Campbell, C., and Glasser, L. Hydrogen bonding structure in alcohols and their solutions. *S. Afr. J. Sci.*, **73** (1977), 11–15.

Brooker, P. J. and Ellison, M. The determination of the water solubility of organic compounds by a rapid turbidimetric method. *Chem. Ind.*, **1974** (1974), 785–787.

Brown, D. J. The monosubstituted pteridines. In *Ciba Foundation Symposium on Chemistry and Biology of Pteridines*, Wolstenholme, G. E. W. and Cameron, M. P., Eds. Little, Brown, Boston, 1954, 63–73.

Brown, H. C., Barbaras, G. K., Berneis, H. L., Bonner, W. H., Johannesen, R. B., Grayson, M., and Nelson, K. L. Strained homomorphs. 14. General summary. *J. Am. Chem. Soc.*, **75** (1953), 1–6.

Brown, I., Fock, W., and Smith, F. Heats of mixing. V. Systems of *n*-alcohols with *n*-hexane. *Aust. J. Chem.*, **17** (1964), 1106–1118.

Brunner, E. Reaktionsgeschwindigkeit in heterogenen systemen. *Z. Phys. Chem.*, **47** (1904), 56–102.

Buchowski, H., Devaure, J., Van Huong, P., and Lascombe, J. Influence des solvants sur les constantes d'association de l'heptyne-1 et du deutériochloroforme avec l'acétone. *Bull. Soc. Chim. Fr.*, **1966** (1966), 2532–2535.

Buckingham, A. D. Discussion. In *Study Week on Molecular Forces* (April 18–23, 1966, Pontificiae Academiae Scientiarum Scripta Varia 31, Vatican City), North Holland, Amsterdam, 1967, pp. 119–120.

Buckingham, A. D. and Orr, B. J. Molecular hyperpolarisabilities. *Q. Rev.*, **21** (1967), 195–212.

Bunick, G. and Voet, D. Crystal and molecular structure of 9-β-D-arabinofuranosyladenine. *Acta Cryst.*, **B30** (1974), 1651–1660.

Burrell, H., Solubility parameter values. In *Polymer Handbook*, 2nd ed., Brandrup, J. and Immergut, E. M., Eds. Wiley, New York, 1975, IV-337–IV-359.

Bury, C. R. and Jenkins, M. O. The freezing point of some benzene solutions. *J. Chem. Soc.* (1934), 688–696.

Butler, J. A. V. The energy and entropy of hydration of organic compounds. *Trans. Faraday Soc.*, **33** (1937), 229–238.

Butler, J. A. V., *Chemical Thermodynamics*, 5th ed., Macmillan, London, 1962, 373–393.

Butler, J. A. V. and Harrower, P. The activities of some aliphatic alcohols and halides in non-polar solvents. *Trans. Faraday Soc.*, **33** (1937), 171–178.

Butler, J. A. V. and Ramchandani, C. N. The solubility of non-electrolytes. II. The influence of the polar group on the free energy of hydration of aliphatic compounds. *J. Chem. Soc.* (1935), 952–955.

Butler, J. A. V. and Reid, W. S. The solubility of non-electrolytes. III. The entropy of hydration. *J. Chem. Soc.* (1936), 1171–1173.

Butler, J. A. V., Thomson, D. W., and Maclennan, W. H. The free energy of the normal aliphatic alcohols in aqueous solution. I. The partial vapour pressures of aqueous solutions of methyl, *n*-propyl, and *n*-butyl alcohols. II. The solubilities of some normal aliphatic alcohols in water. III. The theory of binary solutions, and its application to aqueous alcoholic solutions. *J. Chem. Soc.* (1933), 674–686.

Butler, J. N. *Ionic Equilibrium, A Mathematical Approach*, Addison-Wesley, Reading, MA, 1964.

Byrn, S. R. *Solid State Chemistry of Drugs*. Academic Press, New York, 1982.

Byron, P. R., Notari, R. E., and Tomlinson, E. Calculation of partition coefficient of an unstable compound using kinetic methods. *J. Pharm. Sci.*, **69** (1980), 527–531.

Cabani, S., Conti, G., and Lepori, L. Thermodynamic study on aqueous dilute solutions of organic compounds. Part 1. Cyclic amines. *Trans. Faraday Soc.*, **67** (1971a), 1933–1942.

Cabani, S., Conti, G., and Lepori, L. Thermodynamic study on aqueous dilute solutions of organic compounds. Part 2. Cyclic ethers. *Trans. Faraday Soc.*, **67** (1971b), 1943–1950.

Campbell, C., Brink, G., and Glasser, L. Dielectric studies of molecular association. Concentration dependence of dipole moment of 1-octanol in solution. *J. Phys. Chem.*, **79** (1975), 660–665.

Cantwell, F. F. and Mohammed, H. Y. Photometric acid–base titrations in the presence of an immiscible solvent. *Anal. Chem.*, **51** (1979), 218–223.

Carnahan, B., Luther, H. A., and Wilkes, J. O. *Applied Numerical Methods*, Wiley, New York, 1969, p. 171.

Carstensen, J. T. *Solid Pharmaceutics: Mechanical Properties and Rate Phenomena*, Academic, New York, 1980.

Carter, S., Murrell, J. N., and Rosch, E. J. The evaluation of equilibrium constants and extinction coefficients for weak charge-transfer complexes. *J. Chem. Soc.*, **1965** (1965), 2048–2053.

Cattrall, R. W. and Freiser, H. Coated wire ion selective electrodes. *Anal. Chem.*, **43** (1971), 1905–1906.

Chamberlain, C. S. and Drago, R. S. Comparison of enthalpies of hydrogen bonding determined by a gas–liquid chromatography method with those obtained by calorimetric or frequency shift methods. *J. Am. Chem. Soc.*, **98** (1976), 6142–6144.

Champion, J. V. and Meeten, G. H. Conformation of sodium cromolyn in aqueous solution using light scattering and magnetic birefringence. *J. Pharm. Sci.*, **62** (1973), 1589–1595.

Chapman, D., Williams, R. M., and Ladbrooke, B. D. Physical study of phospholipids. VI. Thermotropic and lyotropic mesomorphism of some 1, 2-diacyl-phosphatidylcholines (lecithins). *Chem. Phys. Lipids*, **1** (1967), 445–475.

Charlton, M. Steric effects. I. Esterification and acid-catalyzed hydrolysis of esters. *J. Am. chem. Soc.*, **97** (1975), 1552–1556.

Cheam, V., Farnham, S. B., and Christian, S. D., Vapor phase association of methanol. Vapor density evidence for trimer formation. *J. Phys. Chem.*, **74** (1970), 4157–4159.

Chiou, C. T., Freed, V. H., Schmedding, D. W., and Kohnert, R. L. Partition coefficient and bioaccumulation of selected organic chemicals. *Environ. Sci. Technol.*, **11** (1977), 475–478.

Chiou, W. L. and Riegelman, S. Pharmaceutical applications of solid dispersion systems. *J. Pharm. Sci.*, **60** (1971), 1281–1302.

Chou, J. T. and Jury, P. C. Computation of partition coefficients from molecular structures by a fragment addition method. In *Physical Chemical Properties of Drugs*, Yalkowsky, S. H., Sinkula, A. A., and Valvani, S. C., Eds. Marcel Dekker, New York, 1980, pp. 163–199.

Chow, A. H.-L., Chow, P. K. K., Wang, Z., and Grant, D. J. W. Modification of acetaminophen crystals: influence of growth in aqueous solutions containing *p*-acetoxyacetanilide on crystal properties. *Int. J. Pharm.*, **24** (1985), 239–258.

Chow, K. Y., Go, J., Mehdizadeh, M., and Grant, D. J. W. Modification of adipic acid crystals: influence of growth in the presence of fatty acid additives on crystal properties. *Int. J. Pharm.*, **20** (1984), 3–24.

Christian, S. D. and Lane, E. H. Solvent effects on molecular complex equilibria. In *Solutions and Solubilities*, Part I. *Techniques in Chemistry*, Vol. III, Dack, M. R. J., Ed., Wiley, New York, 1976.

Christian, S. D., Taha, A. A., and Gash, B. W. Molecular complexes of water in organic solvents and in the vapour phase. *Q. Rev.*, **24** (1970), 20–36.

Chulkaratana, S. Hydrogen Bonding and Solubility in Nonaqueous Systems, Ph.D. Thesis, University of Wisconsin, Madison, 1964.

Clever, H. L. Sechenov salt-effect parameter. *J. Chem. Eng. Data*, **28** (1983), 340–343.

Clever, H. L. and Battino, R. The solubility of gases in liquids. In *Solutions and Solubilities. Techniques of Chemistry*, Vol. VIII, Wiley, New York, 1976, pp. 374–444.

Clint, J. H., Corkill, J. M., Goodman, J. F., and Tate, J. R. Adsorption of *n*-alkanols at the air/aqueous solution interface. *J. Colloid Interface Sci.*, **28** (1968) 522–530.

Coggeshall, N. D. and Saier, E. L. Infrared absorption study of hydrogen bonding equilibria. *J. Am. Chem. Soc.*, **73** (1951), 5414–5418.

Collander, R. The partition of organic compounds between higher alcohols and water. *Acta Chem. Scand.*, **5** (1951), 774–780.

Cook, D. T. A Study of the Wetting Properties of a Hydrophobic Powder, Ph.D. Thesis, University of Nottingham, UK, 1978.

Crank, J. *The Mathematics of Diffusion*, 2nd ed., Clarendon, Oxford, 1975.

Creswell, C. J. and Allred, A. L. Thermodynamic constants for hydrogen bond formation in the chloroform–benzene–cyclohexane system. *J. Phys. Chem.*, **66** (1962), 1469–1472.

Crowley, J. D., Teague, G. S., Jr., and Lowe, J. W., Jr. A three-dimensional approach to solubility. *J. Paint Technol.*, **38** (1966), 269–280.

Crugman, B. Thermodynamic Quantities Associated with the Transfer of the Methylene and Oxygen Functions from Water to Cyclohexane and the Temperature Dependence of Group Contributions to Partition Coefficients. M. S. Thesis, University of Kansas, Lawrence, 1971.

Currie, D. J., Lough, C. E., Silver, R. F., and Holmes, H. L. Partition coefficients of some conjugated heteroenoid compounds and 1,4-naphthoquinones. *Can. J. Chem.*, **44** (1966), 1035–1043.

Dale, A. J. and Gramstad, T. Studies of hydrogen-bonding. XXII. NMR studies of self-association of phenol, mono- and polyfluorophenols. *Spectrochim. Acta*, **28A**, (1972), 639–650.

Danckwerts, P. V. Absorption by simultaneous diffusion and chemical reaction. *Trans. Faraday Soc.*, **46** (1950), 300–304.

Davies, J. T. The mechanism of diffusion of ions across a phase boundary and through cell walls. *J. Phys. Coll. Chem.*, **54** (1950), 185–204.

Davies, M. Hydrogen bond energies. In *Hydrogen Bonding*, Hadzi, D. and Thompson, H. W., Eds. (Symposium on Hydrogen Bonding, Ljubljana, Yugoslavia, July 29–August 3, 1957). Pergamon, New York, 1959, pp. 393–403.

Davis, S. S. Determination of the thermodynamics of the methyl group in solutions of drug molecules. *J. Pharm. Pharmacol.*, **25** (1973a), 1–12.

Davis, S. S. Use of substituent constants in structure–activity relations and the importance of the choice of standard state. *J. Pharm. Pharmacol.*, **25** (1973b), 293–296.

Davis, S. S. Determination of thermodynamics of halogen groups in solution of drug molecules. *J. Pharm. Pharmacol.*, **25** (1973c), 769–778.

Davis, S. S. Determination of the thermodynamics of hydroxyl and carboxyl groups in solutions of drug molecules. *J. Pharm. Pharmacol.*, **25** (1973d), 982–992.

Davis, S. S. and Elson, G. The determination of partition coefficient data using the "AKUFVE" method. *J. Pharm. Pharmacol.*, **26** (Suppl.), (1974), 90P.

Davis, S. S. and Olejnik, O. An ion-selective electrode for the determination of the activity of quaternary ammonium salts in formulated products. *J. Pharm. Pharmacol.*, **31S** (1979), 19P.

Davis, S. S., Higuchi, T., and Rytting, J. H. Determination of thermodynamics of the methylene group in solutions of drug molecules. *J. Pharm. Pharmacol.*, **24** (Suppl.), (1972), 30P–46P.

Davis, S. S., Higuchi, T., and Rytting, J. H. Determination of thermodynamics of functional groups in solutions of drug molecules. In *Advances in Pharmaceutical Sciences*, Vol. 4, Bean, H. S., Beckett, A. H., and Carless, J. E., Eds. Academic, New York, 1974, pp. 73–261.

Davis, S. S., Tomlinson, E., and Wilson, C. G. The effect of ion-association on the transcorneal transport of drugs. *Brit. J. Pharmacol.*, **64** (1978), 444P.

Davis, W. W. and Parke, T. V., Jr. A nephelometric method for determination of solubilities of extremely low order. *J. Am. Chem. Soc.*, **64** (1942), 101–107.

Deal, C. H. and Derr, E. L. Group contributions in mixtures. *Ind. Eng. Chem.*, **60**(4), (1968), 28–38.

Deal, C. H., Derr, E. L., and Papadopoulos, M. N. Activity coefficients and molecular structure. Hydrocarbon solutes in fixed solvent environments. *Ind. Eng. Chem. Fundam.*, **1** (1962), 17–19.

De Meere, A. L. J. and Tomlinson, E. Physicochemical description of the absorption rate of a solute between water and 2,2,4-trimethylpentane. *Int. J. Pharm.*, **22** (1984), 177–196.

Deming, S. N. and Morgan, S. L. Simplex optimization of variables in analytical chemistry. *Anal. Chem.*, **45** (1973), 278A–283A.

Denbigh, K. G., *The Principles of Chemical Equilibrium*, 3rd ed. Cambridge University Press, London, 1971, p. 242.

Denyer, R. L., Gilchrist, A., Pegg, J. A., Smith, J., Tomlinson, T. E., and Sutton, L. E. An investigation of complex formation by partition of a reactant between the vapour phase and solution: The determination of association constants and the heats of formation of some hydrogen bonds. *J. Chem. Soc.*, (1955) 3889–3901.

Derr, E. L. and Deal, C. H., Jr. Analytical solutions of groups: Correlation of activity coefficients through structural group parameters. *Proc. Int. Symp. Distill. l. Chem. E. Symp. Series No. 32*, (1969) 3:40–3:51.

Deutsch E. W. and Hansch, C. Dependence of relative sweetness on hydrophobic bonding. *Nature (London)*, **211** (1966), 75.

De Valera, E., Feakins, D., and Waghorne, W. E. Relationship between the enthalpy of transfer of a solute and the thermodynamic mixing functions of mixed solvents. *J. Chem. Soc., Faraday Trans. I*, **79** (1983), 1061–1071.

De Voe, H. and Wasik, S. P. Aqueous solubilities and enthalpies of solution of adenine and guanine. *J. Soln. Chem.*, **13** (1984), 51–60.

Diamond, J. M. and Katz, Y. Interpretation of non-electrolyte partition coefficients between dimyristoyl-lecithin and water. *J. Memb. Biol.*, **17** (1974), 121–154.

Diamond, J. M. and Wright, E. M. Molecular forces governing non-electrolyte permeation through cell membranes. *Proc. Roy. Soc.*, **B172** (1969a), 273–316.

Diamond, J. M. and Wright, E. M. Biological membranes: The physical basis of ion and non-electrolyte selectivity. *Ann. Rev. Physiol.*, **31** (1969b), 581–646.

Diamond, R. M., The aqueous solution behavior of large univalent ions. A new type of ion-pairing. *J. Phys. Chem.*, **67** (1963), 2513–2517.

Divatia, G. J. and Biles, J. A. Physical chemical study of the distribution of some amine salts between immiscible solvents. *J. Pharm. Sci.*, **50** (1961), 916–922.

Dixon, W. B. Nuclear magnetic resonance study of solvent effects on hydrogen bonding in methanol. *J. Phys. Chem.*, **74** (1970), 1396–1399.

Djerassi, C. *Optical Rotatory Dispersion. Applications to Organic Chemistry*, McGraw-Hill, New York, 1960, p. 60.

Dolezalek, F. Zur theorie der binären gemische und konzentrierten lösungen. *Z. Phys. Chem.*, **64** (1908), 727–747.

Dolezalek, F., Zur theorie der binären gemische und konzentrierten lösungen. II. *Z. Phys. Chem.*, **71** (1910), 191–213.

Doyle, T. D. and Levine, J. Application of ion-pair extraction to partition chromatographic separation of pharmaceutical amines. *Anal. Chem.*, **39** (1967), 1282–1287.

Drago, R. S. and Wayland, B. B. A double-scale equation for correlating enthalpies of Lewis acid–base interactions. *J. Am. Chem. Soc.*, **87** (1965), 3571–3577.

Drago, R. S., Vogel, G. C., and Needham, T. E. A four-parameter equation for predicting enthalpies of adduct formation. *J. Am. Chem. Soc.*, **93** (1971), 6014–6026.

Drago, R. S., Parr, L. B., and Chamberlain, C. S. Solvent effects and their relationship to the E and C equation. *J. Am. Chem. Soc.*, **99** (1977), 3203–3209.

Dreisbach, R. R. *Physical Properties of Chemical Compounds* (Advance in Chemistry Series, Vol. 15). American Chemical Society, Washington, 1955.

Dressman, J. The Effects of Association Equilibria on the Interphase Transport of Phenol. Ph.D. Thesis, University of Kansas, Lawrence, 1981.

Dressman, J. B., Himmelstein, K. J., and Higuchi, T. Simultaneous self-association and diffusion of phenol in isooctane. *J. Pharm. Sci.*, **71** (1982), 1226–1230.

Dressman, J. B., Himmelstein, K. J., and Higuchi, T. Diffusion of phenol in the presence of a complexing agent, tetrahydrofuran. *J. Pharm. Sci.*, **72** (1983), 12–17.

Duer, W. C. and Bertrand, G. L. Calorimetric determination of heats of formation of hydrogen bonds. *J. Am. Chem. Soc.*, **92** (1970), 2587–2588.

Dunn, W. J., III Molar refractivity as an independent variable in quantitative structure–activity studies. *Eur. J. Med. Chem.*, **12** (1977), 109–112.

Dunn, W. J., III, Block, J. H., and Pearlman, R. S., Eds. *Partition Coefficient Determination and Estimation*. Pergamon, New York, 1986.

Durrans, T. H. *Solvents*, 8th ed., revised by Davies, E. H., Chapman and Hall, London, England, 1971.

Eagland, D. Nucleic acids, peptides and proteins. In *Water, A Comprehensive Treatise*, Vol. 4, Franks, F., Ed. Plenum, New York, 1975, pp. 305–518.

Edgell, W. F., Watts, A. T., Lyford, J., IV, and Risen, W. M., Jr. Infrared bands from alkali ion motion in solution. *J. Am. Chem. Soc.*, **88** (1966), 1815.

Eisenberg, H. *Biological Macromolecules and Polyelectrolytes in Solution*, Clarendon, Oxford, 1976.

Eley, D. D. On the solubility of gases. Part I. The inert gases in water. *Trans. Faraday Soc.*, **35** (1939), 1281–1293.

Eley, D. D. On the solubility of gases. Part II. A comparison of organic solvents with water. *Trans. Faraday Soc.*, **35** (1939), 1421–1432.

Elworthy, P. H. and Lipscomb, F. J. A note on the solubility of griseofulvin. *J. Pharm. Pharmacol.*, **20** (1968), 790–792.

Elworthy, P. H., Florence, A. T., and Macfarlane, C. B. *Solubilization by Surface-Active Agents and its Applications in Chemistry and the Biological Sciences*, Chapman and Hall, London, 1968.

Evans, D. F. and Gardam, P. Transport processes in hydrogen-bonding solvents. II. Conductance of tetraalkylammonium salts in 1-butanol and 1-pentanol at 25°. *J. Phys. Chem.*, **73** (1969), 158–163.

Fee, J. V., Grant, D. J. W., and Newton, J. M. Effect of solvent flow Reynolds number on dissolution rate of a nondisintegrating solid (potassium chloride). *J. Pharm. Sci.*, **65** (1976), 48–53.

Fenby, D. V. and Hepler, L. G. Calorimetric investigations of hydrogen bond and charge transfer complexes. *Chem. Soc. Revs.*, **3** (1976), 193–207.

Ferguson, J. The use of chemical potentials as indices of toxicity. *Proc. Roy. Soc., B.*, **127** (1939), 387–404.

Findlay, A., Campbell, A. N., and Smith, N. O. The Phase Rule and Its Applications, 9th ed. Dover, New York, 1951

Fletcher, A. N. The effect of carbon tetrachloride upon the self-association of 1-octanol. *J. Phys. Chem.*, **73** (1969), 2217–2225.

Fletcher, A. N. Molecular structure of ethanol-d_1 solutions. A near-infrared study of hydrogen bonding. *J. Phys. Chem.*, **76**, (1972) 2562–2571.

Fletcher, A. N. and Heller, C. A. Self-association of alcohols in nonpolar solvents. *J. Phys. Chem.*, **71** (1967), 3742–3756.

Florence, A. T. Partial molal volumes of some non-ionic detergents in monomeric and micellar form. *J. Pharm. Pharmacol.*, **18** (1966), 384–389.

Flory, P. J. Thermodynamics of high polymer solutions. *J. Chem. Phys.*, **9** (1941), 660–661.

Flory, P. J. Thermodynamics of high polymer solutions. *J. Chem. Phys.*, **10** (1942), 51–61.

Flory, P. J. *Principles of Polymer Chemistry*, Cornell University Press, Ithaca, New York, 1953.

Flynn, G. L. Structural approach to partitioning: Estimation of steroid partition coefficients based upon molecular constitution. *J. Pharm. Sci.*, **60** (1971), 345–353.

Flynn, G. L., Yalkowsky, S. H., and Roseman, T. J. Mass transport phenomena and models: Theoretical concepts. *J. Pharm. Sci.*, **63** (1974), 479–510.

Fornili, S. L., Sgroi, G., and Izzo, V. Effect of solvent on stacking interactions. A spectrophotometric study of methylene blue dimerization in aqueous solutions of some monohydric alcohols. *J. Chem. Soc., Faraday Trans. I*, **79** (1983), 1085–1090.

Foster, R. and Fyfe, C. A. Interaction of electron acceptors with bases. Part 15. Determination of association constants of organic charge-transfer complexes by NMR spectroscopy. *Trans. Faraday Soc.*, **61** (1965), 1626–1631.

Fowkes, F. M., Additivity of intermolecular forces at interfaces. 1. Determination of the contribution to surface and interfacial tensions of dispersion forces in various liquids. *J. Phys. Chem.*, **67** (1963), 2538–2541.

Frank, H. S. The structure of water. *Fed. Proc. Suppl. 15*, (1965), S-1–S-11.

Frank, H. S. and Evans, M. W. Free volume and entropy in condensed systems. III. Entropy in binary liquid mixtures; partial molal entropy in dilute solutions; structure and thermodynamics in aqueous electrolytes. *J. Chem. Phys.*, **13** (1945), 507–532.

Frank, H. S. and Wen, W.-Y., III. Ion–solvent interaction. Structural aspects of ion–solvent interaction in aqueous solutions: A suggested picture of water structures. *Discuss. Faraday Soc.*, **24** (1957), 133–140.

Frank, S. G. Inclusion compounds. *J. Pharm. Sci.*, **64** (1975), 1585–1604.

Franks, F. *Water*, The Royal Society of Chemistry, London, 1983.

Franzen, J. S. and Franzen, B. C. Solvent dipole competition for interamide hydrogen bonds. *J. Phys. Chem.*, **68** (1964), 3989–3899.

Franzen, J. S. and Stephens, R. E. The effect of a dipolar solvent system on interamide hydrogen bonds. *Biochemistry*, **2** (1963), 1321–1327.

Fredenslund, A. and Rasmussen, P. Correlation of pure component Gibbs energy using UNIFAC group contribution. *Am. Inst. Chem. Eng. J.*, **25** (1979), 203–205.

Fredenslund, A., Jones, R. L., and Prausnitz, J. M. Group-contribution estimation of activity coefficients in nonideal liquid mixtures. *Am. Inst. Chem. Eng. J.*, **21** (1975), 1086–1099.

Fredenslund, A., Gmehling, J., Michelsen, M. L., Rasmussen, P., and Prausnitz, J. M. Computerized design of multicomponent distillation columns using the UNIFAC group contribution method for calculation of activity coefficients. *Ind. Eng. Chem.*, *Process Des. Dev.*, **16** (1977a) 450–462.

Fredenslund, A., Gmehling, J., and Rasmussen, P. *Vapor–Liquid Equilibria Using UNIFAC, a Group Contribution Method*. Elsevier, Amsterdam, 1977b.

Freiser, H. Relevance of solubility parameter in ion association extraction systems. *Anal. Chem.*, **41** (1969), 1354–1355.

Fuchs, R., Young, T. M., and Rodewald, R. F. Enthalpies of transfer of aromatic molecules from the vapor state to polar and nonpolar solvents. *J. Am. Chem. Soc.*, **96** (1974), 4705–4706.

Fung, H.-L. and Higuchi, T. Molecular interactions and solubility of polar nonelectrolytes in nonpolar solvents. *J. Pharm. Sci.*, **60** (1971), 1782–1788.

Fürer, R. and Geiger, M. A simple method of determining the aqueous solubility of organic substances. *Pestic. Sci.*, **8** (1977), 337–344.

Gaginella, T. S., Bass, P., Perrin, J. H., and Vallner, J. J. Effect of bile salts on partitioning behavior and GI absorption of a quaternary ammonium compound, isopropamide iodide. *J. Pharm. Sci.*, **62** (1973), 1121–1125.

Gerrard, W. *Solubility of Gases and Liquids. A Graphic Approach. Data–Causes–Prediction*, Plenum, New York, 1976.

Ghersetti, S. and Lusa, A. Infrared study of hydrogen bonding between various phenols and diphenyl-sulphoxide: Thermodynamic properties. *Spectrochim. Acta*, **21** (1965), 1067–1071.

Gibson, N. A. and Weatherburn, D. C. The distribution of salts of large cations between water and organic solvents. Part II. Factors affecting the magnitude of the distribution ratio. *Anal. Chim. Acta*, **58** (1972), 159–165.

Giles, G. H. and Duff, D. G. Dyestuffs. In *Water, A Comprehensive Treatise*, Vol. 4, Franks, F., Ed., Plenum, New York, 1975, pp. 169–207.

Gill, S. J., Nichols, N. F., and Wadsö, I. Calorimetric determination of enthalpies of solution of slightly soluble liquids. II. Enthalpy of solution of some hydrocarbons in water and their use in establishing the temperature dependence of their solubilities. *J. Chem. Thermodyn.*, **8** (1976), 445–452.

Gmehling, J. and Onken, U. *Vapor–Liquid Equilibrium Data Collection: Aqueous–Organic Systems*. Dechema Chemistry Data Series, Vol. 1, Parts 1–8, Dechema, Frankfurt am Main, F.R.G. 1977–1984.

Gmehling, J., Rasmussen, P., and Fredenslund, A. Vapor–liquid equilibria by UNIFAC group contribution. Revision and extension. 2. *Ind. Eng. Process Des. Dev.*, **21** (1982), 118–127.

Gmehling, J. G., Anderson, T. F., and Prausnitz, J. M. Solid–liquid equilibria using UNIFAC. *Ind. Eng. Chem. Fundam.*, **17** (1978), 269–273.

Goffredi, M., Liszi, J., Nemeth, B., and Liveri, V. T. Free energy of transfer of *n*-nitroalkanes from *n*-octane to water at 25°C. *J. Soln. Chem.*, **12** (1983), 221–231.

Goldammer, E. V. and Hertz, H. G. Molecular motion and structure of aqueous mixtures with nonelectrolytes as studied by nuclear magnetic relaxation methods. *J. Phys. Chem.*, **74** (1970), 3734–3755.

Golderg, A. H. and Higuchi, W. I. Improved method for diffusion coefficient determinations employing the silver membrane filter. *J. Pharm. Sci.*, **57** (1968), 1583–1585.

Goodman, D. S. The distribution of fatty acids between *n*-heptane and aqueous phosphate buffer. *J. Am. Chem. Soc.*, **80** (1958), 3887–3892.

Gould, P. L., Goodman, M., and Hanson, P. A. Investigation of the solubility relationships of polar, semi-polar and non-polar drugs in mixed co-solvent systems. *Int. J. Pharm.*, **19** (1984), 149–159.

Gramstad, T. Studies of hydrogen bonding. Part IX. The effect of solvent interaction on hydrogen bonding. *Spectrochim. Acta*, **19** (1963), 1363–1369.

Gramstad, T. and Mundheim, O. Studies of hydrogen bonding. XXIII. Interaction of chloroform with various phosphoryl compounds. The Higuchi plot–solvent effect. *Spectrochim. Acta*, **28A** (1972), 1405–1413.

Gramstad, T. and van Binst, G. Studies of hydrogen bonding. XVI. The complexing of pentafluorophenol with triphenylphosphine oxide. *Spectrochim. Acta*, **22** (1966), 1681–1696.

Grant, D. J. W. and Abougela, I. K. A. The fatty acid solvates of griseofulvin–desolvation data. *J. Pharm. Pharmacol.*, **33** (1981), 619–620.

Grant, D. J. W. and Abougela, I. K. A. Solubility behaviour of griseofulvin in fatty acids. *J. Pharm. Pharmacol.*, **34** (1982), 766–770.

Grant, D. J. W. and Abougela, I. K. A. Prediction of the solubility of griseofulvin in glycerides and other solvents of relatively low polarity from simple regular solution theory. *Int. J. Pharm.*, **17** (1983), 77–89.

Grant, D. J. W. and York, P. Entropy of processing: A new quantity for comparing the solid-state disorder of pharmaceutical materials. *Int. J. Pharm.*, **30** (1986), 161–180.

Grant, D. J. W., Higuchi, T., Hwang, Y. T., and Rytting, J. H. Partial vapor pressures and solubilities of cyclic polar compounds in iso-octane solutions. *J. Soln. Chem.*, **13** (1984a), 297–311.

Grant, D. J. W., Mehdizadeh, M., Chow, A. H.-L., and Fairbrother, J. E. Non-linear van't-Hoff solubility–temperature plots and their pharmaceutical interpretation. *Int. J. Pharm.*, **18** (1984b), 25–38.

Green, R. D. *Hydrogen Bonding by C–H Groups*, Wiley, New York, 1974.

Groves, M. J. and Alkan, M. H. Column methods—The best way of evaluating drug dissolution? *Manuf. Chem. Aerosol News*, **46** (5), (1975), 37–42.

Grünbauer, H. J. M. and Tomlinson, E. Application of local-composition theory to reversed-phase liquid chromatography. *J. Chromatogr.*, **268** (1983), 277–280.

Grünbauer, H. J. M., Bijloo, G. J., and Bultsma, T. Influence of eluent composition on lipophilicity measurements using reversed-phase thin-layer chromatography. *J. Chromatogr.*, **270** (1983), 87–96.

Grunwald, E. Interpretation of data obtained in nonaqueous media. Proceedings of the 7th Annual Summer Symposium, Developments in titrimetry. *Anal. Chem.*, **26** (1954), 1696–1701.

Grunwald, E. and Ralph, E. K., III. Rate constants for the dissociation of amine–water hydrogen bonds and the effect of nonpolar groups in aqueous solution. *J. Am. Chem. Soc.*, **89** (1967), 4405–4411.

Guggenheim, E. A. *Mixtures*. Oxford University Press, Oxford, 1952.

Gupta, P. A. and Daubert, T. E. Prediction of low-pressure vapor–liquid equilibria of non-hydrocarbon-containing systems—ASOG or UNIFAC. *Ind. Eng. Chem. Process Des. Dev.*, **25** (1986), 481–486.

Gurka, D. and Taft, R. W. Studies of hydrogen-bonded complex formation with *p*-fluorophenol. IV. The fluorine nuclear magnetic resonance method. *J. Am. Chem. Soc.*, **91** (1969), 4794–4801.

Gustin, J.-L. and Renon, H. Équilibres liquide–vapeur de mélanges binaires par la méthode statique. *Bull. Soc. Chim. Fr.*, **12** (1974), 2719–2722.

Gutowsky, H. S. and Saika, A. Dissociation, chemical exchange, and the proton magnetic resonance in some aqueous electrolytes. *J. Chem. Phys.*, **21** (1953), 1688–1694.

Guttman, D. and Higuchi, T. Study of possible complex formation between macromolecules and certain pharmaceuticals. IX. Formation of iodine–iodide complexes with polyethylene glycol. *J. Am. Pharm. Assoc., Sci. Ed.*, **44** (1955), 668–678.

Guttman, D. and Higuchi, T. Possible complex formation between macromolecules and certain pharmaceuticals. X. The interaction of some phenolic compounds with polyethylene glycols, polypropylene glycols, and polyvinylpyrrolidone. *J. Am. Pharm. Assoc., Sci. Ed.*, **45** (1956), 659–664.

Guttman, D. and Higuchi, T. Reversible association of caffeine and of some caffeine homologs in aqueous solution. *J. Am. Pharm. Assoc., Sci. Ed.*, **46** (1957), 4–10.

Guy, R. H. and Honda, D. H. Solute transport resistance at the octanol–water interface. *Int. J. Pharm.*, **19** (1984), 129–137.

Haberfield, P., Kivuls, J., Haddad, M., and Rizzo, T. Enthalpies, free energies, and entropies of transfer of phenols from nonpolar solvents to water. *J. Phys. Chem.*, **88** (1984), 1913–1916.

Hadzi, D. and Sheppard, N., The infrared absorption bands associated with the COOH and COOD groups in dimeric carboxylic acids. I. The region from 1500 to 500 cm^{-1}. *Proc. Roy. Soc., Ser. A*, **216** (1953), 247–266.

Hafkenscheid, T. L. and Tomlinson, E., Estimation of aqueous solubilities of organic non-electrolytes using liquid chromatographic retention data. *J. Chromatogr.*, **218** (1981), 409–425.

Hahn, F.-V.v. Ist hydrotropie eine besondere eigenshaft von organischen salzen? *Kolloid-Z.*, **62** (1933), 202–207.

Hála, E., Wichterle, I., Polák, J., and Boublik, T. *Vapor–Liquid Equilibrium Data at Normal Pressures*, Pergamon Oxford, 1968.

Haleblian, J. K. Characterization of habits and crystalline modification of solids and their pharmaceutical applications. *J. Pharm. Sci.*, **64** (1975), 1269–1288.

Hall, N. A. Solubilization of salicylic acid by polysorbate 80 as determined by solubility titration. *J. Pharm. Sci.*, **52** (1963), 189–191.

Hamlin, W. E. and Higuchi, W. I. Dissolution rate–solubility behavior of 3-(1-methyl-2-pyrrolidinyl)-indole as a function of hydrogen-ion concentration. *J. Pharm. Sci.*, **55** (1966), 205–207.

Hammett, L. P. *Physical Organic Chemistry*, McGraw-Hill, New York, 1940, Chapter 7.

Hanna, M. W. and Ashbaugh, A. L. Nuclear magnetic resonance study of molecular complexes of 7,7,8,8-tetracyanoquinodimethane and aromatic donors. *J. Phys. Chem.*, **68** (1964), 811–816.

Hansch, C. Use of substituent constants in drug modification. *Farm., Ed. Sci.*, **23** (1968), 293–320.

Hansch, C. Quantitative structure–activity relationships in drug design. In *Drug Design*, Vol 1, Ariens, E. J., Ed. Academic, New York, 1971, p. 271.

Hansch, C. and Anderson, S. M. The effect of intramolecular hydrophobic bonding on partition coefficients. *J. Org. Chem.*, **32** (1967), 2583–2586.

Hansch, C. and Fujita, T. ρ-σ-π Analysis. A method for the correlation of biological activity and chemical structure. *J. Am. Chem. Soc.*, **86** (1964), 1616–1626.

Hansch, C. and Leo, A. *Substituent Constants for Correlation Analysis in Chemistry and Biology*, Wiley, New York, 1979.

Hansch, C., Quinlan, J. E., and Lawrence, G. L. The linear free-energy relationship between partition coefficients and the aqueous solubility of organic liquids. *J. Org. Chem.*, **33** (1968), 347–350.

Hansen, C. and Beerbower, A. Solutility parameters. In *Kirk–Othmer Encyclopedia of Chemical Technology*, 2nd ed., Supplement Volume, Standen, A., Exec. Ed. Wiley, New York, 1971, pp. 889–910.

Hanson, W. A. *Handbook of Dissolution Testing.* Pharmaceutical Technology Publication, Springfield, OR, 1982.

Harris, F. W. and Seymour, R. B., Eds., *Structure–Solubility Relationships in Polymers*, Academic, New York, 1977.

Harris, M. J., Higuchi, T., and Rytting, J. H. Thermodynamic group contributions from ion-pair extraction equilibria for use in the prediction of partition coefficients. Correlation of surface area with group contributions. *J. Phys. Chem.*, **77** (1973), 2694–2703.

Hartley, G. S. *Aqueous Solutions of Paraffin-Chain Salts: A Study in Micelle Formation*, Hermann et Cie, Paris, 1936.

Hassel, O. and Hvoslef, J. The structure of bromine 1,4-dioxanate. *Acta Chem. Scand.*, **8** (1954), 873.

Havemeyer, R. N. and Higuchi, T. The complexing tendencies of cyanocobalamin with inorganic compounds. Heteromolybdates and heavy metal chlorides. *J. Am. Pharm. Assoc. Sci. Ed.*, **49** (1960), 356–360.

Hermann, R. B. Theory of hydrophobic bonding. I. The solubility of hydrocarbons in water, within the context of the significant structure theory of liquids. *J. Phys. Chem.*, **75** (1971), 363–368.

Hermann, R. B. Theory of hydrophobic bonding. II. The correlation of hydrocarbon solubility in water with solvent cavity surface area. *J. Phys. Chem.*, **76** (1972), 2754–2759.

Hersey, J. A. prediction of dissolution rates of drugs. *J. Pharm. Sci.*, **62** (1973), 514.

Higuchi, T. Hydrogen bonded complexes in nonpolar solutions—Influence of structure and solvent on formation tendency and stoichiometry. In *Proceedings of the American Association of Colleges of Pharmacy Teachers' Seminar*, Vol. 13, July 9–15, 1961, Lemberger, A. P., Pharmacy Ed. School of Pharmacy, University of Wisconsin, Madison, pp. 107–111.

Higuchi, T. Pro-drug, molecular structure and percutaneous delivery. In *Design of Biopharmaceutical Properties Through Pro-drugs and Analogs*, Roche, E. B., Ed. American Pharmaceutical Association, Washington, D.C., 1977.

Higuchi, T. and Bolton, S. The solubility and complexing properties of oxytetracycline and tetracycline. III. Interactions in aqueous solution with model compounds, biochemicals, metals, chelates, and hexametaphosphate. *J. Am. Pharm. Assoc., Sci. Ed.*, **48** (1959), 557–564.

Higuchi, T. and Chulkaratana, S. The Hydrogen Bond Formation of Various Alcohols with Salicylic Acid, Catechol and Hydroquinone in Nonaqueous Solution. Unpublished observations in Chulkaratana, S., M.S. Thesis, University of Wisconsin, Madison, 1961.

Higuchi, T. and Connors, K. A. Phase-solubility techniques. In *Advances in Analytical Chemistry and Instrumentation*, Vol. 4, Wiley, New York, 1965, pp. 117–212.

Higuchi, T. and Drubulis, A. Complexation of organic substances in aqueous solution by hydroxyaromatic acids and their salts: Relative contributions of several factors to the overall effect. *J. Pharm. Sci.*, **50** (1961), 905–909.

Higuchi, T. and Kato, K. Ion-pair extraction of pharmaceutical amines. II. Extraction profile of chlorpheniramine. *J. Pharm. Sci.*, **55** (1966), 1080–1084.

Higuchi, T. and Kristiansen, H. Binding specificity between small organic solutes in aqueous solution: Classification of some solutes into two groups according to binding tendencies. *J. Pharm. Sci.*, **59** (1970), 1601–1608.

Higuchi, T. and Lachman, L. Inhibition of hydrolysis of esters in solution by formation of complexes. I. Stabilization of benzocaine with caffeine. *J. Am. Pharm. Assoc., Sci. Ed.*, **44** (1955), 521–526.

Higuchi, T. and Michaelis, A. F. Mechanism and kinetics of ion pair extraction. Rate of extraction of dextromethorphanium ion. *Anal. Chem.*, **40** (1968), 1925–1931.

Higuchi, T. and Pisano, F. D. Complexation of organic substances in aqueous solution by hydroxyaromatic acids and their salts II. Influence of halogen, nitro, and other substituents and correlation of binding tendencies toward prednisolone, theophylline and phenacetin. *J. Pharm. Sci.*, **53** (1964), 644–651.

Higuchi, T., Gupta, M., and Busse, L. W. Influence of electrolytes, pH, and alcohol concentration on the solubilities of acidic drugs. *J. Am. Pharm. Assoc., Sci. Ed.*, **42** (1953), 157–161.

Higuchi, T., Michaelis, A., Tan, T., and Hurwitz, A. Ion-pair extraction of pharmaceutical amines. Role of dipolar solvating agents in extraction of dextromethorphan. *Anal. Chem.*, **39** (1967), 974–979.

Higuchi, T., Richards, J. H., Davis, S. S. Kamada, A., Hou, J. P., Nakano, M., Nakano, N. I., and Pitman, I. H., Solvency and hydrogen bonding interactions in nonaqueous systems. *J. Pharm. Sci.*, **58** (1969), 661–671.

Higuchi, T., Lee, H. K., and Pitman, I. H. Analysis of possible accelerating influence of irreversible chemical reactions on dissolution rates of reactive solids. Dissolution behavior of 7-acetyltheophylline. *Farm. Aikak.*, **80** (1971a), 55–90.

Higuchi, T., Michaelis, A., and Rytting, J. H. Role of solvating agents in promoting ion-pair extraction. *Anal. Chem.*, **43** (1971b), 287–289.

Higuchi, T., Dayal, S., and Pitman, I. H. Effects of solute–solvent complexation reactions on dissolution kinetics: Testing of a model by using a concentration jump technique. *J. Pharm. Sci.*, **61** (1972), 695–700.

Higuchi, T., Shih, F.-M. L., Kimura, T., and Rytting, J. H., Solubility determination of barely aqueous–soluble organic solids. *J. Pharm. Sci.*, **68** (1979), 1267–1272.

Higuchi, W. I. Diffusional models useful in biopharmaceutics. Drug release rate processes. *J. Pharm. Sci.*, **56** (1967), 315–324.

Higuchi, W. I. and Misra, J. Solubilization in non-polar solvents. Influence of the chain length of solvent on the solubilization of water by dioctyl sodium sulfosuccinate. *J. Pharm. Sci.*, **51** (1962), 455–458.

Higuchi, W. I., Parrott, E. L., Wurster, D. E., and Higuchi, T. Investigation of drug release from solids II. Theoretical and experimental study of influences of bases and buffers on rates of dissolution of acidic solids. *J. Am. Pharm. Assoc., Sci. Ed.*, **47** (1958), 376–383.

Higuchi, W. I., Lau, P. K., Higuchi, T., and Shell, J. W. Polymorphism and drug availability. Solubility relationships in the methylprednisolone system. *J. Pharm. Sci.*, **52** (1963), 150–153.

Higuchi, W. I., Nelson, E., and Wagner, J. G. Solubility and dissolution rates in reactive media. *J. Pharm. Sci.*, **53** (1964), 333–335.

Hildebrand, J. H. The entropy of vaporization as a means of distinguishing normal liquids. *J. Am. Chem. Soc.*, **37** (1915), 970–978.

Hildebrand, J. H. The vapor pressures of liquid metals. *J. Am. Chem. Soc.*, **40**, (1918), 45–49.

Hildebrand, J. H. Solubility. XII. Regular solutions. *J. Am. Chem. Soc.*, **51** (1929), 66–80.

Hildebrand, J. H. A criticism of the term "hydrophobic bond," *J. Phys. Chem.*, **72** (1968), 1841–1842.

Hildebrand, J. H. and Scott, R. L. *Solubility of Nonelectrolytes*, 3rd ed., Reinhold, New York, 1950, pp. 11–13, 47, 160, 175–197.

Hildebrand, J. H. and Scott, R. L. *Regular Solutions*, Prentice-Hall, Englewood Cliffs, NJ, 1962.

Hildebrand, J. H., Prausnitz, J. M., and Scott, R. L. *Regular and Related Solutions*, Van Nostrand Reinhold, New York, 1970, pp. 64–67.

Hine, J. S. *Structural Effects on Equilibria in Organic Chemistry*, Wiley, New York, 1975, pp. 66, 202–205.

Hirano, E. and Kozima, K. The intermolecular hydrogen bonding between methanol and triethylamine in various states. *Bull. Chem. Soc. Jpn.*, **39** (1966), 1216–1220.

Hirschfelder, J. O. Intermolecular forces. In *Study Week on Molecular Forces* (April 18–23, 1966,

Pontificiae Academiae Scientiarum Scripta Varia 31, Vatican City), North Holland, Amsterdam, 1967, p. 87.

Hirschfelder, J. O., Curtiss, C. F., and Bird, R. B. *Molecular Theory of Gases and Liquids*, Wiley, New York, 1954, p. 280.

Hixon, A. W. and Baum, S. J. Mass transfer and chemical reaction in liquid–solid agitation. *Ind. Eng. Chem.*, **36** (1944), 528–531.

Hixon, A. W. and Crowell, J. *Ind. Eng. Chem.*, **23** (1931) 923–926.

Hofacker, G. L., Marechal, Y., and Ratner, M. A. Dynamic properties of hydrogen bonded systems. In *The Hydrogen Bond, Recent Developments in Theory and Experiments*, Schuster, P., Zundel, G., and Sandorfy, C., Eds. North-Holland, Amsterdam, 1976, pp. 295–357.

Hollenbeck, R. G. Determination of differential heat of solution in real solutions from variation in solubility with temperature. *J. Pharm. Sci.*, **69** (1980), 1241–1242.

Hou, J. P. H. Hydrogen Bond Formation. Phenols with Organophosphorus Compounds in Solution. Ph.D. Thesis, University of Wisconsin, Madison, WI, 1967.

Hovorka, F., Lankelma, H. P., and Axelrod, A. E. Thermodynamic properties of the hexyl alcohols. III. 2-Methylpentanol-3 and 3-methylpentanol-3. *J. Am. Chem. Soc.*, **62** (1940), 187–189.

Howarth, O. W., Reassessment of hydrophobic bonding. *J. Chem. Soc. Faraday Trans. I*, **71** (1975), 2303–2309.

Hoy, K. L. and Martin, R. A. *Tables of Solubility Parameters*, Union Carbide Corp., New York, May 16, 1975. (3rd reprinting as the *Hoy Tables of Solubility Parameters*, Union Carbide, Solvents and Coating Materials, South Charleston, WV, 1985.)

Huggins, M. L. Certain properties of long-chain compounds as functions of chain length. *J. Phys. Chem.*, **43** (1939), 1083–1098.

Huggins, M. L., Solutions of long chain compounds. *J. Chem. Phys.*, **9** (1941), 440.

Huggins, M. L. Thermodynamic properties of solutions of long-chain compounds. *Ann. NY Acad. Sci.*, **43** (1942), 1–32.

Hull, R. L. and Biles, J. A. Physical chemical study of the distribution of some amine salts between immiscible solvents II: Complexation in the organic phase. *J. Pharm. Sci.*, **53** (1964), 869–872.

Hussain, A. Prediction of dissolution rates of slightly water-soluble powders from simple mathematical relationships. *J. Pharm. Sci.*, **61** (1972), 811–813.

Hüttenrauch, R. Molekulargalenic als Grundlage moderner Arzneiformung. (Molecular pharmaceutics as the basis of modern drug formulation). *Acta Pharm. Technol., APV Informationsdienst Suppl.*, **6** (1978a), 55–127.

Hüttenrauch, R. Fortschritte und tendenzen in der entwicklung neuer arzneiformen, *Pharmazie*, **33** (1978b), 481–499.

Inagi, T., Maramatsu, T., Nagai, H., and Terada, H. Influence of vehicle composition on the penetration of indomethacin through guinea pig skin. *Chem. Pharm. Bull.* (*Tokyo*), **29** (1981), 1708–1714.

Irmann, F. Eine einfache Korrelation zwischen Wasserlöslichkeit und Struktur von Kohlenwasserstoffen und Halogenkohlenwasserstoffen. *Chem. Ing. Tech.*, **37** (1965), 789–798.

Irwin, G. M., Kostenbauder, H. B., Dittert, L. W., Staples, R., Misher, A., and Swintosky, J. V. Enhancement of gastrointestinal absorption of a quaternary ammonium compound by trichloroacetate. *J. Pharm. Sci.*, **58** (1969), 313–315.

IUPAC publications on solubility.

IUPAC (International Union of Pure and Applied Chemistry), *Pure Appl. Chem.*, **51** (1979), 1–41.

Iwasa, J., Fujita, T., and Hansch, C. Substituent constants for aliphatic functions obtained from partition coefficients. *J. Med. Chem.*, **8** (1965), 150–153.

Jaffé, H. H. A reexamination of the Hammett equation. *Chem. Rev.*, **53** (1953), 191–261.

James, K. C. *Solubility and Related Properties*. Dekker, New York, 1986.

James, K. C. and Roberts, M. The solubilities of the lower testosterone esters. *J. Pharm. Pharmacol.*, **20** (1968), 709–714.

Jauquet, M. and Laszlo, P. Influence of solvents on spectroscopy. In *Solutions and Solubilities*, Part 1, Dack, M. R. J., Ed. Wiley, New York, 1975, pp. 195–258.

Jencks, W. P. *Catalysis in Chemistry and Enzymology*, McGraw-Hill, New York, 1969, pp. 175–178.

Jentschura, Von U. and Lippert, E. NMR-spektroskopische Untersuchungen über die Eigenassoziation von Karbonsäuren in inerten Lösungsmitteln I. Iterative Berechnung thermodynamischer Grössen für Essigsäure. Part 2. *Ber. Bunsenges. Phys. Chem.*, **75** (1971), 556–564; 782–787.

Joesten, M. D. and Schaad, L. J. *Hydrogen Bonding*. Dekker, New York, 1974.

Johnson, C. D. Linear free-energy relationships and the reactivity–selectivity principle. *Chem. Rev.*, **75** (1975), 755–765.

Johnson, G. C. and Francis, A. W. Ternary liquid system, benzene–heptane–diethylene glycol. *Ind. Eng. Chem.*, **46** (1954), 1662–1668.

Johnson, J. R., Christian, S. D., and Affsprung, H. E. Self-association and hydration of phenol in carbon tetrachloride. *J. Chem. Soc.*, **1965** (1965), 1–6.

Jones, J. R. Proton transfer reactions in highly basic media. *Prog. Phys. Org. Chem.*, **9** (1972), 241–274.

Jones, M. M. *Elementary Coordination Chemistry*. Prentice-Hall, Englewood Cliffs, NJ, 1964, pp. 282–285.

Jones, T. M. The physico technical properties of starting materials used in tablet formulation. *Int. J. Pharm. Technol. Prod. Mfr.*, **2** (1981), 17–24.

Jonkman, J. H. G. Ionpaar–extractie als isoleringsmethode bij de analyse van geneesmiddelen en metabolieten in lichaamsvloeistoffen. II. Praktische uitvoering en toepassingen. *Pharm. Weekbl.*, **110** (1975), 673–689.

Jordan, T. E. *Vapor Pressure of Organic Compounds*, Interscience New York, 1954.

Joris, L., Mitsky, J., and Taft, R. W. The effects of polar aprotic solvents on linear free-energy relationships in hydrogen-bonded complex formation. *J. Am. Chem. Soc.*, **94** (1972), 3438–3442.

Jost, W. *Diffusion in Solids, Liquids, and Gases*. Academic, New York, 1960.

Judy, C. L., Pontikos, N. M., and Acree, W. E. Solubility in binary solvent systems: Comparison of predictive equations derived from the NIBS model. *Phys. Chem. Liq.*, **16** (1987), 179–187.

Kakeya, N., Yata, N., Kamada, A., and Aoki, M. Biological activities of drugs. VIII. Structure–activity relationship of sulfonamide carbonic anhydrase inhibitors. *Chem. Pharm. Bull.*, **17** (1969), 2558–2564.

Kamlet, M. M., Abraham, M. H., Doherty, R. M., and Taft, R. W. Solubility properties in polymers and biological media. 4. Correlation of octanol/water partition coefficients with solvatochromic parameters. *J. Am. Chem. Soc.*, **106** (1984), 464–466.

Karger, B. L., Snyder, L. R., and Eon, C. An expanded solubility parameter treatment for classification and use of chromatographic solvents and adsorbents. Parameters for dispersion, dipole and hydrogen bonding interactions. *J. Chromatogr.*, **125** (1976), 71–88.

Karger, B. L., Snyder, L. R., and Eon, C. Expanded solubility parameter treatment for classification and use of chromatographic solvents and absorbents. *Anal. Chem.*, **50** (1978), 2126–2136.

Kauzmann, W. Some factors in the interpretation of protein denaturation. *Adv. Protein Chem.*, **14** (1959), 1–63.

Keller, R. A., Karger, B. L., and Snyder, L. R. Use of the solubility parameter in predicting chromatographic retention and eluotropic strength. In *Gas Chromatography*, Stock, R., Ed. Institute of Petroleum, London, 1970, pp. 125–140.

Kertes, A. S. The chemistry of solvent extraction. In *Recent Advances in Liquid–Liquid Extraction*, Hanson, C., Ed. Pergamon, Oxford, England, 1971, pp. 15–92.

King, C. V. and Brodie, S. S. The rate of dissolution of benzoic acid in dilute aqueous alkali. *J. Am. Chem. Soc.*, **59** (1937), 1375–1379.

Kinkel, J. F. M., Tomlinson, E., and Smit, P. Thermodynamics and extrathermodynamics of organic solute liquid–liquid distribution between water and 2,2,4-trimethylpentane. *Int. J. Pharm.*, **9** (1981), 121–136.

Kirkwood, J. G. *Theory of Solutions*, Gordon and Breach, New York, 1968.

Kitaigorodsky, A. I. *Molecular Crystals and Molecules*. Academic, New York, 1973.

Kitaigorodsky, A. I. Non-bonded interactions of atoms in organic crystals and molecules. *Chem. Soc. Rev.*, **7** (1978), 133–163.

Kivinen, A., Murto, J., and Kilpi, L. Fluoroalcohols. Part 7. Intermolecular hydrogen bonding of 2,2,2-trifluoroethanol and 1,1,1,3,3,3-hexafluoro-2-propanol with proton acceptors of various types. *Suom. Kemistil. B.*, **40** (1967), 301–313.

Kivinen, A., Murto, J., Korppi-Tommola, J., and Kuopio, R. Fluoroalcohols, Part 15. A near-infrared study of the self-association of trifluoro and hexafluoro substituted tertiary butyl alcohols. A model for the structure of alcohol associates. *Acta Chem. Scand.*, **26** (1972), 904–922.

Klotz, I. M. Protein hydration and behavior. *Science*, **128** (1958), 815–822.

Klotz, I. M. *Chemical Thermodynamics*, rev. ed. W. A. Benjamin, New York, 1964, pp. 348–358.

Klotz, I. M. Role of water structure in macromolecules. *Fed. Proc., Suppl.*, **15** (1965), S-24–S-33.

Kojima, I. and Davis, S. S. The effect of salt concentration on the distribution of phenol between aqueous sodium chloride and carbon tetrachloride. *Int. J. Pharm.*, **20** (1984), 203–207.

Kolarik, Z. and Pankova, H. Acidic organophosphorus extractants. I. Extraction of lanthanides by means of dialkyl phosphoric acids—effect of structure and size of alkyl groups. *J. Inorg. Nucl. Chem.*, **28** (1966), 2325–2333.

Kollman, P. A. Noncovalent interactions. *Acc. Chem. Res.*, **10** (1977), 365–371.

Kollman, P. A. and Allen, L. C. The theory of the hydrogen bond. *Chem. Rev.*, **72** (1972), 283–303.

Korenman, I. M., Gur'ev, I. A., and Gur'eva, Z. M. Solubility of liquid aliphatic compounds in water. *Russ. J. Phys. Chem.*, **45** (1971), 1065–1066. Original article in Russian, *Z. Fiz. Khim.*, **45** (1971), 1866.

Kostenbauder, H. B. and Higuchi, T. Formation of molecular complexes by some water-soluble amides. I. Interaction of several amides with *p*-hydroxybenzoic acid, salicylic acid, chloramphenicol and phenol. *J. Am. Pharm. Assoc., Sci. Ed.*, **45** (1956a), 518–522.

Kostenbauder, H. B. and Higuchi, T. Formation of molecular complexes by some water-soluble amides. II. Effect of decreasing water solubility on degree of complex formation. *J. Am. Pharm. Assoc., Sci. Ed.*, **45** (1956b), 810–813.

Kretschmer, C. B. and Wiebe, R. Thermodynamics of alcohol–hydrocarbon mixtures. *J. Chem. Phys.*, **22** (1954), 1697–1701.

Krishnan, C. V. and Friedman, H. L. Solvation enthalpies of hydrocarbons and normal alcohols in highly polar solvents. *J. Phys. Chem.*, **75** (1971), 3598–3606.

Kristiansen, H., Nakano, M., Nakano, N. I., and Higuchi, T. Effect of solvent composition on association between small organic species. *J. Pharm. Sci.*, **59** (1970), 1103–1106.

Kumar, R. and Prausnitz, J. M. Solvents in chemical technology. In *Solutions and Solubilities*, Part 1, *Techniques of Chemistry*, Vol. VIII, Dack, M. R. J., Ed., Wiley, New York, 1975, pp. 259–326.

Kuntz, I. D., Jr., Gasparro, F. P., Johnston, M. D., Jr., and Taylor, R. P. Molecular interactions and the Benesi–Hildebrand equation. *J. Am. Chem. Soc.*, **90** (1968), 4778–4781.

Lachman, L. and Higuchi, T. Inhibition of hydrolysis of esters in solution by formation of complexes. III. Stabilization of tetracaine with caffeine. *J. Am. Pharm. Assoc., Sci. Ed.*, **46** (1957), 32–36.

Lachman, L., Ravin, L. J., and Higuchi, T. Inhibition of hydrolysis of esters in solution by formation of complexes. II. Stabilization of procaine with caffeine. *J. Am. Pharm. Asoc., Sci. Ed.*, **45** (1956), 290–295.

Lachman, L., Guttman, D., and Higuchi, T. Inhibition of hydrolysis of esters in solution by formation of complexes. IV. Stabilization of benzocaine with 1-ethyltheobromine. *J. Am. Pharm. Assoc., Sci. Ed.*, **46** (1957), 36–38.

Lagerström, P.-O., Borg, K. O., and Westerlund, D. Fluorimetric determinations by ion-pair extraction. 4. Studies on quantitative ion-pair extraction of ammonium compounds with anthracene-2-sulfonate as counterion. *Acta Pharm. Suec.*, **9** (1972), 53–62.

Laiken, N. and Némethy, G. A statistical–thermodynamic model of aqueous solutions of alcohols. *J. Phys. Chem.*, **74** (1970), 3501–3509.

Lamb, D. J. and Harris, L. E. Correlation of the distribution coefficients of various barbituric acids. *J. Am. Pharm. Assoc.*, **49** (1960), 583–585.

Langenbucher, F. *In vitro* assessment of dissolution kinetics: Description and evaluation of a column-type method. *J. Pharm. Sci.*, **58** (1969), 1265–1272.

Langmuir, I. Distribution and orientation of molecules. *Colloid Symp. Monograph*, **3** (1925), 48–75.

Lassettre, E. N. and Dickinson, R. G. A comparative method of measuring vapor pressure lowering with application to solutions of phenol in benzene. *J. Am. Chem. Soc.*, **61** (1939), 54–57.

Latimer, W. M. and Rodebush, W. H. Polarity and ionization from the standpoint of the Lewis theory of valence. *J. Am. Chem. Soc.*, **42** (1920), 1419–1433.

Lauffer, M. A. Polymerization–depolymerization of tobacco mosaic virus protein. VII. A model. *Biochemistry*, **5** (1966), 2440–2446.

Laurence, C. and Wojtkowiak, B. Etude par spectroscopie infrarouge des effets de substituants sur les associations moléculaires par liaison hydrogène. *C. R. Acad. Sci. (Paris)*, **264c** (1967), 1216–1219.

Laurence, C. and Wojtkowiak, B. Relation entre facteurs de Hammett et déplacements de fréquence par liaison hydrogène en spectroscopie infrarouge. *Bull. Soc. Chim. Fr.*, **7** (1968), 2780–2783.

Lawson, D. D. and Ingham, J. D. Estimation of solubility parameters from refractive index data. *Nature (London)*, **223** (1969), 614–615.

Lee, H. K., Lambert, H., Stella, V. J., Wang, D., and Higuchi, T. Hydrolysis and dissolution behavior of a prolonged-release prodrug of theophylline: 7,7'-succinylditheophylline. *J. Pharm. Sci.*, **68** (1979), 288–295.

Le Fèvre, R. J. W. Molecular refractivity and polarizability. *Adv. Phys. Org. Chem.*, **3** (1965), 1–90.

Leo, A., Hansch, C., and Elkins, D. Partition coefficients and their uses. *Chem. Rev.*, **71** (1971), 525–616.

Leo, A., Jow, P. Y. C., Silipo, C., and Hansch, C. Calculation of hydrophobic constant (log P) from π and f constants. *J. Med Chem.*, **18** (1975), 865–868.

Leo, A., Hansch, C., and Jow. P. Y. C. Dependence of hydrophobicity of apolar molecules on their molecular volumes. *J. Med. Chem.*, **19** (1976), 611–615.

Leo, A. J. Relationships between partitioning solvent systems. In *Biological Correlations—The Hansch Approach* (Advances in Chemistry Series, No. 114), American Chemical Society, Washington, D.C., 1972, pp. 51–60.

Lesson, L. J. and Carstensen, J. T. *Dissolution Technology*. American Pharmaceutical Association, Washington, D.C., 1974.

Levich, V. G. *Physicochemical Hydrodynamics*. Prentice-Hall, Englewood Cliffs, NJ, 1962.

Levy, G. and Hayes, B. A. Physicochemical basis of the buffered acetylsalicylic acid controversy. *New Engl. J. Med.*, **262** (1960), 1053–1058.

Levy, O., Markovits, G. Y., and Perry, I. Thermodynamics of aggregation of long chain carboxylic acids in benzene. *J. Phys. Chem.*, **79** (1975), 239–242.

Lewis, G. N. and Randall, M. Thermodynamics, 2nd ed., revised by Pitzer, K. S. and Brewer, L., McGraw-Hill, New York, 1961, pp. 244–249.

Lin, L.-N., Christian, S. D., and Tucker, E. E. Solute activity study of the self-association of phenol in cyclohexane and carbon tetrachloride. *J. Phys. Chem.*, **82** (1978), 1897–1901.

Lindau, G. v. Zur erklärüng der hydrotropie. *Naturwissenschaften*, **20** (1932), 396–401.

Linder, B. Continuum-model treatment of long-range intermolecular forces. I. Pure substances. *J. Chem. Phys.*, **33** (1960), 668–675.

Lippert, E. von. Wasserstoffbrückenbindung und magnetische protonenresonanz. *Ber. Bunsenges. Phys. Chem.*, **67** (1963), 267–280.

Long, F. A. and McDevit, W. F. Activity coefficients of non-electrolyte solutes in aqueous salt solutions. *Chem. Revs.*, **51** (1952), 119–169.

Longsworth, L. G. The diffusion of hydrogen bonding solutes in carbon tetrachloride. *J. Colloid Interface Sci.*, **22** (1966), 3–11.

Longuett-Higgins, H. C. Solutions of chain molecules—A new statistical theory. *Discuss. Faraday Soc.*, **15** (1953), 73–80.

Lönnberg, H., Ylikoski, J., and Vesala, A. Effect of alkyl substituents on the thermodynamics of the self-association of purine in aqueous solution. *J. Chem. Soc., Faraday Trans. I.*, **80** (1984), 2439–2444.

Lovering, E. G. and Black, D. B. Drug permeation through membranes III. Effect of pH and various substances on permeation of phenylbutazone through everted rat intestine and polydimethyl-siloxane. *J. Pharm. Sci.*, **63** (1974a), 671–676.

Lovering, E. G. and Black, D. B. Diffusion layer effects on permeation of phenylbutazone through polydimethylsiloxane. *J. Pharm. Sci.*, **63** (1974b), 1399–1402.

Lumb, E. C. and Winsor, P. A. Solubilization titration. A rapid method for the analysis of binary mixtures of organic liquids. *Analyst*, **77** (1952), 1012–1016.

Lumry, R. and Rajender, S. Enthalpy–entropy compensation phenomena in water solutions of proteins and small molecules: A ubiquitous property of water. *Biopolymers*, **9** (1970), 1125–1227.

MacNicol, D. D., McKendrick, J. J., and Wilson, D. R. Clathrates and molecular inclusion phenomena. *Chem. Soc. Rev.*, **7** (1978), 65–87.

Magnussen, T., Rasmussen, P., and Fredenslund, A. UNIFAC parameter table for prediction of liquid–liquid equilibria. *Ind. Eng. Chem. Process Des. Dev.*, **20** (1981), 331–339.

Maitland, G. C., Rigby, M., Smith, E. B., and Wakeham, E. G. *Intermolecular Forces, Their Origin and Determination.* Clarendon Press, Oxford, 1981, p. 20.

Manzo, R. H. Effects of solvent medium on solubility. A linear free energy relationship treatment. *J. Pharm. Pharmacol.*, **34** (1982), 486–492.

Marsden, R. J. B. and Sutton, L. E. Evidence of wave-mechanical resonance in the carboxylic ester and the lactone group from electric dipole moments. *J. Chem. Soc.*, **139** (1936), 1383–1390.

Martin, A., Newburger, J., and Adjei, A. Extended Hildebrand solubility approach: Solubility of theophylline in polar binary solvents. *J. Pharm. Sci.*, **69** (1980), 487–491.

Martin, A., Paruta, A. N., and Adjei, A. Extended Hildebrand solubility approach: Methylxanthines in mixed solvents. *J. Pharm. Sci.*, **70** (1981), 1115–1120.

Martin, A. J. P. Some theoretical aspects of partition chromatography. *Biochem. Soc. Symposia (Cambridge)*, **3** (1949), 4–20.

Martindale, W. *The Extra Pharmacopoeia*, 27th Ed., Wade, A. and Reynolds, J. E. F., Eds. The Pharmaceutical Press, London, 1977.

Masterton, W. L. Partial molal volumes of hydrocarbons in water solution. *J. Chem. Phys.*, **22** (1954), 1830–1833.

Martire, D. E. Thermodynamics of dilute solutions: Aliphatic solutes in aromatic solvents and various solutes in paraffinic solvents. In *Gas Chromatography*, 1966, Littlewood, A. B., Ed. Proceedings of the Sixth International Symposium on Gas Chromatography and Associated Techniques (Sept. 20–23, 1966, Rome), Institute of Petroleum, London, 1967, pp. 21–340.

Martire, D. E. and Riedl, P. A thermodynamic study of hydrogen bonding by means of gas–liquid chromatography. *J. Phys. Chem.*, **72** (1968), 3478–3488.

Martire, D. E., Sheridan, J. P., King, J. W., and O'Donnell, S. E. Thermodynamics of molecular association. 9. An NMR study of hydrogen bonding of $CHCl_3$ and $CHBr_3$ to di-*n*-octyl ether, di-*n*-octyl thioether and di-*n*-octylmethylamine. *J. Am. Chem. Soc.*, **98** (1976), 3101–3106.

McAuliffe, C. Solubility in water of paraffin, cycloparaffin, olefin, acetylene, cycloolefin and aromatic hydrocarbon. *J. Phys. Chem.*, **70** (1966), 1267–1275.

McClellan, A. L. *Table of Experimental Dipole Moments*, W. H. Freeman, San Francisco, 1963.

McClellan, B. E. and Freiser, H. Kinetics and mechanism of extraction of zinc, nickel, cobalt and cadmium with diphenylthiocarbazone di-*o*-tolylthiocarbazone and di-alpha-naphthylthiocarbazone. *Anal. Chem.*, **36** (1964), 2262–2265.

McGlashan, M. L. Internationally recommended names and symbols for physicochemical quantities and units. *Ann. Rev. Phys. Chem.*, **24** (1973), 51–76.

McGowan, J. C. The physical toxicity of chemicals. II. Factors affecting physical toxicity in aqueous solutions. *J. Appl. Chem.*, **2** (1952), 323–328.

McGowan, J. C. Partition coefficients and biological activities. *Nature* (*London*), **200** (1963), 1317.

McGowan, J. C. and Mellors, A. *Molecular Volumes in Chemistry and Biology. Applications Including Partitioning and Toxicology*. Ellis Harwood, London, 1986.

McGowan, J. C. Atkinson, P. N., and Ruddle, L. H. The physical toxicity of chemicals. V. Interaction terms for solubilities and partition coefficients. *J. Appl. Chem.*, **16** (1966), 99–104.

McHan, D. R., Rytting, J. H., and Higuchi, T. unpublished observations, 1980.

McKee, R. M. Use of hydrotropic solutions in industry. *Ind. Eng. Chem.*, **38** (1946), 382–384.

McLaughlin, E. and Zainal, H. A. The solubility behaviour of aromatic hydrocarbons in benzene. *J. Chem. Soc.*, **1959** (1959), 863–867.

Mehdizadeh, M. and Grant, D. J. W. Solubility and complexation behavior of griseofulvin in fatty acid–isooctane mixtures. *J. Pharm. Sci.*, **73** (1984), 1195–1202.

Meier, J. and Higuchi, T. Determination of stability constants from optical rotatory dispersion measurements. Camphor–phenol system in carbon tetrachloride. *J. Pharm. Sci.*, **54** (1965), 1183–1186.

Mellan, I. *Compatability and Solubility*, Noyes Development Corporation, Park Ridge, NJ, 1968, pp. 1–15.

Mellan, I. *Industrial Solvents Handbook*, 2nd ed., Noyes Data Corporation, Park Ridge, NJ, 1977.

Michaelis, A. F. and Higuchi, T. Ion-pair extraction of pharmaceutical amines. IV: Influence of anion on enthalpic, entropic, and free energy changes resulting from phase transfer. *J. Pharm. Sci.*, **58** (1969), 201–204.

Mikkelson, T. J., Watanabe, S., Rytting, J. H., and Higuchi, T. Effect of self-association of phenol on its transport across polyethylene film. *J. Pharm. Sci.*, **69** (1980), 133–137.

Miller, K. W. and Hildebrand, J. H. Solutions of inert gases in water. *J. Am. Chem. Soc.*, **90** (1968), 3001–3004.

Mirrlees, M. S., Moulton, S. J., Murphy, C. T., and Taylor, P. J. Direct measurement of octanol–water partition coefficients by high-pressure liquid chromatography. *J. Med. Chem.*, **19** (1976), 615–619.

Mitsky, J., Joris, L., and Taft, R. W. Hydrogen-bonded complex formation with 5-fluoroindole. Applications of the pK_{HB} scale. *J. Am. Chem. Soc.*, **94** (1972), 3442–3445.

Modin, R., Persson, B.-A., and Schill, G. Ion-pair extraction in the isolation of ionisable organic substances, Proc. Int. Solvent Extraction Conf., The Hague, 19–23 April, 1971. Vol. II, Soc. Chem. Ind., London, 1971, pp. 1121–1220.

Mooney, K. G., Mintun, M. A., Himmelstein, K. J., and Stella, V. J. Dissolution kinetics of carboxylic acids. I: Effect of pH under unbuffered conditions. *J. Pharm. Sci.*, **70** (1981a), 13–22.

Mooney, K. G., Mintun, M. A., Himmelstein, K. J., and Stella, V. J. Dissolution kinetics of carboxylic acids II: Effect of buffers. *J. Pharm. Sci.*, **70** (1981b), 22–32.

Mooney, K. G., Rodriguez-Gaxiola, M., Mintun, M., Himmelstein, K. J., and Stella, V. J. Dissolution kinetics of phenylbutazone. *J. Pharm. Sci.*, **70** (1981c), 1358–1365.

Morawetz, H. *Macromolecules in Solution*, Wiley, New York, 1975.

Morgan, S. L. and Deming, S. N. Simplex optimization of analytical chemical methods. *Anal. Chem.*, **46** (1974), 1170–1181.

Morokuma, K. Why do molecules interact? The origin of electron donor–acceptor complexes, hydrogen bonding, and proton affinity. *Acc. Chem. Res.*, **10** (1977), 294–300.

Morozowich, W., Oesterling, T. O., Miller, W. L., Lawson, C. F., Weeks, J. R., Stehle, R. G., and Douglas, S. L. Prostaglandin prodrugs. I: Stabilization of dinoprostone (Prostaglandin E_2) in solid state through formation of crystalline C_1-phenyl esters. *J. Pharm. Sci.*, **68** (1979), 833–836.

Mukerjee, P. Use of ionic dyes in the analysis of ionic surfactants and other organic compounds. *Anal. Chem.*, **28** (1956), 870–873.

Mukhayer, G. I. and Davis, S. S. Interactions between large organic ions of opposite charge. I. Stoichiometry of the interactions of sodium dodecylsulfate with benzyltriphenylphosphonium chloride. *J. Colloid Interface Sci.*, **53** (1975), 224–234.

Mukhayer, G. I. and Davis, S. S. Interactions between large organic ions of opposite charge. II. The effect of ionic strength, temperature and urea on the interaction between sodium dodecylsulfate and benzyltriphenylphosphonium chloride. *J. Colloid Interface Sci.*, **56** (1976), 350–359.

Mukhayer, G. I. and Davis, S. S. Interactions between large organic ions of opposite charge. III. The effect of molecular structure on interaction between sodium alkyl sulfates and phosphonium salts. The alkyl chain. *J. Colloid Interface Sci.*, **59** (1977a), 350–359.

Mukhayer, G. I. and Davis, S. S. Interactions between large organic ions of opposite charge. IV. The effect of molecular structure. The interaction between sodium dodecyl sulfate and substituted benzyltriphenylphosphonium chlorides. *J. Colloid Interface Sci.*, **61** (1977b), 582–589.

Mukhayer, G. I. and Davis, S. S. Interactions between large organic ions of opposite charge. VI. Coacervation in mixtures of sodium dodecyl sulfate and benzyltriphenylphosphonium chloride. *J. Colloid Interface Sci.*, **66** (1978a), 110–117.

Mukhayer, G. I. and Davis, S. S. Interactions between large organic ions of opposite charge. V. The effect of substituted ureas on the interaction between sodium dodecyl sulfate and benzyltriphenylphosphonium chloride. *J. Colloid Interface Sci.*, **65** (1978b), 210–215.

Mukhayer, G. I., Davis, S. S., and Tomlinson, E. Automated conductimetric titrimeter: Use in studying ionic solute–solute interactions. *J. Pharm. Sci.*, **64** (1975), 147–151.

Mukhopadhyay, M. and Dongaonkar, K. R. Prediction of liquid–liquid equilibria in multicomponent aromatics extraction systems by use of the UNIFAC group contribution model. *Ind. Eng. Chem. Process Des. Dev.*, **22** (1983), 521–532.

Mukhopadhyay, M. and Pathak, A. S. L-L-E data for aromatics extraction calculations using a modified UNIFAC model. *Ind. Eng. Chem. Process Des. Dev.*, **25** (1986), 733–736.

Müller, W. Löslichkeit der wichtigsten Alkaloide in Wasser, mit Aether gesättigtem Wasser, mit Wasser gesättigtem Aether, Essigäther, Chloroform, Aether, Benzol, Petroläther und Tetrachlorkohlenstoff, *Apoth.-Ztg.*, **18** (1903) No. 25, 208–209; No. 26, 218–219; No. 27, 223–25; No. 28, 232–234; No. 30, 248–250; No. 31, 257–258; No. 32, 266–267.

Muratsugu, M., Kamo, N., Kurihara, K. and Kobatake, Y. Selective electrode for dibenzyl dimethyl ammonium cation as indicator of the membrane potential in biological systems. *Biochim. Biophys. Acta*, **464** (1977), 613–619.

Murthy, K. S. and Zografi, G. Oil–water partitioning of chlorpromazine and other phenothiazine derivatives using dodecane and *n*-octanol. *J. Pharm. Sci.*, **59** (1970), 1281–1285.

Nagakura, S. Studies on the hydrogen bonding by use of the near ultraviolet absorption spectrum. II. On the proton-donating powers of *o*-, *m*-, *p*-chlorophenol, *o*-, *m*- and *p*-cresol and α- and β-naphthol. *J. Chem. Soc. Jpn. Pure Chem. Sect.*, **75** (1954), 734–737 (the original article is in Japanese).

Nakano, M. and Higuchi, T. Determination of molecular binding in aqueous solution from optical activity measurements. Interaction of tryptophan with alkylxanthines. *J. Pharm. Sci.*, **57** (1968), 1865–1868.

Nakano, M., Nakano, N. I., and Higuchi, T. Calculations of stability constants of hydrogen-bonded complexes from proton magnetic resonance data. Interactions of phenol with dimethylacetamide and various ketones. *J. Phys. Chem.*, **71** (1967), 3954–3959.

Nakashima, T. T., Traficante, D. D., and Maciel, G. E. Carbon-13 chemical shifts of the carbonyl carbon. VII. The phenol–acetone system. *J. Phys. Chem.*, **78** (1974), 124–129.

Nancollas, G. H. *Interactions in Electrolyte Solutions*, Elsevier, Amsterdam, 1966.

Nelson, H. D. and De Ligny, C. L. Interpretation of solubility data on the *n*-alkanes and some other substances in the solvent water. *Rec. Trav. Chim. Pays-Bas.*, **87** (1968), 623–640.

Nelson, K. G. and Shah, A. C. Convective diffusion model for a transport-controlled dissolution rate process. *J. Pharm. Sci.*, **64** (1975) 610–614.

Némethy, G. and Scheraga, H. A. Structure of water and hydrophobic bonding in proteins. I. A model for the thermodynamic properties of liquid water. *J. Chem. Phys.*, **36** (1962a), 3382–3400.

Némethy, G. and Scheraga, H. A. Structure of water and hydrophobic bonding in proteins. II. Model for the thermodynamic properties of aqueous solutions of hydrocarbons. *J. Chem. Phys.*, **36** (1962b), 3401–3417.

Némethy, G. and Scheraga, H. A. The structure of water and hydrophobic bonding in proteins. III. The thermodynamic properties of hydrophobic bonds in proteins. *J. Phys. Chem.*, **66** (1962c), 1773–1789.

Némethy, G., Scheraga, H. A., and Kauzmann, W. Comments on the communication "A criticism of the term 'hydrophobic bond'" by Joel H. Hildebrand. *J. Phys. Chem.*, **72** (1968), 1842.

Nernst, W. Theorie der reaktionsgeschwindigkeit in heterogenen systemen. *Z. Phys. Chem.*, **47** (1904) 52–55.

Neuberg, C. Hydrotropische erscheinungen. I. Mitteilung. *Biochem. Z.*, **76** (1916), 107–176.

Nicklasson, M., Brodin, A., and Sundelöf, L.-O. Studies of some characteristics of molecular dissolution kinetics from rotating discs. *Int. J. Pharm.*, **23** (1985), 97–108.

Nogami, H., Nagai, T., and Suzuki, A. Studies on powdered preparations. XVII. Dissolution rate of sulfonamides by rotating disk method. *Chem. Pharm. Bull.*, **14** (1966), 329–338.

Northrop, J. H. and Kunitz, M. Solubility curves of mixtures and solid solutions. *J. Gen. Physiol.*, **13** (1929), 781–791.

Noyes, A. A. and Whitney, W. R. The rate of solution of solid substances in their own solutions. *J. Am. Chem. Soc.*, **19** (1897), 930–934.

Nyburg, S. C. and Faerman, C. H. A revision of van der Waals atomic radii for molecular crystals: N, O, F, S, Cl, Se, Br and I bonded to carbon. *Acta Cryst.*, **B41** (1985), 274–279.

Nys, G. G. and Rekker, R.F. Statistical analysis of a series of partition coefficients with special reference to the predictability of folding of drug molecules. Introduction of hydrophobic fragmental constants (*f* values). *Chim. Ther.*, **8** (1973), 521–535.

Occam, William of. *Dialogus*. Publisher to the Holy Roman Emperor, Ludwig of Bavaria, Munich, 1343.

Ochsner, A. B. and Sokoloski, T. D. Prediction of solubility in nonideal multicomponent systems using the UNIFAC group contribution model. *J. Pharm. Sci.*, **74** (1985), 634–637.

Okada, S., Nakahara, H., Yomota, C., and Mochida, K. Partition behavior of procaine and *p*-aminobenzoic acid in pentanol/water and ethyl acetate/water systems. *Chem. Pharm. Bull.*, **32** (1984), 3287–3290.

Olabisi, O. (Union Carbide Corp.), Polyblends. In *Kirk–Othmer Encyclopedia of Chemical Technology*, 3rd ed., Vol. 18, Grayson, M. and Eckroth, D., Eds. Wiley, New York, 1982, pp. 443–478.

Olander, D. R. Simultaneous mass transfer and equilibrium chemical reaction. *Am. Inst. Chem. Eng. J.*, **6** (1960), 233–239.

Pakula, R., Pichnej, L., Spychala, S., and Butkiewicz, K. Polymorphism of indomethacin. Part I. Preparation of polymorphic forms of indomethacin. *Pol. J. Pharmacol. Pharm.*, **29** (1977), 151–156.

Palit, S. R. and Venkateswarlu, V., Solubilization of water in non-polar solvents by detergent mixtures. *J. Chem. Soc.*, (1954) 2129–2134.

Papadopoulos, M. N. and Derr, E. L. Group interaction. II. A test of the group model on binary solutions of hydrocarbons. *J. Am. Chem. Soc.*, **81** (1959), 2285–2289.

Parrish, C. F. Solvents, industrial. In *Kirk–Othmer Encyclopedia of Chemical Technology*, 3rd ed., Vol. 21, Grayson, M. and Eckroth, D., Eds. Wiley, New York, 1983, pp. 377–401.

Parrott, E. L., Wurster, D. E., and Higuchi, T. Investigation of drug release from solids. I. Some factors influencing the dissolution rate. *J. Am. Pharm. Assoc., Sci. Ed.*, **44** (1954), 269–273.

Paruta, A. N. and Irani, S. A. Dielectric solubility profiles in dioxane–water mixtures for several antipyretic drugs. Effects of substituents. *J. Pharm. Sci.*, **54** (1965), 1334–1338.

Paruta, A. N., Sciarrone, B. J., and Lordi, N. G. Solubility of salicylic acid as a function of dielectric constant. *J. Pharm. Sci.*, **53** (1964), 1349–1353.

Paruta, A. N., Sciarrone, B. J., and Lordi, N. G. Dielectric solubility profiles of acetanilide and several derivatives in dioxane–water mixtures. *J. Pharm. Sci.*, **54** (1965), 1325–1333.

Patel, R. M. and Zografi, G. Factors influencing the surface activity of chlorpromazine at the air–solution interface. Effect of inorganic and organic electrolytes. *J. Pharm. Sci.*, **55** (1966), 1345–1349.

Pauling, L. *The Nature of the Chemical Bond*, 3rd ed., Cornell University Press, Ithaca, NY, 1960, p. 453.

Peterson, S. W. and Levy, H. A. A single-crystal neutron-diffraction study of heavy ice. *Acta Cryst.*, **10** (1957), 70–76.

Philbrick, F. A. The association of phenol in different solvents. *J. Am. Chem. Soc.*, **56** (1934), 2581–2585.

Pierotti, G. J., Deal, C. H., Jr., and Derr, E. L. Activity coefficients and molecular structure. *Ind. Eng. Chem.*, **51** (1959), 95–102.

Pierotti, R. A. A scaled particle theory of aqueous and nonaqueous solutions. *Chem. Rev.*, **76** (1976), 717–726.

Pierotti, R. A. The solubility of gases in liquids. *J. Phys. Chem.*, **67** (1963), 1840–1845.

Pikal, M. J., Lang, J. E., and Shah, S. Desolvation kinetics of cefamandole sodium methanolate: The effect of water vapor. *Int. J. Pharm.*, **17** (1983), 237–262.

Pimentel, G. C. and McClellan, A. L. *The Hydrogen Bond.* Freeman, San Francisco, 1960.

Pimentel, G. C. and McClellan, A. L. Hydrogen bonding. *Ann. Rev. Phys. Chem.*, **22** (1971), 347–385.

Plakogiannis, F. M. Lien, E. J., Harris, C., and Biles, J. A. Partition of alkylsulfates of quaternary ammonium compounds: Structure dependence and transport study. *J. Pharm. Sci.*, **59** (1970), 197–200.

Pohl, H. A., Hobbs, M. E., and Gross, P. M. Electric polarization of carboxylic acids in dilute solutions of nonpolar solvents. I. The relation of electric polarization to the association of carboxylic acids in the hydrocarbon solvents. *J. Chem. Phys.*, **9** (1941) 408–414.

Prakongpan, S., Higuchi, W. I., Kwan, K. H., and Molokhia, A. M. Dissolution rate studies of cholesterol monohydrate in bile acid–lecithin solutions using the rotating disc method. *J. Pharm. Sci.*, **65** (1976), 685–689.

Prausnitz, J. M. *Molecular Thermodynamics of Fluid-Phase Equilibria*, Prentice-Hall, Englewood Cliffs, NJ, 1969, pp. 190–192, 418–422.

Prausnitz, J. M., Anderson, T. F., Grens, E. A., Eckert, C. A., Hsieh, R., and O'Connell, J. P. *Computer Calculations for Multicomponent Vapor–Liquid and Liquid–Liquid Equilibria.* Prentice-Hall, Englewood Cliffs, NJ, 1980.

Prigogine, I. *The Molecular Theory of Solutions.* North-Holland Publishing Company, Amsterdam, 1957.

Prokic, R., Sarunac, V., and Renon, H. Méthodes expérimentales d'évaluation des solvants d'extraction des hydrocarbures aromatiques. *Rev. Inst. Fr. Petrole*, **25** (1970), 327–347.

Redlich, O., Derr, E. L., and Pierotti, G. J. Group interaction. I. A model for interaction in solutions. *J. Am. Chem. Soc.,* **81** (1959), 2283–2285.

Reinhardt, H. and Rydberg, J. Rapid and continuous system for measuring the distribution ratios in solvent extraction. *Chem. Ind. (London),* (1970) 488–491.

Reiser, A. *Symposium on Hydrogen Bonding, July 29–Aug 3, 1954, Ljubljana,* Hadzi, D. and Thompson, H. W., Eds. Pergamon, Oxford, 1959, pp. 443–447.

Reisman, A. *Phase Equilibria: Basic Principles, Applications, Experimental Techniques.* Academic, New York, 1970.

Rekker, R. F. *The Hydrophobic Fragmental Constant: Its Derivation and Application. A Means of Characterizing Membrane Systems,* Elsevier, Amsterdam, 1977.

Restaino, F. A. and Martin, A. N. Solubility of benzoic acid and related compounds in a series of *n*-alkanols. *J. Pharm. Sci.,* **53** (1964), 636–639.

Riddick, J. A. and Bunger, W. B. *Organic Solvents,* 3rd ed. Wiley, New York, 1970.

Riddick, J. A., Bunger, W. B., and Sakano, T. K. *Organic Solvents. Physical Properties and Methods of Purification.* 4th ed. *Techniques of Chemistry,* Vol. II. Weissberger, A. Ed. Wiley, New York, 1986.

Rogers, D. W. and Özsogomonyan, A. Phase titrations IV. New applications including the assay of water in pyridine. *Talanta,* **11** (1964), 652–655.

Rogers, D. W., Özsogomonyan, A., and Sümer, A. Phase titrations III. New applications and the phase titration of binary solutions of chemically similar components. *Talanta,* **11** (1964), 507–514.

Rogers, J. A. and Davis, S. S. Functional group contributions to the partitioning of phenols between liposomes and water. *Biochim. Biophys. Acta,* **598** (1980), 392–404.

Rohrschneider, L. Solvent characterization by gas–liquid partition coefficients of selected solutes. *Anal. Chem.,* **45** (1973) 1241–1247.

Ronc, M. and Ratcliff, G. A. Prediction of excess free energies of liquid mixtures by an analytical group solution model. *Can. J. Chem. Eng.,* **49** (1971), 825–830.

Rossotti, F. J. C. and Rossotti, H. *The Determination of Stability Constants and Other Equilibrium Constants in Solution,* McGraw-Hill, New York, 1961, p. 277.

Roux, A. H., Hétu, D., Perron, G., and Desnoyers, J. E. Chemical equilibrium model for the thermodynamic properties of mixed aqueous micellar systems: Application to thermodynamic functions of transfer. *J. Soln. Chem.,* **13** (1984), 1–25.

Rowlinson, J. S., *Liquids and Liquid Mixtures,* 2nd ed. Butterworths, London, 1969.

Rytting, J. H., Davis, S. S., and Higuchi, T. Suggested thermodynamic standard state for comparing drug molecules in structure–activity studies. *J. Pharm. Sci.,* **61** (1972), 816–818.

Rytting, J. H., Anderson, B. D., and Higuchi, T. Vapor pressure studies of the self-association of alcohols in isooctane. 2. The effect of chain branching. *J. Phys. Chem.,* **82** (1978a), 2240–2245.

Rytting, J. H., Huston, L. P., and Higuchi, T. Thermodynamic groups contributions for hydroxyl, amino, and methylene groups. *J. Pharm. Sci.,* **67** (1978b), 615–618.

Rytting, J. H., McHan, D. R., Higuchi, T., and Grant, D. J. W. Importance of the Debye interaction in organic solutions: Henry's law constants for polar liquids in nonpolar solvents and vice versa. *J. Soln. Chem.,* **15** (1986), 693–703.

Saket, M. M., James, K. C., and Kellaway, I. W. Partitioning of some 21-alkyl esters of hydrocortisone and cortisone. *Int. J. Pharm.,* **21** (1984), 155–166.

Saleh, A. M., Badwan, A. A., and El-Khordagui, L. K. A study of hydrotropic salts, cyclohexanol and water systems. *Int. J. Pharm.,* **17** (1983), 115–119.

Saleh, A. M., Ebian, A. R., and Etman, M. A. Solubilization of water by hydrotropic salts. *J. Pharm. Sci.,* **75** (1986), 644–647.

Saracco, G. and Spaccamela Marchetti, E. Influenza della cartena idrocarburica sulla solubilità in acqua di serie omologhe. Nota 1. *Ann. Chim.,* **48** (1958), 1357–1370.

Sarma, T. S. and Ahluwalia, J. C. Experimental studies on the structure of aqueous solutions of hydrophobic solutes. *Chem. Soc. Rev.*, **2** (1973), 203–232.

Saunders, M. and Hyne, J. B. Study of hydrogen bonding in systems of hydroxylic compounds in carbon tetrachloride through the use of NMR. *J. Chem. Phys.*, **29** (1958), 1319–1323.

Scatchard, G. Equilibria in non-electrolye solutions in relation to the vapor pressures and densities of the components. *Chem. Rev.*, **8** (1931), 321–333.

Schanker, L. S. On the mechanism of absorption of drugs from the gastrointestinal tract. *J. Med. Pharm. Chem.*, **2** (1960), 343–359.

Scholtan, W. Die hydrophobe Bindung der Pharmaka an Humanalbumin and Ribonucleinsäure. *Arzneim.-Forsch. Drug Res.*, **18** (1968), 505–517.

Schreiber, D. R. Thermodynamics of *n*-Alkanols in Solution, Master's Thesis, University of Kansas, Lawrence, 1979.

Seidell, A. *Solubilities of Inorganic and Metal Organic Compounds*, Linke, W. F., Ed. Van Nostrand, Princeton, NJ, 1958a.

Seidell, A. *Solubilities of Organic Compounds*, Linke, W. F., Ed. Van Nostrand, Princeton, NJ, 1958b.

Seiler, P., Bischoff, O, and Wagner, R. Partition coefficients of 5-(substituted benzyl)-2,4-diaminopyrimidines. *Arzneim. Forsch. Drug Res.*, **32**(II) (1982), 711–714.

Setschenow, J. Über die konstitution der salzlösungen auf grund ihres verhaltens zu kohlensäure. *Z. Phys. Chem.*, **4** (1889), 117–125.

Sewell, J. M. Royal Aircraft Establishment Technical Report, No. 66185, Ministry of Aviation, Farnborough, Hampshire, UK, 1966.

Shah, A.C. Automated *in vitro* dissolution-rate technique for acidic and basic drugs. *J. Pharm. Sci.*, (1971) 1564–1567.

Shah, A. C. and Nelson, K. G. Evaluation of a convective diffusion drug dissolution rate model. *J. Pharm. Sci.*, **64** (1975), 1518–1520.

Shami, E. G. Formation of Hydrogen-Bonded Complexes by Amides in Nonaqueous Solvents. Ph.D. Thesis, The University of Wisconsin, Madison, WI, 1964.

Shefter, E. and Higuchi, T. Dissolution behavior of crystalline solvated and nonsolvated forms of some pharmaceuticals. *J. Pharm. Sci.*, **52** (1963), 781–791.

Sheridan, J. P., Martire, D. E., and Tewari, Y. B. Thermodynamics of molecular association by gas–liquid chromatography. II. Haloalkane acceptors with di-*n*-octyl ether and di-*n*-octyl thioether as electron donors. *J. Am. Chem. Soc.*, **94** (1972), 3294–3298.

Sherwood, T. K. and Gordon, K. F. A note on the additivity of diffusional resistances. *Am. Inst. Chem. Eng. J.*, **1** (1955), 129.

Sherwood, T. K. and Pigford, R. L. *Absorption and Extraction* McGraw-Hill, New York, 1952, p. 327.

Shinoda, K. The critical micelle concentration of soap mixtures (two-component mixture). *J. Phys. Chem.*, **58** (1954), 541–544.

Shinda, K. "Iceberg" formation and solubility. *J. Phys. Chem.*, **81** (1977), 1300–1302.

Shinoda, K. *Principles of Solution and Solubility*, translated in collaboration with P. Becker, Marcel Dekker, New York, 1978.

Silcock, H. L., Ed., *Solubilities of Inorganic and Organic Compounds*. Vol. 3, Parts 1, 2, and 3 (1979), *Ternary and Multicomponent Systems of Inorganic Substances*, Pergamon, Oxford. Translated from the Russian Spravochnik po Vastvorunosti, Izdatel'stro Nauka, 1969.

Singleton, W. S. Solution properties. In *Fatty Acids*–Part I, 2nd ed., Markely, K. S., Ed. Wiley–Interscience, New York, 1960, pp. 609–678.

Skjold-Jørgensen, S., Kolbe, B., Gmehling, J. and Rasmussen, P. Vapor–liquid equilibria by UNIFAC group contribution. Revision and extension. *Ind. Eng. Chem. Process Des. Dev.*, **18** (1979), 714–722.

Smith, F. Properties of very dilute alcohol solutions. *Aust. J. Chem.*, **30** (1977a), 23–42.

Smith, F. Properties of *n*-alcohol + *n*-alkane mixtures. *Aust. J. Chem.*, **30** (1977b), 43–69.

Smith, F. and Brown, I. Thermodynamic properties of alcohol + alkane mixtures. II. Contributions to the excess energies other than those due to hydrogen bonding. *Aust. J. Chem.*, **26** (1973), 705–721.

Snyder, L. Solutions to solution problems. 1. *Chemtech*, **9** (1979), 750–755.

Snyder, L. Solutions to solution problems. 2. *Chemtech*, **10** (1980), 188–193.

Snyder, L. R. Solvent selection for separation processes. In *Separation and Purification*, 3rd ed. *Techniques in Chemistry*, Vol. XII, Perry, E. S. and Weissberger, A., Eds. Wiley, New York, 1978.

Sorby, D. L., Bitter, R. G., and Webb, J. G. Dielectric constants of complex pharmaceutical solvent systems I. Water–ethanol–glycerin and water–ethanol–propylene glycol. *J. Pharm. Sci.*, **52** (1963), 1149–1153.

Spaccamela Marchetti, E. and Saracco, G. Influenza della catena idrocarburica sulla solubilità di serie omologhe di composti organici in solventi. Nota II. *Ann. Chim.* **48** (1958), 1371–1394.

Speakman, J. C. *The Hydrogen Bond and Other Intermolecular Forces*, The Chemical Society Monographs for Teachers, No. 27, The Chemical Society, London, 1975.

Spencer, J. N., Harner, R. S., Freed, L. I., and Penturelli, C. Hydrogen bonding of resorcinol to ethers and thioethers. *J. Phys. Chem.*, **79** (1975), 332–335.

Spencer, J. N., Gleim, J. E., Hackman, M. L., Blevins, C. H. and Garrett, R. C. Spectrophotometric and calorometric study of the nitrogen–hydrogen bond. *J. Phys. Chem.*, **82** (1978), 563–566.

Spencer, J. N., Gleim, J. E., Blevins C. H., Garrett, R. C., and Mayer, F. J. Enthalpies of solution and transfer enthalpies. An analysis of the pure base calorimetric method for the determination of hydrogen bond enthalpies. *J. Phys. Chem.*, **83** (1979), 1249–1255.

Staveley, L. A. K. and Milward, G. L. Solutions of alcohols in non-polar solvents. Part IV. Some thermodynamic properties of glycols in benzene, heptane, and cyclohexane. *J. Chem. Soc.*, **1957** (1957), 4369–4375.

Steinberg, W. H., Hutchins, H. H., Pick, P. G., and Lazar, J. S. Automated technique for determining dissolution and reaction rate of antacids. I. Instrumentation and evaluation of antacid raw materials. *J. Pharm. Sci.*, **54** (1965a), 625–633.

Steinberg, W. H., Hutchins, H. H., Pick, P. G., and Lazar, J. S. Automated technique for determining dissolution and reaction rate of antacids. II. Commercial antacid products. *J. Pharm. Sci.*, **54** (1965b), 761–771.

Stella, V. J. Nonclassical phase transfer behavior of phenylbutazone. *J. Pharm. Sci.*, **64** (1975), 706–708.

Stella, V. J. and Pipkin, J.D. Phenylbutazone ionization kinetics. *J. Pharm. Sci.*, **65** (1976), 1161–1165.

Stellner, K. L., Tucker, E. E., and Christian, S. D. Thermodynamic properties of the benzene–phenol dimer in dilute aqueous solution. *J. Sol. Chem.*, **12** (1983), 307–313.

Stephen, H. and Stephen, T., Eds., *Solubilities of Inorganic and Organic Compounds*. Vol. 1, Part 1 (1963), Binary Systems; Vol. 1, Part 2 (1963), Binary Systems; Vol. 2, Part 1 (1964), Ternary Systems; Vol. 2, Part 2 (1964), Ternary and Multicomponent Systems. Pergamon, Oxford. Original Russian work by Kafarov, V. V., Chief Coordinator, Akademii Nauk S.S.S.R., Vsesoiwznyi Institut Naucknoi i Tekhnicheskoi Informatisii, Moscow.

Stillinger, F. Abstracts, Royal Society Discussion Meeting on Water Structure, London, 1976.

Sunwoo, C. and Eisen, H. Solubility parameter of selected sulfonamides. *J. Pharm. Sci.*, **60** (1971), 238–244.

Suryanarayanan, R. and Mitchell, A. G. Evaluation of two concepts of crystallinity using calcium gluceptate as a model compound. *Int. J. Pharm.*, **24** (1985), 1–17.

Swarbrick, J., Ed. *Current Concepts in the Pharmaceutical Sciences: Biopharmaceutics*. Lea and Febiger, Philadelphia, PA, 1970.

Symposium on Ion Pair Partition. Function in analytical and preparative organic chemistry and in membrane transport, Stockholm, October 24–26, 1972. *Acta Pharm. Suec.*, **9** (1972), 609–654.

Szwarc, M. Ions and ion pairs. *Acc. Chem. Res.*, **2** (1969), 87–96.

Szwarc, M. Concept of ion pairs and their distinction from free ions. *Acta Pharm. Suec.*, **9** (1972), 612–614.

Taft, R. W., Jr., Separation of polar, steric, and resonance effects in reactivity. In *Steric Effects in Organic Chemistry*, Newman, M. S., Ed. Wiley, New York, 1956, pp. 556–675.

Taft, R.W., Gurka, D., Joris, L., Schleyer, P. von R., and Rakshys, J. W. Studies of hydrogen-bonded complex formation with *p*-fluorophenol. V. Linear free energy relatonships with OH reference acids. *J. Am. Chem. Soc.*, **91** (1969), 4801–4808.

Taft, R. W., Abraham, M. H., Doherty, R. M., and Kamlet, M. J. The molecular properties governing solubilities of organic nonelectrolytes in water. *Nature*, **313** (1985a), 384–386.

Taft, R. W., Abraham, M. H., Famini, G. R., Doherty, R.M., Abboud, J.-L.M. and Kamlet, M. J. Solubility proerties in polymers and biological media 5: An analysis of the physicochemical properties which influence octanol–water partition coefficients of aliphatic and aromatic solutes. *J. Pharm. Sci.*, **74** (1985b), 807–814.

Takahashi, F., Karoly, W. J., Greenshields, J. B., and Li, N. C. Solvent effects in ultraviolet spectral studies of hydrogen bonding between phenol and *N, N*-dimethylacetamide. *Can. J. Chem.*, **45** (1967), 2033–2038.

Tamres, M. Charge–transfer complexes in the vapour phase. In *Molecular Complexes*, Vol. I, Foster, R., Ed. Elek Science, London, 1973, pp. 49–116.

Tamres, M. and Yarwood, J. Complexes of *n* and π donors with halogens and related σ acceptors. In *Spectroscopy and Structure of Molecular Complexes*, Yarwood, J., Ed. Plenum, New York, 1973, p. 217.

Tanford, C. Contribution of hydrophobic interactions to the stability of the globular conformation of proteins. *J. Am. Chem. Soc.*, **84** (1962), 4240–4247.

Tanford, C. *The Hydrophobic Effect*, Wiley-Interscience, New York, 1973.

Thomas, E. R. and Eckert, C. A. Prediction of limiting activity coefficients by a modified separation of cohesive energy density model and UNIFAC. *Ind. Eng. Chem. Process Des. Dev.*, **23** (1984), 194–209.

Timmermans, J. La théorie des solutions concentrées, Revue historique et critique (The theory of concentrated solutions: A historical and critical review). *J. Chim. Phys.*, **19** (1921). 169–178.

Tomlinson, E. Chromatographic hydrophobic parameters in correlation analysis of structure–activity relationships. *J. Chromatogr.*, **113** (1975), 1–45.

Tomlinson, E. Enthalpy–entropy compensation analysis of pharmaceutical, biochemical and biological systems. *Int. J. Pharm.*, **13** (1983), 115–144.

Tomlinson, E. and Davis, S. S. Increased uptake of an anionic drug by mucous membrane, upon formation of ion-association species with quaternary ammonium salts. *J. Pharm. Pharmacol.*, **28** (Suppl.) (1976), 75P.

Tomlinson, E. and Davis, S. S. Interactions between large organic ions of opposite and unequal charge. I. Complexation between alkylbenzyldimethylammonium chlorides, bischromones, and indigo carmine. *J. Colloid Interface Sci.*, **66** (1978), 335–344.

Tomlinson, E., Poppe, H., and Kraak, J. C. Extrathermodynamic group contribution values determined by high pressure reversed-phase liquid/solid chromatography. *J. Pharm. Pharmacol.*, **28** (1976), 43P.

Townley, E. R. Griseofulvin, In *Analytical Profiles of Drug Substances*, Vol. 8, Florey, K., Ed. Academic, New York, 1979, pp. 219–249.

Treybal, R. E. *Liquid Extraction*, 2nd ed. McGraw-Hill, New York, 1963.

Tsonopoulos, C. and Prausnitz, J. M. Activity coefficients of aromatic solutes in dilute aqueous solutions. *Ind. Eng. Chem. Fundam.*, **10** (1971), 593–600.

Tsuji, A., Nakashima, E., Hamano, S., and Yamana, T. Physiochemical properties of amphoteric β-lactam antibiotics. I: Stability, solubility, and dissolution behavior of amino penicillins as a function of pH. *J. Pharm. Sci.*, **67** (1978), 1059–1066.

Tucker, E. E. and Becker, E. D. Alcohol association studies. II. Vapor pressure, 220-MHz proton magnetic resonance, and infrared investigations of *tert*-butyl alcohol association in hexadecane. *J. Phys. Chem.*, **77** (1973), 1783–1796.

Tucker, E. E. and Christian, S. D. Hydrogen bond cooperativity. The methanol–tri-*n*-octylamine system in *n*-hexadecane. *J. Am. Chem. Soc.*, **97** (1975a), 1296–1271.

Tucker, E. E. and Christian, S. D. Suprabinary hydrogen-bonded complexes. The methanol-*N*, *N*-diethyldodecanamide system in *n*-hexadecane. *J. Phys. Chem.*, **79** (1975b), 2484–2488.

Tucker, E. E. and Christian, S. D. 13th Midwest Regional Meeting, American Chemical Society, Nov. 3–4, 1977a.

Tucker, E. E. and Christian, S. D. Alcohol association studies. 3. Vapor pressure measurements for the ethanol-*n*-hexadecane system. *J. Phys. Chem.*, **81** (1977b), 1295–1299.

Tucker, E. E., Farnham, S. B., and Christian, S. D. Association of methanol in vapor and in *n*-hexadecane. A model for the association of alcohols. *J. Phys. Chem.*, **73** (1969), 3820–3829.

Tucker, E. E., Christian, S. D., and Lin, L.-N. Hydrogen bonding of phenol in carbon tetrachloride. The use of activity data to evaluate association models. *J. Phys. Chem.*, **7** (1974), 1443–1445.

Turi, P., Dauvois, M., and Michaelis, A. F. Continuous dissolution rate determination as a function of the pH of the medium. *J. Pharm. Sci.*, **65** (1976), 806–810.

Ubbelohde, A. R. *Melting and Crystal Structure*, Clarendon Press, Oxford, 1965.

Ueda, S. The mechanism of solubilization of water-insoluble substances with sodium benzoate derivatives. I. The interaction between water-insoluble substances and sodium benzoate derivatives in aqueous solution. *Chem. Pharm. Bull.*, **14** (1966), 22–29.

Underwood, F. L. and Cadwallader, D. E. Automated potentiometric procedure for studying dissolution kinetics of acidic drugs under sink conditions. *J. Pharm. Sci.*, **67** (1978), 1163–1167.

United States Pharmacopeia XXI and National Formulary XVI, United States Pharmacopeial Convention, Inc., Rockville, MD, 1985, pp. 1243, 1244, 1343, 1344, 1483–1490.

Vachon, M. G. and Grant, D. J. W. Enthalpy–entropy compensation in pharmaceutical solids. *Int. J. Pharm.*, **40** (1987), 1–14.

Valvani, S. C., Valkowsky, S. H., and Amidon, G. L. Solubility of nonelectrolytes in polar solvents. VI. Refinements in molecular surface area computations. *J. Phys. Chem.*, **80** (1976), 829–835.

Vanderborgh, N. E., Armstrong, N. R., and Spall, W. D. A cryoscopic study of the association of phenolic compounds in benzene. *J. Phys. Chem.*, **74** (1970), 1734–1741.

Van Krevelen, D. W. and Van Hoftijzer, P. J. Kinetics of gas–liquid reactions. Part I. General theory. *Rec. Trav. Chem.*, **67** (1948) 563–586.

Van Krevelen, D. W. and Van Hooren, C. J. Kinetics of gas–liquid reactions. Part II. Application of general theory to experimental data. *Rec. Trav. Chem.*, **67** (1948), 587–599.

Van Ness, H. C., Van Winkle, J., Richtol, H. H., and Hollinger, H. B. Infrared spectra and the thermodynamics of alcohol–hydrocarbon systems. *J. Phys. Chem.*, **71** (1967), 1483–1494.

Vogel, A. I. Physical properties and chemical constitution. Part XXIII. Miscellaneous compounds. Investigation of the so-called co-ordinate or dative link in esters of oxy-acids and in nitro-paraffins by molecular refractivity determinations. Atomic, structural, and group parachors and refractivities. *J. Chem. Soc.*, (1948), 1833–1855.

Von Hahn, F.-V. Ist Hydrotropie eine besondere Eigenshaft von organischen Salzen? *Kolloid-Zeitschrift*, **62** (1933), 202–207.

Vora, K. R. M., Higuchi, W. I., and Ho, N. F. H. Analysis of human buccal absorption of drugs by physical model approach. *J. Pharm. Sci.*, **61** (1972), 1785–1791.

Wagner, J. G. *Biopharmaceutics and Relevant Pharmacokinetics*. Hamilton Press, Hamilton, IL, 1971.

Wakabayashi, T. Some applications of the regular solution theory to solvent extraction. IV. Oxine–inert solvent system. *Bull. Chem. Soc. (Jpn.)*, **40** (1967), 2836–2839.

Walden, P. Über die schmelzwärme, spezifische kohäsion und molekulargrösse bei der schmelztemperatur. *Z. Eletrochem.*, **14** (1908), 713–724.

Wall, L. A., Flynn, J. H., and Straus, S. Rates of molecular vaporization of linear alkanes. *J. Phys. Chem.*, **74** (1970), 3237–3242.

Ward, H. L. and Cooper, S. C. The system benzoic acid, orthophthalic acid, water. *J. Phys. Chem.*, **34** (1930), 1484–1493.

Warycha, S., Rytting, J. H., and Higuchi, T. Unpublished observations, 1980.

Weast, R. C. and Astle, M. J. *CRC Handbook of Chemistry and Physics*, 61st ed. CRC, Boca Raton, FL, 1980.

Weast, R. C., Lide, D. R., Astle, M. J., and Beyer, W. H. *CRC Handbook of Chemistry and Physics*, 70th ed., CRC, Boca Raton, FL, 1989.

Wells, P. R. Linear free energy relationships. *Chem. Rev.*, **62** (1962), 171–219.

Westerlund, D., Borg, K. O., and Lagerström, P.-O. Fluorimetric determinations by ion-pair extraction. Part 3. Extraction constants of ion pairs between anthracene-2-sulphonate and mono- and divalent amines. *Acta Pharm. Suec.*, **9** (1972), 47–52.

Whetsel, K. B. and Lady, J. H. Self-association of phenol in nonpolar solvents. In *Spectrometry of Fuels*, Friedel, R. A., Ed. Plenum, New York, 1970, pp. 259–279.

Whitman, W. G. The two-film theory of gas absorption. *Chem. Metall. Eng.*, **29** (1923), 146–148.

Wichterle, I., Linek, J., and Hála, E. *Vapor–Liquid Equilibrium Data Bibliography*, Elsevier, Amsterdam, 1973.

Wichterle, I., Linek, J., and Hála, E. *Vapor–Liquid Equilibrium Data Bibliography*, Suppl. I, II, III, IV, Elsevier, Amsterdam, 1976, 1979, 1982, 1985.

Wiley, G. R. and Miller, S. I. Thermodynamic parameters for hydrogen bonding of chloroform with Lewis bases in cyclohexane. A proton magnetic resonance study. *J. Am. Chem. Soc.*, **94** (1972), 3287–3293.

Wilhelm, E. and Battino, R. Thermodynamic functions of the solubilities of gases in liquids at 25°C. *Chem. Rev.*, **73** (1973), 1–9.

Wilhelm, E., Battino, R., and Wilcock, R. J. Low-pressure solubility of gases in liquid water. *Chem. Rev.*, **77** (1977), 219–262.

Williams, N. A. and Amidon, G. L. Excess free energy approach to the estimation of solubility in mixed solvent systems. I: Theory. *J. Pharm. Sci.*, **73** (1984a), 9–13.

Williams, N. A. and Amidon, G. L. Excess free energy approach to the estimation of solubility in mixed solvent systems. II: Ethanol–water mixtures. *J. Pharm. Sci.*, **73** (1984b), 14–18.

Williams, N. A. and Amidon, G. L. Excess free energy approach to the estimation of solubility in mixed solvent systems. III: Ethanol–propylene glycol–water mixtures. *J. Pharm. Sci.*, **73** (1984c), 18–23.

Wilson, G. M. and Deal, C. H. Activity coefficients and molecular structure. Activity coefficients in changing environments–Solutions of groups. *Ind. Eng. Chem. Fundam.*, **1** (1962), 20–23.

Windholz, M., Budavari, S., Blumetti, R. F., and Otterbein, E. S., *The Merck Index: An Encyclopedia of Chemicals, Drugs, and Biologicals*, 10th Ed., Merck, Rahway, NJ, 1983.

Winsor, P. A. Hydrotropy, solubilisation and related emulsification processes. Parts I, II, III, IV. *Trans. Faraday Soc.*, **44** (1948a), 376–398.

Winsor, P. A. Hydrotropy, solubilisation and related emulsification processes. Parts V, VI, VII, VIII. *Trans. Faraday Soc.*, **44** (1948b), 451–471.

Winsor, P. A. Hydrotropy, solubilization and related emulsification processes. Part IX. The electrical conductivity and the water dispersibility of some solubilized systems. *Trans. Faraday Soc.*, **46** (1950), 762–772.

Wood, J. H., Syarto, J. E., and Letterman, H. Improved holder for intrinsic dissolution rate studies. *J. Pharm. Sci.*, **54** (1965), 1068.

Woolley, E. M. and Hepler, L. G. Molecular association of hydrogen-bonding solutes. Phenol in cyclohexane and benzene. *J. Phys. Chem.*, **76** (1972), 3058–3064.

Woolley, E. M., Travers, J. G., Erno, B. P., and Hepler, L. G. Molecular association of hydrogen-bonding solutes. Phenol in carbon tetrachloride. *J. Phys. Chem.*, **75** (1971), 3591–3597.

Wóycicka, M. K. and Rećko, W. M. Heats of mixing of *n*-propanol and *n*-hexanol with *n*-hydrocarbons at high dilutions. Série des sciences chimiques XX. *Bull. Acad. Pol. Sci. Chem.*, **20** (1972), 783–788.

Wright, E. M. and Diamond, J. M. An electrical method of measuring non-electrolyte permeability. *Proc. Roy. Soc.*, **B172** (1969a), 203–225.

Wright, E. M. and Diamond, J. M. Patterns of non-electrolyte permeability. *Proc. Roy. Soc.*, **B172** (1969b), 227–271.

Yaacobi, M. and Ben-Naim, A. Solvophobic interaction. *J. Phys. Chem.*, **78** (1974), 175–178.

Yair, O. B. and Fredenslund, A. Extension of the UNIFAC group-contribution method for the prediction of pure-component vapor pressures. *Ind. Eng. Chem. Process Des. Dev.*, **22** (1983), 433–436.

Yalkowsky, S. H. Estimation of entropies of fusion of organic compounds. *Ind. Eng. Chem. Fundam.*, **18** (1979), 108–111.

Yalkowsky, S. H. Solubility and partitioning. V: Dependence of solubility on melting point. *J. Pharm. Sci.*, **70** (1981), 971–973.

Yalkowsky, S. H. and Valvani, S. C. Precipitation of solubilized drugs due to injection or dilution. *Drug Intell. Clin. Pharm.*, **11** (1977), 417–419.

Yalkowsky, S. H. and Valvani, S. C. Solubility and partitioning. I: Solubility of nonelectrolytes in water. *J. Pharm. Sci.*, **69** (1980), 912–922.

Yalkowsky, S. H., Flynn, G. L., and Slunick, T. G. Importance of chain length on physicochemical and crystalline properties of organic homologs. *J. Pharm. Sci.*, **61** (1972a), 852–857.

Yalkowsky, S. H., Flynn, G. L., and Amidon, G. L. Solubility of nonelectrolytes in polar solvents. *J. Pharm. Sci.*, **61** (1972b), 983–984.

Yalkowsky, S. H., Amidon, G. L., Zografi, G., and Flynn, G. L. Solubility of nonelectrolytes in polar solvents. III: Alkyl *p*-aminobenzoates in polar and mixed solvents. *J. Pharm. Sci.*, **64** (1975), 48–52.

Yalkowsky, S. H., Valvani, S. C., and Amidon, G. L. Solubility of nonelectrolytes in polar solvents IV: Nonpolar drugs in mixed solvents. *J. Pharm. Sci.*, **65** (1976), 1480–1494.

Yalkowsky, S. H., Valvani, S. C., and Roseman, T. J. Solubility and partitioning. VI: Octanol solubility and octanol–water partition coefficients. *J. Pharm. Sci.*, **72** (1983), 866–870.

Yalkowsky, S. H., Valvani, S. C., Kuu, W.-Y., and Dannenfelser, R.-M. *Arizona Database of Aqueous Solubility. An Extensive Compilation of Aqueous Solubility Data for Organic Compounds*, 2nd. ed. Samuel Yalkowsky, Tucson, AZ, 1987.

Yalkowsky, S. H., Pinal, R., and Banerjee, S. Water solubility: a critique of the solvatochromic approach. *J. Pharm. Sci.*, **77** (1988), 74–77.

Yamaoka, Y., Roberts, R. D., and Stella, V. J. Low-melting phenytoin prodrugs as alternative oral delivery modes for phenytoin: A model for other high-melting sparingly water-soluble drugs. *J. Pharm. Sci.*, **72** (1983), 400–405.

York, P. Solid-state properties of powders in the formulation and processing of solid dosage forms. *Int. J. Pharm.*, **14** (1983), 1–28.

Zarkarian, J. A., Anderson, F. E., Boyd, J. A., and Prausnitz, J. M. UNIFAC parameters from gas–liquid chromatographic data. *Ind. Eng. Chem. Process Des. Dev.*, **18** (1979), 657–661.

Zografi, G. and Auslander, D. E. Surface activity of chlorpromazine and chlorpromazine sulfoxide in the presence of insoluble monomolecular films. *J. Pharm. Sci.*, **54** (1965), 1313–1318.

Zografi, G. and Munshi, M. V. Effect of chemical modification on the surface activity of some phenothiazine derivatives. *J. Pharm. Sci.*, **59** (1970), 819–822.

Zografi, G. and Zarender, I. Surface activity of phenothiazine derivatives at the air/solution interface. *Biochem. Pharmacol.*, **15** (1966), 591–598.

Zografi, G., Auslander, D. E., and Lytell, P. L. Interfacial properties of phenothiazine derivatives. *J. Pharm. Sci.*, **53** (1964a), 573–574.

Zografi, G., Patel, P. R., and Weiner, N. D. Interactions between Orange II and selected long-chain quaternary ammonium salts. *J. Pharm. Sci.*, **53** (1964b), 544–549.

Zolotov, Yu. A. On the mechanism of the elementary act in the extraction of inner-complex compounds. *Dokl. Chem., Proc. Acad. Sci. USSR*, **162** (1965), 498–501. Original article in Russian, *Dokl. Akad. Nauk. SSSR*, **162** (1965), 577–580.

AUTHOR INDEX

SUBJECT INDEX